T0221485

A Blue Carbon Primer

Praise for the Book

As the impacts of changing climate are becoming increasingly apparent, the ability of coastal wetlands to continue to absorb carbon dioxide has emerged as a topic of intense importance and interest. Scientists will find that this unique resource provides up-to-date information on the carbon cycle in coastal ecosystems, including their capacity for long-term carbon sequestration. Coastal resource managers and policy makers will benefit from the book's discussion on implications of that science for preserving the ecosystem services of coastal wetlands as seas rise and climate changes.

—*Marcia K. McNutt, President, National Academy of Sciences, Washington, DC, USA*

Our planet's highly biodiverse blue carbon ecosystems play a key role in determining the progression, mitigation, and outcomes of global climate change. Alongside their essential role as carbon sinks, these ecosystems are fundamental to the livelihoods and food security of tens of millions of coastal people, including some of the planet's most vulnerable populations throughout the tropics. Blue carbon habitats are bunkers of biodiversity, fisheries production, coastal protection, and carbon sequestration. Decisive action to protect them is of paramount importance to the fate of both people and nature. Yet for too long, we've known too little about the carbon dynamics of these habitats. This landmark publication brings together leading research from the world's preeminent coastal wetland scientists, and will help guide evidence-based action at a time of unprecedented threats to these ecosystems.

—*Alasdair Harris, Executive Director of Blue Ventures*

I was introduced as a plenary speaker at the 2016 Mangrove and Macrobenthos (MM4) meeting in St. Augustine by my former student, Dr. Ken Krauss, as saying that "Robert Twilley was studying carbon in mangroves before carbon was cool." In support of that statement, a Web of Science search that includes "mangroves and carbon" returns only 20 publications in the year I published my carbon export study of Rookery Bay mangroves in 1985, compared to over 200 papers published last year. It is great to see this global interest associated with the "blue carbon" focus of wetlands as an ecosystem service. However, with that interest comes the challenge of having shared methods that help shape a community of practice that can generate valid comparative information on blue carbon dynamics. The chapters in this edited book, *A Blue Carbon Primer*, sets the bar to achieve this goal. This series of chapters will certainly guide this community of practice for the next generation of blue carbon analyses of wetland ecosystems.

—*Robert R. Twilley, University of Louisiana at Lafayette and Past President, Coastal & Estuarine Research Federation*

In 2009, after 4 years of work, International Union for Conservation of Nature (IUCN) published the landmark report on the management of natural coastal carbon sinks. We wrote that report—the first of its kind and which led to the creation of the phrase "blue carbon"—because of the fundamental need to change the world's views on the importance we should attach to such coastal habitats. The concept of blue carbon captured the attention of policy makers looking to include the biosphere and oceans in climate actions. It is for this reason that I am particularly delighted to see the publication of this book, which will help policy makers, decision takers, and practitioners on the coast to move forward with renewed determination to safeguard these most critical of Earth's ecosystems.

—*Dan Laffoley, Marine Vice Chair, IUCN World Commission on Protected Areas*

Blue carbon is rapidly becoming a major impetus for conservation of tidal marshes, mangrove swamps, and seagrass meadows. These ecosystems are some of the world's most important sinks for organic carbon. With the recent flood of literature on the topic, *A Blue Carbon Primer* is a timely

contribution. The breadth of topics covered, from drivers of storage to mapping and policy, as well as case studies from five different countries, should make this an important resource for both those new and those experienced in the blue carbon field.
—*Gail Chmura, Professor, Centre for Climate and Global Change Research, McGill University*

A Blue Carbon Primer is a comprehensive overview of the nearshore marine ecosystems capable of long-term carbon storage in their sediments—which provide climate mitigation benefits important for the ocean business community to understand and engage in. Managing, protecting, and restoring coastal blue carbon ecosystems are significant means for achieving carbon sequestration services and other benefits to the environment, society, business, and the economy. This timely volume creates a framework for blue carbon research, policy, and project implementation and sets out common blue carbon terminology and concepts for science, policy, and practice necessary for this young field. The book provides examples of blue carbon accounting approaches, management, and projects that should be of interest to the global ocean business community.
—*Paul Holthus, Founding President and CEO, World Ocean Council*

Coastal blue carbon ecosystems, mangroves, tidal salt marshes, and seagrasses are found on every continent of the world, besides Antarctica. They are a source for food and help to protect our coasts. However, a function that might be easily overlooked is that they can be a natural-based solution for climate change, mitigating CO_2 emissions. The Intergovernmental Oceanographic Commission of UNESCO supports scientists and urges policy makers to recognize that when protected or restored, coastal blue carbon ecosystems sequester and store carbon—acting as carbon sinks. However, when degraded or destroyed, these ecosystems emit the carbon they have stored for centuries into the ocean and atmosphere and become sources of greenhouse gases. This book provides a concrete tool to connect the science with action to foster local, national, regional, and finally global activities on coastal blue carbon conservation and restoration.
—*Kirsten Isensee, Program Specialist—Ocean Carbon Sources and Sinks,*
Intergovernmental Oceanographic Commission of UNESCO—Ocean Science Section

Over fifty years ago a report to U.S. President Lyndon B. Johnson cautioned our nation on the effects on climate of the continued use of fossil fuels. This landmark report to policy-makers authored by Roger Revelle and others sparked the beginning of climate sciences and an era of scientific inquiry in a vast array of cutting-edge disciplines and questions, many of which were unforeseen at the time. The field of blue coastal carbon is today at the cutting edge of climate sciences; as it develops, it is providing fundamental and actionable science in support of restoration, adaptation, and mitigation strategies in our most vulnerable geography: the coast. This compendium sets forth the most recent understanding in interdisciplinary coastal and climate science and practice, providing policy-makers and land managers with fundamental knowledge and future direction to address our most urgent environmental priorities.
—*Maria Dolores Wesson, US Environmental Protection Agency, Washington, DC*

CRC Marine Science Series

Michael J. Kennish, Ph.D. and Judith S. Weis, Ph.D.

MARINE SCIENCE SERIES

The CRC Marine Science Series is dedicated to providing state-of-the-art coverage of important topics in marine biology, marine chemistry, marine geology, and physical oceanography. The series includes volumes that focus on the synthesis of recent advances in marine science.

Chemical Oceanography, Third Edition
Frank J. Millero

Coastal Pollution: Effects on Living Resources and Humans
Carl J. Sindermann

Acoustic Fish Reconnaissance
I.L. Kalikhman and K.I. Yudanov

Coastal Lagoons: Critical Habitats of Environmental Change
edited by Michael J. Kennish and Hans W. Paerl

Ecology of Marine Bivalves: An Ecosystem Approach, Second Edition
Richard F. Dame

Climate Change and Coastal Ecosystems: Long-Term Effects of Climate and Nutrient Loading on Trophic Organization
Robert J. Livingston

Habitat, Population Dynamics, and Metal Levels in Colonial Waterbirds: A Food Chain Approach
Joanna Burger and Michael Gochfeld

Living Shorelines: The Science and Management of Nature-Based Coastal Protection
edited by Donna Marie Bilkovic, Molly M. Mitchell, Megan K. La Peyre, and Jason D. Toft

Fishes Out of Water: Biology and Ecology of Mudskippers
edited by Zeehan Jaafar and Edward O. Murdy

A Blue Carbon Primer: The State of Coastal Wetland Carbon Science, Practice, and Policy
edited by Lisamarie Windham-Myers, Stephen Crooks, and Tiffany G. Troxler

For more information about this series, please visit: https://www.crcpress.com/CRC-Marine-Science/book-series/CRCMARINESCI?page=&order=pubdate&size=12&view=list&status=published, forthcoming

A Blue Carbon Primer
The State of Coastal Wetland Carbon Science, Practice, and Policy

Edited by
Lisamarie Windham-Myers, Stephen Crooks, and
Tiffany G. Troxler

CRC Press
Taylor & Francis Group
Boca Raton London New York

CRC Press is an imprint of the
Taylor & Francis Group, an **informa** business

CRC Press
Taylor & Francis Group
6000 Broken Sound Parkway NW, Suite 300
Boca Raton, FL 33487-2742

First issued in paperback 2020

© 2019 by Taylor & Francis Group, LLC
CRC Press is an imprint of Taylor & Francis Group, an Informa business

No claim to original U.S. Government works

ISBN-13: 978-1-4987-6909-9 (hbk)
ISBN-13: 978-0-367-89352-1 (pbk)

This book contains information obtained from authentic and highly regarded sources. Reasonable efforts have been made to publish reliable data and information, but the author and publisher cannot assume responsibility for the validity of all materials or the consequences of their use. The authors and publishers have attempted to trace the copyright holders of all material reproduced in this publication and apologize to copyright holders if permission to publish in this form has not been obtained. If any copyright material has not been acknowledged, please write and let us know so we may rectify in any future reprint.

Except as permitted under U.S. Copyright Law, no part of this book may be reprinted, reproduced, transmitted, or utilized in any form by any electronic, mechanical, or other means, now known or hereafter invented, including photocopying, microfilming, and recording, or in any information storage or retrieval system, without written permission from the publishers.

For permission to photocopy or use material electronically from this work, please access www.copyright.com (http://www.copyright.com/) or contact the Copyright Clearance Center, Inc. (CCC), 222 Rosewood Drive, Danvers, MA 01923, 978-750-8400. CCC is a not-for-profit organization that provides licenses and registration for a variety of users. For organizations that have been granted a photocopy license by the CCC, a separate system of payment has been arranged.

Trademark Notice: Product or corporate names may be trademarks or registered trademarks, and are used only for identification and explanation without intent to infringe.

Visit the Taylor & Francis Web site at
http://www.taylorandfrancis.com

and the CRC Press Web site at
http://www.crcpress.com

Contents

Chapter 1
Defining Blue Carbon: The Emergence of a Climate Context for Coastal Carbon Dynamics 1

Stephen Crooks, Lisamarie Windham-Myers, and Tiffany G. Troxler

Chapter 2
The Importance of Blue Carbon in Coastal Management in the United States 9

Steve Emmett-Mattox and Stefanie Simpson

Chapter 3
Human Impacts on Blue Carbon Ecosystems.. 17

Catherine E. Lovelock, Daniel A. Friess, J. Boone Kauffman, and James W. Fourqurean

Part I
State of Science

Chapter 4
The Fate and Transport of Allochthonous Blue Carbon in Divergent Coastal Systems 27

**Thomas S. Bianchi, Elise Morrison, Savanna Barry, Ana R. Arellano,
Rusty A. Feagin, Audra Hinson, Marian Eriksson, Mead Allison,
Christopher L. Osburn, and Diana Oviedo-Vargas**

Chapter 5
Net Ecosystem Carbon Balance of Coastal Wetland-Dominated Estuaries: Where's the
Blue Carbon?.. 51

Charles S. Hopkinson

Chapter 6
Physical and Biological Regulation of Carbon Sequestration in Tidal Marshes 67

James T. Morris and John C. Callaway

Chapter 7
Accretion: Measurement and Interpretation of Wetland Sediments.. 81

John C. Callaway

Foreword

In November 2010, an intrepid group of coastal scientists ventured into the unknown and, for the first time, entered the bewildering event that is the annual meeting of the United Nations Framework Convention on Climate Change (UNFCCC). We were at COP16—as that meeting was known—in Cancun, Mexico to tell the climate change world about blue carbon—the carbon stored by mangroves, tidal marshes, and seagrass. In particular, we were there to talk about how blue carbon should be an important part of the solution to climate change for many countries. Despite initial naive optimism, our confidence slowly slipped away that day as we presented our case to a largely empty room of skeptical climate negotiators.

"You don't have enough data."

"No country will want blue REDD+."

"Ocean-based climate mitigation won't work."

We were summarily dismissed and probably forgotten by most of those present.

Fast forward to 2018. Over 50 countries have included blue carbon in their commitments to achieving the Paris Climate Agreement. The Intergovernmental Panel on Climate Change has official guidance on how countries should include blue carbon in national inventories greenhouse gas emissions and removals, and the United States, Australia, and other countries now count these ecosystems in their annual climate change reporting. An international partnership of countries exists to promote blue carbon as an important component of the solution to climate change. Over a thousand scientific papers have been written describing the importance and potential of blue carbon globally. More than $100 million is currently being spent on blue carbon research, conservation, and other activities globally. And some of the first blue carbon credits are about to go on sale in carbon markets.

In Cancun 8 years ago, we could not have imagined this astonishing progress toward recognizing conservation and restoration of coastal blue carbon ecosystems as a component of the global response to climate change. In retrospect, by 2010, the science of blue carbon had reached a tipping point that allowed a transition into policy and management. Additionally, the hard work of our forestry colleagues in developing the Reducing Emissions from Deforestation and Forest Degradation (REDD) mechanism provided essential lessons and laid a foundation for blue carbon. From my perspective, however, the successful development of blue carbon is a direct result of two very specific characteristics of the blue carbon community that has come together since 2010.

The rapid development of blue carbon has been the result of an almost unique intellectual openness and willingness to collaborate. From the initial group of scientists and conservationists that first highlighted the value of mangroves, seagrasses, and salt marshes at the UNFCCC and elsewhere, the blue carbon community has rapidly grown to include a global network of scientists, national and regional governments, policy analysts, local coastal communities, conservation organizations, carbon market experts, and supporters spanning every continent except Antarctica. The Blue Carbon Initiative (BCI)—formed in 2011 by Conservation International, IUCN, and the Intergovernmental Oceanographic Commission (UNESCO)—was the first formalization and coordination of this network. The BCI has provided a mechanism to facilitate this network and accelerate the integration of blue carbon science into climate policy and related approaches such as carbon financing and management. However, this success has only been possible because of the unhesitatingly collaborative blue carbon community. We have freely shared knowledge and ideas, solved problems together, collaborated to navigate the labyrinth of international climate institutions, and built tools that are now used globally. Importantly, as a community, we have recognized and celebrated the achievements and contributions of all.

Additionally, the openness of the blue carbon community to new perspectives has been transformational in achieving the integration of science and policy that other disciplines have found elusive.

Not long after our trip to Cancun, the BCI initiated a discussion between scientists and the climate policy community. For a while we talked in circles: The scientists were sure they had more than sufficient science to demonstrate the carbon value of coastal wetlands; the policy community was adamant that there was not enough science to support the integration of blue carbon into climate policy. But with patience and determination, the orbits of the discussion aligned and we began to understand that the questions being asked by policy were not those being answered by science. The question was not how carbon rich these ecosystems were but rather what is the climate change consequence of destroying them. This shift in perspective immediately drove new blue carbon research, which has in turn directly led to the inclusion of blue carbon in national carbon inventories and will soon result in blue carbon projects being included in carbon markets. The blue carbon community has learned to work across previously impervious boundaries—between science and policy, coastal management, and climate regulation—and, in so doing, unified the priorities of mitigating climate change and conserving and restoring the world's rich coastal wetlands.

This book is an important record of what we have learned along the blue carbon journey of the last 8 years. From the basic science of carbon dynamics in coastal ecosystems to the challenges of integrating coastal ecosystems into national climate solutions, this volume provides a key resource for the diverse blue carbon community. Perhaps most importantly, however, this book lays out the challenges and gives us a glimpse of the successes we must work toward in the next 8 years.

Emily Pidgeon, Ph.D.
Senior Director, Blue Climate Program, Conservation International
and Co-Chair, International Blue Carbon Scientific Working Group

Preface

Atmospheric CO_2 is rising globally, as are other common greenhouse gases (GHGs) such as CH_4 and N_2O. Opportunities for reducing and reversing this trend include energy sector adjustments and land management for inhabited regions. A total of 75% of the world's population lives within 20 miles of a coastline. With rising sea levels, coastal land management is critical to climate adaptation, and is poised to provide climate mitigation benefits as well.

Coastal blue carbon (hereafter BC) emerged strongly in 2009 as a means to clarify the climate mitigation benefits of tidally driven ecosystems. The "color" framework—the green carbon of forests, the black carbon of atmospheric particulates—emerged as a way to distinguish carbon fluxes that have different causes and different implications. Tidal wetlands (fresh-to-saline, trees-to-grasses-to microbial mats) and subtidal seagrass meadows are all recognized as Blue Carbon Ecosystems (BCEs), capable of long-term carbon storage in their sediments, and subjected to human management. As of 2018, we now have multiple policies, practices, and scientific datasets in place that recognize the role of coastal wetlands in carbon accounting. We sought to produce this primer in order to set an initial framework of this growing field of research, policy, and project implementation.

We have three primary goals in this edited volume.

1. to set terminology and concepts as understood in 2018 for coastal BC activities in three sectors: science, policy, and practice.
2. to provide examples of BC accounting approaches and management, both historical and proposed, in key regions globally.
3. to engage a larger population of interested parties in this discussion of the future role of coastal BCEs in carbon sequestration services among other human benefits.

We provide a glossary of the language used herein to improve clarity among the 28 chapters. We also provide maps and conceptual diagrams in order to illustrate distributions and drivers of coastal BC, as understood today. The book is laid out in six areas, as described below.

Chapters 1–3 focus on definitions and context of BCEs. Crooks et al. (Chapter 1) identify the conditions and boundaries of BCEs, including their regional and geomorphic context. Emmett-Mattox and Simpson (Chapter 2) review how the emerging recognition of carbon sequestration and storage services is a means to improve communication and support for BCE preservation and restoration of multiple ecosystem services across local-to-global scales. Lovelock et al. (Chapter 3) provide a human context of how management decisions influence net C sequestration and long-term preservation.

Chapters 4–11 focus on processes associated with carbon and GHG accounting in coastal ecosystems. Bianchi et al. (Chapter 4) review the fate and transport of C exchange in BCEs through allochthonous input and export to estuaries. Hopkinson (Chapter 5) reviews the drivers, boundaries, and range of scales that influence the Net Ecosystem Carbon Budget, tracking all C sources in and out of BCEs. Morris and Callaway (Chapter 6) provide a "virtual marsh" analysis to quantify C sequestration model results for all reasonable combinations of tidal marsh conditions. Callaway (Chapter 7) addresses the importance of timescale and method when quantifying and interpreting volumetric changes (vertical accretion/subsidence), the dominant flux that regulates C stock accumulation and loss in BCEs. Keller (Chapter 8) contrasts CO_2 flux assessments from C stock changes with the approaches and interpretations needed to measure other GHG fluxes, namely methane (CH_4) and nitrous oxide (N_2O). Kennedy et al. (Chapter 9) address current knowledge on the carbonate cycle, an especially important flux of C between aqueous and precipitate pools in submerged BCEs, such as seagrass. Christian et al. (Chapter 10) assess the role and variability of sea level rise on wetland accretion and extent from past millennia through to future predictions. Megonigal et al. (Chapter 11) clarify the direct and indirect effects of rising temperatures globally that are expected to influence multiple components of BCE resilience and GHG budgets.

Chapters 12–14 focus on current maps of BCEs globally, including what characteristics are currently and potentially mappable (e.g. biomass, salinity, height, etc.). Oreska et al. (Chapter 12) further identify approaches to detecting changes, both loss and gain, in the extent of seagrass and associated subtidal vegetation communities. Giri (Chapter 13) reports current global distributions for mangroves, and nuances of remote-sensing imagery for interpreting changing extents and structures. Byrd et al. (Chapter 14) summarize current salt marsh extents and illustrates novel approaches to mapping biomass stocks and changes for carbon accounting.

Chapters 15–20 focus on policy developments and their role as incentives for coastal wetland management for C sequestration among other ecosystem services. Herr et al. (Chapter 15) step through the history of international policy development for BCE management. Troxler et al. (Chapter 16) focus on the specific international guidance from UNFCCC that became codified in 2013. Sutton-Grier (Chapter 17) reviews policy incentives for BCEs that have developed within national frameworks. Vegh et al. (Chapter 18) clarify the importance of quantifying additional ecosystem services to complete the portfolio of benefits associated with BCE management. Crooks (Chapter 19) provides examples and pathways for BC interventions with a range of relevant and emerging policy tools. Needelman et al. (Chapter 20) step through the criteria for inclusion of BC within voluntary carbon markets, using a currently approved methodology (VCS VM 0033).

Chapters 21–27 all represent case studies. We report here seven sites across the globe (Indonesia, California, Abu Dhabi, Kenya, Massachusetts, Florida, and Mexico) that have their own unique environmental conditions, ongoing studies, and policy developments (Figure 0.1). The extensive projects and scenarios are described by scientists and practitioners working in these seven settings, namely: Murdiyarso (Chapter 21), Drexler et al. (Chapter 22), Schile et al. (Chapter 23), Kairo et al. (Chapter 24), Surgeon-Rogers et al. (Chapter 25), Sherwood et al. (Chapter 26), and Adame et al. (Chapter 27).

We conclude with a final chapter (Chapter 28) on BC futures, which highlights recent synthesis efforts, using these and earlier chapters to identify significant advances and information needs in science, policy, and practice. We expect that this primer, summarizing the state of knowledge in 2018, will be updatable within 10 years as we see exponential growth in the study of BCEs along all the world's coastlines. We are grateful to all the authors and the large community of BC colleagues with whom these ideas and developments are growing as this goes to print. We look forward to the discussions that develop from this "state of the science, policy and practice."

All errors are our own, and the ideas presented do not reflect official endorsement from any of our respective offices and agencies.

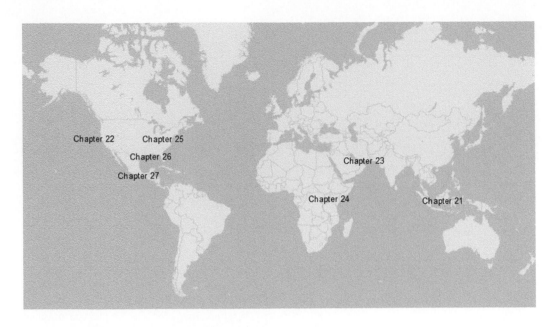

Figure 0.1 Locations of the seven case studies (Indonesia, California (USA), Abu Dhabi, Kenya, Massachusetts (USA), Florida (USA), and Mexico).

Editors

Dr. Lisamarie Windham-Myers has been a wetland ecologist in the USGS National Research Program for 15 years and is currently a lead scientist for the USGS Water Mission Area, based in Menlo Park, California. Her work focuses broadly on biogeochemistry and ecosystem ecology, with approaches that span landscape-to-molecular scales and single-site to synthesis efforts. Across North America, her research has been funded by multiple federal and state agencies, with sites representing a wide range of salinity and management conditions, from rice agriculture to coastal and restored wetlands. Lisa serves in several local, national, and international science advisory efforts to evaluate wetland management and modeling approaches to quantify wetland carbon sequestration, greenhouse gas budgets, and/or mercury methylation and export. She has been a federal U.S. representative in multiple international policy efforts to quantify carbon fluxes in coastal ecosystems, including the Commission for Environmental Cooperation and the IPCC.

Dr. Stephen Crooks is the principal for Wetland Science and Coastal Management at Silvestrum Climate Associates and a wetland restoration practitioner. As a co-founder of the International Blue Carbon Initiative, he has worked with many authors of this volume to pioneer this emerging topic. Steve has been a delegate to UN climate change negotiations since 2010, as a representative of the IPCC in 2013 (Warsaw) and of the French Delegation in 2015 (Paris). He was invited to the UN Climate Summit in 2014 by UN Secretary-General, Ban Ki-Moon. Steve served as a lead author of the IPCC 2013 Supplement to the 2006 IPCC Guidelines on National Greenhouse Gas Inventories: Wetlands. For the United States, he is responsible for translating emerging science on management of coastal wetlands for inclusion within the EPA's National Greenhouse gas Inventory. Steve is co-chair of the Carbon Scientific and Technical Advisory Panel advising the GEF Blue Forests Project assisting country partners in Africa, the Americas, and Asia, to enact blue carbon projects.

Dr. Tiffany G. Troxler is a wetland ecosystem ecologist and research associate professor in the Southeast Environmental Research Center and Department of Biological Sciences at Florida International University (FIU). Her work includes collaborative research that examines the effects of saltwater inundation on Everglades coastal wetlands to inform management actions, monitoring adaptive management actions associated with Everglades restoration and advancing interdisciplinary nature-based solutions in urban coastal environments. She served as co-editor and contributing author on two IPCC methodological reports that guide national greenhouse gas policy support on land use and land-use change and wetlands while based in Japan with the IPCC Task Force on National Greenhouse Gas Inventories. Tiffany is also founding Director and Director of Science of the Sea Level Solutions Center at FIU, a state university center that focuses on advancing knowledge, decision-making, and actions toward mitigating the causes and adapting to the effects of sea-level rise. A near native of Florida, she's lived in Miami for over 20 years.

Editors

List of Contributors

Omar Abdul-Aziz
Department of Civil and Environmental
    Engineering
Ecological and Water Resources Engineering
    Laboratory
West Virginia University
Morgantown, West Virginia

Maria Fernanda Adame
Australian Rivers Institute
Griffith University
Queensland, Australia

Mead Allison
Department of Earth and Environmental
    Sciences
Department of River-Coastal Science and
    Engineering
Tulane University
New Orleans, Louisiana
and
The Water Institute of the Gulf
Baton Rouge, Louisiana

Marcelo Ardón
Department of Forestry and Environmental
    Resources
North Carolina State University
Raleigh, North Carolina

Ana R. Arellano
Department of Geological Sciences
University of Florida
Gainesville, Florida
and
College of Marine Science
University of South Florida
St. Petersburg, Florida

Savanna Barry
Nature Coast Biological Station
University of Florida
Cedar Key, Florida

Thomas S. Bianchi
Department of Geological Sciences
University of Florida
Gainesville, Florida

Richard Birdsey
U.S. Forest Service
Woods Hole Research Center
Falmouth, Massachusetts

Linda K. Blum
Department of Environmental Sciences
University of Virginia
Charlottesville, Virginia

Carly Brody
The National Academies of Sciences,
    Engineering, and Medicine
Washington, District of Columbia

Kristin B. Byrd
U.S. Geological Survey Western Geographic
    Science Center
Menlo Park, California

John C. Callaway
Department of Environmental Science
University of San Francisco
San Francisco, California

Justin Campbell
Climate Change Research Group
Walpole, Massachusetts

Edward Castañeda-Moya
Southeast Environmental Research Center
Florida International University
Miami, Florida

Samantha Chapman
Department of Biology
Villanova University
Villanova, Pennsylvania

Robert R. Christian
Biology Department
East Carolina University
Greenville, North Carolina

Stephen Crooks
Silvestrum Climate Associates, LLC
San Francisco, California

Paul Dijkstra
Center for Ecosystem Science and Society
Northern Arizona University
Flagstaff, Arizona

Bill Dougherty
Climate Change Research Group
Walpole, Massachusetts

Judith Z. Drexler
U.S. Geological Survey
California Water Science Center
Sacramento, California

Igino M. Emmer
Silvestrum Climate Associates, LLC
Jisp, The Netherlands

Steve Emmett-Mattox
Restore America's Estuaries
Arlington, Virginia

Marian Eriksson
Department of Ecosystem Science
  and Management
Texas A&M University
College Station, Texas

Rusty A. Feagin
Department of Ecosystem Science
  and Management
Texas A&M University
College Station, Texas

James W. Fourqurean
Department of Biological Sciences
  and Center for Coastal Oceans
  Research
Institute for Water and Environment
Florida International University
Miami, Florida

Daniel A. Friess
Department of Geography
National University of Singapore
Singapore

Chandra Giri
Office of Research and Development
U.S. Environmental Protection Agency
Research Triangle Park, North Carolina

Jane Glavan
Abu Dhabi Global Environmental Data
  Initiative
Abu Dhabi, United Arab Emirates

Meagan Eagle Gonneea
U.S. Geological Survey Woods Hole Coastal &
  Marine Science Center
Woods Hole, Maryland

Holly S. Greening
Tampa Bay Estuary Program
St. Petersburg, Florida

Greg Guannel
Caribbean Green Technology Center
University of the Virgin Islands
Charlotte Amalie, United States Virgin Islands

A. J. Hamza
Blue Carbon Unit
Kenya Marine and Fisheries
  Research Institute
Mombasa, Kenya

Dorothee Herr
IUCN
Gland, Switzerland

Audra Hinson
Department of Ecosystem Science and
  Management
Texas A&M University
College Station, Texas

James Holmquist
Smithsonian Environmental Research
  Center
Smithsonian Institution
Edgewater, Maryland

Charles S. Hopkinson
Department of Marine Sciences
University of Georgia
Athens, Georgia

James G. Kairo
Blue Carbon Unit
Kenya Marine and Fisheries
    Research Institute
Mombasa, Kenya

J. Boone Kauffman
Department of Fisheries and Wildlife
Oregon State University
Corvallis, Oregon

Jason K. Keller
Schmid College of Science and Technology
Chapman University
Orange, California

Hilary A. Kennedy
Ocean Sciences
Bangor University
Anglesey, United Kingdom

Shruti Khanna
California Department of Fish and Wildlife
Stockton, California

Matt Kirwan
Physical Sciences Department
Virginia Institute of Marine Sciences
Gloucester Point, Virginia

Kevin D. Kroeger
U.S. Geological Survey Woods Hole Coastal &
    Marine Science Center
Woods Hole, Maryland

Michael Kunz
Orizon Consulting LLC
McLean, Virginia

Adam Langley
Department of Biology
Villanova University
Villanova, Pennsylvania

Eduardo Leorri
Geology Department
East Carolina University
Greenville, North Carolina

Elisa López García
Consultant
Cancun, Quintana Roo, Mexico

Catherine E. Lovelock
School of Biological Sciences
The University of Queensland
Brisbane, Australia

Christine May
Silvestrum Climate Associates, LLC
San Francisco, California

Karen J. McGlathery
Department of Environmental Sciences
University of Virginia
Charlottesville, Virginia

Chris Mcowen
United Environment Programme World
    Conservation Monitoring Centre
Cambridge, United Kingdom

J. Patrick Megonigal
Smithsonian Environmental Research
    Center
Smithsonian Institution
Edgewater, Maryland

James T. Morris
Baruch Institute
University of South Carolina
Columbia, South Carolina

Elise Morrison
Department of Soil and Water Sciences
Department of Geological Sciences
University of Florida
Gainesville, Florida

Serena Moseman-Valtierra
Department of Biological Sciences
University of Rhode Island
Kingston, Rhode Island

Ryan P. Moyer
Florida Fish & Wildlife Conservation
    Commission
Fish & Wildlife Research Institute
St. Petersburg, Florida

Daniel Murdiyarso
Center for International Forestry Research
Bogor, Indonesia
Department of Geophysics and Metorologi
Bogor Agricultural University
Bogor, Indonesia

Brian Murray
Nicholas Institute for Environmental Policy
    Solutions
Duke University
Durham, North Carolina

Brian A. Needelman
Department of Environmental Science and
    Technology
University of Maryland
College Park, Maryland

Matthew P. J. Oreska
Department of Environmental Sciences
University of Virginia
Charlottesville, Virginia

James Orlando
U.S. Geological Survey
California Water Science Center
Sacramento, California

Michelle Orr
Environmental Science Associates
San Francisco, California

Robert J. Orth
College of William & Mary
Virginia Institute of Marine Science
Gloucester Point, Virginia

Christopher L. Osburn
Department of Marine, Earth, and
    Atmospheric Sciences
North Carolina State University
Raleigh, North Carolina

Diana Oviedo-Vargas
Department of Marine, Earth, and
    Atmospheric Sciences
North Carolina State University
Raleigh, North Carolina
and
Stroud Water Research Center
Avondale, Pennsylvania

Stathys Papadimitriou
National Oceanographic Centre
Southampton, United Kingdom

Linwood Pendleton
Nicholas Institute for Environmental Policy
    Solutions
Duke University
Durham, North Carolina

Kara R. Radabaugh
Florida Fish & Wildlife Conservation
    Commission
Fish & Wildlife Research Institute
St. Petersburg, Florida

Gary E. Raulerson
Tampa Bay Estuary Program
St. Petersburg, Florida

Doug Robison
Environmental Science Associates
Tampa, Florida

Minerva Rosette
Mexican Center of Environmental Law
Cancun, Quintana Roo, Mexico

Lisa Schile-Beers
Smithsonian Environmental Research
    Center
Smithsonian Institution
Edgewater, Maryland

David H. Schoellhamer
U.S. Geological Survey
California Water Science Center
Sacramento, California

Lindsey Sheehan
Environmental Science Associates
San Diego, California

Edward T. Sherwood
Tampa Bay Estuary Program
St. Petersburg, Florida

Jorge Herrera Silveira
CINVESTAV-IPN Unidad Merida
Merida, Yucatan, Mexico

Stefanie Simpson
Restore America's Estuaries
Arlington, Virginia

Tonna-Marie Surgeon Rogers
Waquoit Bay National Estuarine
    Research Reserve
Waquoit, Maryland

Ariana E. Sutton-Grier
Earth System Science Interdisciplinary Center
University of Maryland
College Park, Maryland
and
MD/DC Nature Conservancy
Bethesda, Maryland

Jianwu Tang
Ecosystems Center
Marine Biological Laboratory
Woods Hole, Maryland

Dave Tomasko
Environmental Science Associates
Tampa, Florida

Tiffany G. Troxler
Sea Level Solutions Center
Institute for Water and Environment
Florida International University
Miami Beach, Florida

Tibor Vegh
Nicholas Institute for Environmental
    Policy Solutions
Duke University
Durham, North Carolina

Moritz Von Unger
Silvestrum Climate Associates, LLC
San Francisco, California

C. Wanjiru
Blue Carbon Unit
Kenya Marine and Fisheries Research Institute
Mombasa, Kenya
and
School of Pure and Applied Sciences
Mombasa Campus
Kenyatta University
Nairobi City, Kenya

Lauren Weatherdon
United Environment Programme World
    Conservation Monitoring Centre
Cambridge, United Kingdom

Ryan Whisnant
PEMSEA
Diliman, Philippines

Dave J. Wilcox
College of William & Mary
Virginia Institute of Marine Science
Gloucester Point, Virginia

Lisamarie Windham-Myers
U.S. Geological Survey
and
Water Mission Area
Menlo Park, California

Lindsay Wylie
Environmental Defense Fund
Washington, District of Columbia

Kimberly Yates
U.S. Geological Survey
St. Petersburg Center for Coastal and
    Marine Science
St. Petersburg, Florida

Keqi Zhang
International Hurricane Research Center
Extreme Events Institute
Florida International University
Miami, Florida

Glossary

Definitions

Coastal blue carbon (BC): the stocks and fluxes of organic carbon and greenhouse gases in tidally influenced coastal ecosystems such as marshes, mangroves, seagrasses, and other wetlands.

> BC stocks include the carbon stored in tidal forests (freshwater or saltwater, such as mangroves), tidal salt marshes, and seagrass meadows within the soil, the living biomass aboveground (e.g., leaves, branches, stems), the living biomass belowground (e.g., roots), and the non-living biomass (litter and dead wood).

> BC Fluxes include net changes in soil pools of carbon and are characterized as:

> Accretion—A volumetric measure of carbon accumulation, usually in reference to sediment accretion
> Accumulation—A mass-based gain in carbon, regardless of source
> Burial—Accumulation over short or long timeframes, all carbon pools documented belowground
> Storage—Long-term preservation of carbon pools in soils (time-dependent)
> Sequestration—Specifically in situ CO_2 uptake and long-term preservation (the most conservative estimate of climate mitigation via carbon dioxide removal)

Activity Data—Geographical data showing the types of land coverage and use in a given area

Allochthonous Carbon—Carbon produced in one location and deposited in another. In the context of BC systems, this type of carbon results from the hydrodynamic environment which controls sediment and associated carbon transport and fate from neighboring ecosystems (offshore and terrestrial).

Allometric Equations—Allometric equations establish quantitative relationships between key characteristics that are easy to measure (i.e., stem height/diameter) and other properties that are often more difficult to assess (i.e., biomass).

Autochthonous Carbon—Carbon produced and deposited in the same location. In the context of BC systems, this type of carbon results from vegetation uptake of CO_2 from the ocean and/or atmosphere that gets converted for use by plant tissue and contributes to the surrounding soil.

Carbon Inventory—A carbon inventory is an accounting of carbon gains and losses emitted to or removed from the atmosphere/ocean over a period of time. Policy makers use inventories to establish a baseline for tracking emission trends, developing mitigation strategies and policies, and assessing progress.

Carbon Pool—Carbon pools refer to carbon reservoirs such as soil, vegetation, water, and the atmosphere that absorb and release carbon. Together carbon pools make up a carbon stock.

Carbon Stock—A carbon stock is the total amount of organic carbon stored in a BC ecosystem of a known size. A carbon stock is the sum of one or more carbon pools.

Coastal Lands—The subset of all terrestrial lands located within the tidal frame, regardless of whether they are connected to tidal hydrology or protected from it by a hydrologic barrier.

Emission Factors—A categorized model for estimating GHG flux rate changes from a predefined area due to change in land coverage and use (i.e., conversion from mangroves to shrimp ponds) or changes within a land use type (i.e., nutrient enrichment of seagrass).

Flux Method—This method estimates the GHG flux between the soil and vegetation and the atmosphere/water column through direct measurements or by modeling and results in Tier 2 and 3 estimates.

Gain-loss Method—This method estimates the difference in carbon stocks based on emissions factors for specific activities (e.g., plantings, drainage, rewetting, deforestation) derived from the scientific literature and country activity data and results in Tier 1 and 2 estimates.

Inorganic Soil Carbon—The term soil inorganic carbon refers to the carbon component of carbonates (i.e., calcium carbonate) and can be found in coastal soils in the form of shells and/or pieces of coral.

IPCC Tiers—The IPCC has identified three tiers of detail in carbon inventories that reflect the degrees of certainty or accuracy of a carbon stock inventory (assessment).

> Tier 1—Tier 1 assessments have the least accuracy and certainty and are based on simplified assumptions and published IPCC default values for activity data and emissions factors. Tier 1 assessments may have a large error range of ±50% for aboveground pools and ±90% for the variable soil carbon pools.
>
> Tier 2—Tier 2 assessments include some country or site-specific data and hence have increased accuracy and resolution. For example, a country may know the mean carbon stock for different ecosystem types within the country.
>
> Tier 3—Tier 3 assessments require highly specific data of the carbon stocks in each component ecosystem or land use area, and repeated measurements of key carbon stocks through time to provide estimates of change or flux of carbon into or out of the area. Estimates of carbon flux can be provided through direct field measurements or by modeling.

Mangrove—A mangrove is a tree, shrub, palm, or ground fern, generally exceeding one-half meter in height that normally grows above mean sea level in the intertidal zone of marine coastal environments and estuarine margins. A mangrove is also the tidal habitat comprising such trees and shrubs.

Remote Sensing—An imaging approach, such as aerial photography, that provides repeated synoptic data for improving the coverage and representativeness of BC datasets.

Resolution—In remote-sensing resolution of an image is an indication of its potential detail, where the smaller the pixel the higher the detail. In other words, 250 m resolution data could identify any earthly feature that is 250 m by 250 m (useful for mapping ecosystem extent). Higher resolution data, such as 30 m can be used to monitor in more detail (useful for identifying encroachment by aquaculture).

Seagrass Meadows—Seagrasses are flowering plants belonging to four plant families, all in the order Alismatales, which grow in marine, fully saline environments. There are 12 genera with some 58 species known.

Soil Organic Carbon—The term soil organic carbon refers to the carbon component of the soil organic matter. The amount of soil organic carbon by weight or volume can vary with hydrology, soil texture, climate, vegetation, and historical and current land cover, use or management.

Soil Organic Matter—The term soil organic matter is used to describe the organic constituents in the soil (undecayed tissues from dead plants and animals, products produced as these decompose and the soil microbial biomass).

Stock Difference Method—This method estimates the difference in carbon stocks measured at two points in time and results in Tier 3 estimates.

Stratification—A technique used to divide large heterogeneous sites (which require many samples to account for variation) into smaller more homogeneous areas (where fewer samples are needed) and is also useful when field conditions, logistical issues, and resource limitations prevent dense sampling regimes.

Tidal Marsh—A tidal marsh is a coastal ecosystem in the upper intertidal zone between land and open water that is regularly flooded by the tides. Salt water or brackish water marshes are dominated by dense stands of salt tolerant plants such as herbs, grasses, or low shrubs, whereas freshwater marshes are dominated by less-salt tolerant species, which may be similar to regional non-tidal wetlands.

Tidal Amplitude—The elevation range of a given wetland, typically mean sea level (MSL) to mean higher high water (MHHW). See Figure 0.2.

Tidal Range—The full range of tidal elevations, typically mean lower low water (MLLW) to mean higher high water (MHHW). See Figure 0.2.

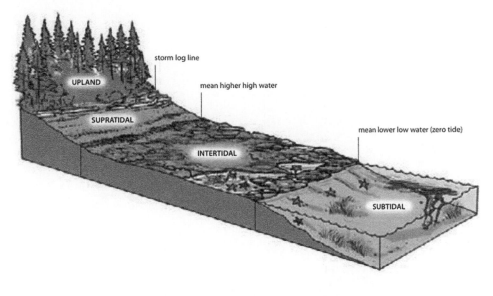

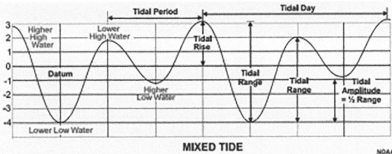

Figure 0.2 Descriptive figures of vertical datums and the tidal frame. (a) Schematic of tidal zones, illustrated by Soren Henrich. (b) Example of tidal data, with averages typically reported over a 19 year period. Figures available from NOAA NOS at https://noaanhc.files.wordpress.com.

CHAPTER 1

Defining Blue Carbon
The Emergence of a Climate Context for Coastal Carbon Dynamics

Stephen Crooks
Silvestrum Climate Associates, LLC

Lisamarie Windham-Myers
U.S. Geological Survey

Tiffany G. Troxler
Florida International University

CONTENTS

HIGHLIGHTS

1. Blue Carbon Ecosystems (BCEs) are defined as coastal wetland ecosystems with manageable and atmospherically significant carbon stocks and fluxes.
2. Policy and management opportunities have promoted the emergence of **blue carbon** as a concept and spurred scientific interest to reduce uncertainties in coastal carbon budgets.
3. The four major BCEs are generally classified by their plant communities: tidal marshes, tidal freshwater forests, mangroves, and seagrass meadows.

1.1 THE GLOBAL CLIMATE CHALLENGE

Global atmospheric carbon dioxide levels are now consistently above 400 ppm, well beyond levels occurring over the past 800,000 years (Lewis and Maslin, 2015). This rise is driven by human activities. In centuries prior to the 20th, land-use change altered carbon stocks through deforestation and soil degradation, but by the 1920s, fossil fuel emissions became the dominant source of increased atmospheric CO_2 concentrations (Lal et al., 2012; Lewis and Marlin, 2015; Le Quéré et al., 2016). These emissions, on top of natural fluctuations, are taken up across the atmosphere, oceans, and biosphere at timescales measured from hours to millennia, and over longer timescales with the lithosphere.

Detailed in the Global Carbon Budget 2016 report (Le Quéré et al., 2016), it is estimated that over the last decade emissions due to fossil fuel burning and land-use change totaled 9.3 ± 0.5 and 1.0 ± 0.5 GtC year^{-1}, respectively. About half of these annual emissions remain in the atmosphere (4.5 ± 0.1 GtC year^{-1}), with significant sinks of carbon partitioned across ocean (2.6 ± 0.5 GtC year^{-1}) and land (3.1 ± 0.9 GtC year^{-1}) ecosystems.

Rising CO_2 levels in the atmosphere are a principle driver of climate change (USGCRP, 2017). Globally annual averaged surface air temperature has increased by 1.8°F (1.0°C) over the past 115 years (1901–2016). This period is now the warmest in the history of modern civilization, with recent years repeatedly breaking records. Rising temperatures also drive the risk of significant unanticipated changes in the Earth's system. At least two types of potential surprises exist: compounded impacts of simultaneous or sequential extreme climate events; and critical thresholds or tipping points, whereby a physical threshold is crossed in the climate system driving large and potentially irreversible impacts over human timescales (USGCRP, 2017). Positive feedbacks (self-reinforcing) have the potential to accelerate climate changes by tripping the Earth's systems into new regimes that are very different from those experienced by current inhabitants of this very crowded world. The list of candidate tipping elements is long and includes: changes to atmospheric-ocean circulations (e.g., disruption to Atlantic Ocean Convection), cryosphere (e.g., accelerated melting of ice sheets), and carbon cycle (e.g., carbon dioxide and methane emission from boreal peatlands). Feedback to earth system warming include associated responses of the hydrosphere, including sea level rise and coastal flooding, changes in rainfall/storm patterns and intensity, and warming and acidification of ocean waters.

Reducing anthropogenic greenhouse gas (GHG) emissions by transitioning energy systems from fossil-fuel dominance is necessary to stabilize atmospheric climate forcing (IDDRI, 2015). Yet, while progress is being made in developing energy alternatives for a growing and increasingly resource-intensive population, globally we have done little to reduce our overall climate impact. With this backdrop, much more can be done to improve the management of the biosphere, recovering terrestrial and ocean sinks as a form of natural climate mitigation (Lal et al., 2012; Lal, 2016; Arneth et al., 2017; Griscom et al., 2017). Expanding our climate mitigation actions to more fully include the biosphere often involves benefits of improved social and environmental conditions despite the complexity associated with natural system fluctuations. Coastal ecosystems are an especially good example of where land-use management can lead to both natural climate mitigation and adaptation.

1.2 THE EMERGENCE OF BLUE CARBON

It is now 10 years since the term **blue carbon** emerged as a recognized concept, and one often associated with coastal carbon management. That is not to say the advent of **blue carbon** generated the first coastal wetland carbon budgets, or that this was even the first time that coastal wetland carbon cycling had been linked to climate change. That is far from the case (e.g., Twilley et al., 1992; Chmura et al., 2003; Duarte et al., 2005). But as a concept, recognizing the opportunity that improved management of certain coastal settings—blue carbon ecosystems (BCEs)—might contribute to a climate response, it captured the imagination of the international policy and conservation community working on climate mitigation and adaptation. For example, likening millions of small tidal wetland losses to **a million little fires** illustrates the climate mitigation potential of coastal management (sensu Kroeger, 2017). Blue carbon as a concept, and thus BCE management as one natural climate solution, has thus arisen at a moment in time when connecting the biosphere to climate mitigation actions is gaining traction within international policy discussions. In Textbox 1, we describe the maturing appreciation of BCE management as a climate mitigation tool.

1.3 DEFINING BCEs IN THE CONTEXT OF CLIMATE RESILIENCE

In the context of climate policy frameworks, **blue carbon** has been taken to represent the carbon accumulating in vegetated, tidally influenced coastal ecosystems such as tidal forests (including mangroves), tidal marshes, and intertidal to subtidal seagrass meadows (International Blue Carbon Science Working Group, 2015). Climate-relevant blue carbon pools and fluxes include physical, chemical, and biological exchanges within and between sediments and soils, waters, living biomass and non-living biomass, as well as GHG exchanges with the atmosphere.

Occupying less than 2% of ocean area (or <5% of global land area) vegetated coastal eco-systems are estimated to be responsible for nearly 50% of carbon burial in marine sediments (Duarte, 2005). This represents a very high concentration of carbon flux from atmosphere and surface waters to long-term sediment storage. More significantly, destruction of these same eco-systems results in fluxes of historical carbon pools back to the atmosphere as carbon dioxide, from soil and biomass stocks that had accumulated over hundreds to thousands of years (Pendleton et al., 2012).

Such a focus on only tidal forests, tidal marshes, and seagrasses reflects perhaps a first step to inclusion of other carbon sequestering, transferring and storing ecosystems in the future. This first step is pragmatic in that it provides for a limited but important set of marine and ocean systems to be evaluated and integrated with developing land-based climate mitigation frameworks under the UNFCCC (Howard et al, 2017; Sutton-Greer and Howard, 2018). By focusing on these three veg-etated ecosystems; however, we may be failing to fully recognize the climate benefits of improved management other coastal and marine ecosystems (e.g., kelp, microalgae, macroalgae, and fish) that are part of the biosphere's process of capturing, fixing, and transferring carbon dioxide to long-term storage of sediment or deep ocean water bodies (Smale et al., 2018).

For the purpose of practicality, recognition as a BCE—specifically one that focuses on connect-ing management of coastal ecosystems to climate policies and finance opportunities associated with land-ownership—requires that the following conditions be met:

TEXTBOX 1

Recognition of blue carbon as a concept of climate policy interest stems back to reports by the IUCN (International Union for Conservation of Nature) and UNEP (United Nations Environment Programme) (Laffoley and Grimsditch, 2009; Nellemann et al., 2009). Occurring against a backdrop of developing international policies for terrestrial forestland management and consideration of peatland soils, these reports invigorated the discussion of coastal and marine ecosystems within the global carbon cycle. The focus of both of these studies is to raise awareness that healthy coastal and marine ecosystems contribute to the removal of carbon from the atmosphere and long-term storage, and also that these ecosystems are threatened and degraded by human actions.

At the same time, foundations and the California Climate Action Reserve (CCAR) were exploring whether tidal wetlands restoration would be a valid inclusion in the carbon market as an offsets mechanism. What emerged was a perspective from the carbon management and project development side (Crooks et al., 2009). This review highlighted that while natural and restoring tidal wetlands do sequester carbon and continuously build soil carbon stocks, these were not the most significant fluxes of relevance to near-term climate mitigation and carbon project development. Rather, the more significant fluxes were driven by destruction of inter-tidal wetlands with conversion of soils from wet and anoxic to dry and oxic conditions, or with remobilization of soils through excavation or erosion. Carbon dioxide emissions from drained organic soils may continue for decades after the disturbance until either the stock is exhausted or the soils are no longer maintained dry (Deverel and Leighton, 2010). Akin to terrestrial peatlands, avoiding emissions, which were occurring at scale around the world, were the most significant near-term climate mitigation opportunity. The study also highlighted the need for coordinated science to improve quantification of drivers, fluxes, and long-term storage potential.

What might have seemed the end of the story at that point was met as a challenge and an opportunity by the conservation community (Chapter 2). In the United States, Restore America's Estuaries have been engaged, establishing a **Blue Ribbon Panel** consisting of the CCAR project team leads, wetland scientists, carbon market, and legal experts to map out a 5-year workplan. This resulted in White House and Federal agency support for science programs, landscape scale project assessments, and the production of methodologies that linked restoration and conservation of tidal marshes, mangroves, and seagrasses to the voluntary carbon markets under the Verified Carbon Standard (Emmer et al., 2015; Emmer et al., 2018a,b). At the global level, Conservation International, IUCN, and UN Intergovernmental Oceanographic Commission launched the Blue Carbon Initiative (BCI) to connect and empower scientists and policy analysts/makers to build capacity around the world and engage with decision makers in senior levels of government at key venues, such as the annual climate change negotiations. The word from the policy community was that they were now curious about coastal and marine ecosystems (remaining uncertain about the term blue carbon) but needed to see demonstration that the science was sound, that interventions could be enacted and that the benefits could be scaled to have meaningful impacts both to benefit local liveli-hoods and have a positive impact on GHG emission reductions.

An early success came at UNFCCC COP 16 (United Nations Framework Convention on Climate Change, Conference of the Parties 16), in Cancun, just prior to the official launch of the BCI. Here a side event was held outlining the early findings of three coordinated reports: a World Bank-funded study that estimated the scale of carbon emissions from a dozen converted large coastal wetland systems from around the world (Crooks et al., 2011); an economic analy-sis funded by the Linden Trust for Conservation that highlighted the potential to apply carbon finance for blue carbon interventions in developing countries (Murray et al., 2011); and, a policy analysis, also funded by Linden, that clarified that blue carbon could be incorporated under existing land-use focused international policy agreements (Climate Focus, 2011). The side event generated some buzz around blue carbon, demonstrating a pathway for inclusion for coastal and marine ecosystems into a policy context (https://blogs.worldbank.org/climat-echange/category/tags/blue-carbon). A particular success of the side event was the subse-quent inclusion of coastal wetlands within guidance for national reporting GHG emissions and removals with land-use change involving wetlands (IPCC, 2014).

At the time of publishing this book, there has been a rapid expansion in the number of scientific, economic, and policy publications on blue carbon. Demonstration projects (such as the GEF, Global Environment Facility Blue Forest Project) and networks (e.g., International Blue Carbon Partnership) are being funded to explore scaling of interventions and facilities are being created to support financing (e.g., Blue Natural Capital Financing Facility).

1. Rates of carbon sequestration and/ or prevention of emissions of GHGs by the ecosystem is cumulatively at sufficient scales to influence climate;
2. Major carbon stocks, change in stock and fluxes of GHGs can be quantified spatially and temporally;
3. Anthropogenic drivers are impacting carbon storage, stock change, or GHG emissions;
4. Management of the ecosystem to improve sequestration or emission reductions is possible and practicable;
5. Interventions can be achieved without causing social or environmental harm; and
6. Management actions can be aligned with existing or developing international policy and national commitments to address climate change.

Those ecosystems currently recognized as BCEs meet these conditions (see discussion in Howard et al., 2017). Drivers of coastal wetland loss globally are resulting in significant and unaccounted for emissions (Pendleton et al., 2012). Approaches for quantifying carbon stocks and stock change exist, including procedures from the IPCC (2014) guiding countries on the inclusion of coastal wetlands in national GHG inventories. Improved management of coastal ecosystems, primarily for benefits beyond those of climate mitigation, has been demonstrated (e.g., Bay Goals, 2016) but remains small-scale despite the critical need in many parts of the world. Being at the edge of the terrestrial zone, and being impacted by dominantly terrestrial human impacts, land-use based climate mitigation policies also can be extended from uplands into the edge of ocean waters (Climate Focus, 2011; Herr et al, 2017).

Other coastal and marine ecosystems meet some of these requirements but are not sufficiently recognized as full BCEs in a climate policy context. For now, we might recognize these as **potential blue carbon ecosystems.** Algal flats in arid regions can have high productivity leading to carbon accumulation (Schile et al., 2016), which through geological time been important sinks transferring carbon to the lithosphere and source rocks for hydrocarbons (El Kammer et al., 2015), as well as in modern settings flushing carbon out to near-shore coastal waters (Adame et al., 2012). Kelps, seagrasses, and man-made floating mariculture, as well as seagrasses may be major sources of laterally transported carbon (Chung, 2011; Krause-Jensen and Duarte, 2016). Offshore, settling algae, mediated by zooplankton and vertebrates may transport some carbon to continental shelf sediments, deep-sea sediments, and deep-ocean waters (e.g., Martin et al., 2016).

One particular challenge facing potential BCEs in open-water is the lack of an appropriate policy context. Policies developed under the UNFCCC under recent decades are focused on stable and spatially fixed carbon pools, and clear ownership of the resource. Trying to integrate open-water potential BCEs into these policies will likely be challenging. Still, we suggest that establishing a separate **Ocean Climate Framework** may have significant climate and management merits (M. von Unger, pers comms, Feb 2018).

1.4 WHAT ARE THE SCALE OF EMISSIONS AND REMOVALS BY BCEs?

The role of coastal wetlands in the global carbon cycle has largely been over looked, perhaps in part because for global carbon budgets, estimated changes in the land sink are derived as the residual of fluxes between the ocean and atmosphere and are not resolved to the level of change in wetland area (Le Quéré et al., 2016). Correctly, the ongoing carbon sequestration from intact wetlands has been seen as a small but positive carbon removal from the atmosphere in the context of near-term global climate mitigation, and at rates for which developing economic instruments was determined challenging (Murray et al., 2011). In contrast with terrestrial carbon sequestration, it should be recognized that tidal wetland carbon stocks grow volumetrically and are thus far less prone to saturation (Morris et al., 2012). Gradual sea level rise and particulate deposition (especially

TEXTBOX 2: 4 MAJOR BCES

TIDAL MARSH

Tidal marshes are communities of grasses, herbs, and low shrubs that are tolerant of flooding and occupy the tidal zone mostly above mean sea level. The composition of tidal marshes is influenced by flooding frequency and salinity. Tidal marshes are found from subpolar regions to subtropics, though at lower latitude they are out competed by mangroves.

MANGROVE

A mangrove is a tree, shrub, palm, or ground fern, generally exceeding one-half meter in height that normally grows above mean sea level in the intertidal zone of marine coastal environments and estuarine margins. The term mangrove also refers to tidal habitat comprised of such trees and shrubs. Mangroves are frost-sensitive and so limited to tropical and subtropical regions.

TIDAL FRESHWATER FOREST

A tidal freshwater forest are communities of trees and associated assemblages on low-lying coastal settings and head of estuaries where freshwater flows fall with the rise and with the tides.

SEAGRASS

Seagrasses are flowering plants belonging to four plant families, all are in the order of Alismatales, which grow in marine, fully saline environments. There are 12 genera with some 58 species known. Seagrasses are found from subpolar to tropical regions.

in large deltaic systems with high sediment loading) foster long term, continuous carbon burial through geomorphic accretion and expansion (Collins et al., 2017).

However, emissions from degradation and destruction of coastal ecosystems has been only recently illuminated, along with the benefits of compounding emissions removals with halting ongoing emissions by restoring wetlands on organic soils of converted wetlands and impoundments (Crooks et al., 2009; Crooks et al., 2011; Donato et al., 2011; Pendleton et al., 2012; Kroeger et al., 2017).

As one component of the biosphere, between terrestrial lands and the ocean, it has been estimated that destruction of coastal wetlands could be contributing emissions of 0.12 GtC year^{-1} (0.45 GtCO$_2$ year^{-1}) (Pendleton et al., 2012). While this amount is relatively small compared to the magnitude of fossil fuel emissions, it is still a sizable emission, compared with the emissions of large economies such as the United Kingdom, France, or California. Though the level of uncertainty is substantial (range 0.04–0.28 GtC year^{-1} (0.15–1.02 GtCO$_2$ year^{-1})), these emissions potentially represent about 10% of those estimated from terrestrial deforestation, but occurring on a much smaller land area (Murray et al., 2011).

Because of the particularly high carbon density per unit area found in coastal wetlands there is scaled benefit to protecting those remaining area and associated carbon stocks. Moreover, the economic damages associated with these emissions are sizable, estimated to be of the order of $18 billion (USD) per year (Pendleton et al., 2012), Halting those emissions would provide a partial but positive contribution to climate mitigation while at the same time reducing the degradation of coastal ecology. Can this be achieved? And in such a way that improves local livelihoods and supports national economies?

1.5 ADVANCING BLUE CARBON INTERVENTIONS

So the big question is, given the reasonable estimates of the scale of emissions, the loss of carbon stocks and ecosystem services and the known drivers of land-use change, can meaningful large-scale actions be taken to reduce or reverse the trend of BCE destruction and degradation?

The information needs to support natural resilience in mitigating and adapting to climate change parallel to those for urban systems (Bai et al., 2018), modified here to include: (i) expanded observations; (ii) understanding ecosystem—climate interactions; (iii) studying landscape responses; (iv) harnessing disruptive technology and policy approaches; (v) supporting transformation; and (vi) recognizing the global sustainability context.

In the chapters that follow, we shall delve into topics that help us explore the depth and limits to our knowledge on these topics. The concept of blue carbon is in itself a potential disruptive technology approach embedded in a broader policy transformation connecting landscapes to climate mitigation and adaptation frameworks, with associated financing to foster more sustainable management practices. The concept is embedded in a world of imperfect, but improving, observations, and understanding of human and natural system response to climate change and disturbance events.

Driving potential transformation are two important accords ratified in 2015 that together provide a basis for scaling up and 'mainstreaming' activities on BCE conservation and restoration. The Paris Climate Agreement and the United Nations 2030 Development Agenda both rest on bottom-up architecture, providing freedom to countries to establish their own pathway toward global common goals. The substantial flow of international finance via a number of transfer mechanisms including the rapidly expanding Green Bonds (including bonds for climate resilience), carbon finance, Global Climate Fund, and World Bank support will be defined by articulation of policy (Herr et al., 2017). Under the Paris Climate Agreement, 58 countries, a high proportion of maritime nations, recognized BCEs in some form under their Nationally Determined Contributions (NDCs), their policy commitments to reduce GHG emissions. Similarly, coastal wetlands and soils, both as a subset of marine and land systems, important cross-cutting components of many of the UNFCCC Sustainable Development Goals (SDGs), but particularly related to decent work and economic growth (SDG 8); industry, innovation, and infrastructure (SDG 9); sustainable cities and communities (SDG 11); responsible consumption and production (SDG 12); climate action (SDG 13); life below water (SDG 14); and life on land (SDG 15). There is continuing potential to improve national articulation of these policies in coming years.

Over the last 50 years, there has also been accelerating growth in experience related to coastal wetlands restoration and conservation. Following establishment of no-net-loss goals under the Clean Water Act, wetlands restoration became a significant economic sector of the U.S. economy spurring science, investment, and a growth of a knowledgeable restoration profession. While basic mistakes in coastal wetland restoration continue to occur around the world, in large part through unintentionally poor project planning, good practice principals for project evaluation and implementation exist and can be applied (Primavera et al., 2013; Crooks et al., 2014; Lewis and Brown, 2014). There is also a rapid growth of marine-protected areas for coastal ecosystem and other coastal management approaches for conservation (Howard et al., 2017), and in developing principles and tools for marine economic development as **blue economies** (Whisnant and Reyes, 2015). To help demonstrate the business opportunities for improved management of BCE's financing facilities are becoming more visible, such as the Government of Luxembourg and IUCN's Blue Natural Capital Financing Facility (https://bluenaturalcapital.org/wp/). Standardized approaches for quantification of blue carbon stocks and stock change (Blue Carbon Initiative, 2015) and accounting of emissions and removals in National GHG Inventories (IPCC, 2014) now exist (see Chapter 16).

To stem the loss of coastal ecosystems there is a need to mainstream good management practices being demonstrated around the world, and scale up these conservation and restoration activities as well as quantification of impacts to carbon stocks. Methodologies that link BCE conservation and restoration are just emerging and while application is still nascent, important carbon market financed examples are appearing from Kenya and Madagascar. Standards for coastal wetlands management do not yet exist under green bond financing initiatives. Maps of BCE extent and loss are well developed for mangroves but poorly developed for marshes and particularly seagrasses.

1.6 CONCLUSION

So at the time of drafting this book, in the post-Paris Climate Agreement world, we are entering a new phase for blue carbon, that being less of awareness raising and more of action and operationalizing. Enacting forest management actions such as under REDD+ policies took several decades to advance to the scale of large-scale projects, but blue carbon, while behind, is on a similar pathway. The challenge over the next few years is to build a portfolio of market-based and non market-based demonstration projects that show successful application in a range of cultural settings and provide a pathway to scaling up. Fortunately, there is an established learning curve and a good number of practice examples in terms of carbon project delivery, and wetland conservation and restoration. In the following chapters, we will explore topics in greater depth that will advance the learning curve and hopefully accelerate implementation of blue carbon interventions.

ACKNOWLEDGMENTS

The contribution by Stephen Crooks was made possible with project funding by NOAA, NASA, USGS and the Global Environment Facility.

CHAPTER **2**

The Importance of Blue Carbon in Coastal Management in the United States

Steve Emmett-Mattox and Stefanie Simpson
Restore America's Estuaries

CONTENTS

HIGHLIGHTS

1. A U.S. network of coastal conservation organizations, Restore America's Estuaries (RAE), has taken the national lead in promoting blue carbon storage and sequestration as an important ecosystem service that could be monetized to support coastal wetland restoration, protection, and management.
2. Blue carbon management is associated with multiple ecosystem services that are valued by U.S. citizens.
3. RAE has elevated blue carbon application potential by improving awareness, eligibility, market tools, and pilot project examples (case studies: see Tampa Bay, Chapter 26).
4. Blue carbon is a new way to communicate about ecosystem services in a common currency that crosses local and global interests, and links past, present, and future conditions.

2.1 INTRODUCTION

It is well established that our nation's estuaries and coastal wetlands are critical for healthy and sustainable coastal communities for people and wildlife, and that they provide tremendous services and benefits. Restore America's Estuaries (RAE) was established in 1995 with a mission to protect and restore the lands and waters essential to the richness and diversity of coastal life. As an alliance of ten regional coastal conservation organizations, it provides a united voice for coastal conservation in the nation's capital and advances the science and practice of protecting and restoring estuaries through on-the ground projects, ground-breaking science, high-level meetings, and the power of convening people.

In 2009, RAE launched a national-scale blue carbon initiative to (i) raise awareness of the climate mitigation and adaptation values of coastal wetlands and (ii) increase public and private investment in the conservation and restoration of these critical natural resources. This chapter explores the strategic importance of blue carbon to RAE's mission, its approach to achieving specific goals through its blue carbon program, and lessons learned from its national leadership of this effort.

2.2 WHAT'S AT STAKE

Estuaries, where rivers meet the sea, are among the most productive ecosystems on the planet. From tidal wetlands to the continental shelf, estuaries provide essential habitat for thousands of species of fish, birds, plants, and animals, including many threatened and endangered species (USFWS 1995).

Nearly 200 million Americans—approximately 70% of the population—visit estuaries and coastal areas every year for vacations, recreation, sport, or sightseeing. And for the 120 million Americans who live near estuaries, they are linchpins in people's quality of life: for their scenic beauty, recreational opportunities, abundance of life and for their mere presence (https://oceanservice.noaa.gov/facts/population.html). Estuaries also improve water quality by filtering pollutants, and they can act as storm buffers and reduce flooding.

Estuaries provide critical habitat for 75% of America's commercial fish catch and 80%–90% of the recreational fish catch (Lellis-Dibble et al. 2008). Healthy coasts and estuaries are essential for protecting more than $800 billion of trade each year, tens of billions of dollars in recreational opportunities annually, and more than 45% of the nation's petroleum refining capacity. Coasts and estuary regions support a disproportionately large share of the nation's economic output and population. And while research shows that environmental damage to coasts and estuaries decreases these values, promoting environmental protection and expanding habitat restoration efforts are likely to increase these values substantially (https://estuaries.org/images/stories/docs/policy-legislation/executive-summary-final.pdf).

However, the United States is losing coastal wetlands faster than it is protecting and restoring them. Between 2004 and 2009, the United States lost an average 80,000 acres of estuarine wetlands, indicating we are still losing wetlands at an unsustainable rate (USFWS 2009). As part of its blue carbon initiative, RAE reviewed tidal wetland restoration progress that is reported by the 28 National Estuary Programs. The analysis showed that between 2009 and 2012, restoration activities occurred in just 27,835 acres of tidal wetlands, out of more than 600,000 acres targeted for restoration (VCS 2014). Unfortunately, a comprehensive and consistent national dataset that tracks restoration carried out by government agencies, as well as other partners, does not exist. Such a database would provide important information to coastal managers and policy makers toward reversing the trend of ongoing habitat losses. The key limiting factor to increasing the level of habitat restoration nationally, and globally, is a shortage of funding, and the funding that is available is subject to political and economic swings, and it cannot be relied upon year after year.

2.3 INTERFACES OF COASTAL WETLANDS AND CLIMATE CHANGE

Blue carbon is a concept that increases the nation's focus on coastal and estuarine habitat restoration and protection. It calls out two distinct marketable and manageable ecosystem services—natural carbon storage and sequestration—that are recognized by voluntary markets as actions that help meet global and national greenhouse gas (GHG) reduction goals. Among RAE's goals for its national blue carbon program is to provide access to funding mechanisms for ongoing climate mitigation and adaptation, quantifying the carbon services provided by these ecosystems, while also supporting all of the other important ecosystem services provided by coastal wetlands.

Our coasts are where the impacts of climate change may be most apparent and recognizable, with increased rates of sea level rise, stronger storms and wave intensity, and changes in precipitation that alter freshwater inputs. Other climate change effects impacting coastal wetlands and estuaries include increased temperatures and higher concentrations of CO_2 (Needelman et al. 2012). At the same time, the coasts are places of hope and actions which protect and restore coastal ecosystems and the surrounding communities, while mitigating climate impacts through carbon sequestration and reduced emissions of CO_2 and other GHGs.

In coordination with the launch of its blue carbon initiative in 2009, RAE began a study to demonstrate the opportunities created by the interconnectedness of coastal habitat restoration, adaptation, and mitigation strategies related to reducing climate change impacts. This study (Needelman et al. 2012) found that coastal communities, climate change, and coastal wetlands are intertwined, and it recommended an integrated approach to managing coastal habitats, providing benefits to coastal communities and their surrounding wetland habitats while addressing climate change through GHG reduction.

Estuaries and coastal wetlands exist at a multitude of interfaces, physically (land/ocean), chemically (fresh/saline water), and also strategically at intersections of history (past/present), climate change responses (adaptation/mitigation), community (people/nature), and management (conservation/restoration).

2.3.1 Climate Change

Restoring wetlands and conserving existing wetlands leads to GHG climate benefits through increased sequestration and prevention of lost carbon stocks. It is the buildup of this organic material (i.e., soil carbon) that allows healthy coastal ecosystems to accrete over time, potentially keeping pace with sea level rise. Increased rates of sea level rise demand that sustainable wetland conservation efforts include adaptation strategies, which could include options such as land acquisition and barrier removal to allow landward migration of coastal wetlands; enhancing a system's ability to sequester carbon (e.g., through water management or removal of invasive species); or providing natural barriers for protecting restored or vulnerable wetland areas (e.g., building oyster reefs to support living shorelines). In this way, recognizing, monitoring, and managing blue carbon ecosystem services can be an added tool for measuring ecosystem resiliency. The potential market value of blue carbon could also be a potential tool to help finance adaptation strategies that utilize both restoration and conservation strategies.

2.3.2 Management

Human infrastructure and growing communities impact much of the natural world, and there are few 'natural' coastal wetlands remaining in the United States. Protecting remaining wetlands is of vital importance. It is also incumbent to plan for the long-term protection of areas being restored to ensure their sustainability. In some coastal wetland environments, such as coastal Louisiana, habitat restoration and conservation are accomplished through the same actions. For example, restoring fringe marshes can provide increased protection for existing marsh and upland habitat, both protecting existing carbon stores and increasing the rate at which carbon is sequestered and stored. There are abundant opportunities for large-scale restoration and conservation throughout the United States, and globally, which would bring substantial blue carbon benefits as well.

2.3.3 Physical and Chemical

The mixing of river water with the ocean leads to an abundance of life. Estuaries are among the most productive ecosystems on the planet, and it is this productivity which also results in high rates of carbon sequestration. Actions which protect and restore estuarine wetlands will increase carbon sequestration, and actions which increase carbon sequestration in estuaries will in turn create and ensure healthier and more productive wetland habitat. Managing for blue carbon and managing for wetland restoration and conservation go hand-in-hand.

2.3.4 Community

The productivity of our coasts and estuaries also makes them attractive places to live, with more than 40% of the nation's population living in coastal communities. Humans bring increased development and pollution pressures to our coastal habitats. Managing the intersection of people and nature successfully is important. Public awareness of the important ecosystem services that our coasts and estuaries provide is critically important, and blue carbon can be a way to bring new and greater attention to them (Resource Media, 2016).

2.3.5 Policy and Action

Across the country and internationally, actions and policies aimed at climate mitigation and adaption will be felt in the coastal environment and can lead to benefits at the local-level. For example, actions that restore and protect coastal wetlands increase carbon capture and storage, thus mitigating the underlying cause of climate change; these same actions also lead to increased resiliency for coastal communities, thus also addressing impacts of climate change such as increased events of storms and flooding. Additionally, the 2013 IPCC Supplemental Guidance for Wetlands was released to encourage nations to include coastal wetlands in national accounting approaches, which can empower nations to improve management of coastal ecosystems for their carbon values.

2.4 GETTING STRATEGIC ABOUT BLUE CARBON AND ACCESS TO THE CARBON MARKETS

Our approach to gaining access to the carbon market, and therefore to new financial resources for restoration and protection projects, has involved establishing eligibility, developing market tools, and initiating pilot projects. In parallel, regular outreach and education have played a vital role in building awareness, capacity, momentum, and support for advancing blue carbon understanding and application.

In 2008, noting the significant GHG offsets potential associated with tidal wetlands projects, the Climate Action Reserve (formerly the California Climate Action Registry) commissioned an issues paper to assess the feasibility of developing a GHG offset protocol for tidal wetlands projects. GHG offset protocols (also called methodologies) provide the necessary guidance to calculate, report, and verify GHG emission reductions associated with offset projects. A wetlands protocol would provide a reliable framework for implementing tidal wetlands restoration and management projects designed to create offset credits that could be recognized as creditable under existing carbon standards and emerging state, regional, and federal climate change laws and regulations.

One of the most intriguing aspects of blue carbon in the late 2000s was the potential to tap into new and substantial sources of funding for carbon offsets, globally and nationally. As previously noted, coastal wetland restoration and conservation are significantly under-funded, despite the myriad of essential environmental, social, and economic services and benefits coastal wetlands provide. The Kerry–Lieberman climate legislation introduced in the U.S. Senate in May 2010 specifically identified "projects to restore or prevent the conversion, loss, or degradation of vegetated marine coastal habitats" (U.S. Senate 2013) as eligible climate offset project types. If this bill had become law, wetland restoration and protection could have received substantial new funding through the sale of carbon offsets under a national compliance market. In the absence of a regulatory approach to climate mitigation, several carbon standards exist in a global volunteer carbon market.

Carbon markets refer to the markets created through the trading of carbon emission offsets and allowances which encourage or help countries and companies limit their GHG emissions. Carbon markets can be regulatory—where a government sets limits on GHG emissions, or voluntary— where companies and others are not required to limit GHG emissions but choose to do so. More than 1 billion metric tons of CO_2e (carbon dioxide equivalents) have been transacted in the voluntary carbon markets since 2000 (Ecosystem Marketplace 2017).

To develop the blue carbon science, practices and policies necessary for coastal wetlands to enter the carbon markets requires access to expertise in a variety of fields. As a convening organization, RAE identified the need for expertise in restoration planning, design, and practice; coastal management; climate finance; wetland GHG science; carbon standards; coastal policy; and community education and advocacy.

RAE's goal was to bring together experts in complex fields to tackle the problem of creating a GHG offsets protocol for one of the more complex and dynamic ecosystems—tidal wetlands, including seagrasses. Establishing a new sector of carbon offsets requires using the best available science to create transparent, credible, and verifiable requirements and procedures.

In 2010, RAE convened the National Blue Ribbon Panel on the Development of a Greenhouse Gas Offset Protocol for Tidal Wetlands Restoration and Management to tackle these questions. The

Panel found that blue carbon presented a significant opportunity to improve wetlands management to achieve GHG reductions and that the methods to quantify and monitor GHGs in coastal wetlands are achievable with then-existing science. The recommendations of the Panel have guided RAE's blue carbon initiative since (Findings of the National Blue Ribbon Panel on the Development of a Greenhouse Gas Offset Protocol for Tidal Wetlands Restoration and Management: Action Plan to guide protocol development. Restore America's Estuaries. August 2010).

Since 2010, RAE has guided partners and experts to enable tidal wetland (including seagrass) restoration and protection actions to receive carbon offsets. This effort followed three paths: **(i) establishing eligibility, (ii) developing market tools, and (iii) initiating pilot projects**.

2.4.1 Eligibility

In 2010, no carbon standard recognized the GHG benefits of protecting and restoring coastal wetlands and seagrass beds. Coordinated by RAE, a technical team worked with the Verified Carbon Standard—the lead issuer of carbon offsets in the voluntary market and the land-use sector—to develop new wetland requirements. The wetland requirements, adopted in 2012 as part of the VCS AFOLU (Agriculture, Forestry and Other Land Uses) Requirements (VCS, 2016), are higher-level requirements governing wetland carbon methodologies and projects. (For more information on carbon offsets, see Chapter 20.)

As a national organization working to protect and restore all coastal wetlands through a variety of restoration and conservation approaches, RAE sought to include as many activities as possible leading to a net GHG benefit in the development of these market tools. The primary activities that provide GHG benefits and are now eligible for carbon offsets are: avoided wetland loss, wetland restoration, wetland management, and wetland creation.

A second question of eligibility relates to the concept of additionality. The purpose of carbon markets is to incentivize new activities that provide GHG benefits that otherwise might not have occurred. This concept is called 'additionality,' and it is required of all carbon offset projects by carbon standards. However, it is difficult to adapt this concept to land-use carbon offset projects such as wetland restoration. A simple case, in which carbon offsets would not be issued, is when a wetland is restored as part of a mitigation requirement. More challenging are individual wetland projects, funded by state and federal grants, which regularly occur throughout the United States and are not required as mitigation or by other laws. RAE's belief was, and remains, that because the need for coastal habitat restoration is so much greater than the funding being provided, any new funding provided by the carbon markets is an important incentive to supporting new projects. The need for restoration and conservation funding is so great that all projects not otherwise required should be considered additional. RAE's inclusive approach to additionality serves the entire coastal community and leads to wetland projects with GHG benefits. This was a critical component of RAE's blue carbon strategy.

Eligible activities that can result in GHG reductions include those that create new wetlands, restore drained or degraded wetlands, and/or protect/avoid the loss of existing wetlands. Below is a list of restoration and management activities that can result in a net GHG benefit:

- Restoring or managing hydrological conditions—e.g., removing tidal barriers, improving hydrological connectivity, restoring tidal flow, lowering water levels on impounded wetlands
- Altering sediment supply—e.g., beneficial use of dredge material, thin-layer spraying, diverting river sediments to sediment-starved areas
- Changing salinity characteristics—e.g., restoring tidal flow to tidally restricted areas
- Improving water quality—e.g., reducing nutrient loads to improve water clarity for seagrass meadows, recovering tidal and other hydrologic flushing and exchange, reducing nutrient residence time
- Reintroducing native plant communities through re-seeding and/or re-planting
- Improved management practices—e.g., removing invasive species, reduced grazing

Priority projects, which can result in higher GHG benefits, include restoration of drained wetlands to reduce emissions, restoring salinity to impounded wetlands to reduce methane emissions, and avoiding conversion of intact wetlands to protect existing carbon stocks.

2.4.2 Market Tools

Carbon methodologies provide project-level criteria and requirements for specific activities. RAE focused on developing two carbon offset methodologies for activities in tidal wetlands: restoration and conservation. Several technical, scientific, and market experts contributed substantially to these two methodologies.

In 2014, the VCS approved VM0033 Methodology for Tidal Wetland and Seagrass Restoration (VCS, 2014), which provided the first comprehensive global GHG offsets methodology for a broad range of restoration activities in tidal wetlands and seagrass beds. This was a significant accomplishment that creates new opportunities for coastal managers through potential carbon offset funding streams. Several locations in the U.S. and globally are considering using this methodology to support restoration and ongoing maintenance and adaptive management.

In 2017, RAE and Silvestrum submitted new methodology modules to the VCS for an existing land-use methodology, which would allow carbon offsets for tidal wetland and seagrass conservation activities, globally. This methodology is expected to achieve final approval from the VCS in early 2018. It also encompasses a broad range of conservation activities and will be eligible for use in all tidal wetland and seagrass habitats worldwide. Many global conservation partners are planning to apply it as soon as it is approved.

2.4.3 Pilot or Regional Case Studies

The National Blue Ribbon Panel recommended regional case studies to identify blue carbon potential in different ecological settings. Several such assessments have been undertaken and completed. In Puget Sound, the Coastal Blue Carbon Opportunity Assessment for the Snohomish Estuary (Crooks et al. 2014) identified the potential GHG benefits of restoring tidal wetlands in this sediment-rich west coast estuary and demonstrated a clear GHG benefit of estuary restoration.

The Tampa Bay Blue Carbon Study (Crooks et al. 2014) assessed the climate mitigation potential of Tampa coastal habitat over the next 100 years and how sea level rise will impact these habitats. The report also provides management recommendations for habitat adaptation (see Chapter 26).

In Cape Cod, the Bringing Wetlands to Market Project is working to develop and validate a model for estimating CO_2 and CH_4 fluxes in salt marsh. RAE is a project partner in this endeavor, working with partners to pilot market concepts for a large planned restoration project at Herring River in Cape Cod National Seashore (see Chapter 25).

In Galveston Bay, (Crooks et al. 2014) RAE led a project with local member group, Galveston Bay Foundation and the Texas A&M University-Galveston lab to compare blue carbon storage and accumulation rates in natural and restored marsh habitat. This data is being incorporated with other ecosystem service values to provide a full-suite view of valuations for the Galveston Bay estuary.

Carbon project feasibility studies are underway or have been conducted in several locations in the United States. These studies lay the ground work for understanding the feasibility of developing habitat restoration projects as carbon offset projects, providing overview of the science and other needs to meeting carbon offset criteria and requirements. One of the most promising is the Herring River Restoration site at Cape Cod National Seashore. The approximately 1,000 acre tidal wetland restoration project will restore salt water tidal flows to restricted areas and is expected to both reduce methane emissions and increase carbon sequestration (see Chapter 25).

Virtually all tidal wetland and seagrass habitat restoration and protection activities are covered by one or both of the new methodologies. This mirrors the RAE mission for both the protection and restoration of estuaries, and will help bring about increased recognition of these ecosystems for their climate values, among other ecosystem services.

Development of the methodologies is also driving new work in modeling GHGs in coastal wetlands and scientific field studies, building capacity for the use and implementation of these methodologies at the project level.

2.4.3.1 Blue Carbon as a Tool for Communication and Education

One of the most important roles that RAE has played is educating stakeholders nationwide, including national programs and agency leadership. Blue carbon provides a new way to communicate the importance of coastal wetlands and can open new doors for partners, collaboration, and support.

Workshops, conferences, and meetings and briefings are all part of RAE's approach to raising awareness of the importance of estuaries to our nation. Blue carbon provides a new opportunity to do this in the context of climate change.

2.5 SUMMARY

Since 2010, recognition of blue carbon as an important ecosystem service provided by seagrasses and tidal wetlands has greatly increased. Correspondingly, support has grown for estuaries as critical ecosystems, especially in the context of climate change mitigation and adaptation. RAE continues to lead the national effort in raising awareness and building capacity for national, state, and local partners to understand this ecosystem service and apply it in various approaches to support restoration and conservation efforts. As funding continues to be a limiting factor to changing the course of habitat loss in the United States and globally, blue carbon offers a new and innovative approach to communicate, measure, and market the benefits provided by coastal wetlands. We look forward to operationalizing blue carbon to help achieve important estuary habitat restoration and conservation goals.

CHAPTER 3

Human Impacts on Blue Carbon Ecosystems

Catherine E. Lovelock
The University of Queensland

Daniel A. Friess
National University of Singapore

J. Boone Kauffman
Oregon State University

James W. Fourqurean
Florida International University

CONTENTS

HIGHLIGHTS

1. Coastal blue carbon ecosystems are often associated with deltas and estuaries; they are highly productive and thus have been sites of intense human activities for thousands of years.
2. Degradation of blue carbon ecosystems has been caused by impoundment, conversion to alternative land uses, over exploitation, and pollution, all of which contribute to CO_2 emissions and reduce carbon sequestration and other ecosystem services offered by blue carbon ecosystems.
3. Emerging threats include damming rivers, reducing sediment supply and increasing nutrient availability, which are likely to exacerbate the effects of climate change and sea-level rise on blue carbon ecosystems.
4. Land-use planning for sustaining blue carbon ecosystems with sea level rise has the potential to increase the area of blue carbon ecosystems and global carbon sequestration.

3.1 INTRODUCTION

Coastal blue carbon ecosystems occur mainly in estuaries and deltas, which is where 20% of the human population of the planet live at densities that are three-fold that in inland areas (Small and Nicholls 2003). This confluence of human settlements and estuaries, deltas and embayments in which blue carbon ecosystems are abundant is no coincidence, as blue carbon ecosystems provide a wide range of services including the provision of food, fuel, fisheries, nutrient processing, and coastal protection (UNEP 2014). However, the colocation of human settlements and blue carbon ecosystems has resulted in losses and degradation of these ecosystems as human settlements have expanded, watersheds have been transformed, economies have become increasingly complex and blue carbon ecosystems have been over-exploited or converted to alternative land-uses (Pendleton et al. 2012).

Human activities on coasts and coastal watersheds have increased with population growth and associated increasing food demand, demands for fuel and timber, increasing urbanization, the extensive use of fertilizers, and extraction of water resources; all of which have led to loss, conversion, and degradation of mangroves, seagrass, and tidal marsh habitats and associated loss of ecosystem services. Emissions of CO_2 after human disturbances of blue carbon ecosystems occur as biomass is removed and often burned and organic matter in soils is oxidized (Table 3.1, Pendleton et al. 2012). Many disturbances enhance the decomposition of the large organic matter stocks held within soils of blue carbon ecosystems, leading to high levels of CO_2 emissions (Figure 3.1), (Mount and Twiss 2005; Marba et al. 2015; Kauffman et al. 2014, 2017; Lovelock et al. 2017). Human activities that result in the loss and degradation of blue carbon ecosystems also result in the loss of carbon sequestration potential as they are often replaced by systems that do not sequester as much carbon as blue carbon ecosystems (Duarte et al. 2013), and are often net sources of CO_2, methane, and nitrous oxide to the atmosphere (Figure 3.2) (Hu et al. 2013; Webb et al. 2016).

The major impacts and timing of human activities on coastal blue carbon ecosystems differ among blue carbon ecosystems and regionally, which has implications for regional strategies for the conservation and restoration of blue carbon stocks. Human activities can have direct effects on blue carbon ecosystems, for example, clearing and conversion of habitat for alternative uses (e.g., conversion to aquaculture, Kauffman et al. 2014, 2017), or indirect, where activities on the land (e.g., fertilizer use, changing hydrology) or in the sea (e.g., overfishing) can have negative effects on blue carbon ecosystems, their carbon stocks and carbon sequestration potential (Deegan et al. 2012; Atwood et al. 2015). While the effects of human impacts are evident on all coastal wetland types, in this chapter we focus on mangrove, seagrass, and tidal marshes (see Chapter 1).

Table 3.1 Global Cover of Blue Carbon Ecosystems, Their Rates of Conversion to Alternative Land-Uses, Estimated CO_2 Emissions due to Human Activities (see Figures 3.1 and 3.2) and Their Estimated Cost

Ecosystem	Global Extent (Mha)	Conversion Rate (% year^{-1})	Organic Carbon in Biomass and the Top Meter of Sediment (Mg CO_2 ha^{-1})	Carbon Emissions (Pg CO_2 year^{-1})	Estimated Cost (Billion US$ year^{-1})
Tidal marsh	2.2–40 (5.1)	1.0–2.0 (1.5)	237–949 (593)	0.02–0.24 (0.06)	0.64–9.7 (2.6)
Mangrove	13.8–15.2 (14.5)	0.7–3.0 (1.9)	373–1,492 (933)	0.09–0.45 (0.24)	3.6–18.5 (9.8)
Seagrass	17.7–60 (30.0)	0.4–2.6 (2.5)	131–522 (326)	0.05–0.33 (0.15)	1.9–13.7 (6.1)
TOTAL	33.7–115.2 (48.9)			0.15–1.02 (0.45)	6.1–41.9 (18.5)

Source: Data are from Pendleton et al. (2012).
Values are minimums and maximums with the central estimate in parentheses.

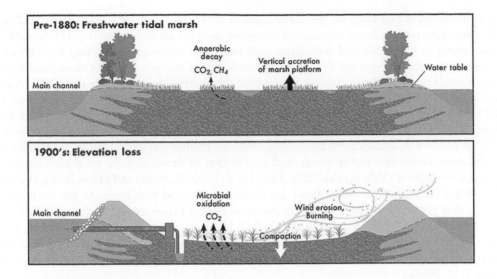

Figure 3.1   Consequences for the soil carbon stock (original condition, upper panel; altered condition, lower panel) when blue carbon ecosystems are impounded, drained, and converted to agricultural lands. (Source: Mount and Twiss (2005) http://escholarship.org/uc/item/4k44725p, originally from Ingebritsen et al. (2000).)

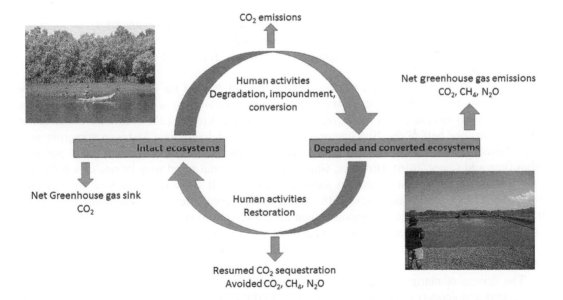

Figure 3.2   **(See color insert following page 266.)** Representation of the effects of human activities on greenhouse gas emissions with degradation, impoundment, and conversion of blue carbon ecosystems, and their restoration.

3.2 MANGROVES

Mangrove ecosystems are distributed over 121 countries along tropical to subtropical coastal zones across the world (Giri et al. 2011; Adame et al. Chapter 27; Giri et al. Chapter 13; Kairo et al. Chapter 24; Murdiyarso Chapter 21). Large-scale human impacts on mangrove forests have been relatively recent, though mangroves were overexploited historically such as the use of mangrove timber to construct fleets of boats and for the construction of buildings on the east coast of Africa (Curtin 1981). Numerous other colonial accounts of mangrove resource extraction also exist (Friess 2016).

Beginning in the 1970s and continuing until today, the expansion of shrimp aquaculture in response to demand for seafood led to widespread losses of mangrove forests throughout central America and Asia as ponds were constructed within the intertidal zone (Valiela et al. 2001; Alongi et al. 2002; Hamilton and Lovett 2015). Many ponds were unsustainable due to low-water quality and disease within the ponds, leading to the need to clear more area to maintain production. In some nations, 80% of mangrove forests have been converted to other uses (e.g., Philippines, Primavera 2005). In addition to aquaculture, agriculture, charcoal production, urbanization, construction of infrastructure, conversion to salt ponds, and exploitation of firewood have been associated with losses in some regions (Valiela et al. 2001; Alongi et al. 2002; Richards and Friess 2016). Diversion and modification of upstream river flows due to dams associated with hydroelectric development and irrigation projects have had global influences on coastal ecosystems (Cloern et al. 2016). For example, in the Chao Phraya Delta in the Gulf of Thailand, reduced sediment accretion within mangrove forests with upstream damming of rivers and low-sediment availability has resulted in losses of forest area as coastlines have retreated under the influence of storms and sea level rise (Goisan et al. 2014).

Indirect negative effects on mangroves due to modifications in hydrology (Jimenez et al. 1985), eutrophication (Lovelock et al. 2009), and pollution (Bayen 2012) have been documented or anticipated. Often indirect effects contribute to vulnerability to extreme events, like intense storms. For example, Feller et al. (2016) showed that fertilized trees had greater susceptibility to hurricanes in Florida, and Lovelock et al. (2009) found that fertilized trees had higher mortality with drought. Human activities that lower the productivity of blue carbon ecosystems are likely to enhanced vulnerability to climate change. For example, reduced root production is likely to lead to vulnerability to sea level rise in forests where roots contribute more than sediment to soil elevation gains (McKee et al. 2011).

Human activities have driven the loss of an estimated 30% of the original global mangrove extent (Alongi 2002), although there is a high level of uncertainty around regional trends (Friess and Webb 2014) and the extent of original mangrove cover. Whatever the exact rate of mangrove loss in the second half of the 20th century, it is clear that global annual losses in mangrove area have slowed in the first decade of the 21st century, averaging at approximately 0.2% per year (Hamilton and Casey 2016). However, while the overall global trend may be one of a reduction in deforestation rates, some countries such as Myanmar still show rapid rates of loss between 2000 and 2012 (Richards and Friess 2016). While in more recent decades aquaculture remains an important threat to mangroves, impoundment for conversion to rice agriculture, cattle pastures, and oil palm have been reported as increasing in impact (Kauffman et al. 2015; Richards and Friess 2016), with losses of soil carbon and other ecosystem services observed (Kauffman et al. 2015; Figure 1).

The drivers of mangrove deforestation vary regionally and often are associated with complex social and economic factors (e.g., Primavera 2006). A focus on enhancing ecosystem services from mangroves and the livelihoods of communities associated with mangrove forests is a recent development (Bosire et al. 2014; Brown et al. 2014) in which blue carbon has a role (e.g., Thompson et al. 2014; Wylie et al. 2016). However, analysis of the CO_2 emissions of shrimp and beef from deforested mangrove lands suggests that a focus on consumers of food sourced from deforested mangrove lands is also an important component of reducing human impacts on mangrove forests (Kauffman et al. 2017). They report that production of one kilogram of beef produces 1,440 kg of CO_2e while a kilogram of shrimp produced 1,603 kg of CO_2e when grown on converted mangrove habitats, highlighting the significant CO_2 emissions associated with these land-use conversions.

3.3 SEAGRASS MEADOWS

Seagrass meadows are suffering huge losses of area largely due to reductions in water quality (Short and Wyllie-Echeverria 1986; Waycott et al. 2009), although local direct effects of boating, dredging, altered hydrology and disease have been important in some locations (Robblee et al. 1991; Erftemelier and Lewis 2006; Serrano et al. 2016). Low-water quality affects: (i) the light available to plants growing on the seafloor, which reduces productivity and eventually results in mortality and (ii) enhances algal growth, which competes with seagrass for light (Duarte et al. 1995; Sherwood et al. Chapter 26; Oreska et al. Chapter 12). In some cases, local extinctions and complete loss of habitat have occurred (van Katwijk et al. 2009). Reductions in water quality have been due to enhanced levels of fine sediment and nutrients delivered to the coast associated with clearing of land in the watershed for agriculture and grazing, which has occurred in the last century in many regions (Morelli et al. 2012) but earlier in Europe and elsewhere (Lopez-Merino et al. 2017).

More recently the intensification and expansion of agriculture has also been an important driver of declining water quality, particularly after the 1940s, when nitrogen fertilizer (from the industrialized Haber Bosch process) became abundant and high loads of nitrogen (urea) were applied to agricultural fields subsequently enriching the coastal oceans (Erisman et al. 2012; Cloern et al. 2001) which has resulted in losses of seagrass globally (Waycott et al. 2009). Losses of tidal marshes and mangroves in these areas are also likely to have increased the negative impacts of nutrient enrichment on seagrass (Valiela and Cole 2002; Jickells et al. 2016). Although reducing the flow of nutrients and other pollutants into the coastal oceans is the target of policies of many countries (e.g., European Union Foden and Brazier 2007; Chesapeake Bay Dennison et al. 1993), many of which have been successful (e.g., Tampa Bay, Greening and Janicke 2006; Greening et al. 2014), much of the fine sediment delivered to the coasts in the past centuries is re-suspended with high levels of wave energy leading to chronically low-water quality in some sites (e.g., Fabricius et al. 2013) which continues to have negative impacts on seagrass meadows (Collier et al. 2016).

Direct conversion of seagrass beds through dredging and filling is rarer than loss of seagrasses as a result of decreased water clarity, but port dredging and filling for development (land reclamation) locally affects seagrasses. The cumulative impacts of unintentional disturbance of seagrasses through accidental groundings or damage from fishing gear can be locally important drivers of seagrass loss as well (Dawes et al. 1997). Further, beach resort developments in areas that support seagrasses have been known to actively remove seagrasses in an attempt to improve the experience of beachgoers (Daby 2003). Of the three blue carbon ecosystems seagrass meadows have the most uncertainty in their current distribution and extent as they are poorly mapped because they occur underwater. As such, they are difficult to map with remote-sensing techniques which have difficulty penetrating water or differentiating seagrass from micro- and macroalgal communities, and mapping by direct observation by divers is expensive and logistically impractical in many regions. As reviewed in Chapter 12, compilations of mapping data worldwide document an extent of mapped area of ca. 175,000 km² (Green and Short 2003); however, that extent is surely an underestimate and the true global extent of seagrasses is likely between 300,000 and 600,000 km² (Duarte et al. 2010). As a result, it is likely that the true extent of seagrasses is underestimated. There are many unknowns about the impact of human activities on the carbon stores of seagrass meadows because besides a few well-studied sites (e.g., Moreton Bay; Shark Bay, Florida Bay; North Sea) there is a lack of data of the distribution and carbon stocks and how these have changed through time with human activities.

3.4 TIDAL MARSHES

Tidal marshes are the dominant blue carbon ecosystem over much of the temperate zone and polar coastal regions of the world but also occur in the high intertidal zone in the tropics. They have been widely used and converted to alternative land-uses for hundreds to thousands of years in the temperate zone (Gedan et al. 2009; Kroeger et al. Chapter 25). Because conversion of tidal marshes occurred so early in the history of human exploitation of the coastal zone, before global mapping, it is difficult to assess the pre-cleared area of tidal marshes and thus the total global amount of marsh converted by human activities is difficult to determine (Gedan et al. 2009). However, in a study of 12 temperate and subtropical estuaries, Lotze et al. (2006) found a 67% loss of coastal wetlands during human history.

Alternative land-uses for tidal marsh lands includes agriculture, grazing lands, urban and industrial development, and mining soils for peat and salt (Gedan et al. 2009; Drexler et al. Chapter 23). Conversion to agriculture, where marshes have been drained and isolated from tidal flows, utilized their often deep, organic matter-rich soils because they supported high levels of plant production with few inputs. This activity stimulates subsidence of the land as organic matter in soils is oxidized (Connor et al. 2001; Turner 2004, Byrd et al. this book, Figure 1). They have been widely filled and reclaimed for development associated with human settlements and have been isolated from tidal flows for conversion to grazing lands (e.g., Ganong 1903). Many of the alternative land-uses for tidal marshes result in losses of carbon sequestration potential, nitrogen removal (Jickells et al. 2016), as well as giving rise to enhanced emissions of CO_2, methane, and nitrous oxides (Deng et al. 2016).

In addition to conversions of tidal marshes to other uses, nutrient enrichment from agricultural pollutants from upstream sources has been significant factor in tidal marsh degradation (Verhoeven et al. 2006; Tuner et al. 2009; Deegan et al. 2012). High levels of nutrients reduce allocation of biomass to roots, making soils less stable and thus more vulnerable to disturbance, erosion, and export of particulate and dissolved carbon (Deegan et al. 2012). The losses of predators of bioturbating crabs, from overfishing, have also been suggested to lead to loss of tidal marsh and significant CO_2 emissions as high levels of bioturbation has facilitated erosion of the marsh edges (Coverdale et al. 2013). Similar to mangrove forests, losses of tidal marshes, for example, within the Mississippi delta, has occurred as a result of reduced sediment inputs from dammed and modified river systems thus diminishing rate of sediment accretion. This makes them more susceptible to submergence and erosion during storms as well as increasing their vulnerability to sea level rise (Day et al. 2011; Mudd 2011). Over many continents, sea level rise in addition to other factors including human activities are contributing to encroachment of mangroves into tidal marshes (Saintilan et al. 2014; Amitage et al. 2015) with consequences for carbon storage (Doughty et al. 2016).

3.5 EMERGING THREATS

While the clearing and conversion of mangroves has slowed in the first part of the 21st century in some areas (e.g., Hamilton and Casey 2016), other human activities are emerging as threats. For mangroves, clearing for oil palm cultivation in Southeast Asia has recently emerged on the research and conservation agenda (Richards and Friess 2016), and is expected to increase substantially in the future as governments set ambitious targets for economic growth and food security. For example, Indonesia has already designated 118,000 ha of mangrove for oil palm concessions (Ilman et al. 2016), many of which will be converted in the future. While the oil palm threat to mangroves has not been systematically assessed in other regions, we would assume that oil palm and the production of other commodities, such as lime, are emerging (Scales et al. 2017).

Blue carbon habitats are affected by human impacts such as land conversion, interactions between human activities and climate change factors. Given the sub- and intertidal nature of blue carbon ecosystems, they are strongly influenced by climate change, experiencing the effects of changes in the oceans, e.g., sea level rise, rising sea surface temperatures as well those on the land, e.g., changes in rainfall, river flows, sediment delivery, air temperature. While the productivity of some species is likely to have been directly enhanced by elevated levels of atmospheric CO_2 (Campbell and Fourqurean 2013; Reef et al. 2015) many climate change factors are likely to interact with human modifications of the land and seascape to have adverse effects on blue carbon ecosystems.

The need for freshwater to support agriculture and human settlements has led to damming of rivers that reduce freshwater and sediment supply to the coast and thus to coastal wetlands (Milliman and Farnsworth 2011). Globally, the building of dams is increasing (Zarfi et al. 2015). While reductions in river flows resulted in salinization of soils and changes in species distributions (Colonnello and Medina 1998), few adverse effects of low-sediment supply were observed. However, evidence for negative effects of these catchment level modifications are becoming more evident for blue carbon ecosystems as rising sea levels and enhanced storm activity result in losses of habitat and erosion of soils (e.g., Leonardi et al. 2016). For seagrass, catchment modifications can lead to hyper-saline conditions which can be devastating (e.g., Florida Bay, Hall et al. 2016). Reduced rainfall with climate change, which is predicted for many regions of the globe, may exacerbate the effects of reduced freshwater riverine inputs, as well as sediment supply, by further suppressing productivity of plants.

A number of stressors on blue carbon also interact with each other, with potentially additive or synergistic effects. For seagrass ecosystems, thermal anomalies are another emerging threat (Diaz-Almela et al. 2007; Nowicki et al. 2017) for which negative consequences maybe increased with human activities such as eutrophication (Unsworth et al. 2015). Similarly, mangroves in the Indo-Pacific are threatened by sea level rise, though their resilience is reduced further due to declining sediment inputs associated with damming of rivers (Lovelock et al. 2015). Low-water quality from eutrophication and legacy sediments remains a major persistent threat to seagrass ecosystems and a challenge to restoration in many locations (van Katwijk et al. 2016).

3.6 SOLUTIONS AND OPPORTUNITIES

While policies based on climate change mitigation are covered elsewhere in this volume (Chapters 15–19), here we focus on human activities that contribute to conserving blue carbon ecosystems as well as contribute to their restoration and maintenance.

Robust and long-term monitoring is urgently required if we are to accurately assess the contribution of habitat loss to blue carbon-related emissions. Currently, we do not have a clear idea of the global distribution of tidal marshes and seagrass. While recent efforts have been made to estimate global tidal marsh extent (McOwen et al. 2017), it is based on a literature review as opposed to a standardized remote-sensing survey, and most probably underestimates the extent of tropical marsh. Additionally, we still lack robust estimates of habitat loss for many blue carbon ecosystems, or use out of date information and sources. For example, previous studies of carbon emissions from mangrove deforestation utilized mangrove loss estimates of 1%–3% per year (e.g., Murdiyarso et al. 2015). More recent studies based on new data suggest substantially lower deforestation rates (Hamilton and Casey 2016).

The targeted protection of blue carbon ecosystems within marine protected areas could reduce losses of habitat and the ecosystem services they provide (Spalding et al. 2014; Thompson et al. 2017). More generally, land-use planning for the persistence of blue carbon ecosystems is critical (Enwright et al. 2016) and highly cost effective (Mills et al. 2016; Runting et al. 2016). Models indicate that landward migration of blue carbon ecosystems with sea level rise could see large increases in the area of blue carbon ecosystems (e.g., Mills et al. 2016), however, changes in habitats, e.g., tidal marsh habitats, converted to mangrove habitats, may also occur (Saintilan et al. 2014).

Restoration (Figure 3.2) provides a clear pathway to redressing many of the losses that have occurred in the past (Lewis 2005; van Katwijk et al. 2016) and may become even more desirable over time as sea level rise progresses and older infrastructure that currently limits tidal flooding is decommissioned (Temmerman et al. 2013). There is a growing knowledge of the role of blue carbon ecosystems in coastal protection (Duarte et al. 2013), which continues to stimulate justifications for restoration (Barbier 2015). The knowledge of processes that can improve the condition of degraded ecosystems is also growing (Lewis et al. 2016). For example, restoration of hydrology has been successful on a large scale in Florida (Howard et al. 2016) and experimental addition of sediments has enhanced marsh condition (Stagg and Mendelssohn 2011). There are developing projects that explore how to encourage restoration within modified aquaculture and agricultural landscapes, such that human livelihoods are maintained and enhanced as blue carbon ecosystems are restored (Lewis et al. 2016).

Reducing pollutants and thereby improving water quality is an opportunity for increasing the area and quality of blue carbon ecosystems. Economic incentives and technological developments have been used to reduce nutrient loads reaching the coastal oceans (Conley et al. 2009). Models have indicated that improving water quality can offset the negative effects of sea-level rise on seagrass ecosystems (Saunders et al. 2015) and is also likely to increase carbon sequestration in many blue carbon ecosystems (Macreadie et al. 2017).

3.7 CONCLUSIONS

Human activities have had enormous impacts on blue carbon ecosystems, but many are reversible and opportunities for increasing the area and quality of blue carbon ecosystems and avoiding greenhouse gas emissions from alternative land uses is available. Currently, the ecosystem services that are derived from blue carbon ecosystems are greatly undervalued (UNEP 2014). Climate change poses an overlay on past and current human activities that adds challenges for sustaining and restoring blue carbon ecosystems. Human activities of damming rivers, reducing sediment supply, and increasing nutrient availability are likely to exacerbate the effects of sea level rise on blue carbon ecosystems. However, land-use planning for sustaining blue carbon ecosystems with sea level rise has the potential to increase the area of blue carbon ecosystems and global carbon sequestration.

State of Science

The Fate and Transport of Allochthonous Blue Carbon in Divergent Coastal Systems

Thomas S. Bianchi, Elise Morrison, and Savanna Barry
University of Florida

Ana R. Arellano
University of Florida
University of South Florida

Rusty A. Feagin, Audra Hinson, and Marian Eriksson
Texas A&M University

Mead Allison
Tulane University
The Water Institute of the Gulf

Christopher L. Osburn
North Carolina State University

Diana Oviedo-Vargas
North Carolina State University
Stroud Water Research Center

CONTENTS

HIGHLIGHTS

1. Because coastal blue carbon ecosystems (BCEs) are one of the largest and most threatened global C reserves, it is imperative to understand coastal transport dynamics, latitudinal gradients, biogeochemical factors, and export rates of allochthonous blue carbon.
2. Several processes remain unconstrained in coastal BCE carbon cycling, like OM respiration, priming effects, and carbon exports.
3. Deciphering C cycling in BCEs depends on many complex factors, such as wetland type, physical drivers, latitude, and microbes.
4. Standardization of methodologies is necessary to have reliable comparisons among depths, timescales, and processes within BCE types (i.e., seagrasses, marches, and mangroves).

4.1 INTRODUCTION

4.1.1 Global Carbon Sequestration and Sinks

Blue carbon ecosystems (BCEs), or the carbon (C) stored in coastal systems such as mangroves, salt marshes, and seagrass meadows, form some of the largest global C reserves (Donato et al., 2011; Pendleton et al., 2012; Beaumont et al., 2014; Howard et al., 2017), and are distributed throughout numerous latitudinal zones. Coastal tidal marshes are largely found in temperate zones but do extend into the tropics, mangroves are confined to tropical and sub-tropical regions due to their freeze intolerance, and seagrasses are broadly distributed from cold polar waters to the tropics (Pendleton et al., 2012). Global coastal ecosystems, particularly BCE that store blue C in the form of peat (Figure 4.1), face numerous threats, including accelerated sea level rise (SLR), land-use conversion, and eutrophication (McLeod et al., 2011; Pendleton et al., 2012). Sea level rise can contribute to peat collapse, shifts in biogeochemical cycling, microbial activity, and C fluxes (including greenhouse gases) in coastal systems (Chambers et al., 2013; Chambers et al., 2014). Tropical peatlands below 5 m elevation contain 8.28 ± 1 Gt of C globally and are particularly vulnerable to SLR (Yu et al., 2010) and to land-use conversion that results in catastrophic peat oxidation and land subsidence (Allison et al., 2016; Whittle and Gallego-Sala, 2016). The threats of SLR on coastal peatlands have been well documented (Yu et al., 2010), and although the downstream fate of terrestrial C in marine systems is an active area of study, few studies have investigated processes. For example, processes such as priming effects, may have the potential to drive degradation of exported blue C transported during open-waters in coastal systems (Bianchi, 2012). Given the global extent of BCE, their role as **hotspot** for the sequestration and storage of atmospheric carbon dioxide (CO_2) continues to be explored (Figure 4.1) (Chmura et al., 2003; Laffoley and Grimsditch, 2009; McLeod et al., 2011; Howard et al., 2017).

Carbon sequestration in BCEs is estimated to be as much as 3–50 times greater than that of similar sized rainforests (Bridgham et al., 2006; Nellemann et al., 2009; Breithaupt et al., 2012). The annual sequestration potential of blue C, not accounting for the current pace of coastal land loss, is between 0.9% and 2.6% of total anthropogenic CO_2 emissions (Murray et al., 2011). Within the United States, coastal wetlands contribute ca. 36% of the total sequestration by all wetlands and 18% of the total sequestration of all ecosystems in the conterminous United States (Bridgham et al., 2006). Moreover, there is ca. 5.3 Pg of total global C stock in BCEs—which provides a net C balance (i.e., the sum of existing wetland C sequestration, former wetland oxidation, and C sequestered in plant biomass) of $\sim$42.6 Tg C year^{-1} (tidal and non-tidal) (Bridgham et al., 2006). Globally, mangroves have been estimated to bury $\sim$218 $\pm$ 72 Tg C year^{-1} (Bouillon et al., 2008). Chmura et al. (2003) and Duarte et al. (2005) have found global burial rates of 4.8 ± 0.5 and 87.2 ± 9.6 Tg C year^{-1} for salt marshes, respectively (McLeod et al., 2011).

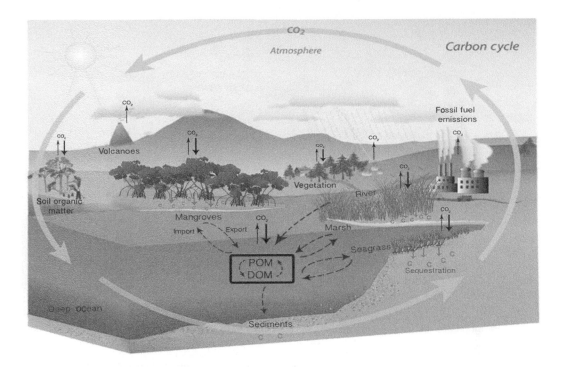

Figure 4.1 **(See color insert following page 266.)** General view of global carbon cycle as it relates to global CO_2 cycling (gray arrows), with blue carbon habitats highlighted with respect to POC and DOC, as modified from from Howard et al. (2017). Carbon dioxide sources (e.g., fossil fuel emissions, fires, volcanoes) and sinks (e.g., ocean, forests, and blue carbon habitats (BCE)) show that in BCEs CO_2 is taken up via photosynthesis (large solid black arrows) and some is respired (smaller solid black arrows). In the case of BCEs, some of this gets sequestered long term (orange arrows) into woody biomass and soils for mangroves and soft tissues and soils for marshes and seagrasses, where it can then be stored in soils/sediments (orange C's). Some of this material in the BCEs can be exchanged as DOC and POC with adjacent deeper shelf waters via import and export processes (dashed arrows), where it ultimately contributes to the global oceanic pool (blue box).

4.1.2 Clarifications and Goals

An issue that has received attention in recent years concerns the use of terms such as **labile** and **recalcitrant** to describe the character of organic carbon (OC) in soil and in aquatic systems (Bianchi, 2011; Schmidt et al., 2011). These terms, which are commonly used to describe differences in the quality of OC as a food resource for heterotrophs, may only be valid in the context of the ambient environment and over the specific timescale of observations (Burdige, 2006). It is increasingly recognized that the reactivity of OC is not a chemically intrinsic property of the material per se, but rather a summation of many ecosystem properties (Schmidt et al., 2011). Therefore, in this review the terms s**table** and **unstable** will be used rather than **recalcitrant** and **labile**, respectively.

Another point of clarification concerns the **novelty**, or lack thereof, of the current blue C research foci from that of older, more classic, dogmatic wetland literature on C export and storage. High burial and export rates of OC in coastal wetland habitats were first documented many years ago (Teal, 1962; Odum and de la Cruz, 1967; Odum, et al., 1973; Valiela, 1995). For example, organic matter exported from coastal wetlands has long been thought to be important for offshore coastal fisheries. More specifically, the **outwelling hypothesis** of Odum (1968) suggested that salt marshes (and rivers) transport biologically available organic matter into near-shore waters, thereby enhancing secondary production in coastal bays and on the shelf (Figure 4.2). While this phenomenon was generally not supported by many of the earlier studies (Teal, 1962; Odum and de la Cruz, 1967;

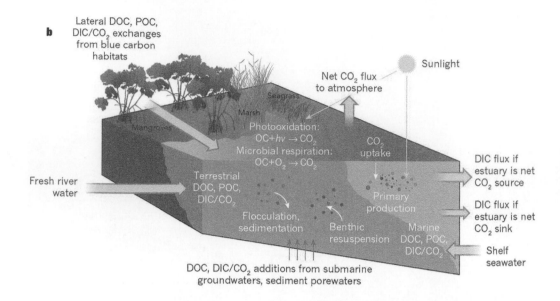

Figure 4.2   Major processes affecting carbon sources and fluxes in estuaries, with a blue-carbon centric emphasis (as modified from Bauer et al., 2013). Many blue carbon habitats are associated with estuaries, which contain a mixture of organic and inorganic carbon sources derived from terrestrial materials, that can be exchanged with the inner shelf. Losses of organic carbon in this coastal exchange can be due to salinity-induced flocculation, sedimentation, microbial respiration, and photooxidation.

Haines, 1977), later work showed (e.g., Moran et al., 1991) that 6%–36% of dissolved organic carbon (DOC) in near-shore waters off the coast of Georgia originates from coastal salt marshes. Other work (Bianchi et al., 1997) demonstrated that a significantly higher fraction of vascular plant materials (both riverine and from coastal marshes) is being transported as particulate organic carbon (POC) further out to the continental shelf and slope (Moran et al., 1991; Trefry et al., 1994; Ward et al., 2017). In tropical regions, previous work also reported significant global export of OC from mangroves (e.g., Alongi, 1998; Dittmar et al., 2006; Alongi, 2014). Having said this, the new research emphasis on BCEs is in fact distinct and timely, simply because of the important new emphasis on the role these communities play in the context of C credits and the natural services they provide in C sequestration and storage in a warming world (Duarte et al., 2013; Siikamäki et al., 2013).

In this chapter, we will discuss the fate of allochthonous POC, either sorbed OC onto clay particles, as detrital organic particles, and DOC in modern coastal BCE. In the seagrass literature, seston, which is allochthonous POC, is usually thought of as phytoplankton and other small organic detritus and sediment that are trapped by seagrass leaves and sedimented out in the bed. The seston can be resuspended and transported out or produced locally and exported with the tide, similarly to the wrack fraction. We will cover a broad spectrum of divergent coastal systems that span across river- and non-river-dominated coastlines.

4.2 IMPACT OF WETLAND TYPE ON COASTAL TRANSPORT DYNAMICS

These BCEs are coastal wetlands and seagrass beds in adjacent estuaries that vary in extent, longevity, and vegetation type with elevation relative to sea level, climate, substrate, exposure to salinity and turbidity, and a host of secondary controls. BCEs are important C sequestration sites for *in situ* autochthonous production and burial, although energetic seagrass settings tend to have lower, but still globally significant C sequestration efficiencies than marsh and mangrove wetlands

(Figure 4.1) (Fourqurean et al., 2012; Howard et al., 2017). Many serve a role as buffers of upland habitats and protect inland infrastructure and populations from storm erosion. One of the key differences between BCEs and non-tidal freshwater wetlands is that sulfate reduction dominates over methanogenesis, thus limiting methane emissions and allowing for greater net sequestration and storage of OC. Furthermore, the greater storage of OC in BCE soil/sediments compared to open ocean sediments is in large part due to the higher primary production rates, greater stability of OC (e.g., lignocellulose in vascular tissues), and lower oxygen exposure time prior to burial (Figure 4.1).

Storm- and tidally-driven inundation of the wetland surface imports and traps allochthonous organic matter and OC particulates that are delivered from adjacent upland areas and from up-current rivers discharging at the land-ocean interface (Allison et al., 2014). Trapping of this allochthonous POC is a complex process that is not only controlled by inundation depth and frequency, but has strong ecogeomorphological control. For instance, trapping efficiency of particulates is strongly controlled by marsh and mangrove stem properties such as density, height, and rigidity (Neph and Vivoni, 2000; Fagherazzi et al., 2004). In seagrass beds, seston adhesion and wrack creation linked to the hydrodynamics, controls the cycling of POC. In general, estuaries serve as temporary (timescales of seconds to decades) storage and mixing points (as well as burying POC themselves) for POC derived from uplands, the shelf, and exported from the wetlands, exchanging with all these environments during energetic hydrodynamic events. Thus, coastal wetlands are sources as well as sinks of blue C, as flooding and wave erosion of the wetland-edge exports wetland dissolved inorganic carbon (DIC), POC, and DOC released from pore waters to adjacent estuaries and the open sea (Bauer et al., 2013) (Figure 4.2).

In the absence of these coastal BCEs, much of the upland POC reaching the land-ocean interface would likely be more rapidly consumed in estuarine and coastal ocean food webs if not exported directly to the deep sea, and thus would not be stored for intervals up to millennia in these lower oxygen and lower energy BCE substrates. The fate of allochthonous blue C released into coastal waters and the open sea or sequestered in the coastal BCEs is highly specific to the coastal margin setting. For example, it depends on a broad spectrum of controlling factors, such as wetland substrate redox conditions, mineral and organic deposition rate, vegetation type and density, and the timing and magnitude of hydrodynamic forcing. The movement of POC and mineral sediment and DOC-laden water is governed by (i) riverine processes that have seasonal hydrographs in large systems, and highly episodic in small; (ii) tidal cycles, which vary from coastal basin-to-basin; and (iii) wind, which drives coastal currents, river plumes, and wave formation. The settings and factors that make the exchange of blue C in coastal basins highly individualistic will form the basis of the subsequent sections of this chapter.

4.3 LATITUDINAL GRADIENTS IN TRANSPORT MECHANISMS OF ALLOCHTHONOUS POC AND DOC

Allochthonous sources of organic matter (those delivered from external sources) are products of continental erosion and weathering as well as upstream production. They are delivered via rivers, the atmosphere, and shoreline erosion. In land-margin ecosystems, it is possible to distinguish between marine and land-derived sources of organic matter using biomarkers, organic compounds that have molecular structures related to their biogenic source. As a result, these compounds can provide detailed information about the origins of particulate and sedimentary organic matter in the environment.

Lignin, a macromolecule found in the cell walls of vascular plants, has proven to be one of the most useful chemical biomarkers tracking terrestrial inputs to continental margins (Prahl et al., 1994; Bianchi et al., 1997; Goni et al., 1998; Bianchi and Canuel, 2011). Oxidation of lignin with CuO yields eight dominant vanillyl, syringyl, and cinnamyl phenols, which can be used to distinguish between woody and non-woody tissues of gymnosperm and angiosperm origin (e.g., seagrasses, mangroves, terrestrial grasses, agricultural plants, trees) (Hedges and Mann, 1979; Bianchi et al., 1999; Bianchi and Canuel, 2011).

Another co-variable affecting the fate and transport of allochthonous POC reaching coastal BCEs is the underlying geology (i.e., siliciclastic versus carbonate) of the coastal system. If the system is siliciclastic, detrital mineral, and POC inputs are primarily controlled by surface hydrology whereas karst (carbonate) systems are often characterized by hydrological exchange with the ocean as subsurface (groundwater) runoff. In the southeastern United States, karst substrates are typically colonized by woody mangrove rather than marsh vegetation typically seen in siliciclastic substrates. The combined effects of carbonate and woody mangroves versus siliciclastic non-woody marshes BCEs need to be considered when making predictions about the vulnerability of stored blue C with potential accelerations in sea-level rise and inundation, and the erosion and release of stored blue C (e.g., mangrove and marsh peat).

Interestingly, in many places around the world coastal marshes are being replaced by mangroves as the warmer BCEs continue to spread to higher latitudes with changing climate. This has been particularly evident in regions of the Gulf of Mexico (GOM). Some work in the GOM as focused on competitive factors controlling zonation at the ecotone between mangroves (*Avicennia germinans*) and salt marshes (*Spartina* spp.) (Patterson and Mendelssohn, 1991; McKee et al., 2002; Stevens et al., 2006). Other more recent work has shown that the advance of mangroves over marshes has resulted in a greater elevation of coastal wetland sites when occupied by mangroves, which may prove beneficial in terms of coastal storm protection (Comeaux et al., 2012). Other work has shown that carbon sequestration rates were greater in *A. germinans* than *S. alterniflora* along the Texas coast, which ranged from 253 to 270 and 101 to 125 g C m^{-2} year^{-1}, respectively (Bianchi et al., 2013). Interestingly, this work also showed that lignin storage rates were also greater for mangrove than marsh site, which ranged from 19.5 to 20.1 and 16.5 to 12.8 g lignin m^{-2} year^{-1}, respectively. Finally, remote-sensing (Light-Detection and Ranging—LiDAR, and aerial images) work combined with core data, has revealed that the amount of C stored in soils of newly established *S. alterniflora* intertidal marshes was significantly lower than in older marshes in the same region (Kulawardhana et al., 2014; Kulawardhana et al., 2015). A transition in soil characteristics also coincided with the timing of transformation in marsh land cover (e.g., vegetation percent cover, plant height, plant density, above-and below-ground biomass, and C) which was driven largely by rapid increases in relative sea-level. Therefore, in order to quantify and predict C sequestration and stocks in coastal wetlands, changes in relative sea-level rise and accretion rates need to be better linked with land-use and vegetation changes (Kulawardhana et al., 2015).

There will be significant differences in the ability of different BCEs to trap POC, depending upon the vegetation type and overall physical energy of the system. Effects of the aboveground biomass will be very different for dense non-woody marsh versus more open woody mangrove habitats. Based on a survey of the BCE literature, it appears that while all mangrove studies reported an export (or no net change) of litter and POC to the coastal ocean, salt marshes were largely exporting POC (with some import), seagrasses were also mostly exporting (wrack) and importing C by trapping sestonic particles (seston) (Figure 4.2). The higher import capability reflected by seagrasses may be best explained by the differences in the hydrodynamics of baffling effects of seagrass blades in submerged habitats compared to salt marshes and mangroves. More specifically, the seagrasses generally have a more **active** surface—per total aboveground plant surface area, due to greater epiphytic-zoic growth (e.g., bryozoans, benthic forams, microalgae/macroalgae) on their leaves than in mangroves and salt marshes. In some studies, the seston import is balanced by the export of wrack and DOC (Ferguson et al., 2017) producing no net change in C exchange.

The highest export of both POC and litter from BCEs occurs in tropical mangroves (Figure 4.3). In contrast, there was no latitudinal gradient in POC export for salt marshes and seagrasses.

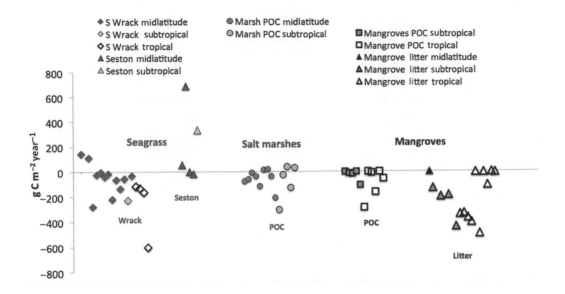

Figure 4.3 Latitudinal comparison of wrack and seston fluxes in seagrasses, POC fluxes in salt marshes, and POC and litter fluxes mangroves. Wrack refers to dead seagrass biomass, while seston denotes the sestonic particles trapped by seagrass meadows. Seston fluxes were from Freguson et al., (2017); Barrón et al., (2004); Cebrian et al., (2000); Barrón and Duarte (2009). Mangrove litter and POC fluxes were obtained from Adame and Lovelock (2011). Estimates for salt marshes are from Mitsch (1994); Nixon (1980); and others. Positive values represent export and negative values import.

The high POC export in tropical mangroves, compared to salt marshes and seagrasses is due to the fact that 50% of C litterfall is exported during tidal inundations in mangroves (Saenger and Snedaker 1993). Additionally, Adame and Lovelock (2011) reported a wide range in mean export (0%–99%) of C litterfall indicating a substantial amount of mangrove production is exported. Adame and Lovelock (2011) also found that variation in mangrove litter export could be explained by precipitation and mean annual temperature. Hypersalinity in dry regions decreases bacterial abundance (Twilley et al., 1997; Ólafsson et al., 2004) and reduces productivity (Cintron et al., 1978; Saenger and Snedaker 1993), which slows down decomposition; therefore, decreasing C recycling within the forest and increasing litter export in comparison to regions with higher precipitation and lower salinities.

In seagrass and mangrove habitats, studies report contrasting DOC fluxes, with some studies showing export and others import as shown in Figure 4.4. Net C import in seagrass communities is due to net uptake from planktonic respiration or seagrass leaves (Vonk et al., 2008). The biological communities that grow on a greater fraction of the total leaf surface area of seagrasses, likely contributes to the uptake of DOC via passive (e.g., sorption) and active uptake (e.g., bacterial biofilms) on the leaf surface (Kirchman et al. 1984; Ziegler and Benner 1999). However, DOC exudation by seagrass leaves and rhizomes is common and accounts for between 1.2% and 30% of gross seagrass primary production (Moriarty et al., 1986; Barrón et al., 2004; Barrón and Duarte, 2009; Kaldy, 2006, 2012). Additionally, the contrasting results in mangrove communities could be due to differences in spatial-scale methodologies (Adame and Lovelock, 2011). In salt marshes and mangroves, DOC export was greater at lower latitudes due to higher temperatures and annual precipitation and thus higher productivity (Saenger and Snedaker, 1993). Although Barrón et al. (2014) reported an increase in DOC fluxes in seagrass beds with warmer temperatures this was not observed across a latitudinal gradient, with the exception of one study in Silaqui Island Bolinao, Philippines, which oddly reported export rates three times higher than other study in the tropics. This could be due to different physical drivers. For example, some riverine mangroves receive more DOC than tidal-dominated mangroves and thus are able to retain DOC (Adame et al., 2010).

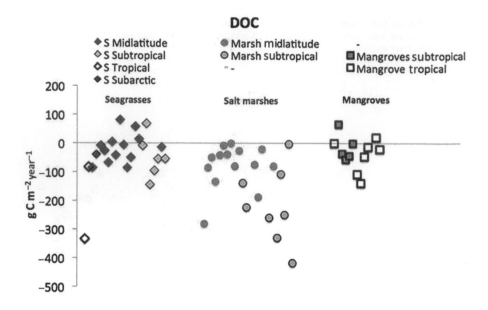

Figure 4.4 Latitudinal (mid-latitude, subtropical, tropical, and subarctic) comparison of DOC fluxes in seagrasses, marshes, and mangroves. Positive values represent export and negative values import. Estimates of seagrass DOC fluxes based on in situ benthic chambers from (Barrón et al., 2014). Estimates for salt marshes are from Mitsch (1994); Nixon (1980); and others. Estimates for mangroves are from Adame and Lovelock (2011). DOC exports in salt marsh and mangrove studies were greater in lower latitudes, while in the seagrass studies, latitudinal trends were not observed.

4.4 BIOGEOCHEMICAL FACTORS CONTROLLING ROLE OF ALLOCHTHONOUS POC VERSUS DOC

Coastal aqueous POC and DOC forms constitute the major pool of organic matter providing C and energy to heterotrophs in estuaries. However, understanding the fate of these materials—i.e., their *recycling* in BCEs—requires a general understanding of the physical and biogeochemical factors that shape the rates at which POC and DOC are metabolized to DIC, as well as the variability among BCEs with respect to these rates (Macreadie et al., 2017) (Figure 4.2). The aim of this section is to provide an overview of approaches to measure these rates, the key biogeochemical controls on those rates, and to conclude with a discussion of the role of the priming effect on allochthonous POC and DOC cycling in BCEs.

4.4.1 Overview of Approaches to Measure OC Recycling

Determining the factors controlling the fates of allochthonous POC and DOC involves quantifying the rates of allochthonous POC and DOC recycling. Rates often are measured as loss of biomass in so-called litter bag experiments in which not only changes in mass of C is tracked (e.g., Newell et al., 1989) but also changes in biochemical composition such as lignin (e.g., Newell et al., 1989; Opsahl and Benner, 1993; 1995). Such studies are focused on mass loss and compositional changes in biomolecules, bulk elemental and isotopic composition of material. In litterbag experiments, a known mass of material is introduced into mesh bags and then suspended in water for a period of time. Harvesting individual bags over time will provide information enabling rate measurements to be computed. For example, the decrease of OC concentration or the increase in CO_2 over time. Recently, a **Tea Bag Index** has been developed to allow for the use of tea bags as litterbags in a global effort to quantify C decomposition in BCEs (Keuskamp et al., 2013). Often results are expressed in percentages to give a wider picture of how organic matter (OM) is recycled into different fluxes. For example, Newell et al. (1989) examined loss of *S. alterniflora* leaf OC on leaves left attached to plants in two cohorts from early autumn through late spring. Roughly equivalent fractions (10%–15%) of C fluxes for dead-leaf carbon flowed to CO_2, dissolved leachate (i.e., DOC), and fungal biomass, respectively. Similarly, Moran and Hodson (1990) found that roughly double the DOC produced by *S. alterniflora* degradation was attributable to lignin rather than cellulose. In fact, more than 40% of DOC in the tidal creek of the salt marsh they studied was due to *S. alterniflora*. These results demonstrated that allochthonous OC has two primary means of export (gas flux and lateral flux) out of BCEs and supports microbial biomass, fueling the microbial loops within BCEs.

Indeed, the primary agents of OM recycling of allochthonous OM in BCEs are fungi and bacteria; in fact, these habitats are where the former class of microorganism play an important role. Bacterial production measurements to quantify OC assimilation can be done using ^{3}H-labeled leucine or thymidine (e.g., Findlay et al., 1991; Findlay, 2002). In the Newell et al. (1989) study, it was determined that OC flow to fungal biomass was roughly five times greater than bacterial biomass. Moreover, rates of fungal productivity were often much greater than bacterial productivity (both expressed as net changes in organic biomass) during the early autumn (2.68 versus 0.17 mg organic mass standard sample^{-1}d^{-1}). In the winter, rates were very similar (0.2 mg organic mass standard sample^{-1}d^{-1}). Flow of *S. alterniflora* OC to fungi and promotion of fungal productivity results in increase in fungal biomass, which also enriches the N content of allochthonous OM.

Work by Opsahl and Benner (1993, 1995) showed the importance of fungal and microbial degradation of BCE leafy material with respect to allochthonous OM quality, which was shown through biochemical changes to the lignin composition of these materials during multi-year degradation studies. These so-called **litter bag** techniques are another key way to assess allochthonous POC degradation. Seagrass blades had the most rapid percentage loss, reaching ca. 10% of initial mass of bulk tissue, lignin, and cutin, as well as other vascular plant biomarkers such as hemicellulose and cellulose, after roughly 1.1 years (Opsahl and Benner, 1993). By contrast, mangrove and smooth cordgrass (*S. alterniflora*) leaves were far more resistant to degradation, taking roughly 2 and 4 years to reach similar degree of mass loss. Lignin often was more resistant to degradation than was cutin (Opsahl and Benner, 1995). These biochemical changes to organic matter quality are important when considering that in situ processes shape the properties of ex situ material transported from BCEs to adjacent estuaries and coastal waters.

One aspect of organic matter quality that is important for its recycling is its size. Extracellular enzymatic activity (EEA) is an important means of quantifying rates of hydrolysis on POC. Considered the first or rate-limiting step in organic matter degradation, EEA can provide information on what biochemical features of POC and high molecular weight (HMW) DOC are being actively utilized (Arnosti, 2010). Major substrates investigated (and associated enzyme hydrolysis rates measured) include polysaccharides (α- and β-glucosidases), peptides (aminopeptidase), lipids (lipase), and lignin (phenol oxidase). Hydrolysis rates can vary across latitudinal gradients (higher at temperate to tropical latitudes and lower at polar latitudes) (Arnosti, 2014) and may be limited by blocking access to substrate, complexing enzymes or blocking their binding sites, absence of an inducer, cofactor, or community member that is required (Arnosti, 2011). Much of the work on EEA in coastal wetlands has been restricted to sediments and more studies have been done for salt marshes (e.g., Ravit et al., 2003; Duarte et al., 2008; Morrissey et al., 2014), than for seagrass meadows (Lopez et al., 1995) or mangroves. A general lack of observations of EEA in the water column of BCEs has been made which can assess lability of POC or DOC specifically.

Coastal regions are generally seen as net heterotrophic and much of the net heterotrophy observed is due to OC **incineration** in BCEs, and leads to lateral fluxes of inorganic C to coastal waters (Cai, 2011; Bauer et al., 2013). Decomposition of OM and its respiration to CO_2 by fungi and bacteria means that quantifying CO_2 dynamics within BCE is an important means of quantifying allochthonous OM recycling. Approaches to measuring respiration via gas consumption or emission in sediments of BCEs include laboratory micro- or mesocosm experiments using chambers on cores collected from the field (e.g., Alongi et al., 2004; Chambers et al., 2013, 2014) and in situ chambers that measure gas fluxes in the BCEs (e.g., N_2, N_2O, CO_2, CH_4, O_2) (e.g., Bowman and Delfino, 1980; Nowicki, 1994; Alongi et al., 2005). The first-order controls on decomposition is oxygen demand (Tobias and Neubauer, 2009); hence, standard water quality protocols for biochemical oxygen demand have been use to quantify respiration, mainly pelagic respiration of DOC.

Rates of decomposition of allochthonous OM in BCE are strongly controlled by tidal inundation. In salt marshes and mangrove stands, baffling by standing stems and leaves creates drag that reduces tidal velocities (Dame et al., 2002). Reduced tidal energy and water velocities promotes sediment deposition, but also low energy quiescent conditions that increase the residence time of water overlying organic-rich sediments during submergence of the tidal wetland platform. These conditions are very conducive to decomposition. Rates of decomposition vary among BCEs. Median rates of decomposition were similar for salt marsh and mangroves and were about 40%–45% of NPP, while median seagrass decomposition was higher (ca. 60% of NPP) (Duarte and Cebrian, 1996). Thus, it can be expected that seagrass OC is recycled more rapidly than is salt marsh or mangrove OC. However, this comparison is a subtle difference and it is notable that a large majority of NPP is actually lost. Much of the present blue carbon research is focused on constraining those losses, including constraining lateral fluxes of POC and DOC from BCEs to adjacent coastal waters (Duarte et al., 2005).

Pelagic organic matter recycling measured as water column respiration is relatively understudied in BCE. For mangroves, Alongi (2014) has reported rates of 0.1–3.5 g C m^{-2} d^{-1}. Apple and del Giorgio (2008) report rates of 0.035 g C m^{-2} d^{-1}, for tidal creeks in salt marshes fringing the Chesapeake Bay. Rates of 0.21–0.68 g C m^{-2} d^{-1} were reported for tidal creeks along the coast of North Carolina (MacPherson et al., 2007). Temperature appears to be a strong control on OC respiration (Shiah and Ducklow, 1995; Apple and Del Giorgio, 2006) whereas Apple and del Giorgio (2008) suggest that OM quality may be more important than nutrients for bacterial growth efficiency, which partitions consumption of organic matter into production of biomass and respiration to CO_2.

4.4.2 Key Biogeochemical Controls on OC Cycling in BCE

As discussed in the Introduction to this chapter, the terms **labile** and **recalcitrant** (or **refractory**) are in need of re-examination, particularly in light of recent evidence that nearly all OC compounds are labile given the right environmental conditions (Marin-Spiotta et al., 2014). We first briefly consider the key concepts with control of OC degradation generally in coastal ecosystems.

Three paradigms currently exist in marine science that contextualizes the degradation of organic matter in aquatic environments. An important point here is that microbes are capable of consuming OM only up to a certain molecular size (ca. 700 Daltons) based on limits of active and transport mechanisms. Thus, a first order control on OM degradation is the size of the material; it must first be broken down in order to be consumed. Arnosti (2004) conceptualized this limitation in terms of **speed bumps** and **barricades** to OM degradation. The focus of this concept is on OM lability based upon EEA for specific biochemical substrates (polysaccharides, proteins, lipids). Limits to enzyme function exert a top-down control on the microbial utilization because: (i) lack of a needed nutrient or trace element; (ii) lack of an organism to produce a specific enzyme; (iii) an inducer for enzyme production may be absent. Thus, inhibiting POM and HWM DOM hydrolysis to smaller substrates will be the rate-limiting step for organic matter recycling.

Indeed, the size of OM and its corresponding geochemical characteristics indicate a **continuum** of reactive states (Benner and Amon, 2015). Under this concept, the larger OM (>1000 Daltons) is the most reactive, linking again to the importance of EEA at starting the degradation process. Smaller (<1000 Daltons) OM have been determined to be more difficult for microorganisms to use and this supposition was made based on the age of OM in different size classes. Low-molecular weight (LMW) DOC makes up a larger fraction in the ocean than HMW DOC (Benner and Amon, 2015). Exponential decay constants for POC were nine and two times higher than LMW and HMW DOC, respectively (Benner and Amon 2015). In essence, the action of extracellular enzymes breaks apart larger organic molecules into smaller fragments until it becomes unavailable to microbes. A possible reason for the low reactivity of LMW DOC is that once substrate size is reduced below the 600–700 Dalton threshold rapid uptake should occur—providing the substrates can be found. Arrieta et al. (2015) conducted experiments with deep ocean marine DOM, which they concentrated and fed to deep ocean microbial communities. Results were consistent with increasing the concentration leading to higher microbial populations and greater rates of degradation. This concept may be intuitive but consider that we often have very little idea of the concentrations of individual substrates and when those are measured, even in BCEs, the concentrations often are very low (pM to μM).

The aforementioned paradigms largely are microbially driven which is sensible because the microbial loop is itself driven by organic matter as a primary resource (especially DOC). These paradigms suggest that allochthonous DOC and POC from BCEs should be reactive, based on this material containing large molecules (e.g., lignin) and evidence of a range of respiration rates, in some cases supporting large amounts of the bacterial C demand of adjacent estuarine and coastal waters. However, the preceding discussion is a reminder that the allochthonous OM exported has qualitative properties distinct from the OM internally recycled in situ (e.g., plant root exudates). Thus, recycling of allochthonous POC and DOC in BCEs should consider the fate of material exported from them to coastal waters.

Aside from microbial degradation previously described, photochemical degradation by sunlight is an important removal process because much of the DOC generated in BCEs is colored. Despite the moniker, **blue carbon** exported from BCEs to coastal waters is brown. This colored organic matter is termed **CDOM** and is detected by light absorption and fluorescence (Stedmon and Nelson, 2015). The brown color is derived from humic-like substances, which absorb strongly in the blue region of the electromagnetic spectrum and deep into the ultraviolet (typically measured from 250 to 750 nm). Compounds such as lignin, which have been noted above to be enriched in marsh, mangrove, and seagrass vegetation, and polyphenols (e.g., tannins) give rise to these optical properties. These compounds are also abundant in the HMW fraction of DOM.

Sediments and soils are where massive amounts of organic matter are stored in BCEs; consequently, these stores of OC are also where much of the BCE CDOM originates. In salt marshes, for example, during flood tide, inundation of the marsh platform with seawater leaches HMW DOM from underlying sediments (Clark et al., 2008; Otero et al., 2010; Osburn et al., 2015). The high rates of microbial activity in BCE sediments is one of the likely mechanisms for generating humic-like compounds that contribute to the HMW DOM and the CDOM. Tidal pumping and advection of marsh sedimentary pore waters creates a constant mechanism of extracting and then exporting this material (Tzortziou et al., 2011). In this study, evidence of the marsh signal was observed at distance of 1 km into the adjacent estuary. This result is important because at such a distance from land, unique BCE signals in CDOM could be detected using satellite-based remote-sensing observations (e.g., Landsat 8: Slonecker et al., 2016; VIIRS: Joshi et al., 2017).

At low tide and during periods of low runoff, internal processing can influence BCE organic matter cycling. Low flow in the North River Farms salt marsh created longer residence time for primary and secondary production to transform CDOM, amending it with fluorescence signals which indicated a greater abundance of plankton-derived organic matter (Osburn et al., 2015). The net effect was that less stable OC was exported from this marsh.

CDOM concentrations in BCEs are often as high if not higher than estuarine CDOM. For example, Tzortziou et al. (2008) found two threefold increases in CDOM concentrations over a tidal cycle in the Kirkpatrick Marsh, Maryland. This flux of CDOM corresponded to a similar magnitude of DOC exported from the marsh (e.g., range of 3 mg L^{-1} at high tide and 12 mg L^{-1} at low tide). Further, this material was quite photoreactive, exhibiting about 50% loss of initial absorbance when exposed to natural sunlight (Tzortziou et al., 2007). Similarly, Osburn et al. (2015) found CDOM concentrations in a restored salt marsh that were two- to sixfold higher than the adjacent estuary. Annual export of HMW, stable, DOC exported from this marsh was 86 g C m^{-2}year^{-1}, roughly 60% of the total DOC export (Osburn et al. 2015).

Inorganic controls on OM recycling in BCEs include the local redox environment and nutrients (Dame et al., 2002; Bouillon et al., 2008; Tobias and Neubauer, 2009). The oxidizing potential, Eh, serves as a quantifiable measure of the degree of oxidation possible in an environment and typically is negative in sediments of salt marshes (Odum, 1988), mangroves, and seagrass meadows (Marbà et al., 2006). The reducing conditions are driven by the large amount of OM produced and buried within these systems, which rapidly consumes available oxygen. In the classic sense of redox zonation, many salt marsh and mangrove ecosystems have plentiful sulfate and iron, both of which are used as electron acceptors after oxygen is depleted. While nitrate is generally low in most coastal environments, BCEs proximal to urbanized coastal ecosystems may have substantial nitrate inputs leading to denitrification in these systems (Tobias and Neubauer, 2009).

Recycling of allochthonous POC and DOC in BCE are strongly controlled by tidal inundation and resupply of key electron acceptors and nutrients. Tidal pumping through pore waters resupplies oxidants to sediments (Howes et al., 1985; Odum, 1988) as does burrowing and other forms of bioturbation (Bouillon, 2008). Benthic-pelagic coupling connects the biogeochemistry of the sedimentary environments to the overlying water column, supplying allochthonous DOC as well as nutrients, much of which is processed by heterotrophic organisms in the seabed (Dame et al., 2002). Tidal action brings in fresh nutrients (and perhaps less stable organic matter) which promotes microbial degradation in the pelagic waters (Kirstensen et al., 2008; Alongi, 2014). Finally, groundwater may be important for OM recycling in BCEs (Valiela et al., 1978). Tidal pumping of nutrients and OC across the sediment-water interface appears to influence the mineralization of OM in mangrove sedimentary pore waters as determined by stable isotope measurements of DOC and DIC (Maher et al., 2013). Rapid recycling of DOC in sedimentary pore waters was proposed by these workers to constitute part of the missing **sink** of carbon in mangroves.

Tidal influences on OM recycling in salt marshes have been shown to have important management and restoration applications. Restoration of tidally restricted marshes resulted in tangible effects to OM cycling, largely driven by vegetative changes occurring from increased salinity as tidal flow is restored (Chambers et al., 1999). Wozinak et al. (2006), for example, used C, N, and S stable isotopes to examine changes to food webs in three New England salt marshes that had tidal flow restored. Notably, they found that the tide-restricted marshes were fresher and dominated by C3 *Phragmites*, whereas after restoration of tidal flow, dominance shifted back to C4 *Spartina*, as salinity increased. Seagrass and benthic and micro- and macroalgae imported from tidal inundation augmented OM sources, which may influence the **priming effect** of organic matter recycling (see below) by adding labile OM from these seagrass and planktonic sources to the marsh.

4.4.3 Potential Priming Effects in BCE

Many BCEs exist at the interface between terrestrial and marine systems, where newly released blue OC can mix with other forms of marine OC (e.g., macroalgae and phytoplankton) in both particulate and dissolved (POC and DOC) forms (Bianchi et al., 2007). In fact, blue C will be released as POC and DOC as well. However, it is the DOC pool that will most likely be the most reactive in terms of consumption by microbes due to its overall higher availability to the microbial community, at least in many cases (McDonald et al., 2007). If this is the case, we posit that the breakdown of stable DOC from blue C sources will experience enhanced conversion to CO_2 in the presence of more unstable algal DOC. Enhanced breakdown of stable DOC in the presence of unstable algal

DOC has been described as "priming" in aquatic systems, and has been recently demonstrated in lab (Bianchi et al., 2015) and field (Ward et al., 2016) studies. These studies found that stable terrestrially derived DOC (TDOC) is more readily degraded when LMW substrates are available. Previous studies have shown that "recalcitrant" C stored in deep peat can be degraded when subject to different environmental controls, suggesting that the concept of "recalcitrant" substrates may need to be reevaluated, and that a substrate's reactivity should be considered in the context of its environmental conditions (Bianchi et al., 2011; Schmidt et al., 2011). Although the C stored in peat is generally considered resistant to microbial degradation, if this "stored" material is transferred to estuarine systems, particularly in the presence of algal DOC, microbial priming is likely to stimulate peat degradation, and subsequently result in greenhouse gas production.

The priming effect has been well described within soil systems, but the dynamics of priming within estuarine systems has only recently been investigated, and inconsistent results have been found (Blanchet et al., 2016). Although several microbially mediated priming mechanisms have been proposed (Bianchi, 2012; Guenet et al., 2010), the specific microbial taxa, and particularly the role of fungi, have not been thoroughly investigated in aquatic priming. Marine fungi can occupy numerous ecological niches (Das et al., 2006; Richards et al., 2012), and can engage in interactions with other members of the microbial community (Richards et al., 2012). These dynamics could have important implications for priming and blue C degradation in estuarine systems, and could potentially transform a previously stable C stock into a CO_2 source. Additionally, increased ocean acidification could promote fungal growth in estuarine systems, and could increase their importance in aquatic systems in the future (Krause et al., 2013).

The leaching, erosion, and export of wetland-derived DOM into estuarine environments greatly alters the potential reactivity of substrates that may have been stored in sediments for hundreds to thousands of years. The combination of an oxic environment with a mixed assemblage of aquatic and marine microbes predisposed to consume wetland DOC likely stimulates the degradation of OC that may have previously been "recalcitrant." Further, the presence of fresh algal exudates and biomass has been shown to stimulate the breakdown of vascular plant-derived DOC as a result of microbial priming effects (Bianchi et al., 2015). The process of priming was discovered by Lohnis (1926), who revealed that rates of soil humus mineralization were enhanced by the addition of fresh organic residues. Since then, and especially in the last decade, a large number of soil studies have confirmed and extended the original observations. Guenet et al. (2010) and Bianchi (2011) have also suggested that priming of TDOC by the presence of a more bioavailable substrate such as algal-derived DOC (ADOC) exudates likely acts as an important factor in regulating carbon transformations in aquatic and marine ecosystems. This process may explain in part recent observations of greater TDOC consumption rates in inland and coastal waters than reported in the past as a result of human influences on nutrient cycles (Cole et al., 2007).

While mention of priming (or in some cases co-metabolism) effects can be found in the aquatic coastal literature, the concept has largely been supported by superficial or equivocal evidence (Bianchi et al., 2011). Poretsky et al. (2010) observed shifts in microbial expression of OC-transporter genes under variable inputs of algal and terrestrially derived OC, providing insight into the biological facilitation of priming processes, but did not address priming per se. Blanchet et al. (2016) and Guenet et al. (2014) made some of the first experimental observations of aquatic priming, but the implications of this effect on globally relevant CO_2 fluxes remains unclear. Recently, Blanchet et al. (2016) investigated the effects of priming on bacterial community activity (BCA) and bacterial community composition (BCC), and found that different types of DOM could influence BCA and BCC, but did not find a clear priming effect within their experiment. However, there is currently no clear consensus on the role of aquatic priming. For example, recent studies have made observations of strong positive priming (Danger et al., 2013; Bianchi et al., 2015; Ward et al., 2016), no priming at all (Attermeyer et al., 2015), and negative priming effects (Gontikaki et al., 2013) in various settings.

The studies described above have established that priming results from bacterial production of hydrolytic enzymes in response to uptake of low molecular mass substrates or substrates derived from readily degradable polymeric organics (e.g., fresh phytoplankton biomass). Hydrolytic enzyme production in turn stimulates formation of additional low molecular mass substrates through the degradation of existing polymer pools that might be compositionally unrelated to the inducing substrates. For example, monomeric sugars and amino acids may stimulate synthesis and excretion of lignin-degrading enzymes, producing simple aromatics and degradation by-products. The net effect is that priming enhances the mineralization of resistant or stable polymers along with those that are more unstable. While the magnitude of priming remains uncertain, it is clear that it can affect soil carbon sequestration, and respond positively or negatively to a variety of disturbances (e.g., Guenet et al., 2014).

Diverse soil studies have addressed multiple aspects of priming, including effects of nitrogen availability, plant-microbe interactions, and the quality/quantity of priming substrates. However, even though priming is clearly a microbial process, remarkably little is known about the microbes that are involved and their specific behaviors. Several studies have used a membrane phospholipid fatty acid (PLFA)-based method to address shifts in microbial community composition during priming events, but this approach has relatively coarse taxonomic resolution and can miss many important changes. Nonetheless, these PLFA-based studies have shown that community composition varies over time during priming, and in response to both priming substrate quality and quantity (Wang et al., 2014). Recently, Blanchet et al. (2016) investigated the effects of riverine DOM and labile DOM on coastal bacterial communities and found that *Actinobacteria* were in greater proportion within control microcosms, while their enriched microcosms exhibited a greater abundance of *Alpha-* and *Betaproteobacteria*. Moreover, BCEs can play a role in coastal food webs via export of DOC. Microorganisms cannot directly utilize POC and must degrade POC first to HMW DOC which then is further degraded to smaller DOC molecules which may be respired to CO_2. Therefore, the role of BCE coastal ocean C cycling is important to understand with respect to the availability of exported DOC to coastal marine microbial communities. The quality of this allochthonous organic matter exported to coastal waters is expected to exert a major control on its utilization by coastal microbes and hence its fate.

4.5 ESTIMATES OF EXPORT RATES OF ALLOCHTHONOUS POC AND DOC IN SELECTED BC ECOSYSTEMS

POC and DOC quantities in the open waters of an estuary are primarily sourced from fringing wetlands or upstream riverine locations (Bauer et al., 2013) (Figure 4.2). A major source of allochthonous POC to coastal waters comes from eroding BCE sediments, and a major source of DOC comes from pore waters that are flushed tidally and allow draining of DOC into receding waters. The "steeping" effect of saturating BCE sediments with water facilitates the solubilization of DOC from accumulating soil organic carbon (SOC). Release of substances from pore waters to overlying waters is not only controlled by diffusion rates but also is influenced by tidal pumping. This export of POC and DOC could be exacerbated by storm events, flushing porewaters as well as eroding sediment and overall enhancing allochthonous inputs of OC from BCEs.

Loss of DOC from POC has been studied in soils and sediments and indicates that substantial C can be released from erosive processes, bioturbation, and resuspension. Other work has shown the importance of sunlight in perhaps facilitating this process. Much of what we know about DOC losses from sediments has come from studies of terrestrial soils (especially peat) so it is anticipated that BCEs are potentially large and unrealized sources of DOC to coastal waters.

Moreover, it can be expected that that BCE that are actively eroding will contribute comparatively larger amounts of allochthonous POC and DOC via these processes than systems less susceptible to erosion. In contrast, prior work has shown that mineral surfaces can resorb OC from solution, creating a potential sink for the desorbed allochthonous DOC.

The proportion of wetland versus riverine OC sources potentially could be identified using a synthetic approach. Our approach is geared toward answering order-of-magnitude questions, and thus should be viewed as an advanced and quantitative, yet "back-of-the-envelope" style of analysis. To accomplish this, we quantify the amount of OC available for export to the water from wetlands. We conduct this analysis for each estuary in the continental United States, and summarize the trends.

4.5.1 Order-of-Magnitude Calculations of Lateral Export Sources to the Estuary

For this order-of-magnitude approach, we begin with knowledge of SOC density in the terrestrial wetland soil. As the depth below the surface increases in wetlands, SOC generally decreases. We obtain depth-explicit SOC data from the CoBluCarb database (Hinson et al., 2017).

To obtain the per year SOC loss rate for a given $1 \, m^2$ of tidal wetland soil surface, T_{soc}, we put Equation 4.1 as:

$$T_{soc}(g \, m^{-2} \, year^{-1}) = -\sum_{n=1}^{4} [(SOC_n - SOC_{n+1}) * a] 10^4 \quad (4.1)$$

where, SOC is the SOC density at a given n increment in a wetland soil profile acquired from CoBluCarb, which details the density of carbon ($g \, cm^{-3}$) at 5 cm increments of depth. The result of Equation 4.1 is negative when carbon is exported from the wetland soil to the adjacent water body. Equation 4.1 also contains a scaling factor for converting the $1 \, cm^2$ in surficial area of the density profile into m^2. The profile increments are divided by the number of years for each to be deposited, such that the SOC losses are converted into time domain units (as shown in reduced equation above). The notation a is the average per year wetland accretion rate, using the sea level rise rate as a rough surrogate for accretion. Accretion and sea level rise were assumed to be linearly related at a 1:1 relationship, which is certainly not always true (e.g., Morris et al., 2002; Mudd et al., 2009). However, it was considered as a rough and first-order guidance. Relative to other potential errors in this analysis, this assumption is small, within an order-of-magnitude.

Importantly, we conducted the analysis to (i) only the upper 20 cm of depth, (ii) 100 cm of depth, and (iii) 300 cm of depth. We chose the 20 cm depth, because it minimized the number of additional assumptions. This choice was based on: (i) the majority of POC exported likely originates from depths of the rooting zone, (ii) differential compaction occurs with greater depth below this zone and accretion measurements available from the literature generally do not assume these greater depths (e.g., the average depth of Cs-137 core-derived accretion data in Chmura et al. (2003), or other similar references, is approximately 20 cm). In many wetland accretion and SOC reviews, soil compaction is either mentioned but not included as an explicit value (e.g., Kirwan et al., 2016), or subsumed in the accretion values themselves (e.g., Chmura et al., 2003). We know that compaction of a given 5 cm increment of soil increases with depth, and thus at greater depths, the 5 cm increments should correspond with a greater number of years—this is not accounted for in Equation 4.1. In the future, one could further refine a with measured literature-derived values, as well as modeling the compaction effect by using the dimensionless ratio of bulk density at one increment relative to another at a greater depth. We also chose the 100 cm depth, because this is the depth most often utilized by international and national synthesis efforts, and carbon at this depth is assumed "stable." We additionally chose the 300 cm depth, because this gave the greatest downward range available from CoBluCarb and SOC appeared to continue to decrease down to this depth in our datasets. To address both the many assumptions required of using greater depths and the limitations of using lesser depths, we present results from all three depths. More work in the future is required on these critical depth issues, relative to POC and DOC transport.

To find the theoretical SOC lateral transport flux from the wetland soil to the open waters of the estuary only, L_{soc}, we note Equation 4.2 as:

$$L_{soc}(g\,m^{-2}\,year^{-1}) = T_{soc} + E_{soc} \tag{4.2}$$

where, E_{soc} is the evasion SOC flux, from the wetland soil to the atmosphere via processes of microbial soil respiration (and E_{soc} is negative when this happens). L_{soc} is thus the difference between the total SOC loss T_{soc} (from Equation 4.1 above) and the amount lost by soil respiration, E_{soc}.

To find E_{soc} we first compiled the literature on soil respiration in wetlands, and then mapped the values across all estuaries in the continental United States. Overall, there are few studies that measuring soil respiration CO_2 efflux in estuaries in the United States (Raich and Schlesinger, 1992; Wigand et al., 2009; Weston et al., 2011; Krauss and Whitbeck, 2012). From these studies, there were six sites from the East Coast, two from the Gulf Coast, and none from the West Coast; one was a review paper that included a total of five locations; and of all studies, only three sites were from obvious tidal wetlands.

We explored differences in methodologies, timing, parameters, salinity, and wetland type among the E_{soc} studies, but there were no clear trends besides latitude and temperature. Thus, we sought a correlation between the measured E_{soc} from each study site to the mean temperature at the study site over a 30-year period using PRISM data (PRISM Climate Group, 2017). We also explored other possibilities, but a linear regression between E_{soc} from each study and the PRISM value at each site produced the best statistical fit model ($r^2 = 0.41$, $p < 0.16$):

$$E_{soc}(g/m^2/year) = -(6.1376 * \text{PRISM mean temperature } °C) - 45.23 \tag{4.3}$$

We then re-mapped this predictive relationship across the continental U.S. as a function of the PRISM data, and averaged the predicted E_{soc} within each estuary.

The lateral export for all wetlands within an entire estuary can then be summarized by multiplying by their areal coverage in m^2, a value obtained from the CoBluCarb database. These individual estuary values can then be summed to provide a rough, order-of-magnitude national-level estimate.

4.5.2 Quantities and Trends of Lateral Export Sources to the Estuary

Table 4.1 shows the predicted lateral flux for all estuaries within the continental United States using this order-of-magnitude approach. Although the wetlands in most estuaries were net exporters of OC, some showed positive L_{SOC} at times suggesting possible influx of OC into the wetlands from the surrounding water column (Table 4.1). In general, T_{SOC} or L_{SOC} quantities that used the entire 0–300 cm of soil data were nearly all negative, suggesting flux of OC from wetlands to the surrounding water column, while the 0–100 cm of soil data were less negative. Interestingly, the 0–20 cm of soil data produced T_{SOC} or L_{SOC} that was often positive, suggesting flux from the water to the wetlands.

It should come as no surprise that the estimate of carbon export from the wetland to the water, or L_{SOC}, varies as a function of the soil depth that one considered when making the calculation. So, what depth should we consider as 'the best' estimate, or perhaps more precisely stated, at what depth are the relevant export processes extinguished? We do not have a clear answer to either of these questions; they would prove fruitful areas for biogeochemistry to explore. We can only submit that (i) the 100 cm of depth is most typically used for national and international carbon accounting efforts, (ii) though the average T_{SOC} depth profile continues the downward trend to 300 cm, the rate slows and the 100 cm versus 300 cm values are relatively similar, (iii) the 300-cm depth amplifies any potential errors made in the assumptions about sedimentary compaction over depth, as described above, (iv) the 20-cm depth misses much of the soil column that lay below, where there

Table 4.1 Order-of-Magnitude Estimates of Wetland Carbon Flux by Estuarine Drainage Area (EDA)

Estuary	Accretion Rate (mm year⁻¹)	T_{SOC} (g m⁻² year⁻¹)			E_{SOC} (g m⁻² year⁻¹)	L_{SOC} (g m⁻² year⁻¹)			Wetland Area (km²)
		0–20 cm	0–100 cm	0–300 cm		0–20 cm	0–100 cm	0–300 cm	
Albemarle Sound	3.96	−0.74	−100.72	−390.15	−53.03	52.29	−47.69	−337.12	357
Alsea River	1.72	25.81	−53.89	−89.12	−22.26	48.08	−31.63	−66.86	3.3
Altamaha River	2.51	−98.90	−104.10	−196.24	−72.56	−26.35	−31.54	−123.68	180.6
Apalachee Bay	2.32	−46.20	−104.74	−171.48	−75.83	29.63	−28.91	−95.65	208
Apalachicola Bay	2.44	−13.41	−37.70	−155.98	−76.24	62.84	38.54	−79.74	135
Aransas Bay	5.26	−9.56	−12.79	−54.30	−88.59	79.02	75.79	34.29	102.9
Atchafalaya/Vermilion Bays	8.92	−26.68	−173.99	−544.54	−76.71	50.03	−97.29	−467.83	2465.5
Barataria Bay	8.83	−4.61	−229.03	−524.89	−81.07	76.46	−147.96	−443.81	1,151
Barnegat Bay	3.63	−33.06	−53.68	−271.42	−29.05	−4.02	−24.63	−242.37	55.5
Biscayne Bay	3.30	−31.00	−101.58	−279.67	−101.73	70.72	0.15	−177.94	131.8
Blue Hill Bay	2.12	−28.78	−87.47	−143.95	4.80	−33.59	−92.28	−148.75	4.5
Bogue Sound	3.18	−64.93	−161.49	−271.07	−59.77	−5.16	−101.72	−211.30	95.1
Brazos River	7.03	−29.70	−258.09	−315.62	−81.44	51.74	−176.64	−234.18	116.6
Breton/Chandeleur Sound	7.49	−66.49	−266.16	−449.97	−81.06	14.58	−185.10	−368.90	1247.1
Broad River	3.03	−2.09	−84.00	−324.86	−70.43	68.35	−13.57	−254.43	308.8
Buzzards Bay	2.72	−33.04	−62.33	−251.19	−18.62	−14.41	−43.71	−232.57	23.8
Calcasieu Lake	7.08	−9.59	−579.62	−659.32	−75.41	65.82	−504.22	−583.91	667.4
Cape Cod Bay	2.96	−16.22	−58.01	−244.43	−17.82	1.60	−40.19	−226.61	52.5
Cape Fear River	2.61	−10.14	−93.55	−232.44	−57.19	47.05	−36.36	−175.25	44.5
Casco Bay	1.87	−11.86	46.40	−71.36	0.80	−12.66	45.59	−72.16	9.5
Charleston Harbor	3.18	−4.89	−60.87	−154.68	−66.92	62.04	6.05	−87.76	150.7
Charlotte Harbor	2.79	−26.63	−159.87	−206.43	−95.12	68.49	−64.74	−111.31	222.5
Chesapeake Bay	3.23	−39.80	−92.99	−249.40	−39.30	−0.50	−53.69	−210.10	1,651
Chincoteague Bay	3.97	−25.27	−130.57	−255.42	−41.43	16.16	−89.13	−213.99	118.6
Choctawhatchee Bay	2.96	−16.87	−30.73	−172.06	−72.70	55.82	41.97	−99.36	38.7
Columbia River	0.66	−4.23	−10.44	−22.22	−14.73	10.49	4.28	−7.50	104.9
Coos Bay	1.09	0.38	−13.50	−63.77	−23.29	23.68	9.79	−40.48	7.5
Coquille River	1.25	−15.07	−19.19	−69.63	−24.99	9.92	5.80	−44.64	1.4

(Continued)

Table 4.1 (Continued) Order-of-Magnitude Estimates of Wetland Carbon Flux by Estuarine Drainage Area (EDA)

Estuary	Accretion Rate (mm year⁻¹)	T_{SOC} (g m⁻² year⁻¹)			E_{SOC} (g m⁻² year⁻¹)	L_{SOC} (g m⁻² year⁻¹)			Wetland Area (km²)
		0–20 cm	0–100 cm	0–300 cm		0–20 cm	0–100 cm	0–300 cm	
Corpus Christi Bay	4.84	−12.53	−14.16	−42.39	−90.17	77.64	76.02	47.78	35.5
Damariscotta River	1.96	−40.70	−57.67	−144.00	0.08	−40.78	−57.75	−144.08	1
Delaware Bay	3.69	−36.30	−12.40	−289.92	−32.82	−3.48	20.42	−257.10	699
Delaware Inland Bays	3.71	−8.77	−93.37	−242.14	−39.13	30.36	−54.23	−203.01	26.2
Drakes Estero	1.72	−0.40	−24.75	−59.01	−32.77	32.37	8.02	−26.23	2.1
East Mississippi Sound	3.67	43.29	−31.28	−231.58	−72.25	115.54	40.97	−159.33	170.2
Eel River	2.74	−76.07	−109.09	−123.03	−32.12	−43.94	−76.97	−90.90	1.6
Englishman/Machias Bay	2.09	−28.09	−57.59	−155.40	7.73	−35.82	−65.33	−163.13	7.3
Florida Bay	4.13	−110.65	−194.38	−389.74	−105.72	−4.93	−88.67	−284.02	330.3
Galveston Bay	7.14	−56.29	−246.90	−299.86	−82.15	25.86	−164.75	−217.71	347
Gardiners Bay	2.57	−91.11	−160.22	−174.44	−22.42	−68.69	−137.80	−152.02	16.5
Grays Harbor	−0.14	1.78	5.23	7.64	−16.07	17.85	21.31	23.72	43.4
Great Bay	2.26	−15.91	−91.70	−175.45	−5.65	−10.26	−86.05	−169.80	10.2
Great South Bay	2.80	−12.50	−33.69	−242.24	−25.55	13.05	−8.14	−216.69	69.8
Hampton Harbor	2.33	−4.71	−21.84	−221.30	−8.30	3.58	−13.55	−213.00	15
Hudson River/Raritan Bay	2.79	−19.47	−63.91	−243.54	−15.88	−3.58	−48.02	−227.65	90.2
Humboldt Bay	3.93	−134.53	−190.45	−212.05	−28.03	−106.50	−162.43	−184.02	4.7
Indian River	2.88	−64.81	−94.26	−153.25	−92.62	27.81	−1.64	−60.63	143.6
Kennebec/Androscoggin	1.46	−6.03	−29.57	−101.48	8.63	−14.66	−38.20	−110.11	31.8
Klamath River	2.11	−82.94	−100.26	−115.76	−26.61	−56.33	−73.64	−89.15	0.4
Long Island Sound	2.24	−13.49	−101.72	−217.30	−15.15	1.66	−86.57	−202.15	72.6
Lower Laguna Madre	3.55	−1.80	−6.67	−18.20	−96.98	95.18	90.31	78.78	60
Maryland Inland Bays	3.93	−117.04	−145.44	−344.11	−39.82	−77.22	−105.62	−304.29	21.2
Massachusetts Bay	2.54	−21.54	−53.60	−236.50	−15.45	−6.09	−38.15	−221.05	27.7
Matagorda Bay	6.26	−11.70	−169.43	−214.72	−83.72	72.02	−85.71	−131.00	255
Mermentau River	7.77	−57.08	−528.15	−639.52	−77.20	20.12	−450.96	−562.32	776.6
Merrimack River	2.39	−30.79	−62.70	−275.64	−4.25	−26.54	−58.45	−271.39	11
Mission Bay	2.10	0.00	−15.06	−15.16	−60.49	60.49	45.44	45.34	0.2
Mississippi River	8.06	−114.56	−216.81	−428.72	−82.78	−31.78	−134.03	−345.94	423.1

(Continued)

Table 4.1 (Continued) Order-of-Magnitude Estimates of Wetland Carbon Flux by Estuarine Drainage Area (EDA)

Estuary	Accretion Rate (mm year⁻¹)	T_{SCC} (g m⁻² year⁻¹)			E_{SOC} (g m⁻² year⁻¹)	L_{SOC} (g m⁻² year⁻¹)			Wetland Area (km²)
		0–20 cm	0–100 cm	0–300 cm		0–20 cm	0–100 cm	0–300 cm	
Mobile Bay	3.14	−22.23	−25.22	−145.11	−73.44	51.22	48.22	−71.67	72.6
Monterey Bay	1.28	−0.49	−1.40	−15.10	−46.23	45.74	44.83	31.12	6.7
Morro Bay	0.96	−0.12	−14.68	−16.55	−43.64	43.51	28.96	27.09	2
Muscongus Bay	2.00	−50.30	−63.59	−157.96	0.76	−51.06	−64.35	−158.72	3.1
Narragansett Bay	2.64	−18.42	−119.84	−232.11	−17.27	−1.15	−102.57	−214.85	15
Narraguagus Bay	2.13	−28.78	−49.15	−165.92	6.37	−35.15	−55.52	−172.29	10.4
Nehalem River	0.44	−1.69	−4.77	−22.51	−14.32	12.63	9.55	−8.19	6.1
Netarts Bay	1.73	−4.41	−14.37	−99.53	−19.15	14.73	4.78	−80.38	0.9
New Jersey Inland Bays	3.95	−17.48	−10.14	−386.32	−31.69	14.21	21.55	−354.63	431
New River	2.74	−6.15	−88.60	−204.35	−57.80	51.65	−30.80	−146.55	27.7
Newport Bay	1.56	−3.12	−6.36	−6.78	−65.94	62.82	59.58	59.16	1.4
North Ten Thousand Islands	3.16	−19.13	−33.91	−266.64	−101.29	82.16	67.38	−165.35	924.9
North/South Santee Rivers	3.11	−18.71	−30.33	−166.62	−65.06	46.35	34.73	−101.56	163.5
Ossabaw Sound	2.72	−4.12	−61.27	−259.28	−70.65	66.53	9.39	−188.63	222.4
Pamlico Sound	3.45	−19.49	−117.17	−315.85	−58.76	39.28	−58.40	−257.09	648.5
Passamaq./St. Croix/Cobs.	2.02	−26.53	−61.46	−124.69	9.56	−36.09	−71.02	−134.25	3.3
Penobscot Bay	1.92	−30.67	−46.92	−146.01	6.09	−36.76	−53.01	−152.10	6.2
Pensacola Bay	2.64	−5.61	−95.36	−257.48	−72.64	67.03	−22.72	−184.85	61.3
Perdido Bay	2.92	−2.31	−38.79	−137.46	−73.79	71.48	35.00	−63.68	14.5
Plum Island Sound	2.42	−5.86	−54.91	−238.99	−12.25	6.39	−42.66	−226.74	40.5
Puget Sound	1.14	−24.67	−31.99	−53.64	−11.85	−12.82	−20.15	−41.80	46
Rio Grande	4.43	−0.50	−13.86	−25.22	−99.59	99.10	85.74	74.37	0.4
Rogue River	0.84	−1.93	−24.15	−30.50	−25.90	23.98	1.75	−4.59	0.6
Rookery Bay	2.82	−13.90	−21.45	−244.11	−99.06	85.15	77.61	−145.05	75.1
Sabine Lake	6.83	−83.55	−483.12	−588.15	−74.49	−9.06	−408.63	−513.66	593.2
Saco Bay	1.81	−19.20	43.09	−95.22	6.39	−25.59	36.70	−101.61	17.9
San Antonio Bay	5.57	−13.17	−28.81	−58.07	−86.18	73.01	57.37	28.11	101.3
San Diego Bay	1.86	−0.04	−32.88	−43.42	−57.61	57.57	24.72	14.19	1.5
San Francisco Bay	1.28	−7.83	−29.90	−102.05	−51.75	43.92	21.86	−50.30	212
San Pedro Bay	1.44	0.00	−0.06	−2.37	−61.23	61.23	61.17	58.86	2.7

(Continued)

Table 4.1 (Continued)   Order-of-Magnitude Estimates of Wetland Carbon Flux by Estuarine Drainage Area (EDA)

Estuary	Accretion Rate (mm year⁻¹)	T_{soc} (g m⁻² year⁻¹) 0–20 cm	T_{soc} 0–100 cm	T_{soc} 0–300 cm	E_{soc} (g m⁻² year⁻¹)	L_{soc} (g m⁻² year⁻¹) 0–20 cm	L_{soc} 0–100 cm	L_{soc} 0–300 cm	Wetland Area (km²)
Santa Monica Bay	1.55	-2.92	-2.93	-71.31	-61.61	58.68	58.68	-9.71	0.6
Sarasota Bay	2.77	-101.52	-158.29	-196.25	-94.71	-6.81	-63.58	-101.54	10.2
Savannah River	2.97	-38.71	-74.87	-237.42	-71.01	32.30	-3.86	-166.41	271.6
Sheepscot Bay	1.91	-31.12	-37.67	-168.33	0.99	-32.12	-38.66	-169.32	9
Siletz Bay	2.15	-13.53	-18.01	-94.12	-18.46	4.93	0.45	-75.66	2.1
Siuslaw River	1.35	-5.15	-30.56	-71.77	-23.59	18.44	-6.98	-48.18	5.9
South Ten Thousand Islands	3.77	NA	NA	NA	-104.03	NA	NA	NA	893.8
St. Andrew Bay	2.90	-31.03	-72.28	-105.72	-76.10	45.08	3.82	-29.61	42.5
St. Andrew/St. Simons Sounds	2.41	-109.86	-113.98	-296.99	-73.26	-36.60	-40.72	-223.73	551.8
St. Catherine's/Sapelo Sounds	2.61	-1.70	-52.62	-280.73	-73.55	71.85	20.93	-207.18	532.3
St. Helena Sound	3.01	-5.98	-76.94	-300.25	-67.89	61.90	-9.05	-232.36	379.9
St. Johns River	2.41	-1.89	-189.77	-230.65	-83.28	81.39	-106.49	-147.37	78.7
St. Mary's River/Cumberland	2.33	-44.01	-48.43	-256.79	-77.97	33.96	29.54	-178.82	184.7
Stono/North Edisto Rivers	3.12	-1.59	-4.38	-11.43	-69.68	68.10	65.30	58.25	177.8
Suwannee River	2.08	-32.17	-93.76	-150.44	-79.48	47.31	-14.28	-70.96	112.4
Tampa Bay	2.69	-42.72	-61.14	-111.35	-93.90	51.19	32.76	-17.45	88.3
Terrebonne/Timbalier Bays	8.94	-30.01	-159.86	-444.75	-81.44	51.43	-78.42	-363.31	1155.6
Tijuana Estuary	1.48	0.00	-14.80	-21.09	-48.02	48.02	33.22	26.93	1.8
Tillamook Bay	1.10	0.17	-5.45	-61.47	-15.23	15.39	9.78	-46.24	5
Tomales Bay	1.70	-13.81	-27.48	-40.76	-41.20	27.39	13.71	0.44	4.1
Umpqua River	1.19	-0.85	-28.12	-58.02	-24.69	23.85	-3.42	-33.32	7.2
Upper Laguna Madre	4.01	-4.46	-12.64	-25.99	-92.33	87.87	79.69	66.34	9.7
Waquoit Bay	2.74	-11.37	-25.64	-189.49	-17.98	6.61	-7.66	-171.51	1.4
Wells Bay	2.80	-9.50	-13.68	-122.98	-3.99	-5.52	-9.69	-119.00	3.5
West Mississippi Sound	7.16	55.94	-94.54	-476.51	-74.87	130.81	-19.67	-401.63	780.3
Willapa Bay	-0.20	1.39	7.23	10.23	-16.74	18.14	23.98	26.97	44.5
Winyah Bay	2.93	-10.94	-32.91	-190.52	-60.10	49.17	27.20	-130.41	339.9
Yaquina Bay	2.15	-10.29	-19.79	-96.57	-21.87	11.58	2.08	-74.70	2.1
National Average	3.05	-23.83	-79.31	-190.74	-48.94	24.63	-30.86	-142.28	197.90

Positive values indicate net-carbon influx going into the wetland from the water column.
T_{soc} flux is total carbon loss from wetlands within each estuary; E_{soc} portion lost to the atmosphere via soil respiration flux; L_{soc} portion lost to the adjacent estuary.

is likely at least some export still occurring, and (v) the 20-cm depth likely amplifies any potential errors made in the assumptions about soil respiration—wherein E_{SOC} is based on literature values that utilized 'CO$_2$ chambers' and assume export occurs through the entire soil profile moving upward to the atmosphere—and thus, when subtracted from T_{SOC} from only a 20-cm-deep column of soil, the calculation will result in negligible or even biased positive L_{SOC} flux.

In addition to the fact that the 100 cm depth is most commonly used by national and international accounting efforts, the range of L_{SOC} among the various estuaries at the 100 cm depth (Figure 4.5) is quite similar to that found from the literature review presented earlier in this chapter (as depicted in Figure 4.4), ranging from −504.22 to 90.31 g cm^2 year^{-1} versus approximately −500 to 100 g cm^2 year^{-1}, respectively. For reasons of minimizing the assumptions outlined in detail above, we consider the 100-cm depth alone from here forward in this chapter for making spatial comparisons across the various estuaries, though the above questions will certainly form the basis for necessary future work. From this type of perspective, the present review allows us to ask questions about the types of processes that may be related to the spatial position of individual estuaries or similar groups of estuaries, and assess trends and patterns.

The estuaries that had the most positive L_{SOC} (import of OC from the water to the wetland) at the 100 cm depth (Figure 4.5) included the Tijuana Estuary and San Diego Bay in Southern California, and the Rio Grande and the Lower Laguna Madre in Texas. Concurrently, these locations had close to zero T_{SOC} as they have little in situ organic materials to begin with in their soils (Figure 4.5a), but strongly negative E_{SOC} as they are in warm locations with high soil respiration rates (Figure 4.5b). Thus, the mass balance of Equation 4.2 forces positive L_{SOC} (Figure 4.5c). These relatively "dry climate" estuaries are also typically high in sandy soils, subject to intermittent yet extreme rainfall events that provide the majority of the OC flux going into the estuarine water from riverine sources, and with some riverine or estuarine-sourced OC material occasionally landing into the wetlands. These hot, dry wetlands are low in OC production and are likely net importers of OC.

Conversely, the estuaries that had the most negative L_{SOC} values at the 100 cm depth (export of OC from the wetland to the water) can be found in highly erosive environments in the Chenier Plain of Louisiana and Texas, and in Southern Louisiana. Ranked in order from the most negative L_{SOC} at the 100-cm depth: Calcasieu Lake, Mermentau River, Sabine Lake, Breton/Chandeleur Sound, Brazos River, and Galveston Bay. At the 300-cm depth, the following can be added: Atchafalaya/Vermillion Bays, Barataria Bay, West Mississippi Sound, and Terrebonne/Timbalier Bays. These estuaries have such strongly negative T_{SOC} values due to large OC loss from the soil (Figure 4.5a), that even though they have negative E_{SOC} as they are in warm locations with high soil respiration rates (Figure 4.5b), the mass balance of Equation 4.2 yields a remaining and great quantity of lateral export into the water column (Figure 4.5c).

Interestingly, the West Mississippi Sound had both the most strongly positive L_{SOC} values at the 20 cm depth (net import) and one of the more strongly negative L_{SOC} values at the 100-cm and 300-cm depths (net export), essentially reflecting high uncertainty in this location. The East Mississippi Sound was somewhat similar, though less pronounced. A possible reason for this behavior is that OC is depositing from riverine sources into the surficial layers of the soil and our dataset has this accounted for appropriately, but large quantities of OC are simultaneously lost at depths lower than 20 cm.

Other interesting features include (i) the relatively low T_{SOC} losses of the U.S. West Coast as compared with the East and Gulf Coasts (Figure 4.5a), perhaps due to the absence of large deltaic plains composed of wetlands high in organic materials; (ii) the positive soil respiration flux in the far northeast, suggesting that this region may be inappropriately modeled by Equation 4.3 and that a threshold may be required at a given temperature (Figure 4.5b); and (iii) the relatively high frequency of positive L_{SOC} in estuaries that contain large rivers or inflow sources such as the Columbia River, portions of the Everglades, and the around South Carolina and Georgia (Figure 4.5c). Also interesting, we found weak linear correlation between estuary size and L_{SOC} export particularly at

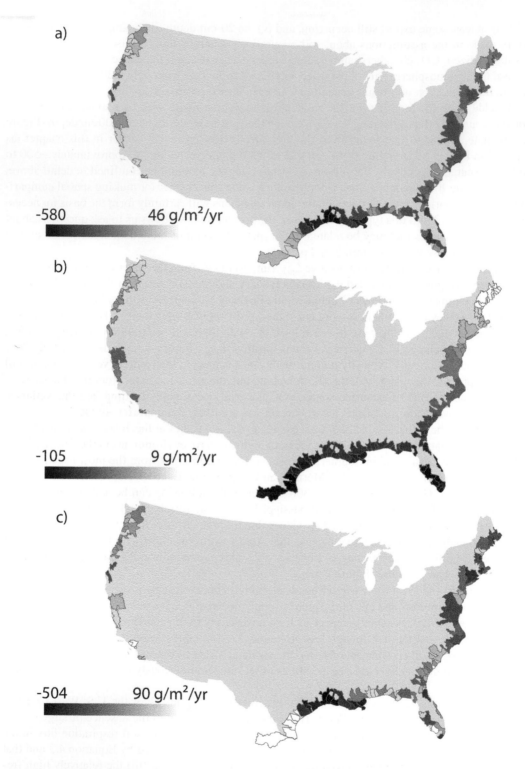

Figure 4.5 Maps of (a) T_{SOC} flux, or total carbon loss from wetland soils within each estuary, (b) E_{SOC}, or the portion lost from wetlands to the atmosphere via soil respiration flux, and (c) L_{SOC}, or the portion lost from wetlands to the adjacent estuarine waters. Positive values indicate net-carbon influx going into the wetland; negative values indicate net export from the wetland.

the 300 cm depth ($r^2 = 0.32$), estuary size and T_{SOC} export ($r^2 = 0.38$), and estuary size and accretion ($r^2 = 0.42$)—suggesting that estuary size may be a surrogate for whether an estuary is deltaic and wetland-dominated, and thus laden with large quantities of wetland OC available for export.

In summary, the total flux of OC from wetlands to the waters of the estuaries of the continental United States was calculated to lie between 1.00 Tg year^{-1} and -5.79 Tg year^{-1}. This range is based on the variability between the estimates of L_{SOC}, using the upper 20 cm of depth versus the full 300 cm of depth. The central value at 100 cm was -1.26 Tg year^{-1}. If we add in the contribution from rivers (13.98 Tg year^{-1}), after summarization across all EDAs from the Coastal Assessment Framework of Bricker et al., 1999), the total flux of OC from wetlands and rivers going into the estuaries of the continental United States ranges from -13.07 to -19.60 Tg year^{-1}, with the central value at 100 cm as -15.27 Tg year^{-1}. These values should be viewed as back-of-the-envelope estimates only, for the purpose of synthetic measurement, to obtain only the most broad and generalized picture of potential OC flux from wetlands and rivers into estuaries.

The greatest value of this synthetic work is not to arrive at a definitive numerical value for a given phenomenon, but rather to expose the areas of science that need further study. The greatest source of potential error that we found concerns OC variability across "depth" and "time." These include the depth to which soil respiration occurs, the compaction rates of soil across depth, and the integration of soil depth across time increments. Choices on these subjects drive the major discrepancies in wetland-to-estuary flux estimates and can even flip their import/export sign. To improve this work in the future, more biogeochemical research will be required on the depth and time limits over which these linked wetland-estuary processes operate.

4.6 SUMMARY

1. Since coastal BCEs are one of the largest global C reserves and have several threats, it is imperative to understand coastal transport dynamics, latitudinal gradients, biogeochemical factors, and export rates of allochthonous blue carbon.
2. The study of BCEs has advanced in the last several decades, but there are still several processes that remain unconstrained, like OM respiration, priming effects, and carbon exports.
3. Deciphering C cycling in BCEs is complex because it depends on many factors like wetland type, physical drivers, latitude, and microbes.
4. Finally, it is important to standardize methodologies in BCEs studies in order to have reliable comparisons across BCE types (i.e., seagrasses, marches, and mangroves), depths and times across which these comparisons are made, and to be able to constrain the processes involved in carbon cycling.

CHAPTER 5

Net Ecosystem Carbon Balance of Coastal Wetland-Dominated Estuaries
Where's the Blue Carbon?

Charles S. Hopkinson
University of Georgia

CONTENTS

HIGHLIGHTS

1. Net ecosystem C budgets are but one tool in the coastal scientist's or manager's toolbox for building a predictive understanding of blue carbon storage in wetland-dominated estuaries. Simulation modeling and incorporating mechanistic controls of key processes identified through field-based experiments is another important tool needed to predict blue carbon storage through the rest of the 21st century.
2. C budgets and conceptual models of C dynamics effectively illustrate some of the complexity and relations between various coastal system habitats, regions, ecosystem processes, and blue carbon storage.
3. Blue carbon storage is a significant component of the net-carbon balance of wetland-dominated estuaries. It represents 9%–29% of the net uptake of CO_2 from the atmosphere by tidal wetlands, or 11%–37% of tidal wetland net ecosystem production (NEP).

4. Blue carbon storage cannot be quantified with mass balance techniques. As has been shown in attempts to define the balance between coastal autotrophy and heterotrophy, the process of interest is too small (blue carbon storage—NEP) relative to the magnitude and high variability and uncertainty in various components that contribute to gross primary production (GPP) and R_{AH}. Blue carbon storage must be directly measured as the product of sediment organic C content, marsh elevation change over time, and the change in wetland areal extent over time.

5. With respect to net-carbon balance and blue carbon storage, the processes which are least understood from a controls perspective are those associated with GPP, R_{AH}, NEP, wetland C export, decomposition of belowground organic matter, and C preservation in wetland soils.

6. A combination of free-water diurnal O_2/CO_2 and eddy covariance tower approaches over long terms (at least 18.6-year-lunar nodal cycle) may provide useful information on how wetland-dominated estuarine metabolism, NEP, and blue carbon storage potential are likely to change over the remainder of the 21st century.

7. Net C balance models seldom incorporate the human dimension, and thus miss critical activities likely to influence blue carbon storage. Critical activities include living shoreline armoring, policies limiting wetland reclamation and changes to hydrology and water circulation, and of course policies to control the emissions of CO_2, which influence global warming and sea level rise (SLR).

Carbon dynamics of wetland-dominated estuaries have been a scientific interest since ecosystem ecology approaches were first initiated in the 1950s. Early interest focused on mechanisms controlling the immense productivity of estuaries (Schelske and Odum 1962), the fate of marsh macrophyte production (Teal 1962; Day et al. 1973) and whether organic carbon (OC) export to tidal waters and incorporation into the detrital food web contributed to enhanced secondary and fisheries production (Nixon 1982). More recently, we have been examining estuaries as **modifiers** of terrestrially derived organic and inorganic carbon (Raymond and Bauer 2001), as to their contribution to overall C dynamics of the global coastal ocean (Bauer et al. 2013) and as to their potential to sequester atmospheric CO_2.

The sequestration of atmospheric CO_2 by coastal and marine ecosystems is referred to as the blue carbon sink. The blue carbon sink for tidal wetlands (salt marshes and mangroves) is attributed to the long-term burial of OC in the wetland sediments. It is generally recognized that C burial rates per unit area are extremely high and when extrapolated to their global area are of comparable magnitude to C burial in all the world's oceans (Nellemann et al. 2009; Hopkinson 2012). Recent research has been focused on how sink strength will respond to decreased rates of sediment supply and increased rates of SLR. These factors play a critical role in wetland elevation gain and areal expansion (Morris and Callaway, Chapter 6). Processes that alter burial rates or global area will directly alter the importance of these systems in mitigating anthropogenic emissions of CO_2 and hence mitigating climate change.

In this chapter, I review blue carbon accumulation by brackish and saline wetland-dominated estuaries from the perspective of estuarine net C balances. I examine the processes or C fluxes that are the basis of C accumulation and the mechanistic controls of those fluxes. Accumulation fluxes are examined in relation to estuarine linkages with adjacent terrestrial and oceanic systems and considered at a variety of scales—from plot to global—and directions—vertical and horizontal. I also discuss how the long-term wetland development trajectories since the last glaciation will likely change direction in the near future. This chapter complements others in this book, several of which focus in more detail on processes such as the import of allochthonous material from adjacent systems, SLR, and wetland accretion and OC burial.

5.1 WETLAND-DOMINATED ESTUARIES

Saline and brackish water wetland-dominated estuaries are relatively simple biotically (compared to temperate and tropical forests or adjacent oceanic regions) with low species diversity of macroflora and macrofauna (excluding prokaryotes), because of their short evolutionary history (past 4,000 years) and stressful environment (e.g., high salt concentration, strong redox gradients, intermittent flooding) (Day et al. 2013). At the same time, however, they are tremendously productive, because of their location at the land-sea intertidal interface (Chapman 1960; Bird 1969; Day et al. 2013). The tidal brackish and saline wetlands within these estuaries are characterized by their rooted, halophytic macrophytes that live in the intertidal region generally between mean sea level (MSL) and slightly above mean higher high water (MHHW). Mangroves (swamps with woody halophytic vegetation) dominate intertidal wetlands in tropical environments, while salt marshes (nonwoody vegetation) dominate in temperate and boreal regions of the world's coastal zone, as mangroves are intolerant of freezing temperatures. Intertidal wetlands, while found worldwide, are most extensive on non-tectonic coasts with wide continental shelves and high sediment delivery from upland runoff (Day et al. 2013). The global extent of mangrove and salt marsh wetlands has high uncertainty, but is in the range of 120,000–200,000 km^2 for mangroves (Giri et al. 2011) and 200,000–400,000 km^2 for salt marshes (Nellemann et al. 2009; Hopkinson et al. 2012). Updated areal extents are shown in Chapter 13 (Giri) and Chapter 14 (Byrd et al.).

Wetland-dominated estuaries are among the most productive ecosystems of the world, often matching that of intensively fertilized agricultural systems (Odum 1971; Day et al. 2013). Several factors contribute to their high productivity, including the variety of producer communities (e.g., seagrasses, wetlands, phytoplankton, benthic micro and macroalgae), the mixing of surface and bottom water that couples benthic nutrient remineralization with surface primary producers, tidal flushing that mixes nutrients throughout the system and flushes toxins from pore waters, and allochthonous material inputs as a result of their location at the land-sea interface, which brings in nutrients and organic matter from terrestrial systems as well as the ocean (Shelske and Odum 1962; Day et al. 2013). Riverine inputs give estuaries the highest C and N loading rates on Earth. Terrestrial OC flux has been estimated to be about 450 Tg C year^{-1} (Bauer et al. 2013). Duarte et al. (2005) estimated areal rates of GPP to be in the range of 100–4,000 gC m^{-2} year^{-1}. While open-water regions of estuaries can vary between being net autotrophic or net heterotrophic systems depending on the balance in inorganic and organic matter loading from watersheds and adjacent tidal wetlands (Smith and Hollibaugh 1993; Hopkinson and Vallino 1995; Kemp et al. 1997), the tidal wetland portion of estuaries is almost exclusively net autotrophic and the basis for blue carbon accumulation.

Any consideration of the blue carbon accumulation in estuarine wetlands of the world must examine the geomorphic development of estuarine wetlands over a range of temporal scales—from their creation (millennia), since the industrial revolution (century), since significant changes in their watershed land use and land cover (decades to century) and into the future in conjunction with climate change and increasing rates of SLR (decades). The reason for this is that blue carbon burial is directly related to the expansion, both vertically and horizontally, of wetlands (with their associated sequestered C) over time, which is largely controlled by the rate of SLR and the availability of sediment (Reed 1995; Morris et al. 2012; Kirwan et al. 2010; Morris and Callaway, Chapter 6). While there was probably always halophytic vegetation in the intertidal zone of all muddy shorelines since the last glacial maximum (~18,000 years ago), the development of deltas and intertidal wetlands as we know them today began only in the past 3,000–6,000 years once SLR decreased to rates less than 1–2 mm year^{-1} (Redfield 1967; McKee et al. 2007). As shoreline transgression slowed and estuaries became more stable, they began to fill with terrigenous sediments. Halophytic vegetation along the estuarine shoreline prograded into shallowing waters, transgressed over uplands as sea level rose, and accreted vertically as vegetation trapped suspended sediments from floodwaters and undecomposed organic matter accumulated in sediments (Bird 1971). The process of wetland development and expansion in estuaries over historical periods has been described based on stratigraphic studies for mangroves and salt marshes, beginning with Mudge (1858) and continued into the 20th century by scientists including Davis, Egler, Chapman, Redfield, Frey, and Woodroffe (Davis 1910; Davis 1940; Chapman 1944; Egler 1952; Redfield 1967; Frey and Basan 1978; Woodroffe 2002). More recently, studies have focused on the impacts of accelerating rates of SLR and decreasing coastal input of terrigenous sediment on continued wetland development or degradation (e.g., Day et al. 1977; Reed 1995, 2002, Morris et al. 2002, 2012; Kirwan 2012; Fagherazzi et al. 2013; Mariotti et al. 2010, 2013; Weston 2014; Morris 2016).

Geomorphic studies show a coastal landscape comprised of wetland and open-water estuarine regions that change over time (Frey and Basan 1978). Both mangroves and salt marshes have fringe or low marsh regions, which are adjacent to water bodies. Further inland is the basin or interior mangroves or the high marsh platform of the salt marsh. It's typically the high marsh or interior mangrove regions that transgress onto the terrestrial uplands as sea level rises. It's the low marsh or fringe mangroves that prograde into open-water areas. Over time, the area of water and the extensiveness of tidal creek systems shrink as a result of sediment infilling. Thus, wetlands accrete vertically and spread horizontally. Another related process, which I will not cover in this chapter, is salt-water intrusion into low-lying, fresh wetlands along the coast and up tidal rivers. Salt-water intrusion can lead to the expansion of salt marshes in a similar fashion to transgression, but it comes with a host of redox controlled biogeochemical processes that add complexity to the blue carbon story. In particular, past freshwater stores of sediment organic matter are susceptible to enhanced decomposition in the presence of sulfate from seawater and increased rates of methane emission C (Weston et al. 2011).

5.2 NET ECOSYSTEM CARBON BALANCE CONCEPTUAL FRAMEWORK

What does blue carbon accumulation mean in terms of the net ecosystem carbon balance (NECB) of coastal wetlands? In the simplest terms, the net carbon balance of a wetland-dominated estuary is the net change in total C stocks of the estuary over time (e.g., Troxler et al. 2013). It could be measured as the slope of a line describing total carbon stocks (y axis) over time (x axis).

$$NECB = dC/dt \qquad (5.1)$$

It is unfeasible to directly measure NECB as the annual change in organic and inorganic stocks of C. Rather NECB is more often determined for a single point in time by balancing all the processes that contribute to change in C stocks over short intervals of time, e.g., GPP, whole system respiration of autotrophs and heterotrophs (R_{AH}), net CO_2 exchange with the atmosphere, input of inorganic and OC from land runoff, OC burial in wetland sediments, etc. The NECB can be determined for tidal wetlands, the estuary, or both regions together. Blue carbon storage within a tidal wetland is a component of the overall wetland-dominated estuarine system NECB. But static measures provide no information on how accumulation changes over decades or longer time frames. For the purpose of this chapter, I consider blue carbon storage to represent only the OC that accumulates and is buried in tidal wetland sediments, which as we will see is related to areal expansion of tidal wetlands in horizontal (area) and vertical space (accretion). Thus, the NECB of a wetland-dominated estuary must be resolved into marsh and open-water estuary regions in order to resolve blue carbon storage.

5.2.1 Blue Carbon Accumulation over Periods of Rising Sea Level

A time-lapse view of wetland development scenarios highlights the accumulation of blue carbon in wetland-dominated estuaries over time (Figure 5.1 left and right). The initial condition (T_0) is a shoreline subject to tides with a tidal amplitude extending between mean low water (MLW) and mean high water (MHW). Here, I simplify the discussion of tidal ranges (e.g., ignoring the difference between MLLW and MLW and MHHW and MHW) and highest elevation attained by a marsh platform. Our understanding of the marsh platform elevation relative to high water levels

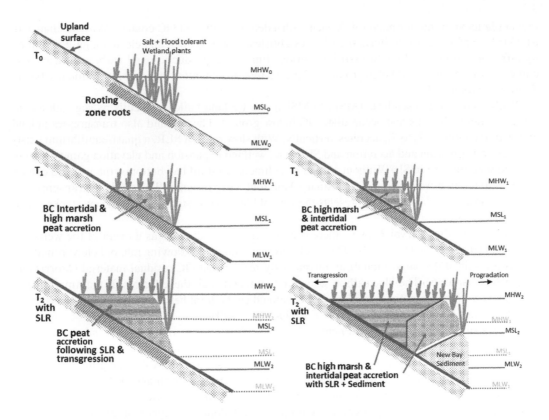

Figure 5.1 **(See color insert following page 266.)** Time series conceptualization of blue carbon storage in tidal wetlands under scenarios of SLR and varying sediment availability (low availability left versus high on right). Only transgression and surface accretion occur under limited sediment availability, while progradation requires high sediment availability. T_0, T_1, T_2 refer to time progressions. MHW, MSL, MLW with subscripts refer to MHW, MSL, and MLW at times T_0, T_1, and T_2. As drawn SL does not change between T_0 and T_1. BC = blue carbon. Three soil organic matter components are shown on left: (1) that associated with roots, which does not change over time and is not included in blue carbon burial, (2) that associated with development of the high marsh platform prior to SLR, and (3) that associated with development of the high marsh platform, plus marsh transgression following a rise in sea level. To the RIGHT, we start with the T_1 stocks from the left, but then superimpose SLR and marsh progradation. In this case, we have blue carbon accumulating as on the left side, plus that associated with progradation. Conceptualization follows Redfield 1967.

is rudimentary, so I simply say that wetland vegetation extends from MSL to MHW. The tidally flooded upland surface between MSL and MHW is vegetated by halophytic vegetation. Sediments contain live and dead roots and rhizomes.

The three-dimensional development of intertidal wetlands is primarily controlled by three factors: availability of sediment, the rate of SLR, and the accumulation of undecomposed wetland vegetation. Consider first the availability of sediment. Without sufficient input of sediments to allow bay infilling (Figure 5.1 left), over time (T_0 to T_1) the halophytic vegetation surface rises as a result of plants trapping sediments from floodwaters and as a result of the accumulation of dead organic matter from roots, rhizomes, dead plant parts (T_1). There is always a surface zone with living roots and rhizomes and recently dead material. It is the dead OC of the accreted marsh sediments that forms the basis of sequestered blue carbon, not the nearly constant or equilibrium mass of live roots and rhizomes at the surface. At this early stage, we can recognize two zones: a creekbank, intertidal zone with robust vegetation growth (equivalent to creekbank low marsh or fringe mangrove) and a high marsh with low stature vegetation (equivalent to high marsh, high marsh platform or interior mangrove), each with a defined sediment OC content. With no change in SL ($MSL_0 = MSL_1$), an equilibrium will be established where growth and elevation gain is limited by infrequent tidal flooding. Stressful conditions limit plant production and carbon storage. Blue carbon in this case is total belowground OC minus that present after initial colonization (carbon at T_1 minus carbon at T_0).

With a slow rise in sea level (MSL_1 to MSL_2), the wetland builds vertically as vegetation continues to trap sediments and accumulate OC below ground. The wetland also transgresses upland and as in the interval T_0 to T_1, accretes vertically. With slow rates of SLR, a quasi-equilibrium exists between transgression and accretion and SLR (i.e., wetland expansion and elevation gain will be in balance with the rise in SL). Without sufficient sediments to infill bay bottoms, the wetland cannot prograde into open-water areas of the estuary. Horizontal expansion will be solely by transgression. Blue carbon accumulation now represents the sum of blue carbon stocks at T_1 and new stocks added between T_1 and T_2.

Throughout the interval T_0 to T_2, there is compaction of sediments as a result of the weight of new material accreting from above. There is likely a continued but slowing rate of belowground OC loss as a result of decomposition of an increasingly recalcitrant OC residual material (Morris and Callaway, Chapter 6). The blue carbon storage rate integrates all these factors. If the timeframe T_0 to T_2 takes a 1,000 years, the rate of storage is defined as the total belowground carbon mass at the end of T_2 minus the amount at the end of T_0 divided by 1,000 years.

With a sufficient supply of sediment and a relatively slow rate of SLR, wetland development also includes a progradation component (Figure 5.1 right) due to bay infilling. Wetland vegetation expands out across infilled portions of bay as bottom elevation reaches MSL. In Figure 5.1 right, we move from T_1 to T_2 and MSL_1 to MSL_2. In addition to the transgression and surface accretion that was associated with the platform or interior wetland region (as in Figure 5.1 left), the low, creek-bank, fringe zone builds out (progrades) and up, as the halophytic vegetation grows on intertidal areas at and above MSL. Blue carbon accumulation includes that associated with transgression, progradation, and accretion. The rate of storage is defined as the change in belowground OC mass during the interval between MSL_1 and MSL_2.

5.2.2 C Dynamics That Contribute to Blue Carbon Accumulation

Numerous C cycling processes contribute to the development of wetland-dominated estuaries and blue carbon accumulation in the coastal zone over time. The clearest way of showing how all the processes relate to each other is to work at multiple scales, showing the (i) greatest detail at the smallest scale, the individual wetland and estuarine subsystems (e.g., a square meter of estuary or a square meter of wetland); (ii) moderate detail at a scale that shows the coupling of wetlands and the estuary and their combined coupling to land and the ocean; and (iii) least detail at a scale that also incorporates a long (decadal and longer) temporal scale (Figure 5.2a–c).

5.2.2.1 Fluxes That Contribute to the Net Carbon Balance of Wetland and Estuarine Subsystems

At the smallest scale, we examine the integrated dynamics of both inorganic and OC that operate at temporal scales from < hourly, in order to capture dynamics such as tidal flooding and draining of a wetland, to annual, in order to capture the annual dynamics and fluxes associated with metabolism (GPP and R), exchange between wetlands and adjacent estuary, and blue carbon storage. A blue carbon assessment requires analysis of both organic and inorganic forms of C, otherwise we might trick ourselves into thinking our system sequesters more C from the atmosphere than it really does. For instance, we might think coral reefs sequester CO_2 from the atmosphere, when in reality there is a net release of CO_2 to the atmosphere during calcification (Kennedy et al. Chapter 9). Seven major fluxes of organic and inorganic C define the time rate of change of OC and inorganic carbon (IC) of wetlands or estuaries (Figure 5.2a).

$$\frac{dOC_{Subsysten}}{dt} = F_{01} + F_{i7} + F_{i01} + F_{02} + F_{03} + F_{i4} + F_{i5} \qquad (5.2)$$

where

dOC/dt—the rate of change in mass of live and dead OC in the subsystem of interest, for example, the combination of live and dead *Spartina alterniflora* leaf, rhizome, and root mass plus any other live and dead autotrophs and heterotrophs in/on a salt marsh plus peat accumulated in sediments since the formation of the wetlands (the blue carbon component of OC).

F_{01}—transport of allochthonous particulate and dissolved OC into the subsystem from adjacent systems. This flux is typically mediated by the flow of water associated with tidal exchange and/or river runoff. This flux could also represent the immigration of larval or juvenile organisms from adjacent systems, e.g., shrimp from the continental shelf, or shrimp from tidal creeks onto the marsh platform during high tide.

F_{02}—transport of allochthonous and or autochthonous DOC and POC to adjacent systems via tidal exchange (an export flux would be designated as a negative flux). An export of autochthonous OC is the result of positive NEP or the depletion of internal C stores, e.g., erosion of creekbank peat into tidal waters. This flux could also include the emigration of organisms to adjacent systems, e.g., adult organisms after growing up in the estuarine nursery

F_{i7}—exchange of CO_2 with the atmosphere in conjunction with GPP and the respiration of autotrophs and heterotrophs (R_{AH}). The balance between GPP and R_{AH} is termed NEP for the subsystem. A positive flux is into the system while a negative flux would be from the system to the atmosphere. For the salt marsh or mangroves, the CO_2 exchange with the atmosphere is increasingly measured with eddy covariance flux tower approaches and is designated as net ecosystem exchange (NEE). NEE in periodically flooded environments, like salt marshes, does not fully account for all GPP and R_{AH} pathways, e.g., when the vertical flux is interrupted by tidal floodwaters and thereby exchanged horizontally with adjacent aquatic systems.

F_{i01}—the net exchange of methane, carbon monoxide, and/or volatile organic compounds with the atmosphere in conjunction with methanogenesis/methane oxidation, UV degradation of organic compounds resulting in the production of carbon monoxide, and the loss of volatile organic compounds, such as dimethyl sulfoxide.

F_{03}—the burial of OC in sediments (**blue carbon**) resulting from a combination of subsystem NEP and the balance of allochthonous *OC* inputs/outputs (F_{01} & F_{02}).

F_{i4}—the respiration of autotrophs and heterotrophs directly into water, resulting in an increase in DIC.

F_{i5}—the uptake of dissolved inorganic carbon (DIC) from water during GPP, e.g., by benthic microalgae, phytoplankton, or intertidal benthic microalgae when inundated.

$$\frac{dIC_{Subsystem}}{dt} = F_{i1} + F_{i2} + F_{i3} + F_{i4} + F_{i5} + F_{i6} \qquad (5.3)$$

where

dIC/dt—the rate of change in the total mass of dissolved inorganic carbon in the subsystem of interest, e.g., the mass of DIC in wetland sediments at low tide or the mass of DIC in wetland sediments and flood waters at high tide.

F_{i1}—the input of DIC from adjacent systems, e.g., the DIC transported onto tidal wetlands at high tide. From the perspective of the estuary, the aquatic system, the import of DIC from water draining off a marsh during ebb tide would be representative of this flux.

F_{i2}—the transport of DIC to adjacent systems, e.g., the DIC transported from the marsh to tidal creeks during ebb tide and creekbank drainage. From the perspective of the wetland subsystem, F_{i1} and F_{i2}

are representative of the DIC fluxes not captured by eddy covariance approaches that measure NEE (see above).

F_{i3}—the net burial or dissolution of carbonate in the subsystem, e.g., burial of bivalve shells.

F_{i4}—the input of DIC in conjunction with the respiration of autotrophs and heterotrophs while submerged in estuarine waters, e.g., phytoplankton or bacterial respiration.

F_{i5}—the uptake of DIC in conjunction with GPP by autotrophs in estuarine waters and sediments or in wetlands while inundated with tidal water.

The units for describing subsystem OC and IC fluxes could be expressed by unit area per unit time (e.g., $gC\ m^{-2}\ year^{-1}$ referring to an average square meter of salt marsh or estuarine water body) or by the total area of wetlands in a particular estuary per unit time (e.g., metric tonnes $C\ year^{-1}$ for a specified area). Note that as these equations are written, for there to be a net increase in a pool of C over time, the sum of the inputs and the outputs to that compartment must be positive. With this convention, flows into a compartment are positive and flows out of a compartment are negative. Thus, the blue carbon flux itself (F_{03}), is a negative flux with respect to the $dOC_{Subsystem}$ box. The signs of fluxes are often confusing, e.g., sometimes net fluxes of CO_2 to the atmosphere are shown as positive and other times negative. One must always check what the flux is in relation to—a flux to the atmosphere box or a flux from the plant or ecosystem box.

5.2.2.2 Fluxes That Contribute to the Net Carbon Balance of a Wetland-Dominated Estuary

Here the system of interest is the integrated estuary and tidal wetlands and their connection to adjacent terrestrial and oceanic systems (Figure 5.2b). Inputs from land via river runoff and exchange with the ocean highlight the **open** nature of wetland-dominated estuaries (Hopkinson 1992). The subsystem dynamics described in Figure 5.2a are simplified at this level, showing only the fluxes associated with NEP and the exchange of allochthonous C between subsystems and adjacent systems. Eight fluxes define the rate of change of total C for wetland-dominated estuaries:

$$\frac{dC_{System}}{dt} = F_{(DOC,POC,DIC)Land-River} + F_{CH_4} + F_{CO} + F_{VOC} + F_{CO_2} + F(IC)_{sed} + F(OC)_{sed} + F_{(DOC,POC,DIC)Ocean}$$

$$(5.4)$$

where

$F_{(DOC,POC,DIC)Land-River}$ represents the riverine input of dissolved and particulate organic and inorganic C,

F_{CO}, F_{CH_4}, and F_{VOC} represent the exchange of these compounds from both wetlands and estuarine waters with the atmosphere. They include fluxes associated with photochemical processes and other abiotic processes as well as methane oxidation and production,

F_{CO_2} represents the exchange of CO_2 with the atmosphere as a result of metabolism (GPP and R_{AH}) and air-sea exchange in conjunction with tidal waters out of equilibrium with atmospheric concentrations of CO_2 (for example, as water temperature varies),

$F(IC)_{sed}$ and $F(OC)_{sed}$ represent blue carbon storage of OC and the balance between carbonate precipitation and dissolution,

$F_{(DOC,POC,DIC)Ocean}$ represents the exchange of organic and inorganic carbon between the wetland dominated estuary and the coastal ocean. It includes the immigration and emigration of living organisms. Compared to the equivalent inputs from land, this flux integrates internal processing of terrestrial carbon (e.g., decomposition) as well as the addition of sources internal to the wetland-dominated estuary, e.g., DOC leached from tidal wetlands, or DIC produced from the oxidation of wetland derived plant material (Wang and Cai 2004).

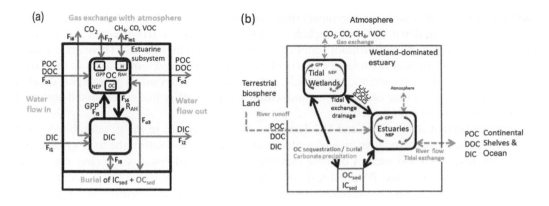

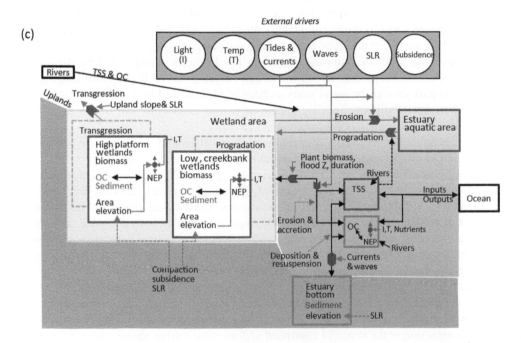

Figure 5.2 **(See color insert following page 266.)** (a–c) Conceptual models of net carbon balance and blue carbon storage for saltwater wetland-dominated estuaries. (a) illustrates the compartments of inorganic carbon and IC and OC and fluxes that contribute to the net C balance for either tidal wetlands or estuaries. Flows are numbered and described in the text. (b) illustrates the relation between tidal wetlands and estuaries and their connection to adjacent upland and ocean systems. Fluxes within the tidal wetland and estuaries boxes are simplified relative to the detail shown in A. (c) a conceptual diagram of the processes associated with the net C balance and blue carbon storage that includes spatial considerations of elevation gain/loss, progradation, and transgression. Key external drivers directly controlling the net C balance are identified. Salt-water intrusion and concomitant saltwater wetland expansion is not considered here. IC implied only.

5.2.2.3 Processes That Contribute to the Net C Balance of Wetland-Dominated Estuaries and Their Controls

For a decadal to millennial scale understanding of the major processes and controls of blue carbon storage in wetland-dominated estuaries, we must examine these systems from a landscape perspective. The landscape captures the vertical and horizontal aspects of tidal wetland OC accumulation. Consider a coastal system comprised of tidal wetlands and adjacent aquatic estuarine waters and bottom sediments that is open to inputs from land and the ocean (Figure 5.2c). In Figure 5.2c, wetlands are broken down further to include the low, creekbank, or fringe zones and the more inland, high platform or interior/basin zones of salt marshes and mangroves. There are six major external drivers of wetland-dominated estuarine C dynamics: light and temperature as they affect metabolism, and tides and currents, waves, SLR, and subsidence as they affect geomorphic development and the storage of blue carbon in wetland sediments. To simplify this conceptual diagram, IC is not illustrated—but it is implied.

Total suspended solids (TSS) and OC dynamics play a major role in the geomorphic development of wetlands and the relative area of wetlands and aquatic areas within the total system landscape. The major external sources of these materials are terrestrial systems, via river runoff, and the ocean, via tidal exchange. Potential sinks of TSS and OC are losses to the ocean, sedimentation on bay bottoms, sedimentation/accretion on wetland surfaces, and metabolism (decomposition). Primary production is an internal source of OC both in the estuary and on wetlands. The major controls of primary production, respiration, and hence NEP of wetland OC are light, temperature, and the depth, duration and frequency of tidal flooding (Morris et al. 2002). In the estuary, the major controls of GPP, R, and NEP are light, temperature, and nutrients.

The elevation or depth of estuarine creek and bay bottoms is controlled by the deposition or resuspension of bottom sediments via waves and currents and SLR (Mariotti and Fagherazzi 2013). Over the past several thousand years as rates of SLR slowed to 2 mm year^{-1} or less, many estuaries experienced infilling from sediments delivered from land and ocean (e.g., Redfield 1967).

Wetland elevation has increased over the past several thousand years in many estuaries as a result of accretion of TSS and OC brought in via tidal flooding and through an excess of wetland plant primary production over wetland respiration (R_{AH})—NEP. Accretion is controlled by several factors including the biomass and density of wetland plants, tidal flooding frequency, depth and duration, TSS and OC concentrations in floodwaters, and distance from tidal creeks and bays. A net effect of wetland trapping of TSS and OC from floodwaters is to decrease concentrations in estuarine source waters. Waves and tides and currents are factors that replace aquatic TSS and OC stores through resuspension of bay bottom sediments and erosion of marsh shorelines. Considering that wetland plant biomass and density are highest in the optimal flooding depths close to creek banks, accretion rates are highest in the same areas, often times resulting in natural levees running parallel to creek/bay shorelines a short distance inland. Three factors contribute to a decline in wetland elevation in opposition to accretionary processes—compaction, subsidence, and SLR. Compaction is the result of continued sedimentation from above which adds weight to the deepest sediments, thereby squeezing them, dewatering them, and effectively increasing their bulk density.

There are two processes that contribute to an expansion of wetland area—transgression and progradation. Recall that I am ignoring saltwater intrusion and the landward migration of salt marshes into brackish and freshwater tidal wetlands. Progradation results in a direct loss of estuarine aquatic system area, while transgression only changes the wetland:water landscape ratio. Transgression is 100% attributable to a rise in the effective rate of SLR (eustatic—minus subsidence or tectonic activity). Halophytic wetland plants will replace upland vegetation as salty water kills plants. The inland extent of transgression reflects the slope of the upland land surface and the elevation of spring tides. Transgression mostly causes an increase in interior/basin or high marsh platform marshes, places generally distant from estuarine bays.

Progradation is the result of bay infilling to a depth above MSL such that wetland plants can colonize, grow, and thereby effectively trap TSS from floodwaters. The rate of progradation is solely a function of the rate of bay infilling relative to SLR. Inadequate sediment inputs can reverse progradation, even in the absence of SLR (Mariotti and Fagherazzi 2013).

5.3 NET CARBON BALANCE OF COASTAL WETLANDS AND ESTUARIES

To the best of my knowledge, the first published analysis of the net carbon balance of an estuary was that of Teal (1962). The study was conducted in the Duplin River salt marsh-dominated estuary adjacent to Sapelo Island, GA. Like several similar aquatic system studies before (e.g., Lindeman 1942; Odum and Odum 1955; Teal 1957), Teal reported the energy budget for the system, even though all fluxes and standing stocks were based on measures of metabolism (measured with either O_2 or CO_2) or changes in OC standing stocks of key ecosystem components. Teal used mass balance to show an imbalance between primary production and ecosystem respiration. He concluded that 45% of the net production of marsh grass was exported to adjacent tidal creeks and bays where it likely contributed to secondary production. There was no mention of the accumulation of OC in wetland soils as another potential fate of the unmetabolized primary production. A decade later, Day et al. (1973) followed suit with a detailed analysis of the community structure and carbon budget of a salt marsh-dominated estuary in Louisiana. Again, mass balance was used to describe the fate of excess primary production of the salt marsh plants—50% exported to estuary and of that 42% exported to the Gulf of Mexico. As with Teal, there was no accounting for OC storage in marsh soils, even though Day et al. (1977) clearly acknowledged long-term trends in SLR and development of extensive areas of marsh and swamp over time. It wasn't until the mid-1970s that measures of OC burial in tidal wetland sediments were directly made and incorporated into net carbon analyses (Woodwell et al. 1979). Over the next decade, with the availability of inexpensive gamma detectors for measuring radioisotopes such as ^{210}Pb and ^{137}Cs, rates of wetland accretion and carbon burial became more common (Pomeroy and Wiegert 1981; Day et al. 1982; Howes et al. 1985).

It is important to note that in all these studies, net carbon balance analyses were not able to define carbon burial by mass balance. Indeed, Hopkinson (1988) showed in a comparative analysis of net carbon balance for six salt marsh-dominated estuaries of the eastern and southern United States, that carbon burial was frequently a trivial component of overall C dynamics. The fluxes that were actually computed by mass balance had huge uncertainty. For example, in the Sapelo Island marsh estuary, carbon burial in the salt marsh amounted to 29 gC m^{-2} year^{-1}, while the mass balance calculation of OC export to the ocean was 673 gC m^{-2} year^{-1}, and this out of a total estimate of net primary production (marsh and estuary) of 1,791 gC m^{-2} year^{-1} (Hopkinson 1988).

5.3.1 Current Global Net Carbon Budget for Wetland-Dominated Estuaries

The global net total carbon balance (inorganic and organic C) for wetland-dominated estuaries clearly illustrates the magnitude and relative importance of blue carbon burial (Figure 5.3). This carbon balance integrates the results of recent global assessments of blue carbon storage (e.g., McLeod et al. 2011; Hopkinson et al. 2012) with analyses of the total C budget for the global coastal ocean (Bauer et al. 2013). In the global C balance, we see the estuary as what would appear to be a straight-through pipe connecting watersheds to the ocean, where the export of inorganic C and organic C (IC and OC) nearly balances watershed inputs (0.85 Pg year^{-1} input and 0.95 Pg year^{-1} output). Numerous studies have shown significant processing of terrestrial OC during passage through estuaries, so this conclusion about a straight-through pipe is clearly a misleading aspect of net carbon analyses. Approximately 0.09–0.18 Pg year^{-1} are accumulated in wetland and estuarine sediments. The excess export and burial is possible only as a result of the high net uptake of CO_2 from the atmosphere by wetland vegetation (0.45 Pg year^{-1}). Much of the estuarine C burial is attributable to riverine sediment runoff, while almost the entirety of tidal wetland burial is due to the accumulation of undecomposed wetland plant organic matter. The net carbon budget shows tidal wetlands to be strongly autotrophic, with a NEP of 0.35 Pg year^{-1}, which is entirely the result of the uptake of CO_2 from the atmosphere. Estuaries on the other hand are strongly heterotrophic, sources

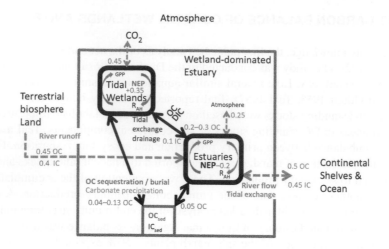

Figure 5.3 Net carbon balance for inorganic and OC for estuaries of the world, specifically identifying the importance of tidal wetlands. Blue carbon storage for the purposes of this chapter is that arising from tidal wetlands. OC storage in estuaries includes that of seagrasses and the burial of terrestrial organic matter. Taken from Bauer et al. (2013) and Hopkinson et al. (2012).

of CO_2 to the atmosphere, and dependent on allochthonous inputs of OC from the ocean, tidal wetlands, and land. While the net C budget does not separate the relative importance of oceanic and land sources to overall heterotrophy, it does show that tidal wetlands contribute both inorganic and OC through a variety of exchange pathways, including tides, creekbank drainage, and storms.

The overall magnitude of wetland blue carbon burial is large. It represents 9%–29% of the net uptake of CO_2 from the atmosphere by tidal wetlands, or 11%–37% of tidal wetland NEP. Considering that tidal wetlands make up less than 0.1% of the surface area of the global ocean, they have a disproportionately large role in C burial, actually comparable to total OC burial in sediments of the entire ocean.

5.4 WHAT DOES THE FUTURE HOLD FOR BLUE CARBON STORAGE?

All is not well for the future of tidal wetland-dominated estuaries. We know that rates of SLR have been increasing since the industrial revolution and the concomitant burning of fossil fuels and deforestation. As a result of global warming, we can expect rates of SLR to approach 20 mm year^{-1} by the close of the century (Kemp et al. 2011; Rahmstorf et al. 2012). We also know that river export of sediments to the world's oceans have declined over the past century as a result of river damming and improved agricultural practices that reduce soil erosion (e.g., Weston et al. 2014). Fortunately, however, for tidal wetlands, we know there is a negative feedback between SLR and wetland flooding depth (Morris 2016; Morris and Callaway, this volume), up to a point. Tidal wetland plants are found most typically at relatively high elevations in the intertidal that are supra-optimal for growth. As SL rises, plant flooding depth and duration increases, which stimulates biomass production, which in turn promotes particle trapping, accretion, and elevation gain. The rise in accretion rate enables wetland surface to maintain elevation relative to the rising rate of SLR. However, there becomes a depth where further increases in the rate of SLR raises the flooding depth to such a point that plant production and standing crop biomass are reduced, which results in a negative feedback on accretion. Unfortunately, for tidal wetlands, we have recently learned that there is a strong relation between marsh sustainability and the balance between rates of SLR and estuarine sediment supply that varies with tidal amplitude. We know little about these relations for mangroves. Further, we have also learned that the recent increase in the rate of SLR relative to declining inputs of sediments has resulted in a reversal of marsh progradation such that marsh shorelines are eroding and leading to an absolute decrease in marsh area (Mariotti et al. 2010, 2013). While none of the empirical erosion studies was done on mangrove shorelines, presumably the processes are identical. We know even less about calcareous surfaces, as in the Everglades, Yucatan peninsula, and elsewhere around the world. On the positive side, increases in sea level should lead to marsh transgression into formerly terrestrial ecosystems. I have seen some estimates that the area of tidal wetlands in the conterminous United States could increase 70% with a 1 m rise in SL. Transgression pre-supposes that transgression will not be blocked by barriers to SLR erected by humans in an attempt to protect private property, however, and that all existing marshes are able to maintain elevation (Morris et al. 2012). Certainly, with the high value of real estate at the marsh/estuary edge in today's industrialized countries, we can expect herculean efforts to halt marsh transgression. From a purely OC accumulation perspective, when transgression occurs, we must also balance the gain in blue carbon storage associated with new transgressive wetlands with the decomposition loss in C stocks of the former land use (e.g., forests).

5.4.1 How Can a Net Ecosystem Carbon Budget Inform Us or Guide Us in Predicting Wetland Blue Carbon Burial over the Remainder of the 21st Century?

There are many interdisciplinary teams developing simulation models to predict the fate and sustainability of tidal wetlands. While there are still some basic physics that remain to be better described, such as the relation between wetland plant biomass and aboveground density and the drag coefficient and the Reynold's number based on stem diameter (Mudd et al. 2004), the current generation of hydrogeomorphic models are fully capable of transporting, eroding, and depositing sediment in 3-D in a wetland-dominated estuary (e.g. Delft 3D). The **physical processes** contained in Figure 5.2c, such as wetland erosion and progradation, transgression up slope, sediment deposition, and resuspension in estuarine aquatic bays, and transport of suspended solids from rivers, through the estuary to offshore, are well known and modeled. The current generation of models can probably also adequately account for compaction and how that varies with organic matter content.

There are several other **ecological processes** illustrated in our conceptual models (Figure 5.2a–c) that are far less understood and perhaps inadequately simulated in the current crop of simulation models (e.g., Mudd et al. 2004; Morris et al. 2002, 2012). Each of the models/balances in this chapter show blue carbon storage to be a component of wetland NEP, which is the balance between GPP and R_{AH} and the balance between NEP and tidal exchange or drainage. While several studies have shown the relation between wetland GPP and flooding depth for marshes, much less exists for mangroves (Morris 2016). Little is known about R_{AH} and flooding and the comparative effect of temperature on GPP and R_{AH}. Theoretically, it has been shown that respiration is more sensitive to temperature change than is primary production, but there are few empirical tests of this for wetlands (see, for example, Yvon-Durocher et al. 2010). Thus, we might expect warming to decrease NEP because respiration will be increased more than GPP. Our diagram also shows one fate of NEP to be export to adjacent tidal waters. How will increased tidal flooding associated with SLR affect export? I am unaware of any studies examining this relation. Likewise, how will increased tidal flooding affect belowground decomposition? What actually controls the C content of soils, a critical parameter in all calculations of blue carbon storage (see Morris and Callaway, Chapter 6)? How does sedimentation of mineral matter (sand, silt, clay) affect C content of soils, other than to **dilute** the C content (being that content is measured as C mass/total soil mass)? Does it protect the organic C from further decomposition, sensu Hedges and Keil (1995)? Increased decomposition will also limit NEP and therefore blue carbon storage and export potential. We know that salinity and sulfides can increase to levels that decrease production and decomposition (Morris 1984; Bradley and Morris 1990; Joye and Hollibaugh 1995). However, increased flooding in saline wetlands will likely minimize increases in porewater salinity and sulfide concentrations as flooding dilutes and flushes out the evaporative or metabolic buildup of these components. Theoretically, we could expect concomitant increases in decomposition, thus potentially lowering NEP. While not specifically identified in any of the conceptual diagrams, it is primarily belowground plant components that contribute to sediment organic content and hence elevation gain and blue carbon storage. Current models simply assume a fixed fraction of belowground production to escape decomposition and thus contribute to NEP and burial (Morris 2016). This certainly requires further research. How do variations in plant production translate to above-belowground production ratios? How does decomposability of belowground plant matter vary with rate of production? How does it vary with temperature, flooding, even nutrient content (a factor we have neglected in this examination)? While there have been studies examining rates of belowground decomposition (e.g., Blum and Christian 2004), and OC decomposition in estuarine sediments in general (e.g., Benner et al. 1991), little to no work has examined controls on belowground decomposition of residual soil OC.

Our conceptual diagrams can also indicate processes unlikely to affect blue carbon storage. For example, we know that the Carbon:Nitrogen:Phosphorous stoichiometry of terrestrial organic matter inputs to the coastal zone plays a central role in estuarine heterotrophy (Bauer et al. 2013), yet estuarine heterotrophy is unlikely to have any effect on adjacent tidal wetlands. The heterotrophy results from metabolism of inputs from land and tidal wetlands; it doesn't **pull** any of the wetland NEP.

One approach to better define overall levels and controls of GPP, R_{AH}, and NEP for tidal wetlands and adjacent estuaries has evolved toward monitoring these processes over time using recently employed technologies that operate at a large spatial scale (e.g., km^2 and larger). Free-water diurnal O_2/CO_2 (e.g., Vallino et al. 2005; Beck et al. 2015; Wang et al. 2017) and eddy covariance flux tower approaches (Kathilankal et al. 2008; Forbrich and Giblin 2015) overcome some of the problems associated with net C balances that are assembled from summations of component measures in bottles and at the plot scale (e.g., ¼ m^2). Conducted over long intervals (decades) such measures can help us resolve how variations in key drivers such as temperature, flooding depth, flooding duration, affect metabolism and the amount of carbon that can potentially be stored as blue carbon (sensu Forbrich et al. 2017). These technologies are not without their own problems, of course. First is the issue of separately resolving what goes on in the water (as measured with free-water techniques) from what is measured with a flux tower whose footprint is flooded during high tides and which often includes open water areas and small tidal creeks (Wang et al. 2016). There's also the issue of accurately converting O_2 measures to C units and accounting for air-sea gas exchange of O_2 and CO_2 during inundation of the wetland surface (Wang et al. 2016).

A major component affecting blue carbon storage is totally missing in net C balance budgets and conceptual models—human actions (Lovelock et al. Chapter 3). Net balances are but one tool to help us better understand the sources of blue carbon and the magnitude of C storage relative to other C transfers and processes in the coastal zone. Controlled experimentation to unravel the controls of key processes is also required to adequately model and predict blue carbon storage through the remainder of the 21st century. But equally important are models that can incorporate the human dimension, perhaps by running various scenarios, like we already do with projections of SLR. Transgression will be largely controlled by human activities. Progradation or marsh edge erosion is also likely to be strongly influenced by human activities. We already see actions attempting to limit edge erosion, such as the installation of living shorelines on or adjacent to existing marsh edges. While such activities may limit loss of private upland property, erosion is to be expected when there is an imbalance in SLR and sediment availability (Mariotti et al. 2013). Indeed, in the Plum Island Sound marshes of the NE United States, we have documented the importance of marsh-edge erosion as a critical source of sediment to help sustain vertical accretion on the remaining marshes of the system (Hopkinson et al. in review). We have observed that shoreline erosion has the potential to meet 31% of marsh platform sediment accretionary needs, which is in stark contrast to rivers that at best can provide 8% of accretionary needs.

ACKNOWLEDGMENTS

This work was partially supported by the Plum Island Ecosystems LTER (NSF grant OCE 1238212). The work benefits from years of discussions with friends and colleagues including J. Vallino, A. Giblin, N. Weston, J. Morris, J. Hobbie, W-J. Cai, S. Fagherazzi, J. Day, and B. Peterson.

CHAPTER 6

Physical and Biological Regulation of Carbon Sequestration in Tidal Marshes

James T. Morris
University of South Carolina

John C. Callaway
University of San Francisco

CONTENTS

HIGHLIGHTS

1. The rate of carbon sequestration in tidal marshes is regulated by complex feedbacks among biological and physical factors including the rate of sea level rise (SLR), biomass production, tidal amplitude, and the concentration of suspended sediment. We used the Marsh Equilibrium Model (MEM) to explore the effects on C-sequestration across a wide range of permutations of these variables.

2. C-sequestration increased with the rate of SLR to a maximum, then decreased down to a vanishing point at higher SLR when marshes convert to mudflats. An acceleration in SLR will increase C-sequestration in marshes that can keep pace, but at high rates of SLR, this is only possible with high biomass and suspended sediment concentrations. We found there were no feasible solutions at SLR >13 mm year[-1] for permutations of variables that characterize the great majority of tidal marshes, i.e., the equilibrium elevation exists below the lower vertical limit for survival of marsh vegetation.

3. The rate of SLR resulting in maximum C-sequestration varies with biomass production. C-sequestration rates at SLR = 1 mm year^{-1} averaged only 36 g C m^{-2} year^{-1}, but at the highest maximum biomass tested (5,000 g m^{-2}) the mean C-sequestration reached 399 g C m^{-2} year^{-1} at SLR = 14 mm year^{-1}.

4. The empirical estimate of C-sequestration in a core dated 50-years overestimates the theoretical long-term rate by 34% for realistic values of decomposition rate and belowground production. The overestimate of the empirical method arises from the live and decaying biomass contained within the carbon inventory above the marker horizon, and overestimates were even greater for shorter surface cores.

6.1 INTRODUCTION

Tidal marshes are one of the blue carbon ecosystems—marine ecosystems that sequester carbon over centuries to millennia. Conservatively, blue carbon ecosystems (marshes, mangroves, other tidal wetlands including forests and seagrasses) store roughly 45 Tg C year^{-1} (Chmura et al. 2003), which is globally significant because of the millennial time scale on which they operate. As long as sea level rises, carbon storage in coastal sediments will continue unabated as marshes either keep pace with SLR or transgress inland. In addition to conservation of extant C pools, continued C accretion creates opportunities for management of greenhouse gas (GHG) offset programs, which have been established by multiple state, regional, and national programs across the world (see Chapters 15–20).

While there is a growing collection of data on tidal marsh carbon dynamics (e.g., see recent papers by Ouyang and Lee 2014; DeLaune et al. 2016; Nahlik and Fennessy 2016; Van de Broek et al. 2016), measured rates have been calculated using a wide range of different approaches; from short-term sediment markers to dated cores using time scales from decades to millennia (see Chapter 7 in this volume for review of methods and related definitions). Most recent estimates have used either ^{137}Cs or ^{210}Pb, along with measurements of soil carbon density, usually by converting estimates of organic matter from loss on ignition (Chmura et al. 2003; Ouyang and Lee 2014). Estimates of sequestration using dated horizons incorporate a variety of carbon pools, including living roots and rhizomes and labile organic matter in modern surface soils; however, the only pool that contributes to long-term carbon sequestration is the refractory one. With dated cores or horizons, it is impossible to separate out these various pools, and better estimates of carbon sequestration can be made by use of models that account for the important, interacting factors that govern the processes contributing to sequestration.

The rate of carbon sequestration in tidal marshes is regulated by complex feedbacks among biological and physical factors. The physical factors include 1) the rate of SLR, 2) the tidal amplitude, and 3) the concentration of suspended sediments in tidal floodwater. Important biological factors include 4) the growth response of vegetation to relative site elevation, 5) the maximum productivity, 6) belowground productivity, and 7) the refractory carbon content of these tissues. In this chapter, we use the MEM to explore interactions of these factors with the goal of generalizing about how feedbacks regulate carbon sequestration. We also use the model to evaluate the contribution of different forms of carbon (labile versus refractory) to measured rates of carbon sequestration using dated sediment cores.

An equilibrium marsh is one that is in balance with sea level, meaning that the elevation of the marsh surface *relative to sea level* is constant through time. The elevation of an equilibrium marsh surface will track the long-term rate of SLR (Redfield and Rubin 1962; Redfield 1972), and by long-term we mean decades to a century. Marshes do not adjust quickly to changes in sea level. Anomalies in annual mean sea level (MSL) change the hydroperiod and productivity of a marsh (Morris and Haskin 1990; Morris et al. 1990; Morris et al. 2002). So, the relative elevation of a marsh in equilibrium is not necessarily constant on annual or shorter time scales, but it is stable over longer time scales. As such, it is a dynamic equilibrium. In equilibrium, the long-term annual productivity of the marsh also should be constant through time, but can and will vary on annual time scales with changes in hydroperiod, climate, and weather. Plant community composition will likely be constant over time and, if not, changes in composition in an equilibrium state will not alter the feedback among tides, plant response, hydroperiod, or the rate of vertical accretion. Additionally, sediment supply and hydrodynamic variables such as tidal amplitude are also assumed to be constant over long-time scales.

Disequilibrium states can arise when there is a step change in one of the forcing variables, like sediment supply, including episodic deposition of sand or sediment from overwash, storms, or thin layer placement of sediment (e.g., Orson et al. 1998). A marsh restoration site can also be in a state of disequilibrium for decades (Craft et al. 1999, 2002; Callaway 2005). MEM simulates marsh responses to multiple environmental factors. The model has been shown to accurately simulate marsh development at multiple sites (e.g. San Francisco Bay, CA and North Inlet, SC) and to synthesize knowledge of key environmental factors driving marsh accretion and sustainability, both abiotic and biotic. It also has been used to predict sustainability of marshes under different climate change scenarios (Kirwan et al. 2010; Schile et al. 2014; Byrd et al. 2016; Alizad et al. 2016a,b). MEM is able to numerically simulate disequilibrium states, but results reported in this chapter apply to equilibrium states calculated for permutations of key parameters.

6.2 MODEL DESCRIPTION

The MEM describes important feedbacks that regulate vertical accretion rate, soil organic matter concentration, and primary production as a function of SLR, tidal range, and suspended sediment concentrations. We report here results of an experiment in which important input variables (maximum biomass, SLR, tidal range, and suspended sediment concentration) were systematically varied to evaluate their effects on marsh resilience and carbon sequestration for virtual marshes in equilibrium with SLR. The current model is the successor to a model published earlier (Morris et al. 2002). We use MEM in an exploratory mode, not calibrated for a single site/location but comparing responses across widely ranging conditions in order to evaluate the determinants of soil organic matter concentration and carbon sequestration rates. We are not using the model here to evaluate marsh dynamics under future scenarios of accelerating SLR. Rather SLR is a constant within an individual model run but is varied among simulations to evaluate the effects of different rates on equilibrium elevation and carbon dynamics.

A feature of MEM is a description of the response of aboveground biomass to the relative elevation of a site and tide range (Morris et al. 2002). For an intertidal species such as *Spartina alterniflora*, the vertical range of growth lies approximately between MSL and mean higher high water (MHHW) (McKee and Patrick 1988). At both the upper and lower extremes of its vertical range, biomass and primary production approach zero (Figure 6.1a). At the lower limit, hypoxia is most important, while osmotic stress sets the upper limit (Mendelssohn and Morris 2000). There is an optimum relative elevation for growth that lies near the middle of the range (Morris et al. 2002, 2013). In other words, the growth response to relative elevation, or alternately depth (D) of the marsh surface below relative MHW, is approximately parabolic:

$$B_s = aD + bD^2 + c : \text{seasonal maximum standing biomass} \, (\mathrm{g\,m^{-2}})$$

The coefficients a, b, and c are calculated after specifying the upper and lower limits of growth, the optimum depth, and biomass at the optimum depth. Biomass at the optimum depth is subsequently referred to as B_{max}. For all simulations reported here we put the lower limit at 10 cm below MSL (positive D) and the upper limit at 20 cm above MHW (negative D), and optimum depth in the middle of the range.

6.2.1 Mass Inputs: Mineral Sediment

The model describes net accretion only; erosion is not explicitly accounted for. Consequently, MEM is not appropriate for marsh edges where erosion is likely to be an important factor. MEM

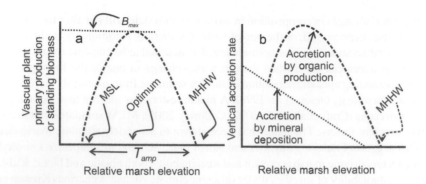

Figure 6.1   Conceptual relationship between marsh productivity (a) and vertical accretion (b) as functions of relative elevation. Marsh production (a) adds organic matter to soil, generating biovolume proportionally and marsh accretion (b). Mineral deposition is proportional to the depth of floodwater and suspended sediment concentration, and decreases as relative elevation increases (b).

decomposes sedimentation dynamics into several contributing processes, reflecting the importance of both mineral and organic matter inputs to sediment accretion (Nyman et al. 1990, 1993; Turner et al. 2000; Morris et al. 2016). Mineral sedimentation resulting from the settling of suspended particles is proportional to the concentration of suspended solids and the amount of time that the surface is flooded. Inorganic sediment load (S_{max}) is calculated as the product of the average depth of water over $1\,cm^2$ of marsh surface during a flood tide, (MHW–Z)/2, the number of tides in a year (704), and the concentration of suspended mineral sediment m ($g\,cm^{-3}$), hereafter denoted SSC. Marsh elevation is Z (cm relative to MSL); MHW is mean high water level.

$$S_{max} = m \times 704 \times 0.5 \times (\text{MHW} - Z) : \text{sediment load}\,(g\,cm^{-2}\,year^{-1})$$

Similar to Krone's settling velocity w_s (Krone 1962), MEM multiplies the sediment load S_{max} by a capture coefficient q, scaled by inundation time (w) to calculate the surface sedimentation rate. Inundation time is calculated as $w = (\text{MHW} - Z)/(\text{MHW} - \text{MLW})$ for $\text{MLW} \leq Z \leq \text{MHW}$. Scaling q by the fractional inundation time accounts for effects of tidal amplitude on sedimentation. For example, the inundation time of 5 cm of water over a marsh in a macrotidal estuary with a 150 cm tidal amplitude is (150–145)/300 = 0.017, which is just a fraction of the inundation time of 5 cm of water over the surface in a microtidal estuary with a 20 cm tidal amplitude, (20–15)/40 = 0.125. The capture coefficient q was set to 2.8 based on work at North Inlet and assumed constant for simulations reported here. Vegetation enhances the deposition of mineral sediment by sorbing suspended sediment directly onto leaf and stem surfaces (Mudd et al. 2010), but this detail is omitted here because of its autocorrelation with the growth of root and rhizome biovolume. With the multi-decadal dataset we have from North Inlet we cannot separate these processes statistically (sediment trapping by vegetation versus biovolume) (Morris, pers. obs.).

6.2.2 Mass Inputs: Organic Matter

Belowground organic matter accumulation has been shown to be a critical component of overall marsh accretion across a wide range of tidal marshes (Nyman et al. 1993, 2006; Turner et al. 2004). Accretion by virtue of primary production is possible because inputs of stable or refractory organic matter ($k_r \varphi \tau B_s$, g dry wt $cm^{-2}\,year^{-1}$) add increments of volume to sediment. Coefficient k_r ($g\,g^{-1}$) is the fraction of belowground production incorporated into the stable fraction of soil carbon, φ is the ratio of belowground production to aboveground standing biomass, and τ is the turnover rate of belowground biomass ($year^{-1}$). Based on the lignin content of *Spartina* (Hodson et al. 1984; Wilson

Table 6.1   Permutations of Parameters That Gave Rise to the Model Results Reported Here

Parameter	Permutations
Tidal amplitude (T_{amp})	20–180 cm in steps of 5 cm
Susp. mineral sediment concentration (SSC)	10–160 mg L^{-1} in steps of 5 mg L^{-1}
Standing biomass at the optimum elevation (B_{max})	1,000–5,000 g m^{-2} in steps of 1,000
Constant rate of SLR	1–40 mm $year^{-1}$ in steps of 1 mm $year^{-1}$

et al. 1986; Buth and Voesenek 1987), we assumed in all simulations here $k_r = 0.1$ and that the ratio of belowground to aboveground standing biomass was 2:1 (Schubauer and Hopkinson 1984). The turnover rate of belowground biomass was assumed to be 1.0 per year. Keep in mind that about half of the belowground biomass is rhizome tissue which is perennial and long-lived. The rhizome fraction of belowground biomass probably turns over much less than 1.0 per year, while the root fraction could turnover several times a year. Because net organic matter inputs are a function of all four parameters above, there is a trade-off in how these parameters affect organic matter inputs. One of the permutations of the simulations reported here includes a range of values of (1,000–5,000 g m^{-2}) for aboveground biomass (Table 6.1), and these can be viewed alternatively as changes in the allocation and turnover of belowground production, i.e., an aboveground production of 1,000 g m^{-2} with root:shoot ratio of 2.0 is equivalent to 2,000 g m^{-2} aboveground production with root:shoot ratio of 1.0, assuming equal turnover rates.

6.2.3 Vertical Accretion, Bulk Density, and Soil Organic Matter

Vertical accretion rate (dz/dt, cm $year^{-1}$) resulting from the settling of particles and biovolume growth is defined by the summation of the inorganic and organic mass inputs divided by their respective self-packing densities:

$$\frac{dz}{dt} = \left[\frac{S_{max}q\omega}{k_2} + \frac{k_r\varphi\tau B_s}{k_1} \right] =$$

$$\left[\frac{qm \times 704 \times \dfrac{0.5D^2}{MHW - MLW}}{k_2} + \frac{k_r\varphi\tau(aD + bD^2 + c)}{k_1} \right] : \text{vertical accretion rate (cm } year^{-1})$$

Coefficients $k_1 = 0.085$ g cm^{-3} and $k_2 = 1.99$ g cm^{-3} are the self-packing densities of organic and mineral sediment respectively (Morris et al. 2016).

In the model, LOI (loss on ignition) is calculated as the concentration of soil organic matter below the root zone at a depth characterized by the stable fraction of organic matter inputs.

$$LOI = k_r\varphi\tau B_s / [k_r\varphi\tau B_s + S_{max}q\omega] : \text{soil organic matter concentration or LOI(g } g^{-1})$$

6.2.4 Equilibrum and Dimensionless Elevation

In equilibrium dz/dt is equal to the rate of SLR. After specifying the values of the constants and variables in the model, dz/dt is determined only by the depth D of the surface below MHW. Finding the equilibrium is a matter of setting a constant rate of SLR and iterating across the range of all possible depths D to identify the depth that results in a vertical accretion rate dz/dt equivalent to

SLR: $f(D) = dz/dt = \text{SLR}$. Relative marsh elevation (Z) at equilibrium was then calculated as tidal amplitude (T_{amp}) minus the equlibrium depth. As a means of standardizing across tidal ranges we computed a dimensionless relative elevation (DimE) at equlibrium as $(Z–E_{min})/(E_{max}–E_{min})$, where E_{max} and E_{min} are the maximum and minimum limits of the vegetation relative to MSL, respectively, and $0 \leq \text{DimE} \leq 1$ irrespective of tide range for a vegetated marsh surface.

6.2.5 Carbon Sequestration

When the marsh is in equilibrium with SLR, the annual rate of carbon sequestration (refractory carbon input) in any year is $0.42 \times k_r \varphi \tau B_s$ where $\varphi \tau B_s$ is the annual production of belowground dry weight, as discussed above. This is the theoretical, long-term rate of carbon sequestration for a marsh in equilibrium with SLR. The constant 0.42 is taken as the elemental carbon fraction of dry *Spartina* tissue based on reports that place it between 0.4 and 0.44 (Osgood and Zieman 1993; Cartaxana and Catarino 1997; Tobias et al. 2014; Byrd et al. in review).

In addition to the theoretical long-term rate of sequestration, we simulated typical field-based measures of carbon sequestration (e.g., Craft et al. 1993; Callaway et al. 2012 and citations in Chmura et al. 2003). Carbon sequestration is often measured empirically by locating a marker horizon or dated soil layer, such as the ^{137}Cs peak corresponding to the peak in fallout from atmospheric nuclear testing in 1963 (DeLaune et al. 1978; Ritchie and McHenry 1990 and see Chapter 7 in this volume). In a typical scenario, one would estimate the total inventory of carbon in the section of core above a dated horizon and divide by time in years. For simplicity we simulated a horizon-derived C-sequestration rate assuming dated horizon of 50 years below the surface, as this is a time frame similar to many measurements made with [137]Cs. In dry weight units (g m^{-2}), the total inventory of organic matter would include the total live root biomass (φB_s), 50 years of refractory inputs ($50 k_r \varphi \tau B_s$), and the

decaying, labile fraction of root inputs given by $\displaystyle\int_1^{50}(1 - k_r)\varphi \tau B_s e^{rt} = \frac{(1 - k_r)\varphi \tau B_s}{r}\left(e^{r50} - e^r\right)$ where r

is the annual decay rate (the fractional loss, $r < 0$), and t is time in years. Since e^{50r} approaches zero, the solution of this equation simplifies to $(1 - k_r)\varphi \tau B_s\left(e^r / -r\right)$. Thus, the total dry weight inventory of organic matter O_i is:

$$O_i = \varphi B_s + 50 k_r \varphi \tau B_s - (1 - k_r)\varphi \tau B_s (e^r / r):$$

soil organic matter inventory in 50 year old sediment column (g dry wt m^{-2})

The estimated annual rate of carbon sequestration (g C m^{-2} year^{-1}) from such a measure of the organic inventory over a 50-year-old marker horizon is $0.42\ O_i\ / 50$.

6.3 MODEL EXPERIMENTS

In order to evaluate constraints on carbon sequestration rates, we exercised the model using a range of permutations encompassing combinations of four key variables: tidal amplitude (1/2 of the tidal range), suspended sediment concentration (m), biomass at the optimum depth (B_{max}), and the rate of SLR (Table 6.1). As above, for each permutation, the model was run iteratively with adjustments to the depth until $dZ/dt = \text{SLR}$, resulting in 838,860 equilibrium solutions. This range of variables (Table 6.1) should define the conditions that describe virtually all tidal marshes. Other model parameters which were held constant across all permutations included the ratio of live belowground biomass to standing biomass ($\varphi = 2\text{g g}^{-1}$), the turnover rate of belowground material ($\tau = 1/\text{year}$), the decay rate of labile organic matter ($-0.4/\text{year}$, Blum 1993; Blum and Christian 2004), the refractory fraction of organic production ($k_r = 0.1$), and the capture coefficient of suspended minerals ($q = 2.8$), as discussed above.

6.4 STATISTICAL ANALYSIS

Outputs from all permutations were grouped first into feasible and non-feasible solutions. A feasible solution is one in which the dimensionless elevation (DimE) at equilibrium ($dz/dt = $ SLR) was greater than zero, i.e., vegetation remained in place within its growth range. Solutions in which the equilibrium elevation was below the lower limit for the vegetation were deemed to be non-feasible since the marsh would be converted to mudflat and no carbon sequestration would be possible. The group of feasible solutions was subsequently analyzed by cross-correlation analysis, grouped by SLR, and by computing the means of select dependent variables (DimE, C-sequestration rate, soil organic matter concentration, mineral accretion rate, and equilibrium biomass).

6.5 RESULTS

Most virtual marshes equilibrated at elevations feasible for vegetation across the range of variables evaluated when SLR $\leq$ 13 mm year^{-1} (Figure 6.2, Table 6.2). For each level of SLR there were 41,943 possible permutations of T_{amp}, SSC, and B_{max}. There was 100% survival at the lowest rates of SLR, 1 and 2 mm year^{-1} (Figure 6.2, Table 6.2). Among all possible permutations, the proportion of feasible combinations started to decline significantly by SLR = 10 mm year^{-1}, and by 23 mm year^{-1} the proportion surviving had dropped to 47% (Figure 6.2 and see Table 6.2). However, marshes first started to drop out at SLR = 3 mm year^{-1} and B_{max} = 1,000 g m^{-2}. When we restricted the permutations to values of B_{max} ($\leq$3,000 g m^{-2}), SSC ($\leq$50 mg L^{-1}), and T_{amp} ($\leq$100 cm)—parameter values that we think characterize the great majority of U.S. tidal marshes, with exceptions found in the Mississippi River and Sacramento–San Joaquin Deltas and other unusual locations—we found no feasible solutions at SLR $\geq$ 13 mm year^{-1}. By SLR = 7 mm year^{-1}, only 43% of the permutations were feasible among this restricted group (Figure 6.2).

The complex feedbacks that characterize the sediment dynamics of an intertidal marsh are illustrated by the correlations among dependent and independent variables, with effects of some factors changing substantially with SLR (Table 6.2). For example, tidal amplitude was negatively correlated with soil organic matter concentration (LOI) and C-sequestration at SLR = 2 mm year^{-1} ($r = -0.08$ and -0.23 respectively), but positively correlated with these variables ($r = 0.49$ and 0.55) at SLR = 20 mm year^{-1}. This can be explained by the fact that elevation capital, which becomes more important as the rate of SLR increases, is affected by tidal amplitude. Suspended sediment

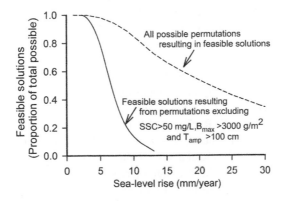

Figure 6.2 The proportion of feasible (DimE > 0) solutions found among 41943 possible permutations of B_{max}, SSC, and T_{amp} as a function of the rate of SLR. Also shown is the proportion of feasible solutions for a restricted group of permutations representing the majority of the universe of existing tidal U.S. saltmarshes ($B_{max} \leq$ 3,000 g m^{-2}, SSC $\leq$ 50 mg L^{-1}, and $T_{amp} \leq$ 100 cm).

Table 6.2　Pearson Correlation Coefficients among Dependent and Independent Variables from the Universe of Simulations Resulting in Feasible (DimE > 0) Virtual Marshes, Each Equilibrated ($dz/dt = \mathrm{SLR}$) with Constant Rates of SLR (2, 10, and 20 mm year^{-1})

| | SLR = 2 mm year^{-1} | | SLR = 10 mm year^{-1} | | SLR = 20 mm year^{-1} | |
	LOI	Carbon Sequestration Rate	LOI	Carbon Sequestration Rate	LOI	Carbon Sequestration Rate
Tidal amplitude (T_{amp})	−0.08	−0.23	−0.07	0.32	0.49	0.55
Suspended Inorg. Sed. Conc. (SSC)	ns	ns	−0.05	0.15	0.40	0.45
Maximum biomass (B_{max})	0.09	0.27	0.78	0.75	0.42	0.37
Means ± 1 SD	87.9 ± 1.8	71.1 ± 0.1	22.4 ± 32.1	189.1 ± 135.7	0.7 ± 1.1	88.4 ± 115.8
	$n = 41,943$		$n = 36,907$		$n = 30,545$	

Also shown are the means ± 1 SD of dependent variables LOI and C-Sequestration and the number of feasible solutions found for each level of SLR. Every correlation is significant at $p < 0.0001$ except where noted.

concentration was not correlated with LOI or C-sequestration at 2 mm year^{-1} SLR, but positively correlated with LOI and C-sequestration ($r = 0.49$ and 0.55) at SLR = 20 mm year^{-1}. At low rates of SLR, marshes equilibrate high in the tidal frame where mineral sediment input is unimportant, but its importance increases at higher SLR because it supports overall accretion and helps to maintain elevation. Consequently, the low input of mineral matter at low SLR results in a high mean LOI (87.9% ± 1.8), while the increasing importance of minerals at high SLR results in a low LOI (0.7% ± 1.1) (Table 6.2).

The most striking example of the importance of these feedbacks is shown by the changes in C-sequestration rates across the range of SLR (Table 6.2). As a consequence of the vertical distribution of biomass productivity and its dependence on equilibrium elevation (Figure 6.1a), C-sequestration shows non-linear behavior with change in SLR. Essentially, it mimics the biomass profile (Figure 6.1), because equilibrium elevation declines with increasing SLR (Morris et al. 2002). Average C-sequestration for marshes in equilibrium ranged from 71 at SLR = 2 mm year^{-1} to 189.1 ± 135.7 g C m^{-2} year^{-1} at SLR = 10 mm year^{-1}, and then dropped to 88.4 ± 115.8 at SLR = 20 mm year^{-1} (Table 6.2). Average C-sequestration at 2 mm year^{-1} is essentially the theoretical maximum of 71.4 g C m^{-2} year^{-1}, calculated as the product of the self-packing density of pure organic matter (Morris et al. 2016), the carbon faction of organic matter, and the rate of SLR (= 0.085 g dry weight cm^{-3} × 0.42 g C/g dry weight × 0.2 cm year^{-1} × 10^4 cm^{2}m^{-2}). As SLR increases, elevation decreases, and biomass productivity and sequestration increase, but at even greater SLR, the elevation decreases to the point where excess flooding inhibits productivity.

Soil organic matter concentration was highly sensitive to SLR, depending on B_{max}. At the lowest SLR, LOI in every case was at the maximum level of 88% (Figure 6.3) corresponding to DimE at or near 1.0 (Figure 6.4). At B_{max} = 1,000 g m^{-2}, mean LOI had declined to 50% at SLR < 3 mm year^{-1}. The LOI was approaching zero by SLR = 5 mm year^{-1} (Figure 6.3). At B_{max} = 3,000 and 5,000 g m^{-2}, LOI declined to less than 50% by SLR <7 and <13 mm year^{-1}, respectively, roughly in line with the decline in DimE (Figure 6.4). In general, at low B_{max}, both LOI and DimE declined rapidly with increasing SLR, which is characteristic of a low fertility marsh. Conversely, LOI and DimE decline less rapidly with SLR at high B_{max}, which is characteristic of a highly fertile marsh or, alternatively, a marsh with a very high production of refractory roots and rhizomes, i.e., very high $k_r \varphi \tau$.

Dimensionless elevation (DimE) is a measure of where the equilibrium elevation occurs within the vertical range of the vegetation. A DimE of 1.0 denotes equilibrium at the highest possible relative elevation for tidal marsh vegetation, assumed here to be MHW + 30 cm, while a DimE of 0 equals the lowest vertical limit of the vegetation, 10 cm below MSL. A DimE of 0.5 is the elevation

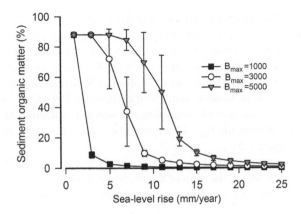

Figure 6.3 Mean dimensionless elevation ± 1 STD for each level of SLR and three levels of maximum biomass (B_{max}). Only feasible solutions (DimE>0) were included. Dimensionless elevation was averaged over all possible SSC and T_{amp} for each combination of SLR and B_{max}.

of maximum biomass along the parabolic response of biomass to elevation. DimE was highly sensitive to SLR and dependent upon the maximum biomass (B_{max}) (Figure 6.4). At the lowest B_{max} (1,000 g m^{-2}), mean DimE (averaged over all feasible solutions by SLR) dropped to 0.48 by 5 mm year^{-1} SLR, indicating that the average marsh in this class was at an elevation below the optimum for the vegetation. For B_{max} = 3,000 g m^{-2} the mean DimE dropped just below 0.5 between 9 and 11 mm year^{-1} SLR, but at B_{max} = 5,000 g m^{-2}, the mean DimE did not drop below 0.5 until SLR exceeded 17 mm year^{-1}. Again, this was for all feasible solutions, including the full range of SSC and T_{amp}.

High rates of mineral sediment input dilute sediment organic matter and reduce sediment LOI. The relative elevation at which a marsh equilibrates, which is inversely related to SLR (Figure 6.4) and SSC determine the input of mineral sediment and biomass production, and consequently, LOI. At low SLR, the high elevations at which these marshes equilibrated (Figure 6.4) resulted in the accretion of soils composed of peat (i.e., very high LOI) (Figure 6.3). Similar peat buildup occurred at 4 mm year^{-1}, but only at higher biomass levels. By 15 mm year^{-1} SLR, all surviving marshes were below the optimum elevation for biomass production and approaching the lower limit of elevation for survival (Figure 6.4) and with mineral inputs dominating sediment accretion (Figure 6.3).

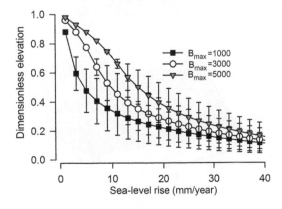

Figure 6.4 Mean sediment organic matter concentration (LOI) ± 1 STD for each level of SLR and three levels of maximum possible B_{max}. Only feasible (DimE>0) solutions were included. LOI at equilibrium was averaged over all possible SSC and T_{amp} for each combination of SLR and B_{max}.

Across all variables, C-sequestration was least at the lowest rate of SLR, and greatest at a combined high SLR and B_{max} (Figure 6.5). Increasing B_{max} led to greater average C-sequestration rates except at the lowest rate of SLR (1 mm year^{-1}). At SLR = 0, marsh elevation will equilibrate at E_{max} where equilibrium biomass and C-sequestration are zero. For every level of B_{max} there is an optimum SLR with respect to C-sequestration, and the optimum SLR increases with increasing B_{max}. At B_{max} = 5,000 g m^{-2}, the maximum, mean C-sequestration reached 399 g C m^{-2} year^{-1} at SLR = 14 mm year^{-1}. At B_{max} = 1,000 g m^{-2}, the mean C-sequestration was maximized at SLR = 4 mm year^{-1}, but only at 77 g C m^{-2} year^{-1}. Across all B_{max}, C-sequestration was uniformly 36 g C m^{-2} year^{-1} when SLR was 1 mm year^{-1} (Figure 6.5).

Dated cores of both the virtual and real varieties include a mix of both labile and refractory carbon. From the derivation above, the ratio of the theoretical, long-term rate of C-sequestration to the simulated horizon-derived rate in a 50-year-core (similar to ^{137}Cs dating methods) is:

$$\frac{0.42 k_r \varphi \tau B_s}{\left\{ 0.42 \left[\varphi B_s + 50 k_r \varphi \tau B_s + (1 - k_r) \varphi \tau B_s \left(\frac{e^r}{-r} \right) \right] \middle/ 50 \right\}}$$

This simplifies to $\left\{ 50 r k_r \tau \middle/ \left[r + \tau e^r (k_r - 1) + 50 r k_r \tau \right] \right\}$ for $r < 0$. Interestingly, the biomass variable and root:shoot ratio (φB_s) drop out, and the quotient depends only on the refractory fraction k_r, the decay rate r, and the turnover rate of belowground biomass τ. Substituting for our assumptions $(k_r = 0.1, r = -0.4,$ and $\tau = 1)$, we found that the theoretical rate of C-sequestration was always 66% of the simulated, horizon-derived rate integrated over 50 years. The difference between the two rates decreases as the decay rate r increases (becomes more negative), due to the fact that there is less labile organic matter remaining in the deeper soil layers, but even when $r = -0.9$, the theoretical rate was still 78% of the horizon-dated rate. The decay rate r does not affect the theoretical, long-term rate of refractory C-sequestration, even though it does affect the measured rate of carbon in dated cores. If the age of the core decreases, as it may in shallower cores, the difference between the theoretical and empirical rates will increase. For a core of length 25 years, the theoretical rate of C-sequestration decreases to 50% of the empirically derived rate using our original assumptions $(k_r = 0.1, r = -0.4,$ and $\tau = 1)$.

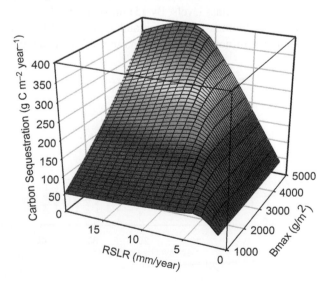

Figure 6.5 **(See color insert following page 266.)** Mean carbon sequestration rate as a function of the relative rate of SLR and maximum biomass (B_{max}). Only feasible (DimE > 0) solutions were included.

6.6 DISCUSSION

6.6.1 Elevation and Resilience

At the lower rates of SLR, virtual marshes equilibrated at an elevation within the vegetated zone of the marsh (Figure 6.1), although the equilibrium elevation tended to decline at increasing rates of SLR. In both natural and modeled conditions, marshes maintain this equilibrium with a mix of mineral and organic matter sediment inputs, and as they gain elevation in the tidal frame there is a shift toward greater domination by organic matter accumulation. Ultimately, for a marsh at the highest elevation, in equilibrium with SLR, it is the input and preservation of refractory organic matter that creates new volume and contributes to vertical accretion. Marshes low in the tidal frame (Figure 6.4) are more dependent on mineral inputs to maintain their equilibrium (Figure 6.3 and Table 6.2). This pattern of peat buildup in high elevation locations within tidal marshes is reflected in the consistent pattern of increasing soil organic matter content from low to high elevation marshes across many different locations (e.g., French et al. 1995; Callaway et al. 2012; Roner et al. 2016). It also is implied in the longstanding concept of marshes tracking SLR through peat accumulation (e.g., Redfield 1972). In cases of high SLR, low SSC, small T_{amp}, or low B_{max}, the equilibrium elevation fell below the lower vertical limit of vegetation (E_{min}), indicating a conversion of vegetated marsh to unvegetated mudflat.

In discussions about marshes, the equilibrium question arises often: how do we know if a marsh is in equilibrium with SLR? We would argue that relative to the low rates of eustatic SLR in the recent past, almost all tidal marshes globally must be in equilibrium, or at least in a dynamic equilibrium. If they were not, they would either grow out of the water and transition to a terrestrial ecosystem outside the tidal frame, or they would lose elevation so rapidly that they would be converted to mudflat ecosystems. There is no evidence of the former occurring, and the latter has occurred only in locations with very high rates of local subsidence. In theory, tidal marshes will always be chasing the equilibrium, and in a perfectly stable climate with constant SLR, they will find it. In a variable world, they will always be in a dynamic equilibrium. However, we know of tidal marshes today that are not in equilibrium. Disequilibrium will occur in modern times when conditions change rapidly, such as a change in sediment supply or a rapid acceleration in subsidence, and hence local SLR. From our sensitivity analysis of the virtual marshes generated by MEM we can project that the first marshes to fail with an acceleration of SLR (those with the least resilience) will be those with low primary productivity, i.e., low fertility, low SSC (e.g., DeLaune et al. 1978; Butzeck et al. 2015), and in microtidal estuaries (e.g., Kearney and Turner 2016). Those that are more resilient and more likely to survive will have the opposite characteristics, namely high productivity, high SSC (e.g., Patrick and DeLaune 1990; DeLaune et al. 1987; 2016), and tides of large amplitude (Cahoon and Guntenspergen 2010).

We have not addressed here how quickly marshes will fail; however, this is a function of the same variables—productivity, SSC, and tidal amplitude, and as well as the acceleration rate of SLR. Much has been written about elevation capital, and how it affects the vulnerability of a marsh to SLR (Reed 2002; Cahoon et al. 2011); in general, marshes at higher elevations, i.e., greater elevation capital, are less vulnerable to SLR because they have more elevation to lose before they reach the threshold elevation for conversion from marsh to mudflat (Cahoon and Guntenspergen 2010). However, it is important to note that elevation capital is limited by tidal amplitude. By definition, tidal marshes in a microtidal estuary will have very little elevation capital to lose before they are converted to mudflats, and hence are highly vulnerable. Whereas, marshes at the same dimensionless elevation in a macrotidal estuary (same relative elevation capital but much greater absolute elevation capital) are much less vulnerable to SLR. A tidal marsh with a meter of elevation capital will survive a 1 m rise in sea level, though it will lose relative elevation over the course of a century and still maintain vegetated.

Marshes can also be in disequilibrium in the opposite direction, i.e., building elevation rapidly. This situation is most likely to relate to newly constructed marshes that are established at relatively low elevations in sediment-rich environments or naturally occurring deltaic marshes. Under these conditions, the new marshes build elevation rapidly through the rapid accumulation of both mineral matter and belowground organic matter (living, decaying, and refractory material that builds soil volume). Prior to vegetation colonization, restoration sites build elevation solely through mineral sediment accumulation (and hence are reliant on relatively high SSC). This evolution has been observed in San Francisco Bay tidal marsh restoration projects (Williams and Orr 2002; Brand et al. 2012), but is likely to occur in any sediment-rich system. While the early development of these marshes is affected by mineral accumulation, over time the gains in elevation are more strongly affected by organic matter accumulation (Figure 6.2), with a gradual gain in relative elevation ($dZ/dt > $ SLR) until the marsh reaches an equilibrium point.

6.6.2 Carbon Sequestration

Although there were complex feedbacks across variables, the modeled rates of carbon sequestration were affected most directly by two variables: B_{max} and SLR. This is not surprising as B_{max} along with elevation determines overall productivity, as well as belowground biomass, which is the primary source of carbon for sequestration, and SLR creates the opportunity for the ongoing accumulation of material. The direct relationship between SLR and carbon sequestration rates over moderate rates of SLR indicates that up to a point, increases in SLR will lead to greater rates of carbon sequestration.

There is an interaction between SLR and B_{max} that determines C-sequestration. B_{max} is really a proxy for maximum potential production of belowground biomass. For any given level of B_{max} there is a SLR that results in maximum C-sequestration, and the SLR that gives maximum sequestration rises as B_{max} increases (Figure 6.5). As you would expect, higher C-sequestration rates are found in productive ecosystems/locations, and many of the high rates from previous compilations of carbon sequestration rates come from locations like Louisiana, with high rates of local SLR (e.g., see data in Chmura et al. 2003; Ouyang and Lee 2014).

At low rates of SLR, marshes tend to maintain an elevation close to MHW (Figure 6.4), and at this high elevation, they build virtual carbon-rich, peat soils, in large part because of the irregular tidal flooding and reduced input of mineral sediment. This has been observed in many natural marshes, from the Everglades (Craft and Richardson 1993) to New England marshes (Bricker-Urso 1980) and high elevation San Francisco Bay marshes (Callaway et al. 2012). In these cases, carbon sequestration is equal to SLR×carbon density. With recent rates of SLR equal to 2–3 mm year^{-1}, and typical carbon density values of 25–35 mg C cm^{-3}, (Gosselink et al. 1984), this would result in carbon sequestration rates ranging from 50 to 105 g C m^{-2} year^{-1}. This pattern also reflects the peat buildup for high elevation marshes discussed above. Stable, low-elevation marshes are more likely to be associated with higher rates of SLR, and they will be maintained by a mix of both mineral and organic matter inputs, resulting in soils with lower organic concentration (Figure 6.3) in low-elevation (Figure 6.4), equilibrium marshes.

At some point, the virtual marsh cannot keep pace with accelerating SLR and begins to lose elevation when $dZ/dt < $ SLR (disequilibrium). Under model conditions with constant inputs, the marsh will continue to lose relative elevation; however, in natural marshes, there could be storm inputs or other shifts that push the marsh back towards equilibrium. As the virtual marsh loses elevation, sequestration will decline due to declining productivity at lower elevations (the declining arm of the elevation-biomass parabola) (Figure 6.1). When equilibrium DimE declines to a level below 0.5, the optimum for biomass production, further increases in the rate of SLR will lead to more rapid loss of relative elevation, and lower biomass production, and eventually to loss of C-sequestration potential as the equilibrium elevation drops below the threshold for plant survival. The virtual system eventually reaches an equilibrium elevation, but this will be below the threshold for plant survival.

Our definition of carbon sequestration focuses on refractory carbon that will be retained in the soil over the long term; however, most field-based measurements of tidal marsh carbon sequestration are based on dated sediment cores with a time frame of ~50 years for ^{137}Cs and ~100 years for ^{210}Pb. Some estimates have also been made using short-term methods, such as marker horizons, as well as longer term methods, e.g., ^{14}C. As indicated above and discussed in the chapter on accretion methods (Chapter 7 in this volume), all of these methods integrate surface soil layers, which inherently include a mix of labile and refractory carbon, and will result in inflated estimates of carbon sequestration. Surface layers have the greatest ratio of labile carbon:refractory carbon, and this ratio is likely to decline consistently with depth. As a result, short-term methods that incorporate more surface layers relative to deep soil layers will result in the largest overestimates of carbon sequestration. This issue can be seen in previous compilations of carbon data using a mix of measurement methods (e.g., Chmura et al. 2003).

Finally, it is important to note that historical rates of carbon sequestration from individual sites are informative, but these rates alone are not necessarily a good predictor of future potential for carbon sequestration. Future rates will be affected by changes in SLR, shifts in marsh elevation that could lead to changes in productivity, as well as changes in suspended sediment inputs (affected by both changes in overall sediment loads and marsh elevation). Given these potential shifts, future carbon sequestration could be higher than historical rates (e.g., if there is a slight increase in SLR, allowing the marsh to keep pace; refer to Figure 6.5) or lower than historical rates (e.g., if the marsh loses substantial elevation, plants become stressed and productivity is reduced). These shifts not only highlight the benefit of using a model that is calibrated with historical accretion and sequestration rates, but also the ability of the model processes to respond to future conditions in order to predict future rates of sequestration. Similarly, a model could be used to predict conditions in a newly developing, restored marsh based on historical conditions in a reference marsh, with shifts in elevation and other critical parameters.

ACKNOWLEDGMENTS

This research was funded by the NASA Carbon Monitoring System (Grant no. NNH14AY671) and by NSF DEB-1654853.

CHAPTER 7

Accretion
Measurement and Interpretation of Wetland Sediments

John C. Callaway
University of San Francisco

CONTENTS

HIGHLIGHTS

1. A range of methods has been used to measure accretion and elevation change in tidal wetlands; these methods cover different time scales and a mix of processes that affect overall sediment and accretion dynamics. These same methods are widely used to estimate carbon sequestration and mass-based sediment accumulation rates.
2. Marker horizons and Sediment Elevation Tables (SETs) have been used most commonly for short-term measurements that focus on wetland sustainability; however, these approaches are not as useful for estimates of carbon sequestration because of their short-term time scale.
3. Sediment cores dated with radioisotopes are widely used for measuring accretion rates on the scale of decades to millennia, with the most common dating methods being ^{137}Cs and ^{210}Pb, as well as ^{14}C.
4. All of these methods can be useful for understanding wetland carbon dynamics, but it is important to be mindful that shorter-term measurements are likely to result in higher rates of measured carbon sequestration because these measurements include a greater proportion of labile carbon that will not be stored over the long term.

7.1 INTRODUCTION

Tidal wetlands are found along the margins of coastal and estuarine areas around the world, and they maintain their elevation relative to sea level by accumulating sediment. The sediment that accumulates within tidal wetlands includes a substantial carbon component, leading to high rates of carbon sequestration in tidal wetlands and other blue carbon ecosystems (e.g., see Ouyang and Lee 2014; Howard et al. 2017). Because sediment accretion is a critical process controlling sequestration rates, measurements of sediment accretion are used directly to estimate carbon sequestration in these blue carbon ecosystems. Accreting sediment in these systems includes a mix of both inorganic (mineral) and organic materials (e.g., see Friedrichs and Perry 2001; Morris 2002). The inorganic sediment can be brought in by daily tidal forces or by less frequent storm inputs, both through tidal channels that dissect the wetland and across the wetland-mudflat interface. Because of the dense vegetation and the location of tidal wetlands in relatively protected bays and enclosures, tidal wetlands tend to be depositional with erosion occurring only under unusual circumstances.

The sediment that is brought in by the tides or storm inputs is primarily inorganic but may include some organic matter that can be deposited and accumulate within the wetland; the carbon associated with organic matter that comes from outside sources of fixation/photosynthesis is referred to as allochthonous carbon. However, the majority of the organic matter and carbon that accumulates within tidal wetlands is a result of local plant production or autochthonous carbon. Both above- and belowground plant biomass can contribute to soil organic matter and eventual carbon sequestration. While extensive data on the relative importance of these two sources is lacking, most evidence points to the fact that much of the aboveground biomass is likely to be removed with the tides or lost to relatively rapid aerobic decomposition on the soil surface or in shallow sediments. Given these dynamics, it is likely that most soil carbon originates from belowground production, which is primarily subject to anaerobic decomposition within wetland soils. This belowground production includes both roots and rhizomes (underground stem tissue); however, there is very little information on the relative contributions of roots versus rhizomes to either accretion or carbon sequestration.

The accumulation of this mineral and organic matter on and in the wetland soil column builds elevation of the wetland soil surface, with organic matter accumulation typically more closely linked to rates of vertical accretion (Turner et al. 2000; Morris et al. 2016); in addition, there are ongoing interactions between accumulation and multiple factors that lead to losses in elevation. The factors that could result in loss of elevation include: eustatic sea level rise (SLR), regional (or deep) subsidence or tectonic activity, shallow compaction of sediment, decomposition, and erosion. Eustatic SLR occurs at the largest scale, as a result of global processes, such as the melting of glaciers and ice sheets, plus the thermal expansion of ocean waters as global temperatures increase. Regional subsidence or deep, subsurface compaction occurs at a moderate scale; most commonly, it is important in river deltas, where large-scale consolidation of deep deltaic sediments can lead to relatively rapid subsidence (e.g., over 10 mm year^{-1} in areas such as the Mississippi River or Nile River deltas (Stanley and Warne 1998; Törnqvist et al. 2008)); these rates of subsidence are in addition to eustatic SLR. Large-scale pumping of groundwater (e.g., South San Francisco Bay; Poland and Ireland 1988) or the oxidation of surface peat soils (e.g., Sacramento and San Joaquin delta; Deverel 1996; Drexler et al. 2009b), can also cause relatively large-scale regional subsidence. Tectonic activity can lead to extremely rapid subsidence or upliftment, with shifts of multiple meters possible from single tectonic events (Thilenius 1990), although in comparison with other processes, this factor is highly unpredictable. The remaining processes (shallow compaction, decomposition, and erosion) are most typically local phenomena. Shallow compaction refers to the consolidation of very light surface sediments as they are buried to greater depths. This results in a reduction of soil volume and an associated increase in soil bulk density. Similarly, decomposition results in the removal of organic matter from the soil through biological activity, and this also results in a loss of soil volume. As noted above, surface erosion of sediments is not common in tidal wetlands; however, in isolated cases (e.g., wave exposure, rapid storm related run-off, or sediment-depleted systems), it can lead to the removal of material and resulting elevation loss. In discussing tidal wetland sediment dynamics, accretion, and elevation changes, it is critical to be explicit about the terms that are of interest and that are being measured by a particular method because of the wide variety of processes that occur simultaneously on and within the soil column. Definitions related to measurement terms are included below.

In addition, it is important to note that there is variation in these processes within a particular tidal wetland. In particular, inputs of suspended sediment to a wetland are likely to be greatest in locations that are flooded more frequently by the tides (typically low elevation areas and/or areas adjacent to tidal creeks; see Krone 1987; French et al. 1995). Plant productivity, and hence organic matter inputs, also vary with elevation, although the relationship is more complex, with a parabolic response (low biomass at extreme elevations and maximum biomass in the middle of the wetland; Morris et al. 2002). Temporal variation can also be important in tidal wetland sediment dynamics, due to storm inputs of sediment, seasonal variations in water levels or suspended sediment concentrations, as well as longer-term, interannual variation (driven by climate/watershed inputs rather than individual storm events).

Over the last few millennia, SLR has been slow enough to allow tidal wetlands to develop and build deep accumulations of sediments around the globe. However, in the early Holocene, SLR was rapid enough that tidal marsh development was precluded (Redfield 1972; Atwater 1979); similarly, there are locations where local processes (primarily regional subsidence) lead to lose of elevation and threaten the long-term survival of coastal wetlands, even with current and recent rates of relatively slow global SLR. For example, many wetlands coastal Louisiana and other abandoned deltaic systems have high rates of subsidence and without very substantial sediment inputs, tidal wetlands in these regions have lost elevation and been converted to lower elevation unvegetated mudflats (Stanley and Warne 1998; Day et al. 2000). These regions give some indication of what could happen on a more global scale as SLR begins to increase more rapidly and more coastal regions shift from a net balance or increase in elevation to net loss of elevation.

In order to understand the long-term viability of tidal wetlands in the face of global SLR and regional and local processes that could lead to loss of elevation, a number of different approaches have been used to measure the rates of sediment accretion, as well as the resulting change in elevation. Most methods focus on vertical accretion rates (mm year^{-1}); however, vertical measurements are easily converted to mass-based rates (g m^{-2} year^{-1}) using data for soil bulk density (g cm^{-3}). Mass-based rates can be separated into mineral and organic rates based on organic matter content (Turner et al. 2000), and they also can be converted into carbon sequestration rates based on soil carbon content.

7.2 METHODS FOR MEASURING ACCRETION

A variety of methods have been used to measure accretion, elevation change, and carbon sequestration rates in tidal wetlands; as above, different methods cover different processes, and also encompass a range of time periods (Table 7.1). It is important to emphasize that the time period of measurement can substantially affect the measured rates of accretion and carbon sequestration. Measurements that cover longer time periods almost always result in slower accretion and sequestration rates compared to short-term measurements. The reduction in the measured accretion rates over time is due to shallow compaction of soil as it is buried and ongoing belowground decomposition of organic matter (i.e., carbon) that also leads to the loss of soil volume. Near the soil surface, wetland soils have very low bulk densities and are subject to compaction as they are buried; this ongoing compaction reduces soil volume, leading to the frequent observation of increasing soil bulk density with depth and the inherent reduction in measured accretion rate over time due to the loss of soil volume.

Simultaneously, belowground decomposition removes organic matter and sediment volume over time. The input of organic material from root and rhizome production, as well as from the burial of aboveground plant matter, occurs at shallow soil depths, typically resulting in higher soil organic content near the surface and a reduction in organic matter with depth, as the buried organic matter is subject to continual decomposition. The loss of organic matter through decomposition results in loss of soil volume and additional compaction of deeper soil layers, again resulting in increasing bulk density with depth in soil cores, and an overall reduction in measured accretion over time. Decomposition and compaction of sediments are inherently linked as organic matter provides much of the structure in any soil, including wetland soils. When organic matter is lost through decomposition, this likely leads to loss of soil strength and increased compaction. The importance of organic matter (carbon) in providing structure in wetland soils is illustrated by the relatively constant nature of carbon density in wetland soils; although bulk densities vary widely and typically are lowest in organic-rich, shallow sediments, profiles of carbon density are less variable.

Similar to the dynamics affecting accretion rates, measured rates of carbon sequestration are also inherently lower over time, and this is reflected in dating methods that cover longer time periods. The reduction in measured carbon sequestration with longer-term methods is caused primarily by the decomposition of labile carbon over time. Shallow, relatively young sediments include a mix of both labile and refractory carbon, with a greater percentage of labile material in more recent sediments. The labile pool of carbon is a major component of short-term measurements but is slowly lost over time. In addition, overall accretion rates are higher for surface sediments compared to older, more deeply buried sediments, although the feedbacks between organic matter, soil structure, and vertical accretion rates complicate the identification of a cause-effect relationship between carbon accumulation and accretion processes.

The mix of simultaneous processes occurring at different rates and different places throughout the sediment column complicates measurement and interpretation of wetland sediment dynamics (Figure 7.1). Input and loss of mineral material occur only at the sediment surface (deposition and erosion), although sediment erosion is uncommon in tidal marshes as they are typically depositional systems. Inputs of organic matter can occur at the surface (deposition of both local aboveground production or allochthonous carbon associated with mineral sediment inputs) or throughout the sediment column via root and rhizome production. Root and rhizome production is usually greatest near the sediment surface and drops rapidly with depth in anaerobic wetland conditions. Similarly, the decomposition of organic matter occurs throughout the sediment column with complex (but poorly studied and understood) patterns with depth and over time due to both differences in decomposition rates under aerobic (more likely near the surface) and anaerobic processes, and the presence of both labile and refractory carbon. Labile carbon can be lost quickly to decomposition, while more recalcitrant material is buried to greater depths but is still subject to ongoing slow rates of decomposition.

A variety of terms has been used to identify these processes and their net results, with varying meanings depending on the author and context. Given this, it is important to clarify terms that are used directly for accretion measurements; additional terms are discussed above. In this chapter, I use the following terms:

accretion: refers to the vertical accumulation of material, and typically is measured using a marker or dated sediment core.
accumulation: the mass-based buildup of material that occurs in and on wetland soils This could include both surface deposition of material as well as the buildup of organic matter belowground. It is typically separated in mineral and organic matter accumulation rates based on the organic content of the soil.
compaction: (also referred to as shallow subsidence): the loss of sediment volume due to the consolidation of light surface sediment as well as the loss of organic matter due to decomposition. This refers to dynamics in shallow sediments (i.e., above-established benchmarks).
elevation change: This refers to the shift in wetland elevation relative to a stable benchmark. Elevation change is measured with SETs and is equal to accretion minus compaction.

Compaction typically is not measured but estimated by subtraction, using estimates of elevation change from SETs and accretion from makers (elevation change = accretion − compaction). Deep

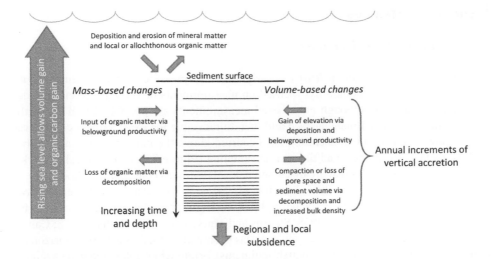

Figure 7.1 Conceptual diagram of processes affecting sedimentation dynamics in tidal wetlands. Processes on the left side of the diagram indicate mass-based changes, while those on the right indicate volume-based changes. Overall elevation change of the sediment surface is a function of accretion, compaction within the shallow sediment column, regional and local subsidence, and eustatic SLR. Factors that lead to increases in sea level (or loss of elevation relative to sea-level) allow for greater accretion (volume gain) and carbon gain.

subsidence is also of interest in determining the net elevation change of a wetland; however, it is not incorporated into SET measurements since it occurs below the SET benchmark. Long-term measurements of water level based on tidal gauges and geodetic leveling give estimates of overall regional subsidence, which includes both shallow and deep subsidence (Turner 1991; Dokka 2011), as does evaluation of sediment stratigraphy (Törnqvist et al. 2008), although there has been debate in how these data are interpreted. Where available, estimates of local SLR and regional subsidence can be incorporated into the interpretation of wetland elevation relative to local water levels.

The terms above focus on sediment accretion processes, but similar terminology is also used for carbon dynamics (see definitions provided in the glossary for details). In a carbon context, accretion relates to volumetric measurements of carbon accumulation; this is directly analogous to the vertical focus of sediment accretion measurements, as vertical sediment accretion is relative to a particular sediment area (height × area = volume). There are slight differences in other terminology usage from a sediment accretion to a carbon context: carbon accumulation is mass-based, similar to the usage for sediment references, but emphasizes the inclusion of both allochthonous and autochthonous carbon. Sequestration is not used in relation to sediment accretion, and with carbon dynamics, this refers specifically to the long-term accumulation of refractory autochthonous carbon. None of the field-based measurements of sediment accretion and carbon sequestration is particularly effective in distinguishing autochthonous versus allochthonous sources of carbon, and in this sense, models of carbon sequestration, calibrated with data from accretion estimates and soil characteristics, are more effective in teasing apart these different contributions (see Chapter 6 in this volume).

The choice of a method for a particular study or location depends on the questions of interest, as well as the conditions at a sampling site (e.g., long-term historical rates at a natural wetland versus current rates at restored wetland). Given the mix of processes that are occurring, there is no single measurement that can provide data on the all dynamics and time scales that may be of interest (Table 7.1). In addition, accretion and sequestration dynamics vary across an individual wetland, so a single measurement within a sampling location will not capture the range of conditions and dynamics that may occur across an entire site, and some consideration should be given to sample locations across the wetland as well as the number of replicate samples.

7.2.1 Short-Term Methods

7.2.1.1 Sediment Marker Horizons

They have been used since at least the early 1900s to measure surface sediment dynamics; most typically, the marker method has focused on measurements of vertical accretion rather than mass-based accumulation, although mass-based measurements are possible. In all cases, the idea is to "mark" the current sediment surface and then to resample the exact same location over time to measure how much sediment accretion has occurred above the marker in the interval between the establishment of the marker and the current sampling (Figure 7.2). Most typically, markers are used in conjunction with SETs that are discussed below (Boumans et al. 1992; Cahoon et al. 2002a, b; see reviews by Callaway et al. 2013; Lynch et al. 2015). A wide variety of materials have been used to mark the sediment surface, including brick dust, glitter, metal plates, resins, and feldspar clay (Carey and Oliver 1918; Steers 1938; Stearns and MacCreary 1957; Richard 1978; Cahoon and Turner 1989). Over the last few decades, feldspar has been widely used as a marker horizon because of its fine texture (similar to most tidal marsh sediments), bright white color, and wide availability at ceramic supply stores. As above, this method is only useful to measure accretion above the marker and does not include deeper processes; however, as the marker is buried deeper and deeper over time, it will integrate root production and other processes. Filter paper also has been used to measure sedimentation measurements over days to weeks (Reed 1989); however, given the extremely short time scale, this approach is not applicable to carbon dynamics. In order for marker horizons to be used successfully, they must remain undisturbed within the sediment column; bioturbation or excessive trampling will reduce their effectiveness. Similarly, the marker material should have the same density and particle size as local sediment to minimize potential movement of the marker within the sediment column. Markers are not likely to be effective in wetland areas exposed to substantial wave action or erosion (e.g., some locations adjacent to mudflats), as this also will lead to marker disturbance or erosion. Last, marker plots must be of sufficient size so that repeated sampling does not disturb them, and they should be well marked so that they can easily be relocated. A number of additional short-term methods for sediment dynamics are available and reviewed in Thomas and Ridd (2004). These methods are of interest for understanding sediment inputs and other short-term issues, but are not likely to be useful for carbon sequestration studies.

Figure 7.2   (See color insert following page 266.) Layer of white feldspar used as a marker horizon in a sediment plug. The depth of sediment above the feldspar is measured with a caliper and indicates the amount of sediment that has accreted since the layer was established, in this case just over 30 mm.

7.2.1.2 Surface Elevation Tables

They also have been used widely in last few decades as a companion to sediment marker horizons because they provide information on elevation change, rather than accretion rates. SETs measure changes in the elevation of the marsh surface relative to a stable benchmark that is established within the marsh; the SET instrument is portable, and it is designed to attach to the benchmark and measure changes with high precision (e.g., ±few mm). SETs were introduced in the United States in the 1990s in Louisiana in large part because of the realization that markers alone were not providing sufficient understand of marsh sediment and elevation dynamics (Boumans and Day 1993; Cahoon et al. 2002a). In this case, many marshes were accreting sediment very rapidly (e.g., over 10 mm year[-1]) but were still converting to unvegetated flats (Day et al. 2000). SET measurements were used to document changes in elevation. Although wetlands were accreting substantial sediment, they were subjected to even higher rates of subsidence and compaction, resulting in the overall loss of elevation (Cahoon et al. 1995). As in this example, when used together, markers and SETs can be used to estimate surface compaction of sediments, because accretion (from marker horizons) − compaction = elevation change (from SETs). In cases where surface sediments are very firm, compaction will be minimal and elevation change will be very similar to accretion; however, in sites with soft, unconsolidated sediments, elevation change may be substantially less than accretion. In addition, it should be noted that the compaction that is measured using the SET is only relative to the bottom of the SET benchmark; any deeper compaction or regional subsidence will not be reflected in estimates of compaction from SET measurements.

As with markers, it is important to be aware of potential site disturbances when establishing SETs. Permanent or removable sampling platforms and access boardwalks are usually established to reduce trampling and other impacts associated with both benchmark installation and ongoing sampling. In order to give meaningful data, SET benchmarks must be stable, which requires pounding the SET rod into the ground until refusal, which varies in depth depending on local substrate and can be 30 m or more. Details on installation, equipment, and sampling methods are provided in Lynch et al. (2015). Typically, one SET benchmark and adjacent marker horizons are established per sampling location, with multiple sampling locations across a marsh to evaluate spatial variation. It is also possible to measure compaction at different depths within the soil column by nesting SET benchmarks at different depths at a single sampling location (Cahoon et al. 2002b). Elevation changes are of critical interest for evaluating marsh sustainability (Webb et al. 2013), and sustainability is essential to maintaining ongoing carbon sequestration. However, the measurement of accretion rates is of more direct interest for carbon sequestration issues since elevation change is not directly convertible to any estimate of carbon sequestration rate.

7.2.2 Longer-Term Methods

Dated sediment cores have been used extensively to evaluate tidal wetland accretion rates over time scales from decades to millennia. Sediments layers within the cores are dated using radioisotopes or other historically deposited materials, such as pollen. In this sense, the dated cores are analogous to the marker studies above; however, the "markers" of radioisotopes or pollen cover much longer time periods. Sediment core dating is effective and widely used in tidal wetlands, because they are consistently depositional and have relatively low rates of bioturbation. In ecosystems that are more dynamic (i.e., may have periods of rapid erosion) or experience substantial bioturbation, the use of dated cores is not likely to be effective because of movement within the sediment column of the radioisotopes or other materials that are used for dating. It is also important to note that dated sediment cores provide information on sediment accretion rates but do not measure changes in surface elevation (as from SET measurements); however, because of the longer-term nature of most sediment core dating, these measurements do incorporate some component of compaction (through the depth

Table 7.1 Summary Comparison of Accretion Methods for Carbon Sequestration Measurements

Method	Time Scale	Strengths/Weaknesses	Additional Notes
SETs	Typically 1–10 years, possibly longer	• Measures elevation change • Does not provide information on accretion	Extremely valuable for evaluating wetland stability, but not useful for estimates of carbon sequestration
Sediment marker horizons	Typically 1–10 years, possibly longer	• Provides very accurate measurements of surface, short-term accretion • Does not incorporate deeper sediment processes • Short time frame leads to overestimates of carbon sequestration	Very widely used in conjunction with SET measurements to evaluation accretion, compaction, and elevation change
^{137}Cs	~50 years	• Provides accretion estimate over one time period (since 1963) • Simple and cheapest of radioisotope methods • Time period is intermediate compared to markers and ^{210}Pb	Very widely used because of low cost; intermediate time scale overestimates long-term sequestration rates
^{210}Pb	~100 years	• Possible to identify different accretion rates over multiple time periods • More expensive than ^{137}Cs because this method requires more samples and longer analysis time • The time scale of 100 years is useful for long-term carbon sequestration estimates and policy issues	Widely used; still incorporates labile carbon in surface sediments, but longer time frame reduces effect of the labile carbon pool on overall sequestration estimate
^{14}C	1,000s years	• Possible to identify different accretion rates over multiple time periods, although time scales are broader than for ^{210}Pb • More expensive than ^{137}Cs because of sample number and analysis time • Long time frame is beyond the scope of evaluating the recent stability of tidal wetlands	Not as widely used, but can provide very valuable information for longer-term processes, including both accretion and carbon sequestration

of the core, but not below). Methods that cover long periods, and therefore greater depths, will incorporate more shallow compaction than shorter-term methods. The three most commonly used methods are discussed below, followed by more general considerations for all types of sediment core dating.

7.2.2.1 ^{137}Cs Dating

This isotope is the result of nuclear fission, and it was spread throughout the northern hemisphere in the 1950s and early 1960s as a result of atmospheric testing of nuclear bombs. The maximum abundance of ^{137}Cs in an undisturbed sediment column reflects the sediment surface in 1963, when testing was at a maximum and just prior to the international agreement that lead to the abolition of atmospheric nuclear bomb testing (Ritchie and McHenry 1990; Reide Corbett and Walsh 2015). The peak of ^{137}Cs activity is directly analogous to a marker horizon that was laid down at that time, and the location of peak within the sediment column provides an average rate of sediment accretion or mass-based accumulation since 1963 (DeLaune et al. 1978; Milan et al. 1995). With a half-life of 30.2 years, ^{137}Cs peaks from 1963 are likely to be found in undisturbed soils for many decades into the future, although they will continue to decline in activity over time (Ritchie and McHenry 1990; Nolte et al. 2013). There also can be more local signals that may provide secondary peaks and additional dating opportunities. For example, the Chernobyl nuclear accident in April 1986 resulted in a plume of ^{137}Cs that has been used to date sediments in northern Europe (Callaway et al. 1996). Spills from power plants or other accidents could also be used to date sediments with local sources (Tsompanoglou et al. 2011). Activity of ^{137}Cs is measured using gamma spectroscopy.

There must be a substantial clay component in the soil to prevent migration of ^{137}Cs within the soil column and to allow for successful dating with [137]Cs. This is not usually an issue in tidal wetlands since they are typically composed of fine sediments; however, it could be an issue in high-energy coastal systems that are likely to contain a large amount of coarse sediments. Bioturbation or erosion also can rework the sediment and remove or mix ^{137}Cs throughout the sediment column and obscure the peak. Milan et al. (1995) identify considerations for evaluating the quality of the 1963 ^{137}Cs peak; if a clear, distinct peak is not observed in a core profile, the core cannot be dated using [137]Cs. Confirmation of the sediment age based on ^{137}Cs using a second method, such as ^{210}Pb dating, is recommended. One drawback of this method is that it typically only provides a single average rate of sediment accumulation based on the 1963 peak, and because there is only one dated layer, there is no way to differentiate changes in sedimentation rates over time. Because of the relatively high activity of ^{137}Cs in most tidal wetland soils and the need to only identify peak activities, this method does not require long measurements periods and is relatively inexpensive compared to ^{210}Pb and [14]C.

7.2.2.2 [210]Pb Dating

Whereas ^{137}Cs is an unnatural isotope that gives a single date related to peak activity, ^{210}Pb is a natural isotope that is part of the [238]U decay chain. It has a relatively long half-life (22.3 years) compared to most other isotopes in the decay chain. As a result, ^{210}Pb accumulates in the atmosphere, is brought to the earth's surface with rain, and eventually can accumulate in tidal wetland soils. Assuming that there is a relatively constant input of ^{210}Pb over time, soil layers can be dated based on the rate of disappearance of ^{210}Pb from the surface to deeper soil layers and its half-life. Two different approaches are used to calculate dates: the Constant Initial Concentration (CIC) and the Constant Rate of Supply (CRS), with slight differences in assumptions regarding surface inputs and corresponding differences in calculations (Appleby and Oldfield 1978; 1983, Nolte et al. 2013; Reide Corbett and Walsh 2015). Activity of ^{210}Pb is measured using gamma spectroscopy. Dating with ^{210}Pb is restricted to about 100 years; after this time, activities of ^{210}Pb are so low that measurements are very difficult. For estimates of carbon sequestration, accretion measurements using ^{210}Pb usually are preferred over ^{137}Cs, as this method covers a longer time period than ^{137}Cs, and the 100-year-time scale of 210Pb dating coincides with some of the policy goals of carbon credits (e.g., the state of California has a 100-year-time horizon for carbon credits). In addition, if there are distinct periods of differing sediment accretion within a sediment core, it is possible to determine these separate accretion rates for these periods based on the differing slopes in ^{210}Pb profiles over the particular time periods. This is advantageous compared to ^{137}Cs dating, which only typically provides a single accretion rate relative to the 1963 peak in ^{137}Cs activity; however, the use of both methods together can provide a useful comparison and verification of accretion rates across slightly different time scales (Kirchner and Ehlers 1998). Reide Corbett and Walsh (2015) recently summarized details for both ^{137}Cs and ^{210}Pb dating methods.

7.2.2.3 [14]C Dating

With radiocarbon techniques, sediments can be dated on the scale of millennia, as ^{14}C has a half-life of almost 5,800 years. While it has been widely used to date SLR (Törnqvist et al. 2015), it has not been applied as often as ^{137}Cs or ^{210}Pb in tidal wetlands (Brevik Homburg 2004; Drexler et al. 2009a; Glaser et al. 2012). Basal peats, shells, seeds, and other plant or animal fragments are analyzed for ^{14}C using accelerator mass spectrometry (AMS), and details on the method are available in Törnqvist et al. (2015). As with ^{210}Pb, if there are substantial shifts in sedimentation rates over time, accretion rates can be established for separate time period using ^{14}C dating. Given the extended time scale, ^{14}C dating is very useful for evaluating carbon sequestration over long time periods, although the data are not as applicable for the 100-year-time scales that are commonly of interest for policy implications. Where available, ^{14}C dating could be very useful for long-term calibration of sediment models as it substantially extends the time scale under consideration.

7.2.3 General Sampling Considerations

Regardless of the analytic method for dating cores, there are a number of important issues to consider in terms of collecting sediment cores. First, it is necessary to select sampling locations carefully, as there can be substantial spatial variation in conditions across a marsh. Locations that have experienced erosion, resuspension of sediments, surface disturbances such as trampling, bioturbation, or similar processes that disturb sediments should be avoided. In addition, to considering current conditions, it is important to determine that the location being sampled sampling must have been relatively free of disturbance over the entire time period for dating (i.e., back to 1963 for ^{137}Cs, back 100 years for ^{210}Pb, and longer for ^{14}C). For this reason, low marsh areas with substantial wave exposure (or that may have been mudflat locations with high rates of resuspension and bioturbation in the not too distant past) may not be useful for historical dating. While intense bioturbation could affect the peak location, ^{137}Cs is more resistant to these impacts than other methods. Slight movement of ^{137}Cs will blur the intensity of the ^{137}Cs peak, but it is unlikely to shift the location of peak within the soil column. Methods that are reliant on relative activities throughout the soil column, such as ^{210}Pb or ^{14}C are more sensitive to bioturbation. If there is interest in determining an average rate for an entire wetland, it is necessary to collect multiple cores to assess spatial variability within the wetland, and as above the choice of sample locations is more challenging for cores than for markers because the location may have changed over time and current conditions may not reflect historical conditions throughout the depth of core, especially for deep sediment cores that cover long time periods.

In collecting cores, it is important to avoid compaction and mixing of sediment layers. Compaction is more likely to be a problem in sediments with very low bulk densities and with very high water content (e.g., low-marsh locations that are frequently inundated). To reduce compaction, it is important to have a sharp cutting head on the core; a variety of different coring devices have been used successfully in tidal wetland soils with minimal compaction, including the Hargis corer (Hargis and Twilley 1994) and the modified Livingstone corer (Wright 1991). Coring tubes with wider diameters area also likely to reduce compaction, and there is some technique involved in carefully inserting the core to cut through roots and rhizomes as it is pushed through the sediment and reduce compaction. In addition, care must be taken to avoid compaction in extruding cores. Cores are commonly extruded in the field and sectioned by hand with a sharp knife. Cores collected in areas with very soft sediments can be frozen prior to sectioning to avoid compactions; frozen cores require sectioning with a power saw. Individual section depths should be fine enough to provide high resolution of isotopic profiles, although more sections will increase costs. Many cores are sectioned at 2-cm intervals, but they may be larger or smaller, depending on the issue at hand (e.g., 1-cm sections were used to identify the shallow depth of Chernobyl peak ^{137}Cs activity (Callaway et al. 1996)). Care should be taken to avoid mixing sediment across sections by cleaning sampling equipment as the core is processed and sectioned; similarly, as core sections are evaluated in the lab, it is critical to avoid mixing of materials across sections. Typically, small representative subsamples from each section are used for isotopic analysis in a gamma counter.

7.2.4 Ancillary Data for Calculation of Carbon Sequestration

In order to calculate carbon sequestration rates for a wetland location, sediment accretion rates must be combined with additional data, in particular depth profiles of soil bulk density and carbon content. These two sediment characteristics are commonly measured in dated sediment cores. In order to estimate soil bulk density (g cm^{-3}), the weight of a known volume of sediment is measured.

Individual core sections determine a specific volume (i.e., core diameter and section depth), so core section dry weights can be directly converted to bulk density estimates. If short-term carbon sequestration rates are required based on marker horizon accretion rates, it is necessary to estimate the bulk density and carbon content of material above the marker horizon. Estimating the volume of samples above the marker horizon can be challenging due to both the unconsolidated nature of surface sediments and shallow depth of relatively recent deposits; care must be taken to avoid compaction and to accurately estimate sediment volume. Extremely sharp coring devices are needed, and samples can be frozen to avoid compaction.

Carbon content of sediment subsamples can be determined using an elemental analyzer (i.e., a CHN or CN analyzer). Often carbon content is only measured for a subset of samples; and a relationship between organic matter content and carbon content is established to convert organic content measurements to carbon content (e.g., Craft et al. 1991). This is done because organic content can be measured very easily and cheaply by loss on ignition (Ball 1964). In some cases, published conversion factors are used; however, carbon:organic matter relationships may vary between different locations, especially if there are significant changes in dominant vegetation. In these cases, a site-specific relationship of carbon:organic matter should be established.

In order to calculate the carbon sequestration rate, it is necessary to combine accretion rates, soil bulk density, and carbon content, following the steps below.

- First, determine the depth of material accumulated over the time horizon of interest (e.g., 10 years with marker horizons, since 1963 for ^{137}Cs cores, or for 100, 1,000, or 5,000 years using other dating methods).
- Second, determine the cumulative carbon mass per unit area (i.e., g C cm^{-2}) to the depth horizon of interest by summing the carbon density for all sections to the depth of interest (equal to sum of the soil bulk density*carbon content*section depth for all sections).
- Finally, divide the cumulative carbon mass accumulated through the depth horizon of interest by the number of years for that particular depth horizon.

7.3 INTERPRETING ACCRETION DATA

Understanding and interpreting accretion rates can be a challenge given the range of approaches to consider accretion dynamics (e.g., vertical accretion, mass-based accumulation, changes in elevation), the many factors that affect accretion processes, and the fact that these dynamics change across both spatial and temporal scales. Where possible, it is useful to use multiple approaches for measuring accretion to confirm and evaluate potential changes in rates over different temporal scales. Many of these issues have been discussed above; however, more consideration of spatial issues is provided here. At the broad spatial scale, relative SLR, typically driven by rapid subsidence, is seen as a major driver affecting accretion rates between different locations, with the highest rates of accretion found in locations with high rates of relative SLR (e.g., Mississippi River Delta and other locations with rapid subsidence, see data summary in Chmura et al. 2003; Ouyang and Lee 2014). In a sense, it is SLR that "accommodates" the opportunity to accumulate large amounts of sediment, although if SLR exceeds accretion, tidal wetlands can rapidly lose elevation and become unsustainable. In addition to SLR, tidal range has also been proposed to affect accretion rates, under the assumption that a larger tidal range results in greater energy to move water and sediment in and out of tidal wetlands. Stevenson et al. (1986) found a strong relationship between tidal range and accretion balance (accretion –relative SLR); however, this was likely due to the fact that many of the sites with small tidal ranges had very high rates of relative SLR. Kirwan and Guntenspergen (2010) confirmed the challenge of testing the relationship of tidal range and accretion rates but found support for the proposed hypothesis from their modeling experiment. Differences across regions or sites in suspended sediment concentrations and marsh productivity also affect overall accretion, as has been shown with both observations and model outputs.

At smaller spatial scales, a number of studies have shown predictable trends in accretion within an estuary, and within a particular wetland. Within an estuary or region, mineral accumulation rates often tend to be greater in salt marshes than in brackish or tidal freshwater marshes, with the opposite trend commonly seen for organic matter accumulation, i.e., greater organic matter accumulation and carbon sequestration at lower salinities (e.g., Craft 2007). Because carbon density is relatively constant in tidal wetland soils (Gosselink et al. 1984; Morris et al. 2016), trends in organic matter accumulation and carbon sequestration within an estuary typically are linked to local accretion patterns (Craft 2007). However, in some cases no consistent patterns in accretion rates or sequestration rates have been observed across salinity gradients within an estuary (e.g., San Francisco Bay, Callaway et al. 2012). Other factors, such as local patterns in subsidence and relative SLR, also could over-ride a trend across the salinity gradient. At the spatial scale within a particular wetland, accretion rates are strongly affected by elevation and location, in particular, adjacency to tidal creeks, with greater accretion rates at lower elevations and near creeks. This has been documented consistently across many wetlands (e.g., Hatton et al. 1983; French et al. 1995; Callaway et al. 2012), and there appears to be a stronger trend in mineral inputs across these gradients than in organic matter accumulation.

In addition to spatial variation in accretion rates across a range of scales, accretion is strongly affected by the time scale of measurement, and this is particularly important for evaluating accretion and carbon sequestration at restoration sites, where quantification of carbon credits are most likely to be need. As discussed above, measured rates of accretion and carbon sequestration are likely to be greater over short time periods owing to the ongoing processes of compaction and decomposition as material is buried to greater depths. At recent restoration sites, the only possibility is to measure short-term rates, and this creates a temporal mismatch between available data and the focus on longer-term processes for carbon credits. The most reliable way to "project" measured short-term rates into a longer time frame is to use simulation models, which incorporate these mix of processes over time (see Chapter 6); however, even these estimates are just projections and future dynamics could be quite different, especially given potential changes in SLR, available sediment, salinity, and other factors. Long-term measurements from appropriate natural reference sites also can provide valuable background information in terms of the range of future rates that may be likely at a restoration site. That said, the multiple dynamics of accretion rates highlight the fact that past accretion rates are not likely to be a direct predictor of future accretion rates; as SLR increases, accretion rates in tidal wetlands are likely to rise, assuming that there is sufficient sediment input to accommodate higher rates of accretion.

ACKNOWLEDGMENTS

This research was funded by the NASA Carbon Monitoring System (Grant no. NNH14AY67I) and by the University of San Francisco Faculty Development Fund. Thanks to Lisamarie Windham-Myers and Jim Morris for ideas and discussions that contributed to the chapter, and especially for feedback on Figure 7.1. Thanks also to additional colleagues who have provided insight on these methods over the years, including Bill Patrick, Ron DeLaune, Andy Nyman, Gene Turner, Charlie Milan, Don Cahoon, Jim Lynch, Judy Drexler, Lisa Schile, and Evyan Borgnis.

CHAPTER 8

Greenhouse Gases

Jason K. Keller
Chapman University

CONTENTS

HIGHLIGHTS

1. Methane and nitrous oxide are potent greenhouse gases. Even small fluxes of these gases from blue carbon ecosystems can significantly offset the uptake of carbon dioxide, which results in the formation of sequestered organic matter.
2. The production and consumption of methane and nitrous oxide are regulated by microbial activities and are influenced by a number of environmental factors, including water level, electron acceptor availability, organic matter, and plants.
3. Measurement of methane and nitrous oxide fluxes can be accomplished by laboratory incubations, modeling and direct field measurements using chambers or eddy covariance approaches. These measurements vary in complexity and appropriateness for defining greenhouse gas fluxes from ecosystems.

8.1 INTRODUCTION

Greenhouse gases are the gaseous constituents of the atmosphere that absorb and emit infrared, long-wave radiation, effectively altering the energy balance of the planet and, all things being equal, increasing the Earth's temperature. These gases have been important drivers of the global climate over geological time and are among the most important anthropogenic climate forcing agents, responsible for a significant fraction of the human-mediated warming since the Industrial Revolution. While there are numerous greenhouse gases present at various concentrations in the atmosphere, this chapter focuses on methane (CH_4) and nitrous oxide (N_2O) which are regulated by biological activities and have important implications in the context of understanding blue carbon in coastal ecosystems.

This chapter begins with a justification for the need to understand CH_4 and N_2O dynamics in blue carbon projects while concomitantly justifying the exclusion of carbon dioxide (CO_2) in this context. Overviews of the ecology of CH_4 and N_2O emissions are included in order to identify environmental factors that are likely to alter the magnitude of the fluxes of these greenhouse gases. The chapter concludes with a brief discussion of some of the methods that can be used to measure CH_4 and N_2O fluxes from coastal ecosystems.

8.2 WHY CH$_4$ AND N$_2$O (AND WHY NOT CO$_2$)?

Atmospheric concentrations of greenhouse gases are typically reported as mole fractions, or mixing ratios, which present the ratio of the number of moles of a greenhouse gas in a given volume to the total number of moles of all gases in that volume. Depending on the amount of the greenhouse gas in question, values are typically expressed as parts per million (ppm), parts per billion (ppb) or in cases of particularly trace gases, parts per trillion (ppt). For example, current CO_2 concentrations in the atmosphere are near 400 ppm, meaning that there are 400 moles of CO_2 for every million moles of gas in a given volume of the atmosphere (or 400 µmol of CO_2 per mole of atmospheric gas, µmol mol^{-1}). Methane and N_2O have lower concentrations in the atmosphere and are expressed as parts per billion (ppb). A value of 1,000 ppb of CH_4 would mean that there are 1,000 moles of CH_4 in a volume that contains a billion total moles of gas (nmol mol^{-1}; 1,000 ppb = 1 ppm).

Globally, the case for considering CH_4 and N_2O as important greenhouse gases is clear. Atmospheric concentrations of CH_4 have increased from pre-industrial levels of 722 ppb to concentrations of 1,803 ppb in 2011 (Hartmann et al. 2013). A number of natural processes regulate atmospheric CH_4 concentrations. Wetlands and other aquatic systems (i.e., permafrost systems and freshwater lakes and rivers) are the largest natural sources of CH_4, although geological activities also contribute to this background flux of CH_4. Methane reacts with a number of compounds in the atmosphere and these chemical losses are the largest global sink for CH_4, with consumption by soils also contributing to the global sink (Ciais et al. 2013). The increase in atmospheric CH_4 concentrations since the Industrial Revolution is the result of anthropogenic activities above and beyond these natural processes. The most significant anthropogenic sources of CH_4 include agriculture and waste (i.e., rice, ruminants, and landfills); biomass burning; and fossil fuel use. Both top-down and bottom-up global estimates suggest that these anthropogenic sources roughly double the amount of CH_4 entering the atmosphere each year (Ciais et al. 2013). The increase in atmospheric CH_4 concentration is responsible for ~17% of the radiative forcing (the added energy that leads to warming) of well-mixed greenhouse gases (Myhre et al. 2013). Methane fluxes from the vegetated coastal systems associated with blue carbon are generally low, and when corrected for the global area of these ecosystems, the fluxes represent only a small fraction (well below 1%) of the average global CH_4 flux (Ciais et al. 2013), (Table 8.1).

Table 8.1   Aerial Extent, CH₄ Flux, and N₂O Flux from Mangrove, Salt Marsh, and Seagrass Ecosystems

Ecosystem	Aerial Extent[a] (Mha) Mean (Range)	CH₄ Flux (μmol CH_4 m^{-2} h^{-1}) Median (Range)	N₂O Flux[b] (μmol N_2O m^{-2} h^{-1}) Median (Range)	Global CH₄ Flux (Tg CH_4-C year^{-1}) Median (Range)	Global N₂O Flux (Tg N_2O-N year^{-1}) Median (Range)
Mangrove	14.5 (13.8–15.2)	5.7 (–0.3–5168.6)[c]	0.95 (0.1–6.0)	0.09 (0.00–78.85)	0.03 (0.003–0.22)
Salt Marsh	5.1 (2.2–40)	2.2 (–1.0–40.6)[d]	0.34 (–2.5–8.9)	0.01 (–0.01–0.22)	0.004 (–0.01-0.87)
Seagrass	30 (17.7–60)	2.6 (0.3–12.8)[e]	0.39 (0.0–5.2)	0.08 (0.01–0.40)	0.03 (0.0–0.77)

Fluxes are representative values from the literature and should not be misconstrued as representing a complete summary of available data. These fluxes should be viewed as a first approximation of fluxes from blue carbon ecosystems.

[a] Aerial extent from Murray et al. (2015).
[b] N₂O flux measurements from Murray et al. (2015).
[c] Mangrove CH₄ flux measurements from summary tables in Nóbrega et al. (2016) and Chen et al. (2010). Duplicate values were removed, and ranges were used.
[d] Salt marsh CH₄ flux measurements from Poffenbarger et al. (2011, used sites with salinity > 18.0 ppt); Moseman-Valtierra (2011); Witte and Gianai (2016, used salt marsh values); and Chmura et al. (2011, 2016).
[e] Seagrass CH₄ flux measurements from Bahlmann et al. (2015) and Oremland (1975, used rates from light and endogenous conditions, ranges were used).

Global concentrations of N₂O have increased from pre-industrial concentrations of 270 ppb to levels of 324 ppb in 2011 (Hartmann et al. 2013). Similar to CH₄, N₂O concentrations are regulated by a variety of natural processes with the increase in overall atmospheric concentrations resulting from additional anthropogenic activities. The most important natural sources of N₂O are soils under natural vegetation, the oceans, and chemical reactions of other nitrogen species in the atmosphere. Atmospheric chemistry in the stratosphere is the largest natural sink for atmospheric N₂O. Anthropogenic activities currently contribute an additional ~60% of natural N₂O sources to the atmosphere, with agriculture being the dominant anthropogenic source. Fossil fuel combustion and industry, biomass burning, human excreta and increased fluxes from atmospheric nitrogen deposition on terrestrial and marine environments are also anthropogenic sources. Particularly relevant in the context of blue carbon, N₂O from rivers, estuaries, and coastal zones is reported to be just under 10% of the annual anthropogenic source of N₂O (Ciais et al. 2013). The increase in atmospheric N₂O concentrations is responsible for another 6% of the radiative forcing of the well-mixed greenhouse gases (Myhre et al. 2013). A recent review (Murray et al. 2015) suggests that median N₂O fluxes from blue carbon ecosystem are 0.95, 0.34, and 0.39 μmol N_2O m^{-2} h^{-1} for mangrove, salt marsh, and seagrass environments, respectively. When corrected for global area of these ecosystems, these fluxes sum to ~0.07 Tg N_2O-N year^{-1}, approximately 0.4% of the total N₂O source to that atmosphere (Ciais et al. 2013) (Table 8.1).

8.2.1 The Case for Considering CH₄ and N₂O in Blue Carbon Projects

Coastal blue carbon is defined as "the stocks and fluxes of organic carbon and greenhouse gases in tidally-influenced coastal ecosystems such as marshes, mangroves, seagrasses and other wetlands" (Crooks et al, this volume). This definition focuses on large stocks of carbon, on the scale of kilograms per square meter, stored in vegetation and soils of coastal ecosystems (Duarte et al. 2013). This large store of carbon is ultimately derived from the uptake and conversion of atmospheric CO_2 to organic matter by plants, with annual carbon burial rates on the scale of hundreds of grams of carbon per square meter (Mcleod et al. 2011; Duarte et al. 2013). However, included in this definition are the fluxes of *all* greenhouse gases. But, why should one be concerned with the comparatively small fluxes of two gases—on the scale of grams of CH_4 m^{-2} year^{-1} (Poffenbarger et al. 2011) and less than a gram of N_2O m^{-2} year^{-1} (Murray et al. 2015)—from these ecosystems? The short answer is that these gases can have disproportionately large effects on the global climate, and thus even relatively small emissions of CH₄ and N₂O can significantly offset the uptake of CO_2, which results in the formation of sequestered organic matter in blue carbon ecosystems.

Defining the relative impact of CH_4 and N_2O (or any other greenhouse gas) on the climate can be expressed through a number of metrics (Myhre et al. 2013), but the global warming potential (GWP) concept is among the most common metrics used. As defined by the Fifth Assessment Report of the Intergovernmental Panel on Climate Change (IPCC 2013), GWP is "An index, based on radiative properties of greenhouse gases, measuring the radiative forcing following a pulse emission of a unit mass of a given greenhouse gas in the present-day atmosphere integrated over a chosen time horizon, relative to that of carbon dioxide. The GWP represents the combined effect of the differing times these gases remain in the atmosphere and their relative effectiveness in causing radiative forcing." This definition highlights that it is both the energy-capturing efficiency of a gas as well as its atmospheric lifetime that determine its impact on the climate.

Carbon dioxide has a radiative efficiency of 1.37×10^{-5} W m^{-2} ppb^{-1}, meaning that for every increase in atmospheric concentration of 1 ppb (i.e., for the addition of one mole of CO_2 for every billion moles of gas in the atmosphere), an additional 1.37×10^{-5} W m^{-2} of energy will be added to the Earth's climate system. Methane and N_2O have radiative efficiencies 26-times and 221-times this value, respectively (Table 8.2). While these radiative efficiencies are governed by the chemistry of CH_4 and N_2O, the effect of a pulse of these gases into the atmosphere changes through time depending on the average lifetime of these gases in the atmosphere. For example, CH_4 has an average atmospheric lifetime of 12.4 years. Over a 20-year time horizon, CH_4 has a GWP of 84 (recall that GWPs are relative to a pulse of CO_2). However, after 100 years, much of the CH_4 in the initial pulse has left the atmosphere and the GWP over this horizon lowers to 28 (Table 8.2). Nitrous oxide is a longer-lived (121-year atmospheric life time) greenhouse gas than CH_4. This longer lifetime and increased radiative efficiency results in GWPs of 264 and 265 for N_2O over the 20- and 100-year time horizons, respectively (Table 8.2). These GWP values are typically used to express fluxes of CH_4 and N_2O as equivalents of CO_2 (CO_2eq) in blue carbon projects (e.g., VCS Methodology 2015).

Careful use of the GWP concept is crucial for understanding the radiative balance of a system and its potential impact on the climate. For example, past work suggested that even freshwater wetlands with relatively high fluxes of CH_4 would have a net cooling effect (net radiative sinks) on the global climate by taking up more CO_2 than releasing CH_4, even after correcting for the higher GWP of CH_4 (Mitsch et al. 2013). However, the model used in this work considered the atmospheric residence time of CH_4 in addition to the GWP of this greenhouse gas. Thus, the model essentially double-counted the loss of CH_4 from the atmosphere over time, significantly underestimating the potential warming effect of CH_4 flux from these wetlands (Bridgham et al. 2014, Neubauer 2014).

More recently, Neubauer and Megonigal (2015) have highlighted the limitations of the GWP concept—which is based on a single pulse of a greenhouse gas—when considering the climate impacts of sustained greenhouse gas fluxes from natural ecosystems. They instead suggest that the sustained-flux global warming potential (SGWP) may be a more appropriate metric to compare the climatic role of ecosystems where greenhouse gas emissions are "persistent-not-one-time-events" (Neubauer and Megonigal 2015, and references cited therein). Sustained-flux global cooling potentials (SGCP) can also be calculated based on sustained uptake of CH_4 or N_2O, which are possible in natural ecosystems. Accounting for this sustained flux, the impact of CH_4 (relative to CO_2) is more

Table 8.2   Radiative Efficiencies, Lifetimes, GWPs and SGWPs of CH$_4$ and N$_2$O

Greenhouse Gas	Radiative Efficiency[a] (W m^{-2} ppb^{-1})	Lifetime[b] (year)	GWP$_{20}$[b]	GWP$_{100}$[b]	SGWP$_{20}$[c]	SGWP$_{100}$[c]
CH_4	3.63×10^{-4}	12.4	84	28	96	45
N_2O	3.03×10^{-3}	121.0	264	265	250	270

[a] Radiative efficiencies from Table 2.1 in Hartmann et al. (2013).
[b] Lifetimes and GWPs from Table 8.7 in Myhre et al. (2013).
[c] SGWPs from Table 8.1 in Neubauer and Megonigal (2015).

dramatic than the pulse-based GWP approach. A sustained emission of CH_4 has a SGWP of 96 (relative to CO_2) over the 20-year time horizon and a SGWP of 45 over the 100-year time horizon (this decrease again reflects that lifetime of CH_4 in the atmosphere; Table 8.2). The impact of this approach compared to the more common GWP approach is smaller for N_2O, which has a SGWP of 250 and 270 over the 20- and 100-year time horizons, respectively (Table 8.2). Values of SGWP are additive, such that over a 100-year time period, a sustained flux of 1 kg of CH_4 m^{-2} year^{-1} and sustained flux of 1 kg of N_2O m^{-2} year^{-1} would balance the sequestration of 315 kg CO_2 m^{-2} year^{-1}. As noted by Neubauer and Megonigal (2015), the pulse-based GWP approach has been used for decades, but more appropriate metrics for assessing the climate impacts of blue carbon projects (e.g., the SGWP metric) should be considered moving forward.

8.2.2 The Case for Ignoring CO_2 Fluxes in Blue Carbon Projects

While a single molecule of CO_2 may not be as potent as CH_4 or N_2O, it is clear that CO_2 plays an important role as a greenhouse gas on the global scale. Atmospheric concentrations of CO_2 have increased from pre-industrial levels of 278 ppm to concentrations of 390.5 ppm in 2011 (Hartmann et al. 2013). As described in more detail below, ecosystems play an important role in the natural exchange of this gas with the atmosphere. On top of these natural exchanges, fossil fuel burning and cement production have added 375 Pg (Pg = 10^{15} g) of carbon to the atmosphere since the Industrial Revolution. An additional 180 Pg of carbon have been added to the atmosphere over this period as a result of anthropogenic land use changes. Less than half of these anthropogenic CO_2 emissions have accumulated in the atmosphere. The remaining CO_2 has been absorbed by the ocean and in terrestrial carbon sinks (Ciais et al. 2013). Despite not being a particularly potent greenhouse gas, the relatively large increase in atmospheric CO_2 concentration is responsible for 64% of the radiative forcing of the well-mixed greenhouse gases on the global scale (Myhre et al. 2013).

On a per mass basis, the exchange of CO_2 from blue carbon ecosystems likely dwarfs the exchange of CH_4 and N_2O, and in many regards, CO_2 is the dominant greenhouse gas in these environments. Indeed, the effective uptake and long-term storage of atmospheric CO_2 is perhaps the most fundamental underpinning of blue carbon projects. As discussed elsewhere, vegetated coastal ecosystems carry out this sequestration of CO_2 more effectively (per area) than virtually any other ecosystem on the planet (Hopkinson, this volume), explaining why their restoration and conservation has generated such interest in the context of global climate change. Completely ignoring CO_2 fluxes in blue carbon projects is thus nonsensical, and that is not the point of this section. Instead, this section highlights that in most cases, the sequestration of CO_2 by coastal ecosystems is more effectively measured as a change in carbon stocks rather than measured as CO_2 fluxes themselves.

Carbon dioxide enters into coastal ecosystems through photosynthesis, the process which converts atmospheric (or dissolved) CO_2 into organic matter. This uptake of CO_2 is termed gross primary productivity (GPP). A fraction of this organic matter is respired by photosynthetic organisms themselves for growth, maintenance, and ion absorption. The organic carbon respired by the photosynthetic autotrophs (autotrophic respiration, A_R) is released as CO_2 and does not enter into an ecosystem. The concept of net primary productivity (NPP) is used to explain the net gain by photosynthetic organisms after accounting for A_R (NPP = GPP$-A_R$). The organic matter that enters an ecosystem through NPP is subsequently consumed and subjected to decomposition by a host of macro- and microorganisms, including invertebrates, fungi, and soil bacteria. These heterotrophic communities consume organic matter to generate cellular energy, and release much of the carbon found in this organic matter back to the atmosphere as CO_2 through heterotrophic respiration (H_R). It is worth noting that these same microorganisms can be responsible for CH_4 and N_2O production, but those processes are discussed in greater detail below. Net ecosystem production (NEP) is used to define the net amount of carbon entering an ecosystem after accounting for total ecosystem respiration (ER) as the sum of A_R and H_R. Often, net ecosystem exchange (NEE) is used as an approximation of NEP. NEE is the net flux of CO_2 from the ecosystem to the atmosphere, and by convention is a negative value when there is a net uptake of CO_2 by an ecosystem (Chapin et al. 2011). Thus,

$$NEP = GPP - A_R - H_R \approx - NEE \qquad (8.1)$$

or,

$$NEP = GPP - ER, \approx - NEE \qquad (8.2)$$

Ultimately, it is the imbalance between CO_2 inputs (via GPP) and CO_2 outputs (via ER) that allow for the accumulation of organic matter in any ecosystem. In coastal environments, this simple mass balance is made more complicated by dissolved inorganic carbon, dissolved organic carbon, and particulate carbon fluxes associated with tidal exchange. Net ecosystem carbon balance (NECB) is used to account for these additional carbon fluxes. The net dissolved flux (F_D) and net dissolved particulate flux (F_P) are positive when dissolved carbon enters the ecosystem. While generally small, the flux of CH_4 (F_M) from an ecosystem is also accounted for by NECB. Positive NECB values represent an increase in carbon storage in an ecosystem (Figure 8.1). Thus,

$$NECB = - NEE - F_M - F_D - F_P \qquad (8.3)$$

It is possible (and defensible) to measure the actual CO_2 flux associated with NPP to calculate the net movement of CO_2 from the atmosphere into plant biomass. However, it is often far easier to simply measure changes in vegetative biomass to integrate NPP over larger time scales in an ecosystem (Craft 2013). Similarly, it is possible to quantify NEE by measurements of CO_2 exchange; and, these measurements coupled to measurements of sediment and dissolved carbon dynamics can quantify the next carbon input to soil of an ecosystem (Neubauer et al. 2001; Rivera-Monroy et al. 2013;

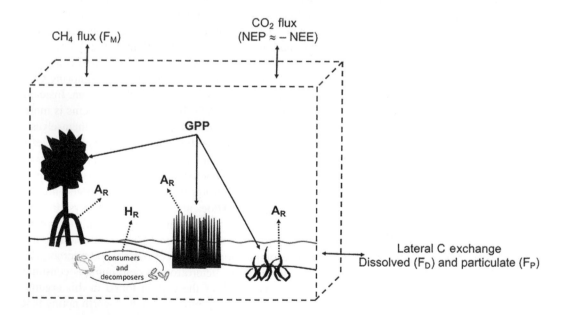

Figure 8.1　Major components of NECB in blue carbon ecosystems. The fluxes that contribute to NEP are shown in bold. In most ecosystems, NEE is used as an approximation of NEP. By convention, negative NEE values reflect CO_2 uptake by an ecosystem. The box represents the ecosystem. Modified from Chapin et al. (2011).

Weston et al. 2014). However, measuring changes in soil carbon stocks over time (e.g., through ^{137}Cs or ^{210}Pb dating) is already integrating many of these measurements and serves as an appropriate proxy for NECB over longer period of time (assuming that vegetation biomass is constant over the time period in question). What is not defensible, however, is to measure changes in stocks of carbon (either in vegetation or in soil pools) and also count direct measures CO_2 inputs as additional sequestration into an ecosystem above and beyond changes in biomass or soil organic matter stocks. Conversely, CO_2 losses as ER (which are easy to measure when measuring CH_4 and N_2O fluxes, see below) should not be deducted from net carbon sequestration values calculated by changes in carbon stock. In short, changes in carbon stocks already account for CO_2 fluxes in coastal ecosystems, and direct measures of CO_2 fluxes into or out of blue carbon ecosystems often represent a form of double counting of this important greenhouse gas. Unfortunately, these changes in carbon stocks do not provide a record of the fluxes of CH_4 and N_2O from the ecosystem, and integrative summaries of the fluxes of these potent greenhouse gases are not possible in blue carbon systems.

8.3 THE ECOLOGY AND CONTROLS OF CH_4 AND N_2O FLUXES

The vast majority of both CH_4 and N_2O cycling in blue carbon ecosystems is regulated by microbial activity. There are excellent reviews of the biogeochemistry and ecology of these gases available (e.g., Megonigal et al. 2004; Bridgham et al. 2013; Murray et al. 2015). This section highlights some of the key ecological aspects of these greenhouse gases with a goal of identifying the environmental factors most likely to control their emissions from blue carbon projects.

8.3.1 Methane Ecology

The flux of CH_4 from an ecosystem is ultimately regulated by the balance between the production of CH_4 (methanogenesis) and the consumption of CH_4 (methanotrophy), although the pathway by which CH_4 leaves an ecosystem can also be important in regulating CH_4 emissions.

Methane is a product of the decomposition of organic matter under anaerobic conditions, such as flooded or saturated soils or sediments, where oxygen in present in low concentrations. In contrast to aerobic decomposition, where a single microorganism can completely degrade organic matter to CO_2 using oxygen as a terminal electron acceptor, anaerobic decomposition relies on a complex microbial consortium that includes both mutualistic and competitive interactions between different microbial groups. Methanogens cannot directly consume complex organic polymers and they rely on "upstream" microbial activities to produce the carbon substrates that they can use for the production of CH_4. Initially complex organic matter is decomposed by microbes using extracellular enzymes to degrade polymers into monomers. Subsequently, these monomers are further degraded by fermenting bacteria into simple substrates, primarily acetate, molecular hydrogen (H_2), and CO_2. These fermentation products are available to methanogens; however, they can also be used be a number of other microbial groups and there is strong competition for these substrates in anaerobic environments. Denitrifiers, manganese reducers, iron reducers, and sulfate reducers all couple the oxidation of these substrates to the reduction of alternative terminal electron acceptors: nitrate (NO_3^-), oxidized manganese (Mn(IV,III)), oxidized iron (Fe(III)) and sulfate (SO_4^{2-}), respectively. These processes are more thermodynamically favorable than methanogenesis and generally win the competition for fermentation substrates, thereby suppressing CH_4 production when they are occurring (Megonigal et al. 2004; Bridgham et al. 2013).

Only after these more favorable alternative terminal electron acceptors have been exhausted, is CH_4 produced through two dominant methanogenic pathways. Acetoclastic methanogens split acetate into CH_4 and CO_2 and hydrogentrophic methanogens oxidize H_2 using CO_2 as an electron acceptor (Megonigal et al. 2004; Bridgham et al. 2013). It should be noted that methanogens can also use a variety of methylated substrates to produce CH_4, including methanol, methylated sulfur compounds (e.g., dimethylsulfate), and methylamines (Boone et al. 1993; Zinder 1993). These methylotrophic methanogenic substrates are non-competitive (i.e., other microbial processes do not compete for them), but are generally thought to play a minor role in natural ecosystems. There is evidence, however, that they can be important in hypersaline systems (e.g., Kelley et al. 2015; Zhuang et al. 2016; Sorokin et al. 2017) and occur in marsh and estuarine environments (King et al. 1983; Franklin et al. 1988; Oremland et al. 1989; Chuang et al. 2016). In systems where these methylotrophic pathways are important, CH_4 can still be produced even in the presence of more favorable terminal electron acceptors.

The production of CH_4 by methanogens in anaerobic environments can be offset by consumption of CH_4 by methanotrophs in the same ecosystem, and this methanotrophy can consume a large fraction of the CH_4 produced within a system (e.g., Giani et al. 1996). Methanotrophs oxidize CH_4 to CO_2 to obtain both the energy and carbon they require (Hanson and Hanson 1996). Traditionally, CH_4 consumption was thought to be limited to aerobic zones in ecosystems—above the water table or in the oxidized root zone—where oxygen is available as an electron acceptor to oxidize CH_4 to CO_2. More recent evidence suggests that CH_4 oxidation can also happen under anaerobic conditions using alternative electron acceptors in place of oxygen (Knittel and Boetius 2009; Gupta et al. 2013). Most notably, it is well established that SO_4^{2-}, present in high concentrations in blue carbon ecosystems, can be used as an electron acceptor under anaerobic conditions and that this process consumes >90% of marine CH_4 production (Hinrichs and Boetius 2002; Reeburgh 2007). Anaerobic oxidation of CH_4 using other alternative electron acceptors including iron, nitrate, and organic matter is also possible (Beal et al. 2009; Segarra et al. 2013; Valenzuela et al. 2017).

Methane can leave an ecosystem through three different pathways: diffusion, plant-mediated transport, and ebullition. The relative role of these pathways can be an important control of CH_4 emissions, in part, by determining how much CH_4 is likely to be oxidized before it leaves an ecosystem. Diffusive fluxes, driven by concentration gradients, are most susceptible to CH_4 oxidation. Wetland plants have developed a variety of physiological adaptations to allow their roots to respire under flooded conditions. The same systems that deliver oxygen to wetland roots frequently allow CH_4 to be vented to the atmosphere. While the effects of plants on CH_4 fluxes are complex (Schutz et al. 1991; Bodelier et al. 2006; Laanbroek 2010), the presence of plants can often increase CH_4 emissions by allowing CH_4 to bypass zones of aerobic CH_4 oxidation. Ebullition, the release of CH_4 through bubbles, also allows CH_4 to bypass aerobic zones and can be in an important, albeit episodic, flux of CH_4 from an ecosystem. Ebullition events are triggered by dissolved CH_4 accumulating to super-saturated conditions and are likely less important in ecosystems with generally low rates of CH_4 production.

8.3.2 Controls of CH_4 Flux

Given the complexity of CH_4 cycling, it is perhaps not surprising that the mechanistic controls of CH_4 flux from an ecosystem are varied and mediated by site-specific conditions. However, there are some key controls of CH_4 emissions that are important when considering CH_4 flux from coastal blue carbon ecosystems (Table 8.3).

The most fundamental constraint on CH_4 flux is arguably water level—the proportion of a soil or sediment that is saturated determines the potential zone of CH_4 production whereas the proportion of a soil above the water level determines the potential zone for aerobic CH_4 oxidation. The tidal nature of blue carbon ecosystems may confound the importance of this factor, and it is important to note that well-mixed surface waters are often well oxygenated. However, Chmura et al. (2011) suggested that lower CH_4 emissions at a microtidal salt marsh relative to a macrotidal salt marsh could have resulted from increased CH_4 oxidation due to more frequent aerobic conditions with lower tidal heights.

Table 8.3 Key Factors Controlling the Flux of CH_4 and N_2O from Blue Carbon Ecosystems

Controlling Factor	CH_4 Flux	N_2O Flux
Water level (anaerobic/ aerobic boundary)	• Methane production is limited to anaerobic zones consistently below water level • Methane consumption is generally highest in aerobic zones above the water level (although anaerobic consumption is also possible)	• Denitrification is limited to anaerobic zones consistently below the water level • Nitrification is limited to aerobic zones above the water level
Alternative Terminal Electron Acceptors	• More favorable electron acceptors (NO_3^-, Mn(IV, III), Fe(III) and most notably SO_3^{2-}) suppress CH_4 production, and can support anaerobic CH_4 oxidation	• Nitrate is required as the electron acceptor for denitrification
Organic Matter	• More electron donors in organic matter will more rapidly consume alternative electron acceptors and allow for increased rates of CH_4 production	• As a heterotrophic process, denitrification generally increases when organic matter is more available • The mineralization of soil organic matter releases NH_4^+, the nitrogen substrate required for nitrification
Plants	• Plant-mediated transport can be an important flux of CH_4 which bypasses zones of aerobic oxidation • Plant release oxygen and/or labile organic substrates in their rooting zones • Over longer timescales, plants are an important control of organic matter quality in ecosystems	• Plant-mediated transport can be an important flux of N_2O, potentially allowing the bypass of zones of N_2O uptake by denitrifiers • Plant release oxygen and/or labile organic substrates in their rooting zones • Over longer timescales, plants are an important control of organic matter quality in ecosystems • Plant compete with nitrifiers and denitrifiers for inorganic nitrogen (NH_4^+ and/or NO_3^-, respectively)

Ultimately, CH_4 emissions require that electrons from organic matter make their way to methanogens that use these electrons to drive CH_4 production. The presence of alternative terminal electron acceptors that intercept these electrons before they reach methanogens will decrease the flux of CH_4 from an ecosystem. In the context of blue carbon, SO_4^{2-} is arguably the most important alternative electron acceptor, and sulfate reduction has long been known to competitively suppress CH_4 production (Bartlett et al. 1987). Indeed, it is this sulfate suppression that has made coastal blue carbon systems more appealing as carbon sinks than their freshwater counterparts despite higher carbon sequestration potential in many freshwater tidal systems (e.g., Craft 2007). Seawater contains on average 28.2 mM SO_4^{2-} (Bianchi 2006) and provides a constant, or at least tidally linked, source of this terminal electron acceptor in blue carbon ecosystems. Frequently, salinity is used as a proxy for sulfate availability. For example, Poffenbarger et al. (2011) demonstrated that saline marshes (those above ~18 ppt) had minimal CH_4 emissions and justified this pattern by increased rates of sulfate reduction in salt marshes. Estimation of SO_4^{2-} based on salinity is reasonable given that seawater has a fairly consistent chloride to sulfate molar ratio of 19.33:1 (Bianchi 2006). However, in cases where microbes can reduce SO_4^{2-} faster than seawater can be exchanged, it is possible to see high salinity systems have low SO_4^{2-} availability. If (generally slow growing) methanogens were able to grow during these periods of relieved competition with sulfate reducers, methanogenesis could play an increasingly important role under these conditions.

Sulfate reducers have received the majority of attention as effective competitors with methanogens in blue carbon ecosystems; however, iron reducers can also suppress methanogens (Neubauer et al. 2005; Tobias and Neubauer 2009). Iron reducers are reliant on oxidized iron (Fe(III)) as an electron acceptor, and while this ferric iron can be an important part of mineral soils and sediments, it is not regularly exchanged with seawater and iron reduction can thus become iron limited. Low tides, oxygen in the plant rooting zone and bioturbation all increase oxygen availability in flooded and saturated soils and sediments and can increase the availability Fe(III). Conceptually, the input of oxidized sediment from the surrounding watersheds could also serve as a source of Fe(III) to coastal ecosystems, leading to the suppression of CH_4 production. Nitrate also serves as a competitively favorable terminal electron acceptor and denitrification suppresses CH_4 production. As discussed below, NO_3^- is present at low levels in all but highly polluted coastal environments and denitrification is generally not thought to play a major role in carbon cycling (e.g., Tobias and Neubauer 2009), despite its importance in N_2O production.

While a large body of literature focuses on the role of electron acceptors in regulating CH_4 production, it is important to remember that it is actually the net flow of electrons to methanogens that matters. Thus, the quantity of electron donors used to reduce the available electron acceptors can also be an important control of CH_4 production (e.g., Sutton-Grier et al. 2011; Vizza et al. 2017). Blue carbon systems with higher organic matter availability or quality (more electron donors) are likely to exhaust competitive terminal electron acceptors more quickly, leading to higher rates of CH_4 production or supporting both sulfate reduction and CH_4 production (Chuang et al. 2016). Methanogenesis has been shown to be directly stimulated by carbon additions in some soils (Irvine et al. 2012), and increased soil carbon content can coincide with higher CH_4 emissions in invaded ecosystems (Tong et al. 2012). Inputs of highly organic waste could also lead to increased CH_4 production in blue carbon ecosystems (e.g., Chen et al. 2011).

Plants are known to be an important mediator of CH_4 dynamics in wetland ecosystems (e.g., Schutz et al. 1991; Laanbroek 2010). As mentioned above, plants can serve as an important conduit for CH_4 to exit an ecosystem without being exposed to zones of CH_4 oxidation. The role of pneumatophores as CH_4 conduits through mangroves may be particularly important in this regard (e.g., Purvaja et al. 2004). In addition to this venting effect, the plant rhizosphere is known to play myriad roles in CH_4 cycling by providing oxygen (and thus other oxidized terminal electron acceptors) to suppress or oxidize CH_4 and/or providing labile carbon substrates (electron donors) to fuel anaerobic metabolism (Bodelier et al. 2006 and references cited therein). Over the longer term, plants are among the strongest determinant of soil organic carbon quality and quantity and alter CH_4 cycling indirectly through this availability of carbon substrates (Tong et al. 2012; Chen et al. 2015).

8.3.3 Nitrous Oxide Ecology

Unlike CH_4, which is the end-product of microbial methanogenesis, N_2O is an intermediate-product, produced during the microbial processes of nitrification and denitrification. While other sources of N_2O, including dissimilatory nitrate reduction to ammonium (DNRA) and abiotic production, are possible, the microbial processes of nitrification and denitrification are thought to be the most important sources of N_2O in many ecosystems (but see Burgin and Hamilton 2007) and will be the focus of this section. The recent review of N_2O in estuarine environments by Murray et al. (2015) provides a more extensive summary of the ecology and controls of N_2O fluxes than what is summarized here.

Nitrification is the microbial process that oxidizes ammonium (NH_4^+) to nitrite (NO_2^-), which is subsequently oxidized to nitrate (NO_3^-). The complete oxidation of NH_4^+ to NO_3^- is typically carried out in two steps by different microbial groups that generally co-exist in soils; so, it is rare to see an accumulation of NO_2^- in natural systems. These microbes are thought to be obligate aerobes (they require oxygen) and use the energy generated by the oxidation of NH_4^+ to reduce CO_2 into the organic carbon required for their growth (i.e., this is an autotrophic process). Heterotrophic nitrification coupled to the decomposition of soil organic matter is also possible in low nitrogen and acidic soils (Chapin et al. 2011), but is not discussed further here. Both nitric oxide (NO) and N_2O are released as by-products of nitrification. Often, this loss is described as analogous to a "leaky pipe," and in general the amount of NO and N_2O released during nitrification is determined by the overall rate of nitrification in a system (Firestone and Davidson 1989). In most cases, only a small percentage of nitrogen "leaks" out during nitrification, and this nitrogen is predominantly in the form of NO as opposed to N_2O (Chapin et al. 2011).

Denitrification describes the reduction of nitrate (NO_3^-) or nitrite (NO_2^-) to di-nitrogen gas (N_2) and is an important pathway of nitrogen removal from ecosystems (Chapin et al. 2011). This process actually progresses through a series of intermediate compounds, in the order: $NO_3^- \rightarrow NO_2^- \rightarrow NO \rightarrow N_2O \rightarrow N_2$. This is a heterotrophic process where denitrifying bacteria are using NO_3^- as an alternative terminal electron acceptor in place of oxygen during anaerobic decomposition (the reduction of NO_3^- is coupled to the oxidation of organic matter to CO_2 in this process). Most microbes responsible for this process are facultative anaerobes, meaning that they have the ability to switch to aerobic respiration when oxygen in present. As discussed above, denitrification is more thermodynamically favorable then methanogenesis and can thus suppress CH_4 production. However, given the generally high carbon to nitrogen ratios in ecosystems (e.g., plants and organic matter), denitrification typically influences a larger fraction of ecosystem nitrogen cycling than carbon cycling (e.g., Tobias and Neubauer 2009). While most bacteria contain the enzymes to fully reduce NO_3^- to N_2, the process of denitrification is often inefficient and a fraction of the nitrogen is released as N_2O. The reasons for this inefficiency, i.e., the reasons for different $N_2O:N_2$ ratios resulting from denitrification are not fully understood. However, the $N_2O:N_2$ ratio is often driven by the relative availabilities of NO_3^- (the electron acceptor) and organic matter (the electron donor). Generally, when NO_3^- is relatively more abundant, denitrification is only partially completed and the resultant $N_2O:N_2$ ratio increases (Chapin et al. 2011; Murray et al. 2015). In addition to bacterial denitrification, there is a growing appreciation for the importance of fungal denitrification and N_2O production (Shoun et al. 1992; Maeda et al. 2015), and recent work highlights the need to consider this process (and chemodenitrification) in coastal environments (Wankel et al. 2017).

The consumption of N_2O is also possible in ecosystems, and most often this consumption is driven by denitrification using N_2O as an electron acceptor instead of NO_3^- or NO_2^-. Essentially, this consumption involves microbes fully reducing the partially reduced N_2O to the end-product of N_2 (recall that N_2O is an intermediate in the denitrification process). In general, this uptake of N_2O by denitrifiers is associated with low NO_3^-, low NO_2^- and low oxygen availability, i.e., only after the more preferred electron acceptors have been exhausted will denitrifiers consume N_2O (Murray et al. 2015 and references cited therein).

8.3.4 Controls of N_2O Flux

Fluxes of N_2O are highly variable in space and time and the controls of this flux are complex and vary by ecosystem. However, similar to CH_4, there are likely to be some key controls of N_2O emissions that are particularly important in the context of blue carbon (Table 8.3). In general, conditions that favor either nitrification or denitrification also favor the flux of N_2O from an ecosystem.

Water level is important in controlling the dominant source of N_2O production in an ecosystem to the extent that it regulates oxygen availability. The production of N_2O by denitrification is limited to anaerobic zones and production by nitrification is limited to aerobic zones. While oxygen availability is linked to water level in sediment and soil environments, surface waters are often well oxygenated and can support nitrification (Murray et al. 2015).

Inorganic nitrogen availability is a key control of N_2O fluxes. In aerobic environments, increased NH_4^+ leads to increased nitrification and associated "leaked" N_2O. Under anaerobic conditions, increased NO_3^- availability provides the necessary terminal electron acceptor for denitrification and increases N_2O flux through this process (Murray et al. 2015 and references cited therein). In most ecosystems, nitrification serves as the primary source of NO_3^- for denitrification. Given that these two processes have different oxygen requirements (i.e., nitrification is aerobic, denitrification is anaerobic), denitrification is often highest at anaerobic/aerobic interfaces, including at the water level, in the oxidized rooting zone or adjacent to burrows of bioturbating organisms (Chapin et al. 2011; Murray et al. 2015). As noted above, increased NO_3^- availability relative to organic matter availability also generally increases the $N_2O:N_2$ ratio produced during denitrification. Increases in either NH_4^+ or NO_3^- associated with eutrophication of coastal water bodies will thus likely increase N_2O fluxes from blue carbon ecosystems. Indeed, nitrogen fertilization frequently increases N_2O flux from blue carbon ecosystems, often causing them to shift from N_2O sinks to N_2O sources (Kreuzwieser et al. 2003; Moseman-Valtierra et al. 2011; Chmura et al. 2016), and nitrogen-rich waste water can also increase N_2O emissions (Chen et al. 2011).

Organic matter directly regulates denitrification by serving as the electron donor for this microbial process. Increased amount, or lability, or organic matter are likely to support higher rates of denitrification assuming that NO_3^- remains available as an electron acceptor (Eyre et al. 2013; He et al. 2016), although the amount of N_2O resulting from this increased denitrification is not well known. Organic matter can also indirectly influence nitrification. In undisturbed ecosystems, the primary source of NH_4^+ oxidized by nitrifiers is often the release of mineralized nitrogen from the decomposition of organic matter. There are generally strong relationships between indexes of soil or sediment quality and rates of nitrogen mineralization; for example, nitrogen mineralization and thus increase NH_4^+ availability are negatively correlated with C:N ratios in soils from a range of ecosystems (e.g., Chapin et al. 2011; although note that this relationship is better studied in terrestrial ecosystems than in coastal ecosystems). Changes in organic matter quality or changes in rates of decomposition (e.g., through increased temperature) are thus likely to increase NH_4^+ within an ecosystem.

Finally, similar to impacts on CH_4 fluxes, plants can regulate N_2O fluxes through changes in oxygen and/or labile carbon availability in the rhizosphere and through longer-term changes in organic matter quality (Chen et al. 2015). Nitrous oxide can also be emitted through vegetation and this plant-mediated flux can be an important source of N_2O to the atmosphere, especially in systems where N_2O subjected to diffusive fluxes would be reduced to N_2 by denitrifiers (Murray et al. 2015). In addition to these impacts, plants are also key regulators of inorganic nitrogen availability and can compete with nitrifiers and denitrifiers for the uptake of NH_4^+ and NO_3^-, respectively.

8.4 MEASUREMENTS OF GREENHOUSE GASES

Given the potential for fluxes of CH_4 and N_2O to negate the climate benefits associated with carbon sequestration in blue carbon projects, measurement of these greenhouse gas dynamics will no doubt continue in marsh, mangrove, and seagrass ecosystems. As protocols for blue carbon projects (e.g., VCS Methodology 2015) continue to develop, the need to accurately quantify these fluxes (or at least justify that they are minimal) will only increase. Detailed methodologies to quantify greenhouse gas fluxes in different ecosystems are well-beyond the scope of this chapter, as are discussions of the sampling approaches necessary to adequately capture the temporal and spatial variability of these fluxes. Instead, this chapter concludes with a brief introduction to the approaches that can be used to measure CH_4 and N_2O fluxes and offers some thoughts on the tradeoffs associated with these approaches. As noted above, many of these methods are equally applicable to fluxes of CO_2, but caution should be used to ensure that CO_2 sequestration is not double counted in blue carbon projects.

8.4.1 Laboratory Incubations

It is relatively straightforward to measure net production of both CH_4 and N_2O by soils, sediments and water in laboratory incubations. In many cases, it is also possible to measure many of the underlying processes responsible for this production, including methanogenesis, methanotrophy, nitrogen mineralization, nitrification, and denitrification. The methods required for quantifying these microbial processes range from straightforward measurements of the accumulation of a greenhouse gas over time (often using gas chromatography) to far more sophisticated approaches using isotopes to track carbon and nitrogen through various pools in the incubations (Bridgham and Ye 2013; Burgin et al. 2013; Inglett et al. 2013a; Inglett et al. 2013b; Inglett et al. 2013c; Roy and White 2013).

The benefit of these incubations is that, assuming the necessary equipment is available, they are relatively straightforward to carry out and allow for the level of control that is only possible in laboratory settings. These approaches are valuable for comparing the potential for a process across sites or through time and for unraveling the mechanistic controls of these biogeochemical processes. Despite their value, controlled laboratory setting rarely match the heterogeneous and dynamic conditions seen in the field; incubations tend to separate soils from vegetation and from hydrological exchange; and sediment and soil structure is frequently altered in incubation vessels. For these and other reasons, scaling up of rates and fluxes measured in these incubations to the ecosystem scale should be done with extreme caution and should be viewed skeptically.

8.4.2 Ecosystem Modeling

A number of models exist to explore the fluxes of CH_4 and N_2O from wetland ecosystems (e.g., Melton et al. 2013; Gilhespy et al. 2014) and there is a growing interest in developing robust and user-friendly models of greenhouse gas dynamics in the context of blue carbon ecosystems (e.g., Surgeon-Rogers et al, this volume).

Ecosystem modeling provides a unique opportunity to estimate greenhouse gas fluxes over larger temporal and spatial scales. Perhaps most appealing, modeling has the potential to project future greenhouse gas fluxes from coastal ecosystems under a range of possible future climate and global change scenarios. An inherent challenge with modeling is to develop models that are robust enough to capture CH_4 and N_2O dynamics in a range of ecosystems and do not need to be calibrated and validated on a site-by-site basis.

8.4.3 Field Measurements

Finally, it is possible to directly measure the fluxes of CH_4 and N_2O from blue carbon ecosystems using a variety of methods. Static chambers remain a common approach to measure greenhouse gas fluxes from soils and sediments (Yu et al. 2013), and floating chambers can be used to measure net flux from open-water areas. Essentially, these methods measure the linear accumulation of CH_4 and N_2O in the headspace of a chamber over time (typically minutes to hours) to calculate the net flux from a given area. Dynamic flux chambers (where air or water is pumped through the chamber at a pre-determined rate) can also over be used to measure greenhouse gas fluxes (e.g., Bahlmann et al. 2015). In most cases, headspace samples are collected and subsequently analyzed using gas chromatography. There are known limitations to these approaches, including changes in temperature, pressure, and concentration gradients over the course of the chamber deployment as well as challenges associated with non-linear fluxes and detection limits for low fluxes. Despite these potential methodological limitations, static chambers remain a generally "low-cost" and "low-tech" approach for quantifying greenhouse gas fluxes. New technologies including, Fourier transform infrared (FTIR) gas analysis, wavelength modulation spectroscopy, and cavity ring-down spectroscopy allow for direct, high precision, in situ measurements of CH_4 and N_2O flux using static chamber over much shorter time periods and that are compatible with auto-chamber approaches (where chambers are deployed automatically to increase the resolution of temporal sampling). While these technologies are generally costlier, they show great promise for quantifying greenhouse gas fluxes.

The benefit of chamber-based approaches is their general simplicity and relative low cost (compared to eddy covariance approaches discussed below). The trade-off for this ease of use is that capturing the spatial (at different elevations, over different vegetation and soil types, etc.) and temporal (tidal cycles, diurnal cycles, seasonal patterns, extreme events, etc.) variability in CH_4 and N_2O fluxes can be challenging.

The eddy covariance approach provides an alternative to chamber-based approaches for measuring greenhouse gas fluxes in the field. In this approach, measurements of upward and downward air movements are coupled with measurements of greenhouse gas concentrations in the air. These measurements are made very rapidly (on the scale of ten measurements per second, or faster) and by looking at the instantaneous covariance between vertical air velocity and gas concentrations, a flux of a greenhouse gas can be calculated (Rivera-Monroy et al. 2013; Yu et al. 2013; Baldocchi 2014). This approach is far more quantitatively intensive than chamber studies and involves computational routines to fill data gaps and further process data. The technology for measuring CO_2 (and water) fluxes using eddy covariance has existed for some time (Baldocchi 2014) and is used in systems around the globe (e.g., Baldocchi et al. 2001; Lu et al. 2017). More recently, sensors capable of measuring CH_4 and N_2O at the high temporal resolution necessary for this approach have been developed, and increasingly eddy covariance is being used to measure the fluxes of these greenhouse gases as well (e.g., Jha et al. 2014; Merbold et al. 2014; Rannik et al. 2015; Holm et al. 2016; Krauss et al. 2016). As the number of these eddy covariance measurements increases, there will be a need to compare them to measurements made with more traditional chamber approaches (e.g., Krauss et al. 2016).

The value of eddy covariance approaches is that they directly measure the flux of greenhouse gases between an ecosystem and the atmosphere with few disturbances to the ecosystem. In contrast to missing spatial and temporal patterns by using chambers, eddy covariance integrates greenhouse gas fluxes over an extended spatial footprint (up to hundreds of meters in length) and collects data over days, weeks, months, or even years. There are, however, limitations of this approach – including issues with integrating spatially patchy greenhouse gases (like CH_4 and N_2O); requirements of certain site conditions (size and horizontal homogeneity); and challenges with measurements at night (Baldocchi 2014; Chuang et al. 2016). In addition to these technical limitations, the use of eddy covariance approaches requires relatively expensive equipment and a particular computational and quantitative skill set that exceeds the requirements of chamber measurements.

8.5 CONCLUSIONS

Methane and N_2O are important greenhouse gases in the global climate and have important implications for understanding the climate impacts of blue carbon ecosystems. Fluxes of CH_4 are likely to be low in these vegetated coastal ecosystems, as a result of SO_4^{2-} in these environments, and fluxes of N_2O are likely limited by low dissolved inorganic nitrogen availability in non-polluted environments. However, given the potency of these greenhouse gases, even small fluxes can negate the climate benefits associated with carbon sequestration in salt marsh, mangrove, and seagrass ecosystems. Thus, the fluxes of these greenhouse gases need to be considered in blue carbon projects.

CHAPTER 9

The Calcium Carbonate Cycle in Seagrass Ecosystems

Hilary A. Kennedy
Bangor University

James W. Fourqurean
Florida International University

Stathys Papadimitriou
National Oceanographic Centre

CONTENTS

HIGHLIGHTS

1. Seagrass meadows provide an important habitat for a wide range of marine organisms, including those that form or secrete skeletal or non-skeletal calcium carbonate (CaCO$_3$), which is deposited on the soil surface along with organic remains.
2. The direction and magnitude of the air-sea CO$_2$ exchange potential in seagrass ecosystems depends on the relative rates of the organic and inorganic carbon cycles within the system.
3. In settings where there is potential for CO$_2$ evasion from seawater to the atmosphere, the soil still provides a long-term store for detrital organic and CaCO$_3$ carbon.

9.1 INTRODUCTION

Seagrass meadows are commonly associated with carbonate-rich soils in the sub-tropics and tropics (Alongi et al. 2008; Mazarrasa et al. 2015), where the functional traits of seagrasses and the ecosystem supported by seagrass meadows facilitate the growth of calcareous organisms (calcifiers) that contribute to the accumulation of detrital carbonate particles ($CaCO_3$) in the soil. The seagrass canopy provides structural complexity for the settlement and growth of calcifiers, and the high productivity of seagrasses, their epiphytes, and associated benthic algae attracts other infaunal and epifaunal sessile and mobile calcifying organisms that benefit from the meadow as an energy source, nursery ground, and a place to avoid predation (Heck et al. 2008). The $CaCO_3$ produced by these organisms is added to the underlying soil and is termed autochthonous. Allochthonous $CaCO_3$ particles can be transported from elsewhere and get trapped in the meadow via the reduction in hydrodynamic flow by the seagrass canopy, which enhances local particle deposition. The soil that accumulates is stabilized by the extensive belowground root and rhizome system of the seagrass beds (Figure 9.1).

Blue carbon is the organic carbon produced and/or trapped by the seagrass meadow and stored in the underlying soil; the ultimate source of the carbon is the atmosphere (Nellemann et al. 2009). This function is dependent on seagrass ecosystems being net autotrophic, that is, where gross primary production (GPP) exceeds respiration (R) and there is a net transfer of carbon dioxide (CO_2) from the atmosphere to seawater. Not all seagrass ecosystems are net autotrophic; Duarte et al. (2005) demonstrated that there is a threshold GPP, below which respiration dominates ecosystem metabolism and there is a net transfer of CO_2 from seawater to the atmosphere. It is now becoming clear that the $CaCO_3$ cycle also has a role to play in determining whether there is a net transfer of CO_2 from the atmosphere to seawater or vice versa in a seagrass ecosystem.

While the organic carbon cycle has been well studied in seagrass ecosystems, the $CaCO_3$ cycle has rarely been evaluated. Here, we outline the processes in seagrass ecosystems that contribute to

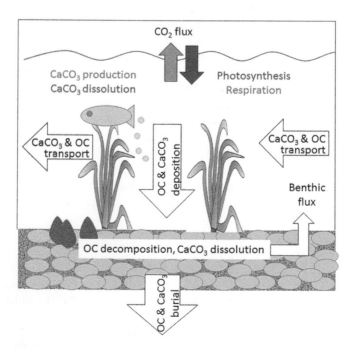

Figure 9.1 (See color insert following page 266.) Conceptual diagram illustrating the major processes affecting the $CaCO_3$ and organic carbon cycle in seagrass ecosystems.

$CaCO_3$ production and accumulation in the underlying soils and how the relative rates of the organic carbon and the $CaCO_3$ cycles affect chemical changes in the water column that determine the direction of the air-sea CO_2 exchange (Figure 9.1). A better understanding of these processes form a basis from which to assess the role of seagrass ecosystems in storing carbon over the short and long time scales, and the vulnerability of the stored carbon to human disturbance and climate change.

9.2 BIOGENIC CaCO₃ PRODUCTION ASSOCIATED WITH SEAGRASS MEADOWS

Calcifying and non-calcifying epiphytic organisms are commonly present on seagrass leaves and stems (Borowitzka et al. 2006), while it has recently been demonstrated that seagrass can also make a direct contribution to $CaCO_3$ production and local $CaCO_3$ deposition by $CaCO_3$ precipitation internally and as coatings externally on their leaves (Enríquez and Schubert 2014). Coralline and articulated red algae, serpulid worms, foraminifera, and bivalves are all regularly observed epiphytic calcifiers (Figures 9.1 and 9.2). Coralline encrusting (non-geniculate) and articulated (geniculate) algae make up a large proportion and are the most prolific $CaCO_3$ producers of the epiphytic community (Land 1970; Patriquin 1972; Nelsen and Ginsberg 1986; Boscence 1989; Frankovich and Zieman 1994; Perry and Beavington–Penney 2005; Corlett and Jones 2007; James et al. 2009; Basso 2012). In southern Australia, for example, corallines made up 53.6% of the taxa, with benthic foraminifera contributing 17.4%, bryozoans 16.4%, spiroids 8%, and bivalves, serpulids, and ostracods each contributing < 0.3% to the epiphyte taxa (James et al. 2009). Many other different calcareous epiphytes have been documented but not quantified (Humm 1964; Wilson 1998).

The non-geniculate epiphytic calcifiers produce thin (≤100 µm) layers of $CaCO_3$ that gradually spread over the leaf surface. The maximum contribution of these epiphytes to the $CaCO_3$ production depends on their growth rate, the surface area available for colonization, the time span of each leaf in a shoot, and the shoot density of the seagrass species (Walker and Woelkerling 1988; Borowitzka et al. 2006). Variation in the epiphyte community is also observed on different structural components of a seagrass. For example, a higher occurrence was observed of non-geniculate corallines on leaf blades and geniculate corallines on the stems of the seagrass *Amphibolis antarctica* in

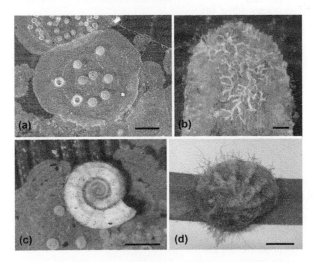

Figure 9.2   (See color insert following page 266.) Calcium carbonate producing organisms: (a) coralline red algae (Rhodophyta), (b) the foraminifer *Cornuspiramia antillarum*, (c) the polychaete *Spirorbis* sp., and (d) the gastropod *Modulus*. Scale bars: (a, b) 1 mm, (c) 500 µm, and (d) 5 mm. Photographs provided courtesy of Tom Frankovich, Marine Education and Research Center, Florida International University.

Australia (James et al. 2009). Species such as *Thalassia, Posidonia,* and *Amphibolis* have large, long-lived, strap-type leaves that facilitate epiphyte colonization and growth. It is these long-lived species that have been most intensively studied (Table 9.1), e.g., *Thalassia testudinum* in Florida Bay, USA, in the Bahamas Bank, and in the Caribbean, *Amphibolis antarctica* in Australia, and *Posidonia oceanica* in the Mediterranean (Walker and Woelkerling 1988; Corlett and Jones 2007; James et al. 2009). The more ephemeral seagrass species with a smaller leaf area available for colonization, such as *Halophila* and *Halodule,* will sustain lower amount of epiphytes and have rarely been studied (Borowitzka et al. 2006).

Beyond these sessile epiphytes, there are more mobile calcifying organisms, which do not always depend directly on the seagrass structural components but are still resident in the seagrass meadow and the surrounding area (Figure 9.2). Mobile gastropods graze on seagrass leaves for food, while ostracods and calcareous mobile and sessile bivalves are also commonly found on the seagrass structures. Fish, which utilize the meadows as nursery grounds and foraging areas, also contribute $CaCO_3$ as they produce and excrete various forms of non-skeletal $CaCO_3$ from their guts as very fine-grained (<2 μm) $CaCO_3$ crystallites. Their overall contribution may be very low (<1%) compared to other calcifiers but they are still considered a measureable contributor to the accumulation of lime mud across the Bahamian archipelago (Perry et al. 2011).

Fish use the meadows as refugia from predation, but the meadows also have the potential to act as refugia for calcifying organisms that are vulnerable to ocean acidification (oceanic pH decrease via absorption of anthropogenically increasing atmospheric CO_2) (Semesi et al. 2009; Manzello et al. 2012; Hendricks et al. 2014). During photosynthesis seagrass removes CO_2 from seawater, which raises

Table 9.1 Seagrass Gross $CaCO_3$ Production Rates Using the Biological Approach

Location	CARB$_{prod}$ (g $CaCO_3$ m^{-2} year^{-1})	Species	Reference
Jamaica	min: 40 mean: 180	*Thalassia testudinum*	Land (1970)
Barbados	max: 5100	*Thalassia testudinum*	Patriquin (1972)
Barbados	134	*Seagrass*	Webster and Parnankap (1977) (in Brown 2005)
Florida Bay (USA)	30–303 (mean: 188 ± 44)	*Thalassia testudinum*	Nelsen and Ginsburgh (1986)
Florida Bay (USA)	55–1042 (site means: 81, 482)	*Thalassia testudinum*	Bosence (1989)
Florida Bay	60[a]	*Seagrass*	Hallock et al. (1986)
Florida Bay (USA)	2–283 (mean: 120)	*Thalassia testudinum*	Frankovitch and Zieman (1994)
Bahamas	0–68 (mean: 28 ± 6)	*Thalassia testudinum*	Armstrong (1989) (in Brown 2005)
Inhaca Island, Mozambique	31–86 (mean: 44)	*Thalassodendron ciliatum*	Perry and Beavington-Penny (2005)
Inhaca Island, Mozambique	8–55 (mean: 33)	*Thalassia hemprichii*	Perry and Beavington-Penny (2005)
Mallorca-Menorca Shelf (NW Mediterranean)	60–70 (mean: 68)	*Posidonia oceanica*	Canals and Ballesteros (1997)
Spain (NW Mediterranean)	82–158	*Posidonia oceanica*	Romero (1988) (in Barrón et al. 2006)
Australia	49–661 (mean: 210 ± 26)	*all species*	James et al. (2009)
Shark bay (Australia)	50–526 36–313	*Amphibolis antarctica*	Walker and Woelkerling (1988)

[a] Only foraminifera.

seawater pH and increases the concentration of carbonate ions (CO_3^{2-}) in seawater (Section 9.4.3). The CO_3^{2-} concentration is a strong driver of the $CaCO_3$ precipitation and dissolution reactions, and its effect can be assessed through the saturation state (Ω) of seawater with respect to a $CaCO_3$ mineral form (e.g., calcite, aragonite) as $\Omega(CaCO_3) = \left[Ca^{2+}\right]\left[CO_3^{2-}\right]/K_{sp}^*$, with brackets denoting concentration and K_{sp}^* = concentration-based (stoichiometric) equilibrium solubility product of the $CaCO_3$ mineral form, a function of salinity, solution composition, and temperature. Because Ca^{2+} is present at relatively constant concentrations in seawater, $\Omega(CaCO_3)$ depends strongly on the variation of CO_3^{2-} concentration. The variability of the CO_3^{2-} concentration is a major control on the ability of calcifying organisms to produce $CaCO_3$ (calcification). Calcification ($CARB_{prod}$) is favorable when the ambient water $\Omega(CaCO_3) > 1$ ($CaCO_3$ supersaturation), and $CaCO_3$ dissolution ($CARB_{diss}$), which destroys biogenic $CaCO_3$, is favorable when $\Omega(CaCO_3) < 1$ ($CaCO_3$ undersaturation). As photosynthesis is highest during daytime and in spring–summer, net calcification ($CARB_{net} = CARB_{prod} - CARB_{diss} > 0$) is often observed during these periods (Yates and Halley 2006; Turk et al. 2015; Muehllehner et al. 2016). And so, by reducing the CO_2 concentration and increasing the pH of seawater, seagrass productivity may also ameliorate ocean acidification stress to calcifying organisms found in the same locality (Manzello et al. 2012).

The effect of $\Omega(CaCO_3)$ on calcifiers will vary as different calcifying organisms secrete a range of structurally and chemically different $CaCO_3$ mineral forms (Scholle 1978). The typical crystalline forms of biogenic $CaCO_3$ are aragonite and calcite, the latter having variable magnesium (Mg^{2+}) content. Magnesium can replace some of the Ca^{2+} in the calcite lattice, and calcite containing $> 4\,wt\%$ of $MgCO_3$ is conventionally defined as high Mg calcite, which is the dominant mineralogy found in the red coralline algae and benthic foraminifera. Bryozoans, foraminifera, echinoderms, arthropods, and brachiopods produce both calcite and high Mg calcite, while green calcareous algae produce aragonite. In tropical regions, aragonite is also a major component of the $CaCO_3$ produced by corals, serpulid worms, and molluscs, while tropical fish excreta consist of low Mg calcite, high Mg calcite, and aragonite. Each of these biogenic $CaCO_3$ mineral forms has a different solubility (K_{sp}^*) in seawater and, hence, a different set of conditions for which seawater favors $CaCO_3$ production or its dissolution. Solubility decreases from aragonite to Mg calcite to pure calcite (e.g., Burdige et al. 2010).

The conditions that lend themselves to $CaCO_3$ dissolution, i.e., seawater CO_2 increase and pH decrease (acidification) are the opposite of those that promote calcification. Organism respiration adds CO_2 to seawater; accumulation of respired CO_2 promotes the decrease of seawater pH, CO_3^{2-} concentration, and, consequently, $\Omega(CaCO_3)$ (Section 9.4.3). Seagrass ecosystem respiration occurs in the water column and underlying soils, and is dominant at night and during the winter months of senescence, thus net $CaCO_3$ dissolution ($CARB_{net} = CARB_{prod} - CARB_{diss} < 0$) is often observed during these times (Barrón et al. 2006; Yates and Halley 2006; Turk et al. 2015; Muehllehner et al. 2016). The overall balance ($CARB_{net}$) between calcification ($CARB_{prod}$) and $CaCO_3$ dissolution ($CARB_{diss}$) depends on a number of factors (Section 9.3).

Calcification has been reported for epiphytic communities, or the whole seagrass ecosystem using different methods, which, in general, divide into two approaches. The first approach (termed biological) uses the epiphytic $CaCO_3$ standing stock and the rates of leaf or shoot production and/or turnover (Table 9.1). Rates of calcification ($CARB_{prod}$) are provided for three seagrass species from six different countries and vary from 2 to $5,100\,g\;CaCO_3\,m^{-2}year^{-1}$. These data are associated, generally, with large coefficients of variation due to the spatial variation in both the epiphytic standing crop and the annual mean rates of seagrass productivity/turnover (Perry and Beavington-Penney 2005). For example, Frankovitch and Zieman (1994) estimated coefficients of variation of epiphyte production between 40% and 136%. The most comprehensive study to date of calcareous epiphyte biomass and $CARB_{prod}$ using the biological approach has been for seagrass species in southern Australia, with $CARB_{prod}$ reported for different locations, seagrass species, epiphytic biota, and seagrass modules, as well as seasonal changes in $CARB_{prod}$ (Brown 2005).

The second approach (termed chemical) uses the seawater alkalinity anomaly generated during calcification and $CaCO_3$ dissolution. This method entails the removal of water samples from an incubation chamber that is inserted into the soil (in situ) or removal of seagrass leaves and their incubation in seawater (ex situ). The in situ method includes changes due to processes occurring in the soil as well as on the seagrass leaves. As the soils are sites of $CaCO_3$ dissolution (Section 9.3.2) they can have a strong influence on the in situ method results. The measurements from the in situ method, therefore, represent net rates of combined calcification ($CARB_{prod}$) and $CaCO_3$ dissolution ($CARB_{diss}$). As mentioned earlier, when $CARB_{prod} > CARB_{diss}$, net ecosystem calcification occurs, with $CARB_{net} = CARB_{prod} - CARB_{diss}$ reported as positive values (>0). In contrast, when $CARB_{prod} < CARB_{diss}$, net ecosystem $CaCO_3$ dissolution occurs, with $CARB_{net}$ reported as negative values (<0).

The biological approach results in $CaCO_3$ production rates only ($CARB_{prod}$) and does not take into account soil $CaCO_3$ dissolution or soil transport (Stockman et al. 1967; Bosence 1989). The chemical method will measure both calcification and dissolution, and, depending on the relative rates of these two processes, can return both negative and positive values (Table 9.2). In the same meadow, $CARB_{net}$ measured by the chemical method cannot only vary numerically, but also between negative and positive values depending on the time interval of measurement. On diurnal scales, Yates and Halley (2006) reported generally positive $CARB_{net}$ during the day when the seagrass ecosystem is net photosynthetic, and negative $CARB_{net}$ during the night when $CaCO_3$ dissolution dominates in the different compartments of the ecosystem (Table 9.2). Estimating annual rates of $CARB_{net}$ by the chemical approach is time consuming and costly, and still only accounts for $CaCO_3$ production and soil $CaCO_3$ dissolution, but not for soil transport (Yates and Halley 2006). Only directly measured soil accumulation rates can account for all three processes.

9.3 THE FATE OF BIOGENIC $CaCO_3$ PRODUCTION

The biogenic production of $CaCO_3$ in seagrass ecosystems generally leads to local $CaCO_3$ deposition and accumulation of $CaCO_3$ stocks. Some part of the biogenic $CaCO_3$ production can be exported by waves and tides out of the meadow to adjacent regions, while some part of the detrital

Table 9.2 [a]$CARB_{net}$ in Seagrass Meadows Determined by the Chemical Approach

Location	$CARB_{net}$ (g $CaCO_3$ m^{-2} year^{-1})	Species	Reference
Florida Bay Dense beds	31–805[b] (mean: 244)	*Thalassia testudinum*	Yates and Halley (2006)
Florida Bay Intermediate beds	−1363–61.9[b] (mean: −84)		
Florida Bay Sparse beds	−328–181[b] (mean: −112)		
Mallorca (NW Mediterranean)	51 ± 110[b]	*Posidonia oceanica*	Barrón et al. (2006)
Shark bay	117[c]	*mixed species*	Smith and Atkinson (1983)
Shark bay (Australia)	35–295[d]	*Amphibolis antarctica*	Walker and Woelkerling (1988)
Caribbean Mexico	590–5,630[d]	*Thalassia testudinum*	Enriquez and Schubert (2014)
Nagura Bay (Japan)	200[b]		Fujita (2002)

[a] $CARB_{net} > 0$ when $CARB_{prod} > CARB_{diss}$ and $CARB_{net} > 0$ when $CARB_{prod} < CARB_{diss}$.
[b] In situ incubation chamber.
[c] Carbon budget.
[d] Ex situ estimate.

CaCO$_3$ accumulated in the soil will be lost due to post-depositional CaCO$_3$ dissolution. The accumulating soils consist not only of CaCO$_3$ but also of the organic carbon produced (and trapped) by the meadow (Figure 9.1).

9.3.1 CaCO$_3$ Soil Stocks and Accumulation Rate

The production of CaCO$_3$ within seagrass meadows (autochthonous CaCO$_3$ production) results in CaCO$_3$ deposition on the soil surface, facilitated by the reduction in current speed within the meadows and the stabilization of the soils by roots and rhizomes (Koch et al. 2006). In quiescent environmental settings, these factors may contribute to the development of CaCO$_3$ mud mounds such as those found in Florida Bay, USA, the Great Bahamas Bank, and in Shark Bay, Australia (Ginsburg and Lowenstam 1958). The particle trapping and soil stability of seagrass beds are also instrumental in the accumulation of CaCO$_3$ transported from adjacent regions (allochthonous CaCO$_3$).

Carbonate soils are not equally distributed around the globe. They have much higher abundance in tropical and sub-tropical regions than in temperate and boreal zones (Mazarrasa et al. 2015). Because the concentration of dissolved CO$_2$ in waters in equilibrium with the atmosphere increases as temperature decreases, the seawater saturation state of carbonate minerals increases with increasing temperature or decreasing latitude (Zeebe and Wolf-Gladrow 2001). Thus, seagrass meadows inhabiting colder waters generally support lower rates of epiphytic CaCO$_3$ production and have lower CaCO$_3$ content in their soils. Reviewing CaCO$_3$ stocks in seagrass soils, Mazarrasa et al. (2015) found a discernible trend of decreasing CaCO$_3$ stocks in seagrass soils with increasing latitude and decreasing temperature, such that stocks tended to decrease polewards at a rate of -67 ± 17 Mg CaCO$_3$ ha^{-1} per degree latitude. Similarly, smaller spatial scale differences in mean water temperature caused by oceanic upwelling can lead to much lower CaCO$_3$ contents of soil in the colder, upwelling-influenced regions of the coastal ocean (Howard et al. 2017). Overall, Mazarrasa et al. (2015) estimated that seagrass meadows store between 25 and 13,833 Mg CaCO$_3$ ha^{-1} (average $\pm 1\sigma$: $5,480 \pm 200$ Mg CaCO$_3$ ha^{-1}) in the upper 1 m of their soils. Soil CaCO$_3$ stocks have also been found to vary across seagrass communities, with the highest stocks found underlying meadows dominated by *Halodule*, *Thalassia*, or *Cymodocea* species, while the soils with the lowest CaCO$_3$ stocks were colonized by *Zostera* and *Halophila* species (Mazarrasa et al. 2015). These observations are not in line with those observations on epiphyte CaCO$_3$ production rates, which are maximal on large, long-lived seagrass species. The lack of a clear effect of the seagrass traits on CaCO$_3$ stocks could be due to other controlling factors on the precipitation, preservation, and accumulation of CaCO$_3$ in the underlying soils. The seagrass species found in warmer seas (e.g., the Caribbean and Mediterranean Seas, and the tropical Indo-Pacific) are the ones associated with higher CaCO$_3$ stocks.

Under conditions of net calcification (CARB$_{prod}$ > CARB$_{diss}$), CaCO$_3$ accumulates in the underlying soils (Yates and Halley 2006). However, the accumulation rate of CaCO$_3$ soils underlying seagrass meadows has rarely been directly measured, with most current estimates derived from calculated rates of epiphyte production or from soil stocks of CaCO$_3$ and soil accumulation rate, or from the alkalinity anomaly method from benthic incubations and seawater transects. Using soil CaCO$_3$ content and estimates of soil accretion in seagrass meadows, Mazarrasa et al. (2015) estimated a mean global CaCO$_3$ accumulation rate of $1,053 \pm 259$ g CaCO$_3$ m^{-2} year^{-1}, while Serrano et al. (2012) measured a soil accretion rate of 483 ± 16 g CaCO$_3$ m^{-2} year^{-1} in a *Posidonia oceanica* meadow.

9.3.2 CaCO$_3$ Dissolution in Seagrass Soils

The detrital CaCO$_3$ that accumulates in the soil alongside the organic matter is subsequently subject to mixing by the infaunal organisms (bioturbation) and dissolution, bathed in the soil pore waters (overlying water buried with the deposited biogenic and lithogenic particles). The dissolution of soil

$CaCO_3$ is driven by the intense modification of the chemical composition of soil pore waters by benthic bacterial oxidation of organic matter (benthic bacterial respiration), which adds CO_2 (metabolic CO_2) and other metabolites to the pore waters, such as ammonium and sulfide in the absence of dissolved oxygen (O_2). It is in the oxic metabolic zone of the surface of marine deposits where bacterial respiration majorly affects the dissolution of detrital $CaCO_3$ as outlined below for seagrass beds.

Seagrasses, rooted in the soil, contribute to $CaCO_3$ dissolution, which can be traced in measurable changes in the chemical composition and in the parameters of the carbonate system (Section 9.4.1) of the pore waters (Rude and Aller 1991; Jensen et al 1998; Ku et al. 1999; Burdige and Zimmerman 2002; Hu and Burdige 2007; Burdige et al. 2008; Jensen et al. 2009; Burdige et al. 2010). Oxic benthic respiration generates acidity in the pore waters via the production of CO_2 from the decomposition and remineralization of soil organic carbon by dissolved molecular oxygen (O_2). The accumulation of metabolic CO_2 in the soil pore waters, buffered by the ambient carbonate system (Section 9.4.1), results in the decline of the pH and $\Omega(CaCO_3)$ (Jensen et al. 1998; Burdige et al. 2010). In addition, at the boundary between the surface oxic metabolic zone and the deeper anoxic zone, where there is no O_2, benthic metabolism can result in conditions favorable to benthic $CaCO_3$ dissolution. The oxidation of anoxic metabolites (e.g., reduced iron and sulfur compounds) at the oxic/anoxic boundary releases acidity and consequently reduces pH and lowers $\Omega(CaCO_3)$ of the pore waters (Ku et al. 1999; Jensen et al. 2009).

The primary source of benthic O_2 is generally the overlying water column via molecular diffusion across the soil-water interface. In seagrass-colonized soils, belowground efflux of photosynthetic O_2 through the root-rhizome system can become the primary source, while, in high-permeability carbonate sands, tidally and wave-driven pore water advective exchange becomes equally important as a source of O_2 to the soil (Burdige et al. 2008; Cyronak et al. 2013). Seagrass metabolism thus supports rates of oxic organic matter remineralization by the benthic bacterial community that are much higher than those that could be supported solely on the O_2 supply to the surface soil pore waters by molecular diffusion alone from the overlying water (Burdige et al. 2008). Seagrass growth also directly adds to benthic metabolism by the net production of organic carbon and its accumulation in the soil, as well as the additional efflux of CO_2 from seagrass roots, both of which lead to a lowered pH in the soil pore waters, facilitating soil $CaCO_3$ dissolution.

In the soils, $CaCO_3$ dissolution is prominent and has been extensively determined in a few locations in the sub-tropics and tropics (Table 9.3). An increasing rate of benthic $CaCO_3$ dissolution with increasing seagrass cover as Leaf Area Index (LAI) has been documented in the most extensively studied carbonate platform, the Bahamas Bank, and was deemed an important loss mechanism for $CaCO_3$ from these ecosystems, equivalent to $CaCO_3$ export by physical transport (Burdige et al. 2010).

The effects of benthic metabolic respiration and $CaCO_3$ dissolution in the soil are large enough to be traceable in the chemistry of the surface waters via solute fluxes across the soil-water interface (Figure 9.1) (Ku et al. 1999; Yates and Halley 2006; Cyronak et al. 2013; Muellehner et al. 2016).

Table 9.3 Benthic CARB$_{diss}$ in Seagrass Soils

Location	[a]CARB$_{diss}$ (g $CaCO_3$ m^{-2}year^{-1})	Species	Reference
Florida Keys (USA)	292–876[b]	*T. testudinum*	Turk et al. (2015)
Indonesia	511–3,979[c]	*mixed species*	Alongi et al. (2008)
Bahamas Bank	51–2,935[d]	*T. testudinum*	Burdige et al. (2010)
Bahamas Bank	281–942[d]	*T. testudinum*	Burdige et al. (2008)

[a] Annual rates estimated as 365 × daily rate.
[b] Maximum night rates, benthic chambers, TA anomaly.
[c] Benthic chambers, Ca^{2+} anomaly.
[d] Transport–reaction modeling of pore water solutes (metabolites).

Benthic metabolism results in accumulation of the end products in the pore waters of the surface soils, such as dissolved inorganic carbon (DIC) and total alkalinity (TA), with TA representing the proton (H^+) buffering capacity of the aquatic system (Zeebe and Wolf-Gladrow 2001; Dickson et al. 2007), and their transport to the overlying water column via molecular diffusion, bioturbation, and, in the case of highly permeable $CaCO_3$ sands, advective exchange by wave action and tidal pumping (Cyronak et al. 2013; Drupp et al. 2016). In seagrass meadows, therefore, the magnitude of changes in seawater chemistry caused by the organisms that produce $CaCO_3$ (i.e., DIC removal and TA increase; Section 9.4.2) can be reduced by the chemically distinct fluxes across the sediment-water interface. The benthic TA flux becomes more influential the shallower the water column, the longer the water residence time, and the higher the $CaCO_3$ dissolution rate. The dissolution rate itself depends on the magnitude of benthic bacterial respiration and the solubility of the deposited $CaCO_3$ mineral, with high Mg calcite being the most soluble. On an even larger scale, the benthic metabolic $CaCO_3$ dissolution rates documented in the Bahamas Bank by Burdige et al. (2010) generated TA fluxes that were an important component of the TA budget of the surface ocean, comparable in this respect to major TA sources to the global ocean such as the riverine TA source.

9.4 EFFECT OF THE CaCO₃ CYCLE ON THE AIR-SEA CO₂ EXCHANGE IN SEAGRASS ECOSYSTEMS

Calcification and $CaCO_3$ dissolution modify the chemistry of seawater in and around seagrass meadows. These changes do not occur in isolation but add to, and modify, those associated with organic matter production and respiration. The seawater is intimately connected to the underlying soils and overlying atmosphere, and fluxes of CO_2 and other relevant dissolved constituents from these distinct reservoirs further modify the seawater chemistry (Figure 9.1). Finally, any net change observed depends on the residence time of the water, i.c., when residence time is long, water mass movement is slow and expressions of chemical change maximal, and vice versa (Frankignoulle and Distèche 1984; Yates and Halley 2006; Tokoro et al. 2014; Turk et al. 2015). Overall, the modification of the seawater chemistry depends on the relative magnitude and rates of hydrodynamic and biogeochemical processes.

9.4.1 The Carbonate System of Seawater

To understand the role of the biogenic $CaCO_3$ reactions (calcification, dissolution) in the exchange of CO_2 across the air-sea interface, first, the dissolved constituents of the seawater carbonate system and the processes that affect them must be outlined. The largest component of the seawater carbonate system consists of the three species of DIC, CO_2^*, HCO_3^- (bicarbonate ion), and CO_3^{2-} (carbonate ion). By definition, CO_2^* includes both the dissolved CO_2 and the trace concentrations of carbonic acid (H_2CO_3), and is proportional to the fugacity of CO_2 (fCO_2) in ambient water as $\left[CO_2^*\right] = K_0 fCO_2$, with $K_0 = CO_2$ solubility (in mol kg^{-1} atm^{-1} or mol L^{-1} atm^{-1}), a function of salinity, temperature, and pressure (Zeebe and Wolf-Gladrow 2001). The (measurable) fCO_2 is related to the partial pressure of CO_2 (pCO_2) via the activity coefficient of gaseous CO_2 as the proportionality factor (Plummer and Busenberg 1982). The concentrations of the DIC species (CO_2^*, HCO_3^-, and CO_3^{2-}) in seawater are typically determined by knowledge of the temperature and salinity (and pressure in deep waters) of seawater, the boron, sulfate, fluorine, phosphate, and silicate concentrations, the acid-base dissociation constants (pK) of water, bisulphate, HF, and the carbonic, boric, phosphoric, and silicic acids, and the measurement of two of the four directly measurable parameters of the seawater carbonate system, pH, fCO_2, TA, and DIC (Dickson et al. 2007; Zeebe and Wolf-Gladrow 2001). These calculations are aided now by the availability of software such as CO2SYS (Lewis and Wallace 1998). Given any two of the four measurable carbonate system parameters, the software calculates the other two, together with the inorganic carbon speciation and the saturation state of seawater with respect to calcite and aragonite. The program also allows the user to select from four different pH scales and several empirical sets of dissociation constants suitable for oceanographic conditions and widely cited in the literature. The changes that photosynthesis, respiration, calcification, and $CaCO_3$ dissolution have on the seawater carbonate system and CO_2 air-sea exchange potential are outlined in Sections 9.4.2–9.4.4.

9.4.2 The Effect of Calcification and $CaCO_3$ Dissolution on the Seawater Carbonate System

Calcification ($CARB_{prod}$) and $CaCO_3$ dissolution ($CARB_{diss}$) affect the aquatic carbonate system. Through the precipitation of CO_3^{2-} as solid $CaCO_3$ by calcifying organisms, calcification leads to a decrease in DIC and the largest relative TA decrease in natural waters in a molar $\Delta TA : \Delta DIC = 2:1$ in a closed system, with consequent decrease in pH and increase in fCO_2 (Figure 9.3). The dissolution of $CaCO_3$, which can be strongly influenced by benthic metabolism results in the opposite DIC, TA, pH, and fCO_2 changes but at the same molar $\Delta TA : \Delta DIC$ in a closed system as calcification (Figure 9.3).

9.4.3 The Effect of Photosynthesis and Respiration on the Seawater Carbonate System

As mentioned in Section 9.2, the biogenic $CaCO_3$ cycle does not occur in isolation but rather occurs in association with the organic carbon cycle of photosynthesis and respiration in seagrass ecosystems with its planktonic and benthic communities. Ecosystem-wide photosynthesis and respiration also modify the parameters of the aquatic carbonate system of ambient waters through the uptake of CO_2 and HCO_3^- by primary producers and the concurrent uptake of inorganic nutrients, such as nitrate and phosphate, and the release of CO_2 via respiration (R) by all organisms, including benthic bacterial decomposition of organic matter. For example, net photosynthesis (net ecosystem production, NEP = GPP − R > 0) results in a decrease in DIC and a relatively weak change in TA due to the biological uptake of inorganic nutrients, depending on the fixed nitrogen species (nitrate, nitrite, ammonium) supporting nitrogen metabolism (Wolf-Gladrow et al. 2007). At constant salinity and temperature in a closed aquatic system, the chemical changes due to net photosynthesis

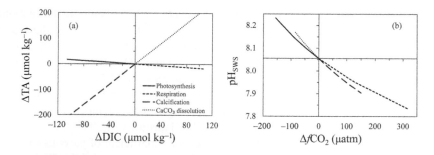

Figure 9.3 Changes in the measurable parameters of the aquatic carbonate system during individual biogeochemical reactions in seawater of salinity 35 at 25°C. The initial conditions for the seawater are equilibrium with the current atmosphere (400 ppmv mol CO_2 fraction in dry air at 1 atm total pressure) and TA = 2,370 µmol kg⁻¹, with derived initial pH_{SWS} = 8.055, fCO_2 = 386 µatm, and DIC = 2049 µmol kg⁻¹. All changes are relative to the initial values and were derived assuming no air-sea exchange during biogeochemical reaction, with photosynthesis and respiration ΔTA and ΔDIC illustrated for nitrate-based metabolism and the mean pelagic phytoplankton C:N:P = 106:16:1 (Redfield C:N:P ratio) (Wolf-Gladrow et al. 2007).

alone lead to an increase in pH and a decrease in fCO$_2$, while net ecosystem respiration (NEP < 0) results in opposite changes in these parameters (Figure 9.3).

9.4.4 Air-Sea CO$_2$ Exchange

As outlined above (Sections 9.4.2 and 9.4.3), the changes in DIC and TA associated with photosynthesis, respiration, calcification, and CaCO$_3$ dissolution result in either an increase or decrease in the seawater pCO$_2$. If the pCO$_2$ of seawater is higher than the atmospheric pCO$_2$, there will be a net CO$_2$ transfer from seawater to the atmosphere (CO$_2$ evasion) (Figure 9.1). If the pCO$_2$ of seawater is lower than the atmospheric pCO$_2$, there will be a net CO$_2$ transfer from the atmosphere to seawater (CO$_2$ invasion) (Figure 9.1). For a given air-seawater pCO$_2$ gradient, the actual magnitude of the exchange is determined by the near-surface turbulence and wind speed conditions.

9.4.5 The Effect of the Aquatic CaCO$_3$ Cycles on the Air-Sea CO$_2$ Exchange Potential of Seagrass Ecosystems

Put together in the ecosystem functioning outlined earlier in Sections 9.4.2 and 9.4.3, net photosynthesis leads to lowered DIC and pCO$_2$, while net calcification leads to lowered DIC but increased pCO$_2$. A similarly competing effect holds when both net respiration and net CaCO$_3$ dissolution dominate ecosystem processes, with net respiration increasing DIC and pCO$_2$, and net CaCO$_3$ dissolution increasing DIC and decreasing pCO$_2$ (Figure 9.3). Generally, the changes in seawater chemistry generated by the combined CaCO$_3$ and organic carbon cycles in the ambient waters will depend on the ratio of the rates of the dominant processes (Gattuso et al. 1995) and the resultant surface seawater pCO$_2$ will then determine the potential of the ecosystem for CO$_2$ exchange with the atmosphere.

Figure 9.4 illustrates the effect of different amounts of calcification or CaCO$_3$ dissolution on the air-sea CO$_2$ exchange potential of an ecosystem depending on its metabolic status. For example, if the rate of calcification equals that of dissolution (ΔDIC$_{CARB\text{-}net}$ = 0) and net photosynthesis occurs (NEP > 0; ΔDIC$_{NEP}$ = −10 µmol kg^{-1} as solid curve in Figure 9.4) in a system originally at air-sea equilibrium, the resulting DIC loss from seawater will cause a reduction in the seawater pCO$_2$ relative to air pCO$_2$ [ΔpCO$_2$ = pCO$_2$(seawater) − pCO$_2$(air) having a negative numerical value], thus generating CO$_2$ invasion potential. An example for a different setting would be that NEP = 0 (dotted line Figure 9.4), DIC is lost from seawater via CaCO$_3$ production (ΔDIC$_{CARB\text{-}net}$ is negative), but the concurrent large change in TA (Section 9.4.2; Figure 9.3) results in a net increase in seawater pCO$_2$ relative to air pCO$_2$ [ΔpCO$_2$ = pCO$_2$(seawater) − pCO$_2$(air) having a positive numerical value] and, therefore, the potential for CO$_2$ evasion from seawater (dashed line in Figure 9.4). Overall, in the examples illustrated in Figure 9.4, a net autotrophic ecosystem (NEP > 0, ΔDIC$_{NEP}$ = −10 µmol kg^{-1} in Figure 9.4) causes CO$_2$ invasion without a CaCO$_3$ cycle (ΔDIC$_{CARB\text{-}net}$ = 0) and when the CaCO$_3$ cycle is driven by net CaCO$_3$ dissolution (ΔDIC$_{CARB\text{-}net}$ is a positive numerical value in Figure 9.4). A net autotrophic ecosystem will result in CO$_2$ evasion when its CaCO$_3$ cycle is driven by net CaCO$_3$ production (ΔDIC$_{CARB\text{-}net}$ is a negative numerical value in Figure 9.4) and its rate is higher than about twice the rate of NEP. In the net heterotrophic ecosystem illustrated in Figure 9.4 (NEP < 0, ΔDIC$_{NEP}$ = +10 µmol kg^{-1} as dotted curve), there is net evasion of CO$_2$ without a CaCO$_3$ cycle (ΔDIC$_{CARB\text{-}net}$ = 0) and when there is calcification (ΔDIC$_{CARB\text{ net}}$ is a negative numerical value in Figure 9.4). A net heterotrophic ecosystem will generate seawater pCO$_2$ conditions favorable to CO$_2$ invasion only when its CaCO$_3$ cycle is driven by net CaCO$_3$ dissolution (ΔDIC$_{CARB\text{-}net}$ is a positive numerical value in Figure 9.4) at a rate higher than about twice the rate of net respiration.

The above examples address the exchange potential of surface seawater with the atmosphere generated by the net ecosystem organic carbon and CaCO$_3$ cycles both in the water column and in the benthic reservoirs. Net autotrophic, $CaCO_3$-producing meadows will maintain their CO_2 invasion potential *from the perspective of air-sea exchange to maintain air-sea equilibrium* when $CARB_{prod} < 2 \times NEP$ but will revert to potential CO_2 evasion status when $CARB_{prod} > 2 \times NEP$. In the same vein, net heterotrophic, $CaCO_3$ dissolving meadows will maintain their CO_2 evasion potential when $CARB_{diss} < 2 \times NEP$, but will revert to CO_2 invasion status when $CARB_{diss} > 2 \times NEP$. Meadows in the Florida reef tract are a good example of a seasonal switch in the metabolic status, the dominant $CaCO_3$ reaction, and the direction of the air-sea CO_2 exchange potential. Based on diurnal rates in Muehllehner et al. (2015), with $CARB_{prod} = 0.3 \times NEP$ (photosynthesis-driven NEP) in spring/summer and $CARB_{diss} = 0.2 \times NEP$ (respiration-driven NEP) in fall/winter, these ecosystems will generally be characterized by CO_2 invasion potential in spring/summer and CO_2 evasion potential in fall/winter. Longer-term studies in *P. oceanica* meadows in the Mediterranean have shown that photosynthesis-driven NEP and CO_2 invasion potential dominate in these ecosystems (Frankignoulle and Distèche 1984; Barrón et al. 2006). It is noted that, *from the perspective of organic carbon and $CaCO_3$-carbon accumulating in soils*, autotrophic, $CaCO_3$ producing meadows can remain carbon stores in the short and long terms of soil accumulation and burial until ecosystem destruction perturbs these stocks. Soil carbon stocks represent longer time scales than the instantaneous exchange potential of surface seawater with air, which, moreover, can be episodic.

There has been no study, to date, that has concurrently measured the rates of all the individual ecosystem processes alongside seawater chemistry. Nonetheless, the numerical results in Figure 9.4 suggest that, in certain settings, seagrass ecosystems in $CaCO_3$-poor soils with modest stocks of soil

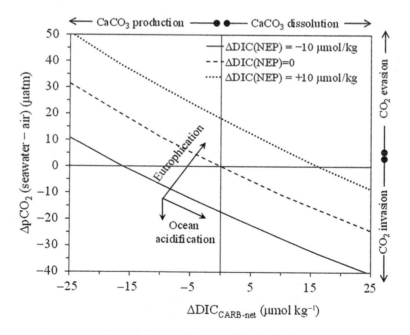

Figure 9.4 The effect of different amounts of calcification ($\Delta DIC_{CARB-net}$ is a negative number) or $CaCO_3$ dissolution ($\Delta DIC_{CARB-net}$ is a positive number) on the air-sea CO_2 exchange potential (positive value represents evasion, negative value invasion) of an ecosystem depending on its metabolic status. The three trends illustrate three specific scenarios for NEP equivalent to $\Delta DIC = -10$, 0, and $+10 \mu mol$ kg^{-1}, while NEP may attain a range of ΔDIC values in natural systems. The results were derived using the numerical approach in Gattuso et al. (1995), assuming negligible TA change due to photosynthesis or respiration ($\Delta TA_{NEP} \approx 0$). The vertical solid line represents an ecosystem with no $CaCO_3$ cycle or where calcification and $CaCO_3$ dissolution are balanced. The horizontal solid line represents no air-sea CO_2 exchange. CO_2 invasion is more frequently observed with $CaCO_3$ dissolution and NEP ≥ 0, while CO_2 evasion is more commonly observed with $CaCO_3$ production and NEP ≤ 0.

organic carbon could be considered as more efficient net sinks for atmospheric CO$_2$ than seagrass systems with higher organic carbon stocks in CaCO$_3$-rich soils (Howard et al. 2017) unless significant benthic CaCO$_3$ dissolution occurs (Figure 9.4).

9.5 POTENTIAL EFFECTS OF EUTROPHICATION AND CLIMATE CHANGE

Whether CO$_2$ invasion or evasion potential predominates in seagrass ecosystems is dependent on the relative rates of NEP and calcification. Field data currently available demonstrate that when seagrass meadows are net autotrophic and support calcifying communities, the overall balance is for CO$_2$ invasion. Increasing atmospheric CO$_2$ concentration in the atmospheric and oceanic reservoirs has been shown to be associated with increased seagrass productivity and, hence, enhanced CO$_2$ invasion potential, although this response maybe species- and site-specific and dependent on the role of changes in other parameters, such as heat stress, increased turbidity, and lowered salinity (Hendricks et al. 2017; Zimmerman et al. 2017). In contrast the effect of increased CO$_2$ in the atmospheric and oceanic reservoirs on calcifiers can be neutral on account of amelioration of local acidification by the chemical effect of primary production of the seagrass system (Cox et al. 2017) or may lead to prevalence of ecosystem-wide net CaCO$_3$ dissolution as already experienced along the Florida Reef Tract (Muehllehner et al. 2015).

The effect of eutrophication on the balance between NEP and calcification is likely to drive toward CO$_2$ evasion by promoting net respiration within the system. Duarte et al. (2005) reported a threshold value for GPP below which seagrass ecosystems changed from net autotrophic to net heterotrophic. Eutrophication leads to a reduction in GPP due to lower light availability, decrease in NEP, and loss of seagrass (Apostolaki et al. 2011). Any change toward a more prevalent and persistent heterotrophic status would enhance CaCO$_3$ dissolution, which may or may not compensate for the increased CO$_2$ depending on the amount of CaCO$_3$ available for dissolution.

9.6 CONCLUSIONS

The interaction of the organic carbon and CaCO$_3$ cycles in seagrass meadows is intricate and directly associated with the air-sea CO$_2$ exchange potential of the ecosystem in the short term. Both cycles contribute to accumulation of detrital carbon stocks in the soil in the longer term. The effects of human disturbance and climate change on biogeochemical reactions and ecosystem functioning generate an additional degree of complexity. Potential scenarios will benefit from a modeling component based on measured process rates and verified by intensive long-term fieldwork studies, so that the role of seagrass ecosystems in storing carbon over the short and long terms can be addressed more fully.

ACKNOWLEDGMENTS

This work was supported by the Ecosystem Services for Poverty Alleviation (ESPA) program [Coastal Ecosystem Services in East Africa (CESEA) NE/L001535/1 research project]. The ESPA program is funded by the Department for International Development (DFID), the Economic and Social Research Council (ESRC), and the Natural Environment Research Council (NERC). Additional support was provided by the U.S. National Science Foundation through the Florida Coastal Everglades Long-Term Ecological Research program under Grant No. DEB-1237517, and this is contribution #47 from the Center for Coastal oceans Research in the Institute for Water and Environment at Florida International University.

CHAPTER 10

Sea Level Change and Its Potential Effects on Coastal Blue Carbon

Robert R. Christian and Eduardo Leorri
East Carolina University

Linda K. Blum
University of Virginia

Marcelo Ardón
North Carolina State University

CONTENTS

HIGHLIGHTS

1. The ability of coastal, sea-level controlled wetlands to sequester carbon is inextricably linked to their respective hydrogeomorphology and ability to maintain themselves in the face of changing sea level.
2. Several geological and archeological approaches are used to observe sea level change in the Holocene; including assessments of basal peats, micro-atolls, paleo-marshes, historical documentation, and for the most recent period tide gauges and satellite altimetry. Each has its own benefits and limitations with modeling providing interpretation and extrapolation.

3. Rates of sea level change vary across time and space with large deviations from global mean sea level change where geographical differences can have opposing trends.

4. The challenge is to understand and quantify how carbon sequestration responds to relative sea level rise (SLR) under the many environmental conditions that promote, maintain, or degrade coastal wetlands.

5. Withstanding high rates of relative SLR by salt marshes and mangroves depends on access to sediment as well as organic matter production for vertical growth. Horizontal expansion is promoted by low-slope, juxtaposed land for transgression. Less is known about the response of seagrass ecosystems, but the importance of adequate light penetration is evident.

10.1 INTRODUCTION: BACKGROUND AND SCOPE

Sea-level controlled wetlands are the primary coastal ecosystems responsible for blue carbon accumulation (Mcleod et al. 2011). They include mangrove, salt marsh, and seagrass ecosystems, and all are addressed within this book (Chapters 12–14). The factors that determine the ability of these ecosystems to sequester carbon are also reviewed in this book (Chapters 3–7, 9 and 11). This chapter describes various aspects of sea level change and how it could affect pools and accumulation rates of blue carbon. While sea level may change near-constantly associated with tides and winds, our focus is on long-term change of decades to millennia. Sea level affects hydroperiod and water quality; it represents one of the most important environmental drivers of carbon processes and budgets of coastal wetlands and controls the very existence of most of these ecosystems.

Eustatic or global sea level has changed throughout geological time, rising and falling in association with climate changes and movements in the Earth's surface (Haq et al. 1987; Miller et al. 2005). Climate change has been controlled, for instance, over the last 140ka, by solar irradiation in response to changes in the Earth's orbit until relatively recently. As a result, coastlines have moved back and forth from their current position. For example, sea level has moved from 3 to 9m (which only equates to ca. 4% of the current global ice volume) above current mean sea level (MSL) to 120m below current MSL over the past 130ka years (Yokohama and Esat, 2011). However, the coastal sea-level controlled wetlands of today are products of Holocene (11,700 years BP; Walker et al. 2009) and Anthropocene (circa 1950 CE; Williams et al. 2016) conditions. Melting of the great ice sheets was largely complete by 6,000 years ago, and it is believed that sea level did not rise significantly again until recently (Woodworth et al. 2011). While global sea level rose ca. 120m since the last glacial maximum (ca. 15,000–20,000 years BP), relative sea level (RSL) (see below) is time and space dependent, reflecting local factors such as glacio-hydro-isostasy (i.e., solid Earth deformation associated with internal buoyancy forces in response to changes in ice and water masses at the surface; Milne and Shennan, 2007) and tectonic land movements. Also, the redistribution of water masses is due to steric (temperature-driven) changes (Johnson and Wijffels, 2011), the gravitational pull (Clark et al. 2002) of the ice masses, and the exchange of waters with the continents. The combination of these factors is reflected locally/regionally (RSL) with rising, falling, or stable sea level. Here we provide a brief review of the science behind sea level change and the trends found that are germane to carbon sequestration of coastal wetlands. This is not meant to be a comprehensive review of sea level change as others already exist (Cazenave and Llovel 2010; Church et al. 2013; IPCC report 2013; Gehrels et al. 2015; Shennan et al. 2015).

We use the terms coastal wetland, sea-level controlled wetland and tidal wetland with distinct intent. Coastal wetland is perhaps the broadest term considered and is based on geography rather than hydrogeomorphology. Note that it is often difficult to define the coast (Christian and Mazzilli 2007). The current book defines coastal lands "as the subset of all terrestrial lands located within the tidal frame, regardless of whether they are connected to tidal hydrology or protected from it by a hydrologic barrier." Thus, for the book, coastal wetlands are any within these coastal lands. Some coastal wetlands are divorced from sea level and tides, such as those isolated on barrier islands, pocosins, or Carolina bays. The other two kinds of wetlands are subsumed under this term. Tidal wetlands are sea-level controlled wetlands influenced by astronomical tides. Salt marshes, mangroves, and seagrasses often are tidal wetlands. Tides along some rivers may extend far upriver, such as the James River in Virginia. These are **tidal** wetlands but freshwater. As SLR occurs, these freshwater tidal systems will become more saline. Other sea level controlled wetlands are connected to large bodies of water but may be non-tidal or minimally tidal. For example, the large sounds of North Carolina have few inlets and thus have minimal flushing by tidal waters. Wind tides dominate the hydroperiod. These wetlands are **sea-level controlled** and **coastal** but not **tidal** in the way most people think. One may find salt marshes, mangroves, and seagrasses within such non-tidal conditions.

Past, present, and future abilities of many coastal wetlands to sequester carbon depend in part on rates of sea level change. They maintain themselves largely by changing their elevation and spatial extent in response to changes in MSL and/or hydroperiod. Many wetlands have maintained themselves for millennia. But MSL is rising at accelerated rates in the past century, with highest rates within the past decades (Chen et al. 2017). Rates in some locales, such as the USA mid-Atlantic region are exceeding the current global SLR by 2–5 times (Church and White 2011 compared to Ezer and Corlett 2012). Future rates are predicted to be even higher as the climate warms (IPCC 2014). These conditions, in conjunction with human alterations to the coast, may exceed the ability of coastal wetlands to maintain both elevation and area. Loss of wetland elevation and area in turn reduce the amount of carbon capable of being stored in the soil. Here we summarize how current and projected rates of SLR affect wetland structure and associated carbon sequestration.

10.2 SEA LEVEL: DEFINITIONS, METHODS, AND MODELS OF STATE AND RATES OF CHANGE

10.2.1 Definitions

We use the definitions from the Fifth Assessment Report by the Intergovernmental Panel on Climate Change (IPCC 2014). Specifically, we cite Chapter 13, "Sea Level Change," by Church et al. (2013). First, RSL is "the height of the ocean surface at any given location, or sea level, is measured ... with respect to the solid Earth" (pg. 1,142). Geocentric sea level relates to satellite altimetry and is measured relative to the reference ellipsoid. Sea level averaged over some time span to remove temporal variability is MSL. Generally, spatial averaging of MSL leads to global mean sea level (GMSL), although this might not be feasible for most of the geological record. For estimates in the geological record, far-field site or isostatic modeling are used (Dutton and Lambeck, 2012; Kopp et al. 2009, 2013), where models are typically tested against field data. This chapter focuses on the aspects of sea level most tied to the ability of coastal wetlands to sequester carbon. Therefore, emphasis is on RSL and its rate of rise. Other climate-related concerns are associated with the rates of SLR (e.g., hydrological cycle changes, intrusion of salt water). Further, while there is a tendency to focus on the rise in sea level, often records show a decrease under some circumstances. We use the word **change** when decreases in sea level may be an important component of the story.

RSL is dependent on numerous processes occurring in the water and on land (Shennan et al. 2015; Wöppelmann and Marcos 2016). The processes affecting sea level change aggregate in two categories. Changes in ocean volume represent what is called eustatic change, while many changes in land and solid Earth position represent isostatic changes in level. Both of these have multiple causes. Simplistically, eustatic changes result from thermal expansion and inputs of water and ice from glaciers, ice sheets, rivers, and groundwater. Isostatic adjustments occur through long-term response to water and glacial redistributions (glacio-hydro-isostatic adjustment) and to local sediment supply and erosion. Further, tectonic activity can cause major alterations to coastal zone elevation and associated wetlands. Other than tectonic activity, groundwater changes, and compaction/subsidence control vertical land movements.

10.2.2 Methods of Measuring Sea Level

The most common method for estimating RSL is by tide gauge. The longest tide gauge record starts in 1711 (Wöppelmann et al. 2008), but tide gauges were not automatic until the 1930s. They became present in most major ports by the end of the 19th century. However, records longer than six decades are from the northern hemisphere; and regardless of location, the majority of tide records are shorter than 60 years long (Woodworth et al. 2011). This spatial and temporal limitation needs to be considered when interpreting global long-term sea level trends. Church and White (2006) used gauge records to make one of the most cited estimates of recent, century-scale changes (1870–2004) in global MSL (1.7 ± 0.3 mm year^{-1}). Others have made similar calculations with different corrections for local isostatic processes and to address more recent accelerations in SLR (Wöppelmann et al. 2007; Jevrejeva et al. 2008; Cazenave and Llovel 2010; Church and White 2011; Wöppelmann and Marcos 2016). For example, Church and White (2011) used 1900 as the reference point (1.7 ± 0.2 mm year^{-1}) (1900–2009) with a second calculation of 1.9 ± 0.4 mm year^{-1} for the period 1961–2009. (More information on rates is given in Section 10.2.3.)

Whereas tide gauges have provided measurements of RSL at tens to hundreds of points for up to the past century or more, satellite altimetry has provided much larger coverage starting in 1992 (Church and White 2006). Altimetry rates of sea level change have been compared to those calculated from tide gauges (Church and White 2006, 2011), even though satellite data are generally not available for near-shore where tide gauges are most commonly installed. After a series of corrections are made to altimetry data, the rates of global sea level change are generally similar between the two methods (Church and White, 2011).

Estimating RSL and relative sea level change rates for periods prior to the tide gauge record present a challenge and require other types of analyses (see Milne et al. 2009 for a review). Perhaps one of the most interesting analyses has been done for Venice by Camuffo and Sturaro (2003). They compared the algal belts on buildings in paintings by Canaletto and his students in the early eighteenth century to those found on the same buildings in 2002. RSL was estimated to rise at 2.3 ± 0.4 mm year^{-1}, a rate similar to estimates using other approaches. But such accurate paintings are rare, and most estimates of relative sea level change prior to the 20th century have been based on the geological record.

Holocene estimates of relative sea level change with errors ranging from few tens of cm to less than 1 m are based mainly on basal peats and paleo-marsh elevation reconstructions based on proxies within saltmarsh soils (Milne et al. 2009). The most accurate proxies to date are foraminiferal (amoeboid protozoa with shells) assemblages found near the soil surface (usually top 1 cm) and at known elevations in the soil profile (Scott and Medioli 1978). Down-core profiles of these assemblages are then linked to a chronological framework derived from ^{14}C dating and other chronological markers. Over the last 25 years, this approach has been combined with multiple regression analyses providing high-resolution sea level reconstructions that extend estimates of RSL hundreds to thousands of years (e.g., Kemp et al. 2011; Barlow et al. 2014).

10.2.3 Rates of Sea Level Change

Rates of sea level change vary across time and space. Estimated rates may also depend on the methods and assumptions of their calculation (Rahmstorf 2007; Vermeer and Rahmstorf 2009; Kemp et al. 2011; Church et al. 2013). Kemp and colleagues (2011, 2017) have provided some of highest resolution records to date, dating to as much as ~3,000 years before present from the mid-Atlantic State of North Carolina, USA. This is an area of currently rapid SLR (Kopp 2013). Kemp et al. (2011) found four stages of rate patterns for one site in the region: from 100 BCE to 950 CE and again from 1350 CE to the late 19th century there was little to no change in sea level: the rate of SLR rose to 0.6 mm year^{-1} from 950 CE until 1350 CE; and then SLR increased to 2.1 mm year^{-1} from the late 1800s until present. However, differences in rate and pattern were noted when multiple sites were analyzed (Kemp et al. 2017). The results highlight the importance of regional and local processes in controlling centennial scales of relative sea level rise (RSLR).

Church et al. (2013) illustrate six sea level versus time gauge records to highlight several aspects of variation (Figure 10.1). First, overall trends, and hence rates, of sea level change differ, even in direction, among locations. Second, patterns of inter-annual change may differ within a location over time (e.g., Manila). Third, there is considerable noise in the signal associated with intra-annual variation. These inter- and intra-annual fluctuations may be associated with several climate and non-climate-related phenomena. Climate-related phenomena include large scale but regional atmospheric oscillations (e.g., El Nino Southern Oscillation), storms, seasonality, and ocean anomalies (e.g., Isla and Bujaleski 2008). Non-climate-related phenomena are related to changes in the tidal amplitude due to astronomical variations (Pugh, 1987; McKinnell and Crawford, 2007), and local alterations in flow from changes in the coastal shape during SLR (e.g., Leorri et al. 2011a).

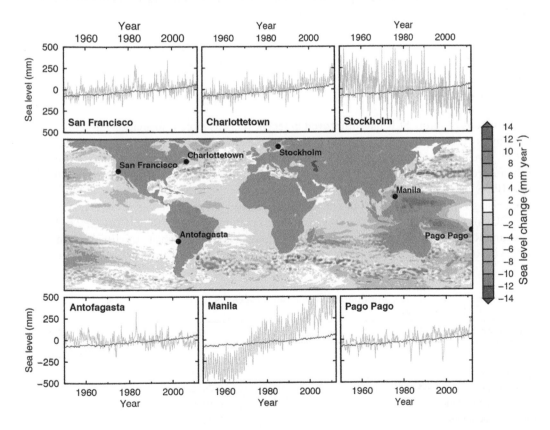

Figure 10.1 (See color insert following page 266.) Differences in trends of sea level change associated with geography. The central map shows rates of sea surface change for 1993–2012 from satellite altimetry. The panels above and below the map show sea levels from 1950 to 2012 recorded by tide gauges at six locations. The red line in each panel represents global sea level change to highlight differences among local trends. (Source: Church et al. (2013); FAQ 13.1, Figure 10.1, p. 1,148.)

The magnitude of differences in relative SL change across the world is large (Figure 10.1 center panel) even when sea level change rates are based on relatively short (1993–2013) altimetry measurements. Note that the Atlantic, Indian, Southern and Western Pacific Oceans have generally shown rates of 2 mm year^{-1} or more. The highest rates of increase were found in the Western Pacific Ocean near the Philippines, Papua New Guinea, and Micronesia. However, the eastern Pacific and northwestern Atlantic Oceans had lower rates with some decreases in sea level (Figure 10.1).

Global mean sea level rate change is estimated from data such as those described thus far and through modeling (Woodworth et al. 2011). Church et al. (2013) summarized GMSL rate change and its components for the Fifth Assessment Report of the IPCC (2014). Both observed and modeled rates were included in the analysis. Three time periods were assessed with a trend of increasing change rates with time. Most recent rates (1993–2010) are highest among the three periods with a mean for observed GMSL rise of 3.2 mm year^{-1} (2.8–3.6 95% CL) and one for modeled GMSL of 2.8 mm year^{-1} (2.1–3.5 95% CL). This period had the most complete dataset for estimating the contributions of different processes (Table 10.1). Thermal or steric expansion of the ocean is seen as the largest, single contributor to increasing sea levels by both observation and modeling. Melting and disruption of glaciers and ice sheets appear less when divided regionally (i.e., Greenland, Antarctica, and others), but combined estimates may rival or exceed thermal expansion. An analysis by Domingues et al. (2008) suggests that the steric (0.7 ± 0.5 mm year^{-1}) and mass (0.8 ± 0.5 mm year^{-1}) contributions are approximately equal within estimated uncertainties for the period 1961–2003. Lastly, change in water storage on land (i.e., natural and man-made freshwater bodies, wetlands, vadose zone, aquifers, and snow pack) contributes to isostatic land surface change relative to sea surface levels. The dominant change in land storage at present appears to be ground water pumping, which would promote relative SLR through land subsidence.

10.2.4 Predictions of Future SLR

Predictions of future SLR have improved in recent years as more information has come to light, especially concerning the contributions of glacial and ice sheet melt waters to the ocean (Church et al. 2013). Predictions depend on which scenarios of human activity and climate change are chosen for the future and which models are used in predictions. The Fifth Assessment Report of the IPCC (2013, IPCC AR5) has focused on four scenarios based on the accumulated radiative forcing from greenhouse gases and the patterns of concentration change (i.e., representative concentration pathways (RCPs)) through 2100. The four scenarios most commonly used (RCP3PD/RCP2.6, RCP4.5, RCP6, and RCP8.5) reflect greater release of greenhouse gases and intensity of climate change from RCP2.6 to RCP8.5. A suite of process models projected that the median GMSL in 2100 would rise

Table 10.1 Estimates of the Contributions of Different Processes to GMSL Rise for the Period 1993–2010

Contribution to GMSL Rise	Observed	Modeled
	Mean mm year^{-1} (± 95% CL)	
Thermal expansion	1.1 (0.8–1.4)	1.49 (0.97–2.02)
Glaciers except Greenland and Antarctica	0.76 (0.39–1.13)	0.78 (0.43–1.13)
Glaciers in Greenland	0.10 (0.07–0.13)	0.14 (0.06–0.23)
Greenland ice sheet	0.33 (0.25–0.41)	
Antarctic ice sheet	0.27 (0.16–0.38)	
Land water storage (observed used for model estimate)	0.38 (0.26–0.49)	
Total GMSL rise	3.2 (2.8–3.6)	2.8 (2.1–3.5)

Source: Adapted from Church et al. (2013).
Estimates from both observations and models are given.

0.44 m (range = 0.28–0.61) for RCP2.6 to as high as 0.74 m (range = 0.52–0.98) for RCP8.5 (Church et al. 2013). These sea levels are at the low end of those from opinions of 90 experts surveyed by Horton et al. (2014). These experts expect between 0.5 m and 1.0 m of SLR by 2100 CE for the lower and upper temperature scenarios, respectively, and a significant probability of exceeding 2.0 m of SLR by AD 2100 under the upper temperature scenario. In agreement with Horton et al. (2014), semi-empirical models also predict higher sea levels than the IPCC AR5 by AD 2100 of 0.5–1.4 m above the 1990 level (e.g., Rahmstorf, 2007). These models are based on the past relationship between SLR rate and temperature change and used for future projections of a scenario of future warming. However, it is unclear if this relationship will hold true in the future.

Ecosystem response may be more impacted by rates of change than actual mean tidal elevation change (i.e., ecosystems might adapt to change unless a threshold in the rate of change is surpassed). Thus, sea level projections can be presented as rates by the end of the century. They range from 4.4 mm year^{-1} (2.0–6.8) in RCP2.6; to 6.1 (3.5–8.8) mm year^{-1} in RCP4.5; to 7.4 (4.7–10.3) mm year^{-1} in RCP6.0; and to 11 (8–16) mm year^{-1} in RCP8.5. These rates could be higher if we consider the expert opinion or the semi-empirical models. In addition, the high stand during the last interglacial (marine isotope stage 5a) could be used as a reference for a potential warming of 2°C, reaching ca. 5 m above present sea level. However, this approach does not allow for the calculation of rates of sea level.

10.3 INTERPLAY OF COASTAL WETLANDS, INUNDATION, AND RATES OF RELATIVE SLR

10.3.1 Environmental Factors Related to Rates of Relative Sea Level Change

The primary effect of changes in RSL is to change the hydroperiod of the wetland. We focus primarily on SLR, rather than sea level fall, as it is the more likely consequence of climate change. The increased flooding of coastal ecosystems is only one of the potential consequences of SLR and rate acceleration (Day et al. 2008). Aspects of water quality may also be affected. Salinity, sulfate concentrations, turbidity, sediment supply, and nutrient concentrations may all be altered by rising sea levels (Herbert et al. 2015). And all of these factors can affect coastal wetland processes that promote carbon sequestration. Other climate-related conditions are expected to change concurrently with accelerated rates of sea level rise. These include increasing temperature and frequency and intensity of major storms (IPPC 2014). Further, human activities that may or may not be related to climate provide further changes to the coast through land-use intensification and changes to water flow and quality (Kirwan and Megonigal, 2013). The interaction of all these various factors creates difficulty in understanding and predicting the efficacy of coastal wetlands to foster blue carbon benefits. Here, we provide a brief introduction to environmental factors specifically related to RSL that may affect carbon sequestration, while other chapters expand on these issues.

Environmental factors directly related to increases in RSL are largely controlled by the position and intensity of the interface between fresh and seawater. All things being equal, as sea level rises seawater pushes farther upstream and becomes a larger proportion of flooding waters. Salinization of coastal wetlands has important consequences to a variety of processes, including plant growth, nutrient cycling, and decomposition (Herbert et al. 2015). Of particular importance to blue carbon is the enhancement of dissimilatory sulfate reduction and resultant increased anaerobic decomposition rates relative to freshwater systems. As nutrient sources to coastal waters are often upstream, seawater intrusion has the potential to alter nutrient availability to coastal wetlands (Weston et al. 2010; Ardón et al. 2013). This in turn may alter plant growth, although the results are not necessarily linear (Turner et al. 2009). Finally, turbidity and sediment load often decrease as fluvial sources become attenuated by seawater intrusion, although this is not necessarily found when tidal energies are high. Reduced sediment availability can limit accretion and jeopardize the capacity for increases in elevation (Lovelock et al. 2015; Morris 2016). Thus, managing the effects of accelerated SLR on blue carbon requires better understanding of these associated and interacting environmental factors.

10.3.2 Hydrogeomorphology of Coastal Wetlands and Carbon Sequestration

Hydrogeomorphology identifies the wetland in the context of its position relative to water. Hydrogeomorphology of and within coastal wetlands affects the interaction between wetlands and sea level (Christian et al. 2000; Day et al. 2008; Kearny and Turner, 2016). Ecosystem functions, including carbon sequestration, and even the very existence of a wetland depends on maintaining its hydrogeomorphic position (Chapters 5 and 6 in this book). Hydrogeomorphological classifications of wetlands have focused on (i) position in the landscape; (ii) water direction; and (iii) water source (ground, surface, and precipitation) (Brinson 1993). The three kinds of coastal wetland ecosystems considered in this book are salt (and brackish) marshes, mangroves, and seagrasses. Because of their low elevations and coastal location, all three are sea-level controlled. Within each ecosystem there may exist zones of community structure and ecosystem function that respond to the factors differently from one another (e.g., low marsh versus high marsh). Water direction and source may also differ among and within wetlands and are the hydrogeomorphic factors most directly affected by sea level. Bidirectional tidal flooding promotes (i) regular flooding deep into an intertidal wetland and control over soil salinity and community structure; (ii) high hydrodynamic energies that entrain sediments available for deposition on wetlands; and (iii) regular and large fluctuations in light regime in seagrass ecosystems. The surface water of coastal wetlands may largely be unidirectional if astronomical tides are not significant or are attenuated by freshwater advection. Ground water may move from lateral or vertical sources, whereas obviously precipitation has an ultimately vertical source. Surface run-off, groundwater, and precipitation can attenuate the salinization by tidal waters, maintain saturated soils, and allow for communities less stressed by tidal energy and salinity. All of these features may be modified if a wetland changes its hydrogeomorphic position (i.e., relative elevation and area).

Overall, carbon sequestration and, thus, blue carbon benefits are promoted when coastal wetlands maintain or grow both relative elevation and area. The coastal wetland ecosystems considered in this book must respond to relative SLR by increasing their elevation and by moving laterally to expand or at least conserve aerial extent (Brinson et al. 1995; Christian et al. 2000; Spencer et al. 2016).

Four major processes contribute to elevation change and associated rate of carbon sequestration: plant growth, sedimentation, decomposition, and compaction. Changes in elevation of these ecosystems are required to maintain the environmental conditions conducive to plant growth (Morris et al. 2002; Chapter 5). In the case of marshes and mangroves, the position is determined relative to the inundation frequencies and durations, which promote or allow growth of specific species. In the case of seagrass ecosystems, elevation increase (i.e., maintaining or decreasing depth) is needed to maintain sufficient light penetration for positive primary production. All of these wetlands increase elevation through deposition of biogenic accretion and sedimentation (Morris et al. 2016). Biogenic accretion involves carbon being translocated below ground through the production of roots and rhizomes and deposition of aboveground plant material. Sedimentation imports both organic and inorganic matter. The increased soil volume from sedimentation provides opportunity for ingrowth of roots near the soil surface. The details of plant-elevation interactions and sedimentation versus autochthonous accretion differ among hydrogeomorphic settings. This is addressed in more detail in other chapters.

The sum of sequestered carbon by a wetland depends on the wetland's area. As sea level rises, that area may be threatened and depends on horizontal movement as well as vertical accretion.

Horizontal movement relies on the interaction between sediment availability and hydrodynamic energy and on the condition of neighboring accommodation space (Brinson et al. 1995). Seawater edges of intertidal wetlands may stay stable, prograde outward, or erode inward. Where the seaward edges are stable, depositional processes are balanced by erosional processes. Expansion of the marsh on the seaward margin, or progradation, requires that the balance between sediment depositional and erosional processes favor deposition. When erosion is greater than deposition, the marsh, seaward edge will move landward. Progradation requires abundant sediment, typically from a landward source, to be carried to the prograding margin (Kirwan et al. 2010). Waves counter marsh expansion by eroding the marsh soil just below the rooting zone (Fagherazzi and Wiberg 2009; Tonelli et al. 2010). It is significant that hurricane-generated waves are not likely to have a disproportionate role in marsh-edge erosion (Leonardi et al. 2016); it is typically lower intensity storm waves that cause erosion and retreat of the marsh edge. Wetland expansion processes in general promote carbon sequestration. Erosion of the seaward edge of a marsh or mangrove without corresponding transgression into uplands decreases carbon sequestration as the aerial extent of the wetland shrinks. Seagrass ecosystems may also contract and expand laterally in response to accretion and erosion from the interplay of sediment availability and hydrodynamic energy. Thus, coastal wetlands can contract in area but build vertically. When the contraction is the result of erosion, the calculation of carbon storage may need special attention. A significant fraction of carbon can be recycled from the front of the wetland itself and be refractory carbon (so-called black carbon for particulate matter). Therefore, the estimate of the carbon burial may need to differentiate between the newly buried carbon and that recycled from previous erosion and redeposition.

Intertidal wetlands often abut uplands or non-tidal wetlands. Maintenance of area in the face of SLR can be accomplished by transgression or movement inland. The speed and extent at which this happens is dependent on the slope from the intertidal wetland into the neighboring ecosystem (Brinson et al. 1995). Low slopes allow greater transgression (Kirwan et al. 2016). Further, human obstructions, such as roads and bulkheads, act as deterrents to transgression. Similarly, coastal wetlands are limited in coasts dominated by wave-cut cliffs. Many neighboring ecosystems without obstructions resist transgression through their own self-maintaining functions. For example, forests shade the ground and prevent marsh plants from establishing themselves. Climate change and accelerated SLR may promote transgression (Craft et al. 2009, 2012). Elevated sea level fosters flooding of the neighboring ecosystems during particularly high tides. This flooding and associated saline waters may stress trees (Krauss et al. 2009). Increased frequency and intensity of storms may exacerbate flooding events and subsequent tree stress. Dying and dead trees allow light penetration through a reduced canopy and hence the invasion of marsh vegetation. The resultant carbon sequestration balance is shifted from that achieved by forest to that of a coastal wetland. Research to date has shown contrasting results regarding which of these ecosystems sequesters more carbon. Working in wetlands along two Georgia rivers, Craft (2012) found that marshes sequestered more carbon than upstream forested wetlands. In contrast, in wetlands along Georgia and South Carolina rivers, Noe et al. (2016) found that forested wetlands could sequester more carbon than downstream marshes. So at this point it is not clear if marsh transgression into forested coastal wetlands will lead to an overall decrease or increase in carbon sequestration at the landscape level.

10.3.3 Threshold Rates of Relative SLR

Accelerated relative SLR has the potential, then, of altering blue carbon by numerous mechanisms. The challenge is to understand and quantify how sequestration responds to relative SLR under the many environmental conditions that promote, maintain, or degrade coastal wetlands. There are some estimates of rates of relative SLR that exceed the ability of wetlands to maintain themselves, mostly for salt marshes. These come from both field measurements and model predictions. Salt marshes in Louisiana, USA, have demonstrated loss when subjected to RSL rise of 10 mm year^{-1} (Boesch et al. 1994; Kearney and Turner 2016). In contrast, sediment supply was found to be highly important in salt marshes in the eastern Cantabrian coast, Spain. These marshes are expected to adapt to ongoing SLR based on the high sedimentation rates (14–18 mm year^{-1}) observed during restoration of reclaimed areas. When surface elevation achieves appropriate position, high marsh plants colonize and sedimentation rates slow down (Garcia-Artola et al. 2016). Morris et al. (2002) used empirical and modeling to predict that mesotidal (2–4 m tidal amplitude) salt marshes with high sediment loads could withstand RSLR of 12 mm year^{-1}. Cahoon et al. (2009) used expert opinion to estimate the impacts of increases in RSLR by 2 and 7 mm year^{-1} above current rate (3–4 mm year^{-1}) on coastal wetlands of the mid-Atlantic states of the USA. The panel suggested that most wetlands would be converted to open water by the higher rate of relative SLR and that many estuarine wetlands within Chesapeake and Delaware Bays would be converted by 2 mm year^{-1} above the current rate. Kirwan et al. (2010) analyzed an ensemble of saltmarsh models to predict threshold relative SLR. They found a wide range of threshold rates from 5 to 100 mm year^{-1} with positive correspondence with tide range and sediment delivery rate. Kirwan et al. (2016) expanded this effort with a meta-analysis of accretion and elevation change measurements. Their work confirmed that a large majority of salt marshes has maintained themselves under current and recent rates of RSL rise as high as 15 mm year^{-1}. However, all but one marsh were subjected to rates less than 10 mm year^{-1} relative SLR, and most SLR rates measured were less than 5 mm year^{-1}. They supported the hypothesis that maintenance ability was highly coupled to availability of sediments and high tidal amplitude. Morris et al. (2016) analyzed the contributions of organic and inorganic components to bulk density of wetlands to predict maximum vertical accretion. If this maximum equals the relative SLR above which the wetland degrades, then the maximum sea level rate is estimated at 5 mm year^{-1} for the **average** wetland of the East coast and Gulf of Mexico coastlines of the USA.

Threshold rates of relative SLR for mangrove and seagrass ecosystems are less available. The responses of intertidal mangroves might be expected to be similar to those of salt marshes. As with marshes, availability of sediment for accretion may be important establishing threshold SLR rates. Long-term rates of accretion in mangroves were reviewed and placed in the context of relative SLR by Ellison and Stoddart in 1991. They proposed that mangroves are often capable of matching modest RSLR rates of 8–9 cm/100 year (0.8–0.9 mm year^{-1}), but not 12 cm/100 year (1.2 mm year^{-1}). More recent studies using surface elevation tables have shown short-term elevation changes up to 6.2 mm year^{-1} with accretion rates reaching 20 mm year^{-1} (Krauss et al. 2014). A follow-up study found that 69% of study sites in the Indo-Pacific had surface elevation gains lower than local SLR (Lovelock et al. 2015). Furthermore, their results suggest that sites with low sediment delivery and low tidal amplitude could be submerged as early as 2070 (Lovelock et al. 2015). In fact, a simplistic classification on mangrove response to SLR rates does not seem adequate. In Seitu, Malaysia, mangroves and coastal lagoons have been accumulating sediment in response to barrier island development controlled by sea level changes. These environments started to develop 300 years ago (Mallinson et al. 2014) and accumulation rates of organic rich mud exceeded more than 1 m/100 years (10 mm year^{-1}). Here mangroves kept up with sea level rates in excess of 3 mm year^{-1} (Culver et al. 2015). On the other hand, mangroves from Seychelles in the western Indian Ocean are dominated by carbonate-rich sands. It is unclear if they can keep up elevation with accelerated sea levels, and their ability to sequester carbon is limited (Leorri et al. 2011b; Woodroffe et al. 2015).

The mechanism of seagrass ecosystem response is more linked to light availability and involves the ability to shift habitat into shallower waters, although increased erosion and uprooting may be at play (Duarte 2002; Waycott et al. 2007). Waycott et al. (2007) suggested that a 1-m increase in RSL around the Great Barrier Reef will increase potential seagrass habitat by 3,000 km^2. Saunders et al. (2013) modeled the response of a large seagrass ecosystem in Queensland, Australia, to future RSLR. They projected that seagrass habitat could decline by 17% under a scenario of a 1.1 m rise in RSLR by 2100. However, decline could be ameliorated by shoreline management and improved water clarity. Further, sediment accretion could increase or decrease habitat area, depending on its rate relative to that of RSLR. Seagrass species now live in an environment with lower temperature and CO_2 than that of their ancestors (Beer and Koch, 1996). Rising sea levels, higher temperatures, and higher dissolved CO_2 could be more conducive for expansion of many seagrass species; however, local human pressures might override any potential benefits (Orth et al. 2006). Thus, maintenance of elevation and area of wetland ecosystems and hence blue carbon benefits do appear to be susceptible to projected rates of RSLR. Some opportunities exist for increases in blue carbon accumulation, while the potential for carbon sequestration in many wetlands may be compromised.

10.4 CONCLUSIONS

We have provided a summary of sea level change with a focus on coastal wetlands. We summarized background on sea level change, factors that control rates of both global and local sea level change, past trajectories, future projections, and mechanisms by which these affect coastal wetlands and their carbon sequestration. The primary tenet of this chapter is that past, present, and future abilities of many coastal wetlands to sequester C depend on rates of sea level change. Wetlands have maintained themselves for millennia, and may or may not do so in the face of projected rates. The ability to withstand higher rates of RSLR appears to depend on access to sediment, associated high tidal amplitudes and low-slope, juxtaposed land for transgression, at least for salt marshes and mangroves. This has been recognized by others. Less is known about the response of seagrass ecosystems. However, the *2013 Supplement to the 2006 IPCC Guidelines for National Greenhouse Gas Inventories: Wetlands* (Hiraishi et al. 2014) did not incorporate SLR into its calculations of carbon sequestration and failed to acknowledge that predictions of success with blue carbon and resulting credits may be jeopardized for wetlands that are negatively affected by accelerated RSLR or in fact enhanced for wetlands with the capacity to respond.

ACKNOWLEDGMENTS

This effort was supported by the National Science Foundation Grant No. 1237733 through the Virginia Coast Reserve Long-term Ecological Research program and the Biology Department of East Carolina University. M. Ardón was also supported by DEB-1452886.

CHAPTER 11

Coastal Wetland Responses to Warming

J. Patrick Megonigal
Smithsonian Institution

Samantha Chapman and Adam Langley
Villanova University

Stephen Crooks
Silvestrum Climate Associates, LLC

Paul Dijkstra
Northern Arizona University

Matt Kirwan
Virginia Institute of Marine Sciences

CONTENTS

HIGHLIGHTS

1. Plant primary production is likely to increase with warming based on evidence from latitudinal gradients of tidal marsh biomass and experimental manipulations.

2. Rates of decomposition are likely to increase based on first principles and lab incubations, but microbial responses to temperature are poorly characterized and confidence in this forecast is low.
3. Models suggest that the net effect of warming on marsh carbon sequestration and the capacity to keep pace with relative sea level rise will be small and related primarily to the indirect effect of temperature on rates of sea level rise.
4. Temperature-driven displacement of tidal marsh plants by mangrove trees will increase carbon sequestration at the boundaries of these ecosystems.
5. Temperature will increase methane emissions through a combination of direct effects on the microorganisms that produce methane and indirect effects on microorganisms that consume methane or compete with methane producing microbes.

11.1 INTRODUCTION AND BACKGROUND

Despite the extraordinary leverage that coastal ecosystems exert over the global carbon cycle, the dynamics of coastal wetland carbon pools are not presently represented in earth system models. Compared to upland soils, the sequestration potential of tidal wetland soils is extremely high because rising sea level gradually increases the potential soil volume, and the rate at which carbon is transferred to deeper, more anoxic soil horizons. Thus, tidal wetlands are not subject to the limits on carbon storage typical of upland soils. Coastal wetlands have only recently been recognized as important carbon sinks, and therefore the response of carbon cycling to climate change in tidal wetlands is largely unexplored. The future sink strength and carbon stock stability of these systems is uncertain because global change drivers such as temperature and elevated atmospheric carbon dioxide (CO_2) perturb the complex biotic and abiotic feedbacks that drive high rates of soil carbon sequestration, and biogeochemical processes such as decomposition, CH_4 emissions, and hydrologic export. Thus, an important reason to study the effects of warming on coastal wetlands is to understand the stability of soil carbon pools, and to determine whether these important ecosystems will continue to gain elevation via carbon sequestration as rates of sea level rise (SLR) accelerate.

Global air temperatures are projected to rise 0.3°C–4.8°C by 2100, a range of outcomes that varies with assumptions about future greenhouse gas emissions (IPCC 2014). Higher air temperatures will cause water bodies and soils to also warm. Over time, this means that soil temperatures will rise to several meters depth (Huang et al. 2000), a possibility that has particularly important implications for tidal marshes with deep, highly organic soils.

The direct effects of warming are accompanied by the direct effects of rising CO_2 levels on plants and a variety of indirect effects such as rising sea level. Carbon dioxide concentrations are projected to continue rising and may exceed 700 ppm by the year 2100 (IPCC 2014). There is evidence that elevated CO_2 acting alone can help stabilize tidal marshes by increasing elevation gain (Langley et al. 2010; Ratliff et al. 2015; Reef et al. 2017), primarily through root production. Warming may either reinforce or negate the effects of elevated CO_2 on elevation, but there is little known about how the two global change factors interact in tidal marshes or tidal freshwater forests.

Rising temperatures are driving accelerated rates of SLR. In addition, many sea coasts are experiencing land subsidence due to natural and anthropogenic phenomena that exacerbate SLR by causing land to sink. The combination of these effects (i.e., relative SLR) tends to increase the frequency of tidal inundation, which in turn changes plant community composition and wildlife habitat value (Krauss et al. 2009; Swanson et al. 2014; Field et al. 2017). When increasing rates of relative SLR cross a poorly understood critical threshold, tidal marshes and forests are threatened with conversion to open water (Cadol et al. 2013; Kirwan et al. 2016b). The capacity of a marsh to respond to SLR depends upon the availability of mineral and organic material to build soil and the availability of space for wetlands to migrate landwards with SLR (Morris et al. 2002; Orr et al. 2003; Kirwan et al. 2010; Spencer et al. 2016).

Temperature and precipitation are coupled through regional and global-scale climate feedbacks that limit the distribution of plant species, and climate will ultimately constrain the responses of foundation plant species and ecosystems to climate change. Because of complex interactions among ecological, hydrological, geomorphological, and biogeochemical processes that tend to maintain coastal wetlands in a state of quasi-equilibrium (Figure 11.1), shifts among functionally different plant community are likely to be non-linear and climate dependent (Gabler et al. 2017). Temperature thresholds are expected to trigger shifts in ecosystem type under humid conditions (e.g., marsh to mangrove), precipitation under arid conditions (e.g., vegetated to unvegetated) (Osland et al. 2016). The influence of macroclimatic drivers on coastal wetlands would benefit from adding coastal processes to Earth System Models (USDOE 2017).

11.2 ECOSYSTEM RESPONSES TO WARMING

Warming is expected to influence the aboveground and belowground feedback loops that regulate soil carbon sequestration, elevation gain, methane emissions, and hydrologic export of carbon and nitrogen (Figure 11.1). Temperature effects on plant production will influence processes that

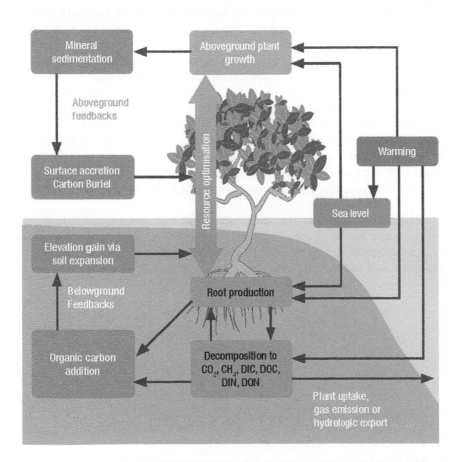

Figure 11.1   (See color insert following page 266.) Conceptual model of temperature effects on key ecosystem processes that regulate carbon input, output, and burial, and export of CH_4 and dissolved forms of carbon and nitrogen. Modified from Kirwan and Megonigal (2013) and reproduced from Megonigal et al. (2016) with permission from the IUCN.

regulate carbon inputs, outputs, and burial, with implications for soil carbon sequestration and marsh stability against SLR. Changes in production will lead to different outcomes for ecosystem function depending on the relative responses of shoots versus roots. Changes in decomposition affect soil organic matter (SOM) pools directly through mass loss, or indirectly through mass gain if increased SOM decomposition increases nitrogen availability and therefore plant growth. Finally, changes in plant and microbial metabolism will have consequences for methane (CH_4) emissions to the atmosphere, and dissolved or particulate forms of carbon and nitrogen to adjacent aquatic ecosystems.

11.2.1 Primary Production

The effects of temperature on plant production have been explored through latitudinal gradients and experimental manipulations (Figure 11.3). All of the experimental manipulations to date have been through passive warming using chambers (Figure 11.2, left panel), which typically causes a 2°C–4°C increase in air temperature, but have negligible effects on soil temperature. In general, warming tends to increase production rates (Gray and Mogg, 2001; Loebl et al. 2006; Charles and Dukes, 2009; Gedan and Bertness, 2009), but the effects tend to be specific to the location and plant species. A latitudinal gradient of *Spartina alterniflora*-dominated salt marshes showed an increase of 27 g m^{-2} year^{-1} per 1°C increase in temperature (Kirwan et al. 2009). Similarly, experimental air warming in a *S. alterniflora* marsh in NE United States suggests an increase in biomass on the order of about 50 g m^{-2} per degree of warming (Gedan et al. 2010). Experimentally warming a *Spartina patens*-dominated NE United States salt marsh also increased biomass (Gedan and Bertness, 2009), while warming a different NE United States salt marsh increased biomass in some species (e.g., *S. alterniflora*), but not others (e.g., *S. patens* and *Distichlis spicata*) (Charles and Dukes, 2009). By contrast, warming increased *D. spicata* biomass by ~250 ± 70 g m^{-2} year^{-1} (mean ± SD, averaged over 2 years) in a SE United States brackish marsh, representing a doubling of salt marsh plant biomass (Coldren et al. 2016). These three in situ warming experiments span a large latitudinal range. The observation that warmer air temperatures resulted in more plant growth in one or more plant species at each site studied indicates that saltmarsh plants may respond positively to warming across their full latitudinal range, and thus may exhibit increased carbon storage in a future, warmer climate.

Warmer air temperatures may also alter salt marsh root biomass and depth distribution, although this response has rarely been measured in the few existing temperature manipulation experiments. In Coldren et al. (2016), wetland plant root mass declined at depths of 40–60 cm in warming chambers and showed a trend toward declines at the surface (0–20 cm). In a greenhouse

Figure 11.2 **(See color insert following page 266.)** Two approaches to experimentally warming tidal marsh ecosystems. Left: Chambers erected on a Florida USA saltmarsh raise air temperature by 2°C–3°C. The chamber has negligible effects on soil temperature. Right: A brackish marsh in Maryland USA is actively heated (1.7°C–5.1°C) aboveground by infrared lamps and belowground by cables to a depth of 1.5 meters. Photos by Samantha Chapman (left) and Roy Rich (right).

experiment, *Spartina anglica* increased in belowground biomass with a rise in temperature (Gray and Mogg 2001). Because belowground plant production plays a dominant role in determining surface elevation (Nyman et al. 2006; Mudd et al. 2009), understanding the influence of temperature on both root growth and decomposition is essential. Root growth responses to soil warming are yet unknown (Figure 11.2, right panel), but may be especially important drivers of long-term carbon stocks in organic soils.

Few experimental manipulations in wetlands have tested hypotheses about the effects of warming on ecological interactions. Bertness and Ewanchuk (2012) studied the role of climate in regulating species interactions by transplanting nine locally common saltmarsh plant species into unvegetated mudflats on the north side (cool) and south side (warm) of Cape Cod, U.S. Soil surface temperature in the two regions differ by 2°C–3°C. Species were planted with and without neighbors. The presence of plant neighbors tended to improve growth rates on the south (warm) side of Cape Cod, but reduce growth rates on the north (cool) side of Cape Cod. This difference in plant interactions seemed to be mediated by the effects of temperature on soil salinity, both of which are greater in the southern sites. The study suggests that interspecific interactions can switch from competitive to facultative in a warming environment, which may help plant communities adapt to climate change. But this conclusion only applies in the aggregate. Another lesson of the study is that predicting the response of such interactions for a given plant species or plant functional group will be exceedingly difficult, limiting forecasts of warming on plant community composition.

Collectively these studies suggest that warming has the potential to affect organismal-level processes such as physiological adaptation to temperature; population-level processes such as reproduction, dispersal, and genetic selection; community-level processes such as interspecific competition and introduction of invasive species; and ecosystem-level functional group shifts such as from herbaceous species to woody shrubs. It is difficult to predict changes in the distributions of species and the consequences of such changes because they are highly dependent on the traits of individual species and on ecological interactions among species in often-novel plant community associations. A great deal of further research on these topics is needed to develop robust forecast models of tidal marsh and forest responses to warming.

11.2.2 Decomposition

Most carbon stored in coastal wetlands lies belowground, yet the critical belowground responses of roots and microbial respiration to warming are far more difficult to assess than the aboveground responses. Though warming-induced increases in plant production are expected to increase carbon inputs to soils, warming may simultaneously increase microbial respiration and decomposition of SOM.

Latitudinal gradient studies suggest that warming will increase decomposition rates at the same time it increases production (Kirwan et al. 2014; Kirwan et al. 2009). The net effect of higher carbon input and output in the short term (decades) will be for warming to enhance carbon storage (Kirwan and Mudd, 2012). However, the long-term effects of warming are more complex than latitudinal patterns suggest, and will be strongly mediated by SLR and gradual increases in total soil-profile decomposition rates as carbon accumulates in the soil profile. Model simulations of a typical *Spartina alterniflora* marsh suggest that carbon storage is enhanced for several decades, followed by weakening and potentially declining carbon storage rates. This pattern is supported by two experimental manipulations of air temperature in New England marshes (Charles and Dukes, 2009; Gedan et al. 2011), but these studies suggest that long-term ecosystem responses to warming will also be controlled by ecological interactions in the form of changes in plant functional groups (Charles and Dukes, 2009; Gedan et al. 2011).

Changes in plant community composition are as important for decomposition rates as for productivity. Plants exert a strong control over SOM decomposition rates in tidal wetlands (Mueller et al. 2015; Ouyang et al. 2017), most likely due to a combination of labile carbon and O_2 additions from plants, and effects vary dramatically across plant species. Bernal et al. (2017) demonstrated that an invasive genotype of *Phragmites australis* accelerates SOM decomposition compared to rates in native North American plant communities. Furthermore, plant stimulation of decomposition differed by soil depth depending on the plant species, with *Schoenoplectus americanus* enhancing SOM decomposition at the soil surface, *Phragmites australis* at greater depths, and *Spartina patens* having no discernable effect on decomposition (Bernal et al. 2017). The potential of warming and ensuing SLR to alter plant community composition represents a large uncertainty for predicting future ecosystem carbon dynamics in tidal marshes and tidal forests.

Initial attempts to measure the sensitivity of organic matter decomposition to warming in marshes range from no responses (Charles and Dukes, 2009) to responses that are larger than those reported in most terrestrial temperature response studies (Kirwan and Blum, 2011). Interestingly, a recent meta-analysis found that decomposition rates in marshes varied with temperature but not latitude (Ouyang et al. 2017). Some models of the process have used a relatively high temperature sensitivity (Q_{10} = 3.44), but more recent analyses suggest a much lower sensitivity to warming (Q_{10} = 1.3–1.5, Kirwan et al. 2014). Because the sensitivity of refractory carbon to warming has never been evaluated in coastal wetlands, the initial models have assumed that both labile and refractory pools respond identically to warming, even though strong differences in temperature effects on labile versus recalcitrant have been observed in upland soils (Frey et al. 2013). Therefore, the precise response of soil carbon decomposition to warming represents a key knowledge gap, and new model experiments informed by field experiments are critical for accurate forecasts of coastal carbon cycling (Figure 11.2). It is important that warming experiments heat both aboveground and belowground portions of the ecosystem.

A largely unexplored question is whether changes in salinity and associated increases in sulfate concentration will influence rates of SOM decomposition (Sutton-Grier et al. 2011; Craft 2007; Stagg et al. 2017). The limited evidence available is equivocal. In many cases, sulfate addition has caused an increase in decomposition rates (Weston et al. 2006), and a meta-analysis suggests that decomposition is slower in mangrove soils with higher salinity (Ouyang et al. 2017). However, in a comparison of ten wetland soils, D'Angelo and Reddy (1999) did not find a difference in decomposition rates under sulfate-reducing or methanogenic conditions. This result may be explained by the fact that the terminal step in microbial respiration where sulfate acts does not necessarily control all of the earlier steps in organic matter degradation, such as depolymerization (Sutton-Grier et al. 2011). There has been limited work on the impacts of different terminal electron acceptors on organic carbon mineralization in soils that differ in organic matter quality.

11.2.3 Microbial Metabolism

Temperature accelerates enzymatic reactions that breakdown organic substrates. Indeed, this is the mechanism by which Arrhenius-based models traditionally forecast changes in soil carbon pools—generally declines—due to warming (Wieder et al. 2013). More recent decomposition models incorporate mechanistic details of microbial population dynamics such as extracellular enzymes, carbon-use efficiency (CUE), and turnover, with a wider range of outcomes from declines to gains in soil carbon (Hagerty et al. 2014; Li et al. 2014). These new models show that predicting changes in soil carbon pools in response to temperature requires detailed knowledge of microbial biomass, physiology, and ecology, and that the understanding of these processes is not presently sufficient to predict whether warming will increase or decrease soil carbon pools.

Changes in microbial metabolism induced by temperature affect the efficiency with which organic matter is converted into microbial biomass, a property known as CUE. Theoretically, warming should reduce CUE by changing the balance between energy production and biosynthesis; at higher temperature, the demand for cell maintenance energy increases, leaving less substrate for biosynthesis of new cell materials (Allison et al. 2010; Manzoni et al. 2012; Sinsabaugh et al. 2013; Cotrufo et al. 2015). However, experimental results are variable with some showing the expected decrease (Steinweg et al. 2008; DeVêvre and Horwáth 2000) and some find no effect of temperature on CUE (Dijkstra et al. 2011; Hagerty et al. 2014). However, estimates of CUE are highly sensitive to the choice of method (Geyer et al. 2019). By comparison, the turnover rate of microbial products does respond strongly to temperature (Hagerty et al. 2014), indicating that it may be important to understand temperature effects on microbial growth as well as death. In freely drained soils dominated by aerobic conditions, microbial mortality is caused by predation and grazing by specialized bacteria and fungi, protozoa, nematodes, and other organisms; the role of such eukaryotic microbes in dominantly anaerobic wetland soils is unexplored. Also relatively less-studied in soils is microbial turnover caused by viruses (Kimura et al. 2008), which may be important as controls on microbial turnover in low and mid-latitude marine ecosystems (Fuhrman, 1999; Mojica et al. 2016). The temperature sensitivity of CUE and its underlying biochemistry in terrestrial ecosystems have mostly been determined under aerobic conditions in mineral soils. Because tidal marshes and forests contain large stores of soil C, it is essential to consider fundamental microbial processes such as CUE and turnover in anaerobic soils.

11.2.4 Methane Emissions

Methane (CH_4) emissions are an important feature of the tidal wetland carbon budget because CH_4 is a powerful greenhouse gas and relatively small rates of release can offset large rates of CO_2 sequestration (Poffenbarger et al. 2011). Each gram of CH_4 released from a marsh into the atmosphere offsets 32–45 g of sequestered CO_2 in terms of the climate impact these gases (Neubauer and Megonigal, 2015). The balance between rates of CO_2 sequestration and CH_4 emissions is important to understand when the goal is to quantify the effects of an activity such as restoration, creation, or management on greenhouse gases. Warming also has the potential to change the balance of CO_2 and CH_4 fluxes.

Subjecting anaerobic soils to a range of temperatures typically shows that warming will increase microbial production of CH_4 (Fung et al. 1991; Meng et al. 2012). Similarly, measuring CH_4 emissions in the field as temperature changes across seasons suggests that warming will increase CH_4 emissions from tidal wetlands (Dunfield et al. 1993; Megonigal and Schlesinger, 2002; Yvon-Durocher et al. 2014). However, the value of such studies is limited either by the absence of real-field conditions (soil incubations) or by the assumption that seasonal variation is entirely due to temperature (field studies). Because of complex interactions between many processes, in situ temperature manipulation experiments are needed to determine whether warming will increase or decrease CH_4 emissions.

The effect of temperature on CH_4 emissions is more complex than the effect on CH_4 production alone for several reasons. One is that emissions of CH_4 are the net outcome of two separate microbial processes—production and oxidation—that can have different temperature responses. Because the amount of CH_4 consumed in oxidation can exceed the amount emitted (Megonigal and Schlesinger, 2002), both must be considered in predictions of future CH_4 emissions. The few Q_{10} values reported for CH_4 oxidation are near 2.0, a value similar to other aerobic biochemical processes, while the Q_{10} values reported for net CH_4 emissions in field and laboratory studies are often much higher (Dunfield et al. 1993). Megonigal (1996) proposed that even a modest difference in the apparent activation energy (equivalent to Q_{10}) of CH_4 production (60 kJ mol⁻¹) versus oxidation (50 kJ mol⁻¹) causes CH_4 oxidation to consume an increasingly smaller fraction of CH_4 production as soils warm (Figure 11.3). The implication is that warming may increase CH_4 emissions from tidal marshes by decreasing percent CH_4 oxidation.

In principle, warming can influence CH_4 emissions indirectly by affecting other coupled aerobic-anaerobic processes that regulate CH_4 emissions. Methanogens compete with a variety of other microorganisms for the organic compounds that support respiration (Megonigal et al. 2004).

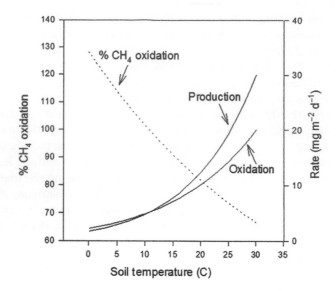

Figure 11.3   A model illustrating the potential for CH_4 oxidation to decline as a fraction of methane production with increasing temperature (Megonigal, 1996).

These organisms generally outcompete the methanogens and suppress CH_4 production wherever there is an abundance of nitrate (denitrifying bacteria), oxidized iron (iron-reducing bacteria) or sulfate (sulfate-reducing bacteria) (Neubauer et al. 2005). Bullock et al. (2013) demonstrated that rising temperature increases iron reduction rates faster than iron oxidation rates, with the result that the abundance of iron oxides declines with warming. Thus, warming may cause increased rates of CH_4 production by releasing methanogens from competition with iron-reducing bacteria.

Climate warming will also change CH_4 emissions through other indirect mechanisms that regulate the supply of metabolism-regulating compounds. Methane production is sensitive to the supply and quality of organic carbon (Megonigal and Schlesinger, 1997; Vann and Megonigal, 2003; Mozdzer and Megonigal, 2012), which can vary due to temperature-driven changes in plant growth or plant species replacement (Gough and Grace, 1998; Baldwin et al. 2001; Langley and Megonigal 2010; Mueller et al. 2016). SLR will increase the supply of sulfate—a critical substrate for the respiration of sulfate-reducing bacteria—which typically outcompete methanogens for organic compounds, thereby suppressing CH_4 production (Neubauer et al. 2005; Weston et al. 2006).

Finally, warming will influence CH_4 emissions by increasing the diffusion rate of gases through soil water or plant tissue, which exhibits a temperature sensitivity similar to biological processes (Kirwan et al. 2014). The net effect of increased diffusion rates on CH_4 emissions is complex, and will depend on the relative change in CH_4 versus O_2 diffusion rates, and whether rates of aerobic methanotrophy are limited by CH_4 versus O_2. For example, a scenario in which CH_4 emissions would increase is where CH_4 oxidation is CH_4-limited (e.g., Megonigal and Schlesinger 2002), and the Q_{10} of CH_4 diffusion exceeds the Q_{10} of oxidation.

11.2.5 Model Forecasts

Numerical models are one approach to forecasting temperature effects on tidal marsh elevation and carbon sequestration. There are several robust tidal marsh elevation models (Fagherazzi et al. 2012), but only a few attempt to mechanistically model SOM accumulation. The Callaway model (Callaway et al. 1996; Callaway and Takekawa, 2013) and the modification named WARMER (Swanson et al. 2014) simulate burial by varying SOM decay rates as a function of age and soil depth, but there are no feedbacks of temperature or flood duration on decomposition rate. The Marsh Equilibrium Model (MEM) of Morris et al. (2002) and similar models add a constant fraction of annual primary production to the SOM pool, which means that SOM storage (the inverse of decomposition) effectively varies only indirectly as a function of flooding on plant production. The Kirwan and Mudd (2012) marsh elevation-carbon model is the only model in this group to date in which SOM decomposition rates respond to temperature.

Kirwan and Mudd (2012) simulated the response of a tidal marsh to a step change in air and soil temperature (Figure 11.4). The model was parameterized for a tidal marsh dominated by *Spartina alterniflora* in a setting with low-suspended sediment concentrations and a constant rate of SLR. The model found that warming increased soil carbon accumulation rates in the years immediately following a sudden increase in temperature. Warming increased plant productivity, which led to enhanced mineral deposition rates, soil elevation gain, and SOM accumulation. However, several factors caused the initial increase in accumulation to decline over time. Warming increased the total pool of SOM over time, which in turn increased the total amount of carbon lost to decomposition. At the same time, gains in marsh surface elevation became too high for optimum plant growth. The net result of these changes was that warming had little impact on net carbon gain after a century in model runs with both plant and decomposition effects (Figure 11.4, black line, Kirwan and Mudd, 2012). Warming increased organic matter accumulation more when the decomposition response was taken out of the model, highlighting the need for research on decomposition responses to temperature, which are poorly understood (Kirwan and Megonigal, 2013).

The Kirwan and Mudd (2012) model forecasts that the positive impact of temperature on *S. alterniflora* production increases with the rate of SLR. This behavior arises for three reasons. First, plant productivity increases with inundation frequency, so proportional increases in growth caused by warming are larger in absolute terms when the rate of sea level is faster. Second, faster rates of SLR tend to offset gains in surface elevation that would otherwise decrease inundation frequency and eventually limit primary production. Third, SLR enhances sediment deposition, so that the carbon concentration in the soil profile and its impact on decomposition is reduced (Mudd et al. 2009; Kirwan and Mudd, 2012). However, these model results are based on simple parameterizations that apply to a specific wetland type, and they remain untested in natural environments.

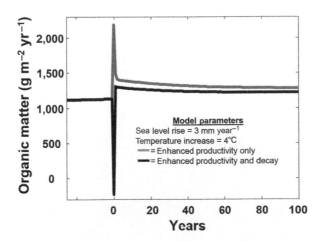

Figure 11.4 Modeled response of organic matter sequestration to an instantaneous 4°C increase at year 0 and a constant rate of SLR. The black line is the effect of warming on both plant growth and microbial decomposition; the green line is with warming of plant growth only (i.e., no warming of decomposition). The sharp change in accumulation at year 0 is a model artifact arising from an immediate increase in decomposition, but a 1-year lag before the enhanced plant growth reaches the soil profile in the model. Adapted from Kirwan and Mudd (2012).

11.2.6 Mangrove Invasion of Tidal Marsh

Mangroves are invading marsh-dominated ecosystems around the world (Figure 11.5), representing one of the most dramatic plant range shifts occurring today (Perry and Mendelssohn, 2009; Doyle et al. 2010; Record et al. 2013; Saintalin et al. 2014). This shift indicates that warming-driven changes in carbon storage are already occurring (Kelleway et al. 2016) are imminent (Doughty et al. 2015). The expansion of mangroves into higher latitudes on a global scale is driven predominantly by declining frequency of severe freeze events (Osland et al. 2013; Cavanaugh et al. 2014). For example, in North America, the black mangrove, *Avicennia germinans*, intermingles with salt marsh species and can expand its distribution during freeze-free intervals (Stevens et al. 2006) or when saltmarsh vegetation is stressed or killed (McKee et al. 2004). Though air temperatures drive global patterns of mangrove expansion, finer-scale changes in mangrove extent respond to many other environmental factors such as erosion, land subsidence, and accretion (Giri and Long, 2014). Thus, some mangrove expansion may be due to re-emergence from previous populations that had disappeared due to disturbance. Regardless, the declining frequency of freeze-related disturbances has generated an overall trend of poleward mangrove range expansion (Saintilan et al. 2014). Woody mangroves encroaching into herbaceous marshes will likely significantly increase wetland carbon storage at these ecotones between temperate marshes and tropical mangroves.

Mangrove area in southeastern Florida (U.S.) has doubled over the past three decades at the northern edge of the historical mangrove range, with a corresponding loss of salt marsh (Cavanaugh et al. 2014). Mangroves are also expanding into vast areas of marshland on the Louisiana and Texas U.S. coasts (Comeaux et al. 2012; Osland et al. 2013), rendering a large portion of the U.S. coastline subject to dramatic change in the next few decades. At a site in northeastern Florida, Doughty et al. (2015) showed that mangrove area increased by 69% in only 7 years, driving a 25% increase in carbon storage over the landscape in less than a decade (Doughty et al. 2015). Kelleway et al. (2016) quantified the lateral encroachment of mangroves into two south-eastern Australian salt marshes over a period of 70 years, increasing soil carbon sequestration rates by as much as 230 Mg C km^{2}year^{-1}. This is one example of a mechanism by which warming may increase carbon storage in a tidal wetland, thereby counteracting global warming.

Figure 11.5 Young mangrove seedlings establish in a tidal marsh previously dominated by grasses. Photo by Adam Langley.

Though extreme climatic events (i.e., reduction of freezes) are driving wetland plant community composition changes, chronic warming will likely continue to alter the carbon sequestration potential of these rapidly shifting ecotones. Root dynamics in transitional wetlands are of particular interest because mangrove roots can potentially oxygenate the soil more than marsh roots, increasing microbial decomposition of organic matter and increasing carbon release. Understanding these dynamics is an important challenge where organic matter is a driver of surface elevation maintenance to SLR (Krauss et al. 2017). As coastal ecosystems provide some of the highest carbon sequestration rates on earth, understanding their shifting carbon storage capacity holds particular importance (Chmura et al. 2003; Donato et al. 2011; Atwood et al. 2017).

11.2.7 Sea Level Rise

Temperature is a major driver of accelerated SLR. The impact of SLR on tidal wetlands is to shift spatial boundaries, the vertical and horizontal ranges in the coastal zone that wetlands occupy (Woodroffe et al. 2016; Lovelock et al. 2017). The response of an individual wetland to SLR is nested within the larger landscape response as estuaries, open coasts, barrier systems, and other large landforms adjust to the changing balance between increased wave energy (including tides) and the redistribution of sediment (Pethick and Crooks, 2000). If tidal wetlands are unable to adjust to this progressive and accelerating shift, they will drown out and transition to subtidal habitat such as mud flat or seagrass meadows.

The persistence of tidal wetlands in the face of accelerated SLR is favored by a high mineral sediment supply and low slopes (Woodroffe et al. 2016). Settings with low sediment supply and limited space for marshes to migrate will see a progressive conversion of vegetated wetland area to mudflat and open water (Spencer et al. 2016; Kirwan et al. 2016b). Field observations and numerical models suggest that tidal range also regulates the transition from a stable to unstable marsh by setting the elevation range over which plants can grow (McKee and Patrick, 1988; Kirwan et al. 2010; Kearney ant Turner, 2016). Relative SLR is most rapid in large river deltas and other areas of rapid subsidence (Kirwan and Megonigal, 2013). Thus, the most sensitive marshes to SLR are those with rapid relative SLR, and small tidal ranges and sediment inputs. Though there is a great deal known about the processes by which tidal marshes gain elevation, large gaps remain in our understanding about how these interact with hydrogeomorphic phenomenon and land-use patterns at large spatial scales. As a result, current forecasts of marsh loss due to SLR may be overestimated (Kirwan et al. 2016). Tidal freshwater marshes and forests as a rule occur at the head of tidal rivers where the nearby uplands have relatively steep slopes. As a result of this geomorphic setting, these systems may be particularly vulnerable to loss of area with accelerated SLR.

11.3 CONCLUSIONS

Warming is one of many global changes that affect the structure and function of tidal wetland ecosystems, and forecasting the effects of warming is hampered by limitations in observational data, experimental evidence, and models. Key among these is information on how plant growth and interspecies competition are influenced by temperature. Temperature-driven shifts in species composition will dramatically alter tidal marsh ecology, biogeochemical cycling, and ecosystem services. Such shifts will arise from species-specific responses of which very little is known. Changes in plant production caused directly by temperature or indirectly by species shifts will fundamentally alter carbon cycling and storage. Forecasting these effects will require information on changes in belowground production and depth distributions, which are not commonly measured.

The processes that regulate decomposition rates are poorly understood. Perhaps the most important consequence of this gap is the challenge of forecasting the effects of warming on marsh stability against SLR. Because soil carbon is a large fraction of tidal wetland soil volume (Morris et al. 2016), an increase in decomposition rate caused directly by warming or indirectly by increased plant production (Wolf et al. 2007; Mueller et al. 2015), has the potential to destabilize existing SOM pools. The feedbacks between plant activity, microbial activity, and carbon storage are complex and will remain a research challenge for decades.

Ultimately, forecasts of tidal marsh responses to warming will require improved models informed by manipulative experiments and process-level observations. Presently, there are few wetland warming experiments, and none of these manipulate soil temperature deeper than a few centimeters. Manipulative experiments should focus on observations that inform models (and vice versa) and seek to elucidate feedbacks between plant responses, microbial responses, nutrient cycling, and elevation change.

State of Mapping

PART II

State of Mapping

CHAPTER 12

Seagrass Mapping
A Survey of the Recent Seagrass Distribution Literature

Matthew P. J. Oreska and Karen J. McGlathery
University of Virginia

Robert J. Orth and Dave J. Wilcox
Virginia Institute of Marine Science

CONTENTS

HIGHLIGHTS

1. Despite extensive recent work on seagrass biogeography, comprehensive meadow distribution data is still lacking for particular species and in particular bioregions, especially the Temperate Southern Oceans.
2. Advances in aerial and satellite remote sensing allow researchers to map changes in meadow coverage over large areas in optically shallow waters; however, other methods are typically needed to map deeper waters, areas with optically similar substrate, and individual seagrass species in mixed species assemblages.
3. Declines in seagrass coverage are likely indicators of vulnerable blue carbon stocks in subaqueous soils.
4. Existing data provide meadow coverage baselines in many regions; however, assessing blue carbon stock changes will often require additional surveys to quantify changes in meadow extent.

12.1 INTRODUCTION

Information on meadow distribution is essential for calculating the blue carbon benefit provided by seagrasses. Blue carbon studies typically estimate seagrass carbon stocks by multiplying the average organic carbon (C_{org}) concentration in individual carbon pools—including above- and belowground biomass and bed sediment C_{org}—by meadow extent (e.g., Russell and Greening 2015). Seagrass habitats represent a significant sink in the global carbon cycle, because individual meadows rapidly bury C_{org} and because meadows cover an estimated 300,000 to 600,000 km^2 (Duarte et al. 2005; Kennedy et al. 2010; Mcleod et al. 2011; Fourqurean et al. 2012). At the global scale, seagrasses potentially sequester 4.2–19.9 Pg C_{org} (Fourqurean et al. 2012). Assuming a global meadow areal loss rate of 0.4%–2.6% year^{-1}, remineralization of this stock may release up to 330 Tg CO_2 year^{-1} back to the atmosphere (Duarte et al. 2005; Waycott et al. 2009; Pendleton et al. 2012). However, seagrass areal surveys remain incomplete (Duarte et al. 2008), despite recent advancements in seagrass remote-sensing techniques (Hossain et al. 2014). Knowledge about distribution varies by species, and many areas that host seagrasses are rapidly changing (Orth et al. 2006a; Walker et al. 2006; Short et al. 2011). Successfully managing the seagrass blue carbon stored in bed sediments and sequestered in plant biomass will, therefore, require accurate maps of meadow distribution, extent, species identity, and configuration changes over time.

Past studies have compiled information on seagrass biogeography (Short et al. 2001; Green and Short 2003; Short et al. 2007; Duarte et al. 2008; and references therein), but many assessments are incomplete with respect to particular species distributions (Short et al. 2001; 2011). Seagrasses include at least 12 genera representing four marine families (den Hertog and Kuo 2006), which are grouped into four temperate and two tropical bioregions (Short et al. 2007; 2011). Predominantly, marine species in two euryhaline families are also typically considered seagrasses, bringing total species diversity to 72 (Short et al. 2011). The Tropical Indo-Pacific is the most diverse bioregion (25 species), and the Temperate North Atlantic is the least diverse (five species) (Short et al. 2011). Indo-Pacific meadows often contain mixed species assemblages, whereas many meadows in the North Atlantic and temperate Pacific contain a single species, *Zostera marina*, which is also the only species to occur above the Article Circle; no seagrasses occur in Antarctica (den Hartog and Kuo 2006; Short et al. 2007; 2011). Tidal areas may host both euryhaline seagrasses and other, predominantly freshwater grass species, which are collectively referred to as submersed aquatic vegetation (SAV) (Orth et al. 2017). Recent advancements in airborne and space-based remote sensing are now providing detailed, updated maps of seagrass location and extent over large areas (Hossain et al. 2014); however, global distribution maps are still incomplete—even for well-studied species (Telesca et al. 2015).

The lack of detailed, baseline distribution maps for many species, coupled with dynamic fluctuations in natural meadow extent, complicate efforts to quantify seagrass loss caused by specific anthropogenic impacts (Spalding et al. 2003). Only a fraction of the world's coastlines has actually been surveyed for seagrasses (Green and Short 2003; Duarte et al. 2008), and many extent changes, including natural recovery, are not documented (Walker et al. 2006). Data syntheses, nevertheless, agree that global meadow loss far exceeds natural expansion (Orth et al. 2006a; Duarte et al. 2008). A recent study, based primarily on data from North America, Europe, and Australia, estimated that 29% of global seagrass meadow extent has been lost since initial observations in 1879, and loss rates appear to be accelerating (Waycott et al. 2009). Fifteen species have been identified as threatened or near-threatened according to the International Union for the Conservation of Nature's (IUCN) Red List criteria (Short et al. 2011). Many of these species occur in South Asia, Southeastern Africa, and around the Korean Peninsula, including two of the three listed as endangered due to their limited distributions, *Phyllospadix japonicas* and *Z. geojeensis*; the third, *Z. chilensis*, occurs in South

America (Short et al. 2011). Several widely distributed species categorized as least concern also showed marked population declines, including the Mediterranean species *Posidonia oceanica* and the temperate species *Z. marina* and *Z. noltii* (Short et al. 2011).

Managing seagrass meadows for blue carbon storage and sequestration will require regularly monitoring meadow extent, which will benefit efforts aimed at conserving seagrass habitats and the services they provide as ecosystem engineers (Maxwell et al. 2016; Nordlund et al. 2016). However, different seagrass mapping approaches may be more or less appropriate, depending on the spatial scale of interest. At the individual project level, managers must regularly account for small changes in meadow area (i.e., <1 ha) within a project area to quantify blue carbon stock changes over time (Emmer et al. 2015). In comparison, national greenhouse gas (GHG) inventory assessments may need to survey entire coastlines, but lower spatial resolution may be sufficient (IPCC 2014). All of these efforts will aid future IUCN seagrass status assessments, but is there enough existing information to establish current seagrass distribution baselines for prospective blue carbon managers?

This chapter provides an overview of recent information on seagrass distribution, including mapping efforts, factors that impact meadow distribution, and mapping techniques. We focus primarily on reports published since the global IUCN seagrass assessment (Short et al. 2011) to address two practical questions: (i) are geographic areas and species that have not received much past research focus being mapped? and (ii) are current mapping techniques able to capture meadow changes that affect blue carbon stocks? In addition, we evaluate the strengths, weaknesses, and potential applications of different mapping techniques for blue carbon projects.

12.2 RECENT SEAGRASS DISTRIBUTION STUDIES

Studies since Short et al. (2011) supply additional seagrass distribution data, obtained using a variety of methods (Table 12.1). Almost half (48%) of these studies use imagery acquired from satellite remote sensing to map meadow coverage (e.g., Kim et al. 2015; Bakirman et al. 2016). Another 10% mapped meadows using airborne multi- or hyperspectral imagery. Only two of these remote-sensing studies attempted to map multiple seagrass species separately within the same study area (Roelfsema et al. 2014; Koedsin et al. 2016). Others focused on mapping specific meadow characteristics, including density (Misbari and Hashim 2016b), biomass (Hashim et al. 2014), and leaf-area index (LAI) (Wicaksono and Hafizt 2013), by relating image pixel values to in situ measurements (e.g., Wicaksono and Hafizt 2013; Roelfsema et al. 2014; Lyons et al. 2015). In comparison, only five of the studies surveyed in Table 12.1 relied primarily on compiled literature data to generate distribution maps (e.g., Boström et al. 2014; Chefaoui et al. 2016).

Despite recent technological advancements that facilitate meadow mapping over large areas, knowledge gaps remain (Hossain et al. 2014). The United Nations Environment Programme-World Conservation Monitoring Centre (UNEP-WCMC) periodically updates its Global Distribution of Seagrasses dataset as new data become available (Figure 12.1: UNEP-WCMC and Short 2016), but a concerted effort to systematically map seagrasses is still lacking. Multiple, recent studies report seagrass distributions in previously under-surveyed areas in the Indian and Pacific Oceans, including Malaysia (e.g., Ooi et al. 2014; Hossain et al. 2015a,b), the Philippines (Fortes 2013; Mizuno et al. 2017), and Thailand (Reungsorn et al. 2015; Kakuta et al. 2016; Koedsin et al. 2016); however, studies are needed in parts of the Pacific and in the Temperate Southern Oceans, especially in South America and Africa (Table 12.1; Figure 12.1). Even among well-studied species, such as *P. oceanica*, surveys are often local in scale and only cover a small part of the species' total range (e.g., Bonacorsi et al. 2013; Vasilijevic et al. 2014). Telesca et al. (2015) identified long stretches of coastline in the Mediterranean where *P. oceancia* distribution data were lacking, including most

Table 12.1 Recent Seagrass Distribution Studies by Bioregion (cf. Short et al. 2007) and Survey Method

Bioregion	Country/Region	Location	Species	Technique	Reference
Temperate North Atlantic	Canada	Richibucto estuary, New Brunswick	Zostera marina	Satellite RS; Acoustic survey	Barrell et al. (2015)
	Scandinavia	Baltic Sea to Atlantic Ocean	Zostera marina	Literature review (including image database)	Boström et al. (2014)
	Spain	Bay of Biscay	Zostera noltii	Airborne RS	Valle et al. (2015)
	UK	Calshot; Ryde	Zostera marina, Zostera noltii, Posidonia oceanica	Acoustic survey	Paul et al. (2011)
	USA	Albemarle-Pamlico Sound, North Carolina	Zostera marina, Halodule wrightii, Ruppia maritima	Aerial photography	Uhrin and Townsend (2016)
	USA	Chesapeake Bay	Zostera marina	Aerial photography	Orth et al. (2017)
	USA	Delmarva coastal bays	Zostera marina	Aerial photography	Orth et al. (2012)
	USA	Great Bay Estuary, New Hampshire	Zostera marina	Airborne RS	Pe'eri et al. (2016)
Tropical Atlantic	Mauritania	Banc d'Arguin	Zostera noltii	Satellite RS	de Fouw et al. (2016)
	USA	Florida Bay, Florida	Syringodium filiforme (wrack)	Airborne RS	Dierssen et al. (2015)
	USA	Florida Bay, Florida	Thalassia testudinum	Airborne RS	Hedley et al. (2016)
	USA	Springs Coast, Florida	(not specified)	Satellite RS	Baumstark et al. 2016
	USA	Saint Joseph Sound and Clearwater Harbor, Florida	Syringodium filiforme, Thalassia testudinum, Halodule wrightii (some Halophila engelmannii)	Satellite RS	Pu and Bell (2017)
	USA	Saint Joseph's Bay, Florida	Thalassia testudinum, Halodule wrightii, Syringodium filiforme, Ruppia maritima, Halophila engelmannii	Airborne RS	Hill et al. (2014)
	USA	Redfish Bay, Texas	Thalassia sp., Halodule sp., Syringodium sp.	LiDAR	Pan et al. 2014
Mediterranean	Croatia	Murter Island	Posidonia oceanica	Underwater photography; Acoustic survey	Vasilijevic et al. (2014)
	France	Cap Corse, Corsica	Posidonia oceanica	Underwater photography; Aerial Photography; Acoustic survey	Bonacorsi et al. (2013)
	Italy	Oristano	Zostera marina, Zostera noltii, Posidonia oceanica	Acoustic survey	Paul et al. (2011)
	Mediterranean	Atlantic Ocean to the Black Sea	Cymodocea nodosa	Literature review and spatial modeling	Chefaoui et al. (2016)

(Continued)

Table 12.1 (Continued) Recent Seagrass Distribution Studies by Bioregion (cf. Short et al. 2007) and Survey Method

Bioregion	Country/Region	Location	Species	Technique	Reference
	Europe-North Africa	(regional distribution)	*Zostera noltii*	Habitat suitability modeling	Valle et al. (2014)
	Mediterranean	Mediterranean	*Posidonia oceanica*	Literature review	Telesca et al. (2015)
	Tunisia	Gulf of Tunis	*Posidonia oceanica, Cymodocea nodosa*	Satellite RS	Hachani et al. (2016)
	Turkey	Gulluk Gulf	*Posidonia oceanica*	Satellite RS	Bakirman et al. (2016)
Temperate North Pacific	Japan	Shinkawa-Kasugagawa Estuary	*Zostera marina*	Acoustic survey	Ruengsorn et al. (2015)
	Japan	Shizugawa Bay	*Zostera caulescens*	Acoustic survey	Hamana and Komatsu (2016)
	Japan	Funakoshi Bay	*Zostera caulescens, Zostera asiatica*	Satellite RS	Sagawa and Komatsu (2015)
	Korea	Jangheung Bay	*Zostera marina*	Satellite RS; Acoustic survey	Kim et al. (2015)
	USA	Willapa Bay, Sequim Bay, and Clinton, Washington	*Zostera marina*	Site surveys	Thom et al. (2014)
Tropical Indo-Pacific	Australia	Moreton Bay	*Halophila ovalis, Halophila spinulosa, Halodule uninervis, Zostera muelleri, Cymodocea serrulata, Syringodium isoetifolium*	Underwater photography; Satellite RS	Roelfsema et al. (2013; 2014)
	Australia	Moreton Bay	*Halophila ovalis, Halophila spinulosa, Halodule uninervis, Zostera muelleri, Cymodocea serrulata, Syringodium isoetifolium*	Satellite RS	Lyons et al. (2015)
	Australia	Great Barrier Reef	*Halophila, Halodule, Cymodocea, Zostera*	Literature; dispersal modeling	Grech et al. (2016)
	Australia	Shark Bay	*Amphibolis antarctica* (also *Posidonia australis, Halodule uninervis*)	Site surveys	Fraser et al. (2014)
	India/Myanmar	Andaman and Nicobar archipelago	(not specified)	Satellite RS	Paulose et al. (2012)
	Indonesia	Karimunjawa Islands	*Enhalus acoroides, Thalassia hemprichii, Cymodocea rotundata, Halodule uninervis, Syringodium isoetifolium, Halophila ovalis*	Satellite RS	Wicaksono and Hafizt (2013)
	Israel	Gulf of Aqaba	*Halophila stipulacea*	Underwater photography	Winters et al. (2016)
	Malaysia	Sungai Pulai Estuary	*Enhalus acoroides, Halophila ovalis, Cymodocea serrulata, Halodule uninervis*	Underwater photography; Satellite RS	Misbari and Hashim (2016a)
	Malaysia	Sungai Pulai Estuary	*Enhalus acoroides, Halophila ovalis* (also 8 other species)	Satellite RS	Hossain et al. (2015a)
	Malaysia	Pulau Tinggi	*Halophila spp., Halodule uninervis, Syringodium isoetifolium, Cymododea serrulata*	Underwater photography	Ooi et al. (2014)

(Continued)

Table 12.1 (Continued) Recent Seagrass Distribution Studies by Bioregion (cf. Short et al. 2007) and Survey Method

Bioregion	Country/Region	Location	Species	Technique	Reference
	Malaysia	Sarawak; Kelantan; Terengganu	*Thalassia hemprichii, Halophila minor, Halophila ovalis, Cymodocea rotundata, Halodule pinifolia*	Satellite RS	Hossain et al. (2015b)
	Malaysia	Merambong Shoal and Tinggi Island	(not specified)	Satellite RS	Misbari and Hashim (2016b)
	Malaysia	Merambong Shoals	*Enhalus acoroides, Halophila ovalis, Cymodocea serrulata, Halodule uninervis*	Satellite RS	Hashim et al. (2014)
	Philippines	Philippines	(multispecies)	Literature review	Fortes (2013)
	Philippines	Mayo Bay	*Halophila decipiens, Halophila minor, Halodule uninervis, Cymodocea serrulata, Syringodium isoetifolium*	Underwater photography	Mizuno et al. (2017)
	Singapore	Singapore	(multispecies)	Satellite RS	Yaakub et al. (2013)
	Thailand	Patong Beach; Tang Ken Bay	(not specified)	Satellite RS	Kakuta et al. 2016
	Thailand	Paklok Bay	*Halodule uninervis, Halophila beccarii, Halophila ovalis, Cymodocea rotundata, Cymodocea serrulata, Thalassia hemprichii, Enhalus acoroides*	Underwater photography; Satellite RS	Koedsin et al. (2016)
	Thailand	Prasare Estuary	*Halodule pinifolia*	Site surveys	Ruengsorn et al. (2015)
	Vietnam	Cam Ranh Bay	*Enhalus acoroides, Halophila ovalis, Halophila minor, Thalassia hemprichii, Halodule pinifolia, Halodule uninervis, Ruppia maritima*	Satellite RS	Chen et al. (2016)
	Western Pacific	Western Pacific	*Thalassia hemprichii, Cymodocea rotundata, Halophila ovalis, Halophila uninervis, Enhalus acoroides, Cymodocea serrulata, Syringodium isoetifolium*	Site surveys	Short et al. (2014)
	Western Pacific	Coral Triangle	(not specified)	Satellite RS	Torres-Pulliza et al. (2013)
Temperate Southern Oceans	Australia	Port Phillip Bay	*Zostera nigricaulis, Zostera muelleri, Halophila australis, Amphibolis antarctica*	Aerial Photography	Ball et al. (2014)

RS, remote sensing.

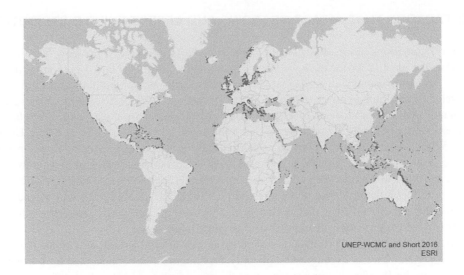

Figure 12.1 **(See color insert following page 266.)** Global seagrass distribution data (green) compiled by the United Nations Environment Programme-World Conservation Monitoring Centre (UNEP-WCWM and Short 2016; the data can be accessed via the Ocean Data Viewer: http://data.unep-wcmc. org/datasets/7; Mediterranean data collection supported by the European Commission, EUR 568,996; data used with permission).

of North Africa and parts of the Greek isles. None of the identified studies in Table 12.1 provided distribution maps for the data deficient species that Short et al. (2011) were not able to characterize in their IUCN assessment. Occurrences for a few of these species have been noted but not mapped, for example, *Lepilaena* spp. (Nordlund et al. 2016).

Approximately one-fifth of the studies in Table 12.1 reported meadow distribution changes; the remainder presented static meadow assessments. The meadow changes were determined from remote imagery (e.g., Kim et al. 2015), repeated site surveys (Fraser et al. 2014; Short et al. 2014; Thom et al. 2014), or compiled from existing reports (Boström et al. 2014; Telesca et al. 2015). Assessing past change over decadal timescales can be accomplished by comparing images collected by multiple satellite sensors (e.g., Hossain et al. 2015b; Kim et al. 2015). A few aerial photograph datasets go back much further, for example, the dataset that Ball et al. (2014) used to assess seagrass meadow change since 1939 in Port Phillip Bay, Australia. Although complete distribution information is still lacking, the aforementioned studies suggest that meadow declines are common, geographically widespread, and not restricted to particular species. For example, Telesca et al. (2015) recently confirmed meadow declines of approximately 34% over the past 50 years for *P. oceanica*. Short et al. (2014) documented multispecies declines over a large area in the Western Pacific from 2001–2009.

Seagrass distribution datasets created by individual studies often become publically available and can aid prospective blue carbon managers. Data compiled by the Global Distribution of Seagrasses dataset can be accessed via the Ocean Data Viewer on the UNEP-WCMC webpage (Figure 12.1: UNEP-WCMC and Short 2016). Additional information on seagrass monitoring is available from the Global Seagrass Monitoring Network (SeagrassNet 2010) and the *Zostera* Experimental Network (ZEN 2017). Detailed local and regional distribution maps can also be obtained from regional monitoring networks, often as downloadable geographic information system data layers (e.g., NOAA 2014), but note that some of the available datasets have not been updated in recent years (e.g., SeagrassNet 2010).

12.3 FACTORS THAT IMPACT MEADOWS AND AFFECT THEIR DISTRIBUTION

Natural fluctuations in meadow extent complicate efforts to establish meadow baseline distributions (Walker et al. 2006). Meadows often expand and contract around a core area, reflecting propagule survival dynamics in areas that vary with respect to water depth, temperature, salinity, current velocity, and tidal exposure (Short et al. 2001; Valle et al. 2013). Clonal expansion and seed recruitment also allow meadows to recolonize natural disturbances from ice scour, storm blowouts, pore water chemistry, and grazing (Olesen and Sand-Jensen 1994; Borum et al. 2014; Unsworth et al. 2015). However, some species spread slowly, making them susceptible to large, infrequent disturbance events (Duarte et al. 2006; Walker et al. 2006). In many areas, multiple anthropogenic stressors contribute to disturbance and complicate recovery (Spalding et al. 2003). Seagrass meadows are prone to sudden collapse when stressed (van derHeide et al. 2007), which leads to bed erosion and the remineralization of sediment C_{org} (Marbà et al. 2015). Consequently, blue carbon managers may need to monitor meadows over short time intervals.

Grech et al. (2012) recently ranked urban/industrial run-off and coastal development as the primary anthropogenic threats to seagrasses, globally. Development has contributed to severe population declines in the Pacific; in comparison, eutrophication from agricultural runoff was ranked as a bigger threat in the North Atlantic (Grech et al. 2012). Many species become light limited below 10 m, making them particularly susceptible to increased turbidity (Short et al. 2011). Other major, ongoing threats include aquaculture, dredging, and warming oceans (Orth et al. 2006a; 2017; Valle et al. 2014). Although several studies suggest that seagrasses may actually benefit from ocean CO_2 enrichment resulting from global change (Russell et al. 2013; Zimmerman et al. 2015), this is far from certain (cf. Arnold et al. 2012). Direct seagrass harvesting is not a major global concern but remains common in parts of the Asia-Pacific region (Yamamuro et al. 2006; Grech et al. 2012). Particular meadow declines have also been attributed to invasive species and disease outbreaks, including wasting disease (Muehlstein et al. 1991; Peirano et al. 2005). Blue carbon projects should adopt a monitoring program capable of capturing meadow extent changes resulting from specific threats within their project area. As noted in the following section, some seagrass mapping techniques are better suited to different spatial and temporal scales than others.

12.4 MAPPING TECHNIQUES

12.4.1 Ground Surveys

Several of the parameters necessary for blue carbon stock mapping must be obtained directly from the meadow(s) of interest, including sediment carbon concentrations. Fourqurean et al. (2014) outline methods for collecting sediment cores for carbon assessment. Although recent studies attempt to map seagrass biomass at large spatial scales using satellite imagery (Hashim et al. 2014; Lyons et al. 2015), blue carbon projects should also collect in situ above- and belowground biomass samples for biomass stock estimation. Determining appropriate average values for each of these parameters for scaling purposes may require direct sampling over large areas, given possible non-random spatial variability within and among meadows in the project area (Oreska et al. 2017). Several other variables that factor into stock estimation are also relatively easy to acquire through direct, repeated measurements taken within a meadow, including shoot density, LAI, and productivity (Short and Duarte 2001). Blue carbon projects seeking VCS offset-credits for seagrass restoration must also account for CH_4 release when calculating the net GHG benefit (Emmer et al. 2015), which will likely entail using in situ chambers or collecting bed sediment cores for ex situ incubations.

In particular cases, managers may also use ground surveys to determine meadow extent for stock scaling. Historically, managers often relied on multiple direct observations, sometimes collected over a period of years, to generate seagrass distribution maps (e.g., Orth and Moore 1983). More recent studies also report data compiled using direct observation methods, ranging from SCUBA to towed video cameras to cameras mounted on remotely operated vehicles (e.g., MacKenzie et al. 2001; Bonacorsi et al. 2013). Although many studies now rely on remote sensing to generate meadow distribution maps (Table 12.2), direct survey methods may be necessary to map meadows in deep waters and in optically deep waters, where incident light does not reflect back to a camera or remote sensor from the bottom, irrespective of actual water depth (Dekker et al. 2006).

12.4.2 Aerial Photography

Film-based aerial photography has traditionally been the preferred method for mapping seagrass meadows, given the high spatial resolution and broad coverage potentially available from images (MacKenzie et al. 2001). Cameras can be mounted on a variety of aerial platforms, ranging from balloons to drones to fixed-wing aircraft. Balloons and drones provide low-cost options but have limited range, which may necessitate multiple surveys to cover all of the meadows in a region. The spectral properties and intensity of images acquired over multiple time periods will differ, given different levels of ambient light, weather conditions, and lack of calibration, which complicates composite image analysis. Balloons and drones are also subject to stability issues. A level flying aircraft can acquire a series of images with minimal geometric distortion. By traversing a flight path that travels back and forth over a project area, airplanes can also capture images that overlap slightly in a single survey, facilitating image compilation. McKenzie et al. (2001) recommend taking images from an appropriate elevation, for example, 1,800 m to provide 1:12,000 map resolution. The specific resolution should be linked to the questions being addressed by a particular project.

Aerial photography has some limitations. Common issues affecting aerial images include sun glare, turbidity, waves, and wind (Ball et al. 2014). Small seagrass patches can be hard to distinguish in aerial images, particularly when macroalgae and diatomaceous mats also appear dark in aerial photos (Uhrin and Townsend 2016). Film-based aerial cameras are being replaced by digital versions that generate images comprised of blue, green, red, and near-infrared bands, which allow some classification but provide less information than most digital remote sensors. Some new analyses for aerial photographs are also available. Uhrin and Townsend (2016) recently used linear spectral unmixing to map seagrass bed coverage. Despite these advancements, seagrass density remains difficult to estimate from aerial photographs, which are not radiometrically calibrated (Hill et al. 2014).

12.4.3 Airborne Remote Sensing

Over the past three decades, seagrass researchers have attempted to adapt remotely acquired multispectral and hyperspectral images for seagrass mapping purposes (McKenzie et al. 2001; Hossain et al. 2014). Multispectral images typically contain several 20–60 μm bands, whereas bands in hyperspectral images are frequently 2–20 μm in width (Dekker et al. 2006). Subtidal environments present several unique challenges for remote sensing, including rapid attenuation of red-edge wavelengths (within approximately 2 m depth), variable light reflectance, scattering, and absorption within the water column, and variable bathymetry (Dekker et al. 2006; Pe'eri et al. 2016). The radiometric properties of water, including radiance and irradiance, can be modeled using radiative transfer theory (Dekker et al. 2006). Zimmerman and Dekker (2006) explain how the loss of radiant energy from light scattering and absorption in both air and water, along with potential energy gains from scattering along the signal detection path, can be approximated linearly. However, dynamic coastal environments present several additional challenges that complicate this process. Correcting for Fresnel reflectance at the water surface and light refraction across the air-water interface is relatively straightforward for calm, clear water but challenging for rough waters (Dekker et al. 2006). Observers must also account for phytoplankton, sediment, and high concentrations of colored, dissolved organic matter in the water column, which are

Table 12.2 Scanners and their Recent Applications for Seagrass Mapping; Coverage Refers to Seagrass Spatial Distribution, Habitat Types Refers to Seagrass and Other Benthic Habitats

	Sensor	Mapping Application	References
Airborne RS	PRISM	Wrack coverage; LAI	Dierssen et al. (2015); Hedley et al. (2016)
	SAMSON	LAI; habitat types	Hill et al. (2014)
	CASI	Habitat types (including two seagrass coverage classes)	Valle et al. (2015)
	AISA Eagle	Habitat types	Pe'eri et al. (2016)
Satellite RS	AVNIR-2	LAI	Wicaksono and Hafizt (2013)
	ASTER	LAI; coverage; coverage change	Torres-Pulliza et al. (2013); Wicaksono and Hafizt (2013); Kim et al. (2015)
	Quickbird-2	Coverage; biomass	Roelfsema et al. (2014); Barrell et al. (2015); Lyons et al. (2015)
	Landsat 5 TM	Percent cover; coverage change	Hossain et al. (2015b); Kim et al. (2015); Chen et al. (2016); Misbari and Hashim (2016a)
	Landsat 7 ETM+ (with and without Scan Line Corrector)	Coverage; coverage change; habitat types	Hossain et al. (2015a); Hossain et al. (2015b); Kim et al. (2015); Chen et al. (2016)
	Landsat 8 OLI	Percent cover; coverage; coverage change; density categories; habitat types; biomass	Hashim et al. (2014); Hossain et al. (2015b); Bakirman et al. (2016); Chen et al. (2016); Kakuta et al. (2016); Misbari and Hashim (2016a); Misbari and Hashim (2016b)
	IKONOS	Coverage; percent cover	Roelfsema et al. (2014); Sagawa and Komatsu (2015); Pu and Bell (2017)
	Worldview-2	Biomass; habitat types; percent coverage categories; species maps	Yaakub et al. (2013); Roelfsema et al. (2014); Lyons et al. (2015); Baumstark et al. (2016); Koedsin et al. (2016)
	IRS IC/ID, I	Coverage change	Paulose et al. (2012)
	IRS P6	Coverage change	Paulose et al. (2012)
	SPOT 4,	Coverage change	Kim et al. (2015)
	SPOT 5 (Google Earth)	Habitat types	Hachani et al. (2016)
	Kompsat-2;	Coverage change	Kim et al. (2015)
Acoustic survey	BioSonics DE-X 430 kHz, single-beam sonar	Present cover (track lines)	Barrell et al. (2015)
	Sonic 2024 narrow multibeam sonar system	Relative abundance categories	Hamana and Komatsu (2016)
	EM 1000 multibeam echosounder; Klein 3000 side-scan sonar	Habitat types	Bonacorsi et al. (2013)
	SportScan-Imagenex sidescan sonar	Coverage	Vasilijevic et al. (2014)
	Fish finder echo-sounder	Small patch distribution	Ruengsorn et al. (2015)
	DT-X digital echosounder	Coverage change	Kim et al. (2015)
	Sediment Imaging Sonar	Habitat types (coarse seagrass density categorization); Coverage estimates (track lines)	Paul et al. (2011)

RS, remote sensing; LAI, leaf area index.

heterogeneous in composition and have different effects on different light wavelengths (Dekker et al. 2006; Hill et al. 2014). Additional background on remote sensing considerations in complex coastal environments can be found in a special issue of *Limnology & Oceanography* (Vol. 48, issue 1 part 2) and in a recent review (Hossain et al. 2014).

Seagrass canopy structural complexity presents additional challenges that must be addressed when interpreting hyperspectral and multispectral images. Shoot density, canopy shading, leaf orientation, and epiphyte coverage hinder simple linear solutions when calculating the expected scatter and absorption of light encountering a seagrass bed and reflecting back to a sensor (Zimmerman 2006). This complexity can be constrained by measuring upwelling and downwelling irradiance and a variety of specific meadow parameters in situ, which allow projects to model expected light reflectance from a canopy (Zimmerman 2003; Dekker et al. 2006). Zimmerman and Dekker (2006) provide a more detailed review of hydrologic optics, including apparent and inherent optical properties of natural waters that affect sunlight reflectance by submerged grass beds.

Recent advances make airborne remote sensing of seagrass meadows increasingly practical (Hossain et al. 2014). Many airborne remote sensors now include instruments for radiometric and geometric registration to facilitate image correction using post-flight correction software (Dekker et al. 2006). Common hyperspectral sensors include CASI and HyMap (Dekker et al. 2006; Pu and Bell 2017). Several sensors, including AVIRIS, SAMSON, HICO, and PHILLS, are radiometrically calibrated (Hill et al. 2014). Spectra in hyperspectral images can be related to particular benthic habitat types, including seagrass meadows, macroalgae-covered areas, and benthic microalgae, provided the images capture a brightness range with sufficient low-end resolution (Stephens et al. 2003; Dekker et al. 2006; Pe'eri et al. 2016). Hyperspectral images with enough individual bands can sometimes be used to identify individual seagrass species using spectral signatures (Fyfe 2003).

In addition to species presence and absence mapping, high-resolution hyperspectral images can potentially provide additional information about meadow structure and health, which may be useful to managers. Several studies use hyperspectral images to calculate seagrass LAI (Dierssen et al. 2003; Hedley et al. 2016) or parse meadows into coverage classes. Valle et al. (2015) explain how combining red and infrared reflectance from CASI images into a Normalized Difference Vegetation Index (NDVI) can be used to discriminate between dense and sparse seagrass patches.

Pan et al. (2014) suggest that it may also be possible to use airborne, full waveform LiDAR to map seagrass canopies in very shallow systems, but this remains uncommon, because of difficulties distinguishing meadow canopies from background echoes. LiDAR is however used to generate bathymetry maps, which are essential for radiometric correction of remotely sensed images.

12.4.4 Satellite Remote Sensing

Satellite-based remote sensors regularly capture images over much larger areas but often with lower spatial resolution. Multiple existing satellite image datasets, nevertheless, provide seagrass managers with images going back more than three decades (Hossain et al. 2014). The Landsat Multispectral Scanner and SPOT satellites began capturing images in the 1970s and 1980s, respectively. More recent satellite images are available from IKONOS, QuickBird, Landsat 7 Thematic Mapper, Landsat 8 OLI/TIRS, SPOT 5, and ASTER (Dekker et al. 2006). Seagrass studies have developed analytical techniques to facilitate image categorization from these sensors. For example, Hossain et al. (2015b) employed several mapping indices, including NDVI, the Normalized Difference Water Index (NDWI), and Normalized Difference Turbidity Index (NDTI), which helped them identify seagrass coverage in Landsat images going back to 1988. Barillé et al. (2010) generated seagrass biomass maps from satellite imagery by relating NDVI to measurements of leaf dry weight.

Satellites map areas with relatively high frequency, which allows for assessment of seasonal changes, but the image resolution available from satellite sensors may be insufficient for identifying small seagrass patches. The broad bandwidth of multispectral bands captured by particular sensors are typically not able to resolve differences between seagrass and other types of underwater vegetation, including macroalgae (Pe'eri et al. 2016). In such cases, airborne hyperspectral mapping may be necessary (Dekker et al. 2006). Chen et al. (2016) compared several methods and determined that aerial photography and remote sensing (with CASI) provided more accurate seagrass maps than Landsat TM/ETM but that the SPOT XS/5 images provided the highest mapping accuracy.

Despite recent advancements, studies note persistent difficulties using remotely sensed imagery from both airborne and space-based platforms. Turbidity remains a challenge (Kim et al. 2015), and epiphytes on seagrass leaves complicate image analysis, albeit in ways that are predictable and can be constrained (Drake et al. 2003; Hedley et al. 2016). Identifying the meadow edge in deeper areas is also difficult, because the change in bathymetry necessitates the use of different attenuation coefficients for image analysis (Misbari and Hashim 2016b).

12.4.5 Acoustic Surveys

A variety of side-scan sonar and multibeam echosounders facilitate seagrass bed mapping in deeper waters not suitable for aerial or space-based mapping (Table 12.2). Although multibeam transducers can scan a fairly wide (e.g., 150°) swath of seafloor (Hamana and Komatsu 2016), most acoustic sensors have a restricted sweep angle that can only scan a narrow area, requiring multiple passes through a meadow to generate a distribution map (Lefebvre et al. 2009; Paul et al. 2011; Barrell et al. 2015). Sonar and other acoustic techniques can, nevertheless, retrieve aspects of meadow structure that may interest managers that are difficult to obtain from aerial and satellite images, including canopy height (Ruengsorn et al. 2015) and finer-scale estimates of percent cover (Paul et al. 2011; Barrell et al. 2015). Studies that use sidescan sonar also commonly use underwater photography (e.g., Vasilijevic et al. 2014).

12.5 CASE STUDY: MAPPING RESTORED *ZOSTERA MARINA* BEDS IN VIRGINIA

As the following case study illustrates, aerial imagery is often necessary to accurately quantify changes in meadow extent over areas of interest to environmental managers, given the potential complexity of regional meadow distribution changes over short timescales. This case study focuses on aerial photography, but note that remotely sensed images can provide many of the same benefits for prospective blue carbon managers.

Z. marina in the southern coastal bays along the Atlantic side of the Delmarva Peninsula, USA, disappeared during the North Atlantic Eelgrass Pandemic circa 1932 (Orth et al. 2006b; 2010). By the mid-20th century, SAV, including *Z. marina* and *Ruppia maritima* meadows, was disappearing throughout the Chesapeake Bay and its tidal tributaries due to eutrophication resulting from urban and agricultural runoff (Orth and Moore 1983). Comparing in situ and aerial surveys conducted in the 1960s and 1970s with older aerial photographs revealed the scope and severity of the SAV decline (Orth and Moore 1984). Beginning in 1983, policies aimed at restoring Chesapeake Bay water quality were enacted, which helped facilitate *Z. marina* conservation and restoration efforts (Orth et al. 2010; 2017). Reseeding was also conducted in the Atlantic coastal bays to restore the meadows lost to wasting disease during the 1932 pandemic (Orth et al. 2006b; 2012).

These restoration efforts have achieved notable success, particularly in the Atlantic coastal bays, where more than 21 km² of eelgrass have been restored (Orth et al. 2012). These restored coastal bay meadows now represent the single largest, successful seagrass restoration effort on the planet. However, the total meadow area in the Chesapeake region exhibits annual variability and remains well below the restoration goal, 748 km² (Figure 12.2).

Annual aerial surveys throughout the Chesapeake and Delmarva coastal bays allow researchers and managers to track the SAV recovery, including fluctuations in areal extent (Moore et al. 2009; Orth et al. 2010). Aerial imagery documenting seagrass distribution has been acquired annually since

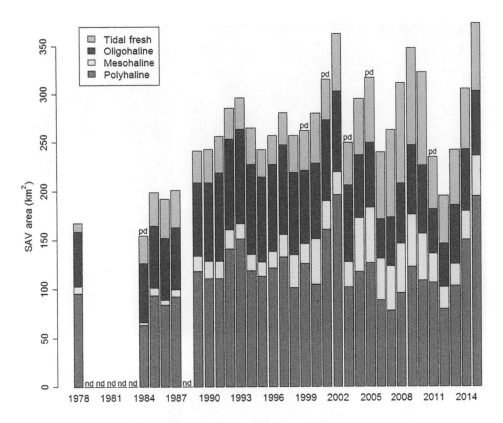

Figure 12.2 Annual fluctuations in SAV area in the Chesapeake Bay by salinity zone, determined from aerial photographs taken by the Virginia Institute of Marine Science (VIMS) SAV Monitoring Program (the regional SAV restoration goal = 748 km²).

1984 to quantify SAV area, with the exception of 1988 (Figure 12.2), along flight lines covering shallow water areas throughout the region (Figure 12.3, inset). In 2014, traditional panchromatic film photography, flown at 3,658 m, was partially replaced with a digital solution and then fully replaced in 2016. The new digital imagery was acquired using a ZI DMC-II 230 multispectral (RGB,NIR) digital mapping camera and IMU at an approximate altitude of 4,023 m, yielding a ground sample distance of approximately 24 cm (Orth et al. 2016). Manual interpretation onto transparent mylar sheets and subsequent heads-down digitizing was used to monitor seagrass distribution until 2001 when a switch was made to in-house orthorectification and on-screen interpretation. Additional low-level imagery of the restoration sites in the Atlantic coastal bays was acquired during the first years (2001, 2003, 2004, & 2006–2010) of the restoration to capture a detailed record of seagrass expansion.

Comparing images over successive years shows how the single largest restored *Z. marina* meadow, located in South Bay, Virginia, expanded naturally in a rapid and irregular fashion from the areas that were reseeded between 2001–2004 (Figure 12.3). Most of this expansion occurred to the southwest, suggesting a dominant role for current-driven seed dispersal. By identifying meadow presence/absence in each successive year of imagery, one can ascertain the blue carbon accumulation time at individual meadow sites with 1-year precision. Oreska et al. (2017) used this information to show that sediment blue carbon concentrations vary with site location relative to the meadow edge and that this spatial variability overshadows concentration differences attributable simply to accumulation time.

Aerial images also provide indications of local meadow changes that could threaten meadow stability or influence blue carbon pools that are not readily apparent from individual ground surveys. Examples include boat propeller scars, blowouts, unexpected distribution changes, and algal

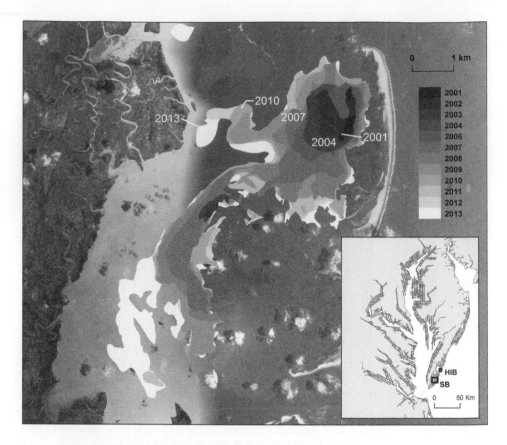

Figure 12.3 (See color insert following page 266.) Annual expansion of the restored *Zostera marina* meadow in South Bay (SB), VA, determined from aerial photographs; the inset map shows the SB and Hog Island Bay (HIB: Figure 12.4) study areas relative to the VIMS SAV Monitoring Program flight-lines (red lines) in the Chesapeake Bay and Delmarva coastal bays.

occurrences. Images taken in 2011 showed an extensive dark area that covered most of the restored South Bay meadow, which ground surveys identified as a benthic diatom mat. The fact that diatoms sometimes occur in abundance throughout the meadow may explain why most of the sediment carbon at a mid-meadow site appeared to derive from benthic microalgae (Greiner et al. 2016). Aerial imagery also confirmed that the restored *Z. marina* meadow in Hog Island Bay, has migrated northeastward over time and no longer encompasses many of the original restoration plots (Figure 12.4).

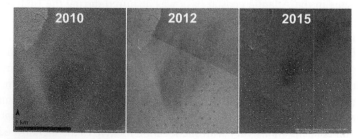

Figure 12.4 (See color insert following page 266.) The northeastward migration of the restored *Zostera marina* meadow in Hog Island Bay (HIB), VA, is discernable from aerial imagery but not ground survey sites (white circles) located in original restoration seed plots (see Figure 12.3 inset for HIB meadow location).

Consequently, the carbon stock cannot be estimated simply by scaling the current meadow area by a single carbon burial flux dating back to the restoration start date.

12.6 SEAGRASS MAPPING FOR BLUE CARBON ASSESSMENT

12.6.1 Seagrass Distribution Changes and Blue Carbon

Most blue carbon projects will focus on preserving or increasing meadow extent to mitigate CO_2 emissions resulting from seagrass habitat conversion (Pendleton et al. 2012). Studies that document past extent changes over time permit assessment of meadow loss (or expansion) rates, which can inform prospective blue carbon projects. For example, Kim et al. (2005) documented a 23-year seagrass decline using satellite imagery in a bay in South Korea, an area that Short et al. (2011) identified as an area of seagrass conservation concern. Existing studies that present a static map of meadow extent in an area provide a starting point for blue carbon monitoring, but projects will need to resurvey these meadows if they want to conduct a stock change assessment.

Uncertainty about the current seagrass distribution in under-surveyed regions places potential blue carbon projects at an initial disadvantage. In the absence of distribution maps, managers must either attempt to delineate meadows from existing satellite or aerial imagery or conduct their own surveys to establish a meadow status baseline. This is particularly important for projects that wish to generate voluntary offset-credits for meadow conservation. These credits will be allocated for preventing meadow declines relative to an extrapolated, business-as-usual, baseline scenario. Project managers must determine meadow areal loss rates within their project area prior to undertaking project activities.

Seagrass restoration projects do not need to establish a pattern of meadow decline to receive voluntary offset-credits, but these projects must stratify the project area to exclude existing meadows when calculating the GHG benefit from the project (Emmer et al. 2015). Tracking meadow recovery requires repeated regional surveys to identify changes in meadow extent over the course of a project. Managers seeking to document blue carbon accumulation over time must verify that the restored meadow(s) remain in the same location and general configuration. If meadows regularly change configuration or migrate, the sediment carbon stock in one area may be remineralized, while the stock in another area may be recent and, therefore, insignificant—even though total meadow area remains unchanged. These same considerations hold for managers seeking to track changes in blue carbon stocks for GHG inventory assessments.

Anticipating future meadow extent using habitat suitability modeling can aid managers. Several studies assess possible seagrass extent for conservation purposes (e.g., Adams et al. 2015). Refer to Valle et al. (2013) for a comparison of different distribution techniques for seagrass mapping, including both machine learning and regression-based models. Most studies apply habitat models to analyze potential seagrass distribution at regional scales (e.g., Downie et al. 2013), but these models can potentially be applied to much larger spatial scales, provided that the environmental parameters used to define the species' niche can be mapped at the larger spatial scale (Valle et al. 2014).

12.6.2 Selecting the Appropriate Mapping Technique to Assess Blue Carbon Stock Changes

Projects attempting to assess seagrass blue carbon stocks will likely need to consider tradeoffs between mapping cost and image resolution, especially those projects that expect to receive financing from carbon offset-credits. According to Dekker et al. (2006), airborne remote sensing using hyperspectral sensors potentially provides the highest level of information about a seascape, including relatively high-resolution images suitable for spectral discrimination among seagrasses, macroalgae, and benthic microalgae. However, projects must anticipate flight costs and acquire the expertise necessary to accurately correct and analyze images (Dekker et al. 2006). Expertise in radiometric correction is not necessary to utilize aerial photographs, but they record less information with fewer bands. Satellite image datasets can be acquired at little or no cost to projects and may provide sufficient resolution to map contiguous meadows; however, some satellite datasets have a minimum resolution of $2.5\,m^2$ or greater, which may be too coarse to monitor low-density species or identify species in multi-species beds and mixed seagrass-macroalgal assemblages (Dekker et al. 2006). A few recent studies attempt to map seagrass distribution using imagery compiled by Google Earth, which Amran (2017) suggests may be of sufficiently high resolution to distinguish meadows from corals and other bottom types. However, Torres-Pulliza et al. (2013) also used Google Earth images for seagrass mapping and noted that the images varied in quality.

The requisite mapping method for a given project will depend on the spatial-scale of the project area (MacKenzie et al. 2001) and the frequency of monitoring necessary to account for seasonal patterns, meadow expansion rates, and recovery rates from episodic disturbance (Dekker et al. 2006; Walker et al. 2006). Managers interested in meadow conservation should establish a monitoring regime based on the frequency of disturbances, both natural and anthropogenic, relative to the pace of natural meadow recovery. Less frequent monitoring may be sufficient for slow growing species. If managers need to assess changes on temporal scales shorter than one year, such as seasonal loss of aboveground biomass, managers may opt to use satellite images together with occasional ground surveys to calibrate biomass distribution maps. Managers who do not need to monitor frequent or fine-scale meadow extent changes may be able to conduct less frequent surveys and rely more heavily on imagery from Google Earth or another freely accessible source. Many blue carbon projects will likely employ a combination of methods (e.g., Kim et al. 2015; Lyon et al. 2015; Ruengsorn et al. 2015), depending on the information needed at different points over the life of the project and the availability of resources. Detailed meadow distribution maps collected repeatedly to aid blue carbon assessments will, in turn, benefit seagrass biologists interested in global seagrass biogeography and the conservation status of individual species.

ACKNOWLEDGMENTS

The authors wish to thank the European Commission and the UNEP-WCMC for permission to use seagrass distribution data and John Porter for helpful feedback on this study. This work was supported by National Science Foundation Virginia Coast Reserve Long Term Ecological Research Grant DEP-1237733, the University of Virginia Jefferson Scholars Foundation, and by Virginia Sea Grant. This is Contribution No. 3725 of the Virginia Institute of Marine Science, The College of William and Mary.

CHAPTER 13

Mapping and Monitoring of Mangrove Forests of the World Using Remote Sensing

Chandra Giri
U.S. Environmental Protection Agency

CONTENTS

HIGHLIGHTS

1. Mangrove forests are found in the intertidal zone of the tropical and subtropical regions of the world and are in constant flux due to both natural and anthropogenic forces.
2. At present, conversion of mangroves to other cover types is the dominant factor responsible for the change; however, loss to climate change (e.g., sea level rise) is becoming increasingly dominant.
3. Mapping and monitoring of the distribution and dynamics of mangroves is central to a wide range of scientific investigations conducted in both terrestrial and marine ecosystems.
4. Recent advancement in remote-sensing data availability, image-processing methodologies, computing and information technology, and human resources development have provided an opportunity to observe and monitor mangroves from local to global scales on a regular basis.
5. The spectral and spatial resolution of remote-sensing data and their availability has improved making it possible to observe and monitor mangroves at unprecedented spatial and thematic details.

13.1 INTRODUCTION

Mangrove forests provide important ecosystem goods and services to the world's dense coastal population and support important biosphere functions. Deforestation and degradation of mangrove forests can lead to the reduction of important ecosystem goods and services and impair critical biosphere functions (e.g., coastal protection, carbon sequestration, and biodiversity conservation) at both local and global scales. However, the forests are under threat from both natural and anthropogenic forces, thus threatening the resilience and vitality of global coastal social-ecological systems.

Mangrove forests are highly productive ecosystems sequestering more carbon per unit area than any other tropical systems (Donato et al. 2011; Mcleod et al. 2011). They have a mean whole-ecosystem carbon stock of 956 t C ha^{-1}, compared with 241 t C ha^{-1} for tropical rain forests, 408 t C ha^{-1} for peat swamps, 593 t C ha^{-1} for salt marshes, and 142.2 t C ha^{-1} for seagrasses. The soil carbon constitutes ~75% of the carbon pool. Although mangroves occupy less than 1% of the global coastal area, they contribute 10%–15% (24 Tg C year^{-1}) to coastal (Mcleod et al. 2011) sediment carbon storage and export 10%–11% of the particulate terrestrial carbon to the ocean. According to the best available estimate, 0.15–1.02 Pg (billion tons) of carbon dioxide are being released annually, resulting in economic damages of $US 6–42 billion annually (Pendleton et al. 2012). Their disproportionate contribution to carbon sequestration is now perceived as a means for conservation and restoration to help ameliorate greenhouse gas emissions (adaptation) and enhance the delivery of other ecosystems goods and services (Duarte et al. 2013). However, uncertainty in the estimation of deforestation, forest degradation, and carbon stock change in the mangrove forests is very high (Mcleod et al. 2011).

Despite the importance of mangrove forests, reliable, accurate, and timely information on mangrove forests cover change of the world is not available. Remote Sensing could play an important role in providing this information. Recent advancement in remote-sensing data availability, image-processing methodologies, computing and information technology, and human resources development have provided an opportunity to observe and monitor mangroves from local to global scales on a regular basis. Spectral and spatial resolution of remote-sensing data and their availability has improved making it possible to observe and monitor mangroves with unprecedented spatial and thematic detail. Novel remote-sensing platforms such as unmanned aerial vehicles, and emerging sensors such as Fourier transform infrared spectroscopy and Lidar can now be used for mangrove monitoring. Furthermore, it is now possible to store and analyze large volume of data using cloud computing.

To increase our scientific understanding of the distribution and dynamics of mangrove forests, scientists are employing new and novel image interpretation and classification techniques. Space agencies such as the National Aeronautics and Space Agency (NASA), the European Space Agency (ESA), Indian Remote Sensing (IRS), and the Japan Aerospace Exploration Agency (JAXTA) are interested in how the remote-sensing technologies are being utilized and their impact on solving complex environmental problems. Country-specific needs include the availability and accessibility of a timely and accurate database of mangrove changes, needed for resources planning, management, and reporting to international treaty and conventions. In some cases, data generation at sub-regional (e.g., Mekong region), national, or sub-national levels is facilitated by donor agencies such as USAID, and NORAD.

Mangrove forests are distributed in the inter-tidal region between sea and land in the tropical and subtropical regions of the world largely between 30°N and 30°S latitude. A satellite-based global inventory determined the total mangrove forest area of the world in the year 2000 was 137,760 km^2 in 118 countries and territories, accounting for less than 1% of total tropical forests of the world (Figure 13.1) (Giri et al. 2011b). This database represents the first reliable, comprehensive, and

Figure 13.1 **(See color insert following page 266.)** Mangrove forest distributions of the world for circa 2000 based on Landsat 30 m data.

globally consistent yet locally relevant high-resolution (30 m) database ever created. Previous estimates of global mangrove extent, based on the compilation of disparate geospatial data sources and national statistics, ranged from ~110,000 to 240,000 km^2 (Giri et al. 2011b). This estimate uncertainty can be attributed to data spatial and temporal variability.

Mangroves possess a very distinct spectral signature in remotely sensed data, particularly in the spectral range corresponding visible red, near infrared, and mid infrared, thus making it easier to classify compared to other land cover types (Figure 13.2).

13.2 GLOBAL DISTRIBUTION

Mangrove forests, consisting of multiple taxa of tropical macrophytes, are distributed mainly in tropical and subtropical regions of the world. The mangroves grow in river deltas, lagoons, and estuarine complexes; they also occur on colonized shorelines and islands in sheltered coastal area with locally variable topography and hydrology. The largest extent of mangroves is found in Asia (42%) followed by Africa (20%), North and Central America (15%), Oceania (12%), and South America (11%) (Giri et al. 2011b). Approximately 75% of mangroves are concentrated in just 15 countries, and Indonesia has the highest percentage of mangroves in the world (Table 13.1).

Our study confirmed earlier findings that mangroves are generally confined to the tropical and subtropical regions of the world, with a few exceptions (Giri et al. 2011b). Mangroves extend to 31° 22′ N in Japan and 32° 20′ N in Bermuda, and to 38° 45′ S in Australia, 38° 59′ S New Zealand, and 32° 59′ S on the eastern coast of South Africa. The upper latitudinal limits of global distribution,

Figure 13.2 **(See color insert following page 266.)** Mangroves (dark red) in the Sundarbans (Bangladesh and India) in a Landsat false color composite (band combinations 4R, 3G, 2B).

Table 13.1   Fifteen Most Mangrove-Rich Countries and Their Cumulative Percentages

SN	Country	Area (m²)	Cumulative (%)	Region
1	Indonesia	3112989.48	22.60	Asia
2	Australia	977975.46	29.70	Oceania
3	Brazil	962683	36.68	South America
4	Mexico	741917.00	42.07	North & Central America
5	Nigeria	653669.10	46.82	Africa
6	Malaysia	505386.00	50.48	Asia
7	Myanmar	494584.00	54.07	Asia
8	Papua New Guinea	480121.00	57.56	Oceania
9	Bangladesh	436570.00	60.73	Asia
10	Cuba	421538.00	63.79	North & Central America
11	India	368276.00	66.46	Asia
12	Guinea Bissau	338652.09	68.92	Africa
13	Mozambique	318851.1	71.23	Africa
14	Madagascar	278078.13	73.25	Africa
15	Philippines	263137.41	75.16	Asia

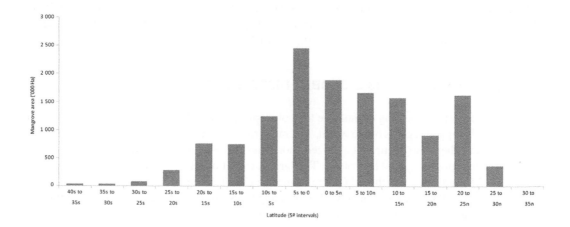

Figure 13.3   Latitudinal distribution of global mangroves (Giri et al. 2011b).

extending into the temperate regions, are characterized by decreased abundance, reduced species diversity, and decreased tree vigor, growth, and biomass (Figure 13.3). The geographic range of mangrove forests is highly dynamic, often expanding and contracting over time. The range mobility could serve as an indicator of climate change reflecting changing environmental conditions or gradual niche evolution over time. What causes areas of mangrove habitat to expand or contract is an important research question with many science and policy implications. Researchers are interested in whether mangroves can adapt to relative sea level rise or "can mangroves withstand and recover from more frequent and extreme tropical storms" have been of research interest for many years.

13.3 MAPPING AND MONITORING USING REMOTE SENSING

Remote-sensing data are increasingly available from multiple platforms including space, air, and ground. Earth observation satellites equipped with instruments for monitoring the earth's surface have been launched into orbit by a host of Nations since the 1970s creating a huge archive of data. However, this wealth of data can be overwhelming and not all data are freely available to the public. In addition to the satellite data, aerial photography has also been used for mangrove mapping and monitoring, especially after natural or anthropogenic disasters (such as hurricanes or oil spills), but those datasets are very limited, research oriented, and not readily available. Currently, Landsat, MODIS, SRTM, PALSAR and ICESat/GLAS datasets are appropriate and freely available for the operational purpose of coastal ecosystem studies at global scales. A typical approach of remote-sensing data analysis is presented in Figure 13.4.

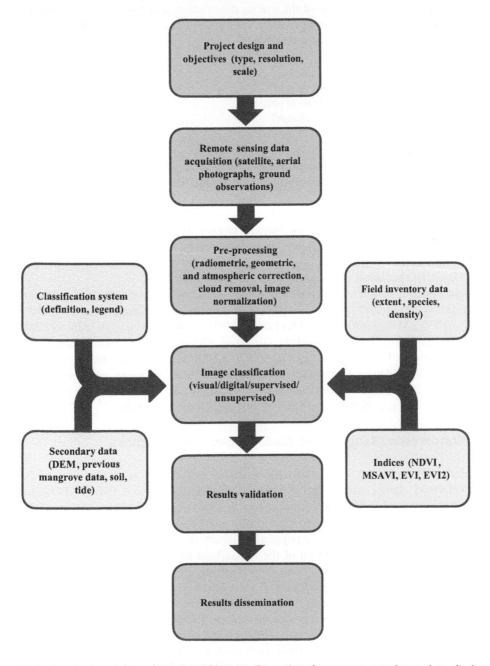

Figure 13.4   A typical work-flow of mangrove Mapping Examples of mangrove mapping and monitoring form various parts of the world are presented below.

13.3.1 Tsunami-Impacted Regions of Asia

Time-series remote-sensing data can be used to regularly monitor mangrove forests. Giri et al. (2008) monitored the regional tsunami-impacted countries of Indonesia, Malaysia, Thailand, Myanmar, Bangladesh, India, and Sri Lanka mangrove forest cover dynamics and identified rates and causes of change from 1975 to 2005 using Landsat data. Results were derived using multi-temporal satellite data and field observations. The repetitive coverage of satellite data provides an up-to-date and consistent overview of the extent, distribution, and dynamics of mangrove forests with better spatial and thematic details than the existing coarse resolution and field inventory data. The analysis addressed the following research questions.

- How much mangrove forests remain?
- Where are these mangrove forests located?
- What is the spatial and temporal rate of change?
- What are the main reasons for the change?
- What are potential areas for rehabilitation/regeneration?

Landsat satellite data were pre-processed and classified for 1975, 1990, 2000, and 2005. Post-classification change analysis was performed subtracting the classification maps, 1975s–1990s, 1975s–2000s, 1975s–2005s, 1990s–2000s, 1990s–2005s, and 2000s–2005s. The change areas were interpreted visually with the help of secondary data and ground data information to identify the factors responsible for the change. Once the mangrove/non-mangrove areas were calculated for each period, the annual rate of change for the region and for each country was calculated. The time-series analysis revealed a net loss of 12% of mangrove forests in the region from 1975 to 2005 (Giri et al. 2008b). Major hot-spot areas were also identified (Figure 13.5).

The rate of deforestation was not uniform in both spatial and temporal domains. The annual rate of deforestation during 1975–2005 was highest (~1%) in Myanmar compared to Thailand (0.73%), Indonesia (0.33%), Malaysia (0.2%), and Sri Lanka (0.08%). Mangrove areas in India and Bangladesh remained unchanged or increased slightly. Giri et al. (2008b) identified the major deforestation fronts in Ayeyarwady Delta (Figure 13.6), Rakhine, and Tanintharyi in Myanmar, Sweetenham and Bagan in Malaysia; Belawan, Pangkalanbrandan, and Langsa in Indonesia; and southern Krabi and Ranong in Thailand. Major reforestation and afforestation areas are located on the southeastern coast of Bangladesh, and Pichavaram, Devi Mouth, and Godavari in India.

13.3.2 Sundarbans (Bangladesh), Madagascar, and Philippines

Similarly, Giri et al. (2007), Giri and Muhlhausen (2008), Long et al. (2013a), and Giri et al. (2014) performed change studies in the Sundarbans (Bangladesh and India), Madagascar, and Philippines. In the Sundarbans, multi-temporal satellite data from the 1970s, 1990s, and 2000s were used to monitor deforestation and degradation of mangrove forests. The spatiotemporal analysis showed that despite having the highest population density in the world in its immediate periphery, the areal extent of the mangrove forest of the Sundarbans has not changed substantially over the last ~25 years. However, the forest is constantly changing due to erosion, aggradation, deforestation, and mangrove rehabilitation programs. The net forest area increased by 1.4% from the 1970s to 1990 and decreased by 2.5% from 1990 to 2000. The change is insignificant in the context of classification errors and the dynamic nature of mangrove forests. The Sundarbans is an excellent example of the coexistence of humans with terrestrial and aquatic plant and animal life. The strong commitment of governments under various protection measures such as forest reserves, wildlife sanctuaries, national parks, and international designations is believed to be responsible for keeping this forest relatively intact (at least in terms of area). Nevertheless, the forest is under threat from

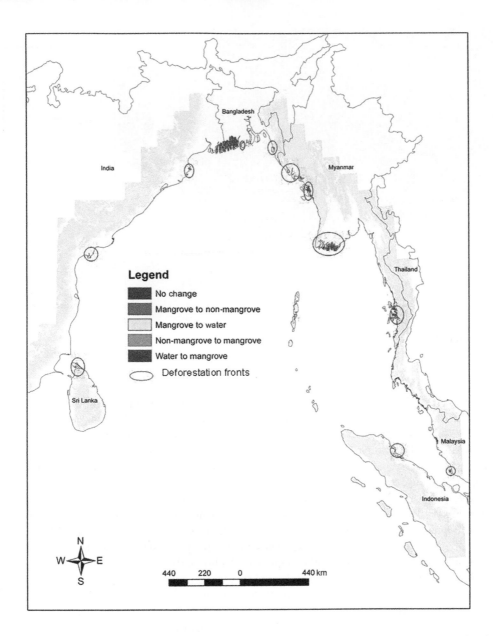

Figure 13.5 Major mangrove forest cover change areas from 1975 to 2005 in the tsunami-impacted region of South and Southeast Asia.

natural and anthropogenic forces that could lead to forest degradation, primarily due to top-dying and overexploitation of forest resources.

Time-series analysis revealed that the mangrove forests of Madagascar are declining, albeit at a much slower rate (~1.5% per year) than the global average. The forests are declining due to logging, overexploitation, clear cutting, degradation, and conversion to other land uses. In this research, Giri and Muhlhausen (2008) interpreted time-series Landsat data from 1975, 1990, 2000, and 2005 using a hybrid supervised and unsupervised classification approach. Landsat data were geometrically corrected to an accuracy of plus-or-minus one-half pixel, an accuracy necessary for change analysis. The results showed that Madagascar lost 7% of mangrove forests from 1975 to 2005, to a

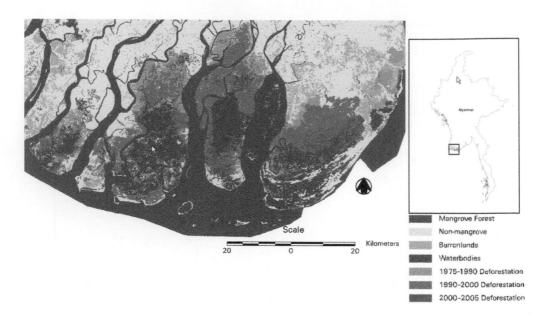

Figure 13.6 (See color insert following page 266.) Spatial distribution of mangrove deforestation in Ayeyarwady Delta, Myanmar, from 1975 to 1990 (cyan), 1990 to 2000 (red), and 2000 to 2005 (purple).

present extent of ~27.97 km². Deforestation rates and causes varied both spatially and temporally. The forests increased by 5.6% (212 km²) from 1975 to 1990, decreased by 14.3% (455 km²) from 1990 to 2000, and decreased by 2.6% (73 km) from 2000 to 2005. Major changes occurred in Bombekota Bay, Mahajamba Bay, the coast of Ambanja, the Tsiribihina River, and Cap St. Vincent. The main factors responsible for mangrove deforestation include conversion to agriculture (35%), logging (16%), conversion to aquaculture (3%), and urban development (1%).

In the Philippines, several national mangrove estimates existed; however, information was unavailable at sufficient spatial and thematic detail for change analysis. Long et al. (2013a) prepared a historical and contemporary mangrove distribution database of the Philippines for 1990 and 2010 at nominal 30-m spatial resolution using Landsat data. Image classification was performed using a supervised decision tree classification approach. Additionally, decadal land cover change maps from 1990 to 2010 were prepared to depict changes in mangrove area using a post-classification technique. Total mangrove area decreased 10.5% from 1990 to 2010. A comparison of estimates produced from this study with selected historical mangrove area estimates revealed that total mangrove area in the Philippines decreased by approximately half (51.8%) from 1918 to 2010.

13.3.3 South Asia

Mangrove forests in South Asia occur along the tidal sea edge of Bangladesh, India, Pakistan, and Sri Lanka. These forests provide important ecosystem goods and services to the region's dense coastal populations and support important functions of the biosphere. Mangroves are under threat from both natural and anthropogenic stressors; however, the current status and dynamics of the region's mangroves is poorly understood. Giri et al. (2014) mapped the current extent of mangrove forests in South Asia and identified mangrove forest cover change (gain and loss) from 2000 to 2012 using Landsat data. Three case studies were also conducted in the Indus Delta (Pakistan), Goa (India), and Sundarbans (Bangladesh and India) to identify rates, patterns, and causes of change in greater spatial and thematic detail than a regional assessment of mangrove forests.

Giri et al. (2014) found that the areal extent of mangrove forests in South Asia is approximately 11,874.76 km^2 representing ~7% of the global total. Approximately 921.35 km^2 of mangroves were deforested and 804.61 km^2 were reforested with a net loss of 116.73 km^2 from 2000 to 2012. In all three case studies, mangrove areas have remained unchanged or increased slightly; however, the turnover was greater than the net change. Both natural and anthropogenic factors are responsible for the change.

Giri et al. (2014) found that although the major causes of forest cover change are similar throughout the region, specific factors are dominant in specific areas. The major causes of deforestation in South Asia include (i) conversion to other land use (e.g., conversion to agriculture, shrimp farms, development, and human settlement); (ii) over-harvesting (e.g., grazing, browsing and lopping, and fishing); (iii) pollution; (iv) decline in freshwater availability; (v) flooding; (vi) reduction of silt deposition; (vii) coastal erosion; and (viii) disturbances from tropical cyclones and tsunamis. The forests are changing due to distinct reasons in some locations, including sea salt extraction in the Indus Delta in Pakistan, over-harvesting of fruits in the Sundarbans, and garbage disposal in Mumbai, India. Conversely, mangrove areas are increasing in some regions because of aggradation, plantation efforts, and natural regrowth. Protection of existing mangrove areas is facilitating regrowth. The region's diverse socioeconomic and environmental conditions highlight complex patterns of mangrove distribution and change. Results from this study provide important insight to the conservation and management of the important and threatened South Asian mangrove ecosystem.

13.3.4 Continental United States

Changes in the distribution and abundance of mangrove species within and outside of their historical geographic range can have profound consequences in the provision of ecosystem goods and services. Mangroves in the conterminous United States (CONUS) are believed to be expanding poleward (north) due to decreases in the frequency and severity of extreme cold events, while sea level rise is a factor often implicated in the landward expansion of mangroves locally. Giri and Long (2016) used ~35 years of satellite imagery and in situ observations for CONUS and report that: (i) poleward expansion of mangrove forest is inconclusive, and may have stalled for now, and (ii) landward expansion is actively occurring within the historical northernmost limit. They revealed that the northernmost latitudinal limit of mangrove forests along the east and west coasts of Florida, in addition to Louisiana and Texas has not systematically expanded toward the pole. Mangrove area, however, expanded by 4.3% from 1980 to 2015 within the historical northernmost boundary, with the highest percentage of change in Texas and southern Florida (Figure 13.7). Several confounding factors such as sea level rise, absence or presence of sub-freezing temperatures, changes in land use activities, impoundment/dredging, changing hydrology, fire, storm, sedimentation and erosion, and mangrove planting are responsible for the change. Besides, sea level rise, relatively milder winters and the absence of sub-freezing temperatures in recent decades may be enabling the expansion locally. The results highlight the complex set of forces acting on the northerly extent of mangroves and emphasize the need for long-term monitoring as this system increases in importance as a means to adapt to rising oceans and mitigate the effects of increased atmospheric CO_2. They also highlighted the limitations of periodic monitoring and importance of annual monitoring (Giri and Long 2016).

13.4 SPECIES DISCRIMINATION

Accurate and reliable information on the spatial distribution of mangrove species is needed for a wide variety of applications, including sustainable management of mangrove forests,

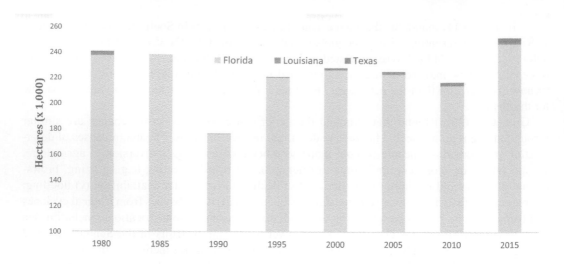

Figure 13.7   Distribution of mangrove areas in Continental United States from 1980 to 2015.

conservation and reserve planning, ecological and biogeographical studies, and invasive species management. Remotely sensed data have been used for such purposes with mixed results. Myint et al. (2008) employed an object-oriented approach with the use of a lacunarity technique to identify different mangrove species and their surrounding land use and land cover classes in southern Thailand using Landsat data. Results from the study showed that the mangrove zonation could be mapped using Landsat (Figure 13.8). It was also found that the object-oriented approach with lacunarity-transformed bands is more accurate (overall accuracy 94.2%; kappa coefficient = 0.91) than traditional per-pixel classifiers (overall accuracy 62.8%; and kappa coefficient = 0.57). Besides, multispectral images, hyperspectral data have been used to discriminate mangrove species (Chakravortty et al. 2014).

13.5 IMPACT/DAMAGE ASSESSMENT FROM NATURAL DISASTERS

Information regarding the present condition, historical status, and dynamics of mangrove forests was needed to study the impacts of the Gulf of Mexico oil spill of 2010. Data was not initially available for Louisiana at sufficient spatial and thematic detail to support a detailed analysis. Giri et al. (2011a) prepared mangrove forest distribution maps of Louisiana (before and after the oil spill) at 1.0 m and 30 m spatial resolution using aerial photographs and Landsat data, respectively. Image classification was performed using a decision-tree classification approach. Maps were prepared of mangrove forest cover change pairs for 1983, 1984, and every 2 years from 1984 to 2010 depicting **ecosystem shifts** (e.g., expansion, retraction, and disappearance).

Direct damage to mangroves from the oil spill was minimal, but long-term impacts need to be monitored. This new spatiotemporal information can be used to assess long-term impacts of the oil spill on mangroves. The study also proposed an operational methodology based on remote sensing using Landsat, Advanced Spaceborne Thermal Emission and Reflection Radiometer (ASTER), hyperspectral, light detection and ranging (LIDAR), aerial photographs, and field inventory data, to monitor the existing and emerging mangrove areas and their disturbance and regrowth patterns. Several parameters such as spatial distribution, ecosystem shifts, species composition, and tree height/biomass can be measured to assess the impact of the oil spill and mangrove recovery and restoration. Future research priorities will be to quantify the impacts and

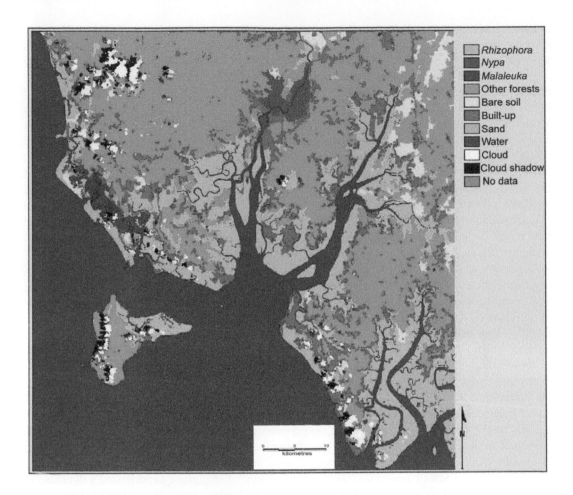

Figure 13.8 **(See color insert following page 266.)** Species zonation map generated by the object-oriented approach using lacunarity-transformed bands in southern Thailand.

recovery of mangroves considering multiple stressors and perturbations, including the oil spill, winter freeze, sea level rise, land subsidence, and land use/land cover change for the entire U.S. Gulf Coast.

Similarly, Landsat imagery was used to map mangrove damage caused by Typhoon Haiyan in November 8, 2013, in the Philippines (Long et al. manuscript underdevelopment). The Normalized Difference Vegetation Index (NDVI) was used as a standardized measure. NDVI is one of the most widely used vegetation indexes (Tucker 1979) to measure and monitor plant growth, vegetation cover, and biomass production. NDVI values range from 0.1 to 1.0 with dense vegetated areas (e.g., closed canopy tropical forest) generally yielding high NDVI values (0.6–0.8), sparsely vegetated areas (e.g., open shrub and grasslands) yielding moderate values (0.2–0.3), and non-vegetated (e.g., rock, sand, and snow) yielding low NDVI values (0.1 and below). Numerous studies have employed repeated measures of NDVI to monitor mangrove vegetation response from varying disturbances (Giri et al. 2011a), but few studies have applied this approach to monitor mangrove disturbance from typhoons (Wang 2012) and still fewer at a 30-m spatial resolution.

Typhoon Haiyan transected five Landsat path/row footprints in the Philippines. Landsat 7 and Landsat 8 (ETM+and OLI) imagery was used for the study. Image pre-processing for all imagery included converting from digital numbers to top-of-atmosphere (TOA) reflectance, stacking, masking for atmospheric contamination, and NDVI transformation. Mangrove areas were mapped prior to Haiyan for the year 2013 using a supervised decision tree classification approach applied to the Landsat imagery captured before November 8, 2013 (Long et al. 2013b). Independent variables used for mangrove classification included Landsat TOA imagery captured before November 8, 2013, a 30-m Shuttle Radar Topography Mission (SRTM) Digital Elevation Model (DEM), a slope index derived from the 30-m DEM, NDVI index transformed from Landsat TOA imagery, and a 2010 mangrove land cover map. Next, the 2013 mangrove land cover map was applied to mask values from all NDVI indexes (i.e., before and after) for mangrove areas. The NDVI images before and after the storm were differenced to create NDVI change maps for all five Landsat footprints. These NDVI change maps illustrate where substantial changes occurred in NDVI following Typhoon Haiyan. Larger decreases in NDVI values indicate greater damage in mangrove canopy. Last, mangrove NDVI change maps were quality checked with field photos and field notes. Field data collection was independent from mapping and useful for quality checking mangrove damage maps.

Differing spatial patterns of mangrove damage resulted from variations in storm surge intensity, bathymetry and topography, wind speed and direction, and type and condition of the existing mangrove vegetation (Figure 13.9). Overall, a general pattern of mangrove damage with the greatest decreases in NDVI values, indicating greater mangrove disturbance, occurring in proximity nearest Typhoon Haiyan's eye transect, especially north of the eye-wall path, and decreasing with greater distance north and south from the eye path. Mangrove damage was also generally highest in the eastern Philippines and generally lessened westward corresponding with decreasing storm intensity from east to west. Average NDVI mangrove values prior to Haiyan were approximately 0.8, indicating vigorous vegetation growth. In this analysis, mangrove damage was considered significant if NDVI values reduced more than 0.5.

The total mangrove area affected (i.e., experiencing some decrease in NDVI) by Haiyan was estimated to be 214.45 km^2 about 9% of the Philippines' total mangrove area. Not all mangrove areas were included in the analysis because of insufficient data resulting from cloud cover in a few areas. Of the total affected mangrove area, only 6.53 km^2 of mangrove experienced a decrease in NDVI $\geq$ 0.5, indicating substantial damage to mangrove health. Areas of substantial damage are considerably smaller than the mangrove areas experiencing minor damage.

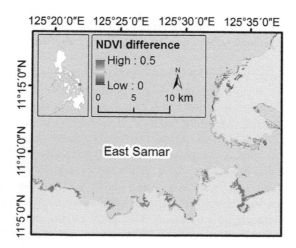

Figure 13.9 NDVI difference before and after Typhoon Haiyan in the Philippines.

13.6 HIGH-RESOLUTION MANGROVE MAPPING

As mentioned in the previous section, Landsat-based monitoring can capture the dynamic nature of mangrove forests. However, higher resolution (<5 m spatial resolution) remotely sensed data are required to capture newly colonized individual stands or relatively small patches of mangrove stands that cannot be captured with 30 m Landsat imagery. At this resolution, it is possible to create precise information on individual land cover features such as species composition and alliances, ecosystem shifts, watercourses, infrastructure, and developed areas.

High-resolution monitoring, deemed impossible until recently, is now becoming a reality due to increasing availability of high-resolution (<5 m) multispectral satellite data, improvement in image processing algorithms, and advancement in computing resources. Processing **big geo data** under this emerging paradigm is a challenge because of the data volume, software constraints, and computing needs. It is difficult, time consuming, and expensive to store, manage, visualize, and analyze these datasets using conventional image processing tools and computing resources. The United States Geological Survey (USGS) mangrove research team proposes to develop a cloud-computing environment to visualize and analyze high-resolution satellite data for land change monitoring focusing on mangrove mapping and monitoring of Florida. This new methodology will develop the next generation of tools to analyze and visualize **big geo data** and improve our scientific understanding of mangrove forest cover change needed for decision-making. Through the National Geospatial-Intelligence Agency (NGA), USGS, and commercial vendors, the research team has access to all available (unclassified) high-resolution satellite data (e.g., WorldView-2, GeoEye, IKONOS) from 2000 onward.

The USGS research team proposes a web-based mapping and monitoring platform and application to pre-process, analyze, and visualize the data. This application environment, based mainly on open source tools, will be deployed in a cloud-based system for improved performance, effectively unlimited storage capacity, reduced software costs, scalability, elasticity, device independence, and increased data reliability. Using this platform, the team plans to map and monitor mangrove distribution and change (area, density, and species composition), and species expansion or squeeze due to climate change. The team also plans to quantify the impact and recovery from natural disasters such as hurricanes. High-resolution data are suitable to capture newly colonized individual mangrove stands, small island mangroves, or relatively small, fragmented, and linear patches of mangrove stands that are difficult to capture using 30-m Landsat imagery (Figure 13.10).

It is anticipated that this system will improve the availability and usability of high-resolution data, algorithms, analysis tools, and scientific results through a centralized environment that fosters knowledge sharing, collaboration, innovation, and direct access to computing resources. Additional data such as Landsat and LIDAR can be added for similar analysis. The information generated from this study will improve our scientific understanding of distribution and dynamics of mangroves in Florida. The innovative tools and methods developed from this study will pave the way for future land change monitoring at high spatial resolution from local to national scales.

13.7 ANNUAL VERSUS PERIODIC MONITORING

Periodic mangrove monitoring (e.g., every 5 years) has its own limitations by not capturing seasonal and annual changes. Increasing interested is being paid to adopt annual monitoring to capture these changes. Annual land cover is now feasible due to the availability of multi-temporal satellite data (e.g., Landsat 7 and 8, sentinel and other data), improved pre-processing capacity, advancement in image classification methodology, and increased availability of resources (e.g., brain-power and computing resources). Information need for annual monitoring has also increased. For certain changes, annual monitoring may not be needed. For example, species composition change may not

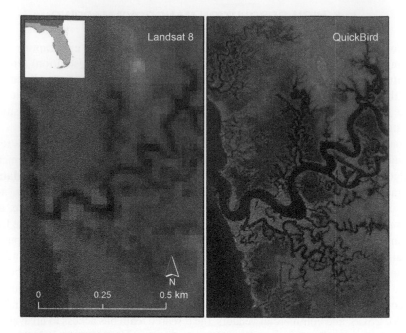

Figure 13.10 Comparison of mangrove distribution in Landsat 8 (lower resolution) and QuickBird images (higher resolution).

happen year to year—for such an application periodic monitoring is better suited. Similarly, vegetation structure, height, canopy may not change that much unless disturbances. Other parameters such as Net Primary Productivity and biomass could change annually. Moreover, certain phenomenon such as fire, floods, storms, may need daily or weekly or monthly monitoring and reporting.

13.8 CONCLUSIONS AND RECOMMENDATIONS

This chapter demonstrates how Landsat and high-resolution satellite data remote-sensing data can be used for mangrove forest cover mapping and monitoring. These studies show that global mangroves can be mapped and monitored using Landsat; however, for specific use in localized areas, high-resolution satellite data or aerial photographs are needed. High-resolution data are useful to detect early colonization, small patches of mangrove forests, and changes. With the availability of high-resolution satellite data, advancement in computer technology, and improvement in classification algorithms, it will be possible to map and monitor global mangroves using high-resolution satellite data in the future. Remote sensing can help answer the following five major research topics.

1. Observed and predicted changes in relative sea level rise (RSLR), temperature, precipitation, atmospheric CO_2 concentration, hydrology, more frequent and destructive tropical storms, land-use change in neighboring ecosystems, and human response to climate change are major components of environmental change that threaten the distribution and health of mangrove ecosystems (Alongi 2008; Gilman et al. 2008; Saintilan et al. 2014). Monitoring mangrove forests using remote sensing could serve as an indicator of global climate change. For example, poleward migration beyond their historical range could indicate a changing climate.
2. Our scientific understanding of the impact of climate change on mangrove ecosystems and adaptation options is limited. RSLR is expected to be the major climate change component affecting mangrove ecosystems globally (Krauss et al. 2013; McKee et al. 2007). Mangroves may respond

to RSLR by migrating upslope or maintaining their surface elevation through vertical accretion or sedimentation. However, several case studies around the world show that mangrove systems do not have the ability to keep pace with predicted RSLR rates (McKee et al. 2007), resulting in contraction or squeeze in extent at the shoreline from erosion and/or submergence, suffocation, and eventual death. RSLR could also impact the species composition of mangrove stands and related floral and faunal biodiversity. In contrast, "mangroves could be able to keep pace with sea level rise in some places" (McIvor et al. 2013). Therefore, global impact will likely be site-specific, driven by the rate and degree of RSLR and other biophysical factors. Remote-sensing observations could be used to quantify these changes spatially and temporally.

3. Mangrove forests are among the most carbon-rich habitats on the planet (Alongi 2014; Donato et al. 2011) compared to other terrestrial forests, owing to the ecosystems' high productivity, rich organic input in long-term soil carbon storage, and low emissions of methane in the saltwater environment. These inter-tidal ecosystems are strong carbon sinks. Therefore, the forests could contribute to climate change mitigation efforts through carbon financing mechanisms such as Reduced Emissions from Deforestation and Degradation (REDD+) and other coastal marine conservation and mitigation initiatives (a.k.a. **Blue Carbon**). Remote sensing could be used to quantify carbon stock and carbon stock change in support of Monitoring, Reporting, and Verification (MRV) of REDD+.

4. It has been suggested that mangrove ecosystems have expanded poleward in the last three decades primarily due to an increase in temperature or decrease in the frequency of winter freeze (Cavanaugh et al. 2013; Saintilan et al. 2014). However, other reports contradict these findings stating that the observation period is inadequate (Giri et al. 2011a). For example, Cavanaugh et al. (2014) used 1984–2011 Landsat data coupled with climate data and concluded that mangrove extent was expanding poleward along the northeast coast of Florida beyond its northern historical range. Giri and Long (2014) and (Giri et al. 2011a) conducted a similar study but included one additional year of Landsat data (i.e., 1983–2011) and concluded that although mangroves expanded from 1984 to 2010, the area has not reached the 1983 extent. The dramatic decrease in mangrove extent was caused by a severe winter freeze in December 1983 that reduced mangrove extent in the northern boundary of Florida and Louisiana by approximately 90%. Historical Landsat data dating back to 1972 and high-resolution satellite data could be used to answer whether mangrove is expanding poleward or not.

5. The Indian Ocean tsunami of December 2004, Hurricane Katrina along the U.S. Gulf Coast in 2005, Cyclone Nargis in Myanmar in 2008, and similar natural disasters in the past decade have highlighted the importance of the mangrove ecosystem as a **bio-shield** or **natural barrier** in protecting vulnerable coastal communities from natural disasters. Several studies after the 2004 tsunami concluded that mangrove ecosystems provided protection to life and property (Barbier 2006; Dahdouh-Guebas et al. 2005; Danielsen et al. 2005; Kathiresan and Rajendran 2005; UNEP 2005). Subsequent publications argued against this claim and concluded that mangrove ecosystems did not provide protection during the tsunami (Kerr and Baird 2007; Vermaat and Thampanya 2006). Studies conducted in India showed that villages located behind mangrove forests suffered less damage than those directly exposed to the coast (Kathiresan and Rajendran 2005). In addition, Kathiresan (2012) suggested that the destructive power of the storm surge was exacerbated during Cyclone Nargis by recent loss of mangroves in Myanmar, although no primary evidence to support these statements was presented. Some researchers who are skeptical about the ability of mangroves to protect against tsunamis have noted that mangroves might be more capable of protecting against tropical storm surges. Storm surges differ from tsunamis in having shorter wavelengths and relatively more of their energy near the water surface. The current consensus is that mangroves provide protection to a certain extent, but quantifying magnitude remains is a scientific challenge. Increased frequency and intensity of extreme events could have profound impacts to coastal development and human safety in the future. Remote-sensing data can provide unbiased and transparent information to help address some of these issues.

CHAPTER 14

Status of Tidal Marsh Mapping for Blue Carbon Inventories

Kristin B. Byrd
U.S. Geological Survey Western Geographic Science Center

Chris Mcowen and Lauren Weatherdon
United Environment Programme World Conservation Monitoring Centre

James Holmquist
Smithsonian Institution

Stephen Crooks
Silvestrum Climate Associates, LLC

CONTENTS

HIGHLIGHTS

1. Remote-sensing-based maps of tidal marshes, both of their extents and carbon stocks, will play a key role in conducting greenhouse gas (GHG) inventories.
2. The U.N. Environment Programme World Conservation Monitoring Centre has produced a new Global Distribution of Salt Marsh dataset that estimates global salt marsh area at 5.5 Mha.
3. A Tier 1–2 GHG Inventory of U.S. Coastal Wetlands has been developed using the NOAA Coastal-Change Analysis Program Landsat-based land cover maps as a primary dataset.

4. Successful mapping of tidal marsh biomass with optical satellite images provides opportunity to improve GHG Inventories.
5. Further work is needed to map tidal marsh salinity gradients, the extent of tidal vs. non-tidal marshes, methane emissions, and high-resolution elevation.

14.1 REQUIREMENTS FOR IPCC GHG INVENTORIES OF TIDAL MARSHES

The purpose of national GHG inventories is to enable countries to consistently estimate anthropogenic emissions by all sources and removals by all sinks. The level of detail required depends upon the importance of the activity as a source of emissions or removals. The Intergovernmental Panel on Climate Change (IPCC) 2003 Good Practice Guidelines offer a hierarchical three-tiered approach to constructing GHG Inventories in the Agriculture, Forestry and Other Land Use (AFOLU) Sector (IPCC 2003). These approaches range from Tier 1 methods that use IPCC—provided default emission factors for activities, to Tier 2 methods that apply emission and stock change factors based on country or region-specific data, higher temporal and spatial resolution, and more disaggregated activity data, to Tier 3 methods that include models and repeated inventory measurement systems driven by high resolution, disaggregated activity data (IPCC 2006).

One of the main steps in preparing an inventory gain-loss estimate for the AFOLU Sector is to compile data on the change in area of land in each land-use category for at least two points in time since 1990, and to categorize the land area by specific management systems. The 2006 IPCC Guidelines provide three generic approaches for obtaining land use and land cover data. The most complex approach, Approach 3, includes tracking land use conversion on a spatially explicit basis (IPCC 2006). In 2013, the IPCC adopted the Wetlands Supplement to the 2006 IPCC Guidelines for National GHG Inventories. This document provides guidance on accounting for emissions and removals associated with wetland management. Chapter 4 covers "Coastal Wetlands", which include tidal freshwater and salt marshes, mangroves, and seagrasses (IPCC 2014). Remote-sensing-based maps of tidal marshes, both of their extents and carbon stocks, will play a key role in conducting these inventories.

14.2 THE STATE OF GLOBAL TIDAL MARSH MAPPING

Mapping change in land area is particularly important for tidal marshes, as land conversion generates the greatest carbon emissions for these ecosystems (Crooks et al. 2011; Pendleton et al. 2012). Among Blue Carbon ecosystems, information on the distribution and extent of tidal marshes globally is lacking compared to mangrove forest (e.g., Giri et al. 2011). One of the principal challenges is that data are often collected at local, national, or regional scales, generally as part of coastal wetland inventories, but are not synthesized to provide an integrated overview. Reasons for these data gaps include technical barriers (e.g., the data are not preserved or only exists as a hard copy or in an antiquated format), motivational barriers (e.g., there is no incentive to share the data), economic barriers (e.g., data sharing requires human and technical resources), and political and legal barriers (e.g., agencies may have developed official policy guidelines that restrict data sharing).

To address this discontinuity of spatial data, the United Nations Environment Programme World Conservation Monitoring Centre (UNEP-WCMC) collated and integrated salt marsh datasets from 50 data providers globally (Table 14.1), with support from Conservation International and The Nature Conservancy. The "Global Distribution of Saltmarsh" (Mcowen et al. 2017) dataset provides a baseline inventory of the extent of these ecosystems, with data obtained from peer-reviewed articles, reports and databases created by non-governmental and governmental organizations, universities, research institutes, and independent researchers globally. The dataset captures 5,495,089 ha of salt marsh across 43 countries in a Geographic Information System polygon shapefile (Table 14.2),

Table 14.1 List of 50 Data Sources Used to Compile "The Global Distribution of Saltmarsh Dataset" (Mcowen et al. 2017)

Source Type	Source	Country
Intergovernmental organizations	European Environment Agency	Denmark
	The Ramsar Convention Secretariat	Switzerland
	UK Environment Agency	United Kingdom
Governmental organizations	Rijkswaterstaat Dienst Zeeland (Ministry of Infrastructure and the Environment)	The Netherlands
	National Parks and Wildlife Service Ireland	Ireland
	Servicio de Conservación e Inventariación de Humedale, Subdirección General de Medio Natural, Dirección General de Calidad y Evaluación Ambiental y Medio Natural, Ministerio de Agricultura, Alimentación y Medio Ambiente	Spain
	Department of Commerce (DOC), NOAA, National Ocean Service (NOS), and the Coastal Services Center (CSC)	USA
	Wasser- und Schifffahrtsamt Bremerhaven	Germany
	Federal Institute of Hydrology	Germany
	Landesamt für Landwirtschaft, Umwelt und ländliche Räume	Germany
	Florida Fish and Wildlife Conservation Commission - Fish and Wildlife Research Institute, Center for Spatial Analysis	USA
	Cawthron Institute (and various regional and city councils)	New Zealand
	National Office of the Environment (ONE)	Madagascar
	Naturvårdsverket (Swedish EPA)	Sweden
	Abu Dhabi Environment Agency	United Arab Emirates
	U.S. Fish and Wildlife Service	USA
	Geoscience Australia	Australia
	Prince Edward Island Department Of Agriculture and Forestry	Canada
	Nova Scotia Department of Natural Resources, Wildlife Division	Canada
	New Brunswick Department of Natural Resources, Fish and Wildlife Branch	Canada
	Ministerio de Agricultura	Peru
	Consorzio Venezia Nuova	Italy
	Division of Water Resources, Hydraulic and Maritime Engineering	Greece
	European Topic Centre on Biological Diversity, Museum National d'Histoire Naturelle	Latvia

(Continued)

Table 14.1 (*Continued*) List of 50 Data Sources Used to Compile "The Global Distribution of Saltmarsh Dataset" (Mcowen et al. 2017)

Source Type	Source	Country
Non-governmental organizations	Estonian Environmental Information Centre (EEIC)	Estonia
	Comisión Nacional de Áreas Naturales Protegidas (CONANP)	Mexico
	The Nature Conservancy	USA/United Kingdom
	Johannes Burmeister (Manfred-Hermsen-Stiftung) and Eiler Tabilo Valdivieso (Centro Neotropical de Entrenamiento en Humedales)	Peru/Germany
	BirdLife International	United Kingdom
Research institutes	Institute of Nature Conservation	Poland
	Komarov Botanical Institute Russian Academy of Sciences	Russia
	Gerasimov YN, Pacific Institute of Geography, Russian Academy of Sciences, Petropavlovsk, Kamchatka	Russia
	Institute of Remote Sensing and Digital Earth	China
Academic or independent researcher(s)	Environmental Hydraulic Institute "IH Cantabria," Universidad de Cantabria	Spain
	Instituto de Investigaciones Marinas y Costeras (IIMyC), Universidad Nacional de Mar del Plata	Argentina
	Ping Zuo, Nanjing University	China
	João M. Neto (IMAR-Univ. Coimbra) and Isabel Caçador (CO-FFCUL)	Portugal
	Pilar García, INDUROT	Spain
	Botany Department, Nelson Mandela Metropolitan University	South Africa
	Neil Saintilan, Macquarie University	Australia
	Olga A. Mochalova, Institute of the Biological Problems of the North (pers.comms)	Russia
	Tomkovich, P. S., Syroechkovski, E. E. Jr., Lappo, E. G. & Zöckler, C. (2002) First indications of a sharp population decline in the globally threatened Spoon-billed Sandpiper Eurynorhynchus pygmeus. Bird Conserv. Internatn. 12: 1–18.	Russia
	Syroechkovski, E. E., Zöckler, C. & Lappo, E. (1998) Status of Brent Goose in northwest Yakutia British Birds 93(2):94–97.	Russia
	Sandring, S., Kratzsch, G., Loele, A. & Haese, D. (2001) Vegetation of the Malkachan Bay area. In: Andreev, A.V., Bergmann, H.-H. (Hrsg.), Biodiversity and Ecological Status Along the Northern Coast of the Sea of Okhotsk. Russian Academy of Sciences—Far Eastern Branch, Institute of Biological Problems of the North: 134–145, Vladivostok.	Russia
	Kruckenberg, H., Kondratyev, A., Zöckler, C., Zaynagutdinova, E. & Mooij, J.H. (2012). Breeding waders on Kolguev Island, Barents Sea, N Russia, 2006–2008. Wader Study Group Bull. 119(2): 102–113.	Russia

(Continued)

Table 14.1 (*Continued*) List of 50 Data Sources Used to Compile "The Global Distribution of Saltmarsh Dataset" (Mcowen et al. 2017)

Source Type	Source	Country
	Tolvanen, P. (1998) Lesser White-fronted Goose Anser erythropus expedition to the Kanin Peninsula in 26 August–12 September, 1996, and the establishment of the Shoininsky Reserve. In: Tolvanen, P., Ruokolainen, K., Markkola, J. & Karvonen, R. (eds.): Finnish Lesser White-fronted Goose conservation project. Annual Report 1997. WWF Finland Report No 9: 33–35.	Russia
	Spilling (pers.comms)	Russia
	Mineev, Y.N. & Mineev, O.Y. (2009). Birds of Malozemelskaka Tundra and Pechora River delta. Nauka, St. Petersburg, Russia [in Russian].	
	Syroechkovski, E. E. Jr. (1995): Current status of Asian Population of Pacific Black Brant Branta bernicla nigricans. Bull of the Goose Study Group of Eastern Europe and North Asia 1:57–67.	Russia
Legacy project	The British Columbia Marine Conservation Analysis (BCMCA)	Canada

Table 14.2 Recorded Salt Marsh Extent by Region (in ha), Number of Polygons, and Time Frame and High-Level Methodology of Data Collection, as Derived from the "Global Distribution of Saltmarsh" Polygon Dataset (Mcowen et al. 2017)

Region		Area (ha)	Number	Start Date	End Date	Methodology
North and Central America	USA (Mainland, Hawai'i)	1,723,410	123,697	01/01/1977	31/12/2012	Remote sensing, field survey
	Mexico	272,527	437	N/A	N/A	Remote sensing
	USA (Alaska)	161,483	8,949	01/01/1977	31/12/2012	Remote sensing, field survey
	Canada	111,274	10,502	01/01/1995	31/12/2002	Field survey
South America	Argentina	118,870	3,086	01/03/2000	28/02/2003	Remote sensing, field survey
	Brazil, Uruguay, Chile, Peru	37,858	385	01/03/2000	31/12/2011	Remote sensing, field survey
Africa and the Middle East	South Africa	6,147	1,561	01/01/2008	31/12/2008	Remote sensing
	Madagascar	5,810	4	01/01/2011	31/12/2011	Remote sensing, field survey
	United Arab Emirates	4,797	174	07/10/2011	20/05/2014	Remote sensing
Europe	Mainland Europe[a]	356,947	31,830	01/01/1999	13/02/2015	Remote sensing, field survey, ground-truthed
	Great Britain	81,842	117,052	01/01/2006	31/12/2009	Remote sensing, ground-truthed
	Ireland (Republic of)	9,889	13,127	01/01/2006	31/12/2008	Remote sensing, field survey
	Iceland	2,617	32	01/01/2006	31/12/2006	Remote sensing
Russian Federation		700,719	50	01/07/1973	31/07/2011	Field survey
China		549,506		01/01/1999	31/12/2008	Remote sensing
Oceania	Australia	1,325,854	14,145	01/01/2001	31/12/2001	Remote sensing, ground-truthed
	New Zealand	19,650	2,482	01/01/2007	31/12/2008	Remote sensing
Small-Island Developing States	Puerto Rico and the U.S. Virgin Islands	5,879	1,498	01/01/1977	31/12/2012	Remote sensing, field survey
	Guam and the Commonwealth of Northern Marianas	8.2	13	01/01/1977	31/12/2012	Remote sensing, field survey
	American Samoa	0.1	3	15/12/2003	15/12/2003	Remote sensing, field survey
	TOTAL	5,495,089	350,985			

Area calculated after dissolving to remove overlapping polygons.
a Coverage in mainland Europe includes 20 countries: Spain, France, Sweden, Italy, Netherlands, Albania, Cyprus, Germany, Denmark, Belgium, Finland, Estonia, Croatia, Montenegro, Portugal, Romania, Turkey, Slovenia, Latvia, and Bulgaria.

with salt marsh presence in an additional 56 countries* documented in a point shapefile (Figure 14.1). In addition, the dataset includes a global point shapefile linked to a Microsoft Access database with information on species found at each location.

The time frames for data collection ranged from 1973 to 2015, and scale of the data is variable, ranging from 1:10,000 to 1:4,000,000, with most falling within the 1:10,000 to 1:100,000 range. Where feasible, detailed descriptions of the datasets were obtained and recorded within the associated metadata (e.g., time of data collection, source, resolution, and methods of processing). Salt marshes were identified in accordance with definitions provided by data sources, where available, or were determined using expert opinion. Plans to update the global map are guided by data availability and vary by data provider.

All data were originally collected using remote sensing and field-based survey methods, with data quality ranging from high-resolution maps to low-resolution representations. Sources of country data range from coordinated national mapping programs to academic studies. For example, Australia has assembled a national dataset of tidal marsh maps, which may be obtained via the Internet on the Australian Online Coastal Information (OzCoasts) website (http://www.ozcoasts. gov.au/nrm_rpt/habitat_extent.jsp). Salt marshes are represented as high spatial resolution polygons as mapped by state agencies, and have been attributed with the National Intertidal/Sub-tidal Benthic Habitat Classification Scheme. High spatial resolution habitat maps are scaled to 10 and 50 km grid cells to indicate habitat presence, absence or if unknown to enable display of data at national and state scales (Mount and Bricher 2008). Also, the United Kingdom Environment Agency produced a map delineating the extent of salt marsh in England and Wales during the period from 2006 to 2009 using high-resolution aerial photography using a consistent and repeatable mapping approach (Phelan et al. 2011). The final map serves as a national baseline from which to compare future salt marsh maps produced using the same methodology.

The UNEP-WCMC composite dataset is indicative of current knowledge of the extent of salt marshes globally based on the data obtained through the inventory, and is not intended to reflect a comprehensive assessment of salt marsh presence and absence. While mapped using a conservative approach, the

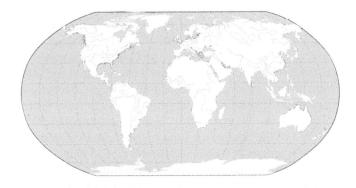

Figure 14.1   (See color insert following page 266.) Global map of digitally recorded extent of salt marsh ecosystems, representing datasets collated between 1973 and 2015 (Mcowen et al. 2017).

---

* The additional 56 countries represented in the point shapefile (but not the polygon shapefile) include: Angola; Antigua and Barbuda; Azerbaijan; Bahrain; Bahamas; Cook Islands; Costa Rica; Cuba; Djibouti; Algeria; Ecuador; Egypt; Western Sahara; Fiji; Georgia; Ghana; Gambia; Greece; Greenland; India; Iran; Iraq; Japan; Kenya; Cambodia; Republic of Korea; Kuwait; Libyan Arab Jamahiriya; Sri Lanka; Lithuania; Morocco; Malta; Mauritania; Namibia; New Caledonia; Nicaragua; Norway; Oman; Pakistan; Panama; Philippines; Poland; North Korea (People's Republic of Korea); Qatar; Saudi Arabia; Sudan; Somalia; Suriname; Tonga; Tuvalu; Taiwan; Tanzania; Ukraine; Venezuela; Viet Nam; and Vanuatu.

UNEP-WCMC estimate of salt marsh area (~5.5 Mha) supersedes a lower estimate of 2.2 Mha that did not include data from Asia, South America, and Australia (Chmura et al. 2003). Notably, there are areas in Canada, Northern Russia, South America, and Africa where salt marshes are known to occur that require additional spatial data on salt marsh extent. Nevertheless, the most extensive saltmarsh world-wide are found outside the tropics, including the low-lying, ice-free coasts, bays, and estuaries of the North Atlantic, which are well represented in the global polygon dataset. Therefore, despite the gaps, the data covers many of the important areas in Europe, the United States, and Australia.

While some datasets used within the global dataset offer local or regional time series, the data-set as a whole cannot be used for temporal analyses of change given an incomplete systematic survey of salt marsh extent globally over time. In addition, the dataset may contain other habitat types such as freshwater marshes, mudflats, and salt pannes due to varying collection and mapping methodologies, and variation between definitions of salt marsh by geographical region.

14.3 THE STATE OF TIDAL MARSH MAPPING IN THE UNITED STATES FOR GHG INVENTORIES

In the United States, the primary dataset being used for tidal marsh GHG inventories is the National Oceanic and Atmospheric Administration (NOAA) Coastal Change Analysis Program (C-CAP) data-set (NOAA Office for Coastal Management 2015). Established in the 1990s, C-CAP is a set of nation-ally consistent, raster-based inventories that cover coastal intertidal areas, wetlands, and adjacent uplands for the coastal U.S. Wall to wall 30-m resolution coastal maps with up to 25 land use and land cover classifications, as well as maps of land cover change between two dates from and to each class, are updated every 5 years using Landsat satellite imagery. Data may be downloaded on a Web Interface at https://coast.noaa.gov/ccapftp/#/. These data products use standardized data format and processing methods to ensure consistency through time (since 1975) and across geographies. Key coastal wetland classes include palustrine forested, palustrine scrub/shrub, and palustrine emergent, where palustrine represents freshwater wetlands with salinity less than 0.5%, and estuarine forested, estuarine scrub/shrub, and estuarine emergent, where estuarine represents brackish and saline wetlands with salinity equal to or greater than 0.5%. In a recent accuracy assessment of the 2010 C-CAP land cover dataset, overall accuracy was 84.0%. An assessment of the 2006–2010 change dataset identified that the overall change/no-change accuracy was 88.7% (NOAA Office for Coastal Management 2014).

According to the C-CAP land cover change dataset, between 1996 and 2010, 44.3% of the U.S. total land use-land cover change occurred in coastal areas (e.g., "coastal zone"), which include coastal watersheds and counties extending inland of tides, though these areas accounted for 25% of the contiguous U.S. land mass (NOAA Office for Coastal Management 2016). Slightly more than 8% of the coastal zone of the contiguous United States changed from 1996 to 2010, mainly due to timber management practices in the Pacific Northwest and Southeastern United States and increased development. Overall, wetlands (primarily palustrine-forested wetlands) lost 3,976 km^2 mainly due to development (NOAA Office for Coastal Management 2016). While national net loss in estuarine and palustrine emergent and scrub/shrub marsh was approximately 84 km^2, in the Gulf Coast, these wetland types experienced approximately 453 km^2 of net loss primarily to open water (NOAA Office for Coastal Management 2015). Here loss from hurricane Katrina damage and ongo-ing subsidence and sediment deprivation outpacing new wetland gains from sediment diversions in the Atchafalaya and Wax Lake Deltas.

To support the U.S. Environmental Protection Agency's (EPA) national reporting on GHG emis-sions and removals (see Chapter 16, Troxler et al), a NOAA-funded team has developed a Tier 1–2 GHG Inventory of U.S. coastal wetlands, using C-CAP as a primary dataset (USEPA 2017). There are two approaches to accounting; this inventory was being conducted using a "managed land proxy" approach rather than an activities-based approach. Similar to AFOLU guidelines for U.S. forest and cropland accounting, all wetlands were considered as managed lands, and any quantifiable gains or losses of wetland carbon stocks were counted in the national GHG inventory regardless of the direct or proximate causes of those fluxes. Under this managed lands approach, carbon accumulated in wetlands that remain wetlands can also be counted as a GHG sink. As a result, monitoring change in emissions and removals within wetlands over time is important for conducting accurate inventories.

As part of the inventory, the C-CAP derived land use change data will be accompanied with a literature review database describing soil carbon stock, carbon accumulation rates, aboveground biomass and methane emissions. Broken down by climatic zone, these values provide a first assessment of emissions and removals associated with changes in wetlands. According to the literature, emissions result primarily from erosion of salt marshes in the Mississippi Delta (Crooks et al. 2009; DeLaune and White 2012; Couvillion et al. 2013), ongoing emissions from legacy organic soils on drained former wetlands, as well as methane emissions from impounded waters behind coastal barriers (Kroeger et al. 2017).

14.4 LIMITATIONS TO TIDAL MARSH MAPPING FOR GHG INVENTORIES

14.4.1 Salinity Gradients

Currently in the United States, NOAA C-CAP datasets provide the most useful wetland maps at the national scale for GHG accounting due to their consistency, vegetation classes, and change products. However, in these datasets, brackish marshes are not distinguished from salt marsh, and tidal and non-tidal wetlands are not mapped. Methane emissions can vary strongly based on salinity gradients, and a key threshold for methane emissions is the presence of year-round salinity greater than 18 parts per thousand (ppt) (Poffenbarger et al. 2011). Because of this trend, the 2013 IPCC Wetlands Supplement calls for the need to distinguish between tidal salt marsh (>18 ppt) and tidal brackish/freshwater marsh (≤18 ppt) to accurately assign Tier 1 methane emission factors. In C-CAP, whereas palustrine boundaries are sufficient in delineating freshwater systems (<5 ppt), there is a need to distinguish between brackish and fully saline marshes within the estuarine wetland categories. NOAA has produced a three-zone average annual salinity map for marshes in the United States with marine, mixed, and freshwater classes that correspond to marine, brackish, and fresh categories (https://catalog.data.gov/dataset/noaa-average-annual-salinity-3-zone452ce). However, this product is only available for certain U.S. estuaries, and extends only to near-shore environments, omitting wetlands inland of the shoreline as mapped by the NOAA Coastal Assessment Framework. This limits its use for adjusting C-CAP estuarine classifications into saline and brackish categories.

Distinguishing brackish and saline tidal marshes with remote-sensing data has been challenging. However recently, Mo et al. (2015) succeeded in identifying significantly different satellite-based phenological signals between freshwater, brackish, and salt tidal marshes in Louisiana using moderate resolution remote sensing, which could potentially be used to map the three wetland types. The researchers found that a Gaussian function was the best phenological model for Louisiana coastal marshes, where freshwater intermediate, brackish and saline marshes could be distinguished by peak normalized difference vegetation index (NDVI), day of peak NDVI, and growth duration. These phenology differences remained significant both within a drought year and during the year of Hurricane Katrina.

14.4.2 Tidal versus Non-Tidal Marshes

In addition to challenges with mapping brackish wetlands, C-CAP datasets cannot distinguish between tidal and non-tidal palustrine wetlands. However, the U.S. Fish and Wildlife Service

National Wetlands Inventory (NWI) data (Dahl and Stedman 2013; U.S. Fish and Wildlife Service 2014) can be used to supplement information on the extent of tidal wetlands. NWI maps are vector-based and primarily produced using aerial photographs, photointerpretation techniques, and field verification. The spatial coverage of these maps encompasses the entire contiguous United States. However, the temporal coverage is variable, and there is no associated accuracy assessment available with this product.

NWI maps contain detailed wetland classes based on vegetation, hydrologic regime and salinity, and sub-classification "modifiers" in NWI can help to delineate the boundary of tidal extent. For example, NWI correctly classifies impounded brackish wetlands in Suisun Marsh, the largest brackish marsh in California (Moyle et al. 2014), located on the eastern edge of San Francisco Bay, while C-CAP maps this area as estuarine emergent marsh (Figure 14.2a). Despite this improved classification, the detail and consistency of coding can vary depending on the time and location of the air photo interpretation. In addition to variable hierarchical classification, conditions can change over time because of sea level change, or the removal of dikes and impoundments. Also, the change to the wetland sub-classes in NWI lags the actual land use change, by however long it takes for the change to be reported, recorded, and updated in a new NWI version. This lag time can lead to misclassification of restored sites, for example, in the Nisqually National Wildlife Refuge in Puget Sound, WA, where a dike was removed in 2009 to restore 308 ha of tidal salt marsh (Figure 14.2b).

Relying solely on the NWI "tidal modifiers" likely underestimates the area of tidal wetlands in the United States, in part, because wetland delineation and classification in NWI are based on air photo interpretation, and can vary by survey. In terms of the NOAA-led U.S. National Tier 1 GHG Inventory for Coastal Wetlands, the approach submitted to review is to recognize the suite of all wetlands falling at or below a Mean Higher High Water Spring Line elevation whether they are otherwise classified as tidal and non-tidal, collectively as coastal wetlands. This approach could also allow for the inclusion of other land-use categories (such as cropland) that were originally coastal wetlands before conversion (Figure 14.3).

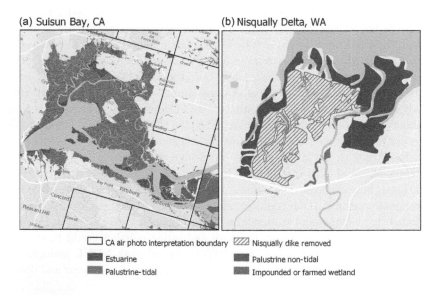

(a) Suisun Bay, CA   (b) Nisqually Delta, WA

☐ CA air photo interpretation boundary   ▨ Nisqually dike removed
■ Estuarine   ■ Palustrine non-tidal
■ Palustrine-tidal   ■ Impounded or farmed wetland

Figure 14.2 **(See color insert following page 266.)** (a) USFWS National Wetland Inventory (NWI) map of Suisun Marsh, California. Here NWI distinguishes between tidal and impounded wetlands, but these distinctions can vary based on the survey (bottom right corner). (b) USFWS NWI map of the Billy Frank Jr. Nisqually National Wildlife Refuge, Washington. Here, NWI does not indicate the presence of tidal wetland restored in 2009 from a dike removal, illustrating that NWI can be out of date and that tidal restorations are not included in the classification.

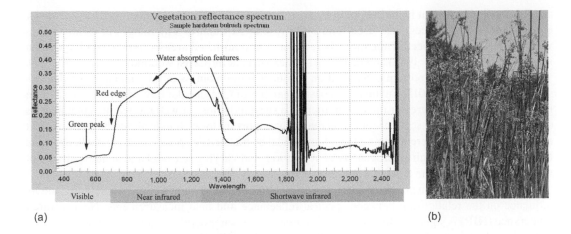

(a)   (b)

Figure 14.3   **(See color insert following page 266.)** (a) Typical hardstem bulrush (*Schoenoplectus acutus*) reflectance spectrum. (b) Hardstem bulrush. Plants absorb light in the red wavelengths and reflect light in the near-infrared wavelengths. Vegetation indices—combinations of surface reflectance at two or more wavelengths—are used to estimate aboveground biomass in tidal marsh. The commonly used NDVI is calculated as: $(R_{NIR} - R_{Red}/R_{NIR} + R_{Red})$, with R_{NIR} = reflectance in the near infrared wavelengths and R_{Red} = reflectance in the red wavelengths.

14.5 REMOTE SENSING OF TIDAL MARSH BIOMASS

The IPCC inventory approach estimates carbon emissions based on change in carbon pools. For tidal marshes, these pools include aboveground biomass, belowground biomass, dead wood, litter, and soil organic matter. In a Tier 1 inventory, aboveground biomass of tidal marshes is not included; however, the opportunity is open for Parties to apply nationally specific data should that be available. Combined, above- and belowground biomass carbon stocks may be of the order of 15–30 TC ha^{-1}, a small pool compared to soil carbon. However, aboveground biomass, through its physical interactions with suspended sediments and elevation, has a disproportionate positive influence on tidal marsh response to sea level rise (Morris et al. 2002; Byrd et al. 2016). Furthermore, peak biomass provides a threshold for projecting potential tidal marsh C sequestration (Morris and Callaway Chapter 6; Megonigal et al. Chapter 11) and methane emissions (Keller, Chapter 8).

It is good practice for Parties to account for emissions or removals with changes in the extent of biomass carbon pool, if data is available. For tidal marsh, these data are typically available through published academic studies or through long-term monitoring programs, such as with the U.S. National Science Foundation Long-Term Ecological Research (LTER) Network. These data are increasingly being made available online such as through the LTER Network Data Portal (https://portal.lternet.edu/nis/simpleSearch).

In addition to mapping land cover change, remote sensing has become a proven technology for mapping aboveground biomass, and emerging research is demonstrating the potential to estimate belowground biomass with remote sensing as well (O'Connell et al. 2014; O'Connell et al. 2015). One of the main benefits of remote sensing to measure biophysical vegetation features is the capacity to capture the extreme heterogeneity of marshes at multiple spatial scales that plot based sampling could not capture without large numbers of spatially distributed samples (Schalles et al. 2013). Remote-sensing data may be evaluated with uncertainty analyses (Congalton and Green 2009), and repeat satellite and airborne images provide the potential for frequent updates.

Optical remote sensing of the biomass of tidal marsh vegetation began in the 1980s, primarily in *Spartina alterniflora*-dominated marshes in the eastern United States, using vegetation indices such as NDVI calculated from hand-held radiometer reflectance data (Hardisky et al. 1983ab; Hardisky et al. 1984). These indices were used with 30-m resolution Landsat TM images to estimate site-wide biomass in monospecific marshes where vegetation-dominated areas several times larger than image pixels (Gross et al. 1987) and in a more species-rich Pacific Coast wetland (Zhang et al. 1997).

Image scale is an important consideration for quantifying biomass in tidal marshes. As demonstrated in Turpie et al. (2015), in the large coastal marshes of Chesapeake Bay, Maryland, in satellite images with pixel sizes greater than 30 m, there is a substantial loss of wetland features that may be delineated. This issue of scale becomes increasingly important for wetlands along the Pacific Coast of the United States, where mapping plant biophysical characteristics poses multiple challenges due to small wetland size, high heterogeneity of vegetation, and varying seasonal and annual patterns of wetland hydrology (Phinn et al. 1996; Phinn 1998).

Optical remote sensing of wetland vegetation is particularly limited by water inundation and its influence on the relationship between spectral reflectance and field measurements. Near-infrared reflectance (NIR) is significantly reduced with progressive water depth, which can highly influence NDVI-based estimates for marsh biomass (Kearney et al. 2009). The vertical stem structure of tidal marsh vegetation canopies such as those dominated by rushes, sedges, and grasses poses additional challenges; while it contributes to high photosynthetic rates by increasing light penetration within canopies, vertical stem morphology also increases light scattering and absorption in spaces between vegetation, leading to lower overall canopy reflectance (Mutanga and Skidmore 2004; Ollinger 2011). Another challenge of NDVI-type indices, particularly those based on the red and NIR portion of the spectrum, is that they reach a saturation level at high biomass density (Tucker 1977), and do not estimate biomass well at 100% vegetation cover [a common feature of freshwater marshes (Rocchio 2005)].

14.5.1 New Approaches to Aboveground Biomass Mapping

One approach for addressing the saturation effect of NIR reflectance is the use of the wide dynamic range vegetation index (WDRVI), which is an NDVI type index that scales NIR reflectance by a factor ranging from 0.1 to 0.5 (Table 14.3). The WDRVI increases the linearity between biomass and NIR reflectance, thus reducing sensor saturation (Gitelson 2004). Recently in the Gulf Coast of Louisiana, Mishra and others (2012) produced an aboveground biomass model based on this index with a percent-normalized root mean square error [RMSE/(max observed − min observed)] of 7% (Mishra et al. 2012). In a recent publication, Ghosh et al. (2016) mapped biophysical characteristics of tidal marsh vegetation in the northern Gulf of Mexico, including vegetation fraction and aboveground green biomass using field surveys combined with NASA's Moderate Resolution Imaging Spectroradiometer (MODIS) 250 m and 500 m images. The researchers used the WDRVI ($\alpha = 0.1$) for MODIS 250 to map biomass and achieved a percent normalized RMSE of 19.6% ($R^2 = 0.85$) for pixels without a water signal.

Also, in Suisun Marsh, in San Francisco Bay, California, Landsat OLI reflectance data were used to produce a model of biomass in order to estimate peak biomass. Use of the WDRVI ($\alpha = 0.2$) produced the best model for fully vegetated pixels, and best-predicted high biomass values in fully vegetated pixels, while the red/green simple ratio index (R_{Red}/R_{Green}) produced the best model for all pixels including mixed vegetation/water pixels, generating a hybrid model with percent normalized RMSE of 16% (Byrd et al. 2016).

Hyperspectral data such as from the 224 band Airborne Visible/Infrared Imaging Spectrometer (AVIRIS) has the potential to improve biomass mapping (Turpie et al. 2015). With a large number of spectral bands that extend into the shortwave infrared region of the spectrum, hyperspectral data can remove saturation problems, and potentially increase the dynamic range of predicted biomass values (Mutanga and Skidmore 2004; Thenkabail et al. 2004). There have been limited applications of mapping tidal marsh biomass with hyperspectral imagery, and many have relied on vegetation

Table 14.3 Optical and Lidar-Based Indices for Mapping Biomass in Recent Publications

Index	Formula	Sensor	Spatial Resolution	Locations Where Demonstrated	Citation
WDRVI	$(\alpha \times R_{NIR} - R_{red})/$ $(\alpha \times R_{NIR} + R_{red})$ $0.1 <= \alpha <= 0.5$	Landsat TM (5) MODIS 250 Landsat OLI (8)	30 m 250 m 30 m	Louisiana Coast Northern Gulf of Mexico Suisun Marsh, California	Mishra et al. (2012), Byrd et al. (2016), Ghosh et al. (2016)
Narrowband NDVI	$(R_{857} - R_{675})/$ $(R_{857} + R_{675})$	AISA Eagle imaging spectrometer	1 m	Sapelo Island, Georgia	Schalles et al. (2013)
Narrowband NDVI	$(R_{857} - R_{675})/$ $(R_{857} + R_{675})$	Field spectrometer data	2 m	Bahia Blanca Estuary, Argentina	Gonzalez Trilla et al. (2013)
Visible atmospherically resistant index	$(R_{green} - R_{red})/$ $(R_{green} + R_{red})$	MODIS 500	500 m	Northern Gulf of Mexico	Ghosh et al. (2016)
Simple ratio red/green	R_{red}/R_{green}	Landsat OLI (8)	30 m	Suisun Marsh, California	Byrd et al. (2016)
Red-edge NDVI	$(R_{NIR} - R_{red-edge})/$ $(R_{NIR} + R_{red-edge})$	Digital Globe World View-2	2 m	KwaZulu-Natal Province, South Africa	Mutanga et al. (2012)
Soil-adjusted vegetation index	$(R_{NIR} - R_{red}) \times$ $(1 + L)/(R_{NIR} +$ $R_{red} + L)$ $L = 0.5$	National Agriculture Imagery Program aerial imagery	0.5 m	Galveston Island, Texas	Kulawardhana et al. (2014)
Maximum vegetation height (L_{max})		Discrete return LIDAR		Galveston Island, Texas	Kulawardhana et al. (2014)
Waveform amplitude and width		Full waveform LIDAR		Cape Cod, MA	Rogers et al. (2015)

R, spectral reflectance at a given satellite band or wavelength.

indices for model development (Gonzalez Trilla et al. 2013; Schalles et al. 2013). In contrast, use of all information in the reflectance spectrum, instead of a limited number of bands, with statistical approaches such as partial least squares regression, can reduce model error. In a recent analysis of first derivative-transformed full spectrum field spectrometer data with PLS, biomass model accuracy was higher than when using simulated broadband spectra (e.g., Landsat ETM+) for values captured under low inundations conditions (Byrd et al. 2014) (12% versus 16% normalized RMSE).

In addition to optical data, in recent years, lidar data, which can identify vegetation physical structure and characteristics such as plant height, have also been used to map biomass in tidal marshes. Kulawardhana et al. (2014) studied fusion of discrete return lidar and 0.5 m multispectral aerial imagery from the USDA National Agriculture Imagery Program (NAIP) to quantify salt marsh carbon stocks in Texas, United States. The lidar metric, L_{max} (lidar derived maximum vegetation height), provided the best agreement with field measured height, but explained only 41% of variance in vegetation height and underestimated field measured values. Fusion of lidar with NAIP data using L_{max} and the soil adjusted vegetation index, explained 47% of biomass data variance, while the best models using lidar (L_{max}) or spectral data alone explained 37% and 28% variance respectively.

Using more advanced techniques, Rogers et al. (2015) used full waveform lidar and field techniques to assess the distribution of biomass throughout its height and its light blocking properties in *Spartina alterniflora* marsh in Cape Cod. It was expected that the shape of the received pulse would vary across the marsh as a function of terrain and marsh characteristics. Researchers used simple metrics from lidar waveform to estimate salt marsh parameters including height and biomass. Regression of vegetation height on waveform width and waveform amplitude returned an R^2 of 0.72. They estimated quadrat biomass density with $R^2 = 0.32$, using waveform amplitude and width.

14.5.2 Mapping Belowground Biomass

In addition, new emerging remote-sensing techniques are demonstrating the potential to estimate belowground biomass with optical satellite data. Quantifying belowground biomass is important because roots and rhizomes form soil organic carbon (SOC), and SOC represents more than two-thirds of terrestrial organic carbon, with soil residence times exceeding thousands of years (Nyman et al. 1993). O'Connell et al. (2014) tested the theory that leaf nitrogen (N) concentration, which can be remotely sensed, may be an indicator of root:shoot ratios and belowground growth, given that nitrogen inputs often increase leaf N (Kao et al. 2003; Larkin et al. 2012) and sometimes reduce belowground biomass (Darby and Turner 2008; Turner et al. 2009; Deegan et al. 2012). In a nitrogen addition experimental study of *Schoenoplectus acutus* in the Sacramento San Joaquin River Delta, California, O'Connell et al. (2014) tested whether remotely sensed vegetation indices derived from a full spectrum field spectrometer could predict leaf N concentration, root:shoot ratios and belowground biomass of *S. acutus* measured after 5 months of summer growth. A first-order derivative simple ratio index using wavelength 2,184 nm and 1,780 nm was negatively correlated with leaf N concentration ($R^2 = 0.88$, $p < 0.01$) and also was positively correlated with root:shoot ratio ($R^2 = 0.54$, $p = 0.02$), and belowground biomass ($R^2 = 0.44$, $p = 0.04$).

In a follow-up field-based study, O'Connell et al. (2015) used remotely sensed estimates of two aboveground plant characteristics, aboveground biomass, and leaf N concentration to explore biomass allocation in an impounded freshwater wetland in the Sacramento-San Joaquin River Delta. They developed a hybrid modeling approach to relate Landsat ETM+ estimates of leaf N and aboveground biomass to belowground biomass measured in situ with root ingrowth cores for species specific and mixed species models. Landsat reflectance data explained up to 70% of leaf N and 67% of aboveground biomass. Spectrally estimated foliar N or aboveground biomass had negative relationships with belowground biomass and root:shoot ratio in both *Schoenoplectus acutus* and *Typha* spp. Hybrid models explained up to 76% of variation in belowground biomass and 86% of variation in root:shoot ratio. Across sites, the average field measured root:shoot ratio for *S. acutus* was 2.5 (standard deviation: 2.2), while the Landsat-based average was 2.6 (standard deviation: 2.5). For *Typha* spp., average field measured root:shoot ratio equaled the average Landsat-based estimate (mean: 0.3, standard deviation: 0.2). This modeling approach provided the first method using optical remote-sensing data for developing maps of spatial variation in wetland belowground biomass.

14.5.3 Considerations for Mapping Tidal Marsh Biomass

Multiple sources of error can contribute to the challenges of mapping biomass in tidal marsh, including field measurements, allometric estimates of biomass, errors around scaling biomass estimates, and spatial and temporal plot co-location errors (Asner et al. 2013). The presence of background site characteristics such as water, litter, and aquatic vegetation can also affect biomass model accuracy. The accuracy and representativeness of in situ data is particularly important when scaling biomass measurements with moderate or coarse resolution satellite data (>30 m) (Mishra and Ghosh 2015). To reduce error, the number of sampling points averaged to represent a pixel needs to be sufficient and cover all species present. High-resolution imagery and site knowledge can be used to develop a detailed ground sampling design that includes a set of homogenous pixels representing a wide range in biomass values (Mishra and Ghosh 2015). In addition, use of satellite sensors with a high return frequency may be desirable for generating a robust model of biomass, as a multi-temporal dataset can capture a greater range in plant growth (Byrd et al. 2014; Byrd et al. 2016). Tidal marsh biomass can be highly variable from year to year (Morris and Haskin 1990), so multi-temporal datasets can also help to build models representative of this inter-annual variability. Species-specific models tend to have higher accuracy than multispecies models, though they require the classification of wetland plant communities, which takes significant ground data and effort (Mishra and Ghosh 2015). However, the new Landsat OLI has shown potential for generating species-invariant models (Byrd et al. 2016). Finally, estimates of biomass are scale dependent; while the highest values are typically found infrequently at the $1\,m^2$ scale, averaging biomass at the 30 m scale results in substantially lower values. This discrepancy can be reduced by scaling satellite-measured biomass by fraction vegetation cover in mixed pixels, which can help to obtain values comparable to those measured in the field. As mentioned earlier, sensor saturation can reduce the potential to estimate biomass, though hyperspectral data shows potential for improvement with more advanced analysis of spectra [e.g., use of first derivative reflectance spectra with partial lease squares regression (Byrd et al. 2014)]. In the end, managers need to consider the tradeoffs between cost, additional spectral information, and the high spatial resolution of most commercial satellite imagery, which can identify within-site variability particularly for smaller marshes.

14.6 MAPPING TIDAL MARSH DEGRADATION AND RECOVERY

Given the decision by the United States to use a "managed land proxy" approach for IPCC reporting, any quantifiable gains or losses of wetland carbon stocks will be counted in the national GHG inventory regardless of the direct or proximate causes of those fluxes. Because this approach recognizes the influence of management and alterations in affecting wetland quality, there is also substantial growing interest to quantify carbon stock changes due to wetland degradation. The forms of wetland degradation are widespread, but are known to have some form of impact on carbon stocks, emissions, and removals.

At the time the 2013 Wetland Supplement was developed, there was insufficient observational data to provide emissions factors specific to degradation. Recently though, new studies have reported on methods to use remote-sensing data to track degradation of tidal marsh that results in changes in carbon stocks. In a study of tidal marsh dieback, loss and recovery in Louisiana, Ramsey et al. (2014) documented site-specific changes in biomass associated with *Spartina alterniflora* marsh dieback from hurricanes and recovery using ground-based methods. The researchers measured a ground-based metric, the live/dead biomass ratio, as an indicator of marsh degradation caused by elevated salinity from hurricane storm surges. They determined the best radar indicator of dieback and recovery from NASA Uninhabited Aerial Vehicle synthetic aperture radar (UAVSAR) polarimetric SAR (PolSAR) data, which was then compared to an optical NIR/Red ratio vegetation index from SPOT XS4 and Landsat ETM+. With both the radar and optical metrics, they mapped degradation and recovery across the region over time, and related regional changes in these metrics within marshes that experienced dieback and those that did not. PolSAR data transformed to cross-polarization HV (horizontal send and vertical receive) backscatter were negatively correlated with the live/dead biomass ratio, and was the best indicator of degradation ($R^2 = 0.85$, $n = 6$, $p < 0.01$). The vegetation index was positively correlated with the biomass ratio ($R^2 = 0.94$, $n = 5$, $p < 0.01$). Both indices detected and mapped a sizable and previously undetected marsh dieback event between 2010 and 2012, and were also related to marsh recovery. While optical and radar results were similar, the PolSAR backscatter data is directly relatable to the three dimensional distribution of water contained in marsh leaves and stalks, and underlying sediments, has higher sensitivity to canopy structure than optical imaging, and can be collected day or night, under all weather conditions.

During the same time period of 2010–2011 in Louisiana, researchers used high resolution (< 10 m) AVIRIS data flown over Barataria Bay to study the effects of oil contamination from the Deepwater Horizon Oil Spill on salt marsh vegetation (Khanna et al. 2013). Researchers mapped oil contamination with oil absorption features in pixel spectra. Plant stress from oil contamination (or other disturbances) results in loss of pigments, water and ultimately leaves that result in changes to the reflectance spectrum, particularly in the red edge region. Researchers found the impact from oil to extend 14 m from shore, as indicated by differences in vegetation indices, in particular, the normalized difference infrared index, sensitive to plant water content ($R^2 = 0.42$) and the angle at Red, an angle index sensitive to plant pigments ($R^2 = 0.34$). Indices indicated that tidal marsh communities began to recover after one year.

14.7 FUTURE MAPPING NEEDS AND OPPORTUNITIES

A considerable body of science in the form of observational data and model development exists describing the state of salt marsh carbon stocks and carbon sequestration rate [see review by Ouyang and Lee (2014)]. Far less observational data exists describing the emissions from salt marshes when converted to other land uses, or lost to open water, or those wetlands in a degraded state. Moreover, understanding methane emissions across the salinity and elevation gradient is a newly developing science, and as described earlier, the capacity to map this gradient is just emerging (e.g., Mo et al. 2015).

Reintroducing tidal flow to drained or impounded tidal marsh is a potential method for reducing methane emissions from degraded freshwater wetlands (Kroeger et al. 2017). This activity has been highlighted as a cost-effective way for stakeholders to participate in carbon markets through wetland restoration (Rogers 2015). For reporting this activity at the national scale in GHG inventories, two major datasets are needed: a nationally consistent map of formerly (or potentially) tidal wetland soils, and a spatial database of in-progress or planned restoration activities including restoration of tidal flow. A map of drained and impounded formerly tidal soils could be built partially from NWI; however, as discussed before and displayed in Figure 14.2, tidal and impoundment coding of NWI polygons are not always consistent across surveys. Alternative data sources could be used to supplement NWI such as USDA's Soil Survey Geographic (SSURGO) Database (Natural Resources Conservation Service 2016), which maps hydric soils and differentiates between histosols and mineral soils. Both NWI and SSURGO were used in the development of C-CAP's wetland potential layer, which provides the likelihood of a wetland being present at a given location (https://data.noaa.gov/dataset/noaa-c-cap-national-wetland-potential). As demonstrated by Figure 14.2b, wetland restorations are typically not documented by NWI, and there is no national database, spatial or tabular, for tracking these activities. Planned and in-progress restoration project information is available locally through sources such as South Bay Salt Pond Restoration Project and Callaway et al. (2011) for the San Francisco Bay, and The Louisiana Coastal Master Plan (http://coastal.la.gov/2017-coastal-master-plan/); however, the mapping of diked former wetlands, and in-progress or planned restoration of tides, represent a major data need and an opportunity for future research.

Carbon fluxes, especially soil carbon accumulation rates, can vary along elevation gradients in tidal marshes (Kirwan et al. 2016). Soil accretion is likely the most important variable for measuring soil carbon accumulation in wetlands, and soil accretion is significantly greater in low marsh environments than high marsh environments (Ouyang and Lee 2014; Kirwan et al. 2016). Generally in tidal marshes, low elevation marsh and high elevation marsh are distinguished by distinct vegetation zonation patterns; for example, tall *Spartina alterniflora* dominates low marsh and short *S. alterniflora*, *S. patens*, *Batis maritima* and *Salicornia virginica* dominating high marsh in the Southeastern United States. (Schalles et al. 2013). Because these marsh elevation zones are correlated with distinct plant communities, maps of these communities generated from high-resolution imagery can provide proxies for elevation (Schalles et al. 2013; Hladik and Alber 2014).

The use of Unmanned Aircraft Systems (UAS) is creating new opportunities for identifying plant communities and other fine-scale characteristics of coastal wetlands, though challenges exist with this new technology (Zweig et al. 2015). UASs can be used to collect hyper-spatial resolution data with timing and frequency that are tailored to the application need, such as timing of tide cycles. Some of the benefits of UASs include data collection on cloudy days, flexible mission planning, lower cost, and variable spatial resolution. However, given that it is an emerging technology, regulatory and technical challenges exist. These include the need to obtain Certificates of Airworthiness for every geographic area, restrictions of flights within 30 miles of major airports without difficult to obtain permissions, and notification of local Federal Aviation Administration towers, decisions on the type of airframe and the type of payload, such as the sensor, physical dimensions, weight, power consumption, heat production, resolution, and resilience (Zweig et al. 2015).

Precise and accurate elevation data is also essential for tidal marshes; minor changes in elevation in the order of centimeters can drive key processes such as accretion (Gesch 2009). Bare earth digital elevation models (DEMs) derived from LIDAR typically have high vertical error in tidal marshes, due to the interference of dense marsh vegetation with the LIDAR signal (Chassereau et al. 2011). In addition to LIDAR-based uncertainty, there are also associated errors introduced when converting NAVD88 vertical data to local tidal data, introduced both by the distance from an associated the tide gauge data sources and uncertainty in the data themselves (VDATUM; NOAA Datum Uncertainty Report, http://vdatum.noaa.gov/docs/est_uncertainties.html). This uncertainty in elevation data can limit their usefulness for management and planning (Runting et al. 2013), so multiple approaches are being developed to correct this error. In particular, the use of vegetation-specific correction factors with real-time kinematic GPS and high-resolution vegetation maps have been found to reduce RMSE in a Georgia salt marsh from 16 cm to 10 cm (Hladik and Alber 2012), and reduce RMSE from 21.2 cm to 9.8 cm in a San Francisco Bay salt marsh (McClure et al. 2016). Statistical treatment of individual returns on full waveform LIDAR is another new technique for correcting coastal marshes for vegetation interference (Parrish et al. 2014). A third new statistical method for correcting DEMs relies on NDVI from NAIP imagery and real-time kinematic GPS (Buffington et al. 2016). The number of GPS points needed depends on the spatial variation of vegetation height and density, but with national coverage of NAIP imagery, this method has the potential to be feasibly transferable across sites.

14.8 CONCLUSION

In the United States, most of the National Inventory of GHG Emissions and Sinks for the AFOLU sector such as the USDA Forest Inventory Assessment and the Natural Resources Assessment have been derived from plot-based data. With the increasing application of spatial datasets, there is a growing interest in adopting wall-to-wall accounting procedures. This process is already being applied for the coastal wetlands analysis included in the 2017 GHG Inventory of Coastal Wetlands Report. With land conversion being one of the greatest sources of carbon emissions for tidal marsh, nationally consistent spatial datasets of marsh area, which are updated regularly, provide a key resource for GHG Inventories. Likewise, freely available optical remote-sensing data provide the capacity to track aboveground biomass, degradation and recovery, and potentially belowground biomass, enabling estimates of carbon stock changes over large areas. However, development of these remote-sensing models requires careful planning and effort in field data collection. Despite these capabilities, more accurate GHG Inventory reporting will rely on improved mapping of tidal wetlands, salinity gradients, and plant communities, as well as reduced error in bare earth DEMs. Emerging technologies, including the ability to process large datasets in Google Earth Engine, capture high-resolution imagery with UASs, and development of new hyperspectral and radar sensors, will further push our capacity to meet these objectives.

State of Policy

PART III

State of Policy

CHAPTER 15

State of International Policy for Blue Carbon Actions

Dorothee Herr
IUCN

Tibor Vegh
Duke University

Moritz Von Unger
Silvestrum Climate Associates, LLC

CONTENTS

HIGHLIGHTS

1. Since **blue carbon** was first talked about on the international policy stage in 2010, policy development itself has come a long way.
2. Blue carbon has found its footing in the relevant process of the UNFCCC, as well as what might lie ahead for incentivizing better management of blue carbon ecosystems via the implementation of the 2015 Paris Agreement.
3. Multiple opportunities exist for integration of coastal carbon management activities via the CBD and the implementation of the Sustainable Development Goals.

15.1 INTRODUCTION

The sustainable management and conservation of coastal ecosystems are goals enshrined across a diverse set of multilateral environment agreements (MEAs), such as the Convention on Biological Diversity (CBD), the Ramsar Convention, the United Nations Framework Convention on Climate Change (UNFCCC), and the recently adopted Paris Agreement. The UNFCCC in particular has since its adoption in 1992 put a specific emphasis on the sustainable management of and the need for increased cooperation for the conservation and enhancement of sinks and reservoirs of all greenhouse gases (GHGs) in natural environments, "including biomass, forests and oceans as well as other terrestrial, coastal and marine ecosystems" (Art. 4.1(d)).

However, despite early work under the Clean Development Mechanism (CDM; small- and large-scale afforestation/reforestation methodologies available for mangroves), it took more than a decade to create specific tools for implementation. The development of and growing international interest in REDD+* as a promising incentive scheme for forest restoration, conservation, and overall sustainable management materialized and brought Art 4.1(d) eventually to "full fruition." It was then, around 2007, that also the marine community started looking into the other "important missing sinks" (Pidgeon 2009) with respect to climate mitigation, which were coined as "Blue Carbon" (Nelleman et al. 2009). The year 2009 marked a turning point in the recognition of blue carbon as an important sink as two key reports by IUCN on **The Management of Natural Coastal Carbon Sinks** (Laffoley and Grimsditch 2009) and by a UNEP lead report called **Blue Carbon** as they drew together the scientific understanding of coastal and marine systems for climate change mitigation.

These reports highlighted that there was a clear gap, as well as an opportunity to strengthened climate mitigation actions. They showed that national and international carbon accounting processes under the UNFCCC had largely failed to calculate the removals by sinks and the emissions by sources from coastal wetland management and to provide a mechanism for incentivizing climate change mitigation efforts in this area.

This article will provide an overview of how the UNFCCC, as primary international, intergovernmental forum for negotiating the global response to climate change, has started to incorporate mitigation activities from nature-based solutions in coastal ecosystems into its efforts. The article will further show the linkages with other international MEAs such as the CBD or Ramsar, and outline some of their synergistic activities with the UNFCCC. The history of blue carbon in international policy fora shows how this topic can be further addressed in international fora. Outlooks on what the next priorities for policy-making are included in each section.

---

* Reducing Emissions from Deforestation and forest Degradation in developing countries, and the role of conservation, sustainable management of forests, and enhancement of forest carbon stocks in developing countries.

15.2 TERMINOLOGY AND THE COASTAL FOCUS

The term and concept of blue carbon—or coastal blue carbon—is still used in many informal and conceptual discussions, in side events or academic circles—such as this book—to summarize a set of opportunities for specific ecosystems. Typically referring to mangrove forests, seagrass meadows, and salt marshes, blue carbon has been coined primarily to access UNFCCC and related climate change mitigation incentive and financing strategies to enhance coastal and marine management efforts (Howard et al. 2017). While the term can in some minor instances be found in UNFCCC documents (e.g., SBSTA agenda, see Section 15.3.1.1), the term as such has no UNFCCC-consolidated definition or refers to a specific activity nor does to date any international decisions of the Conference of the Parties (COPs) to the UNFCCC or related conclusions from subsidiary bodies, such as the UNFCCC SBSTA, exist on the matter.

The Blue Carbon Policy Framework, developed in 2011, underscored that it uses the term Blue Carbon "for the exclusive purpose of easy identification and conceptualization." The use of this terminology does not imply intent to create new or separate policy or financing schemes. Rather, the framework is designed to allow for smooth inclusion of Blue Carbon activities into existing international policy and financing processes whenever possible, broadening the definitions and terminologies as appropriate and necessary (Herr et al. 2011). Possible examples could be the explicit and comprehensive inclusion of blue carbon ecosystems in REDD+ or the establishment of "wetlands" at the basis of a targeted adaptation-*cum*-mitigation intervention instrument (Crooks et al. 2017).

There are ongoing efforts to investigate open-ocean carbon systems, such as coral reefs, phytoplankton, kelp, and marine fauna with respect to their role for climate change mitigation as part of the UNFCCC. Given that they do not store significant amounts carbon for long periods of time and that the practical management of coral, phytoplankton, kelp, and marine fauna for carbon is hindered by jurisdictional issues and scientific knowledge gaps (Howard et al. 2017) blue carbon in this article, and within the context of the UNFCCC, focuses on mangrove, tidal marsh, and close-shore seagrass systems,

15.3 UNFCCC

The UNFCCC is one of three 'Rio Conventions' adopted at the 'Rio Earth Summit' in 1992 before entering into force on 21st March 1994. There are 197 Parties to the Convention. The ultimate objective of the Convention is the "...stabilization of GHG concentrations in the atmosphere at a level that would prevent dangerous anthropogenic interference with the climate system...."

In Art 4.1(d) the UNFCCC calls for Parties to "promote sustainable management, and promote and cooperate in the conservation and enhancement, as appropriate, of sinks and reservoirs of all greenhouse gases not controlled by the Montreal Protocol, including […] oceans as well as […] other coastal and marine ecosystems."

This section starts with a brief description of how the Blue Carbon issue found its way into UNFCCC, including the recently agreed Paris Agreement (3.1), followed by a discussion about the opportunities and limitations the UNFCCC framework offers for enhanced coastal management. Note that the UNFCCC provides for few specific mechanisms; yet it presents indirect incentives (through reporting and accounting), and it does set the scene for specific mechanisms created under the Kyoto Protocol, the Paris Agreement (with its NDC approach), or separately as in REDD+ or NAMAs (National Appropriate Mitigation Actions)—further detailed below.

15.3.1 The Political Process

15.3.1.1 Developments under SBSTA

Immediate international "uptake" of Blue Carbon within the UNFCCC was complicated by the "lack of detailed knowledge about their potential for climate change mitigation, and absence of applicable carbon accounting methodologies" (Crooks et al. 2011). The years 2010 and 2011 were marked by many efforts focused on increased engagement with the scientific community to provide additional technical information not only on the role of coastal and marine ecosystems as carbon sinks, but importantly as carbon sources that can be destroyed or degraded (see Chapter 1).

As these became more evident, Papua New Guinea requested "Blue Carbon: Coastal Marine Systems" to be added to the agenda of the 34th session of the Subsidiary Body for Scientific and Technological Advice (SBSTA) UNFCCC in June 2011 (UNFCCC 2011; UNFCCC 2011a). The SBSTA provides scientific and technological information to the COPs, the decision-making body of the UNFCCC.

Immature scientific understanding (Brazil) and the view that this was a "'underhanded' way to include new market mechanisms on the agenda under the guise of a research item" (Akanle et al. 2011) initially kept blue carbon from being considered in a proposed workshop for SBSTA 36 (Murray and Vegh 2012).

Every inter-sessional SBSTA holds a so-called Research Dialogue, a science-policy exchange without negotiations. As requested by SBSTA in its 35th session, in advance of SBSTA 36, eight parties and eight research organizations submitted their views on specific research themes to be addressed. Half of these submissions specifically mention blue carbon or related topics—see Murray and Vegh (2012) for details. Several presentations given during this Research Dialogue provided information on technical and scientific aspects of emissions and removals of all GHGs from coastal and marine ecosystems (Murray and Vegh 2012; UNFCCC 2012b). That year's participants left with an increased understanding and knowledge on scientific issues—especially the role of mangroves, tidal salt marshes, wetlands, and seagrass meadows as carbon stocks, and the threats to and potential carbon releases from these ecosystems (UNFCCC/SBSTA 2012).

In a separate move, Papua New Guinea tried to raise the issue of blue carbon also during the Research and Observation agenda item, one of the official negotiation sessions under SBSTA 36. While the attempt by Papua New Guinea that intended to bring up the idea of a blue carbon workshop failed that year, a specific workshop was held before SBSTA 39.

In October 2013, scientific experts met with Party representatives on the request of SBSTA 37 (UNFCCC/SBSTA 2012a) for a workshop to consider information on the technical and scientific aspects of ecosystems with high-carbon reservoirs not covered by other agenda items under the Convention, such as coastal marine ecosystems, in the context of wider mitigation and adaptation efforts (UNFCCC 2013a). A set of presentations addressed in detail methodological aspects and the ecology of seagrasses, mangroves, and tidal marshes (UNFCCC/SBSTA 2014). Examples of current conservation and restoration approaches in coastal marine ecosystems were also highlighted.

While no discussion around the use of incentive and financial mechanisms like REDD+ for coastal marine systems was held, the report provided recommendations, including potential roles of SBSTA to guide further work on the technical and scientific aspects of high-carbon ecosystems and provide guidance on the use of carbon accounting methodologies. For example, the report encouraged the development of common terminology, data sharing and knowledge platforms, acknowledged methodological developments by the IPCC on high-carbon ecosystems, and considered possible recommendations to the IPCC for further work (UNFCCC/SBSTA 2014).

The report notes that, "while methodological advances have been made, there are still large data gaps, especially in developing countries. Furthermore, for carbon in soils, remaining uncertainties are still high" (UNFCCC/SBSTA 2014). Despite these scientific and technical challenges, an increased confidence on the use of coastal blue carbon as part of a nature-based solution to climate change miti-gation evolved. This growth in confidence could be attributed, in part, to analysis that showed how coastal carbon management activities was, or was ready to be, reflected in existing incentive and finan-cial mechanisms established under the UNFCCC. Therefore, no new market mechanisms needed to be developed to be able to act on blue carbon, as had been feared by some Parties (Murray et al. 2012).

15.3.1.2 Blue Carbon Outside SBSTA

The UNFCCC related discussions held during 2012–2014 showed two particular traits. First, there was a move toward the use of Coastal Marine Ecosystems (CME) instead of Blue Carbon. As suggested by Murray et al. (2012): "Instead, it may be appropriate to modify the language to be more logically connected to the language already used by the UNFCCC (article 4.1[d]). Terms such as carbon pools and soil carbon are already used within the IPCC and the UNFCCC." The most recent UNFCCC documents however refer more broadly to natural sinks and sources, or the "oceans" and "forest" and do not reference CMEs particularly.

Secondly, many efforts went into advancing CEM/blue carbon by indirect means. As high-lighted amongst others by Murray and Vegh (2012), management activities in coastal carbon eco-systems can fall, or be accounted for, under already existing negotiations and mechanisms, such as REDD+, or the development NAMAs—see details in Section 15.3.2.

As an example, UNFCCC COP Decision 12/CP.17 (2011) highlights the possibility of a step-wise approach to national forest reference emission level and/or forest reference level development and also notes that significant carbon pools and/or activities should not be excluded (UNFCCC 2011c).* The step-wise approach would allow Parties to improve their calculations of forest refer-ence emission levels and/or forest reference levels over time by using updated data, improved meth-odologies and information as they become available—a process in favor of including mangrove information in a step-wise manner. This was put forward by several Parties and NGOs that favor the inclusion of mangroves under REDD+.

15.3.1.3 The Paris Agreement and Beyond

The Paris Agreement was adopted by all 196 Parties to the UNFCCC at COP21 in December 2015. This landmark agreement sets out an action plan to limit global warming **well below** 2°C, marking a turning point with nations now striving for a low-carbon economy using innovation in the technology, energy, finance, and conservation sectors. Less than 1 year since its adoption, the Agreement entered into force on November 4, 2016. This is considered extremely fast for an interna-tional treaty and signals the overall sense of urgency to address climate change (Schleussner 2016). With the second commitment period of the Kyoto Protocol not yet entered into force (and coming to an end on December 31, 2020, regardless of its ultimate adoption), the Paris Agreement shapes international actions on climate change for all countries in a long-term perspective.

"Forests and other carbon-absorbing ecosystems are at the heart of the (imperfect) Paris Agreement," says Leonard (2015). Several references to all sinks and reservoirs/emissions by sources and removals by sinks of GHG, also quoting Art 4.1(d), which specifically lists coastal and marine ecosystems, are included in the Agreement. Such references further provide ample opportunity to continue the work being undertaken by many Parties and NGOs in implement-ing various mitigation related efforts such as REDD+, NAMAs, CDM, IPCC 2013 Wetlands Supplement, among others.

---

* Decision 12/CP.17 provides guidance on systems for providing information on how all the safeguards referred to in deci-sion 1/CP.16, Appendix I are being addressed and respected. The decision also elaborates modalities relating to forest reference emission levels and/or forest reference levels as referred to in decision 1/CP.16, paragraph 71(b).

> ### TEXT BOX: RELEVANT TEXT PASSAGES FROM THE PARIS AGREEMENT FOR BLUE CARBON
>
> Preamble "Recognizing the importance of the conservation and enhancement, as appropriate, *of sinks and reservoirs of the greenhouse gases referred to in the Convention,*"
>
> Art 5 "Parties should take action to *conserve and enhance, as appropriate, sinks and reservoirs of greenhouse gases as referred to in Art. 4.1(d),* of the Convention, including forests."
>
> Art 13. "Each Party shall regularly provide the following information:
>
> (a) A *national inventory report* of anthropogenic *emissions by sources and removals by sinks of greenhouse gases*, prepared using good practice methodologies accepted by the *Intergovernmental Panel on Climate Change* and agreed upon by the Conference of the Parties serving as the meeting of the Parties to the Paris Agreement;

Blue carbon was not specifically discussed during the negotiations at COP 22 (2016) in Marrakech, Morocco. COP 22, in its attempt to lay out detailed rules for implementation, did not get as far as discussing sectoral (UNFCCC 2016).

Prior to COP 22 there was however one submission from Chile (Chile 2016), broadly on the ocean and its global response to climate change, which also made references to coastal carbon ecosystems. While Chile warned to adopt wrong incentives related to ocean mitigation solutions, it did mention "coastal ecosystems that capture CO_2 as the result of biological processes" as an exception, for which specific incentives could be granted.

Coastal blue carbon ecosystems and their related value for climate mitigation operate in the nexus between land and ocean. While there is not (yet) a separate agenda item for "The Ocean," at COP 22, for the first time "The Ocean" was part of the official Global Climate Action Agenda of the UNFCCC—as Oceans Action Day (IISD 2016). The Strategic Action Roadmap on Oceans and Climate: 2016 to 2021 also covers a section on actions on Blue Carbon (Cicin-Sain et al. 2016). The topic has also been prominently featured in many side events (GEF Blue Forest Project 2016), including one hosted by the Australian government, which, a year prior, launched the International Blue Carbon Partnership (IBCP 2016) during COP 21 in Paris. It certainly allows for addressing those blue carbon elements which—despite a range of mechanisms already open to accommodate coastal environments (see below)—would still fall outside their scope due to limitations in land-based national carbon accounting. And it permits to venture into a new area of trans-boundary action and financing (see Chapters 16 and 19).

15.3.2 Financial and Incentive Mechanisms

A suite of UNFCCC related policy, financial and incentive mechanisms and interventions (Climate Focus 2011) can be accessed to finance mitigation projects or programs in coastal blue carbon ecosystems (Herr et al. 2011, 2012). The sections below briefly describe them, as well as their opportunities and/or limitations for enhancing coastal carbon management practices.

15.3.2.1 National Reporting and IPCC 2013 Wetlands Supplement

All Parties to the UNFCCC are required to submit national reports on the implementation of the Convention to the COPs (UNFCCC Art. 4.1 and 12) which include information on emissions by sources and removals by sinks of GHGs. Reporting requirements differ between Annex I and non-Annex I countries and are detailed, for example, in Crooks et al. (2011).

The IPCC has a set of good practice guidance and guidelines available for countries to use for these communications, generally referred to as Good Practice Guidelines (GPG) (Murray and Vegh 2012).

In 2013, the IPCC published a landmark *2013 Supplement to the 2006 IPCC Guidelines for National Greenhouse Gas Inventories: Wetlands ('2013 Wetland Supplement').* This Wetland Supplement provides national-level inventory methodological guidance on wetlands, including a full chapter on coastal wetlands with default emission factor values. The Wetland Supplement is an addition to and aims to fill gaps in the coverage of wetlands and organic soils in the 2006 IPCC Guidelines (Crooks et al. 2011). The guidance covers the estimation of carbon stock (and changes in stock) in drained inland organic soils, rewetted organic soils, coastal wetlands, inland wetland mineral soils, and constructed wetlands for wastewater treatments. It adds to the 2006 IPCC Guidelines by including off-site carbon dioxide emissions via water borne losses, guidance on CH_4 emissions from rewetting of organic soils, ditch/open water emissions and CO_2, CH_4, and CO_2 emissions from peat fires (Herr at al. 2014).

The 2013 Wetlands Supplement is now being tested by several countries, including the USA and the United Arab Emirates (UAE). Countries are asked to report back to SBSTA 46 (May 2017).

15.3.2.2 National LULUCF Accounting, Annex-I Countries

The Kyoto framework introduced into GHG 'accounting', i.e., the calculation of particular carbon budgets for a range of nations (mostly industrialized countries). However, as the coverage of land use, land-use change and forestry (LULUCF) emissions was severely restricted, the accounting for blue carbon ecosystems (which do exist even in the colder countries of the North) was entirely neglected.

It took until 2011—the Conference of Durban—when countries finally agreed to allow countries—on a voluntary basis (that no country had picked up prior to the end of the first commitment period)—to account for wetland drainage and rewetting (WDR). Emissions arising from fresh drainage actions of tidal saltmarshes, for example, would be an eligible activity, so would the subsidence reversal by gradually raising water levels and building soil surfaces to intertidal elevation. Legacy issues (previous drainage with ongoing emissions) would remain outside the scope of the accounting rule, however. Other coastal restoration practices are also not admitted, even though some—such as the restoration or creation of tidal wetlands by removing barriers or adding/removing fill and planting native plants—could fall under existing activity items (such as **revegetation**). The restoration or creation of seagrass meadows by planting seeds or shoots arguably could fall under this category, as no land-restriction has been specified in the definition. However, additional guidance from technical and scientific bodies such as SBSTA or the IPCC on the inclusion of fully submerged systems into the GHG accounting process would be useful (Herr et al. 2011).

15.3.2.3 Clean Development Mechanism

The other exposure to blue carbon ecosystems of the Kyoto Protocol was indirectly through the incentive scheme of the Clean Development Mechanism (CDM), which allows industrialized countries to earn certified emission reductions (CERs) credits for investing in emission-reduction projects in developing countries. CERs can be traded and sold to meet a part of emission reduction targets under the Kyoto Protocol. The scope of eligible projects remains limited, but it includes afforestation and reforestation activities. A site-level methodology for afforestation and reforestation of degraded mangrove habitats under the CDM was approved in 2011.

Avoided deforestation or forest degradation, or enhancing forest carbon stocks are not eligible for crediting under the CDM.

The discussions are ongoing, with SBSTA considering and developing modalities and procedures for possible additional land use, land-use change and forestry activities, including revegetation, under the CDM (UNFCCC 2011b; UNFCCC/SBSTA 2015; UNFCCC/SBSTA 2016) particular
concern for many countries relates to the concept of permanence of CERs. A forest, once planted,
may be logged or burned down reversing the positive impact on the climate by releasing the carbon
dioxide sequestered. The issue is technically limited to A/R activities only, not conservation activities, for which the baseline does not include sequestration. Conceptually, however, the discussion
has burdened LULUCF interventions in all forms for many years (Joosten et al. 2016).

15.3.2.4 REDD+

REDD+ aims to mitigate climate change by reducing and removing GHG emissions through
enhanced forest management in developing countries. The '+' in REDD+ refers to the conservation,
sustainable management, and the enhancement of forest carbon stocks for the purpose of increased
removals of GHGs.

Over recent years, REDD+ has emerged as one of the main negotiation blocks between UNFCCC
Parties (La Viña 2016). A comprehensive framework—the Warsaw Framework for REDD+—(UNFCCC
2013) has been established, and the Paris Agreement (Art. 5.2) specifically encourages Parties to take
actions on **results-based payments** with relation to REDD+. Important progress has been made over
the years on many fronts and many technical details including the calculation of forest emission reference levels, REDD+ safeguards, REDD+ implementation phases, and jurisdictional approaches.

Each country, however, applies their own definition of **forest** (within the limits prescribed by
international definitions from either FAO or the UNFCCC), and not all countries consistently include
mangrove forests in their reference level calculations (baseline emissions). Therefore, coastal wetland activities sometimes qualify as REDD+, sometimes it does not. Governments increasingly
design REDD+ projects and toolkits for coastal environments, in general, and mangrove forests, in
particular (e.g., King 2012). The concept of **results-based** (or **performance-based**) support, in this
context, is particularly helpful for both the promotion of transparent impact evaluation (on the basis
of measuring–reporting–verification (**MRV**)) guidelines and the installation of community-focused
benefit systems (carbon benefits as well as non-carbon benefits). REDD+ policy development has
also advanced land tenure discussions and participatory engagement actions (including recourse
mechanisms), also benefiting indigenous populations (Savaresi 2013). Nevertheless, REDD+ development remains a slow process, and most countries have not yet tapped into the scheme's full potential. Various governments and NGOs remain skeptical as to the social, if not moral compatibility
of REDD+—seen, in this view, as a foreign (market-speculative) governance tool dispossessing
communities, rather than empowering them—and traditional (forest) land management (Bayrak
and Marafa 2016). It is noted, however, that all of the negative case study narratives known to the
authors concern terrestrial REDD+ projects, not coastal ones. This may ultimately be explained by
the unique set of characteristics—historically higher level of regulation (and thus control), looser
accessibility including for communities, and the dual habitat pressure (sea-sides and land-sides),
among others—which sets terrestrial REDD+ and coastal REDD+ apart.

15.3.2.5 NAMA

NAMAs refer to any action that reduces emissions in developing countries and is mostly prepared under the umbrella of a national governmental initiative in the context of sustainable development. NAMAs were designed to emphasize financial assistance from developed to developing
countries and to be flexible, allowing developing countries to foster activities based on their priorities and capabilities (UNFCCC 2016a). In other words, national carbon wetland programs or individual NAMAs may qualify for support if the program or activity not only reduces GHG emissions

but also contributes to economic and social development and/or poverty eradication. NAMAs can also be self-supported by a country with additional financial assistance. The UNFCCC provides a matchmaking platform to match NAMAs with financing opportunities (UNFCCC 2016b).

The NAMA structure provides an opportunity for countries to tailor them to their specific needs and mitigation potential. Countries could use NAMA readiness activities to increase the understanding of the sink capacity of Blue Carbon ecosystems and of the emissions resulting from conversion and degradation of mangroves, saltmarshes and/or seagrasses, identify drivers of these emissions, and activities needed to address those drivers.

The Dominican Republic registered the first Blue Carbon NAMA under the UNFCCC's NAMA mechanism. The first of its kind, the submission is effectively a declaration of intent by the government to mitigate GHG emissions in a manner commensurate with capacity and in line with national development goals. The experience and lessons learned in the Dominican Republic can serve as pilot and facilitate the development of blue carbon programs globally (see Chapter 17).

15.3.2.6 Nationally Determined Contributions

Nationally Determined Contributions (NDCs), as part of the UNFCCC Paris Agreement (see Section 15.3.4.1), are a means for countries to independently decide how to lower their emissions. These national level climate action and emissions reduction plans are prepared to reflect countries economic and environmental differences. Each successive NDC is to represent a progression from the previous one, representing the highest possible ambition (Art. 4.3 of the Paris Agreement) and each party shall communicate a revised NDC every five years (Art 4.9 of the Paris Agreement).

An analysis of the 163 submitted Intended Nationally Determined Contributions (INDCs)[*] showed that 28 countries have included a reference to coastal wetlands in terms of mitigation, either as part of LULUCF or other forest commitments, or general mitigation aims (Herr and Landis 2016)—see Table 15.1.

Herr and Landis (2016) also show the dual value of coastal ecosystems for mitigation and adaptation and that we need a more nuanced and refined approach for addressing those in tandem. There is no one-strategy-fits- all available, as these three systems operate between definitions and categories construed by the UNFCCC. Acknowledging the variety of synergistic mitigation and adaptation blue carbon opportunities for countries is important when considering future revised commitments (Table 15.2).

While single projects can lead to much bigger programmatic efforts, and are useful to pilot ideas, future NDCs would benefit from outlining coastal adaptation and mitigation efforts in the broader policy contexts of, for example, coastal zone management or, in the case of mangroves, including efforts within national REDD+ strategies (Herr and Landis 2016).

15.3.2.7 Article 6 of the Paris Agreement

During the last days of the Paris conference, the firm negotiation of Art. 6 of the Paris Agreement was a reminder that emissions trading did not die with the demise of the CDM. Art. 6 provides for a variety of offset or reward mechanisms (see also Streck et al. 2016) namely:

- *Cooperative Approaches*: Parties may engage in "voluntary cooperation" (Art. 6.1) and "cooperative approaches," using "internationally transferred mitigation outcomes" or "ITMOs" in climate parlance (Art. 6.2) to achieve their consolidated NDC targets. Market enthusiasts have been quick to refer to **ITMOs** as a new carbon commodity.

---

[*] Intended Nationally Determined Contributions were submitted prior and during UNFCCC COP21. Upon ratification, INDCs turned into the first NDCs.

Table 15.1 Different NDC Categories of Blue Carbon Efforts

Action Type	Countries			
Mitigation				
LULUCF and Forestry Countries that include coastal wetlands as part of LULUCF and other forest commitments	Angola Australia Bahamas Bangladesh Brunei		El Salvador Guinea Haiti Iceland Philippines	Senegal Sri Lanka Suriname USA
General Mitigation Countries that include coastal wetlands as part of general mitigation aims	Antigua and Barbuda Bahrain Belize China		Cook Islands Comoros Ecuador Guyana Kiribati	Marshall Islands Mexico Saudi Arabia Seychelles UAE
Adaptation				
Conservation, Protection, and Reforestation Countries that include coastal wetlands adaptation solutions, with references to conservation and management, protection, and reforestation measures	Bahamas Bahrain Bangladesh Belize Benin Cameroon Cape Verde Congo, the Republic of Cook Islands Cote d'Ivoire Cuba Djibouti Dominican Republic Ecuador Egypt Fiji Gabon	Gambia Grenada Guinea-Bissau Guyana Haiti Honduras India Lebanon Liberia Madagascar Marshall Islands Mauritius Mexico Morocco Myanmar	Nauru Niue Oman Philippines Saint Lucia Saint Vincent and the Grenadines Saudi Arabia Senegal Seychelles Singapore Somalia South Africa Sri Lanka Sudan Suriname Tanzania	Thailand Togo Uruguay Venezuela Vietnam
Coastal Zone Management Countries that include information and make specific references to planning tools, such as ICZM	Bangladesh Belize Cambodia Cameroon Egypt Eritrea	Gabon Gambia Georgia Grenada Guatemala Haiti	India Liberia Malaysia Morocco Myanmar	Niue Saint Lucia Saudi Arabia Sudan Vanuatu
Fisheries Countries that include information and/or see the need to prioritize adaptation in job-generating sectors using coastal and marine resources (e.g., fisheries)	Bahamas Belize Brunei Cambodia Cameroon	Cape Verde Costa Rica Djibouti Haiti	Jamaica Mauritius Nigeria Oman	Saint Lucia Sierra Leone Tanzania Vietnam
Other				
Countries that use the term blue carbon	Bahrain Philippines		Saudi Arabia Seychelles	UAE

(Continued)

Table 15.1 (*Continued*)   Different NDC Categories of Blue Carbon Efforts

Action Type	Countries			
Countries that specifically recognize both the mitigation and adaptation benefits of coastal wetlands	Antigua and Barbuda Bahrain Belize Cook Islands	Madagascar Marshall Mexico Philippines St Lucia	Saudi Arabia Islands Togo	Suriname

Source: Herr and Landis 2016.

Table 15.2   A Schematic Overview of How to Address, and Scale Up the Mitigation Contribution of Coastal Blue Carbon Efforts, While Aligning with Existing or Planned Adaptation Efforts for Future NDCs

NDC	Mitigation	Adaptation
List of potential actions for enhanced ambition of future submissions	*mangroves, seagrasses and saltmarshes*	*mangroves, seagrasses and saltmarshes, and other coastal and marine ecosystems*
1. Basic mitigation and adaptation information		
1.1 Scope and coverage	Detail inclusion of mangroves, seagrasses, and saltmarshes as mitigation solutions	Detail inclusion of mangroves, seagrasses, and saltmarshes as adaptation solutions along with relevant other coastal systems (e.g., coral reefs, bivalves) and coastal and marine resource activities (e.g., fisheries) as adaptation solutions
1.2 Methodological processes	National GHG Inventory: Detail the use of IPCC Wetlands Supplement LULUCF: Detail whether wetlands are included, and if so, how coastal wetlands are included.	
2. Examples of programmatic opportunities for coastal nature-based mitigation and adaptation		
2.1 Outline relevant policies and plans, or reforms needed	National Climate Change Plans	
	National GHG Mitigation Plans REDD+ (mangroves only) NAMAs Use of market mechanisms that are aligned with decisions and modalities of the UNFCCC	-NAPs/NAP As
2.2 Outline the national legal and regulatory context for blue carbon ecosystems	Environmental Law Coastal Management Law Fisheries Law	
2.3 Detail steps on efforts toward a more synergistic and programmatic approach for coastal areas	ICZM Detailed mangrove management plan which include the appropriate linkages of relevant mitigation (e.g., REDD+, NAMA) and adaptation (NAP/NAPAs) activities, laws and other plans, e.g., National Biodiversity Plans or other; Recognizing the synergies and co-benefits of coastal blue carbon management for both mitigation and adaptation of climate change	
2.3 Outline relevant opportunities and pilot case studies that can help promote the integration of blue carbon into NDCs	With clear mitigation (blue carbon) benefits and primary purpose, contributing to GHG emission reductions	With clear adaptation benefits and primary purpose, with mitigation as a co-benefit
2.4 Outline Funding	Identify funding opportunities and means to fast-track effective implementation while upholding environmental integrity	

Source: Herr and Landis 2016.

- *Sustainable Development Mechanism*: Based on an intervention by Brazil and the E.U., the PA also defines a sustainable development mechanism that allows private and public entities to support mitigation projects that generate transferrable GHG emissions (Art. 6.4). Programs and projects—the PA avoids using either term alone by itself—developed under this new mechanism can generate "emission reductions" which may be used by another Party to fulfill its NDC. The mechanism is implemented under the "authority and guidance" of the CPA, which, according to the Paris Agreement is to develop relevant "modalities and procedures." The provision in the Paris Decision links back to the mechanisms of the Kyoto Protocol, namely the CDM and Joint Implementation (JI), when requesting that the new mechanism be modeled after them (para. 38.f). Similar to the CDM, the mechanism addresses subnational public and private entities, and it foresees a **share of proceeds** to cover both administrative costs and adaptation needs for nations most vulnerable to climate change (Art. 6.6). This opens a future for the Adaptation Fund, created under the Kyoto Protocol, which Parties decided should be retained and transferred into the framework of the Paris Agreement.

  However, unlike the CDM, the new mechanism must 'deliver an overall mitigation in global emissions' (Art. 6.4(d)), that is, it must go beyond offsetting and have a net positive mitigation effect. Also, emission reductions may be accounted for only once in the context of NDCs, either by the host Party or by another Party (Art. 6.5). An early country submission on further implementation suggested that in the context of REDD+, only country-wide implementation (as opposed to project-based one) should be eligible. The issue can be expected to remain contentious for some time to come.

- *Framework for non-market approaches*: The Paris Agreement recognizes "the importance of integrated, holistic and balanced non-market approaches" (Art. 6.8) to assist Parties with implementing their NDCs, in the context of sustainable development and poverty eradication. It aims at both mitigation and adaptation, "enhance[s] public and private sector participation" and seeks opportunities for coordination "across instruments and relevant institutional arrangements." The conceptual scope and meaning of non-market approaches—as opposed to the kind of instruments, which are seen (though no longer called) market mechanisms, for which we find precedence in the Kyoto mechanisms—is hard to gauge. In a technical paper published in 2014, the UNFCCC secretariat summarized non-market approaches as "any actions that drive cost-effective mitigation without relying on market-based approaches or mechanisms (i.e., without resulting in transferable or tradable units)." The technical paper listed as examples from country experience fiscal instruments (such as carbon taxes) and regulation, but also voluntary agreements on mitigation action, and results-based payments for REDD+. The concept, in this interpretation, is very wide, indeed, and there will be much work ahead for SBSTA, which is charged with preparing a draft work program for the 2017 session.

Eventually, all three mechanisms may provide important incentives for blue carbon activities. It may take years, however, for detailed rules to materialize. These technical challenges aside, market (or non-market) demand will be needed to stimulate action. The CDM was able to rely on the market demand from Europe's emissions trading system, before that source largely closed in 2012 (World Bank et al. 2016). To what extent national emissions trading activity will be prepared to drive future emissions reduction and sequestration action remains, at this stage, uncertain.

The basis for fresh demand from other sources, in the meantime, was recently set out by the International Civil Aviation Organization (ICAO). During its 39th assembly session in 2016, ICAO's Member States voted to install a global carbon-offsetting scheme for the international aviation sector. The "Carbon Offset and Reduction Scheme for International Aviation" (CORSIA) will start in 2021 on a voluntary basis, with 65 states having committed to join from the start. It is still unclear what kind of offsets will be allowed and who will decide on it, but blue carbon proponents should prepare to advocate their cause.

15.4 CBD

While the UNFCCC is the main international convention dealing directly with climate change and related policy and financial instruments, different other international regimes have a certain degree of intersection with the climate change agenda. The CBD is one of them. It has three main objectives: 1. The conservation of biological diversity; 2. The sustainable use of the components of biological diversity; 3. The fair and equitable sharing of the benefits arising out of the utilization of genetic resources (CBD 1992).

While the CBD does not address national GHG reduction strategies (and blue carbon being one of them) it creates other policy and financial instruments which can be beneficial to conservation of coastal and marine ecosystems—for climate change mitigation, adaptation, and biodiversity. Countries adequately managing their coastal blue carbon ecosystems can not only contribute to climate mitigation (and adaptation) and report back to the UNFCCC but similarly contribute to their commitments under the CBD (and others, see www.ramsar.org).

In 2010, the CBD in its decisions from COP10 in Nagoya invited Parties to incorporate marine and coastal biodiversity into national climate change strategies and action plans and to promote ecosystem-based approaches to climate change mitigation and adaptation (CBD 2010). This would support the conservation, sustainable use, and restoration of marine and coastal habitats while contributing to climate change mitigation. Given the increased attention given to the mitigation value of coastal carbon ecosystem around 2009/2010, the CBD also called at the time for increased effort to identify the current scientific and policy gaps in order to enhance the sustainable management of the natural carbon sequestration services of marine and coastal biodiversity (CBD 2010a). The CBD includes sustainable wetland management, restoration of degraded wetlands and conservation of mangroves, salt marshes and seagrass beds as part of ecosystem-based mitigation approaches contributing to UNFCCC, the Ramsar Convention on Wetlands and the CBD (CBD 2010). The CBD also made aware its Parties the need to conserve soil biodiversity, especially in regard to conserving and restoring organic carbon in soil and biomass, including peatlands and other wetlands (CBD 2010a).

While COP 11 (2012) and COP 12 (2014) did not go into specific recommendations on blue carbon issues like COP 10, the need for integrating biodiversity considerations into climate change-related activities and the need to build synergies between national biodiversity strategies and action plans and national climate change strategies or action plans, including on coastal carbon issues, has been a topic ever since (CBD 2012; CBD 2014).

Ecosystem restoration, one option to mitigate climate change through coastal management, received greater attention since 2012 (CBD 2012a). A short-term action plan on ecosystem restoration was adopted at the CBD COP 23, detailing a timeline for action. Nations' effort undertaking coastal and marine restoration activities can thus help to achieving their CBD commitments while at the same time contribute to climate mitigation (and adaptation).

15.4.1 Aichi Targets

CBD COP 11 also produced the Aichi Targets, with Targets 11, 14, and 15 being the most relevant to blue carbon ecosystems:

Target 11   By 2020, at least 17% of terrestrial and inland water, and 10% of coastal and marine areas, especially areas of particular importance for biodiversity and ecosystem services, are conserved through effectively and equitably managed, ecologically representative and well-connected systems of protected areas and other effective area-based conservation measures, and integrated into the wider landscapes and seascapes.

Target 14   By 2020, ecosystems that provide essential services, including services related to water, and contribute to health, livelihoods, and well-being, are restored and safeguarded, taking into account the needs of women, indigenous and local communities, and the poor and vulnerable.

Target 15   By 2020, ecosystem resilience and the contribution of biodiversity to carbon stocks has been enhanced, through conservation and restoration, including restoration of at least 15% of degraded ecosystems, thereby contributing to climate change mitigation and adaptation and to combating desertification.

COP 12 further emphasized the need for information on experiences, lessons learned, and best practices on ecosystem-based approaches to climate change adaptation, mitigation, and disaster-risk reduction (CBD 2014), a topic still currently being promoted (CBD/SBSTA 2016).

The years 2015/2016 marked again a higher, more dedicated effort on blue carbon issues. A report *Managing ecosystems in the context of climate change mitigation. A review of current knowledge and recommendations to support ecosystem-based mitigation actions that look **beyond terrestrial forests*** (Epple et al. 2016) has been issued by the CBD. The report provides an overview of the state of knowledge on the global distribution of organic carbon stocks and rates of GHG flows to and from ecosystems under different land use intensities and in different ecological settings. It also sets out areas for future research, including the need for scenario analysis of possible impacts on ecosystems based on different socio-economic development trajectories and related changes in drivers of ecosystem degradation and conversion, as well as their implications for the feasibility and long-term likelihood of success of ecosystem-based approaches to mitigation.

15.4.2 CBD and the Post-Paris world

Following the UNFCCC Paris Agreement and the submissions of Nationally Determined Contributions, action from within the CBD focuses on the need "fully take into account the importance of ensuring the integrity of all ecosystems, including oceans, and the protection of biodiversity, and to integrate ecosystem-based approaches therein, involving the national focal points to the Convention on Biological Diversity in this work and ensuring that information, tools and guidance developed under the Convention on Biological Diversity are used" (CBD 2016a).

The potential for synergies between climate change adaptation and mitigation measures in the conservation of biological diversity and disaster risk reduction in all ecosystems is being recognized by the CBD. The CBD also encourages using platforms, such as those established under the UNFCCC, for the exchange of experiences and sharing of best practices on ecosystem-based approaches to climate change adaptation and mitigation (CBD 2016a).

Similar to what has been witnessed in terms of official UNFCCC side-events as well as beyond, the CBD also notices an increase of dedicated blue carbon events, such as the one organized during COP 13 in Cancun, Mexico: Mainstreaming Blue Carbon for conservation and sustainable development organized by the Fondo Mexicano para la Conservacion de la Naturaleza.

As mentioned prior, different international agreements can unleash different policy incentives and financial instruments to better manage coastal and marine ecosystems with various co-benefits. A revised guide by IUCN and partners discusses climate finance and other financial mechanisms to support coastal wetland programs and projects and identifies between sources, including from the CBD (Herr et al. 2015). With the new National Blue Carbon Policy Assessment Framework countries can now also better identify which policy and financing mechanisms most suits their national context (Herr et al. 2016). By applying the National Blue Carbon Policy Assessment Framework countries can achieve a first-order analysis leading to a more comprehensive and integrated approach to coastal management, with clear answers of whether and when climate and carbon-related policies and mechanisms make sense for them, and how they can be aligned with existing coastal regulation and policies.

15.5 RAMSAR CONVENTION

The Convention on Wetlands of International Importance, called the Ramsar Convention, also has a degree of intersection with the climate change agenda and blue carbon and is another international tool to foster better coastal wetland management. Ramsar is a global intergovernmental treaty that promotes the conservation and sustainable use of all wetlands through local and national actions and international cooperation. The designation of Ramsar Sites together with their effective management, as well as sustainable use of other wetlands, can, in some regions, play a vital role in carbon sequestration and storage and therefore in the mitigation of climate change (Ramsar 2015).

The Convention uses a broad definition of the types of wetlands covered in its mission, including lakes and rivers, swamps and marshes, wet grasslands and peatlands, oases, estuaries, deltas and tidal flats, near-shore marine areas, mangroves and coral reefs, and human-made sites such as fish ponds, rice paddies, reservoirs, and salt pans (Ramsar Convention Secretariat 2016).

One of the priorities of the Ramsar Strategic Plan 2016–2024 on climate change and wetlands is that the critical importance of wetlands for climate change mitigation and adaptation is understood (Ramsar Convention Secretariat 2016). Its Target 12 focuses on restoration in degraded wetlands, with priority to wetlands that are relevant for biodiversity conservation, disaster-risk reduction, livelihoods, and/or climate change mitigation and adaptation. Ramsar also seeks to align its work with the Aichi Targets (see also Section 15.4.1) as well as the Sustainable Development Goals (SDGs) (see Section 15.6).

Ramsar has also urged its Contracting Parties that are also Annex I Parties to the Kyoto Protocol to consider the wise use of wetlands for accounting of GHG emissions from wetlands under a second commitment period under the Kyoto Protocol (Ramsar 2012).

15.5.1 Technical Advice

The Convention's Scientific and Technical Review Panel (STRP) is continuously working on different tasks related to climate change mitigation and wetlands (Herr et al. 2012). The most recent Work Plan of the STRP envisions technical guidance on the implications of REDD+ for the sustainable use of wetlands as well as best practices in wetland restoration (Ramsar/STRP 2016), both with implications to foster implementation of coastal projects for climate change mitigation.

STRP has continually been asked by the COP to see its work on climate change as a high priority and, in conjunction with the Ramsar Secretariat, to collaborate with relevant international conventions and agencies in the development of a multi-institutional coordinated program of work to investigate the potential contribution of wetland ecosystems to climate change mitigation and adaptation, in particular, for reducing vulnerability and increasing resilience to climate change (Ramsar 2008).

The Contracting Parties of the Ramsar Convention have adopted a number of Resolutions that have relevance to coastal carbon management. Ramsar COP further insisted back in 2009 for urgent action to encourage expansion of demonstration sites on peatland restoration and wise-use management in relation to climate change mitigation and adaptation activities and to undertake studies of the role of wetlands in carbon storage and sequestration, in adaptation to climate change. (Ramsar 2008).

Ramsar COP further encouraged Contracting Parties and relevant organizations to undertake studies of the role of the conservation and/or restoration of both forested and non-forested wetlands in relation to climate change mitigation, including the role of wetlands in carbon storage and sequestration, GHG emissions from degraded wetlands, avoidance of GHG emissions through removals of wetland carbon sinks (Ramsar 2012).

15.5.2 Increased Cooperation and Information Exchange

Given that there is a only limited amount information exchange, the Ramsar Convention keeps recalling the need to initiate and foster greater information exchange on the actual and potential roles of wetland conservation, management, and restoration activities in implementing relevant strategies, as appropriate, in mitigating GHG emissions through enhancing carbon sequestration and storage in wetlands (Ramsar 2012, 2015).

15.6 SUSTAINABLE DEVELOPMENT GOALS

Last, but definitely not least, the SDGs are guiding the above international conventions and national policy-making. Adopted in 2015, the United Nation's 2030 Agenda for Sustainable Development includes a set of 17 SDGs to end poverty, fight inequality and injustice, and tackle climate change by 2030. They are meant to guide policy development and implementation for the next 15 years.

Of the SDG's, Goal 14 is the most relevant for blue carbon. This goal, focused on **Life below water** aims to "conserve and sustainably use the oceans, seas and marine resources for sustainable development." While other goals and targets are of relevance to coastal and marine carbon ecosystems (IUCN 2017) the below targets are key. If properly implemented, these targets directly contribute to the mitigation of climate change, beyond the provision, conservation, and restoration of other ecosystem services from coastal and marine carbon ecosystems.

14.2 By 2020, sustainably manage and protect marine and coastal ecosystems to avoid significant adverse impacts, including by strengthening their resilience, and take action for their restoration in order to achieve healthy and productive oceans

14.5 By 2020, conserve at least 10 per cent of coastal and marine areas, consistent with national and international law and based on the best available scientific information

14.c Ensure the full implementation of international law, as reflected in the United Nations Convention on the Law of the Sea for States parties thereto, including, where applicable, existing regional and international regimes for the conservation and sustainable use of oceans and their resources by their parties

Coastal carbon management also supports targets under SDG Goal 13, focused on climate action, which calls to "take urgent action to combat climate change and its impacts." For example, Target 13.1 aims to "strengthen resilience and adaptive capacity to climate-related hazards and natural disasters in all countries." Also, action around the timely implementation of Target 13.2 to "integrate climate change measures into national policies, strategies and planning" can already be witnessed in many countries (see Case Studies, Chapters 21–27). In these efforts, the need to ensure cross-sector planning—from development policies to climate change, forestry, and coastal marine resource management—is being recognized. As a particularly useful framework for cross-sector planning, Integrated Coastal Zone Management (ICZM) can help countries toward effective climate change-related planning and management, another target promoted by the SDGs (13.b).[*]

The outcome document from The Ocean Conference (2017) called Our Ocean, Our Future: Call for Action (subparagraph (k)) calls on all stakeholders to "develop and implement effective adaptation and mitigation measures that contribute to increasing and supporting resilience to ocean and coastal acidification, sea-level rise, and increase in ocean temperatures, and to addressing the other harmful impacts of climate change on the ocean as well as coastal and blue carbon ecosystems such as mangroves, tidal marshes, seagrass, …."

15.7 KEY CHALLENGES AND OPPORTUNITIES

The main objective of the international policy negotiations and developments is to establish and maintain the proper channels for the effective use of financial support to flow to blue carbon programs and projects. In this context, some key take away points from the policy and scientific developments over the last decades are:

---

[*] 3.b Promote mechanisms for raising capacity for effective *t* in least developed countries and small island developing states, including focusing on women, youth and local and marginalized communities.

- The unique nature of blue carbon (coastal areas in the interface between terrestrial and marine environments, high soil carbon content, land tenure issues, jurisdictional issues) make it a particularly challenging ecosystem to include in effective international policy-making processes, but developments over the past couple of decades created policy mileposts particularly within the UNFCCC, CBD, and Ramsar Convention. Implementing blue carbon projects and programs helps countries to reach international commitments and pledges, including the SDGs.
- There is no **one size fits all** solution or mechanism, suitable for every country, program, or project, for the inclusion of blue carbon in mitigation or adaptation projects, actions, and activities, or for the development of programs around these ecosystems within the contexts of sustainable development goals, or ecosystem sustainability initiatives. Currently, several countries are undertaking national policy assessments to determine what type of **blue carbon** policy and financial incentives best fit their national circumstances alongside the coastal management policies and practices already in place. The National Blue Carbon Policy Assessment Framework (Herr et al. 2016) helps countries identify which policy and financing mechanisms most suits their national context.
- There is a clear need to link mitigation with adaptation and other coastal management polices during national implementation. It is also important to put coastal carbon management in sync with what is already happening with regard to coastal management overall, via marine resources polices, biodiversity strategies, climate change adaptation efforts, or overall development goals. The management of blue carbon cannot happen in isolation, and the inclusion of relevant stakeholders and local communities in finding the right approach is therefore key.

CHAPTER 16

Introduction of Coastal Wetlands into the IPCC Greenhouse Gas Inventory Methodological Guidance

Tiffany G. Troxler
Florida International University

Hilary A. Kennedy
Bangor University

Stephen Crooks
Silvestrum Climate Associates, LLC

Ariana E. Sutton-Grier
University of Maryland
MD/DC Nature Conservancy

CONTENTS

HIGHLIGHTS

1. Recognizing the problem of global climate change, one of the IPCC's activities is to support the UN Framework Convention on Climate Change (UNFCCC) through its work on methodologies for national greenhouse gas (GHG) inventories.
2. In 2014, the IPCC released the 2013 Supplement to the IPCC National Greenhouse Gas Inventory: Wetlands (*Wetlands Supplement*), which[*] provides methods for estimating anthropogenic emissions

---

[*] *Wetlands Supplement*, Overview Chapter.

and removals of GHGs (CO_2, CH_4, and N_2O) associated with specific activities including aquaculture, salt production, extraction, drainage, rewetting, revegetation and creation, and forest management practice in mangroves (IPCC 2014).

3. A U.S. case study, applying the guidance in the U.S., illustrates the effect of accounting for management activities directly rather than stock changes associated with Land Use, Land-use Change and Forestry whereby all lands are defined as managed.

4. The IPCC guidance provided and U.S. case study illustrate how any country can include coastal wetlands as part of their national GHG inventory prepared for the UNFCCC.

16.1 INTRODUCTION

16.1.1 Global GHG Mitigation—Urgency and Priorities for Policy

Through nationally determined contributions, The Paris Agreement (Decision 1/CP.21,) aims to strengthen the response of countries around the world to combat climate change. In all sectors, aggressive changes in greenhouse gas (GHG) emissions will be necessary to achieve the temperature goal of limiting the increase well below $2°C$. The United Nations Framework Convention on Climate Change (UNFCCC) is essential for addressing climate change as a primary mandate is to: "...to achieve stabilization of GHG concentrations in the atmosphere at a level that would prevent dangerous anthropogenic interference with the climate system" (http://unfccc.int).

The IPCC Task Force on National Greenhouse Gas Inventories produces guidelines for estimating national GHG emissions and removals associated with anthropogenic activities. Those guidelines are intended to be used by all Parties (countries) to: A) provide generic, default data and methods; B) provide standardized methodology and QA/QC guidance to ensure transparency, accuracy, completeness, consistency, and comparability among countries (TCCCA); C) allow the use of more sophisticated methods if countries wish to use them and they are consistent with the guidelines; D) satisfy the requirements of international climate change policy; and E) contribute to national mitigation and emission reduction plans. These general guidelines provide methodological procedures to estimate GHG emissions and removals for every sector of a nation's economy, including Energy, Industrial Processes and Product Use (IPPU), Agriculture, Forestry and Other Land Use (AFOLU) and Waste sectors. The methodological approaches are presented to estimate GHG emissions and removals from the product of an emission factor and activity data following increasing robustness and reduced uncertainties in a three-tiered hierarchy, with Tier 1—a simple, first-order approach with default values provided, based on globally available data via expert review of the available scientific literature, Tier 2—replacing the Tier 1 default values with country- or region-specific values, greater stratification, and/or more disaggregated activity data, and Tier 3—the highest-order approach, using detailed modeling and inventory measurement systems with greater data resolution (IPCC 2006). National circumstances include the availability of data and knowledge, and contribution made by the category to total national emissions and removals and to their trend over time. The most important categories, in terms of total national emissions and the trend, are called key categories, which generally require Tier 2 or Tier 3 methods.

16.1.2 Coastal Wetlands as C Sinks Contributing to Negative Emissions

Initially, coastal wetlands capture carbon dioxide (CO_2) from the atmosphere and convert it to organic carbon that is stored in their vegetation. The incorporation of the biomass (both above- and belowground) into the soil and its burial contributes to the long-term accumulation of organic carbon that is termed autochthonous. Furthermore, particles derived from other sources such as the adjacent land masses and oceanic regions, thus termed allochthonous, are transported via seawater currents and trapped as they flow through the vegetation, further contributing both mineral sediments and organic matter to the soil. As a consequence, the elevation of coastal wetland soils increases at a pace similar to that of sea level rise and can result in organic carbon deposits several meters thick (Chmura et al. 2003; Donato et al. 2011; Duarte et al. 2013). The storage of organic carbon in the soils is aided by the low rates of decomposition that occur in oxygen-poor saturated soils. Overall, the burial rate of organic carbon in vegetated coastal soils is at least an order of magnitude greater than that of terrestrial forests. Although coastal vegetated habitats only represent about 3% of the area covered by terrestrial forests, their contribution to long-term organic carbon burial is similar (Mcleod et al. 2011; Duarte et al. 2013). In comparison with carbon burial in marine soils, coastal vegetated habitats occupy only 0.2% of the ocean surface but are estimated to account for approximately 50% of total carbon burial rate in the oceans (Duarte et al. 2005; Nelleman et al. 2009).

About one-third of the area occupied by coastal vegetated habitats has been lost over the past half century as a result of management activities concerned with deforestation, urbanization, aquaculture, reclamation, and drainage (McLeod et al. 2011). The loss of coastal vegetation due to these activities compromises the continued accumulation of carbon and increasing elevation of the soils. The removal of biomass can also result in erosion of the soil and potential release of stored carbon through oxidation of organic carbon back to gaseous CO_2. The largest impact occurs when the soil is also disturbed and particularly when the previously water-logged, oxygen-poor soil is drained or moved to ground above the local water table, when the soil organic carbon is exposed to oxygen and microbial decomposition of the organic carbon results in enhanced fluxes of CO_2 back to the atmosphere (Lovelock et al. 2011). Pendleton et al. (2012), using available data on land use conversion rates, estimated that between 0.15 and 1.02 Pg of CO_2 was being released annually to the atmosphere, equivalent to 3%–19% of the CO_2 assigned to deforestation globally.

16.1.3 An Important Part of a Portfolio for Achieving Negative Emissions with Significant Ecosystem Services and Other Co-benefits

It is important to recognize that there are also important additional benefits to society of achieving negative emissions, particularly when natural climate solutions are part of a country's portfolio of activities. Natural climate solutions is a term that means methods of reducing emissions and also adapting to climate change impacts such as sea level rise and coastal flooding, by using natural or nature-based options. These options include activities such as ecosystem restoration or creation and the use of nature-based (or hybrid) approaches such as living shorelines (Sutton-Grier et al. 2015). When these natural solutions, such as ecosystem restoration, are implemented, society gets many benefits from these projects because coastal ecosystems provide many co-benefits like important recreational and tourism opportunities, key fishery habitat, water quality improvements, and flood and erosion mitigation which can reduce risks to life, property, and economies (Barbier et al. 2011, Vegh et al. Chapter 19).

16.2 HISTORY

"Recognizing the problem of global climate change, the WMO and UNEP co-established the IPCC in 1988. One of the IPCC's activities is to support the UNFCCC through its work on methodologies for national greenhouse gas inventories" (IPCC 2014). Under the UNFCCC, all Parties have agreed to report anthropogenic emissions by sources and removals by sinks, including from the Land Use, Land-use Change and Forestry* (LULUCF). Reporting is accomplished through the submission of national reports (National Communications and National GHG Inventories, biennial reports or biennial update reports), with different requirements for Annex I and non-Annex I countries (Iverson

---

* UNFCCC Article 4.1 (a). http://unfccc.int/files/essential_background/background_publications_htmlpdf/application/pdf/conveng.pdf.

et al. 2014). The primary directive of inventory reporting is to "...reduce catastrophic interference with the climate system...." Thus, the focus of GHG inventory reporting is to improve the management of GHG at the national level, understanding that the primary sources and sinks of GHG emissions and removals are anthropogenic in nature.

Notably, significant CO_2 emissions have been reported for wetlands across the world associated with impacts through anthropogenic activities. A Florida example is the conversion of Everglades wetlands for agricultural production of sugarcane (Figure 16.1). In the Florida Everglades, more than 283,280 ha (700,000 acres) in area, the Everglades Agricultural Area was established in the Central and South Florida Project for Flood Control and Water Supply. By draining this area for agriculture (mostly sugar), former Everglades peat soils have experienced significant oxidation—losing as much as 6 ft. in elevation across the area over the last 80 years (Aich 2013).

The IPCC GHG reports provide the methodological guidance documents countries are encouraged to use to report their national level GHG estimates. The UNFCCC requests updates to these reports as directed by countries that are party to the Convention. The IPCC focal point for each country nominates experts to serve on IPCC panels to develop these updates. The country reports, developed using the IPCC Guidance documents, are reviewed annually by a UNFCCC Expert Review Team (ERT), each expert satisfactorily completing examinations focused in general inventory reporting and specific to their sectors and policies of expertise.

As part of the comprehensive GHG estimation methodologies, the IPCC GHG documents have developed and updated significant guidance on the LULUCF sector [i.e., Good Practice Guidance for Land Use, Land-use Change and Forestry (GPG-LULUCF) and the 2006 IPCC Guidelines for National GHG Inventories, Volume 4, Agriculture Forestry and Other Land Use (2006 IPCC Guidelines)]. The 2006 IPCC guidance maintains six land-use categories: Forest Land, Cropland, Grassland, Wetlands, Settlements and Other Lands (e.g., bare soil, rock, ice, etc.). The 2006 IPCC Guidelines use these categories for the purposes of estimating anthropogenic emissions and removals from LULUCF (i.e., managed lands). Emissions and removals are not reported for unmanaged lands, but the area for those lands is tracked over time. The same six

Figure 16.1   More than 700,000 acres in area, the Everglades Agricultural Area was established in the Central and South Florida Project for Flood Control and Water Supply. By draining this area for agriculture (mostly sugar), former Everglades peat soils have experienced significant oxidation—losing as much as 6 ft. in elevation across the area over the last 80 years (photo credit: Dr. Ramesh Reddy, University of Florida).

categories are used in the agreed UNFCCC Common Reporting Format (CRF) for submission of developed country (Annex I) national GHG inventories (Iverson et al. 2014). For each of the six land-use categories, emissions and removals from the following pools are estimated: Living biomass (separate above- and belowground pools); Dead organic matter (deadwood and litter); and soil organic carbon (mineral- and organic-rich soils). In addition, wood products such as timber used in construction or furniture, as well as wood products put in landfills, referred to as harvested wood products (HWPs) are reported as an additional pool under LULUCF. A general assumption of AFOLU approaches with regard to estimation of CO_2 emissions and removals is that the sum of C gains and losses is equivalent to net stock changes and equivalent to total emissions and removals. Thus, the inventory compiler is not restricted to the use of C stock data or CO_2 flux data, as long as the general equations used (stock-difference or gain loss) are complete, accurate, and consistent, avoiding over- and under-estimates (e.g., double-counting).

Notably, however, the coverage of the 2006 IPCC Guidelines on wetlands was restricted to peatlands drained and managed for peat extraction, conversion to flooded lands, and limited guidance for drained organic soils(IPCC 2014). Agreeing to address the former gaps in methodological guidance, IPCC released the 2013 Supplement to the 2006 IPCC Guidelines for National Greenhouse Gas Inventories: Wetlands (herein **Wetlands Supplement**, IPCC 2014). The Wetlands Supplement "extends the content of the 2006 IPCC Guidelines by filling gaps in coverage and providing updated information reflecting scientific advances, including updating emission factors." It covers inland organic soils and wetlands on mineral soils, coastal wetlands including mangrove forests, tidal marshes and seagrass meadows, and constructed wetlands for wastewater treatment.

The Wetlands Supplement, like the 2006 IPCC Guidelines, generally provides guidance, with a series of decision trees to help guide the inventory compiler to produce national inventory estimates that are transparent, consistent, credible, complete, and accurate (TCCCA). Estimation methods are similarly at three levels of detail, from Tier 1 (the default method) to Tier 3 (the most detailed method; IPCC 2014). The provision of different tiers enables inventory compilers to use methods consistent with their resources and to focus their efforts on those categories of emissions and removals that contribute most significantly to national emission totals and trends.

It is also worth noting that the IPCC is currently in the process of refining the guidance in the 2006 IPCC Guidelines and the 2013 Supplement to improve the methodologies and default emission factors. This guidance, to be called the 2019 Refinement to the 2006 IPCC Guidelines for National GHG Inventories will be presented to the IPCC Plenary in 2019 for approval.

16.3 INTRODUCTION OF COASTAL WETLANDS IN GHG INVENTORY REPORTING

Significant land-use change in wetlands, and increasing data availability with which to estimate GHG emissions and removals associated with it, brought an invitation from the UNFCCC to the IPCC to "undertake further methodological work on wetlands, focusing on the rewetting and restoration of peatland."* In the case of coastal wetlands, a synthesis provided by Valiela et al. (2001) reported that the most important human activities contributing to the loss of mangrove were shrimp culture (38%), forestry uses (26%), fish culture (14%), diversion of freshwater (11%), land reclamation (5%), and another 5% collectively to herbicides, agriculture, salt ponds, and coastal development. Significant land-use change combined with the significant carbon stock contained in soils and biomass of coastal wetlands subject to those land-use changes can produce CO_2 emissions of large magnitude. Alternatively, management that rewets or restores coastal

---

* Paragraph 72 of SBSTA 40 report (FCCC/SBSTA/2010/13), available at: http://unfccc.int/resource/docs/2010/sbsta/eng/13.pdf-insert.

wetlands can result in active sequestration and long-terms inks of CO_2, with increasing areas of intact coastal wetlands contributing to a portfolio of anthropogenically derived atmospheric C reductions. The guidance provides estimation methodologies covering management activities with the largest global influence on GHG emissions and removals on managed coastal wetlands (IPCC 2014). Annex I countries, which includes the U.S., are encouraged to use the Wetlands Supplement in preparing their annual inventories under the Convention from 2015 (Decision 24/CP.19 paragraph 4).

Regardless of whether a land-use change occurs or not, it is *good practice* to quantify and report significant emissions and removals (see Table 4.1 of the Coastal Wetland Chapter) resulting from management activities on coastal wetlands in line with their country-specific definition (Wetlands Supplement, Chapter 4, p. 4–6). Chapter 4 of the Wetlands Supplement provides guidance on estimating emission and removals of GHGs (CO_2, CH_4, and N_2O) associated with specific activities on managed coastal wetlands (Wetlands Supplement, Overview Chapter, p. 0–8). Emissions and removals may be quantified as a rate or a stock change. For example, agricultural lands with organic soils that were formerly coastal wetlands will continuously emit CO_2 once drained (and CH_4 if extensive ditching accompanies the drainage; see Chapter 2, IPCC 2014), and would be reported under the management activity of drained, wetland soil. Previously, terrestrial land-use conversions rarely considered the persistent nature of CO_2 emissions associated with drained lands that were formerly wetland. As long as organic soils persist, CO_2 emissions continue, particularly when drainage depths are lowered to maintain agricultural production, allowing those emissions to continue at significant rates (Chapter 2, IPCC 2014). Conversion to Settlements is reported as causing CO_2 emissions when development practices remove anaerobic wetland soils, on average to 1 m depth, and remove or otherwise expose those soils to aerobic conditions. In this case, emissions are quantified at the time of the conversion activity (i.e., instantaneous within the year that settlements in coastal wetlands were constructed). A similar assumption is applied for other coastal development activities like construction of aquaculture and salt production ponds. Regardless of activity, the methodological guidance for coastal wetlands provide that the area upon which the activity occurred in a year or changed to that activity within a year was the basis for applying the fundamental equation (GHG flux = emission factor * activity data).

The IPCC source categories that comprise new sources/sinks fall under four categories:

A. New CO_2 sources/sinks associated with specific management activities that can be reported using the guidelines in the IPCC Wetland Supplement Forest Land, Cropland, Grassland, Wetlands, Settlements, and Other Land categories (CO_2).

1. *Forest management in mangroves*: Removal of wood occurs to different extents throughout the tropics where mangrove forests are harvested for fuel wood, charcoal, and construction (Ellison and Farnsworth 1996; Walters et al. 2008). Natural disturbances are another form of biomass carbon stock loss. There may also be conversion to forest land where mangrove replanting can take place on rewetted, or already saturated, soils. The guidance provides updates for data used to estimate C stock change specifically in mangrove living biomass and dead wood pools, relevant to IPCC CO_2 source/sink categories that may fall under any land-use category, especially pertaining to mangrove forests, including: aboveground biomass, aboveground biomass growth, ratio of belowground to aboveground biomass, C fraction of aboveground biomass (Table 16.1), wood density and litter and dead wood C stocks.

2. *Extraction in mangroves, tidal marshes, and seagrass meadows (including excavation generally, and construction for aquaculture and salt production specifically)*: Extraction collectively refers to: Excavation of saturated soils above the local groundwater level, leading to unsaturated soils (Table 16.1) and removal of biomass and dead organic matter. Activities that lead to the excavation of soil often lead to loss of coastal wetlands. The excavated or dredged soil is also commonly used to help develop coastal infrastructure where there is a need to

Table 16.1 C stocks and Fluxes (i.e., Emission Factors) Presented in 2013 Wetlands Supplement (Reproduced from 2014 IPCC)

Stock/Flux	Pool	Vegetation Type	Climate Sub-Region	Value (95% CI)
C stock (Tonnes dry mass ha⁻¹)	AG biomass	Mangrove	Tropical Wet	192 (187–204)
			Tropical Dry	92 (88–97)
			Subtropical	75 (66–84)
	BG biomass[a]	Mangrove	Tropical Wet	94 (90–98)
			Tropical Dry	27 (26–28)
			Subtropical	72 (68–75)
C stock (Tonnes C ha⁻¹)	Soil[b]	Mangrove		386 (351–424)
		Tidal Marsh		255 (254–297)
		Seagrass Meadow[d]		108 (84–139)
C flux (Tonnes C ha⁻¹year⁻¹)	Soil[c]	Mangrove		−1.62 (1.3–2.0)
		Tidal Marsh		−0.91 (0.7–1.1)
		Seagrass Meadow		−0.43 (0.2–0.7)
C flux (Tonnes C ha⁻¹year⁻¹)	Soil[e]	Mangrove Tidal Marsh		7.9 (5.2, 11.8)

[a] Based on ratio of BG/AG; see IPCC 2014 for values for tidal marsh and seagrass meadow.
[b] Aggregated organic and mineral soils; see IPCC 2014 for organic and mineral soil C stocks disaggregated by vegetation type; see Holmquist et al. (2018a) for discussion on soil C stock parsimony among vegetation types and climate sub-regions.
[c] Soil C sequestration (removal from atmosphere); aggregated organic and mineral soils.
[d] Mineral soils only.
[e] Aggregated organic and mineral soils.

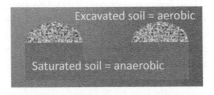

Figure 16.2 Methodological approach for estimating soil C removal and CO_2 emissions associated with excavation and the extraction of soil C with the construction phase of the management activity of aquaculture.

raise the elevation of land in low-lying areas and/or contribute to new land areas for settlement (Figure 16.2). Aquaculture and salt production are also common activities in the coastal zone and similarly require excavation of soil and removal of biomass and dead organic matter to facilitate their construction (Table 16.1). Global default C stock values (Tier 1) that can be applied for this management activity include soil C stocks for mangroves, tidal marshes (in both mineral- and organic-rich soils), and seagrass meadows in mineral soils (Table 16.1). All of these default factors lead to significantly higher emissions than a loss of mangrove biomass C stocks from the same area. Excavation can result in significant emissions, for example, applying a point-change analysis for activity data derived from MODIS imagery, 21,000 ha were converted from mangrove forest to aquaculture in Kalimantan, Indonesia (Rahman et al. 2013). Applying methodological guidance for conversion of mangrove forest to aquaculture, approximately 47 million metric tons (MMT) of CO_2 were emitted from 2000 to 2010, and approximately 13MMT in 2002 alone. As an example of the magnitude of these emissions, it is equivalent to 70% of all GHG emissions in Southeast Florida region in 2009, roughly 6 million people. (B) Construction is only the first phase in aquaculture and salt production. The second phase, termed use is when fish ponds, cages, or pens are stocked and fish production occurs and salt production ponds in mangroves and tidal marshes is the use of these facilities.

3. *Soil drainage in mangroves and tidal marshes (CO_2)*: Mangroves and tidal marshes have been diked and drained to create pastures, croplands, and settlements since before the 11th century (Gedan et al. 2009). The practice continues today on many coastlines. On some diked coasts, groundwater of reclaimed former wetlands is pumped out to maintain the water table at the required level below a dry soil surface, while on other coasts drainage is achieved through a system of ditches and tidal gates. Due to the substantial carbon reservoirs of coastal wetlands, drainage can lead to large CO_2 emissions (Table 16.1).

4. *Rewetting, revegetation and creation in mangroves, tidal marshes and seagrass meadows*: Rewetting is a pre-requisite for vegetation reestablishment and/or creation of conditions conducive to revegetation. Revegetation can occur by natural recolonization, direct seeding, and purposeful planting. This activity is also used to describe the management activities designed to reestablish vegetation on undrained soils in seagrass meadows. Also included in this activity are mangroves and tidal marshes that have been created, typically by raising soil elevation or removing the upper layer of upland soil or dredge spoil, and grading the site until the appropriate tidal elevation is reached to facilitate reestablishment of the original vegetation. Alternatively, created wetlands with mangroves can be found where high riverine sediment loads lead to rapid sediment accumulation, so that previously subaqueous soils can be elevated above tidal influence. This naturally created land can be reseeded or purposefully vegetated. Once the natural vegetation is established, soil carbon accumulation is initiated at rates commensurate to those found in natural settings (Craft et al. 2002, 2003; Osland et al. 2012; Table 16.1).

B. *New category*: CH_4 Emissions from Rewetting of Mangroves and Tidal Marshes (CH_4)
    Rewetting of drained soils, through reconnection of hydrology, shifts microbial decomposition from aerobic to anaerobic conditions, increasing the potential for CH_4 emissions (Harris et al. 2010). A strong inverse relationship between CH_4 emissions and salinity of mangrove and tidal marsh soils exists (Purvaja and Ramesh 2001; Poffenbarger et al. 2011). The global default CH_4 emission factors are $193.7 \pm 100\,kg\ CH_4\,ha^{-1}\,year^{-1}$ for coastal wetlands with salinity less than 18 psu (assumed to be tidal freshwater and brackish marsh and mangrove wetlands) and $0\,kg\ CH_4\,ha^{-1}\,year^{-1}$ for coastal wetlands with salinity that exceeded 18 psu (assumed to be tidal saline water marsh and mangrove).

C. *New category*: N_2O Emissions from Aquaculture (N_2O)

The most significant activity contributing to N_2O emissions from managed coastal wetlands is aquaculture. One-third of global anthropogenic N_2O emissions are from aquatic ecosystems, and nearly 6% of anthropogenic N_2O–N emission is anticipated to result from aquaculture by 2030 at its current annual rate of growth (Hu et al. 2012). In seagrass meadows, this direct N_2O source arises from N added to fish cages (e.g., off-shore installations). N_2O is emitted from aquaculture systems primarily as a byproduct of the conversion of ammonia (contained in fish urea) to nitrate through nitrification and nitrate to N_2 gas through denitrification. The N_2O emissions are related to the fish production (Hu et al. 2012). The global default emission factor is based on a synthesis developed by Hu et al. (2012) as kg N_2O-N per kg fish produced (0.00169 ± 0.001). This EF would be applied when the aquaculture was in use and would be additional to any CO_2 emission related to excavation associated with the construction of an aquaculture facility.

16.4 METHODOLOGICAL APPROACHES FOR COASTAL WETLANDS GHG ESTIMATION

The general methodological approach provided in the Coastal Wetlands chapter was an activity-based approach. The intention of this was to reduce any potential additional burden on countries so that, at a minimum, GHG emissions and removals for managed coastal wetlands could be provided for the management activities that were expected to result in the largest sources and sinks. Also important was that countries may have more available data on management activities rather than comprehensive coverage of all managed and unmanaged coastal wetlands to use areas of land use and land-use change for the entire country in a way that follows *good practice*. Starting a new sub-sectoral GHG contribution also presents several opportunities for identifying additional sources and sinks of emissions to those outlined in the guidelines and which can potentially contribute significantly to meeting a country's nationally determined contribution (NDC). Several challenges are also presented, related to identifying new sources of data, and ensuring consistency with the existing inventory, applying additional, sector-specific quality assurance, including: 1) the difficulties in distinguishing managed from unmanaged coastal wetlands (as is common to Forest and Grassland categories of LULUCF), 2) the significant area of coastal wetlands under management within and outside of the land base, 3) ensuring consistency with the existing, land (terrestrial)-based inventory, and 3) using limited resources most efficiently.

16.4.1 Generic Step-by-Step Guide to Implementation

Implementation of methodologies to develop coastal wetland GHG estimates for national inventories is not unlike developing an inventory for the first time, but at much smaller scale and required effort. Given the focus of the IPCC Wetlands Supplement Chapter 4 on coastal wetland management activities, additional considerations concerning whether to apply a land-based (i.e., managed land proxy) or activity-based approach are warranted. The following generic implementation steps describe a possible land-based approach. A specific applied example based on the U.S. case study follows.

1. Define coastal wetlands and identify national area of coastal wetlands
   a. Apply consistently through time. For example, U.S. used mean spring high water (MHHWS) line to identify an upper boundary for coastal lands (Crooks et al. in review). If including seagrass meadows, the lower boundary offered is as the extent of vegetated submerged lands. Given that this line will change with time, a buffer or transitional area could be delineated for focusing some additional resources to ensure that over or underestimates are avoided where the terrestrial inventory meets the coastal wetland inventory—an area of greatest uncertainty in terms of GHG flux estimates (freshwater-brackish water transition) and where certain flux rates may be greatest relative to their area (DOE report) and scaled to Earth system models, as inventories and technologies applied to develop them, improve.
2. Identify areas managed and unmanaged coastal wetlands
   a. Identify best available activity data for identifying and tracking coastal wetland area. Refer to Chapters 1 and 4 of the IPCC Wetlands Supplement (IPCC 2014) for further guidance.
   b. Define protocols for consistent tracking of lands. Refer to Volume 4, IPCC 2006, and Chapter 7 of the IPCC Wetlands Supplement (IPCC 2014) for further guidance.
   c. Develop approach for harmonizing existing tracking of terrestrial lands with coastal wetlands, with special attention to the intertidal transition area and areas where coastal wetlands have been converted to other uses (i.e., drainage for agriculture, conversion to settlements)
3. Develop a cross-walk of types of land-use change to coastal wetland management activities described in the IPCC Wetlands Supplement (IPCC 2014). An example approach can be found in Table 16.1.
4. Determine available, carbon stock and emission and removal factor data following *good practice*, and applying default Tier 1, Tier 2 or Tier 3 data, as appropriate to national circumstances. Refer to Chapters 4 and 7 of the IPCC Wetlands Supplement for further guidance.
   a. Determine level of disaggregation to apply, based on soil type (organic and mineral), climate zone, vegetation type, salinity level, etc. as appropriate to national circumstances. Note: appropriate level of disaggregation can be first assessed based on available data. If there is significant variation among different categories of disaggregation, consider applying carbon stock and emission/removal factor data to that level of disaggregation. See Holmquist et al. (in review) for further considerations.

5. Apply appropriate equations found in the IPCC Wetlands Supplement consistent with the type of land-use change (management activity). Apply expert judgment to verify assumptions and as appropriate to address national circumstances.
6. Sum emission estimates across levels of disaggregation for Land Remaining and Land Converted sub-categories.
7. Follow IPCC guidance (IPCC 2006, IPCC 2014) for TCCCA (defined in section 16.2) including full documentation, uncertainty assessment, time series consistency, quality assurance, recalculations, and future improvements.

A potential approach for cross-walking the change in land-use available from a land-use change database that approximates GHG methodologies for management activities found in the IPCC Wetlands Supplement, each land-use change might be aggregated by comparable activity types that are appropriate to a respective methodology (i.e., EPA 2017), following either the Supplement guidance or concurrence through expert judgment (Table 16.2).

16.5 U.S. CASE STUDY

The United States is among those countries with significant coastal wetland area change through conversion. For example, between 1922 and 1954, 6.5% of coastal wetlands were lost. The annual rate of loss in the 1980's was nearly 0.2% year^{-1}, with the highest overall rate of coastal wetland loss observed in Louisiana during the 1958–1974 period (Valiela et al. 2006). The U.S. Environmental Protection Agency (EPA) completed their first GHG inventory to include coverage of coastal wetlands in 2017 (EPA 2017). In previous reports of the U.S. Inventory, inclusion of GHG emissions and removals for managed Wetlands were restricted to those lands where the water table is artificially changed (i.e., lowered or raised) and was currently limited to emissions and removals from managed peatlands. Based on expert judgment, in applying the Wetlands Supplement, those coastal (intertidal) wetlands, which fall within the U.S. Land Representation, because of accessibility and level of regulation oversight,[*] were also considered Managed Lands. This simplifies reporting, as databases for some activities (notably restoration of wetlands) are absent, enabling estimation of emissions and removals by tracking changes in land cover. Greater than 99.5% of the conterminous U.S. falls under Managed Lands Proxy and this extends to U.S. coastal wetlands.

The U.S. defined coastal wetlands as "a wetland at or near the coast that is influenced by brackish/saline waters and/or astronomical tides. Coastal wetlands may occur on both organic and mineral soils. Brackish/saline water is water that normally contains more than 0.5 or more parts per thousand of dissolved salts." The landward extent of coastal land is defined by the extent of mean spring high water line and includes all lands (wetlands and other land uses) at or below this elevation seaward to the extent of offshore waters (United States Coast Guard Lawson Line; Crooks et al. *in review*). In order to determine the inland boundary of U.S. Coastal Land and approximate the coastal area most likely influenced by tidal fluxes along the U.S. coasts we established a Mean Higher-High Water Spring (MHHWS) surface. This area was determined by extrapolating the values of those tides that accounted for 95% (2 standard deviations) of all values that exceeded MHHW over the last three full years (Crooks et al. in review). Emissions and removals on seagrasses, which fall beyond the area included in the U.S. Land base, were not included at this stage and are a planned

---

[*] Wetlands in the U.S. are managed for a number of ecosystem services, and are covered by regulatory permits as well as several laws. For example, the U.S. Clean Water Act regulates discharges of pollutants into the waters of the United States, which includes many wetlands, in order to ensure that U.S. freshwater resources are protected. The U.S. also has the **No Net Loss** policy which was first adopted as a national goal in 1988. This policy uses multiple policy tools, including the Clean Water Act, the Coastal Zone Management Act, and conservation easements to stem the loss of wetlands in the U.S. by working to offset impacts to wetlands with created or restored wetlands that serve a similar function.

Table 16.2 2015 Summary of U.S. Coastal Wetland CO_2 and Non-CO_2 Emissions and Removals (EPA 2017 Unless Otherwise Noted)

CO_2 IPCC C Pool	LULUCF category (2015)	Sub-category[a]	IPCC Wetlands Supplement Coastal Wetlands Management activity	IPCC Equation	CO_2 Eq. emission (MMT year^{-1})	CO_2 Eq. 95% CI
Soil	Land Remaining	Wetland remaining Wetland	Rewetting, Revegetation, and Creation[c]	Eq. 4.7: $CO_{2SO-RE} = \Sigma_{v,s,c} (A_{RE} \times EF_{RE})_{v,s,c}$	(12.2)	±29.5%
		Wetland converted to open water[b]	Extraction	Equation 4.6: $\Delta C_{SO-CONVERSION} = \Sigma_{v,s}(SO_{AFTER} - SO_{BEFORE})_{v,s} \times A_{CONVERTED\ v,s}$	3.5	±41.7%
		Open water to wetland	Rewetting, Revegetation, and Creation	Eq. 4.7: $CO_{2SO-RE} = \Sigma_{v,s,c} (A_{RE} \times EF_{RE})_{v,s,c}$	(0.0)	±29.5%
	Land Converted[e]	To Wetland	Rewetting, Revegetation, and Creation	Eq. 4.7: $CO_{2SO-RE} = \Sigma_{v,s,c} (A_{RE} \bullet EF_{RE})_{v,s,c}$	(0.02)	±29.5%
		To Settlement[d]	Extraction	Eq. 4.6: $\Delta C_{SO-CONVERSION} = \Sigma_{v,s}(SO_{AFTER} - SO_{BEFORE})_{v,s} \bullet A_{CONVERTED,v,s}$	0.1	±41.7%
Non-CO_2 IPCC Gas						
CH_4	Land Remaining	Wetland remaining Wetland	Rewetting, Revegetation, and Creation[c]	Eq. 4.9: $CH_{4-SO-REWET} = \Sigma_v (A_{REWET} \bullet EF_{REWET})_v$	3.5	±29.8%
	Land Converted	To wetlands	Rewetting, Revegetation, and Creation	Eq. 4.9: $CH_{4-SO-REWET} = \Sigma_v (A_{REWET} \bullet EF_{REWET})_v$	0.01	±29.8%
N_2O	Land Remaining	Wetland remaining Wetland	Aquaculture	Eq. 4.10: $N_2O\text{-}N_{AQ} = F_F \bullet EF_F$ (based on fish production)	0.14	±116%

Wetlands always refers to coastal wetlands, following U.S. definition. See EPA 2017 and Crooks et al. (in review) for full list of references.

a Derived using NOAA C-CAP; see Crooks et al. Wetlands, in review.

b Because of the former wetland classification, open water coastal wetlands are considered coastal wetlands.

c The U.S. definition of managed as pertaining to coastal wetlands is consistent with terrestrial-based reporting (see footnote 3, above). Because maintaining wetland character through management is an active and persistent sink of atmospheric CO_2, the U.S. rejected the assumption of EF = 0 and applied the methodological approach for the soil C pool consistent with EF_{RE} (Section 4.2.3.3, Eq. 4.7; IPCC 2014)

d These are reported as part of other land-use categories and therefore not explicitly provided in EPA (2017). See Crooks et al. (in review) for a more detailed treatment.

e Drainage is addressed in other land use categories of Cropland and Grassland.

improvement. Following the existing U.S. inventory, GHG emissions and removals for managed coastal wetlands do not include Alaska, Hawaii, or U.S. territories at this time.

The inventory approach applied for U.S. coastal wetlands follows the general methodology of: (1) defining the coastal land base (disaggregated and classified by palustrine/estuarine and forest, scrub/shrub and marsh); (2) quantifying land use within the coastal land area; (3) quantifying land-use change for the 1990–2015 time series; (4) ascribing an appropriate equation or set of equations to quantify emissions and removals associated with the type of land-use change; (5) applying to respective classified (and at level of disaggregation available) land areas; and (6) summing to respective sub-categories to determine respective emissions and removals. Activity data were derived primarily from the Coastal Change Analysis Program (C-CAP) data from NOAA (https://coast.noaa.gov/digitalcoast/tools/lca) (EPA 2017). The C-CAP data provide nationally standardized wall-to-wall land cover and land change data for the coastal region of the Conterminous United States and Hawaii. The data are derived from Landsat-derived 30 m resolution products. C-CAP land-cover data are available for the years 1996, 2001, 2006, and 2010, with a 2015 update in production. Additional activity data for N_2O emission from aquaculture were obtained from the annual NOAA fisheries report (https://www.st.nmfs.noaa.gov/st1/publications.html) (EPA 2017).

The U.S. expanded on specified management activities identified in the Wetlands Supplement to describe and include additional likely major anthropogenic GHG sources and sinks of managed coastal wetlands using Tier 2, country-specific data to estimate emissions associated with human-induced subsidence and erosion (e.g., conversion to open water). Other activities considered important, but not yet included are: CH_4 emissions associated with impaired tidal drainage and forestry activities on tidally influenced forests. GHG emissions and removals associated with these additional activities are consistent with the methodologies developed in Wetlands Supplement guidance on managed coastal wetlands. To cross-walk the change in land-use available from our land cover database (NOAA CCAP) with GHG methodologies (Table 16.3), each land-use change was aggregated by comparable activity types that were appropriate to a respective methodology, following either the Supplement or concurrence through expert judgment. Additional to those land-use changes in Table 16.2, the case study presented below estimated GHG emissions for the following land conversions Wetland to Cropland, Wetland to Grassland, and Wetland to Settlement.

For each category of source or sink, managed coastal wetlands were disaggregated into classes by soil type (organic and mineral), climate zone (cold temperate, warm temperate, subtropical and Mediterranean), and vegetation types (palustrine scrub/shrub, palustrine emergent, estuarine emergent, estuarine scrub/shrub, estuarine emergent and water). See EPA 2017 and Crooks et al. (in review) for a full description of the following approaches and estimation methods.

A. *Carbon Dioxide*: Carbon stock and flux estimates are based on the change in soil pools only. Tier 2 country-specific soil C stock estimates were synthesized from available U.S. literature (EPA 2017). Whereas soil CO_2 flux estimate for drainage was based on the IPCC default value. Lands on mineral and organic soils were summed and average stock was applied. When soil C data were not available for a specific climate zone or vegetation class, C stock data were estimated from other classes. The elements covered below are fundamental to a country's GHG inventory report submitted to the UNFCCC.

   Vegetated Coastal Wetlands Remaining Vegetated Coastal Wetlands (VRV): This source category includes managed coastal wetlands that fall under management protection (see footnote 4). Within VRV is also an Open Water sub-category, which is recognized as Coastal Wetlands within the Inventory. Therefore, sub-categories included are Vegetated Coastal Wetlands to Unvegetated Open Water Coastal Wetlands (VCO) and Unvegetated Open Water Coastal Wetlands converted to Vegetated Coastal Wetlands (OCV).

   a. *Summary of fluxes*: Soils are the largest pool of C in Vegetated Coastal Wetlands Remaining Vegetated Coastal Wetlands (VRV). Nationally aggregated, intact Vegetated Coastal Wetlands Remaining Vegetated Coastal Wetlands are considered using the Tier 1 EF for rewetting as C is continuously accumulating under management at a CO_2 removal rate of 12.2 (± 29.5% C.I.)

Table 16.3 Sample Approach for Developing a Cross-Walk between Land-Based CO$_2$ Reporting in LULUCF and Activity-Based CO$_2$ Reporting Specific to Chapter 4, Coastal Wetlands of the IPCC Wetlands Supplement (IPCC 2014)

LULUCF Category	Sub-Category	IPCC C Pool	IPCC Wetlands Supplement Coastal Wetlands Management Activity	Vegetation Types Affected	IPCC Tier 1 Methodology	IPCC Equations	Notes
Land Remaining	Wetlands remaining Wetlands	Biomass	Forest Management	Mangrove (and tidal forest)	Gain-Loss/Stock-Difference for biomass; BCEF and biomass can be computed from BEF and density	Section 2.2.1, Chapter 2, Volume 4 of the 2006 IPCC Guidelines; Eq. 4.1, Chapter 4, 2013 IPCC Wetlands Supplement	Forest management activities provide guidance for mangrove and tidal forest biomass management
			Extraction (Aquaculture)	Mangrove (and tidal forest), coastal marsh,	Stock-Difference for mangrove biomass (BCEF and biomass can be computed from BEF and density; BG/AG can be used to estimate marsh biomass	Section 2.2.1, Chapter 2, Volume 4 of the 2006 IPCC Guidelines; Eqs. 4.2–4.4, Chapter 4, 2013 IPCC Wetlands Supplement	The Tier 1 methodology assumes that the biomass, dead organic matter, and soil are all removed and disposed of under aerobic conditions where all carbon in these pools is emitted as CO$_2$ during the year of the extraction with no subsequent changes.
Land Converted	To Wetlands	Biomass	Rewetting, Revegetation and Creation	Mangrove (and tidal forest)	Gain-Loss/Stock-Difference for biomass; BCEF and biomass can be computed from BEF and density	Section 2.2.1, Chapter 2, Volume 4 of the 2006 IPCC Guidelines; Eq. 4.1, Chapter 4, 2013 IPCC Wetlands Supplement	If former land use is non-forest, good practice to consider change in emissions/removals with biomass C
	To Cropland[a]		Drainage	Mangrove (and tidal forest), coastal marsh	Gain-Loss/Stock-Difference for mangrove biomass; BCEF and biomass can be computed from BEF and density	Section 2.2.1, Chapter 2, Volume 4 of the 2006 IPCC Guidelines; Eq. 4.1, Chapter 4, 2013 IPCC Wetlands Supplement	Tier 1 = 0 as assumption for herbaceous (marsh) biomass gains = losses; national circumstances may not satisfy this assumption

(Continued)

Table 16.3 (*Continued*) Sample Approach for Developing a Cross-Walk between Land-Based CO$_2$ Reporting in LULUCF and Activity-Based CO$_2$ Reporting Specific to Chapter 4, Coastal Wetlands of the IPCC Wetlands Supplement (IPCC 2014)

LULUCF Category	Sub-Category	IPCC C Pool	IPCC Wetlands Supplement Coastal Wetlands Management Activity	Vegetation Types Affected	IPCC Tier 1 Methodology	IPCC Equations	Notes
	To Settlement[b]		Extraction	Mangrove (and tidal forest), coastal marsh, seagrass meadow	Stock-Difference for mangrove biomass (BCEF and biomass can be computed from BEF and density); BG/AG can be used to estimate marsh and seagrass biomass	Section 2.2.1, Chapter 2, Volume 4 of the 2006 IPCC Guidelines; Eqs. 4.2–4.4, Chapter 4, 2013 IPCC Wetlands Supplement	The Tier 1 methodology assumes that the biomass, dead organic matter, and soil are all removed and disposed of under aerobic conditions where all carbon in these pools is emitted as CO$_2$ during the year of the extraction with no subsequent changes.
Land Remaining	Wetlands remaining Wetlands	Soil	Forest Management	Mangrove (and tidal forest)	Soil removal factor is applied to managed area of vegetated coastal wetlands	Eq. 4.7, Chapter 4, 2013 IPCC Wetlands Supplement	Forest management could be extended to marsh and seagrass meadow management when considering the soil pool. The EF$_{RE}$ may be considered since maintaining the wetland character through management (see footnote 4) is an active and persistent sink of CO$_2$ (Citation or elsewhere).
			Extraction (Aquaculture)	Mangrove (and tidal forest), coastal marsh	Stock-Difference for soil is applied to specific area where management activity occurring	Eqs. 4.2, 4.3, and 4.6, Chapter 4, 2013 IPCC Wetlands Supplement	The Tier 1 methodology provides guidance for differentiating between mineral and organic soils where other activity data are not available.

(Continued)

Table 16.3 (Continued) Sample Approach for Developing a Cross-Walk between Land-Based CO$_2$ Reporting in LULUCF and Activity-Based CO$_2$ Reporting Specific to Chapter 4, Coastal Wetlands of the IPCC Wetlands Supplement (IPCC 2014)

LULUCF Category	Sub-Category	IPCC C Pool	IPCC Wetlands Supplement Coastal Wetlands Management Activity	Vegetation Types Affected	IPCC Tier 1 Methodology	IPCC Equations	Notes
Land Converted	To Wetlands		Rewetting, Revegetation, and Creation	Mangrove (and tidal forest), coastal marsh, seagrass meadow	Soil removal factor is applied to relevant area of vegetated coastal wetlands	Eq. 4.7, Chapter 4, 2013 IPCC Wetlands Supplement	Forest management could be extended to marsh and seagrass meadow management when considering the soil pool. Apply EF$_{RE}$ when vegetation has been established through replanting or reseeding or otherwise verified.
	To Cropland[a]		Drainage	Mangrove (and tidal forest), coastal marsh	Soil drainage (emission) factor is applied to area of cropland with wet organic or mineral soils.	Eq. 4.8, Chapter 4, 2013 IPCC Wetlands Supplement	
	To Settlement[b]		Extraction	Mangrove (and tidal forest), coastal marsh, seagrass meadow	Stock-Difference for soil	Eqs. 4.2, 4.3 and 4.6, Chapter 4, 2013 IPCC Wetlands Supplement	The Tier 1 methodology provides guidance for differentiating between mineral and organic soils where other activity data are not available.

Only biomass and soil pools are considered in this example.
a Reported in Cropland.
b Reported in Settlement.

MMT CO_2 eq. for the United States in 2015. Within the sub-category of VRV, although both soil and biomass C stocks are lost associated with VCO, only CO_2 flux associated with soil C stock change are used to compute VCO CO_2 emissions and these emissions are reported in the year of conversion. Including biomass is an area of planned improvement. In 2015, it is estimated that conversion to VCO corresponded to a CO_2 emissions rate of 3.5 ($\pm$ 41.7% C.I.) MMT CO_2 eq. OCV reflects either the building of new vegetated marsh through sediment accumulation or the transition from other lands uses through an intermediary open water stage as flooding intolerant plants are displaced and then replaced by wetland plants. Biomass and soil C accumulation when unvegetated open-water coastal wetlands are converted to vegetated coastal wetlands begins with vegetation establishment, but only soil C stock changes are reported. It is estimated that conversion to OCV corresponded to a CO_2 removal of 0.01 ($\pm$ 29.5% C.I.) MMT CO_2 eq. in 2015.

b. *Methodology*:

  i. VRV. Following Section 4.2.3.3 (i.e., rewetting and revegetation, following assumption ii, p. 4.27), Chapter 4 of the Wetland Supplement, an annual removal factor was applied to each climate zone/vegetation class subcategory and summed to estimate annual CO_2 removals. Carbon fluxes for mineral and organic soils were not differentiated (i.e., areas of land on mineral and organic soils were summed prior to applying removal factor).

  ii. *VCO*: Following Section 4.2.2.3 (i.e., excavation), Chapter 4 of the Wetland Supplement, the C stock conversion for each climate zone/vegetation class was estimated and summed. Carbon stocks for mineral and organic soils were not differentiated (i.e., areas of land on mineral and organic soils were summed prior to applying removal factor).

c. *Uncertainty Analyses and Time Series Consistency*: Uncertainties are estimated based on the 95% CI of each stock or flux estimate. The time series is consistent to the extent that the C-CAP land representation data have not changed over the 1990–2015 period. Carbon stocks and stock changes are estimated from country-specific data whereby uncertainties should be reduced with higher resolution and climate zone/vegetation class disaggregation.

d. *QA/QC and Verification*: Soil C stocks and CO_2 fluxes (removals) are synthesized from published U.S. literature.

e. *Recalculations*: None required at this time.

f. *Planned Improvements*: See Crooks et al. (in review).

  Other Land-Use Categories Converted to Vegetated Coastal Wetlands: Land Converted to Vegetated Coastal Wetlands occur as a result of inundation of unprotected low-lying coastal areas with gradual sea level rise, flooding of previously drained land behind hydrological barriers, and through active restoration and creation of coastal wetlands through removal of hydrological barriers. All other land categories are identified has having some area converting to Vegetated Coastal Wetlands.

a. *Summary of fluxes*: Between 1990 and 2015, the rate of annual transition for Lands Converted to Vegetated Coastal Wetlands ranged from 2,619 ha year^{-1} to 5,316 ha year^{-1}. Conversion rates were higher during the period 2010–2015 than during the earlier part of the time series, driven an increase in the extent unvegetated lands bear ground in Other Lands converted to wetlands. In 2015, C removal rate was 0.02 ($\pm$ 29.5% C.I.) MMT CO_2 eq. and CH_4 emission was 0.01 ($\pm$ 29.8% C.I.) MMT CO_2 eq. (EPA 2017).

b. *Methodology*: Following Section 4.2.3.3 (i.e., rewetting and revegetation), Chapter 4 of the Wetland Supplement, an annual removal factor was applied to each climate zone/vegetation class subcategory and summed to estimate annual CO_2 removals. Carbon fluxes for mineral and organic soils were not differentiated.

c. *Uncertainty Analyses and Time Series Consistency*: Uncertainties are estimated based on the 95% CI of each stock or flux estimate. The time series is consistent to the extent that the C-CAP land representation data have not changed over the 1990–2015 period. Carbon stocks and stock changes are estimated from country-specific data whereby uncertainties should be reduced with higher resolution and climate zone/vegetation class disaggregation.

d. *QA/QC and Verification*: Soil C stocks and CO_2 fluxes (removals) are synthesized from published U.S. literature.

e. *Recalculations* None required at this time.

    f. *Planned Improvements*: See Crooks et al. (in review).

B. *Methane*: Methane emissions estimates were developed using IPCC default values applied to lands differentiated by palustrine and estuarine C-CAP classes. A CH_4 emission value of zero was applied to estuarine classes.

    Vegetated Coastal Wetlands Remaining Vegetated Coastal Wetlands: As with CO_2 sequestration, management of coastal wetlands for conservation or other purposes can also result in release of CH_4.

    a. *Summary of fluxes*: Methane emissions of 3.5 (± 29.8% C.I.) MMT CO_2 eq. (Table 16.3) offset C removals resulting in an annual net C removal rate of 8.7 MMT CO_2 eq.

    b. *Methodology*: Following Section 4.3.1.1 (i.e., CH_4 emissions from rewetted soils), Chapter 4 of the Wetland Supplement, the same annual emission factor was applied to each climate zone/vegetation class subcategory and summed to estimate annual CH_4 emissions. CH_4 fluxes for mineral and organic soils were not differentiated.

    c. *Uncertainty Analyses and Time Series Consistency*: Uncertainties are estimated based on the 95% CI of CH_4 flux estimate for the IPCC default value. The time series is consistent to the extent that the C-CAP land representation data have not changed over the 1990–2015 period. Greater uncertainty as compared with country-specific values as there is no disaggregation by climate zone. There is also significant uncertainty in applying an assumption of zero flux to estuarine vegetation classes.

    d. *QA/QC and Verification*: The assumption of zero average CH_4 flux is reported for U.S. marshes when salinity exceeds approximately 18 psu. However, it is highly unlikely nor can we verify whether estuarine classes are continuously 18 psu or above.

    e. *Recalculations*: None required at this time.

    f. *Planned Improvements*: See Crooks et al. (in review).

C. *Nitrous Oxide*: Nitrous oxide emissions estimates were developed using IPCC default values applied to areas identified as aquaculture. Shrimp and fish cultivation in coastal areas increases nitrogen loads resulting in direct emissions of N_2O.

    Vegetated Coastal Wetlands Remaining Vegetated Coastal Wetlands: These are areas maintained as wetlands but for aquaculture use.

    a. *Summary of fluxes*: Overall, aquaculture production in the U.S. has fluctuated slightly from year to year though it is essentially at a similar level since 2011 as in baseline year of 1990. Data for 2015 were not yet available thus we used emissions from 2014 as a proxy (0.14 ± 116% C.I. MMT CO_2 eq; EPA 2017).

    b. *Methodology*: Following Section 4.3.2.1 (i.e., N_2O emissions during aquaculture use), Chapter 4 of the Wetland Supplement, the same annual emission factor was applied to activity data of annual fish production to estimate annual N_2O emissions. All annual aquaculture production is included in estimate with the exception of Clams, Mussels, and Oysters.

    c. *Uncertainty Analyses and Time Series Consistency*: Uncertainty estimates are based upon the Tier 1 default 95% confidence interval provided in the Wetlands Supplement. The time series is consistent to the extent that activity data have not changed over the 1990–2014 period (2015 was estimated from 2014 because 2015 data were not available).

    d. *QA/QC and Verification*: NOAA provide internal QA/QC review of reported fisheries data. N_2O emissions estimates were applied to any fish production to which food supplement is supplied be they pond or open water (EPA 2017).

    e. *Recalculations*: None required at this time.

    f. *Planned Improvements*: See Crooks et al. (in review).

16.6 FINAL THOUGHTS

Wetland ecosystem science is increasingly vital to addressing global policy issues. The IPCC Task Force on National Greenhouse Gas Inventories (TFI) is supported by the best available science to develop methodological guidance for use by countries to quantify and report national-scale net GHG emissions associated with anthropogenic activities. The TFI maintains an "Emission Factor Database" http://www.ipcc-nggip.iges.or.jp/EFDB/) that provides updated and country-specific data for improved data quality and reduced uncertainties in estimates. Sustained research on multiple sites, particularly through experiments, is imperative to increase data availability to reduce uncertainties in ecosystem carbon budgets, minimize cost, and provide new methodologies and techniques for country-specific assessments and data syntheses i.e., coupled ecosystem models, aquatic C transport, attribution of changes in near-shore water quality (boundary issues). Methodologies, data synthesis and emission estimates associated with other current and emerging anthropogenic impacts are needed to address data and methodological gaps—for example, estimates of organic and inorganic C export, attribution of nutrient enrichment to sources of pollution, improved estimates of seagrass meadow area, and areas of conversion, and differentiating upland from wetland nutrient sources—are all important areas of improvement that can improved inventory estimates. As the intersection between science and policy strengthens, the scientific issues of incorporating better coverage of land-use GHG emissions and removals will improve, providing policy makers with more robust tools and information to achieve the primary goal of reducing potential for catastrophic impacts of global GHG emissions.

ACKNOWLEDGMENTS

Many scientists and policy experts contributed to the eventual approval of the IPCC Wetlands Supplement. We owe a debt of gratitude to Sir Jim Penman for his guidance, keen awareness, and dedication to advancing GHG inventory methodologies. He is greatly missed. Partial support for this work was provided by Restore America's Estuaries and NOAA. This is publication #10 of the Sea Level Solutions Center in the Institute of Water and Environment at Florida International University, Miami, Florida.

CHAPTER 17

National Policy Opportunities to Support Blue Carbon Conservation

Ariana E. Sutton-Grier
University of Maryland
MD/DC Nature Conservancy

Carly Brody
National Academeis of Science, Engineering and Medicine

Michael Kunz
Orizon Consulting LLC

Dorothee Herr
IUCN

Elisa López García and Lindsay Wylie
Independent

Minerva Rosette
Mexican Center of Environmental Law

James G. Kairo
Kenya Marine and Fisheries Institute

CONTENTS

HIGHLIGHTS

1. There are many options for implementing blue carbon projects and for leveraging policy opportunities to restore wetlands and improve ecosystem service valuation for climate change mitigation.
2. Voluntary carbon markets are one mechanism that has been successful in Kenya, but other market mechanisms, such as sustainable shrimp labeling that requires mangrove restoration, is another that is being used in Vietnam.
3. Countries are also incorporating coastal wetlands into national activities such as the greenhouse gas reporting efforts in the United States, and climate mitigation commitments under the Paris agreement in Mexico.
4. Challenges to blue carbon efforts include the low price of carbon in the market and a lack of mapping and carbon data in many places but these can be overcome with targeted investments.
5. The efforts described here are some of the first successful blue carbon projects and can serve as models for other countries with the hope that we can build a global database of data and projects to help facilitate implementation of projects around the world.

17.1 INTRODUCTION

The international climate community is increasingly recognizing the role of natural systems in climate change mitigation, including the role of coastal wetlands as effective long-term carbon sinks (IPCC (Intergovernmental Panel on Climate Change) 2014; Howard et al. 2017). The scientific evidence continues to build in demonstrating the importance of marshes, mangroves, and seagrasses—coastal wetland blue carbon ecosystems—in helping to mitigate climate change. In addition, these ecosystems provide a number of additional benefits that support coastal human communities and a diversity of marine and coastal species (see Figure 17.1).

And yet these ecosystems are some of the most threatened on the planet as a result of conversion to coastal aquaculture, agriculture, or urban development; over-harvesting and deforestation in the case of mangroves specifically; and pollution from agricultural, urban and industrial runoff (McLeod et al. 2011; Pendleton et al. 2012). As a result, it is critical that nations rich in blue carbon ecosystems: (1) maintain an accurate inventory of each type of their blue carbon ecosystems; (2) assess the threats to these ecosystems and the loss rates; and (3) understand the legal options for protecting coastal wetlands in order to determine how best to protect or restore these ecosystems.

Here we examine case studies in six countries, including both developed and developing, detailing the different policy strategies countries are taking to protect and/or restore coastal blue carbon wetland ecosystems. The case studies include: the Dominican Republic, Indonesia, Kenya, Vietnam, the United States, and Mexico. These case studies examine the different policy strategies taken by countries to better protect or restore coastal ecosystems. They include two mechanisms supported by the United Nations Framework Convention on Climate Change (UNFCCC), namely Nationally Appropriate Mitigation Actions (NAMAs) and Reducing Emissions from Deforestation and Degradation (REDD), carbon offsets via the voluntary carbon market, sustainable shrimp labeling to support mangrove conservation, and incorporating coastal wetlands into national greenhouse gas (GHG) inventories and into existing national environmental policies. Figure 17.2 summarizes these different types of policy opportunities that countries are using to support blue carbon conservation and management, either via specific projects or via the incorporation of carbon services into broader policy frameworks.

We conclude with discussions of the challenges to successfully implementing coastal blue carbon projects and to coastal blue ecosystem conservation, and we suggest potential ways to overcome these challenges.

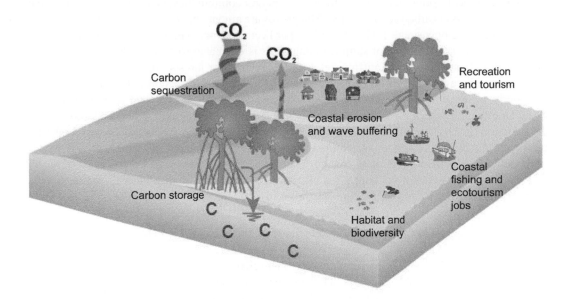

Figure 17.1 **(See color insert following page 266.)** Coastal wetlands provide many additional ecosystem services including fish and bird habitat, storm and erosion-risk reduction, recreational opportunities, coastal jobs, and water quality improvements. (Adapted from NOAA figure.) Also, see the chapter ecosystem services, this book.

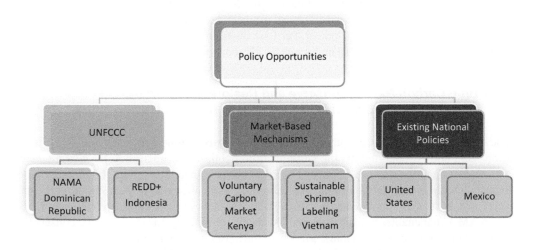

Figure 17.2 Policies for coastal management and/or climate mitigation.

17.2 CASE STUDIES

17.2.1 Project-Based Case Studies Using UNFCCC Mechanisms

17.2.1.1 NAMA Mangrove Project: Dominican Republic

Mangrove ecosystems provide many important benefits to human communities; they serve as buffers against tropical storms, generate livelihoods for poorer communities, and are more effective than tropical rainforests at carbon sequestration. Mangroves are among the most carbon-rich forests in the tropics, containing on average 1,023 Mg carbon per hectare (Donato et al. 2011). In the Dominican Republic, carbon sequestration of sampled mangroves is estimated to be roughly ten times higher than the Brazilian rainforest (Kauffman et al. 2014). Mangrove forests are severely degraded and declining rapidly as a result of overharvesting and deforestation, conversion to coastal aquaculture, and poorly managed dredging and coastal development. As a result, the total area of mangroves in the Dominican Republic has declined by as much as 50% since 1980 (Food and Agricultural Organization (FAO) 2007). This high loss rate, well above global averages and among the highest in the Americas, threatens important ecosystem services that underpin coastal livelihoods and support mitigation and adaptation to climate change. The destruction of these blue carbon ecosystems not only reduces climate change resilience, but also accelerates climate change because the carbon that has been stored in the healthy ecosystem gets released when the ecosystem is destroyed.

17.2.1.2 The National Climate Policy Context in the Dominican Republic

The Dominican Republic's current national climate change policy is aligned with national development strategies and sector strategies. The existing legal framework in the Dominican Republic incorporates environmental protection and the need for adaptation to climate change through the Constitution of the Republic and laws on environment and natural resources. Legislation is based on the international conventions and agreements ratified by the Dominican Government, such as the UN Convention on Climate Change Framework and the Kyoto Protocol.

The Dominican Republic is one of the few countries that have a low emissions development strategy. The Climate Compatible Development Plan (CCDP) is focused mainly on the three key sectors of energy, transport and forestry, and a cross-sectoral group called **quick wins** which includes the cement, waste, and tourism sectors. The CCDP is linked to the National Strategy, which is meant to create a national framework for coordinated and enhanced action to strengthen learning relevant for green, low emission and climate-resilient development. Per Law 01–12 of the National Strategy, the Dominican Republic is committed to cut its GHG emissions by 25% compared with the 2010 level by 2030.

17.2.1.3 Blue Carbon NAMA: Conserve and Restore Mangroves in the Dominican Republic and Commitment to the Blue Carbon NAMA

The blue carbon project concept, **Conserve and Restore Mangroves,** was developed in consultation with, and is supported by, the Dominican Republic's leading agency for climate change, the National Council for Climate Change. The Project is designed to facilitate the preparation of a Blue Carbon NAMA. Capacity building and technology assistance will be provided to policy makers, scientists, as well as community and private sector leaders to support the preparation of a Blue Carbon NAMA that enables transformational change of community-based mangrove conservation, sustainable management, and economic growth in the agriculture and tourism sectors.

The first of its kind, this DR Blue Carbon NAMA is designed to address climate change mitigation and adaptive capacity by improving livelihoods through the implementation of community-based actions. To promote the institutionalization of sustainable services, the Blue Carbon NAMA is designed to establish a community-based financial mechanism with funds from carbon credits. This co-benefit approach will be the first Blue Carbon Mangrove NAMA in the world and serve as a model for transformational change that can be replicated in the region and around the globe for countries with significant mangrove stocks.

The Government of the Dominican Republic is strongly committed to the Blue Carbon NAMA support project as it directly addresses the CCDP and National Development Strategy objectives. The government's commitment is reflected in a memorandum of understanding between the National Council for Climate Change and the International NGO Counterpart International. Interestingly, recent reforms in the marine sector mean the Dominican Republic now has up-to-date environmental legislation. The Protected Areas Law of 2004 and its related regulations and decrees provide the framework and authority for a National System of Protected Areas. However, it is not fully implemented due to a lack of capacity and financial resources. Therefore, the main driver for the Blue Carbon NAMA is the climate change legal framework and not the marine protection legal framework.

17.2.1.4 Institutional Arrangements: Governance and MRV of "Conserve and Restore Mangroves"

A National Steering Committee will govern the **Blue Carbon Conserve and Restore Mangroves** project with support from an Advisory Committee that has international expertise in NAMA design and blue carbon technology.

The Blue Carbon Support NAMA will be designed through a Steering Committee made up of members of the National Development Strategy (NDS), which was developed by the Ministry of Economy, Planning, and Development. Members of the Council are the Office of the President, as well as the Ministries of Environment, which serves as secretariat, Energy, Agriculture, Industry and Trade, Economy, Planning and Development, and other key actors. The roles and responsibilities of each member are articulated in the NDS organizational structure.

The Blue Carbon Steering Committee will bring together all sectors and stakeholders with the goal of fostering action in a decentralized and coordinated way. The Platform will facilitate the mobilization of resources, coordinate implementation of priority actions, and foster regular dissemination of information and learning materials among key partners. In order to facilitate effective coordination with donors, line ministries and sub-national organizations, the Steering Committee will serve as a single-window that can coordinate with donors and demonstrate to the international community both commitment and the capacity to conduct projects in an organized manner. To coordinate operational activities, the Blue Carbon Steering Committee will be supported by a small Secretariat located in the National Council on Climate Change.

The Blue Carbon Advisory Committee will provide advice and consultation to the Steering Committee, including technical assistance, international best practices, and capacity development. Members of the committee will represent organizations with international experience, including members of the Blue Carbon International Scientific Working Group* and the Task Force on National Greenhouse Gas Inventories.

The National Council for Climate Change will manage the Measurement, Reporting, and Verification (MRV) mechanism for blue carbon. The MRV mechanism is used in international agreements on climate change and refers to states' actions to collect data on emissions and mitigation

---

* Blue Carbon International Scientific Working Group convened by CI, IUCN and the Intergovernmental Oceanographic Commission (IOC) of the UN Educational, Scientific and Cultural Organization (UNESCO). The group was formed to determine the role of coastal vegetated ecosystems, such as mangroves, seagrasses, and saltmarshes, in carbon storage and sequestration; develop accounting methodologies and standards to encourage the inclusion of these systems in national carbon accounting; and identify critical scientific information needs and data gaps.

actions and publish this information in reports and inventories.* The Blue Carbon Support project will assist in the development and articulation of methodologies, sectors, and gases covered, global warming potential values used, as well as provide support for implementation. Support will be provided to identify an appropriate voluntary GHG program Verified Carbon Standard (VCS) methodology applicable to the proposed project. The VCS Registry System will be designed to serve as a secure platform where credits can be assigned unique serial numbers, allowing any project and any credit to be searched and tracked online. In order to maintain quality assurance, the VCS registries will adhere to strict conflict of interest policies.

The Dominican Republic's National Council for Climate Change has registered the Blue Carbon Conserve and Restore Mangroves with the UNFCCC NAMA Registry to attract potential support and also to share its ideas and techniques with other countries. The Climate Change Council estimates an 18-month time frame for preparation and is seeking financing from potential investors, including multilateral development banks, bilateral assistance programs, and the private sector.

17.2.1.5 REDD Mangrove Project: Indonesia

Indonesia, an archipelagic country, located along the equator has both mangrove and seagrass ecosystems. Although there are conflicting reports and data on the total area of mangrove forests, recent estimates place the coverage at 1.5 million hectares, storing up to 3.14 billion tons of CO_2 equivalent (tCO_2eq) in coastal carbon storage (D. Herr, Personal Communication). National seagrass coverage is estimated at 3,100,000 ha, which translates into roughly 1.35 billion tCO_2eq.

Indonesia's blue carbon ecosystems are among the most threatened, with high levels of mangrove deforestation, cleared to make room for aquaculture or converted into plantations, often for palm oil. Other destructive activities include land conversion for housing and industrial projects, and intense logging for timber and charcoal sourcing. Seagrass depletion also is happening at an alarming rate due to a variety of causes including increased boat traffic and anchoring, and widespread destruction for port and industrial estate development.

As much as 85% of Indonesia's GHG emissions are caused by land-use activities (National Council on Climate Change 2010). As a result, addressing land-use change and deforestation is a priority for Indonesia in the fight against climate change (Indonesian Ministry of National Development Planning/National Development Planning Agency 2010) and this includes developing Reducing Emissions from Deforestation and Forest Degradation in Developing Countries (REDD+) pilot projects (Indonesian Minister of Forestry 2008 p.68). In September 2012, the REDD+ Task Force launched the National REDD+ Strategy. The Strategy advocates for a strategic rehabilitation program, for example, recommending the replanting and rehabilitation of mangrove forests as well as the expansion of community-managed areas (The REDD Desk 2017). The National Strategy looks to support the commitment made by the Indonesian President to achieve an emissions reduction target of 26% below the 2020 projection based on a business-as-usual scenario, using the forestry sector as a way to influence change (Indarto et al. 2012). More than 60 REDD+ activities that are either active or in the preparation phase currently exist (The REDD Desk 2017).

While the government of Indonesia has substantially improved its national system for data collection, processing, and even real-time monitoring of the country's forests and wetlands, the inventory assessment of blue carbon ecosystems remains patchy. The ONE Map initiative, a government-led effort, aims to combine all conflicting maps into one comprehensive map to allow for ease of understanding data and improve coordination for monitoring and conservation. Mangroves are planned to be included in this ONE Map effort in Indonesia, which will improve the national data on mangroves. For more information, also see the Murdiyarso (Chapter 21).

---

* https://mitigationpartnership.net/measuring-reporting-and-verification-mrv-0.

17.2.2 Using the Voluntary Carbon Market to Support Mangrove Restoration: Kenya

Kenya has approximately 50,000–60,000 ha of mangroves, having lost about one fifth of its mangroves since 1985 (Food and Agricultural Organization (FAO) 2016). These mangroves are threatened by degradation for human uses, mainly for the extraction of wood products, but also for salt extraction and pond aquaculture. Additionally, land use changes upstream have led to sedimentation, resulting in smothering of mangroves (Lang'at and Kairo 2013). In an effort to restore mangrove forests and involve local communities in the care and sustainable use of mangroves, one community has explored the use of carbon financing with voluntary carbon credits to fund mangrove restoration. This project, called Mikoko Pamoja, is a mangrove restoration and reforestation project being implemented in Gazi Bay, Kenya (Crooks et al. 2014). The project protects 117 hectares (about 20% of the total mangrove cover in Gazi bay) of nationally owned mangroves, with the potential to grow. Eighty percent of the community makes their living from fishing-related activities, and therefore the Gazi Bay community heavily depends on the mangroves for their important role as a nursery for inshore and offshore fisheries. Mangroves also provide building materials, tourism, and coastal protection, but the harvest of wood for building materials and firewood has led to loss and degradation of mangroves across the country (Kirui et al. 2013).

The Mikoko Pamoja project is community-led, and is financed by the sale of voluntary carbon credits in the voluntary market. The main objectives of the project are to facilitate development in the area, restore mangrove ecosystems, enhance ecosystem services (including carbon sequestration), promote sustainable mangrove-related income, and act as a model for future projects (Crooks et al. 2014). Five years of research on carbon storage potential was completed in order to develop this project (Lang'at and Kairo 2013). According to the Project Document submitted to Plan Vivo, the contracting period of the Mikoko Pamoja is 20 years starting 2013 (Huxham 2013). Annually, 3,000 tons of carbon credits are generating an income of approximately US$15,000 per annum. The first two years of profits have earned the community about US$30,000 (Obiria 2016) and have exceeded their original target (Abdalla et al. 2015). The revenues collected from the sale of credits have been invested in the project implementation (one full-time staff member, mangrove planting and conservation) and have supported community development projects including rehabilitating the local schools and buying text books, as well as providing piped water to the community (Obiria 2016). They have also received support in 2015 through the World Bank's Kenya Coastal Development Project (KCDP). The funding will be used to construct a watchtower in the community-managed mangrove forest for improved surveillance of the project area (see Abdalla et al. 2015, Annex 6 for a technical drawing of the approved watch tower). More details on the financing of the project and the specifics about credits are available in Abdalla et al. (2015). The project has many additional benefits including supporting the onshore and offshore fishery, biodiversity conservation, and coastal protection.

There have been challenges with the project including dealing with the fluctuating carbon market and lacking similar mangrove projects for reference. The small scale of the project makes it hard to achieve economies of scale and to find buyers for the carbon (Abdalla et al. 2014; Wylie et al. 2016).

Despite these and other challenges, the project has been successful in both completing the sale of carbon credits, as well as meeting targets for mangrove planting and conservation (Wylie et al., 2016). Simultaneously, the project has benefited the local community by diversifying mangrove sources of income, such as beekeeping, aquaculture, and ecotourism related to the **Gazi Women's Mangrove Boardwalk.** To reduce deforestation pressure on the mangroves, fast growing terrestrial wood has been cultivated near the project site, which provides alternative sources of wood products, particularly for Gazi women (Plan Vivo Foundation 2015). The project's success is likely due to the active participation of the community in the development and implementation, a strong scientific foundation of published literature on the mangroves of the area, and key support from the government (Crooks et al. 2014, see more discussion on the project successes in Wylie et al.

2016). This project demonstrates that the voluntary carbon market can sustainably fund small-scale community-based blue carbon projects, and that these projects can support local community development in developing countries. The success of Mikoko Pamoja has encouraged the establishment of the East African Forum for Payments for Ecosystem Services (EAFPES), a networking body, to support and promote and upscale mangrove PES projects throughout Western Indian Ocean (WIO) region (Wylie et al. 2016).

17.2.3 Sustainable Shrimp Labeling Leading to Mangrove Restoration: Vietnam

Seafood and aquaculture are a major source of revenue in Vietnam comprising a six billion dollar industry, with shrimp farming making up one-third of that revenue (McEwin 2014). Accompanying this large amount of revenue is an impressive amount of environmental impact, mostly from the destruction of mangroves. Over half of Vietnam's mangroves have been lost in the last 30 years (McEwin 2014). National law in Vietnam requires 60% mangrove cover in aquaculture areas, but this is not typically enforced such that in practice cover is much less (McEwin 2014). One project that is seeking to decrease the loss of mangroves is the Markets and Mangroves (MAM) project, which was initiated in 2012 and is located in Ca Mau, one of the 12 provinces in the Mekong Delta. The project site encompasses 3,371 hectares of land (1,715 hectares of mangroves).

The innovation in mangrove conservation of the MAM project is helping shrimp farmers get organic certification for their shrimp farming activities by ensuring the farmers are planting or maintaining at least 50% mangrove cover in their farms (Wylie et al. 2016). If farmers can demonstrate that they have achieved this cover, then the sustainability label can apply to their shrimp, which gives farmers a premium price for their shrimp. At the same time, the certification bans additional mangrove destruction for the construction of shrimp ponds thereby halting the loss of ~23.5 ha of mangrove loss per year (Wylie et al. 2016).

It is interesting to note that carbon financing was initially expected to play an important role in the MAM project, but organic certification was a more lucrative and expedient financial mechanism (Wylie et al. 2016). This mechanism works by having a global seafood export company, Minh Phu, buy the organic shrimp at a price that is 10% more for the organic shrimp, and then sell them in markets in Europe, the United States, and Canada (SNV World 2014). An added benefit of this arrangement is that the increased mangrove cover has been linked to increased shrimp production, giving farmers an additional reason to comply with the sustainability label (McEwin 2014). The project has been fairly successful with 1,150 farmers certified with Naturland,—a recognized international organic aquaculture and agriculture standard (Wylie et al., 2016), and they are hoping to continue to expand to as many as 6,000 participating farming households (SNV World 2014). Thus, climate mitigation goals for reducing the loss of mangroves are being met, but using a completely different, innovative mechanism that is not based on carbon accounting. The project successfully assessed and integrated the economic needs of local communities, and linked conservation, climate mitigation, and economic growth through the organic shrimp market. It is likely that the diversity of stakeholders involved in MAM has also contributed to the success of the project (Wylie et al. 2016).

17.2.4 National Policy Analyses to Determine Opportunities to Incorporate Coastal Blue Carbon: United States and Mexico

17.2.4.1 United States: National Policy Analysis

The United States has about 41 million acres of wetlands in coastal watersheds of which about 6.3 million are saltwater wetlands (Dahl and Stedman 2013). Despite the U.S. environmental policies meant to protect ecosystems including wetlands, the U.S. is losing about 80,000 acres of wetlands in coastal counties annually, with approximately 70% of the loss occurring in the Gulf of

Mexico, much of it due to the conversion of estuarine wetlands to open water (Dahl and Stedman 2013). These losses suggest that there is still a need for better understanding wetland ecosystem processes in order to better mitigate losses and effectively restore coastal wetlands.

There has been growing interest in the U.S. federal government for the past two decades in better understanding the links between healthy ecosystems and healthy human societies and economies. Beginning in 1998 under the Clinton administration, the President's Council of Advisors on Science and Technology (PCAST) Panel on Biodiversity and Ecosystems issued a report that emphasized the importance of sustaining healthy ecosystems for economic growth (PCAST 1998). Then in 2006 under the Bush administration, the Committee on Environment and Natural Resources created an Interagency Ecosystem Services working group to coordinate activities across agencies in ecosystem service applications to policy (Schaefer et al. 2015). The Obama administration continued these efforts with a 2011 report issued by PCAST that called on the federal government to play an essential role in protecting the nation's environmental capital by better valuing those services and using that information to inform its planning and management decisions (PCAST 2011). Additional executive guidance for federal agencies to incorporate ecosystem services into planning and decision-making is included in the 2014 *Priority Agenda for Enhancing the Climate Resilience of America's Natural Resources* and in the Memorandum for Executive Departments and Agencies called "Incorporating Ecosystem Services into Federal Decision Making" (Council on Climate Preparedness and Resilience Climate and Natural Resources Working Group 2014, White House Office of Management and Budget et al. 2015).

This growing interest in better accounting and valuing of ecosystem services in federal planning and decision-making has occurred in parallel with a growing understanding of coastal blue carbon and the importance of the climate mitigation benefits of coastal wetlands. As a result, there has been increasing federal interest in understanding and accounting for the carbon in coastal wetlands and particularly in the potential for wetland restoration as a new carbon sink, which has led to some innovative policy analyses and efforts (Sutton-Grier and Moore 2016).

Some of the first analyses of opportunities for incorporating carbon benefits from coastal wetlands into existing national policies were completed in the United States and focused on the most important U.S. environmental legislation that impacts the management and protection of coastal ecosystems, including the Clean Water Act, the Coastal Zone Management Act, the Oil Pollution Act, and the National Environmental Policy Act (NEPA) (Pendleton et al. 2013; Sutton-Grier et al. 2014). These analyses determined that carbon benefits were not being incorporated into the implementation of these federal policies but that they could be. This was an important finding because it suggested that the United States did not need any new environmental legislation in order to begin accounting for carbon benefits, but that implementation of existing policies would need to be modified in order to incorporate carbon benefits. More recently, the U.S. National Oceanic and Atmospheric Administration (NOAA) has incorporated the carbon benefits associated with salt marsh restoration into one NEPA Environmental Assessment and into the NOAA Restoration Center's Programmatic Environmental Impact Statement (PEIS), which will guide all of the Restoration Center's future NEPA assessments for ecosystem restoration (Sutton-Grier and Moore 2016). Thus, there is now a precedent and guidance for including carbon benefits in NOAA's restoration project design process. These are initial steps to help institutionalize the incorporation of carbon benefits into the implementation of U.S. federal policies.

In October 2014, the U.S. White House's interagency Council on Climate Preparedness and Resilience (Resilience Council) Climate and Natural Resources Working Group (CNRWG) released their Priority Agenda for Enhancing the Climate Resilience of America's Natural Resources (Priority Agenda).* The second priority in this document was to manage and enhance natural carbon sinks and the language specifically includes wetlands and coastal ecosystems. One of the actions in

---

* https://www.whitehouse.gov/sites/default/files/docs/enhancing_climate_resilience_of_americas_natural_resources.pdf.

the Priority Agenda was to assess, restore, and protect coastal habitats in order to understand and enhance the storage of blue carbon. Stemming from this action was a specific task to do a baseline study of coastal wetlands and carbon benefits in the U.S. NOAA has been leading this effort for the past two years. The result of this effort is that coastal wetlands are included in the U.S. GHG inventory for the first time in the 2017 submission to the UNFCCC.[*] The United States is one of the first countries to include coastal wetlands in the national GHG inventory[†] and is writing a report examining the process and challenges of doing so to help inform future greenhouse inventory efforts.

In addition, the United States has been focusing on opportunities to incorporate wetlands into the voluntary market. There is no national carbon market in the United States at this time, but the voluntary carbon market is available in the United States and globally. NOAA and other agencies have supported Restore America's Estuaries (RAE) in the development of the VCS "Methodology for Tidal Wetland and Seagrass Restoration" (VM0033)[‡] that would provide carbon credits to global coastal managers and voluntary carbon market project developers of tidal wetland restoration projects (Sutton-Grier and Moore 2016). Other important studies include the RAE Snohomish Estuary assessment and the Tampa Bay assessment, which examine the climate benefits of watershed-scale restoration.

17.2.4.1.1 Mexico: National Policy Analysis

With 775,555 km^2 of mangrove coverage, Mexico is the fourth country in the world for mangrove area; mangroves extend along its three coasts, Pacific, Gulf and Caribbean, 72.5% of which are located within protected areas (Troche-Souza et al. 2016). In addition to this, Mexico has 9193 km^2 of mapped coast covered by sea grass, although little information is available for this ecosystem (CEC 2016). Considering only mangroves, the benefits provided annually to Mexico in ecosystem services have been valued at more than U.S. $100,000 per ha.[§] Despite the goods and services provided by blue carbon ecosystems in Mexico, their rates of degradation and loss continue to increase with an average loss of 0.3% per year since 1980 (Troche-Souza et al. 2016). This destruction is mainly caused by poorly managed human activities driven by short-term economic gain, such as massive tourism, agriculture and aquaculture, and vulnerability to climate-related changes.

Mexico is a developing country highly vulnerable to the effects of climate change despite the fact that its national emissions of GHG represent only 1.4% of global emissions. Mexico's location between two oceans, as well as its latitude and topography, significantly increases Mexico's exposure to extreme hydro-meteorological events.[¶] In 2013, Mexico moved from position 48th to 4th in the Climate Risk Index due to the arrival of two simultaneous storms: Hurricane Manuel and Hurricane Ingrid on the Pacific coast. The strength of the storms caused economic damage worth U.S. $5.7 billion, for events that took place over less than 24 hours (Global Climate Risk Index 2014). These events created the opportunity for Mexico at the COP20 in Lima in 2014 to take leadership in the Latin American region for the design and implementation of domestic strategies to mitigate and adapt to climate change. This leadership was confirmed on April 22, 2016, when Mexico signed its adhesion to the Paris Agreement and ratified it on September 21, 2016. Mexico unconditionally committed to reducing its greenhouse gas emissions by 25% and short-lived climate pollutants below **the business as usual** (BAU) benchmark by 2030. This commitment involves both mitigation and adaptation actions, and in both cases, the conservation and restoration of blue carbon ecosystems is the most cost-effective solution. At the same time, these actions may help the Mexican government

---

[*] https://www.epa.gov/ghgemissions/inventory-us-greenhouse-gas-emissions-and-sinks-1990-2015.
[†] https://www.epa.gov/ghgemissions/inventory-us-greenhouse-gas-emissions-and-sinks-1990-2016.
[‡] http://database.v-c-s.org/methodologies/methodology-tidal-wetland-and-seagrass-restoration-v10.
[§] <http://thenaturalnumbers.org/mangroves.html>.
[¶] Mexico's Intended Nationally Determined Contribution http://www4.unfccc.int/submissions/INDC/Published%20 Documents/Mexico/1/MEXICO%20INDC%2003.30.2015.pdf.

to achieve the goals of international agreements on biodiversity such as the Aichi Targets and the Sustainable Development Goals (SDGs) (CCA 2017).

An analysis examined opportunities to include coastal blue carbon within Mexico's international commitments and national legal framework, as well as financial mechanisms (CCA 2017). The analysis highlighted the wide array of Mexican public policy that will enable blue carbon strategies to contribute to the country's ambitious international commitments. In fact, the international treaties signed and ratified by Mexico on environmental protection, biodiversity, and climate change have set the basis for establishing guidelines in Mexico's national laws and relevant public policies, such as the General National Property Act, the General Climate Change Act, the General Wildlife Act, and the General Ecological Balance and Environmental Protection Act.

The Mexican legislative framework addresses blue carbon issues from assorted viewpoints, making it necessary to address conservation of blue carbon ecosystems from an inclusive approach focused on four lines of action: (i) the right of ownership is granted to the federal government, as these ecosystems are mainly located within the Federal Maritime Land Zone; (ii) the recognition of the payment of environmental services and offset mechanisms; (iii) the recognition of blue carbon ecosystems as carbon sinks, and their conservation and restoration as potential mitigation matters accordingly, especially within protected natural areas, and iv) the recognition of mangroves as priority protection species (CCA 2017).

Although Mexico has a legal framework in which blue carbon projects could be incorporated, the need for further knowledge on carbon storage, the financial schemes most suited to attracting investors, and the best strategies to engage communities remain. In this regard, it is imperative to draft, propose, and implement an Official Mexican Standard that specifically regulates blue carbon matters. At the same time, considering that most of the coastal ecosystems that collect and store blue carbon are located in areas under federal jurisdiction, it is necessary to ensure a federal land concession system that allows for the correct allocation of benefits and establishes strict conditions for the conservation of those ecosystems. Finally, more research is needed to achieve a reliable estimate of the coverage of these ecosystems in Mexico. This will allow for a more accurate quantification of carbon reserves (in surface plant matter and in the soil) and of the country's true natural capital, which will help secure financial support for projects (CCA 2017).

Once these elements are available, emissions from coastal degradation can be included in the national emissions inventory, triggering the government to regulate them and use conservation as part of its strategy to fulfill its international commitments. The best resource for this purpose would be to create a regulated national emissions trading scheme similar to and supported by the California Emission Trading System (ETS) (CEC 2016). This would include blue carbon in the guidelines imposed by the specific Official Mexican Standard and a system of allowances to allocate carbon credits. Nonetheless, even if Mexico considers the possibility of creating a national ETS of its own, it may take years until it becomes a reality. It is also essential to be cautious to ensure that there are no additionality conflicts between the ETS and previous projects, and that the benefits from the sale of blue carbon credits will not be dissipated in a centralized system that addresses national forest priorities away from the coastal marine area.

With these considerations in mind, the design of a national blue carbon strategy for Mexico will be supported by the lessons learned and good practices carried out by local voluntary markets (Figure 17.3). It should rely on a nested approach across all levels of government, where financing and capacities flow not only from international levels to community levels, but also the other way, escaping from the paradigm of traditional subsidies, and where socio-environmental costs are included as part of the ecosystem's actual value. To develop this platform, the Nationally Appropriate Mitigation Actions (NAMAs) are of fundamental importance, given their potential to secure financial support and technical training, and contribute to Mexico's NDC (Martin et al. 2016; CCA 2017). Also see Adame et al (Chapter 27).

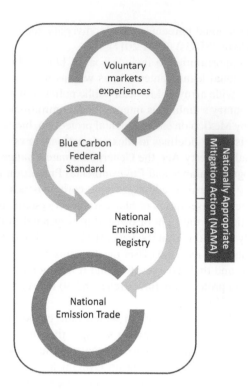

Figure 17.3 Mexico's national blue carbon strategy as proposed by CCA (2016).

17.3 DISCUSSION

Based on these national policy analyses, it becomes clear that there are many current options for implementing blue carbon projects using different funding mechanisms, and also many opportunities to leverage different policy opportunities to restore wetlands and improve ecosystem service valuation for climate change mitigation. Capitalizing on market mechanisms is one option that has been widely adopted by several countries. For example, the Blue Carbon NAMA in the Dominican Republic created a community-based financial mechanism that is funded by carbon credits and which aims to account for these credits through a VCS methodology. Kenya's project to fund mangrove restoration also is financed by the sale of voluntary carbon credits in the voluntary market. And the United States and Mexico are exploring opportunities to capitalize on the voluntary carbon market for wetland restoration and conservation.

Carbon market mechanisms are not always the solution; however, as seen in countries like Vietnam where organic shrimp certification has been a more successful market opportunity for protecting mangrove forests than carbon financing. In Indonesia, blue carbon projects are addressing land-use change and deforestation, which are the country's largest contributors to climate change.

Another important note about successful blue carbon projects is that community engagement is key (Wylie et al. 2016). An integral component of the Dominican Republic's Blue Carbon NAMA is community participation in mangrove conservation and a community-based financial mechanism funded by carbon credits. In Kenya, the success of the Mikoko Pamoja mangrove restoration and reforestation project is likely because it is community-led, and the community is invested in the health of mangroves for their fishing-dominated economy (see Kairo et al., Chapter 24). Furthermore, the project addresses the community's needs because profits from the carbon credit project are funding education and water projects in the community. Sustainable shrimp labeling in Vietnam also directly involves the shrimp farmers in mangrove restoration. Community involvement in the design and implementation of projects seems to be a key to project success.

Many countries have existing legislation that can be used to support coastal wetland conservation. For example, the Dominican Republic has policies in place that address climate change mitigation which can be used to support coastal wetland conservation. The United States has federal legislation in place to protect clean water and reduce the impacts of oil spills and other disasters that help to protect and restore wetlands. In addition, there has been a growing interest in incorporating ecosystem services, including carbon benefits, into federal planning and decision-making. Mexico has a legal framework that could include blue carbon policies. In addition to these national policies, international commitments to conventions on climate change mitigation have an important role in influencing existing legislation on environmental protection in some countries including the Dominican Republic and Mexico.

Nevertheless, several hurdles exist when implementing blue carbon projects. For example, in many countries there is a need for more carbon data and habitat mapping to support the incorporation of carbon benefits into policies and markets. The fluctuating and relatively low price of carbon in carbon markets has affected Kenya's project, which relies on the voluntary carbon market, and may affect other projects as well. Furthermore, because most of these projects are the first of their kind, they lack similar projects to use as references in their development.

We recommend that governments develop frameworks for incorporating wetland protection and restoration into their climate change mitigation efforts to meet national and international goals. Coastal wetland conservation also should be part of national climate change adaptation efforts because these ecosystems provide important storm and erosion-risk reduction (Sutton-Grier et al. 2015). Furthermore, we suggest that the governments support research efforts in mapping and carbon data collection, and support the inclusion of wetlands in national GHG inventories and other ecosystem inventories. This will help ensure that these data are available for use in carbon markets and other financial mechanisms for blue carbon projects, while also improving tracking and management of blue carbon ecosystems. When developing specific blue carbon projects, we recommend that project developers work to understand local community and indigenous needs and to involve the community in the development and implementation of a project since this tends to lead to more successful projects. Finally, as more countries begin to incorporate coastal ecosystems into their national inventories and policies, and as more blue carbon projects are developed, we recommend that information about these efforts are shared digitally and in reports and publications, and ideally in an international project database, in order to inform the international community and national decision makers worldwide about the possibilities and opportunities related to blue carbon conservation.

ACKNOWLEDGMENTS

We thank Lucie Robidoux and Gail Chmura for comments that improved this chapter. A. Sutton-Grier was supported by NOAA grant NA14NES4320003 (Cooperative Institute for Climate and Satellites—CICS) at the University of Maryland/ESSIC.

CHAPTER 18

Ecosystem Services and Economic Valuation
Co-Benefits of Coastal Wetlands

Tibor Vegh, Linwood Pendleton, and Brian Murray
Duke University

Tiffany Troxler, Keqi Zhang, and Edward Castañeda-Moya
Florida International University

Greg Guannel
University of the Virgin Islands

Ariana E. Sutton-Grier
University of Maryland
MD/DC Nature Conservancy

CONTENTS

HIGHLIGHTS

1. Blue carbon ecosystems (BCEs) provide many ecosystem services as "the benefits people obtain from ecosystems." These include important recreational and tourism opportunities, key fishery habitat, water quality improvements, and flood and erosion mitigation. These are co-benefits of BCEs in addition to carbon sequestration.
2. The monetary value of these services are best known for coastal protection and carbon sequestration, with monetary value of coastal protection services estimated at \$23 billion $year^{-1}$ between 1980 and 2008.
3. Environmental degradation of coastal areas reduces the capacity of BCE to confer these co-benefits.
4. While some uncertainties persist, given what is known about the aggregate monetary and non-monetary value of BCE, the co-benefits of BCE may outweigh the cost of managing, restoring, and creating BCE as nature-based features in coastal adaptation projects.

18.1 INTRODUCTION

The deterioration and loss of coastal ecosystems worldwide has resulted from intense and increasing human activities in the coastal zone. Barbier et al. (2011) report that 50% of salt marshes, 35% of mangroves, and 29% of seagrass meadows have been either lost or degraded during the last 20 years. These values presume we have inventoried or mapped them well enough to accurately document these rates of deterioration, which in many places may not be the case. During the 19th century, coastal wetlands often were considered a nuisance environment and perceived as competing against other human uses that serve immediate economic needs. However, during the last few decades, the literature is replete with examples describing how coastal ecosystems provide many co-benefits including important recreational and tourism opportunities, key fishery habitat, water quality improvements, flood and erosion mitigation, and mitigation of greenhouse gases through carbon storage and sequestration in vegetation (above- and belowground) and soil, each with monetary and non-monetary value, which can reduce risks to life, property, and economies (Barbier et al. 2011; Mcleod et al. 2011; Lovelock et al. 2017). These ecosystem services (ESs) have been broadly defined as "the benefits people obtain from ecosystems" (MEA 2005). In 2005, the Millennium Ecosystem Assessment (MEA) highlighted what were considered as the most important wetland ESs affecting human well-being: fisheries and water availability, and identified others in which coastal wetlands are particularly relevant including water purification, climate regulation, and mitigation of climate impacts through physical buffering (MEA 2005). This chapter provides a general description of benefits and consideration of economic valuation of those benefits that can be derived from coastal habitats that store and sequester carbon: seagrass meadows and wetlands, which include marshes and mangrove forests.

18.2 COASTAL ESs

When estimating ES it is important to recognize that not all coastal wetlands are structurally and functionally equal (Ewel et al. 1998). There are a wide variety of wetland species and eco-types even within a coastal setting that differ in their functional capacity in providing specific ESs (Ewel et al. 1998; Barbier et al. 2008). Estimating ecological production functions that translate the structure and function of coastal wetlands into their capacity to provide ES is difficult because the functions are nonlinear. Taking mangrove forests as an example, mangrove forest structural complexity (i.e., above- and belowground biomass, basal area, tree density) varies considerably around the globe, with regional and local variations caused by many factors including forest age, species composition, and response to environmental conditions. Mangrove vegetation patterns result from the interaction of complex environmental gradients and natural disturbances (i.e., hurricanes) that operate at different spatial and temporal scales, creating an array of different ecotypes (sensu Lugo and Snedaker 1974; Castañeda-Moya et al. 2013) across the continuum landscape. On a global perspective, the distribution of mangroves species and aboveground biomass is correlated with gradients in climatic factors including temperature, precipitation, and solar radiation (Blasco 1984; Twilley et al. 1992; Saenger and Snedaker 1993). Regional variability in geophysical processes (river, tide, and waves) within a coastal landform (e.g., delta, estuary, carbonate) control the basic patterns in mangrove forest structure and productivity and determine in large part the network in energy flow and material cycling (Thom 1982; Twilley 1995; Woodroffe 2002; Twilley and Rivera-Monroy 2009). At the local scale, ecological processes of resource competition (light and nutrients) and tolerance to regulator gradients (salinity and sulfide) control vegetation patterns depending on site-specific conditions of topography and hydrology, resulting in distinct ecotypes such as riverine, fringe, basin, scrub, and overwash (Lugo and Snedaker 1974). The interaction and degree of these resources and regulators along with hydroperiod gradients across the intertidal zone define a constraint envelope for determining the structure and productivity of mangrove wetlands within a coastal setting (Twilley and Rivera-Monroy 2009). Further, external drivers that impact mangroves including climate change, nutrient availability, and stream discharge are very dynamic and complex. Thus, before assessing an economic value to these ESs at different spatial scales, it is critical to obtain accurate estimates of mangrove biomass and C stocks for mangrove ecotypes (Alongi 2014).

However, there are a significant number of literature publications that help to resolve these uncertainties and improve economic valuation of these important ecosystems, including their resilience to stressors. Mangroves, for example, have been shown to exhibit a high degree of ecological stability and are highly resilience to natural disturbances such as tropical storms and hurricanes due to their life history traits that allow trajectories of recovery in ecosystem structure and function at decadal time scales depending on the frequency and intensity of disturbance (Alongi 2008; Danielson et al. 2017).

And there are many other ES that coastal wetlands provide. For example, coastal wetlands have long been recognized as providing primary habitat for a number of species and supporting great biodiversity. Given their continued destruction and degradation, one can argue that preserving biodiversity for biodiversity sake has not provided enough incentives to lead to the protection of coastal wetlands. Changes started to occur as these natural systems were more widely recognized for providing a suite of benefits (ES) vital for human health and well-being. A growing body of literature provides estimates of ES values derived from mangroves, the most studied of the three BCEs, which include mangroves, seagrass meadows, and salt marshes—referenced in this chapter. In BCEs, benefits other than C storage and sequestration have been identified, and, in some cases, valued including cultural (e.g., recreational), provisioning (e.g., fish stocks), regulating (e.g., water quality and hazard mitigation), and supporting (e.g., soil formation) ESs (Table 18.1). Coastal management approaches are now taking a more integrated approach, looking for beneficial tradeoffs and development schemes that maintain flow of ES to society and nature. The extent to which these services are maintained is determined largely by drivers of ecosystem change including human activities and anthropogenic climate change (Sandifer and Sutton-Grier 2014). Anthropogenic perturbations including localized impacts of vegetation disturbance, altered hydrological regimes, biological invasions, dredging, eutrophication, overharvesting, coastal development, pollution, and regional to global impacts of sea level rise (SLR) and climate change all affect the capacity for coastal ecosystems to deliver services to humans (Barbier et al. 2011; Sandifer and Sutton-Grier 2014).

18.2.1 Coastal Ecosystem Cultural Services

Coastal ecosystems provide significant non-material benefits—spiritual and inspirational, recreational, aesthetic, and education benefits (MEA 2005; Pleasant et al. 2014). For example, values

Table 18.1 Ecosystem Service Categories, Types and Examples

ES Category	Example ES Types	Ecosystem Process or Function	Relative Magnitude of known or Perceived Ecosystem Services Derived per Unit Area (H,M,L)[a]		
			Coastal Marshes	Mangroves	Seagrass Meadows[b]
Cultural	Recreation, spiritual	Tourism and recreational activities; Personal feelings and well-being	H	L	L
	Aesthetic	Appreciation of natural features	M	L	NA
Provisioning	Food	Fish, algae, invertebrates	H	H	L
	Freshwater	Storage and retention of water; provision of drinking and irrigation water	L	L	NA
	Other raw materials and products	Production of timber, fuel wood, peat, fodder, aggregates; extraction of materials from biota; medicines, genes for pathogen resistance, ornamental species, etc.	L	L	NA
Regulating	Climate and Biology	Regulation of GHG, temperature, precipitation; resistance to species invasions, regulating trophic interactions	H	H	L
	Natural hazards	Attenuates and/or dissipates waves; flood control	H	H	M
	Pollution control and detoxification	Provides nutrient and pollution uptake, as well as retention, particle deposition	H	H	L
	Erosion control	Provides sediment stabilization and soil retention in vegetation root structure	M	H	L
Supporting	Maintenance of fisheries	Provides sustainable reproductive habitat and nursery grounds, sheltered living space	M	M	L
	Soil formation; nutrient cycling	Sediment retention and organic matter accumulation; storage, recycling, processing and acquisition of nutrients	M	M	NA

Sources: MEA (2005); Barbier et al. (2011).
[a] Of inland and coastal wetland ecosystem types.
[b] Recent research illustrates higher magnitude of ESs derived from seagrass meadows that what is reported in MEA (2005).

among rural and indigenous communities may have particular importance, as passing down their culture, heritage, and traditional knowledge is part of a **way of life** that is closely interconnected with the environment. People may also benefit just from knowing that an ecosystem exists or that it can be enjoyed by future generations (Barbier 2017). While for some it is more difficult to quantify the benefits these services provide, like spiritual and cultural benefits, recreational fishing and local income derived from tourism produce significant economic benefits. For mangrove, marsh and seagrass meadow ecosystems, some important components controlling ecosystem processes and functions include species, plant density and productivity, habitat quality and area, prey species availability and health of predator species, as well as external forcings like storm events (Barbier et al. 2011). Although it is difficult to assess the economic and biophysical value in part because they are interrelated with other services and there are few indicators to monitor the contributions of cultural ESs to social systems (Atkinson et al. 2012; Daniel et al. 2012), they are reported to significantly contribute to people's well-being (Chan et al. 2012) and human health (Sandifer and Sutton-Grier 2014), both of which have been suggested to be the "ultimate ecosystem service" (Sandifer et al. 2015). A case study assessing the importance of cultural ESs for Hawaii suggested that, compared to regulatory and provisioning services, cultural ESs are an important management priority in Hawaii, especially for maintaining community security, in part, due to threats to native Hawaiian cultural security (Pleasant et al. 2014). Despite this and the benefits that cultural ESs were found to generate both locally and internationally, there were few policies that clearly guided their management (Pleasant et al. 2014).

18.2.2 Coastal Ecosystem Provisioning Services

In many areas around the world, coastal ecosystems provide food, freshwater, fiber and fuel, and biochemical and genetic materials (MEA 2005). Foods can include fish, game, fruits and grains, freshwater for domestic, industrial and agricultural use, fiber and fuel including production of logs, fuel wood and charcoal, peat and fodder, and genetic and biochemical materials of all sorts (from medicines and genes for resistance to plant pathogens; MEA 2005). The magnitude by which coastal wetlands provide these provisioning services depends on ecosystem type (Table 18.1). Food and other provisioned services within a wetland ecosystem type largely depend on the type and density of vegetation and habitat quality, while food and other materials from marshes also depend on inundation depth, and healthy prey and predator populations (Barbier et al. 2011). The economic benefit derived from these services is due to their role in food production. For example, MEA (2005) estimated that capture fisheries from coastal waters contributes $34 billion to gross world product every year (MEA 2005). Coastal wetland degradation can lead to both water scarcity and a decline of food provisioning both of which can be a significant risk in both urban and rural communities worldwide. Of note, human demand for these provisioning services is only expected to increase as with population growth and changing consumption patterns (MEA 2005). Further, trade-offs among ESs may produce unintended consequences that could be otherwise avoided with sound science and robust scenario development. For instance, agricultural production remains a significant driver of wetland loss pitting one source of food (crop production) against another (seafood production) (MEA 2005).

18.2.3 Coastal Ecosystem Regulating Services

Coastal wetlands provide a number of regulating services, including regulation of biology, regulation of natural hazards to confer coastal protection, nutrient and pollution control to regulate water quality and control of coastal erosion (MEA 2005; Barbier et al. 2011).

18.2.3.1 Climate Regulation

Coastal wetlands regulate global climate change through the carbon they sequester from the atmosphere and store in living biomass and the soil, and methane and nitrous oxide they release to the atmosphere (MEA 2005), each an important greenhouse gas with respectively increasing global warming potentials (Mitsch et al. 2013; Holm et al. 2016; Megonigal et al. this volume). BCE sequester and store large amounts of carbon (C) which have been shown to have negative impacts on the climate system through the greenhouse gas effect of carbon dioxide (CO_2; Nellemann et al. 2009; Barbier 2011). Recent whole C storage estimates in mangrove wetlands of the Indo-Pacific region have estimated an average storage of 1,023 MgC ha^{-1}±88, which exceeds 2.5–5 times the mean C stock in tropical upland, temperate, and boreal biomes (200–400 MgC ha^{-1}) (Donato et al. 2011). Thus, this biomass value suggests that mangrove wetlands are "among the most C-rich forests in the tropics" (Donato et al. 2011). Such rates of C sequestration underscore their potential

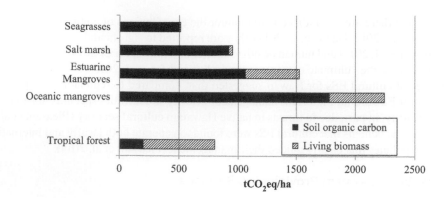

Figure 18.1 Global averages for soil organic carbon and living biomass carbon pools of selected coastal vegetated habitats. (Source: Pendleton, Murray et al. 2014; Note: Only the top meter of soil is included in the soil carbon estimates. Tropical forests are included for comparison.)

for global climate change adaptation and mitigation (Barbier et al. 2011; Siikamaki et al. 2012; Murdiyarso et al. 2015; Lovelock et al. 2017; i.e., Murdiyarso, Chapter 21).

Stored carbon can be released to the atmosphere upon disturbance and can exacerbate negative climate impacts. In fact, globally significant levels of carbon emissions result from mangrove deforestation due to coastal population growth, urbanization, and conversion to aquaculture/agricultural uses (Nellemann et al. 2009; Pendleton et al. 2012; Lovelock et al. 2017; Thomas et al. 2017). To better understand the role BCEs play in the global carbon cycle, a large body of literature examines the coverage and C storage of BCEs.* This body of literature finds that C storage and sequestration rates of BCEs is high relative to rates observed in tropical forests, which have been widely discussed in climate change discussions (Figure 18.1). Furthermore, the conservation and restoration of BCEs have been on the forefront of ecological, economic, and policy research recently, not only because of their connection to climate mitigation both as a source of C emissions, but also as a provider of other ESs, such as coastal protection, fishery support, among others.

Understanding the climate regulation role of wetlands at the Earth system scale has become an area of increasing interest, despite the relatively small area they occupy (US DOE 2017). Mangroves are forested wetlands that dominate tropical and subtropical coastlines and are among the most productive ecosystems on Earth, ranking second in terms of net primary productivity only to coral reefs (Duarte and Cebrian 1996). Global estimates indicate that mangrove coverage is approximately 137,760 km², which represents 0.7% of total tropical forests of the world (Giri et al. 2011; Figure 18.2). However, despite their disproportionate area compared to tropical forests, mangrove wetlands provide a wealth of ESs to coastal communities and industries (Nagelkerken et al. 2008; Alongi 2011; Barbier et al. 2011; Zhang et al. 2012; Lee et al. 2014).

18.2.3.2 Biological, Water, and Nutrient Regulation

Biological regulation is also considered significant, especially for mangrove ecosystems. Examples include resistance to species invasions, regulating interactions of species across trophic levels in food webs, and regulation of functional diversity (MEA 2005). Ecosystem processes and functions include nutrient and pollution uptake as well as retention of nutrients and particles or contaminants (Barbier

---

* Provide key references, if needed, for mangroves: e.g., (Hamilton and Casey 2016), (Jardine and Siikamaki 2014), seagrasses: (Chmura 2013, Duarte, Kennedy et al. 2013, Lavery, Mateo et al. 2013), saltmarshes: (Chmura 2013); (Duarte, Middelburg et al. 2005, Mcleod, Chmura et al. 2011, Pendleton, Donato et al. 2012, Duarte, Losada et al. 2013, Grimsditch, Alder et al. 2013, Pendleton, Murray et al. 2014, Thomas 2014).

Global Distribution of Mangroves USGS (2011)

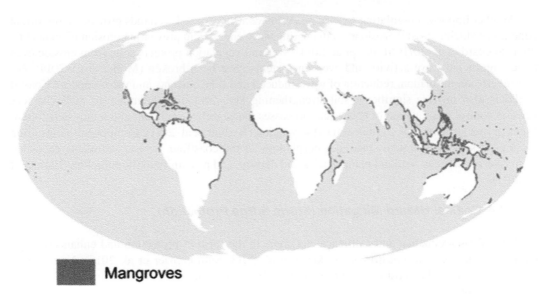

Mangroves

Figure 18.2 Global distribution of mangrove forests (Giri et al. 2011; http://data.unep-wcmc.org/pdfs/4/WCMC-010-MangrovesUSGS2011-ver1.3.pdf?1435856641).

et al. 2011). For example, water purification within a wetland ecosystem type largely depends on plant species and density, wetland quality and area, nutrient and sediment loads to the receiving wetland, supply and quality of water, and the extent to which healthy predator population are present (Barbier et al. 2011). Wetlands have long been known for their nutrient regulation functions (Mitsch and Gosselink 1986). Wetlands can reduce nutrient through sedimentation, nutrient sorption to sediments, nutrient uptake by plants, and through denitrification (Fisher and Acreman 2004). In a meta-analysis of 57 wetlands, Fisher and Acreman (2004) found that 80% and 84% of wetlands studied for N and P functioning, respectively, functioned to reduce loads of those nutrients. The mean change for those wetlands reducing N and P loads was 67 ± 27 and 58 ± 23, respectively. For N, there was greater reduction of N loads at lower N loading to those wetlands (Fisher and Akerman 2004). For instance, mangroves purify water by retaining nutrients and pollutants passing through them (Barbier et al. 2011), burying heavy metals and other contaminants adsorbed onto fine sediment particles and leaf matter (Lewis et al. 2011). Wetlands have also been shown to be effective at reducing nutrients or transforming nutrients from organic to inorganic form and vice versa, depending on wetland condition and structure.

Interestingly, water regulation as an ES links ES supply with ES demand by delivering goods and services from providing to benefiting areas (Syrbe and Walz 2012), and this spatial connection defines the benefit (Villamagna et al. 2013). With increasing demands for water inland, the supply of water that sustains coastal services (i.e., intact mangrove forests) can be weakened, with diminished benefits that feedback to the inland area (Van Oudenhoven et al. 2012; Villamagna et al. 2013). This leads not only to competition between demands for different services, but also increases the vulnerability of coastal communities by weakening coastal defense services (Arkema et al. 2013; Wolff et al. 2015) when natural coastal infrastructure is degraded or destroyed (i.e., decreases in freshwater inputs can decrease mangrove forest productivity and biomass which decreases their capacity for storm surge attenuation (Zhang et al. 2012)).

Coastal wetlands can also influence local and regional temperature, humidity, precipitation, and other climatic processes, and regulating chemical composition of the atmosphere (MEA 2005).

18.2.3.3 Erosion Control

Another important regulating service is erosion control. Coastal wetlands provide an important benefit in reducing coastal erosion by attenuating wave energy and preventing erosion (Gedan et al. 2011; Shepard et al. 2011; Möller et al. 2014). Wetland vegetation protects soils from erosion even under significant wave activity and even once stems have been broken (Möller et al. 2014). For coastal erosion mitigation, reduction of wave-induced and tidal currents and wave impacts. Coastal wetlands also provide significant soil strengthening and formation (reference to SLR chapter). Erosion control is a function of ecosystem processes that stabilize sediment and retain soils, often-times in root structures of coastal wetland vegetation (Barbier et al. 2011). Components controlling these processes include tidal stage, geomorphology of the coastline, load of sediments deposited from riverine sources, wetland plant species and density, and proximity to the shoreline (Barbier et al. 2011).

18.2.3.4 Coastal Hazard Mitigation (Storm Surge Protection)

Coastal risk reduction and resilience can come in the form of protection and enhancement of natural features on our coastlines (Arekema et al. 2013; Sutton-Grier et al. 2015). Natural features can provide a low-cost means of reducing vulnerability, allowing cost savings to be invested into more costly, but alternative risk reduction measures, thereby increasing overall resilience for equivalent cost. Vegetated coastal wetlands, such as mangrove forests and coastal marshes, are natural features that contribute to coastal risk reduction performance (USACE 2013). Mangroves and marshes reduce impacts of wave action along the shoreline by stabilizing sediment, increasing intertidal elevation, and providing obstructions that increase friction to reduce wave velocity, height, and duration (Barbier et al. 2011). Performance factors include elevation and continuity of the vegetated feature, and density and type of vegetation (USACE 2013; Sutton-Grier et al. 2015). However, significant uncertainties exist in the protective function of coastal wetlands, primarily due to high variability in coastal wetland features and limited, systematic field data, specifically for factors that determine coastal risk reduction performance of vegetated features (Loder et al. 2009; Zhang et al. 2012; Bricker et al. 2015).

In terms of coastal flooding mitigation, coastal wetlands provide reduction in wave run-up and storm surge elevation, resulting in fewer damages; however, strong storms and the increasing rates of SLR limit their capacity under certain conditions. For example, the magnitude of energy absorption coastal protection through hazard mitigation strongly depends on tree density, stem and root diameter; Alongi 2008). A recent study in south Florida Everglades mangroves have underscored their role in attenuating the storm surge from Hurricane Wilma using a combination of field observations and numerical simulation models (Zhang et al. 2012). Results showed that a mangrove zone of 7–8 km width reduced storm surge amplitude by 80%, and provided protection to the wetlands inland of the mangrove zone. Models also indicated a decrease in the surge attenuation rate from 40 to 50 cm km^{-1} when mangroves were present, to 20 cm km^{-1} with a mixture of mangrove islands and open water. In comparison, without the presence of a mangrove zone, the surge attenuation decreased inland at a rate of only 6–10 cm km^{-1} (Zhang et al. 2012). These results highlight the buffering capacity of mangroves against storm surges and their significant role in shoreline protection. This is particularly significant given the projected increases in sea level in the next half-century. Moreover, there has been some debate as to the hazard mitigation service offered by marshes as compared with mangroves and **bare land** as compared with marshes. In a review, Engle (2011) found that the range in surge reduction for estuarine marsh and shrub communities of the Gulf of Mexico ranges from 4.2 to 14.6 cm km^{-1}). A review and linear regression analysis of USGS surge data from mobile gauges and FEMA high water marks for the 2005 Hurricane Rita along the Galveston, TX coast (Figure 18.3) showed good agreement and modeled slopes of 0.13 and 0.14 m km^{-1} ($r = 0.94$ and 0.92, respectively) in a 6 km buffer zone around two coastal-inland transects. It is unclear whether this storm surge attenuation capacity is any different from **bare land** as bottom friction might actually be reduced depending on the structure of marsh vegetation with a smoother boundary layer between long-stemmed marsh vegetation and surge water. More high-quality measurement campaigns of this nature are needed to provide better understanding of the comparative storm surge attenuation capacity of different coastal wetland vegetation types in response to storms of different wind strength, storm speed and track direction, and ultimately surge height.

However, coastal wetlands have been found to be some of the highest potential for reducing wave heights (Narayan et al. 2016a). The type of storm can also affect the wave reduction potential of natural ecosystems with natural ecosystems better able to reduce wave energy for faster moving storms (Sheng et al. 2012; Zhang et al. 2012). Also, depending on non-stationary factors like accelerating SLR (Zhang et al. 2013), these benefits vary. Mangroves block incoming water and wave energy from storm surges (Zhang et al. 2012). Mangrove trunks, along with the complex root systems provided by species as *Rhizophora mangle*, create a physical impediment to incoming waves, generating drag that dissipates wave energy and protects inland areas (Mazda et al. 1997). This

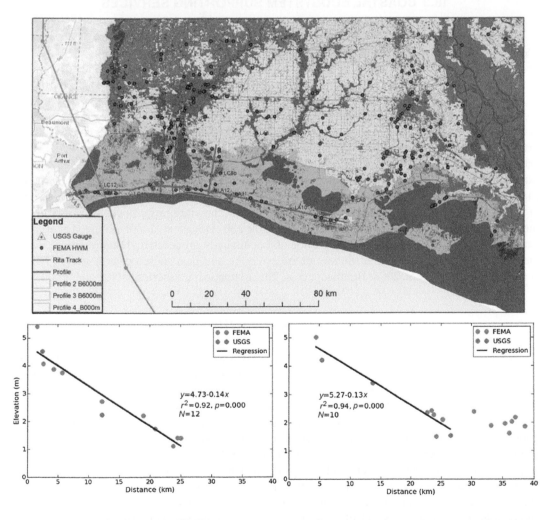

Figure 18.3　**(See color insert following page 266.)** Storm tide profile locations with Hurricane Rita. USGS mobile gauges and FEMA high water marks indicate approximately 13–14 cm reduction in surge height 1 km over vegetated land.

coastal protection service is significant as many cities are located along shorelines, placing cities at risk of flooding and wave damage that is exacerbated by SLR (Temmerman et al. 2013). Planting mangrove forests to restore or create new mangrove ecosystems is one way to stabilize shorelines and adapt to changing sea level (Sutton-Grier et al. 2015).

Regardless, there is growing interest in protecting and restoring coastal wetlands to provide storm protection benefits and risk reduction as an alternative to traditional built approaches, including seawalls, levees, and dikes, thereby improving the resilience of coastal communities to climate change (Arkema et al. 2013; Spalding et al. 2014; Sutton-Grier et al. 2015). In fact, by understanding how to engineer these ecosystems for the multitude of services they can provide, these can be cost-effective measures to achieve multiple ecosystem benefits through coastal adaptation. However, more research is needed to determine optimal designs for coastal adaptation projects that achieve not only coastal protection but to optimize these other ESs we know natural and many restored coastal systems to provide.

18.3 COASTAL ECOSYSTEM SUPPORTING SERVICES

Coastal wetlands support an array of benefits. Examples of supporting services include fisheries maintenance and soil formation (and elevation) and nutrient recycling. Coastal wetlands can provide reproductive habitat, nursery grounds and sheltered living space for aquatic species, and can also store significant organic matter under anaerobic soil conditions to sequester atmospheric CO_2 as organic matter that provides the foundation for wetland habitat and ecosystem structure. Two mechanisms maintain coastal peat soil elevations: net organic matter (OM) accumulation from plant production, and mineral sediment accretion via tidal transport (Smoak et al. 2013). In mangrove and marsh ecosystems that receive little sediment input (i.e., karstic environments), root production is the primary driver of vertical peat accretion and soil C accumulation (Nyman et al. 2006; McKee 2011; Baustian et al. 2012). Shoreline morphology and coastal elevation have great socio-economic significance, supporting recreation, flood control, and protection of water resources. Mangrove ecosystems provide an excellent example. Mangroves are ecosystems that store large stocks of carbon (C) in their living above- and belowground biomass and in the soil (Donato et al. 2011; Mcleod et al. 2011; Lovelock et al. 2017). The loss of some portion of that annually produced and sequestered C makes these ecosystems significant contributors to the estuarine and near-shore ocean C cycle (Twilley et al. 1992; Bouillon et al. 2008; Breithaupt et al. 2012), supporting fisheries habitat and production.

18.4 ECONOMIC VALUATION

While ES in BCEs are well documented, there are many fewer data available on their monetary and non-monetary values, and where monetary values are available, it has proven difficult to bring these services into the marketplace, with the exception of carbon in mangroves (Jerath et al. 2016). Mangroves provide coastal protection, fishery support, filtration, fuel wood/timber, tourism/recreation, biodiversity support, and various cultural benefits (Barbier et al. 2011; Vo et al. 2012; Duarte 2000). Seagrass and salt marsh ES values have also been extensively studied but we know much less about the value of ES these ecosystems provide (Table 18.2; For a review, see Barbier et al. 2011): seagrass (Orth et al. 2006; Waycott et al. 2009), salt marsh (Craft et al. 2008; Gedan et al. 2009).

Several attempts have been made to calculate economic values for these ESs globally, but economic valuation studies have estimated more credible results when focused on a smaller area where marginal values of some of these ESs can be calculated.* For example, bio-economic

---

* For a review, see Barbier et al. 2011, or for an example, see Das and Vincent 2009.

Table 18.2 Ecosystem Service Value Examples for BCEs

Ecosystem Service	Ecosystem Process or Function	Ecosystem Service Value Example		
		Mangrove	Seagrass Meadow	Coastal Marsh
Raw materials and food provisioning	Generates biological productivity and diversity	$484–595 ha^{-1} year^{-1} (2007 USD)	N/A	GBP 15.27 ha^{-1} year^{-1} (1995 GBP)
Natural hazard regulation	Attenuates and/or dissipates waves	$8,966–10,821 ha^{-1} (2007 USD)	N/A	$8,236 ha^{-1} year^{-1} (2008 USD)
Regulation of erosion	Provides sediment stabilization and soil retention in vegetation root structure	$3,679 ha^{-1} year^{-1} (2001 USD)	N/A	N/A
Regulation of pollution and detoxification	Provides nutrient and pollution uptake, as well as retention, particle deposition	N/A	N/A	$785–15,000/acre (1995 USD)
Maintenance of fisheries	Provides sustainable reproductive habitat and nursery grounds, sheltered living space	$708–987/ha (2007 USD)	$18.50/ha (2006 AUD)	$981–6,471/acre (1997 USD)
Organic matter accumulation	Generates biogeochemical activity, sedimentation, biological productivity	$30.5 ha^{-1} year^{-1} (2011 USD)	N/A	$30.5 ha^{-1} year^{-1} (2011 USD)
Recreation and Aesthetics	Provides unique and aesthetic submerged vegetated landscape, suitable habitat for diverse flora and fauna	N/A	N/A	GBP 32.80/ person (2007 GBP)

Source: Barbier et al. (2011).

models for the fisheries support function of mangroves have been used to calculate the economic value of mangrove fishery support ES in some areas in the world.[*] The coverage of BCE ES valuation studies to date has been globally patchy and focused mostly on mangroves (Table 18.2). A more recent estimate for coastal mangroves of the Florida Everglades compared methodologies for assessing value based on social costs of carbon (SCC), marginal abatement costs (MACs), and market prices (Jerath et al. 2016). The authors concluded that abatement costs were most appropriate for estimating value of stored carbon, based on the Comprehensive Everglades Restoration Plan costs ($56/tC; 2015 USD) and estimated the value at $18,794/ha or $2.7 billion. They also concluded that SCC would be a more appropriate valuation method for carbon sequestration.

Many studies have specifically evaluated the value of the protective services provided by coastal wetlands. One study estimated that coastal wetlands in the United States provide $23.2 billion per year in storm protection services based on a regression model of 34 major hurricanes to hit the United States since 1980 (Costanza et al. 2008). That same model determined that a loss of 1 ha of wetland corresponded with increased average storm damages of $33,000 from specific storms (Costanza et al. 2008). Another estimate for southeast Louisiana determined that coastal wetlands demonstrably reduced storm surge and that a 0.1 increase in the ratio of wetland to open water resulted in saving three to five properties—avoiding damages estimated between $590,000 and $792,000—for a given storm (Barbier et al. 2013). A recent analysis after Hurricane Sandy determined that, insurance industry models estimate that wetlands saved more than $625 million in avoided flooding damages and communities behind marshes experienced 20% less property loss during Hurricane Sandy (Narayan et al. 2016b).

[*] For a review, see Barbier et al. 2011.

Table 18.3   Geographic Focus Area of Mangrove Ecosystem Service Studies

	Percent of World's Mangroves (%)	Percent of Studies (%)
Africa	22	7
Americas	30	19
Asia	38	63
Pacific	9	10

Source: Vegh, Jungwiwattanaporn et al. (2014).

However, while there are a handful of studies about the value of protective services and carbon sequestration, valuation studies are lacking for many other coastal wetland ES. Consequently, this makes it much more difficult to develop ES marketplaces for these other coastal ES (Table 18.3).

18.4.1 Economic Benefits of Clearing BCEs

Of the BCEs we know the most about mangroves which are currently being lost at an average unweighted global rate of 0.164% per year (2000–2012), with some areas experiencing much higher loss rates (Hamilton and Casey 2016). In fact, a recent study found that seagrasses and salt marshes globally are both converted at a rate of 1.5% per year, while mangrove loss is estimated at 1.9% per year (Pendleton et al. 2012). To tie loss of BCEs back to the climate effect of the released CO_2, it has been estimated that the released greenhouse gas amounts to 240 million tons of CO_2–equivalent to 1.08 billion barrels of oil, 118 coal fired power plants, or 94.6 million passenger vehicles per year (Herr et al. 2016).

Commercial and industrial activities such as coastal development or conversion to shrimp aquaculture tend to drive mangrove conversion because the economic returns per hectare of converted land are higher than returns from traditional uses. As an illustration, shrimp farming has been reported to achieve returns of up to thousands of USD per hectare in Southeast Asia, and agriculture returns can be as high as a few hundred USD per hectare in West Africa.[*] These returns are much higher than economic returns from traditional mangrove uses. However, it has been shown that sustainable shrimp farming that includes mangrove protection and restoration can be very profitable to the shrimp farmer and can lead to more mangrove conservation suggesting that there are solutions that can help maintain economic productivity and mangrove ecosystems (Wylie et al. 2015).

In the case of urban, or urbanizing, areas, such as those around some large African coastal cities, the main loss driver is development for residential, commercial, and recreational uses. Urban development can achieve even higher returns on a per hectare basis than rural land uses, providing strong incentives to cut down mangrove forest and build on previously forest-covered land.

From a microeconomic perspective, conversion decisions are made by the landowners and reflect market price of goods that can be produced on a given area of land. Market price refers to the economic price for which a good or service (e.g., shrimp produced on a converted mangrove forest land) is offered on the marketplace. Market value and market price are equal only under conditions of market efficiency, equilibrium, and rational expectations. However, the market or economic value of most of the ESs blue carbon (BC) ecosystems produce are not reflected in the marketplace. In the case of BC, as of the time of writing, the value of ESs (e.g., C) is not reflected in the market price of C and is therefore not a part of rational landowner decision-making. If it were, C offsets would be trading at the level of the **social cost of carbon** value and revenue from the sale of C credits would raise the value of intact BCE habitats.

SCC is an estimate of the economic damages associated with a small increase in C emissions, in a given year (IWG-SCC 2015). Alternatively, SCC represents the value of damages avoided for

[*] Agriculture returns per hectare based on IFAD (2001).

(i.e., benefit of) a small quantity of emission reduction. The SCC attempts to estimate a comprehensive set of climate change damages such as agricultural productivity, human health, property damages from increased flood risk, and changes in energy system costs, such as reduced costs for heating and increased costs for air conditioning. However, due to modeling and data limitations, it does not include all-important damages. The IPCC Fifth Assessment Report (AR5, IPCC 2013) observed that SCC estimates omit important physical, ecological, and economic impacts of climate change, that likely underestimate damages.

If C were trading at SCC, a more accurate value of damages due to BC conversion would result in lower levels of conversion worldwide. According to some studies done to date, a socially more optimal (i.e., lower) level of conversion would be achieved with higher C prices that more accurately reflect the damages of C emissions that occur when these ecosystems are degraded or destroyed. When BCEs are disturbed or converted to an alternative use coastal vegetation stops removing CO_2 from the atmosphere through photosynthesis, and C stored on site in the soils and biomass begins to be released back into the atmosphere as CO_2.

18.4.2 BCE Loss as a Market Failure

Clearing BCEs or allowing their degradation results in C emissions into the atmosphere. Despite this link to the climate system, market signals often favor clearing over the protection of mangroves because externalities (unpriced benefits and costs) distort decisions. In other words, the loss drivers discussed above only consider the private (e.g., developers') economic benefit as opposed to wider societal costs of mangrove conversion, which would be more accurately valued using the SCC. Currently, forest C is trading at around \$4–5 per ton CO_2e on various C markets worldwide (Forest Trends 2014). For carbon values much lower than the SCC, it has been shown that BCEs offer a cost effective way of meeting emissions goals through pricing C stored in these systems (Mcleod, Chmura et al. 2011; Murray, Pendleton et al. 2011; Pendleton, Donato et al. 2012; Pendleton, Murray et al. 2014). Specifically, research has shown that BC conservation can be economically feasible at low to moderate prices of \$2–11 per ton CO_2e, not accounting for transaction costs (Murray et al. 2011). Other research has shown that the majority of potential emissions from mangroves could be avoided at less than \$10 per ton CO_2e (Siikamäki 2012).

18.4.3 Fixing the Market Failure

Market failure from the perspective of BCE conversion occurs when the private and social benefits and costs are not accurately represented in decisions. More specifically, BCE conversion is carried out to the point where the private benefits (i.e., marginal returns) equal the private cost of conversion (Figure 18.4). However, the social benefits of converting BCEs are lower, and the costs higher than from the private point of view. At the socially optimal level of conversion, due to C prices at the SCC as well as payments for ES (PES; higher opportunity cost of conversion), conservation would be reduced to the point where the marginal social benefit of conservation would equal the marginal social cost. At the social optimum level of conversion, CO_2 emissions from BCE conversion would be lower than the current level.

When attempting to fix this market failure, distributional issues (**winners and losers** local versus global, tenure arrangements) must be taken into consideration by way of a socio-economic impact assessment. This assessment should take into consideration the potential for differentiated impacts on different groups of participants as well as other vulnerable stakeholders. Just like with many other types of interventions, there will be winners and losers. However, the negative externalities caused by higher than socially optimal level of BCE conversion are certainly going to be outweighed by the increase in social welfare due to a reduction of CO_2 emissions.

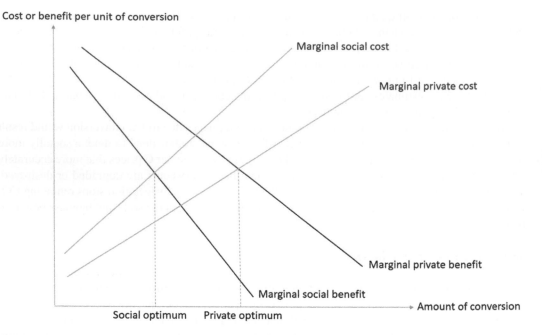

Figure 18.4   Illustration of BCE conservation as a market failure.

18.4.4 Policy Options

The market failures discussed above could be addressed by several approaches, each having its own policy design and implementation challenges. The most commonly analyzed, developed, and tested approach to date has been through C markets where the monetized C benefits (ES) of BCEs, measured in terms of tons of C per hectare, are traded. Carbon markets can be compliance driven (e.g., California, European Emissions Trading Scheme) or voluntary (e.g., Regional Greenhouse Gas Initiative) and could be regional, national, or international, either within or outside of the United Nations Framework Convention on Climate Change (UNFCCC) framework.

Carbon markets—There are examples for regional, national, and international C markets that are currently operating. However, no global market has been established either within or outside of the UNFCCC, even though CO_2 is a globally mixed pollutant. Moreover, trading volumes to date have been low, and prices even in the compliance markets do not reach levels close to the SCC. Projections typically indicate rising C prices, but 2015 C prices have been very low (Forest Trends 2014). For BC to be included in C market transactions at scale, steps for the policies around C market development are expected to target the definition of mangroves as forests. At the time of writing, it is less clear how seagrass meadows and salt marshes would be eligible to participate in C market transactions. That is because most of the C in these ecosystems is stored in the soil as opposed to biomass (e.g., Fourqurean et al. 2012). A major issue with forest C to date has been that even in tropical forests, much of the C is stored in the soils and not the biomass. This is especially true for BCEs where as much as 90% of the C is stored as soil C. This needs to be assessed and appropriate MRV protocols developed specifically for BCEs, in addition to the recently published Verified Carbon Standard Tidal Wetland Restoration Protocol.

Defined as **forest**, tall mangrove BC appears to be a good candidate for inclusion under the Reducing Emissions from Deforestation and Forest Degradation (REDD+) market mechanism. REDD+ prices greenhouse gas emission reductions from forest conservation. Additional international finance mechanisms that could be relevant BCEs include several UNFCCC specific funds, bi- and multilateral, as well as national climate funds (Herr et al. 2016). Also, see Chapter 17 (Sutton-Grier et al. this volume) on "National Blue Carbon Projects" where this is a UNFCCC Nationally Appropriate Mitigation Action (NAMA) example. Additionally, novel financing options such as debt-for-nature swaps or payments for biodiversity have also been considered recently for BC conservation. However, an analysis of four mangrove restoration projects that attempted to use carbon finance to fund the restoration determined that half of them were able to use carbon financing, but not through any UNFCCC mechanism; instead, they were using the voluntary carbon market due to ease of access to the market, primarily, and the other projects opted for completely different funding mechanisms after exploring carbon options (Wylie et al. 2015). This suggests there remain significant challenges to BC projects that need to be overcome to facilitate more restoration and protection of coastal wetlands.

Allowing for and designing payment mechanisms for ESs can generate new or additional fund flows for BCE conservation (Vo, Kuenzer et al. 2012; Lau 2013). Currently, the policy framework is inadequately developed to support PES for BCE conservation. For example, there are not mechanisms through which those undertaking BC conservation programs or activities could receive financing. However, progress has been made on the fronts of assessing the monetary and non-monetary values of these ESs, which can in turn lead to the development of appropriate policies or regulations. In this setup, the beneficiaries of BCE ESs could be the ones paying for the continued provision, thus the conservation of, the ESs. Some of the most promising initiatives to date have been national PES mechanisms. These mechanisms combine the C, biodiversity, or water provisioning ESs of BCEs.

Improved valuation of a more complete suite of ESs provided by BCEs may lead to additional markets to pay for these ESs. Since pricing all benefits and costs—practically speaking, as many as is feasible—may not be realistic in every conversion decision, several pathways, other than C finance, or payments to ESs, to fix the market failure are being deployed. In addition to these pathways, other policy options also exist to curtail BCE conversion, including sustainable supply chain efforts. For instance, governments could mandate that no shrimp could be sold from farms created by mangrove clearing. Also, a business case could be made to the private sector, investors, and donors regarding the sourcing of products from sustainable supply chain systems. This was the case for a sustainable shrimp mangrove restoration project in Vietnam where the shrimp **sustainability** label ensured that at least 50% mangrove cover was maintained and the farmers were guaranteed a higher price for their labeled shrimp (Wylie et al. 2015 and Chapter 17 this volume). This model could be more widely applied with sustainability certification of coastal wetland products, and effective legal enforcement could lead to increased funding for the conservation of BCEs, further reducing their conservation toward the socially optimal level.

18.5 DISCUSSION AND CONCLUSION

Coastal ecosystems provide a wealth of services important to human health and well-being, along with biodiversity benefits. In this chapter, we presented examples of important ESs and their valuation. It is also important to know that there are limits to services, which can be modeled and tested. But well-managed ecosystems and resources can minimize those limitations and improve flow of services. Market and policy constraints remain significant barriers.

Importantly, ecosystems are complex and resilient, but also fragile to human and natural disturbances, and seascape or watershed scale conservation or management approach to maintain supply and delivery of services is critical. Historically, rates of vertical soil accretion of mangrove and tidal marsh wetlands have kept pace with rates of past SLR (e.g., McKee 2011). However, as the relative areas and functions of coastal elevation and vegetation types change, the ESs they provide are also likely to change (i.e., Jerath 2012) without intervention. As such, coastal stressors drive landscape change, and landscape change drives either increases or decreases in ESs. For example, in the coastal Everglades, vegetation modeling has shown that mangroves will continue to transgress inland, but freshwater marshes inland of the mangrove forests are at risk of being converted to open water systems if their peat soil elevation is compromised by peat soil collapse before mangroves can transgress inland (Karamperidou et al. 2013; Figure 18.5). The term **peat collapse** has been used to describe a relatively dramatic shift in soil C balance in response to saltwater intrusion. It results from a net loss of organic C, is manifest as a rapid loss of soil elevation, and culminates in a conversion of vegetated freshwater marsh to open water. Peat collapse has been documented to varying degrees across the United States (Cahoon et al. 2003; Nyman et al. 2006; Voss et al. 2013), and has been attributed to increased sulfate reduction, increases in other avenues of soil microbial respiration, sulfide accumulation, reduced root production, fire history, and vegetation damage from tropical storms. All can synergistically contribute to the instability of freshwater marsh soils. Additionally, recent work by Chambers et al. (2014) showed that SLR—through its effect on increasing salinity and inundation—may also influence C cycling rates in mangrove peat soils. If less salt-tolerant wetlands are unable to adapt quickly enough to the salinity changes associated with accelerating SLR, then significant coastal wetland loss will likely occur (CISRERP 2014), dramatically altering and increasing the vulnerability of the south Florida coastline (Troxler et al. in press). This peat collapse phenomenon has already happened in areas of the Everglades (i.e., Cape Sable) where Wanless and Vlaswinkel (2005) attributed soil elevation loss to canals dug in the 1920s and subsequent saltwater intrusion into freshwater marshes.

Notably, coastal wetland management can stop, reverse, or advance ecosystem restoration, recovery, and/or creation (i.e., where coastal wetlands may have occurred but historical management practices degraded them) and coastal ESs. For instance, freshwater flow restoration in the Everglades is projected to halt the loss of coastal peat marshes (Figure 18.5; although uncertainties in these projects are the subject of intense study; see Chambers et al. 2014; Troxler et al. in press; Wilson et al. submitted; Servais et al. submitted). Importantly, coastal adaptation that also emphasizes ESs in addition to coastal protection and flood mitigation has large potential to yield multiple short- and long-term benefits.

Recent global political events, such as the United Nations Framework Convention on Climate Change Conference of the Parties (COP) 21 in Paris in December 2015, sent a strong signal to the global environmental community that carbon pollution is to be reduced such that the predicted

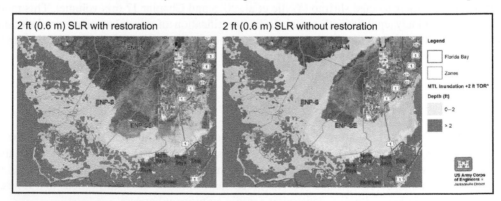

Figure 18.5 (See color insert following page 266.) Everglades National Park with 2 ft of SLR with and without the tentatively selected plan (Alternative 4R) of the Central Everglades Planning Project (CEPP; adapted from CISRERP 2014). The scenario without restoration assumes total peat loss due to saltwater intrusion, illustrating a key area of uncertainty for the fate of Everglades wetlands related to persistence of peat soils under projected SLR.

global average temperature increase remains **well below** 2°C above pre-industrial levels. The extent to which BC will be able to play a role in achieving this target will depend on the development of market-specific methodologies to credibly measure, report, and verify (MRV) greenhouse gas emissions from BCEs and, to some degree, the price of carbon which is currently far below the SCC and low enough that long-term funding of BC projects exclusively through carbon finance can be challenging. Moving forward, the two biggest uncertainties for BC projects to access carbon markets at scale remain: (1) the uncertainty of whether policies will be enacted to create carbon markets of global scale and breadth, and (2) whether such markets will accept BC conservation or restoration as credible activities, particularly for marshes and seagrasses.

Currently, the economics of conversion decisions in BCEs favor conversion as evidenced by a total cumulative global mangrove loss rate of 1.97% between 2000 and 2012. If this trend continues, we will have lost 25% of global mangrove coverage by the end of the next century. To halt the conversion and degradation of BCEs the ESs they provide must be accurately valued and represented in management decisions. But unless ES values are monetized and estimated for at least the key BCE areas, and these data are considered as part of the decision-making process of coastal land-use decisions, private profits as opposed to socially optimal decision-making will continue to drive coastal land-use decisions putting pressure on BC resources worldwide.

18.5.1 Ecosystem-Based Adaptation as an Opportunity and Interdisciplinary Approaches as Integral

There is a critical need for interdisciplinary science and decision-support framework for coastal environments that integrates and meets both social and ecological needs for water in the long-term, while maximizing the benefit provided by coastal environments to mitigate the impacts of increasing risk to coastal stressors. The risks of planning coastal adaptation strategies that are based on short-term return will exacerbate the impact of coastal stressors and reduce our capacity to sustain our livelihoods long-term. In populated and developing regions of the world, existing human water security infrastructure will require different levels and approaches to integrated water resource management, from new management systems to re-engineering of existing systems to protect flows of ES for both provisioning and regulation needs of society. While studies illustrate that society has failed to institute the principle that human water security relies on balancing needs of humans and nature (Vorosmarty et al. 2010), there are promising, cost-effective approaches to preserve and rehabilitate ecosystems (Palmer and Filoso 2009). Engineers, for instance, can re-work dam operating rules to maintain economic benefits while simultaneously conveying adaptive environmental flows for ecosystems (i.e., Arthington et al. 2006). Protecting catchments reduces costs for drinking water treatment, whereas preserving river floodplains sustains valuable flood protection and rural livelihoods (UNESCO 2009). Where resources are few, protection of coastal communities will be increasingly dependent on natural coastal defenses. Such options offer developing nations the opportunity to avoid the high environmental, economic, and social costs that heavily engineered water development systems have produced elsewhere (Gleick 2003).

Further, in coastal systems, inland water use determines the strength of feedbacks between coastal stressors and delivery of coastal ES to the inland natural-built environment. This framing, as a flow of ES, highlights services that sustain health and welfare through protection of property, freshwater supply, and benefits. These benefits are brought by adaptation toward sustained ES rather than economics, and emphasize how ecosystem-based adaptation measures can be the new norm for decision-making. This calls for an improved treatment of co-benefits and feedbacks to the built environment (Geneletti and Zardo 2015). Of critical concern is that short-term solutions without a long-term strategy considering the suite of potential system vulnerabilities will threaten long-term sustainability through increased fossil fuel use (and GHG emissions), loss of ES and reduced coastal hazard mitigation. Further, climate adaptation approaches include actions undertaken in order to reduce harm or exploit benefits from actual or expected climatic stimuli or their effects, in order to reduce harm or exploit benefits. Although historically adaptation to climate change has received less attention than mitigation (Füssel 2007), there has been a recent surge of interest in adaptation interventions, which are already a necessity in many contexts, particularly until greenhouse gases emissions are stabilized (Pickett et al. 2013). Adaptation to climate change may be attained in different ways. One way that is attracting increasing attention is through ecosystem-based approaches. Ecosystem-based adaptation (EbA) is defined as the use of biodiversity and ES to help people to adapt to the adverse effects of climate change (CBD 2009). Recognizing and better valuing the climate mitigation AND adaptation benefits of coastal wetlands brings us closer to realizing their true value for planning, decision-making, or policy choices to consider these multiple benefits provided by ecosystems for long-term coastal sustainability.

Adaptation to extreme weather and climate-related hazards may be attained in different ways and it is an emerging coastal urban science. The use of Natural and Nature-based Features (NNBF) has gained traction as a means to enhance coastal resilience and mitigate the potential impacts of extreme events, SLR and inundation to coastal communities (Sutton-Grier 2015). NNBF refer to a spectrum of features from natural coastal ecosystems (e.g., marshes and dunes) to nature-based features that utilize a combination of natural and human-engineered features to create a 'hybrid' shoreline (Sutton-Grier 2015; SAGE 2015). Specifically, NCCOS/CSCOR defines NNBF as: "existing ecosystems including forests, wetlands, floodplains, dune systems, seagrasses, barrier islands and reefs that provide multiple benefits to communities, such as storm protection through wave attenuation or flood storage capacity and enhanced water services and security." For example, while seawalls were shown to reduce biodiversity and numbers of organisms along shorelines by 23% and 45%, respectively, as compared with natural shorelines, riprap or breakwater shorelines were not different in this regard (Gittman et al. 2016a). Design and performance assessment of coastal adaptation projects in the short- and long-term for multiple benefits under various conditions is an important area of research (Gittman et al. 2016b).

Importantly, coastal ecosystems are not a one-size-fits-all solution. Complementary approaches, both in terms of an integrated coastal ecosystem model, one that optimizes ecosystem processes and functions along the freshwater-marine gradient, to include forested wetlands, marshes, and seagrass meadows, but also an integration of gray and green coastal adaptation strategies, can confer that greatest overall benefits. These are spatial approaches, but temporal approaches are also critical—maintaining **baseline** ecosystem functioning and levels of service while incorporating components that can withstand extreme events. For example, while seagrass meadows may not provide significant storm surge protection, the ESs they provide for maintaining local fisheries is critical for recreation and fisheries-based economies (MEA 2005). Science-based approaches to coastal engineering projects with strong inclusion of ecological features are a promising direction for coastal adaptation and building coastal resilience. Overall, BC approaches to improving ESs and bringing more value to coastal ecosystems is exciting, but we need healthy ecosystems, smart management, and more comprehensive coastal adaptation projects implemented at a large scale to take advantage of new opportunities for applying and creating strategies to maintain the services they provide in the long term.

ACKNOWLEDGMENTS

This is publication #11 of the Sea Level Solutions Center in the Institute of Water and Environment at Florida International University. Partial support for this work was provided by the NASA Carbon Monitoring System Project NNH14AY671.

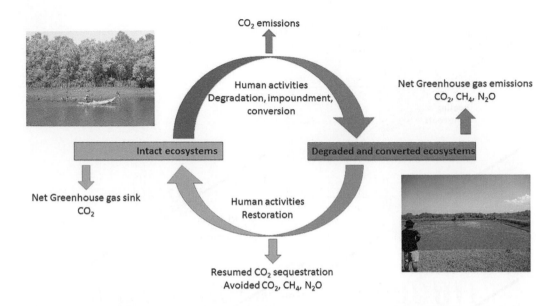

Figure 3.2 Representation of the effects of human activities on greenhouse gas emissions with degradation, impoundment, and conversion of blue carbon ecosystems, and their restoration.

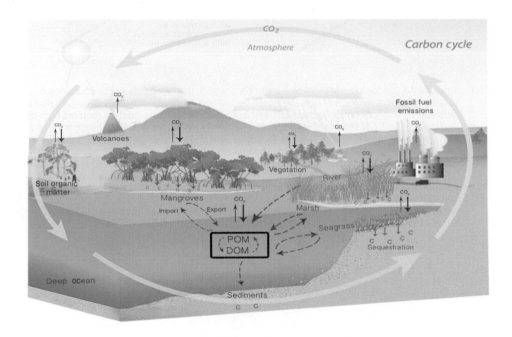

Figure 4.1 General view of global carbon cycle as it relates to global CO_2 cycling (gray arrows), with blue carbon habitats highlighted with respect to POC and DOC, as modified from from Howard et al. (2017). Carbon dioxide sources (e.g., fossil fuel emissions, fires, volcanoes) and sinks (e.g., ocean, forests, and blue carbon habitats (BCE)) show that in BCEs CO_2 is taken up via photosynthesis (large solid black arrows) and some is respired (smaller solid black arrows). In the case of BCEs, some of this gets sequestered long term (orange arrows) into woody biomass and soils for mangroves and soft tissues and soils for marshes and seagrasses, where it can then be stored in soils/sediments (orange C's). Some of this material in the BCEs can be exchanged as DOC and POC with adjacent deeper shelf waters via import and export processes (dashed arrows), where it ultimately contributes to the global oceanic pool (blue box).

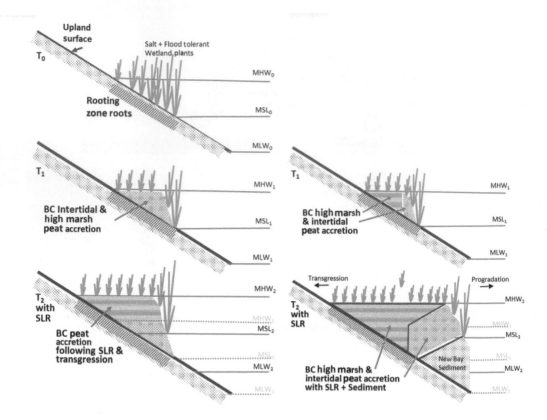

Figure 5.1 Time series conceptualization of blue carbon storage in tidal wetlands under scenarios of SLR and varying sediment availability (low availability left versus high on right). Only transgression and surface accretion occur under limited sediment availability, while progradation requires high sediment availability. T_0, T_1, T_2 refer to time progressions. MHW, MSL, MLW with subscripts refer to MHW, MSL, and MLW at times T_0, T_1, and T_2. As drawn SL does not change between T_0 and T_1. BC = blue carbon. Three soil organic matter components are shown on left: (1) that associated with roots, which does not change over time and is not included in blue carbon burial, (2) that associated with development of the high marsh platform prior to SLR, and (3) that associated with development of the high marsh platform, plus marsh transgression following a rise in sea level. To the RIGHT, we start with the T_1 stocks from the left, but then superimpose SLR and marsh progradation. In this case, we have blue carbon accumulating as on the left side, plus that associated with progradation. Conceptualization follows Redfield 1967.

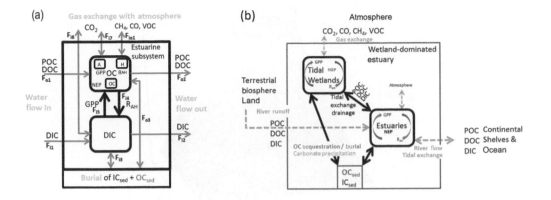

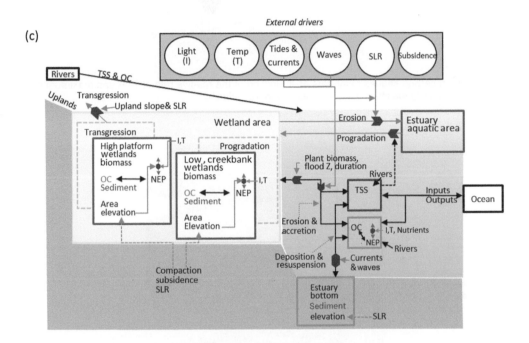

Figure 5.2 (a–c) Conceptual models of net carbon balance and blue carbon storage for saltwater wetland-dominated estuaries. (a) illustrates the compartments of inorganic carbon and IC and OC and fluxes that contribute to the net C balance for either tidal wetlands or estuaries. Flows are numbered and described in the text. (b) illustrates the relation between tidal wetlands and estuaries and their connection to adjacent upland and ocean systems. Fluxes within the tidal wetland and estuaries boxes are simplified relative to the detail shown in A. (c) a conceptual diagram of the processes associated with the net C balance and blue carbon storage that includes spatial considerations of elevation gain/loss, progradation, and transgression. Key external drivers directly controlling the net C balance are identified. Salt-water intrusion and concomitant saltwater wetland expansion is not considered here. IC implied only.

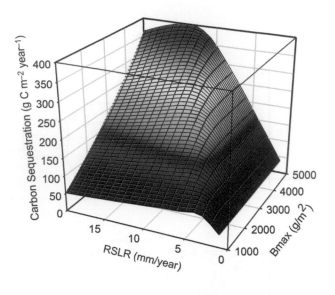

Figure 6.5 Mean carbon sequestration rate as a function of the relative rate of SLR and maximum biomass (B_{max}). Only feasible (DimE > 0) solutions were included.

Figure 7.2 Layer of white feldspar used as a marker horizon in a sediment plug. The depth of sediment above the feldspar is measured with a caliper and indicates the amount of sediment that has accreted since the layer was established, in this case just over 30 mm.

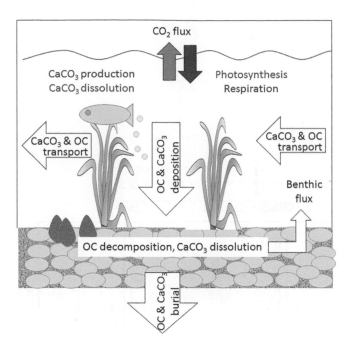

Figure 9.1 Conceptual diagram illustrating the major processes affecting the $CaCO_3$ and organic carbon cycle in seagrass ecosystems.

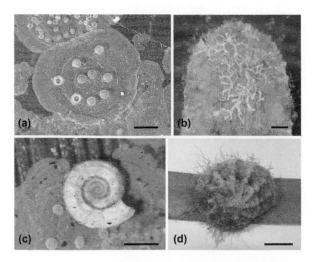

Figure 9.2 Calcium carbonate producing organisms: (a) coralline red algae (Rhodophyta), (b) the foraminifer *Cornuspiramia antillarum*, (c) the polychaete *Spirorbis* sp., and (d) the gastropod *Modulus*. Scale bars: (a, b) 1 mm, (c) 500 μm, and (d) 5 mm. Photographs provided courtesy of Tom Frankovich, Marine Education and Research Center, Florida International University.

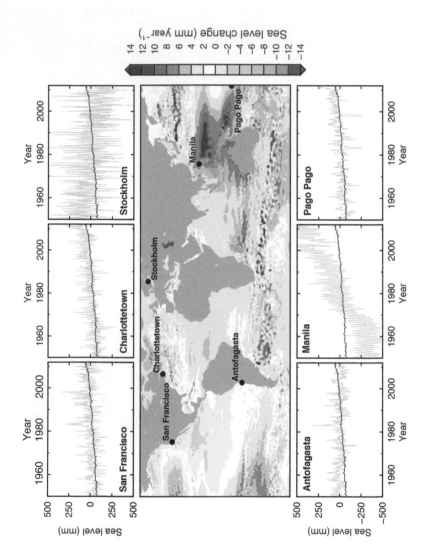

Figure 10.1 Differences in trends of sea level change associated with geography. The central map shows rates of sea surface change for 1993–2012 from satellite altimetry. The panels above and below the map show sea levels from 1950 to 2012 recorded by tide gauges at six locations. The red line in each panel represents global sea level change to highlight differences among local trends.

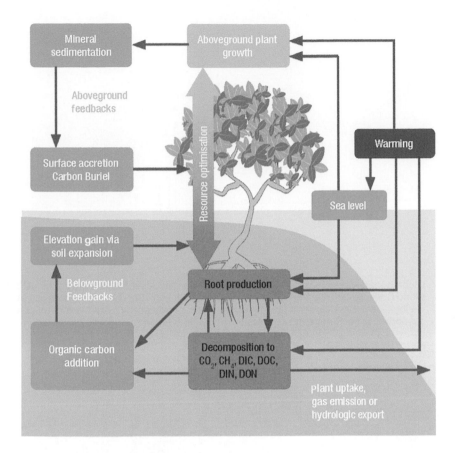

Figure 11.1 Conceptual model of temperature effects on key ecosystem processes that regulate carbon input, output, and burial, and export of CH_4 and dissolved forms of carbon and nitrogen. Modified from Kirwan and Megonigal (2013) and reproduced from Megonigal et al. (2016) with permission from the IUCN.

Figure 11.2 Two approaches to experimentally warming tidal marsh ecosystems. Left: Chambers erected on a Florida USA saltmarsh raise air temperature by 2°C–3°C. The chamber has negligible effects on soil temperature. Right: A brackish marsh in Maryland USA is actively heated (1.7°C–5.1°C) aboveground by infrared lamps and belowground by cables to a depth of 1.5 meters. Photos by Samantha Chapman (left) and Roy Rich (right).

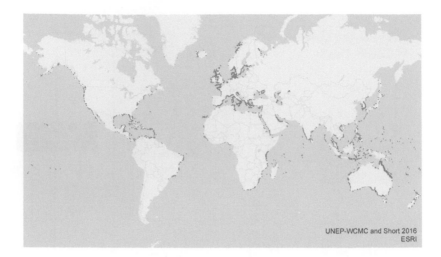

UNEP-WCMC and Short 2016
ESRI

Figure 12.1   Global seagrass distribution data (green) compiled by the United Nations Environment Programme-World Conservation Monitoring Centre (UNEP-WCWM and Short 2016; the data can be accessed via the Ocean Data Viewer: http://data.unep-wcmc.org/datasets/7; Mediterranean data collection supported by the European Commission, EUR 568,996; data used with permission).

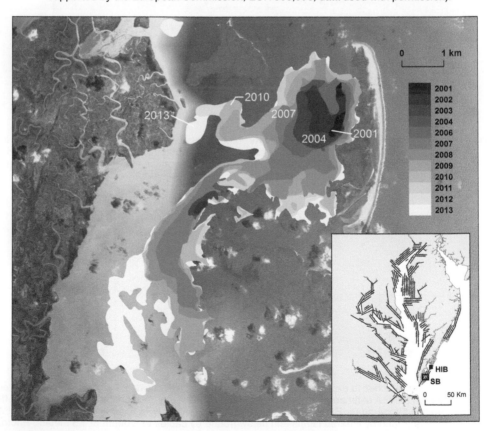

Figure 12.3   Annual expansion of the restored *Zostera marina* meadow in South Bay (SB), VA, determined from aerial photographs; the inset map shows the SB and Hog Island Bay (HIB: Figure 12.4) study areas relative to the VIMS SAV Monitoring Program flight-lines (red lines) in the Chesapeake Bay and Delmarva coastal bays.

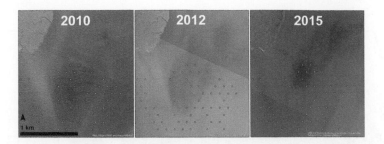

Figure 12.4 The northeastward migration of the restored *Zostera marina* meadow in Hog Island Bay (HIB), VA, is discernable from aerial imagery but not ground survey sites (white circles) located in original restoration seed plots (see Figure 12.3 inset for HIB meadow location).

Figure 13.1 Mangrove forest distributions of the world for circa 2000 based on Landsat 30 m data.

Figure 13.2 Mangroves (dark red) in the Sundarbans (Bangladesh and India) in a Landsat false color composite (band combinations 4R, 3G, 2B).

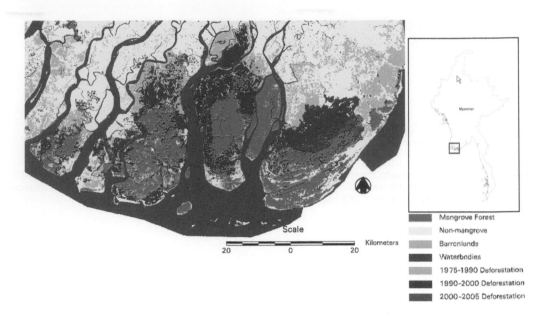

Mangrove Forest
Non-mangrove
Barrenlands
Waterbodies
1975-1990 Deforestation
1990-2000 Deforestation
2000-2005 Deforestation

Figure 13.6 Spatial distribution of mangrove deforestation in Ayeyarwady Delta, Myanmar, from 1975 to 1990 (cyan), 1990 to 2000 (red), and 2000 to 2005 (purple).

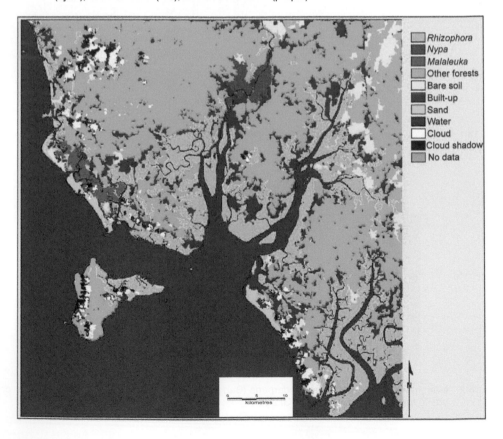

Rhizophora
Nypa
Malaleuka
Other forests
Bare soil
Built-up
Sand
Water
Cloud
Cloud shadow
No data

Figure 13.8 Species zonation map generated by the object-oriented approach using lacunarity-transformed bands in southern Thailand.

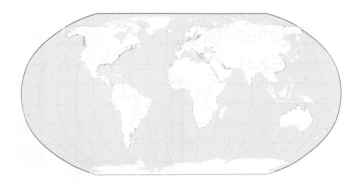

Figure 14.1 Global map of digitally recorded extent of salt marsh ecosystems, representing datasets collated between 1973 and 2015 (Mcowen et al. 2017).

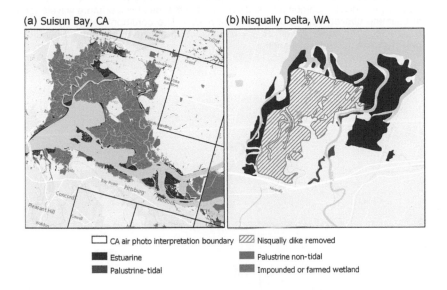

Figure 14.2 (a) USFWS National Wetland Inventory (NWI) map of Suisun Marsh, California. Here NWI distinguishes between tidal and impounded wetlands, but these distinctions can vary based on the survey (bottom right corner). (b) USFWS NWI map of the Nisqually National Wildlife Refuge, Washington. Here, NWI does not indicate the presence of tidal wetland restored in 2009 from a dike removal, illustrating that NWI can be out of date and that tidal restorations are not included in the classification.

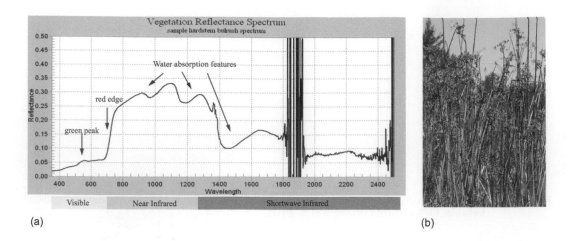

(a)   (b)

Figure 14.3 (a) Typical hardstem bulrush (*Schoenoplectus acutus*) reflectance spectrum. (b) Hardstem bulrush. Plants absorb light in the red wavelengths and reflect light in the near-infrared wavelengths. Vegetation indices—combinations of surface reflectance at two or more wavelengths—are used to estimate aboveground biomass in tidal marsh. The commonly used NDVI is calculated as: $(R_{NIR} - R_{Red}/R_{NIR} + R_{Red})$, with R_{NIR} = reflectance in the near infrared wavelengths and R_{Red} = reflectance in the red wavelengths.

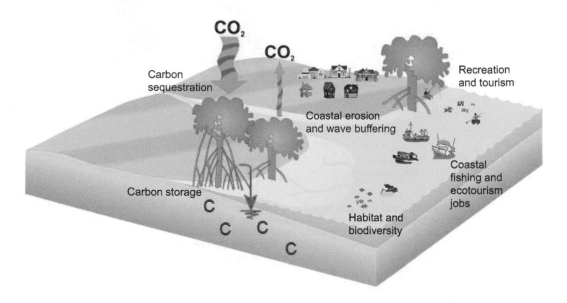

Figure 17.1 Coastal wetlands provide many additional ecosystem services including fish and bird habitat, storm and erosion-risk reduction, recreational opportunities, coastal jobs, and water quality improvements. (Adapted from NOAA figure.) Also, see the chapter ecosystem services, this book.

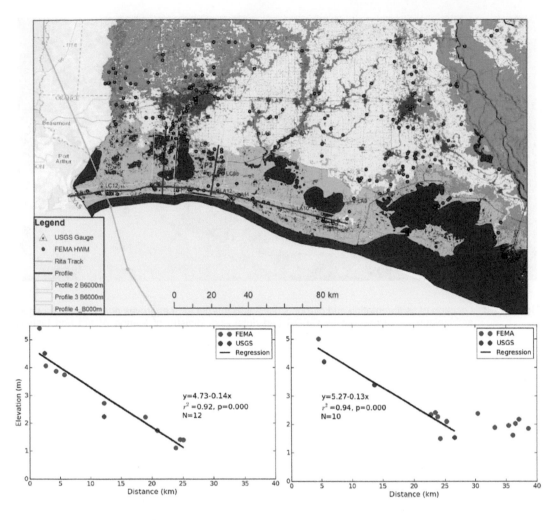

Figure 18.3 Storm tide profile locations with Hurricane Rita. USGS mobile gauges and FEMA high water marks indicate approximately 13–14 cm reduction in surge height 1 km over vegetated land.

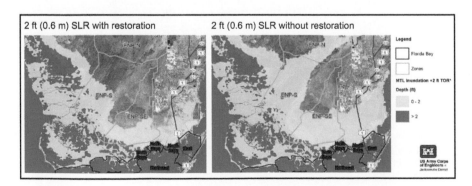

Figure 18.5 Everglades National Park with 2 ft of SLR with and without the tentatively selected plan (Alternative 4R) of the Central Everglades Planning Project (CEPP; adapted from CISRERP 2014). The scenario without restoration assumes total peat loss due to saltwater intrusion, illustrating a key area of uncertainty for the fate of Everglades wetlands related to persistence of peat soils under projected SLR.

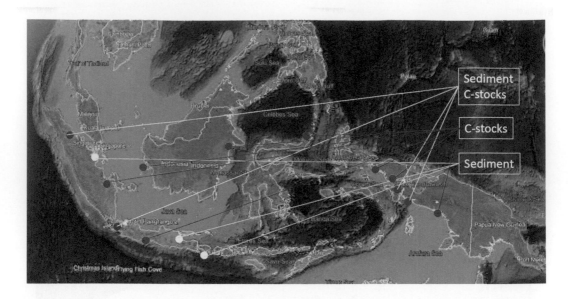

Figure 21.3 Location of sedimentation and carbon stock assessments for mangroves across the Indonesian archipelago as carried out by CIFOR (red dots) and Ministry of Marine Affairs and Fishery (yellow dots).

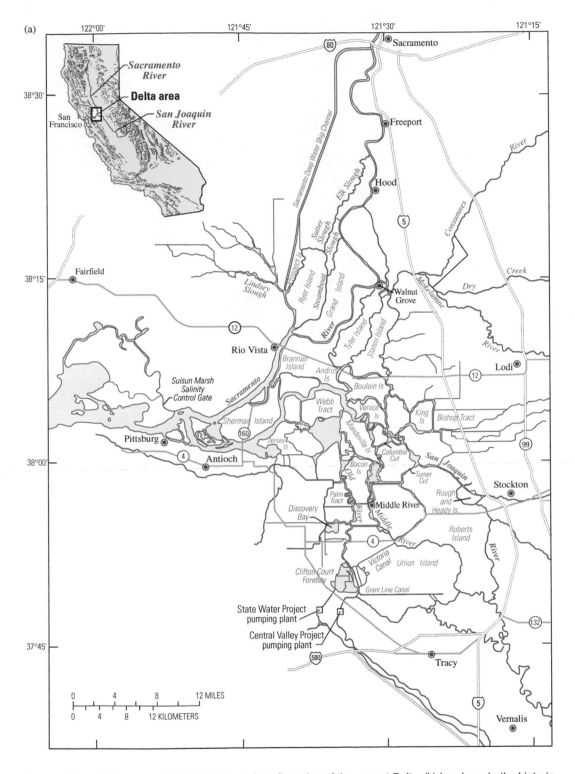

Figure 22.1 (a) Map showing the location and configuration of the current Delta. (b) Land use in the historic and current Delta (Adapted from SFEI-ASC (2014).).

(*Continued*)

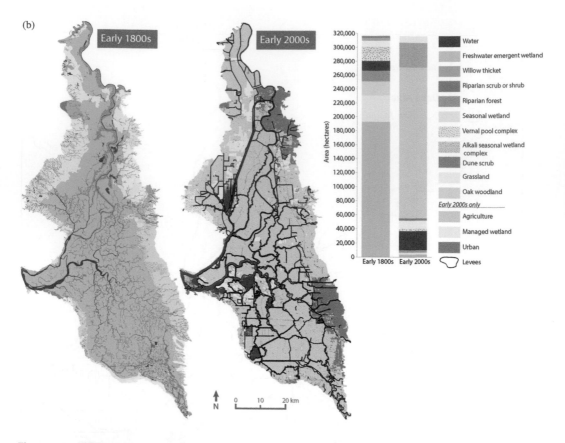

Figure 22.1 (CONTINUED) (a) Map showing the location and configuration of the current Delta. (b) Land use in the historic and current Delta (Adapted from SFEI-ASC (2014).).

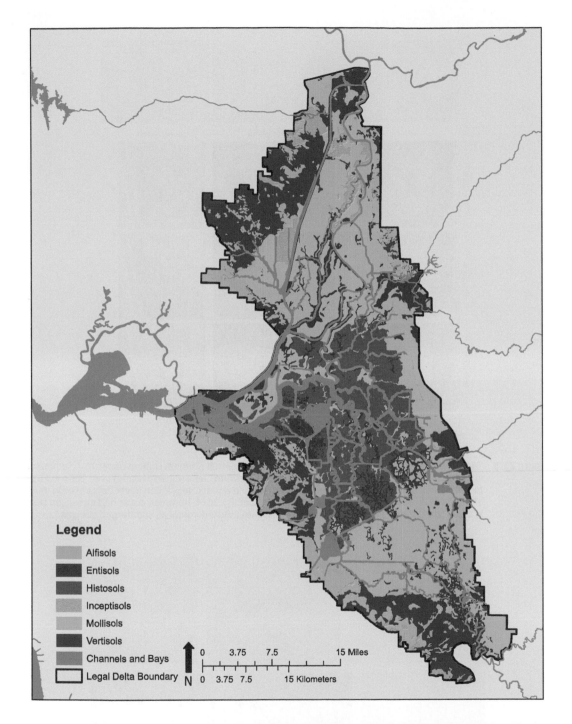

Figure 22.3 Taxonomic soil order map of the Delta showing the central area covered by organic soils (histosols), and the large areas in the north and south covered by mineral soils including mollisols, which contain a thick organic surface horizon, and vertisols, which contain a high proportion of clay (Sprecher 2001). Data from Soil Survey Geographic Database (SSURGO) (Soil Survey Staff 2016).

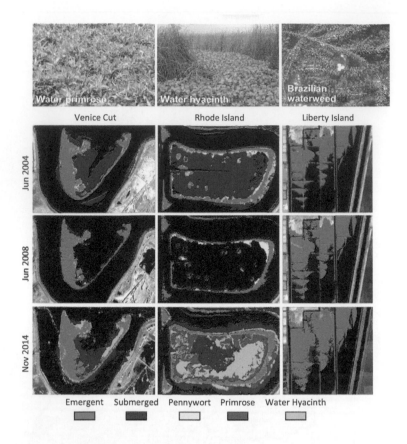

Figure 22.5 Top panel: Dominant IAV species in the Delta. Credit: Shruti Khanna. Bottom panel: IAV distribution in the central (Venice Cut and Rhode Island) and northern Delta (Liberty Island) from 2004 to 2014. In June 2008, after a rigorous spray effort, submerged aquatic vegetation cover decreased in much of the Delta, including Rhode Island and Venice Cut. By 2014, however, cover had increased back to 2004 levels. Emergent=all emergent macrophytes, Submerged=all submerged species.

Figure 23.2 Images of United Arab Emirate blue carbon ecosystems: (a) *Avicennia marina*, (b) *Arthrocnemum macrostachyum*, (c) microbial mat, (d) coastal sabkha, (e) *Halophila ovalis*, and (f) mix of *Halodule uninervis* (thin ribbons) and *Halophila stipulacea*.

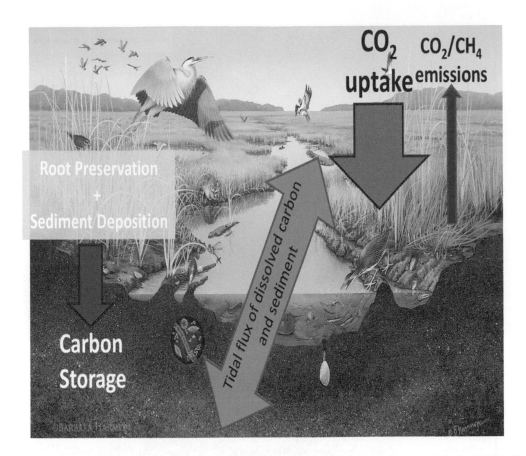

Figure 25.2 Schematic diagram of the major carbon and GHG fluxes of interest in the BWM field research program. Net photosynthetic uptake of carbon dioxide supplies root biomass that contributes to soil carbon stock increase and soil elevation accretion. Tidal exchange with adjoining coastal waters carries lateral fluxes of respiratory and detrital carbon, as well as net import or export of sediment.

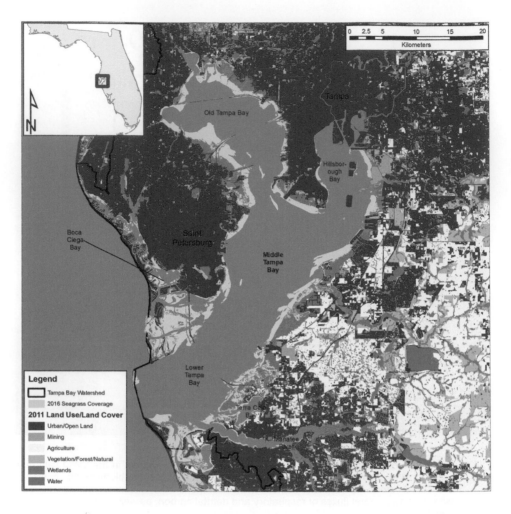

Figure 26.1 Current (as of 2011 for emergent areas and 2016 for subtidal seagrass meadows), generalized land use, and land cover along coastal Tampa Bay. (Sources: SWFWMD 2011, 2016.)

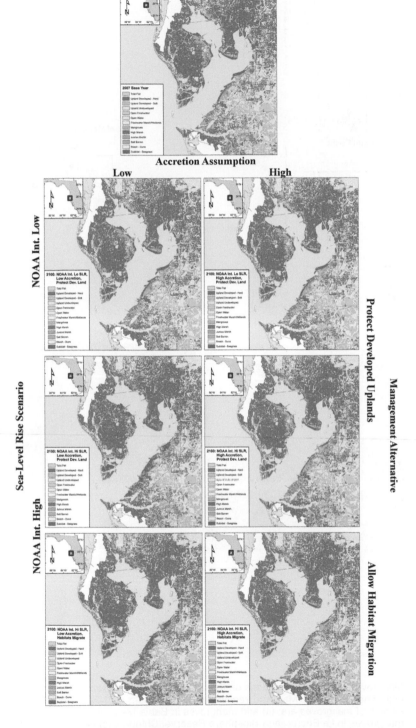

Figure 26.3 Comparison of 2007 blue carbon ecosystem extent and distribution to modeled changes in ecosystems relative to two SLR scenarios (low versus high), two accretion assumptions (low vs. high), and two management alternatives (protect currently developed land vs. habitats allowed to migrate to undeveloped and "softly" developed lands) (Adapted from Crooks and Sheehan (2016), Sources: SWFWMD, TBEP.).

Figure 27.1 Mangrove extension and regions on the basis of climate, hydrology, geologic and geomorphologic characteristics. Map and regionalization are from CONABIO (2009). Key locations for studies of mangroves are: (a) arid mangroves in Bahia Magdalena, Baja California; (b) riverine mangroves in La Encrucijada Biosphere Reserve, Chiapas; (c) karstic mangroves in Celestun, Yucatan; and (d) karstic mangroves in Sian Kaan, the Mexican Caribbean. Photos from MF Adame.

CHAPTER 19

Integrating Blue Carbon into Sustainable and Resilient Coastal Development

Stephen Crooks and Christine May
Silvestrum Climate Associates, LLC

Ryan Whisnant
PEMSEA

Michelle Orr
Environmental Science Associates

CONTENTS

19.1 INTRODUCTION

Maintaining healthy coastal wetlands, as components of the wider terrestrial and marine ecosystems that they bridge, are important for many reasons from supporting local economies to buffering climate change. Ecosystem services, derived from storage, biogeochemical cycling, and flux functions, underpin benefits ranging from marine fisheries production to coastal flood protection (Barbier 2011; Costanza et al. 1997, 2014). Yet, the extent and quality of coastal ecosystems has been in steep decline over the past century (Millennium Ecosystem Assessment 2005). A grand challenge facing humanity right now is to change this trend, finding a way of deriving benefit from coastal ecosystems while at the same time conserving and restoring them to provide for generations to come. The long-term health of human, economic, and environmental systems are intertwined and co-evolutionary (Pearce and Turner 1990; Crooks and Turner 1999). Challenges of sustainable management must be met in the face of changing climate and a growing human population, both of which are directly threatening the very existence of coastal ecosystems (Neumann et al. 2015). Despite this gloomy outlook, there are reasons for hope that sustainable management can be achieved, as global policies are agreed and, on the ground, new partnerships are developed, demonstrating examples of best practice and application of new tools.

Here we shall explore the actions that countries can take by including blue carbon ecosystems (BCEs) in management plans and strategies in order to operationalize commitments under international treaties on sustainable development and climate mitigation along with national and local needs to reduce vulnerability of coastal populations and infrastructure to changing environmental conditions and weather events. While practical experience in this domain continues to grow, there is a need for demonstration activities to highlight best practice, along with technical guidance that supports risk analysis, illustrating assessment of trade-offs in land management decisions involving the destruction, conservation, and restoration of BCEs. As an example, we shall discuss here the status of knowledge about the capacity of BCEs as natural infrastructure to provide coastal flooding risk reduction. We also highlight a range of tools through which countries might integrate BCEs in to land use, climate response, and coastal management decision-making.

19.2 MOVING BEYOND THE FAILURE OF SOCIETY TO PROTECT BCE

If it is the case that BCEs provide so many benefits, it is reasonable to ask the question: "How can this loss and degradation continue?" Clearly, to some degree the desirability of flat areas, fertile lands, and ease of access has drawn agriculture, industry, and urban settlements to coastal areas, consequently displacing natural areas (Crooks and Turner 1999). When large areas of wetlands exist, the marginal cost to society of conversion of these areas is relatively low, but the penalty increases as the total area of habitat declines and there is less and less remaining (Pearce and Turner 1990). Some of the land-use changes are in society's best interest, when the returns from the land-use are relatively high. However, coastal and estuarine ecosystems have frequently been lost to activities that result in limited gain or even net loss to society. This loss has been attributed to interrelated market and policy failures, through a lack of information and understanding of the full value of the multitude of functions that coastal and estuarine ecosystems provide (Turner and Jones 1991).

Market failures reflect the inability of the economy to properly value a range of functions from which the ecosystem and society draw benefits but do not pay the full cost. This is further complicated by inappropriate or absent property rights regimes, and the broader conflict between public and private interests. Many human activities undertaken to maximize private gain (such as pollution related to agricultural fertilizer use or aquaculture waste water, or urban/industrial expansion) can produce adverse effects. Due to the lack of enforceable public rights in some places, no compensation is provided to those who are adversely affected, resulting in a broad-based public loss. In addition, where coastal and estuarine ecosystems are privately owned, financial incentives are often insufficient to convince land-owners to conserve those ecosystems (maximizing the benefit to society overall). Such is the situation with the carbon market today, as the market value of carbon falls well below the social cost of emissions, with revenue derived from credits often uncompetitive with other land-use options. The result is that land is converted to intensive agriculture, housing, etc., in order to maximize private benefits.

Policy intervention failures stem from government or public policies that induce a detrimental impact on environmental ecosystems, which society might otherwise have chosen to conserve. Often, the failure results from a lack of co-ordination between sectoral interests, for instance, subsidies to enhance aquaculture production at the expense of mangrove conservation goals or simply lack of coordination among government agencies with differing and overlapping mandates for managing coastal areas.

Despite the recognition that BCEs provide a wide range of life-sustaining services, and the advancement of environmental economics as a research discipline capable of recognizing the value of these systems, the data to support valuation of services is often sparse, unreliable, or unavailable (Himes-Cornell et al. 2018). This paucity of information, particularly at larger scale, recent (many studies are more than 10 years old) and within a relevant geographic scope, serves as a barrier to planning and management actions. Nevertheless, it is increasingly recognized that while valuation datasets might be sparse, the ecological science is clear that important services such as fisheries production, shoreline stabilization and protection, and water quality maintenance are tied to the presence and health of these ecosystems. This thinking is permeating the decision-making of governments, donors and financing organizations, for instance, in the case of the, e.g., Sustainable Development Strategy for the Seas of East Asia (SDS-SEA), a shared marine strategy among 14 countries in East Asia coordinated by Partnerships in Environmental Management for the Seas of East Asia (PEMSEA). The SDS-SEA includes provisions in its climate strategy for employing and optimizing a range of new and innovative financial mechanisms available in the United Nations Framework Convention on Climate Change (UNFCCC) and other international agreements, including payment schemes involving blue carbon.

Significant progress has been made over the last 30 years to address these market and policy failures, supported by a growing environmental awareness, international agreements, national initiatives, and actions of non-state actors. In wealthier countries, this has been achieved by taking an integrated, multi-sectoral approach, providing strong legal protection, establishing management based on the best available science and clear planning, and engaging and investing in communities. In places such as the United States and Europe, billions of dollars and euros are spent each year recovering ecosystems that were degraded or destroyed or including wetlands as components of water management infrastructure projects (Bayraktarov et al. 2016). For less wealthy, economically emerging nations, international financial mechanisms are focusing on providing support with greater social and environmental impact than typical development finance of the past, particularly those enacting global sustainability and climate-change policy commitments (Herr et al. 2017).

19.3 GLOBAL POLICY COMMITMENTS

19.3.1 Sustainable Development

It is now over 30 years since the Brundtland Commission of the World Commissions on Environment and Development released the report *Our Common Future* and set in motion the concept of sustainable development.

Sustainable development is development that meets the needs of the present without compromising the ability of future generations to meet their own needs.

Brundtland Commissions 1987 (WCED 1987)

This concept and discussions that followed encapsulate the importance of recognizing that the world's global economy and the sustenance of human populations are bounded by the limits of the environment and its carrying capacity. This definition also speaks to intergenerational equity, maintaining development approaches that will enable future generations to achieve a level of well-being comparable to or better than the current generation. In the years since *Our Common Future* was published, approaches for delivering sustainable development have ranged from those considering the balance between natural and technological mechanisms, to those that achieve a supply of goods and services that recognize sustainable resource management (Pearce and Turner 1990; Hammer and Pivo 2017).

Continuing progress has been made, though at times haltingly. Decades of policy work have culminated in two major sustainable development policy frameworks set forth in 2015—the 2030 Agenda for Sustainable Development and the Paris Climate Agreement. Both represent ambitious, and critically needed, global agreements. Sustainable management of coastal ecosystems, involving conservation of intact coastal resources and restoration of degraded systems, is fundamental to achieving commitments under both agreements.

19.3.2 2030 Agenda for Sustainable Development

In September 2015, 194 countries of the UN General Assembly adopted the 2030 Development Agenda entitled "Transforming Our World: the 2030 Agenda for Sustainable Development".[*] The adoption by countries of the 2030 Development Agenda builds on progress from the 1972 UN Conference on Human Environment (Stockholm), the 1992 UN Conference on Environment and Development (Rio), the Rio+20 Conference and the Millennium Development Goals.

The 2030 Agenda commits each country to take an array of actions to tackle the root causes of poverty and to increases economic growth and prosperity through improving human health, education, and social needs while protecting the environment. Speaking at the 2030 Development Agenda Summit, the Secretary General noted, "The true test of commitment to Agenda 2030 will be implementation. We need action from everyone, everywhere. Seventeen Sustainable Development Goals are our guide. They are a to-do list for people and planet, and a blueprint for success."

At the 2016 G20 Leaders' Summit, global leaders endorsed the G20 Action plan on the 2030 Agenda for Sustainable Development (Rickels et al. 2017). The Action Plan states that countries will "integrate sustainable development in domestic policies and plans and international development efforts." The document outlines a set of 17 Sustainable Development Goals (SDGs) and 169 associated targets.[†] Coastal wetlands and soils, both as a subset of marine and land systems, are important cross-cutting components of many of the SDGs, but particularly ensuring availability and sustainable management of water and sanitation for all (SDG 6); decent work and economic growth (SDG 8); industry, innovation, and infrastructure (SDG 9); sustainable cities and communities (SDG 11); responsible consumption and production (SDG 12); climate action (SDG 13); life below water (SDG 14); and life on land (SDG 15).

One important concept emerging from the dialogue on sustainable development is that of **green economy**. The **Rio+20** United Nations Conference on Sustainable Development focused on two key themes—the further development of the institutional framework for sustainable development and the advancement of the **green economy**. While the concept of green economy focuses on the sustainable use of natural resources, and corollary concept of **blue economy** has emerged, capturing the economic importance of maintaining healthy marine resources (Golden et al. 2017).

Blue economy is defined as a practical ocean-based economic model using natural infrastructure and technologies, innovative financing mechanisms and proactive institutional arrangements for meeting the twin goals of protecting our coasts and oceans, and enhancing their potential contribution to sustain development, including improving human well-being, and reducing environmental risk and ecological scarcity (U.N Habitat 2010). From an integrated coastal management perspective, BCEs are valued for their role as vital resources supporting the development of blue economies.

[*] https://sustainabledevelopment.un.org/post2015/transformingourworld
[†] The Goals are as follows: 1. No poverty, 2. Zero hunger, 3. Good health and well-being, 4. Quality education, 5. Gender equality, 6. Clean water and sanitation, 7. Affordable and clean energy, 8. Decent work and economic growth, 9. Industry, innovation, and infrastructure, 10. Reduce inequalities, 11. Sustainable cities and communities, 12. Responsible consumption and production, 13. Climate action, 14. Life below water, 15. Life on land, 16. Peace, justice, and strong institutions, and 17. Partnership for the goals.

Protecting, restoring, and sustaining healthy coastal and marine ecosystem services are important elements of supporting blue economy development. Equally important is assessing the trade-offs between engineered and natural approaches to managing coastal areas. While natural systems require more space, they often provide greater benefits when the total economic value is assessed.

19.3.3 UNFCCC

The UNFCCC, adopted in 1992 at the Earth Summit in Rio, is an international environmental treaty established in response to a growing awareness of the threats of climate change. The objective of the framework is to stabilize atmospheric greenhouse gas (GHG) concentration at levels that would prevent dangerous anthropogenic interference with the climate system. The framework places no binding limits on GHG emissions for individual countries and holds no enforcement mechanisms. Rather, it hosts specific international treaties negotiated by countries toward achieving the goals of the framework.

In 2015, all but two countries globally adopted the Paris Climate Agreement, an ambitious agenda under the UNFCC to tackle climate change through emissions reductions.[*] Under the Agreement, each country commits to stepping up actions on emissions reductions, and planning and regularly reporting on progress. The prior 1997 Kyoto Protocol, widely regarded as a failure, focused almost exclusively on non-biosphere climate-change mitigation, without focusing on emissions from deforestation and ecosystem degradation (Joosten et al. 2016). Eighteen years later, adoption of the Paris Agreement addressed these failings by providing nations with the option of including land management as a component to GHG reductions or increased sequestration.

Though many components of the biosphere, including coastal wetlands, were not specifically called out in the final text, in its preamble, the Paris Agreement notes the "importance of ensuring the integrity of all ecosystems, including oceans, and the protection of biodiversity." A full article is reserved for the Parties' commitment to "conserve and enhance, as appropriate, sinks and reservoirs of greenhouse gases," where coastal wetlands play a critical role.

An important component of the Agreement is the enactment of Nationally Determined Contributions (NDCs), though which countries set out their national strategy for GHG reductions. In preparation for the Paris Climate Change Conference, 58 countries recognized BCEs within their strategy documents (i.e., Intended NDCs) as one approach for tackling climate change. For the most part, countries focused on the importance of BCEs in the context of climate change adaptation and resilience (e.g., the need for flood protection or food provision services), while a few early movers and adopters committed to conservation and restoration of these ecosystems for their climate change mitigation benefits (Herr et al. 2017). An increasing number of governments and non-government institutions have also started to conceptualize specific opportunities presented by linking blue carbon interventions with conservation finance (Huwyler et al. 2014) and payment-for-ecosystem services (PES) schemes (Locatelli et al. 2014), on the one hand, and new climate finance tools—such as results-based finance, blue bonds, and debt-swap-for-nature agreements—on the other hand (Hannam et al. 2015; Thiele and Gerber 2017). Several country-specific efforts demonstrate a growing interest in including the carbon value of these ecosystems into national policy, planning, and decision-making.

19.4 A NEED FOR CLIMATE RESILIENT COASTAL DEVELOPMENT

Coastal areas bear the brunt of climate change impacts (Neumann et al. 2015). More than half of the world's population lives within 200 km of the coast, with a good portion of those living at or

[*] At time of writing Syria and Nicaragua have joined the Accord, with the position of the United States now being uncertain.

below sea level protected by flood defenses. There is a need to increase both the resilience of coastal communities and infrastructure and environmental systems to the changing conditions brought by climate change and the major disturbance events that occur along the way.

The concept of resilience refers to the capacity of social and environmental systems to bounce back from a shock or be sustained under long-term changing conditions (Adger 2000). Maintaining resilience is an important component of helping coastal communities to achieve sustainable development (under by SDGs) and adapt to climate change. Adaptation and resilience have gained traction as key components of disaster risk reduction and management (DRRM) for coastal development (Renaud et al. 2013). Increasing sea level compounds storm events (which have been growing in magnitude and frequency) threatening coastal populations and infrastructure once thought safe from loss and damage.

Failure to address these risks comes with enormous human and economic costs. A June 2016 storm that battered eastern Australia, sweeping away property and infrastructure, and eroding shorelines across a 200-km stretch caused AUS$75 billion in damage (Mitchell et al. 2017). One review of natural hazard costs worldwide from 1995–2015 found that floods accounted for 46% of damage (Perelman et al. 2017). The total rises to 71% when including other aspects of storm-related damage that often accompany flooding. Such damages were apparent in the Caribbean and United States in 2017, a year when 17 major storms were generated in the Atlantic, of which Harvey, Irma and Maria devastated coastal and island communities with estimated economic damages of $290 billion (Aster 2017). According to the United Nations, annual flooding increased by 34% to 171 flood events per year from 2005 to 2015, compared with the prior decade (Perelman et al. 2017). In China, flooding in the more economically developed coastal provinces already accounts for more than 60% of the country's economic losses due to flooding (World Economic Forum 2017). Sea level rise will only exacerbate these impacts, as storm surges occurring once every 100 years are projected to occur at 5–10 times that frequency with only 30 cm of sea level rise (Neumann et al. 2015).

An increasing number of coastal states and municipalities are beginning to invest in damage risk reduction and resilience strategies and infrastructure to limit economic damages. The Rockefeller Foundation-funded 100 Resilient Cities program, for instance, provides resources for cities to hire a Chief Resilience Officer and implement a Resilience Strategy. But disaster-prone cities and public utilities will require significant financing and mobilization of effort to construct the expensive infrastructure needed.

At the same time, we are experiencing a worldwide biodiversity crisis. Southeast Asia is an example of particularly high biodiversity and coastal pressures. According to the Association of Southeast Asian Nations (ASEAN) Center for Biodiversity, the ASEAN region is poised to lose 70%–90% of its habitats and 13%–42% of species by 2100. According to the Convention on Biological diversity, the global need for financing resources to fulfill its 2020 strategy plan is estimated at between $150 and 440 billion per year, a value 3–8 times current available funds (UNDP 2016).

19.5 INCLUDING NATURAL INFRASTRUCTURE IN COASTAL RESILIENCE AND RISK REDUCTION

The terms **green** or **natural** infrastructure covers a wide range of practices, but in essence refers to the application of natural systems as components of the landscape that meet human infrastructure needs. Coastal ecosystems can provide substantial coastal flood defense benefits. A growing body of evidence characterizes the conditions for which these ecosystems provide wave sheltering, shoreline stabilization, and coastal storm surge reduction (e.g., Shepard et al. 2011; Gedin et al. 2011; Temmerman et al. 2013; Spaulding et al. 2014; Narayan et al. 2016; Currin et al. 2017; Morris et al. 2018). Given their ability to adapt with sea level rise and provide co-benefits, natural coastal infrastructure can provide significant benefits over traditional **hard** infrastructure in many circumstances. In the context of managing water resources, natural infrastructure is recognized for its role in storing, filtering, and purifying water bodies.

In terms of scale, it is common to find natural infrastructure applied in small-scale projects, such as installation of a **living shoreline** or individual wetland restoration projects. But natural infrastructure can scale across the landscape. We consider large-scale floodplain reactivation (removing levees to allow water to reach river or coastal floodplains), the cumulative impact of multiple wetland restoration projects, and large natural reefs and wetlands as examples of system-scale natural infrastructure. Coordinated projects and natural systems bring benefits that accrue across the landscape such as hydrologic and ecological connectivity, habitat mosaics, species refugia, flow dissipation, sediment supply, and carbon sequestration.

In considering the landscape context, there is also a need to consider linkages between ecosystems utilized as natural infrastructure. Coral reefs, for example, are highly effective in attenuating wave energy (Ferrario et al. 2014) and providing sheltered conditions for coastal residents, mangroves and seagrass beds. Mangroves and seagrasses help stabilize sediment from upland areas, thus protecting coral reefs from harmful sedimentation. A sequence of habitats such as reefs, seagrasses, and mangroves provide cumulative risk reduction benefits. These linkages and the shoreline protection, food security, and climate ecosystem benefits provided are dependent on maintaining integrated healthy ecosystems.

The appropriate use of natural infrastructure and its benefits relative to traditional **hard** approaches depend on the landscape setting and planning context. Significantly, natural infrastructure requires space. While meaningful wave attenuation can occur within the first few meters of the wetland margin, large areas (kilometers rather than meters) of mangroves and coastal marshes are required to reduce surging flood water levels, with the magnitude of reduction dependent on the strength and duration of a given storm (Wamsley et al. 2009; Wamsley et al. 2010; Zhang et al. 2012). Traditional hard infrastructure typically requires a smaller footprint. Because hard infrastructure is static, fixing the shoreline in place, the shore protection benefits and limits of hard infrastructure are more readily quantifiable from an engineering perspective, which provides a level of comfort to decision-makers even when use of natural infrastructure may be more appropriate. However, because of its static nature, hard infrastructure can be **brittle** when thresholds are exceeded (Gittman et al. 2014). Natural systems are adaptable to highly dynamic conditions and can often recover following damage (e.g., Paling et al. 2008; Gittman et al. 2014). Over time, coastal wetlands accumulate sediments, building in elevation and thus naturally maintaining their benefits with sea level rise, up to a point. Under conditions of high rates of sea level rise, or other forms of stress, coastal wetlands can drown and convert from intact vegetated ecosystems to unvegetated flats and open water (Morris et al. 2012; Kerwin and Megonigal 2013). But, it is important to create space for wetlands to migrate landwards as part of a resilient response to sea level rise (Pethick and Crooks 2000).

In the right context, natural infrastructure can be more cost effective when compared to traditional infrastructure, such as submerged breakwaters (Narayan et al. 2016), and more so when co-benefits are factored in (Costanza et al. 2014; Gittman 2016). Natural systems can be combined with **hard** engineering components along a **soft-hard** or **green-gray** continuum, to provide shoreline protection with ecosystem benefits within site-specific constraints. Of course, some extreme events can overwhelm both natural and hard infrastructure. Just as in hurricane prone areas of United States, where residents behind hard infrastructure are evacuated to high ground and shelters, such approaches will be required for communities living behind natural infrastructure (Figure 19.1).

The opportunities to include natural infrastructure in land-use planning vary greatly around the world depending on the legacy of land use and geomorphic setting (Figure 19.2). Historically, major diking of coastal wetlands in Europe occurred for agrarian use and many of these areas have remained as open space. Similarly, wetland loss in North America between 1800 and 1970 (marking

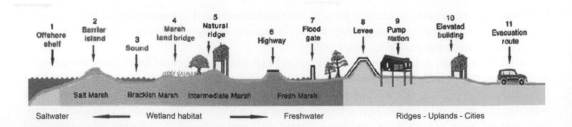

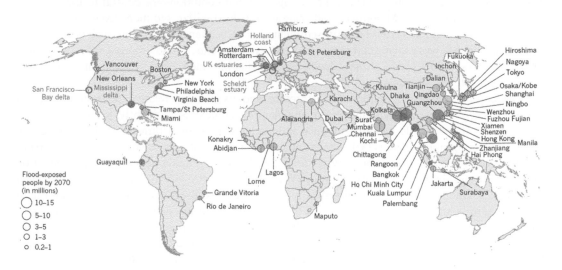

Figure 19.1   Examples of multiple lines of defense strategy. This hybrid approach in Louisiana applies a combination of natural and built lines of flood defense along with maintaining an escape route to higher ground. (Source: Lake Pontchartrain Basin Foundation.)

Figure 19.2   Global need for flood protection with example locations for potential application of nature-based approaches (Temmerman et al. 2013).

enactment of wetland protection regulations) largely occurred away from major urban settings and most land-use conversion was to open space uses. In both settings, opportunities are being taken to recover wetlands, recognizing their natural infrastructure and ecosystem benefits. In Asia, dramatic land-use change has occurred largely post-1950, fueled by population growth, migration to rapidly growing urban settings, the intensification of aquaculture in coastal wetlands, and industrial development (Neumann et al. 2015). Many cities have extended to the very edge of rivers and the sea, limiting risk reduction options mostly to hard infrastructure, even as concepts of greener city design and natural infrastructure awareness are starting to take hold. In some places, such as coastal Vietnam and southern India, former aquaculture ponds remain as open space and have been planted with mangroves to recover natural buffering from storms and other benefits. Some countries in Africa and South America have relatively intact coastal wetland systems, where perhaps there is an opportunity to build coastal development in a climate-smart and environmentally resilient manner, though institutional, regulatory, and economic constraints abound.

Large and growing sources of funding are becoming available for coastal development incorporating natural infrastructure (such as green bonds), but barriers to investment are hindered by a lack of technical guidance for designing and evaluating such projects and lack of ability to adequately quantify ecological and economic benefits. Guidance is needed in: (1) identifying locations where natural infrastructure can play a significant role in coastal resilience; (2) developing the experience and standards to overcome institutional biases in favor of **proven** hard infrastructure; and (3) developing institutional arrangements capable of matching available funding with the needs of individual situations (Colgan 2017). According to a report from the U.S. government, general recognition is that the services provided by natural infrastructure are necessary, but not sufficient, prerequisites for its integration into mainstream coastal resilience and risk reduction strategies. To justify use, planners and decision makers need specific, quantitative information demonstrating that the benefits of natural infrastructure (including co-benefits) are wise investments (White House NSTC 2015).

Yet, though the capacity to quantify all ecosystem services is often lacking, increasingly funds are being provided to support natural infrastructure approaches. In states such as California, for example, the *Water Quality, Supply and Infrastructure Improvement Act* of *2014 (Proposition 1)* authorized $1.5 billion of $7.545 billion in general obligation bonds to fund ecosystem and watershed protection and restoration and water supply infrastructure projects, including coastal wetlands restoration. There have been a number of similar bond financing products, and with a growing focus on climate resilient urban community development, the linkages are being made between land use, climate adaptation and mitigation, with requisite funding. In terms of small scale but environmentally meaningful approaches, the State of Maryland has shifted the burden of proof in permit applications for shoreline protection, requiring that living shoreline (small-scale natural infrastructure) approaches be considered first, with hard infrastructure approved only where living shorelines cannot be used (Pace 2017).

While knowledge gaps currently impede more widespread global adoption, there are a growing number of examples and reviews that we can draw upon to inform the evaluation of natural infrastructure (e.g., Shepard et al. 2011; Temmerman et al. 2013; Spaulding et al. 2014; Narayan et al. 2016; Bilkovic et al. 2017; Morris et al. 2018). The literature addresses natural and restored systems and, in some cases, compares natural infrastructure to conventional engineering approaches, such as seawalls (Gittman et al. 2014) and breakwaters (Narayan et al. 2016). A limit of much of the data collected quantifying the natural infrastructure benefits of BCEs has been collected during everyday conditions, and therefore data representing extreme events are sparse (Sheppard et al. 2011; Narayan et al. 2016). It is at the time of extreme events that communities are at risk of high wave energy and flooding. Detailed data are needed to develop metrics that allow planners and engineers to quantify risk reduction considering location-specific conditions (Spaulding et al. 2014) and capacity is needed within the planning and engineering community to plan and design natural infrastructure solutions.

19.6 EXPERIENCE OF COASTAL WETLAND CONSERVATION AND RESTORATION

To help inform recovery of BCEs and advance the application of natural infrastructure, there are a good number of lessons to be drawn from over four decades of coastal wetlands restoration experience in places such as the urban setting of San Francisco Bay (Williams and Faber 2001; Brew and Williams 2010; Primavera et al. 2012; Crooks et al. 2014; Lewis and Brown 2014; Bay Goals 2015). This experience is not evenly distributed around the globe. The industry is substantial in the United States, but by factoring in geomorphic setting, the principles are transferable to other geographies.

1. **Have a clear and coherent project planning approach**: Successful conservation and restoration is a more likely outcome when projects have a coherent planning process that identifies goals and objectives, opportunities and constraints, adopts best available conceptual models, and sets performance metrics that track project performance relative to achievable success criteria

(Clewell and Aronson 2013). Investing in planning and design is sometimes seen as an unnecessary expense (Williams and Faber 2001). However, an appropriate level of planning can lower risk, save construction costs, increase certainty of project outcomes, reduce the scale of post-project adaptive management or remediation, and greatly improve the ecological value of the project. Given that land availability is often a constraint, there are positive cost-benefit outcomes to restoring or conserving higher value wetlands on a per unit area basis. The success of the project will be judged on its outcome and this, in turn, will influence public support and funding for future projects.

2. **Plan conservation and restoration projects in the wider landscape context**: When planning to conserve and restore coastal ecosystems it is important to consider how individual projects fit within the wider landscape context (Goals Project 2015). Ecosystem benefits are derived from maintaining and recovering expansive and connected areas rather than a patchwork of isolated fragments. Providing space offers resilience to gradual changes and the capacity to respond to disturbance events such as storms (Pethick and Crooks 2000). Landscape mosaics also offer a degree of ecosystem redundancy, which is important to maintaining resilient populations of species. In urban settings, which do not offer large-scale restoration potential, strategic location, and connection of projects can make the most of space available to maximize ecological benefits (Simenstad et al. 2006) and integration with natural infrastructure approaches for flood risk reduction in channels that transition uplands to coastal waters.

3. **Prioritize to enhance sustainability**: Not all coastal areas will respond resiliently to climate change and sea level rise (Orr et al. 2003). Given scarce resources, coastal planners may need to prioritize conservation and restoration actions (Goals Project 2015). At higher rates of sea level rise, some coastal ecosystems will cease to respond resiliently in their current position and will migrate or convert to other ecosystem found at deeper water depth. Planners can reduce risk by locating projects in a way that accounts for landscape evolution and targets locations that will be resilient to future conditions. This may involve phasing projects at different stages of restoration maturity to provide for habitat complexity and taking opportunities of sediment availability.

4. **Restore physical processes and ecosystem dynamics**: Natural processes and dynamics underlie coastal ecosystems' delivery of goods and services. Natural adjustment in the structure and composition of coastal ecosystems results from natural environmental fluctuations and disturbance dynamics. Attempts to control these natural processes may increase fragility and reduce resilience to climate change (Simenstad et al. 2006).

5. **Understand the restoration trajectory and ecological thresholds**: Restoration of coastal ecosystems is a process that may take decades to be fully achieved. Through good science-based planning, projects can be set along a trajectory towards a successful final outcome (Stralberg et al. 2011; Goals Project 2015). Ecological and geomorphic thresholds are perhaps the most challenging aspect of habitat restoration to predict, but can be illuminated through science, trial projects, and modeling (Kirwin et al. 2010). Project planners should consider the likelihood of critical thresholds impacting a project (e.g., insufficient inputs to build intertidal sediment in a wave exposed location) and plan accordingly to reduce risks (e.g., add sediment, select site at elevations high in tidal range or include features that will dissipate wave energy as the site recovers) (Williams and Orr 2002). Landscape scale thresholds may also exist, heightening the importance of considering the project beyond the site scale (e.g., loss of a protective dune barrier due to disruption to longshore sediment supply).

6. **Conserve and restore BCEs sooner rather than later**: The magnitude of climate change impacts will likely increase over time as rate of change increases. Moreover, the capacity of coastal ecosystems to be resilient to change is also a function of system maturity (e.g., building of soils to high intertidal elevations and strengthening with root material). Therefore, restoring wetlands sooner rather than later, as well as conserving intact wetlands will heighten overall system resilience (Stralberg et al. 2011).

7. **Restoration of historic conditions is not always possible**: In many cases, landscapes have been altered to such a degree that full restoration of resilient historic conditions cannot be achieved. Restoration should plan for the future, in these cases (Simenstad et al. 2006). Where restoration of resilient historic ecosystems is feasible, prior conditions can provide a positive and clear restoration

Table 19.1 Geomorphic Metrics for Projecting Resilience and Restoration Potential of Intertidal Wetlands to Sea Level Rise (Crooks et al. 2011)

Parameter	Description
Accommodation Space Index	The area below wetland plain (mean high water spring tide elevation) divided by the volume below that plain on subsided lands. A high value indicates relatively low levels of mineral sediment required to fill volumes above subsided lands to rebuild wetlands.
Readily Restorable Area Index	The area of land that would support vegetated wetlands upon tidal reconnection divided by the total subsided area. A high value suggests higher rates of wetland revegetation and wetland building by organic sedimentation with restoration.
Sediment Supply Index	The index is derived from two values for subsided volume: (1) full subsided volume below marsh plain and (2) subsided volume within the Readily Restorable Area. The index is calculated as the annual sediment supply divided by the subsided volume and represents the amount of mineral sediment required to build subsided lands back to marsh plain elevations. A high value indicates good amounts of sediment availability to rebuild wetlands.
Transitional Area Index	This index is derived from the footprint of upland areas just above the tides (approx. 1 m above) that would support further tidal wetlands divided by the subsided area below the marsh plain. A high value of the index indicates that there is good opportunity for wetlands to migrate with sea level rise (assuming barriers are removed).
Sea Level Rise Vulnerability index (Syvitski et al. 2009)	A comparison of current and projected sediment supply against rate of relative sea level rise (including land subsidence). A high value suggests higher resilience to sea level rise provided by mineral sediment contribution to wetland building.

target, in particular to support endemic species. Often, the goal of estuary or shoreline scale restoration is to return a balance in habitat types and connectivity across the landscape mosaic, where restoration of historic conditions is not feasible, establishing optimal conditions for other BCEs may be the target (e.g., recovery of emergent tidal marshes on lands that once supported forested tidal wetlands prior agriculture induced subsidence).

8. **Ecosystem restoration takes time**: Depending on the extent of disturbance, a system may take decades or longer to fully recover. Good project planning will focus on establishing the restoration trajectory that a recovering ecosystem will track against (Lewis and Brown 2014). Protection of intact ecosystems avoids risks and costs of restoration actions.

9. **Avoid transplantation of non-indigenous and nuisance species**: There are numerous examples of introduced species bringing cascading negative ecological and biodiversity impacts on a region. While levels of awareness are now much higher, care should be taken to minimize such risks (Primavera, et al. 2012; Lewis and Brown 2014).

For further information on modeling and planning approaches for siting wetlands projects in coastal landscapes see reports such Needelman et al. (2018) and Goals Report (2015). Table 19.1 provides a set of geomorphic metrics for quantitatively exploring the resilience and restoration potential of intertidal coastal wetlands areas (derived from Crooks et al. 2011, which provides examples of application and additional metrics for carbon project site selection). These metrics can be derived from topographical data (e.g., LIDAR and SRTM) and information or proxies for sediment supply.

19.7 PRACTICAL STEPS TO ADVANCING BLUE CARBON INTERVENTIONS

Each coastal country across the world has the opportunity to enact or strengthen actions that improve management of BCEs either to support climate change policy responses, as described in prior sections, or for other purposes. Conservation and restoration of BCEs (and other coastal and marine ecosystems) cuts across and underpins many aspects of marine economic sectors. Improved management also benefits from regional and cross-border collaboration.

Developed for the study Understanding Blue Carbon Opportunities in the Seas of East Asia (Crooks et al. 2017), Table 19.2 below highlights a framework of actions that countries can take to advance management of BCEs and climate response planning, in support of blue economy growth. The Framework is based upon three main pillars: (1) awareness building, (2) knowledge exchange, and (3) acceleration of practical action. These pillars are not necessarily seen as being sequential, rather, steps can be taken in parallel as a country may be more advanced or may be still advancing under other coastal management activities, such an inclusion of coastal wetlands in marine-protected areas or as part of natural infrastructure for flood risk reduction. At the most basic level, there is a need for awareness building through assessments of status and trends in coastal land use change, to inform understanding of drivers, pressures and state change. Such information then provides the basis for accounting for blue carbon stocks and stock change, which may be included in national reports on GHG emissions and removals. Illuminating states and trends and the scale of impacts to blue carbon ecosystem services provides knowledge supporting decision-making for practical action.

While some countries may already be advanced in some of these actions, there is value in regional coordination, through intergovernmental bodies like PEMSEA in East Asia and other donor-funded large marine ecosystem (LME) bodies in regions around the world. Many impacts (e.g., sediment diversion) and interventions (international financial agreements) affecting BCEs and their management are trans-boundary in nature. NDCs, which convey the intent of countries to improve management of BCEs, can support international and regional coordination efforts.

19.8 CONCLUDING REMARKS: TIME AND TIDE

Fundamentally, goals of sustainable management of coastal resources, including BCEs and climate change adaptation, are fully compatible. As highlighted in Table 19.2, there are steps that countries can take to build awareness, facilitate knowledge exchange, and accelerate practical actions. BCE restoration and conservation, natural infrastructure approaches, and other management actions are occurring around the world at an accelerated pace and at growing scales. There is a need to connect, share and mainstream these activities into coastal and climate resilience planning. Likewise, there is a need to better promote the many benefits of blue carbon and natural infrastructure, not only in general terms, but in ways that are meaningful to both scientists and communities and decision makers. While more research is needed to improve the quantification of ecosystem services providing by BCEs and natural infrastructure, best practice guidance can already be developed, allowing for steps forward in managing uncertainty and adaptive learning. Finally, and critically, there will be a need for increased financial investment in activities that support nature-based climate resilience and divestment away from unsustainable practices.

Over the last 30 years, concepts around sustainable development, climate change mitigation and adaptation, and integrated coastal zone management have emerged and evolved. Consideration for the biosphere—lately, including blue carbon and the value of natural infrastructure—and including communities as an integral part of these concepts has been recognized more fully and continues to grow in importance. Yet, we are up against the mounting pressures of a growing population and rising global middle class, and the demands for space and resources they bring, as well as accelerating sea level rise. Taken together, these compress the timeline when proactive action is still possible. Time and tide, wait for no man (Chaucer, Franklin's Tale), and this is truer now, more so than ever.

Table 19.2 Opportunities for Countries to Incorporate Blue Carbon Ecosystems into Integrated Coastal Management, Climate Response, Biodiversity Maintenance, and Blue Economy Planning

Action	Benefit	Actor
Build Awareness		
Include blue carbon in policy dialogue.	Supports development of national and subnational policies, cooperation between governments and intra-government agencies and inclusion of private sector and community groups.	National government; International agencies; International NGOs; Academic community.
Apply 2013 IPCC Wetland Supplement and include BCEs in GHG National Inventory and Communications.	Improved quantification of emissions and removals due to land management. Enables setting of goals and benchmarks for management plans.	National Government
Report status and trends of coastal ecosystem, including improved mapping of BCEs, their change through time, threats, and status.	Supports management planning and inclusion of BCEs in GHG national Inventories and communications.	National government; International agencies; Academic community.
Facilitate Knowledge Exchange		
Join networks such as the International Partnership for Blue Carbon[a] and the International Blue Carbon Initiative.[b]	Bring together key organizations to coordinate international activities.	National government; International NGOs, Academic Community
Facilitate/contribute to technical and policy workshops (e.g., The Blue Carbon Initiative[b]).	Enable communication between technical experts and shared science, policy and implementation experience.	National government; International NGOs; Private sector; Academic community.
Support science programs and technical analysis.	Improved quantification of blue carbon benefits and understanding of intervention opportunities.	National government; International NGO; Private sector; Academic community.
Develop knowledge products and demonstration activities (for example, see activities under GEF Blue Forest Project[c] and by Restore America's Estuaries[d]).	Demonstration and communication of experience and good practice to support mainstreaming and upscaling of blue carbon interventions.	National government; International NGOs; Private sector; Academic community.
Accelerate Practical Action		
Investigate appropriate policy frameworks for including BCEs within national commitments to the Paris Agreement.	Including BCEs within NDCs and related plans provides guidance to coastal planners and assists in securing international funding for climate adaptation and mitigation.	National government.
Including management of BCEs within integrated coastal management plans.	Integrated coastal management plans help to steer on-the-ground climate response and blue economy development. Including the status of and goals of for, BCEs can provide a foundation for broader coastal management.	National and local government.
Assess and promote national opportunities for conservation and restoration of BCEs, including quantification of GHG benefits.	BCEs are being lost across East Asia at a high rate. Reversing these losses support components of NDCs, the UN SDGs and blue economy growth.	National and local government.

(Continued)

TABLE 19.2 *(Continued)* Opportunities for Countries to Incorporate Blue Carbon Ecosystems into Integrated Coastal Management, Climate Response, Biodiversity Maintenance, and Blue Economy Planning

Action	Benefit	Actor
Accelerate Practical Action		
Provide training and technical support to local and national government agencies, field schools, and communities on the value of BCEs and good practice for conservation and restoration.	Experience in restoring BCEs exists, but success rates are still relatively low. Training and improved planning can support more successful delivery.	International development organizations; National government; International NGOs Private sector.
Develop climate change adaptation strategies that consider migration of BCEs with sea level rise and human impacts (such as dam construction) on sediment supply to coastal regions.	Space is one of the scarcest resources in coastal areas. Adapting to climate change requires that plans incorporate landward movement of coastal assets including BCEs. There is an opportunity to plan buffer areas of no or low development that will both create space for coastal wetlands to migrate landwards in the future as well as reduce risk of coastal communities to climate change.	National and local government.
Include BCEs in coastal vulnerability assessments.	Along with hard infrastructure, natural infrastructure, including BCEs, is an important element in reducing ecosystem and human vulnerability to climate change. Developing blue carbon vulnerability assessments will empower governments and communities to manage natural resources in to the future.	International development organizations; National government; International NGOs; Private sector.
Include BCEs in national economic development plans.	Recognizing the natural capital value of intact and restored coastal wetlands in economic development plans can support development of sustainable blue economies.	National and local government.
Include BCEs as a component of natural infrastructure.	Coastal and river wetlands provide valuable flood risk reduction services. Including wetlands in development plans provides additional levels of protection during storm and high-low events, along with additional ecosystem services not provided by hard infrastructure.	International development organizations; National government; International NGOs; Private sector.
Include BCEs within marine protected areas.	BCEs are important elements of marine protected areas, supporting biodiversity, providing fish nurseries and other services underpinning marine ecology and productivity. Agreements established to support MPAs provide a basis for other blue carbon interventions.	International development organizations; National government; International NGOs.
Include BCEs as part of marine spatial planning and other tools for managing multi-use coastal landscapes.	Marine spatial planning offers the opportunity to map and track changes in BCEs through time and to support alignment of management approaches for their conservation.	International development organizations; National government; International NGO; Private sector.

(Continued)

Table 19.2 (*Continued*)   Opportunities for Countries to Incorporate Blue Carbon Ecosystems into Integrated Coastal Management, Climate Response, Biodiversity Maintenance, and Blue Economy Planning

Action	Benefit	Actor
Accelerate Practical Action		
Develop/apply soil management plans for watershed and coastal regions.	Improved soil management results in reduced release of carbon either through erosion or directly to the atmosphere in the form of carbon dioxide or methane.	National and local government.
Correlate health of BCEs with industry inputs and outputs of blue economy	Clarify the interdependency of blue economy industries with function of coastal ecosystems. Minimize industry environmental liabilities and maximize benefits.	National and Local government; Private sector.

[a] http://bluecarbonpartnership.org/.
[b] http://thebluecarboninitiative.org.
[c] http://www.gefblueforests.org/.
[d] https://www.estuaries.org/bluecarbon-resources.

CHAPTER 20

Blue Carbon Accounting for Carbon Markets

Brian A. Needelman
University of Maryland

Igino M. Emmer
Silvestrum Climate Associates, LLC

Matthew P. J. Oreska
University of Virginia

J. Patrick Megonigal
Smithsonian Institution

CONTENTS

HIGHLIGHTS

1. The release of global greenhouse gas accounting methodologies has removed a barrier disconnecting blue carbon ecosystems from carbon finance for conservation and restoration.
2. The methodologies provide robust and simplified procedures for quantification of greenhouse gas emissions and removals in blue carbon ecosystems.

3. The methodologies are most readily applied to large projects and support advances toward landscape-scale greenhouse gas management.
4. Applying the methodologies to small, individual parcels remains a challenge because of high project transaction costs.

20.1 INTRODUCTION

Coastal wetlands sequester carbon dioxide, and this greenhouse gas mitigation benefit has a financial value on carbon offset-credit markets. Offset markets are a potential source of funding for tidal wetland conservation and restoration projects; however, coastal managers must first quantify the magnitude of the greenhouse gas offset that results from the project in order to monetize this benefit. Several new standards provide greenhouse gas flux accounting rules for wetlands, which specify how projects can determine the net greenhouse gas benefit that results from conserving or restoring a coastal wetland habitat. In both cases, this benefit equates to enhanced greenhouse gas sequestration or emissions reductions directly attributable to the project, relative to a baseline (i.e., business-as-usual) scenario. These new accounting procedures must be rigorous enough to generate credible offset-credits, yet flexible enough to be applied to a diverse range of coastal wetland habitat types and conditions.

Estimating greenhouse gas fluxes and projecting the baseline scenario is a technically complex step in a carbon project, and project developers need to rely on one or more approved methodologies to do so. In this chapter, we discuss essential science and policy components of greenhouse gas accounting methodologies, captured in the VCS VM0033 Methodology for Tidal Wetland and Seagrass Restoration (Emmer et al. 2015a; Emmer et al. 2015b) and modules in the VM0007 REDD+ methodology Framework (Emmer et al. 2018a; Emmer et al. 2018b). See Needelman et al. (2018) for a more in-depth description and critique of the restoration methodology. For project developers, we refer to Emmer et al. (2015b), which answers questions on how to set-up, implement, and organize a blue carbon project on the ground.

The VM0033 Tidal Wetland and Seagrass Restoration Methodology is the first globally applicable methodology for coastal wetland restoration activities and provides project developers with the protocol needed to generate wetland carbon credits. It outlines procedures to estimate net greenhouse gas emission reductions and removals resulting from restoration of coastal wetlands along the entire salinity range. The scope of VM0033 is global and includes all tidal wetland systems, including mangroves, tidal marshes, tidal forested wetlands, and seagrass meadows. It incorporates best practices and principles in restoration and carbon management, while leaving the flexibility necessary to enable projects to emerge in diverse coastal settings. VM0033 also provides the basis for the tidal wetland greenhouse gas accounting modules (Emmer et al. 2018a; Emmer et al. 2018b) incorporated into the VCS VM0007 REDD+ Methodology Framework, which is a modular methodology covering both conservation and restoration in the land-use sector.

We also address concerns about potential offset-credit misallocation in seagrass blue carbon projects, which may result from estimating long-term carbon accumulation by extrapolating sediment carbon burial rates (e.g., Johannessen and Macdonald 2016; Oreska et al. 2018). Credit overallocation to seagrass meadows or other blue carbon systems would devalue legitimate offset credits.

The Verified Carbon Standard (VCS) is the largest carbon standard in the agriculture, forestry, and other land use (AFOLU) sector. In its decade of existence, the standard registered over 150 projects and introduced methodologies for forest conservation, improved forest management, agricultural land management, and wetlands (VCS 2018). The Wetlands Restoration and Conservation (WRC) category offers guidance for the accounting of greenhouse gas removals and emission reductions across 'blue carbon' ecosystems, covering subjects such as eligible project categories, greenhouse gas sources and carbon pools, baseline determination, leakage calculation, and greenhouse gas emission reductions and removals calculation. So far, the standard approved three methodologies under this category—the Methodology for Coastal Wetland Creation (VM0024), the Methodology for Tidal Wetland and Seagrass Restoration (VM0033), and the REDD+ Methodology Framework (VM0007).

Since 2018, the VCS is, together with the Climate, Community & Biodiversity (CCB) Program and the Verra California Offset Project Registry (OPR), managed by Verra (Verra 2018).

20.2 OVERVIEW OF THE ACCOUNTING PROCEDURES FOR TIDAL WETLAND AND SEAGRASS RESTORATION AND CONSERVATION

The VCS methodologies for tidal wetland and seagrass restoration and conservation provide greenhouse gas accounting procedures for restoration, creation, and conservation of marshes, mangroves, seagrasses, and forested tidal wetlands.

The methodologies consider emissions of carbon dioxide (including carbon stock changes), methane, and nitrous oxide and fulfill the requirements for the VCS Wetland Restoration and Conservation and Afforestation, Reforestation and Revegetation project categories (VCS 2017). The restoration methodology covers the variety of restoration practices that may be used to restore degraded tidal wetland systems (Perillo et al. 2009). Restoration activities must have a net greenhouse gas benefit and fall under some combination of the following practices: creating, restoring, and/ or managing hydrological conditions; altering sediment supply; changing salinity characteristics; improving water quality; (re-)introducing native plant communities; and improving management practices. The procedures for conservation cover protecting at-risk wetlands (e.g., establishing conservation easements, establishing community supported management agreements, establishing protective government regulations, and preventing disruption of water and/or sediment supply to wetland areas), improving water management on drained wetlands, maintaining or improving water quality for seagrass meadows, recharging sediment to avoid drowning of coastal wetlands, and creating accommodation space for wetlands to migrate with sea level rise.

Greenhouse gas emissions are estimated for both a most-likely baseline scenario and a with-project scenario; accounting is then done by subtraction. This basic principle of project greenhouse gas accounting is shown in Figure 20.1. Therefore, mere burial or sequestration rates of carbon do not translate directly in to carbon credits. Emissions may be either estimated or set to a conservative value. Accounting methods for each greenhouse gas include measured data, default values, published values, or models.

20.3 KEY SCIENTIFIC COMPONENTS OF THE ACCOUNTING PROCEDURES

20.3.1 Soil Carbon Sequestration Default Values

Allowing the project to use default values for soil carbon sequestration and other greenhouse gas flux rates greatly increases project feasibility but must be scientifically credible. There has been extensive data collection on soil carbon sequestration in marsh and mangrove systems; we derived a default value of 1.46 t C ha^{-1} year^{-1} from Chmura et al. (2003). A single, general default value for

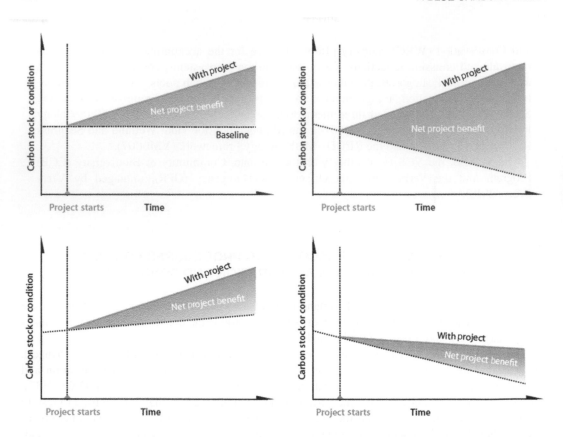

Figure 20.1   Various hypothetical scenarios for net project benefits of carbon projects, based on the difference between the baseline and the project scenario (After: Olander and Ebeling 2011). For example, salt pond restoration may be represented by a flat baseline carbon stock and an increasing project carbon stock (top left). Restoration projects that accumulate carbon may have a declining baseline (top right) and therefore have an avoided emissions component as in conservation projects. A project may improve on both an increasing baseline and a declining baseline (bottom left and right, respectively). Conservation projects are typically represented by the top-right and bottom-right graphs, where the project scenario may be decreasing, flat, or increasing.

soil carbon sequestration in seagrass systems cannot yet be justified, given continued uncertainty about long-term seagrass carbon sequestration (Belshe et al. 2017). However, the procedures allow projects to justify that the use of external default values (emission factors) is appropriate for project conditions, including the IPCC emission factor for seagrasses of 0.43 t C ha^{-1} year^{-1} (IPCC 2014).

20.3.2 Accounting for Mineral-Protected Allochthonous Carbon

Allochthonous carbon is carbon that was removed from the atmosphere outside of the project area and transported into the project area. Such carbon may only be counted as a project benefit if it would have been returned to the atmosphere in the baseline scenario (i.e., in the absence of the project), but preserved in the project scenario (VCS 2017). However, there is a portion of the allochthonous carbon that is resistant to decomposition in both the baseline and project scenarios on project timescales (e.g., 100 years), and therefore should not be counted as a project benefit. This portion has extremely slow rates of decomposition even in aerobic environments due to a close association of carbon compounds with mineral particles (Shields et al. 2016). We assumed that mineral-protected allochthonous carbon is equally retained in both the baseline scenario and the project scenario. Therefore, projects are required to estimate mineral-protected allochthonous soil carbon and subtract this from their carbon sequestration estimates. It is conservative to set this value to zero in the baseline scenario. The methodologies provide procedures to estimate this mineral-protected allochthonous carbon fraction based on organic carbon percentage of tidal wetland soils and deposited sediments.

The methodologies do not currently account for possible seagrass wrack accumulation in certain regions where seagrass beds are abundant. Seagrass aboveground biomass is buoyant and often exported hydrologically from the meadow area when it is dislodged to form floating wracks. For simplicity, most seagrass projects do not count the aboveground biomass as sequestered carbon on the assumption that it will be exported rather than be buried within the seagrass project area. This assumption is conservative in cases where the wrack decomposes outside the project area but may not be conservative when regional hydrodynamics cause seagrass wrack to be exported to adjacent blue carbon habitats. In such situations, this allochthonous carbon may accumulate in the project area where it would also have accumulated in the baseline scenario, so that the above assumptions about mineral-protected carbon do not hold. This special case warrants validator discretion until both methodologies are updated to address such conditions.

20.3.3 Soil Carbon Fate Following Erosion

Coastal wetland soil carbon pools are vulnerable to enhanced rates of oxidation to carbon dioxide when disturbed through erosion and conversion to open water. The possible fates of soil carbon following erosion or conversion to open water depend on the hydrological and geomorphic setting of the tidal wetland or seagrass system through its influence on integrated molecular oxygen exposure time Blair and Aller (2012). Soil carbon oxidation rates are the greatest when the soil is eroded into geomorphic systems that expose the carbon to aerobic conditions. This occurs when eroded carbon is entrained in river-estuary systems that transport materials seaward by continual resuspension; coastal margins and embayments with sufficient wave energy to continually resuspend sediments into an aerobic water column; or subaquatic settings with low sediment organic carbon content and course-grained sediments that act to maintain aerobic conditions in the upper soil profile. Exposure to oxygen is far less when eroded soil carbon is deposited in a low-oxygen environment, or rapidly buried by sediment, separating it from aerobic overlying water. In cases where wetlands are eroded but there is no hydrologic connectivity between the site and a river-estuary system, soil carbon loss may be minimal if it remains submerged and undisturbed in a low-oxygen environment (Lane et al. 2016).

The influence of hydrologic connectivity and depositional environment on preservation of eroded soil carbon was captured in the VCS conservation methodology (VM0007 REDD+ Methodology Framework and associated modules) by defining carbon preservation depositional environments (CPDEs) as "sub-aquatic sediment deposition environments that impact the amount of deposited organic carbon that is preserved. Carbon preservation is affected by mineral grain size, sediment accumulation and burial rates, oxygen availability in the overlying water column and sediment hydraulic conductivity." The carbon preservation in each of four CPDEs (see Table 20.1) were based on a literature review by Blair and Aller (2012).

Recent work has confirmed that erosion of subtidal seagrass beds also contributes to the loss of sediment organic carbon (Macreadie et al. 2015; Marbà et al. 2015).

20.3.4 Avoided Drainage or Excavation of Wetland Soils

Carbon in wetland soils is maintained due to anaerobic conditions; rapid carbon loss to the atmosphere can occur when these soil materials are exposed to an aerobic environment. The term

Table 20.1   Carbon Preservation in Each of Four Carbon Preservation Depositional Environments

Hydrologic Setting	Geomorphic Setting	Fraction Preserved (%)	Fraction Lost (%)
Hydrologic Connectivity	Normal Marine or Deltaic Fluidized Mud	20	80
Hydrologic Connectivity	Depleted O_2 at Sediment Surface	53	47
Hydrologic Connectivity	Transport in Small Mountainous Rivers	39	61
Hydrologic Connectivity	Extreme Sedimentation Rates	49	51
No Hydrologic Connectivity		100	0

avoided losses refers to projects that avoid such soil organic matter oxidation in the baseline scenario; this benefit can be substantially larger than the other greenhouse gas benefits in many conservation and restoration projects. In the conservation methodology, procedures are available for projects that prevent the drainage of wetlands or the excavation of wetland soils and subsequent placement into aerobic conditions. A variety of methods are available to estimate these losses, including historical data collected from the project area or time series (chronosequence) data collected at similar sites. Estimates may be made directly based on changing soil volume, density, and carbon concentrations or indirectly based on initial carbon mass and projected oxidation rates.

20.3.5 Methane Emissions

Accounting for methane emissions is critical for blue carbon projects because of the high global warming potential of methane relative to carbon dioxide and the large variation in methane emissions rates. Due the general trend of decreasing methane emissions with increasing salinity, methane emissions are consistently low in systems with salinities greater than 18 ppt (Poffenbarger et al. 2011; Holm et al. 2016; see Figure 20.2). The VCS methodologies include two default values for use in these systems: 0.011 Mg CH_4 ha^{-1} year^{-1} (0.374 Mg CO_2eq ha^{-1} year^{-1}) for systems with salinity >18 ppt and 0.0056 Mg CH_4 ha^{-1} year^{-1} (0.19 Mg CO_2eq ha^{-1} year^{-1}) for salinities >20 ppt. Default values for tidal wetlands with salinities <18 ppt were not included due to limited data availability and the high variation that has been observed in these systems (Poffenbarger et al. 2011). Without a default value available for brackish and freshwater tidal wetlands, projects will need to use more expensive and labor-intensive quantification methods such as field-data collection, modeling, or proxies. The development of cost-effective methods to estimate methane emissions from brackish and freshwater tidal wetlands is among the greatest research needs in the field of blue carbon accounting.

Ponded areas can act as methane emission hotspots even in high salinity systems, particularly if they are not tidal flushed regularly to replenish the sulfate in seawater. For this reason, the methodologies require that areas of ponds, ditches, or similar bodies of water within the project area that do not have surface tidal water connectivity be treated as separate strata for the estimation of methane emissions.

20.3.6 Nitrous Oxide Emissions

Nitrous oxide emissions are generally low from tidal wetland systems because of low-oxygen availability in anaerobic saturated soils, which favors complete denitrification (reduction of NO_3^- and N_2O to N_2). For most projects, nitrous oxide emissions should be lower in the project scenario than in the baseline scenario. However, some projects involve the lowering of water levels, leading to increased oxygen availability and potentially increased nitrous oxide emissions—these projects are required to account for nitrous oxide emissions in the VCS methodologies. An example of such a project is an impoundment breaching in which the water level in a ponded system is lowered

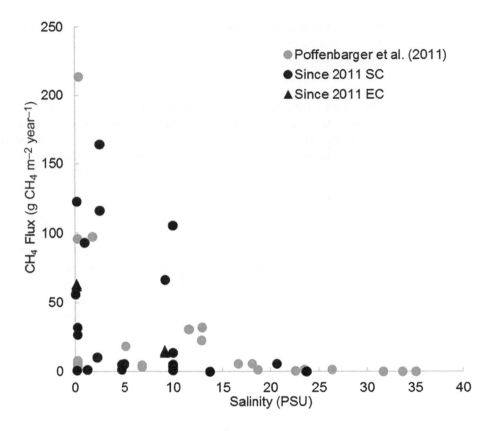

Figure 20.2   Revised version of Poffenbarger et al. (2011) data, based on the updated data available from the Second State of the Carbon Cycle Report (2018). SC refers to Static Chamber approaches and EC refers to Eddy Covariance approaches for methane fluxes. See Keller (Chapter 8) for more detail.

to create a wetland system. Projects that create wetlands from initially open-water systems also function to lower water levels relative to the soil surface.

The methodologies provide a full suite of methods to estimate nitrous oxide emissions including default values, proxies, field data collection, published data, and modeling. The default values were derived from a study in the Barataria Basin in the Gulf of Mexico in which nitrous oxide emissions from three marsh sites (fresh, brackish, and salt) and adjacent open water areas were compared (Smith et al. 1983). High nitrogen inputs can substantially increase nitrous oxide emissions; therefore, these default values may not be used in systems that receive direct inputs of nitrogen (Needelman et al. 2018).

In the methodologies, seagrass projects are not required to account for nitrous oxide emissions, because emissions from seagrasses are expected to be lower than the baseline scenario (Purvaja et al. 2008). This is partly due to the nitrogen limitation generally found in seagrass communities, such that nitrogen release is low (Welsh et al. 2000).

20.3.7 Soil Profile Sampling Methods

Field sampling of soil carbon stocks over time is a widely used technique to estimate soil carbon sequestration rates. In the VCS methodologies, projects have several alternatives to direct measurement of soil carbon sequestration—including default values—but many will opt to use field values in order to capture potential high rates of sequestration. Field sampling of soil carbon stocks generally involves collection of soil cores and laboratory analysis for bulk density and either carbon or organic matter. The key additional step that must be taken to estimate carbon sequestration rates over time is to establish a consistent reference plane within the soil profile and then truncating soil cores at this plane. This is needed because the depth (volume) of tidal wetland soils often changes over time, such that samples collected to the same depth in different years are not directly comparable. The change in soil organic carbon (SOC) content above the reference may be used to estimate carbon sequestration rates based either on the age of the reference plane or the start of project activities. There are many options to establish a reference plane, including marker horizons (most commonly using feldspar) (Cahoon and Turner 1989), a strongly contrasting soil layer (such as the boundary between organic and mineral soil materials), an installed reference plane (such as the shallow marker in a surface elevation table) (Cahoon et al. 2002), a layer identified biogeochemically (such as through radionuclide, heavy metal, or biological tracers) (DeLaune et al. 1978), and a layer with SOC indistinguishable from the baseline SOC concentration (Greiner et al. 2013).

The method of establishing a reference plane using a layer with SOC indistinguishable from the baseline SOC concentration is particularly relevant for seagrass projects. Most seagrass projects occur in dynamic, subtidal environments, where establishing a reference plane may not be possible. In this method, a seagrass soil core profile is compared with a depth-calibrated, background carbon concentration profile (e.g., the soil carbon profile prior to the project start date or at bare control sites). Subtracting the bare concentration from the meadow concentration along the calibrated cores down to the depth where the profiles intersect gives the sediment carbon pool enhancement attributable to the meadow. Repeated calibrated core comparisons can facilitate stock-change accounting in a seagrass bed over time, which should avoid issues with burial flux estimation that may over-estimate seagrass carbon sequestration rates (Oreska et al. 2018). Johannessen and Macdonald (2016) note that most seagrass studies simply extrapolate a single burial flux rate from a dated sediment core into the future to estimate carbon accumulation. This approach does not account for within-bed organic carbon remineralization over long time-scales, sediment mixing, lateral movement, and other factors that may diminish long-term carbon accumulation rates (Johannessen and Macdonald 2016).

20.3.8 Sample Size

Collecting a sufficient number of field samples for laboratory analysis to achieve statistically confident estimates can be a significant project cost and may in some cases render a project infeasible. The methodologies include alternative estimation strategies that may be less expensive; in particular, default values were provided whenever they are scientifically valid. Nonetheless, field sampling may be necessary in some cases and may be preferable to alternative methods to achieve the maximum estimate of greenhouse gas benefits.

The methodologies use a dedicated VCS tool to determine sampling size requirements. The equations in this tool determine sample size requirements as a power function of the coefficient of variation of the quantity being estimated. Projects are instructed to target a confidence interval of 95% with a 30% allowable error (a 90% confidence interval with 20% allowable error is also allowed but requires a greater number of samples).

The sample size requirements in the methodologies should be reasonable for most projects for variables with moderate levels of variation such as soil carbon stocks. Most projects analyzed in the review by Chmura et al. (2003) had a coefficient of variation less than 0.5 for carbon sequestration rates, which would translate to a sample size requirement of about ten per stratum. However, variables with greater variation will require a substantially greater number of samples; for example, methane emissions could require about 40 samples per stratum per sampling event (Needelman et al. 2018). This represents a substantial burden to project implementation due to the high cost of methane flux sampling.

20.4 POLICY COMPONENTS OF THE METHODOLOGIES

20.4.1 Additionality

Carbon markets provide an incentive for new projects and activities that result in net greenhouse gas benefits. To demonstrate this incentive, a project must meet the additionality requirement in order to be awarded carbon offsets. The VCS established two approaches to demonstrate additionality: an individual approach at the project level and a standardized approach for a class of project activities. The VM0033 utilizes the standardized approach for projects within the United States and the individual approach for projects outside of the United States. The VM0007 REDD+ methodology utilizes the standardized approach as developed for the United States; it has, however, been extended to a global scope. Using the standardized approach, it has been demonstrated that all tidal wetland and seagrass restoration and conservation projects are additional. For a further description, see Needelman et al. (2018). The standardized method removes the significant burden for projects to demonstrate additionality. VM0033 will be updated to apply the standardized approach globally.

20.4.2 Leakage

An increase in emissions or a decrease in removals of greenhouse gases outside of the project area as a result of the project intervention is called leakage. Leakage is traditionally broken down into a) activity-shifting leakage related to shifting an activity such as agriculture from the project site to some other location; or b) market-effect leakage, when a project reduces the local supply of a product increasing production elsewhere (Aukland et al. 2003). Specific to wetlands, an additional type of leakage is ecological leakage, i.e., an increase in emissions or decrease in removals in an ecosystem outside the project boundary that is hydrologically connected to the project area (VCS 2017). VM0033 requires projects to avoid leakage by setting specific limitations to projects pertaining to the kind of pre-project land use permitted, and a careful establishment of project boundaries.

Avoiding activity-shifting and market leakage can be achieved if one of the following conditions is met:

a. Demonstrate that prior to the start of the project the land is free of land use that could be displaced outside the project area,
b. Require that a land use that could be displaced outside the project area (e.g., timber harvesting) is not accounted for in the baseline scenario, or
c. Require a pre-project land use that will continue at a similar level of service or production during the project crediting period (e.g., reed or hay harvesting, collection of fuel wood, subsistence harvesting).

For example, project developers may demonstrate that farmers have abandoned the project area prior to project start or that the land has already become unproductive (e.g., due to salinity intrusion). The methodologies do not currently allow projects to demonstrate the lack of activity-shifting leakage except through the absence of a displaceable land use. The methodology could be improved if it allowed projects to demonstrate the absence of activity shifting.

VM0007, however, also allows for quantifying leakage emissions in the with-project scenario. It allows for a variety of approaches since it already includes leakage accounting modules developed for forest conservation projects as well as a module for ecological leakage originally developed for peatlands, where ecological connectivity is of similar importance as in tidal wetlands.

Ecological leakage in tidal wetland projects is avoided in both the restoration and conservation methodology by a project design which manages hydrological connectivity with adjacent areas so as to avoid a significant increase in net greenhouse gas emissions outside the project area, for example, by establishing a project boundary wide enough to capture expected water level changes that are linked to project activities.

20.5 CONCLUDING REMARKS

The VCS VM0033 Methodology for Tidal Wetland and Seagrass Restoration and the tidal wetlands modules in the VCS VM0007 REDD+ Methodology Framework allow the diversity of tidal wetland restoration and conservation projects to receive VCS-approved carbon credits. These and other reputable methodologies are designed to underestimate the net greenhouse gas benefit unless applicant projects take thorough, rigorous, direct measurements that convince validators that the actual project benefit is higher than the conservative, estimated benefit. The procedures are designed to be feasible to implement and highly flexible, while maintaining scientific rigor. The science and policy of greenhouse gas emissions and carbon storage in tidal wetlands is evolving— as evidenced by several innovative approaches in the methodologies—yet it remains limited by knowledge gaps. The accounting tools provided have a broader applicability and may be used to complement currently available systems of national to project-level greenhouse gas accounting for tidal wetland systems.

The existence of approved methodologies is one less barrier to market entry. Tidal wetlands restoration and conservation projects are now served with their own dedicated methodologies. However, appropriate greenhouse gas accounting is—under any carbon standard—a great burden for offset projects in any category, requiring a resourceful team and sufficient funding from the onset. Small-scale projects are unlikely to benefit from carbon finance, unless methodologies are further simplified or unless projects are grouped to realize economies of scale.

Isolated single-category restoration or conservation projects in the coastal zone are likely to face a significant risk of failure. This is because in most coastal settings, sea level rise will require projects to accommodate a landward shift of coastal ecosystems. With greenhouse gas accounting methodologies now ready to assist, it is time to explore landscape-scale interventions including the entire sub to supra-tidal sequence, considering—where relevant—restoration or conservation of wetland and vegetation, or combinations of those. This would have to occur at an appropriate scale to become part of regional land-use planning and to reduce development and transaction costs.

PART **IV**

Case Studies

CHAPTER 21

The Nexus between Conservation and Development in Indonesian Mangroves

Daniel Murdiyarso
Center for International Forestry Research
Bogor Agricultural University

CONTENTS

HIGHLIGHTS

1. With around 3 million hectares, Indonesia houses almost a quarter of the world's mangroves, larger than any continent.
2. To date, they face tremendous pressures from aquaculture and agriculture development with current loss of 1.2% annually.
3. Mainstreaming mangrove sustainable management through a national regulatory framework and linking it with global framework such as Paris Agreement and Sustainable Development Goal is very timely.

21.1 BACKGROUND

It is a common phenomenon that people have different perspectives on a particular subject or issue. Often these perspectives are contrasting or even contradictory or dichotomous. Development and conservation are among the most common themes when one is confronted with the challenges of natural resources. Decision-making entities face difficult choices between short-term economic objectives and long-term resource sustainability. This dilemma is particularly prominent in many developing countries, Indonesia included.

Mangrove ecosystems are not spared in this regard. As most of the problems stem from governance systems, especially in the fishery and forestry sectors, the solution should come from the same systems. The unique ecosystems that lie in the coastal zone are not necessarily considered in an integrated fashion. Rather, they are treated as a production system and often for a single commodity.

Mangrove conversion following deforestation is often practiced based on only minimum levels of information about ecosystem functioning. The main drivers of the development of land (settlement and other infrastructure), food (aquaculture and agriculture), and energy (fuel wood and charcoal) are often transboundary and market forces.

The rate of conversion is alarming; for example, in 20 years (1985–2005), Indonesia has lost more than 1 million hectares (Mha) of its mangroves (FAO 2007). Most of the converted mangroves were developed into shrimp farms (Giri et al. 2011). However, more recent studies show that oil palm expansion has become a new threat to Indonesian mangroves (Richards and Friess 2016).

Only very recently have the significant roles of mangroves in mitigating and adapting to climate change been recognized as part of the coastal **blue carbon** (Donato et al. 2011; UNEP 2014; Murdiyarso et al. 2015). The term blue carbon is used widely to attract policymakers to bring these issues into the global climate arena, such as via the United Nations Framework Convention on Climate Change (UNFCCC), the Bonn Challenge, and Sustainable Development Goals (SDGs).

New scientific information around coastal blue carbon has become available to support decision-making processes and implementations by land managers, practitioners, and the private sector. This information supports the conservation of existing mangroves and the restoration of degraded ones.

21.2 INDONESIAN MANGROVES: EXTENT, DISTRIBUTION, AND TRENDS

Indonesia houses the largest areas of mangroves in the world and was ranked as the most mangrove-rich country, as almost a quarter of the world's mangroves are found in Indonesia (Giri et al. 2011). However, the extent has decreased very rapidly in the past three decades. In 1980, there were 4.20 Mha of mangrove forests along Indonesia's 95,000 km of coastline (FAO 2007). Over two decades, mangrove forest cover had declined about 26%, to an estimated 3.11 Mha (Giri et al. 2011). In 2005, mangrove forest cover had further decreased to 2.90 Mha (FAO 2007). Five years later, based on government data (MoF 2013), our recent estimate shows that mangrove cover is around 2.56 Mha.

Based on FAO and MoF data, cumulatively, Indonesia lost 40% of its mangrove forests between 1980 and 2010. This is equivalent to an annual deforestation rate of 52,000 ha year^{-1} (1.24%). Compared with the national deforestation of 0.84 Mha year^{-1} (Margono et al. 2014), mangroves contribute 6% of national forest loss.

As shown in Table 21.1, mangroves are mainly (86%) distributed around three main islands—Papua, Kalimantan, and Sumatra—and the remaining 14% are located in Maluku, Sulawesi, Java, and Nusa Tenggara. Based on the land-use intensity as indicated by canopy cover, Indonesia's primary mangrove forests (protected and undisturbed) constitute 58% of mangrove cover, while the remaining 42% are considered as secondary mangrove forests.

Figure 21.1 shows that most of the primary Indonesian mangrove forests are located in Papua, followed by Maluku and Sulawesi. The distribution of mangroves based on canopy cover is closely related to mangrove conversion, which was started in Java as early as 1,800 during the Hindu Majapahit Kingdom era. Expansion of aquaculture to Sumatra took place in the early 1900s, in Sulawesi in the mid-1950s and finally in Kalimantan in 1970 (Ilman et al. 2016).

Such a generalized grouping based on Landsat-TM images may be used to guide mangrove conservation activities in protected areas as well as restoration priorities in degraded mangroves. If these conservation activities are related to efforts that reduce greenhouse gas emissions at their

Table 21.1   Mangrove Areas in Descending Order and Distribution in the Main Islands of Indonesia

Island	Primary Mangroves (ha)	Primary Mangroves (%)	Secondary Mangroves (ha)	Secondary Mangroves (%)	Total (ha)	Total (%)
Papua	1,166,406	45.6	78,682	3.1	1,245,088	48.6
Kalimantan	57,532	2.2	442,119	17.3	499,651	19.5
Sumatra	138,431	5.4	314,826	12.3	453,257	17.7
Maluku	66,759	2.6	90,685	3.3	157,444	6.1
Sulawesi	32,929	1.3	112,166	4.4	145,094	5.7
Java	8,865	0.3	23.339	0.9	32,205	1.3
Nusa Tenggara	16,196	0.6	111,753	0.5	27,950	1.1
Total	1,487,118	58.1	1,073,570	41.9	2,560,688	100.0

Source: MoF (2013).

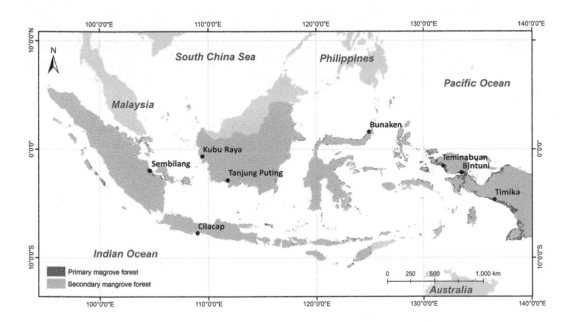

Figure 21.1   Distribution of Indonesian mangroves following classes of canopy cover (primary and secondary forests) and eight sampling sites for carbon stock assessments.

source and/or enhance sequestration by their sinks, they may be categorized as climate change mitigation. In contrast, adaptation to climate change requires measures that respond to the impacts of the changing climate, including sea level rise, to which coastal ecosystems and communities are vulnerable.

21.2.1 Mangroves and Climate Change Mitigation

A comprehensive assessment on Indonesian mangroves has been carried out and reported for the first time in the context of their potential for climate change mitigation (Murdiyarso et al. 2015). The study involved eight sites and 39 plots across the archipelago as shown in Figure 21.1. Except for the Cilacap site on Java, these sites are mainly protected areas, about which one can expect to find background information dating from before the ecosystems were disturbed.

Each site consisted of 5–7 plots that were established perpendicular to the coastlines or creeks to ensure that properties related to plot gradient were considered. Each plot consisted of six subplots within which the density of carbon of the living biomass above the ground (trees, saplings, and seedlings), dead necromass (woody debris and litter) and soil carbon was systematically sampled.

Following Kauffman and Donato (2010), the plots were selected and laid out, where the samples were collected and analyzed. Results were summarized and the ecosystem C stocks across the sites are shown in Figure 21.2.

It was reported that the average C stocks of mangrove ecosystems across Indonesia was $1{,}083 + 378$ MgC ha^{-1} (Murdiyarso et al. 2015). This is slightly higher than the ecosystem C stocks of mangroves in the Asia Pacific reported earlier of 1,023 MgC ha^{-1} (Donato et al. 2011). If this is extrapolated to the country-level mangrove extent of 2.90 Mha, Indonesia's mangroves contain as much as 3.14 PgC (Murdiyarso et al. 2015).

Figure 21.2 also indicates that the highest C stocks were found in Papua and Sulawesi mangroves. More interestingly, most of the carbon (90%) was stored in the soil. It is very important to note that mangrove deforestation followed by excavation for pond development will jeopardize large amounts of carbon storage in the soil, which will then be exposed and oxidized. In the last three decades, Indonesia has lost 40% of its mangroves, with more than 60% of this loss being due to aquaculture development (Giri et al. 2008). Sustainable shrimp farming is a rarity and in many cases only leaves 10–15% of the original stores of carbon when the shrimp farms are abandoned (Kauffman et al. 2014). This has resulted in annual emissions of 0.07–0.21 Pg CO_2e (Murdiyarso et al. 2015).

Avoiding further mangrove conversion, Indonesia could reduce up to 30% of national emissions from the land-use sector. Such action may be included in the nationally determined contribution as stipulated under the Paris Agreement. It is still not clear, however, what kind of mitigation measures would be adopted by the government to meet an emissions reduction target of 830 Gg CO_2e in 2030.

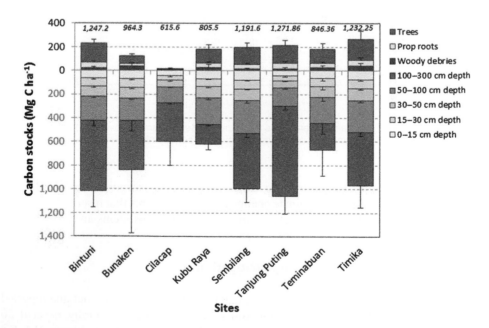

Figure 21.2   Ecosystem carbon stocks in eight mangrove sites in Indonesia. The number above zero indicates aboveground carbon stocks and the number below zero indicates soil carbon stocks up to 3 m below the surface.

21.2.2 Mangroves and Climate Change Adaptation

As unique coastal vegetation that lives in a very harsh environment (salinity, sea waves, and current and irregular tidal range), mangroves offer enormous and sometimes intangible ecosystem services. Some of the most prominent and relevant to climate change adaptation are the ability of mangroves to trap sediment, build the land, and cope with increasing sea levels. This provisioning service is not commonly monetized and hardly captures the interests of the private sector in developing countries. As a result, and as for many other adaptation measures, public funding is spent to adapt to climate change.

Among other factors affecting sedimentation rates are: (i) hydro-geomorphic settings, (ii) types and sources of sediments, and (iii) disturbance regimes, both man-made and natural. Elevation deficit or surplus may be assessed from long-term measurements.

In Indonesia, recent studies have generally combined carbon stock assessments and sedimentation, as summarized in Figure 21.3 (Murdiyarso et al. 2015). By doing so, the studies could address both climate change mitigation and adaptation. Studies ranged from the riverine, estuarine and the fringing mangroves around the coastlines, examined from different perspectives.

They involve the understanding of contemporary sediment accretion and surface elevation changes, which are measured using the rod surface elevation tables (RSETs) combined with marker horizons (MHs) introduced by Cahoon and Lynch (1997), and promoted by Webb et al. (2013). To trace historical sedimentation processes, a radionuclide technique using ^{210}Pb tracer (Marchio et al. 2016) was also applied.

Table 21.2 shows the location coordinates, the institutions involved and the starting date of the campaign. Most of these sites were newly established and had begun to generate information on accretion rates and the age of sediments from the radionuclide analysis. Yet they will have to be verified with the contemporary sediment (and carbon) accumulation using the RSET-MH approach.

Figure 21.3   **(See color insert following page 266.)** Location of sedimentation and carbon stock assessments for mangroves across the Indonesian archipelago as carried out by CIFOR (red dots) and Ministry of Marine Affairs and Fishery (yellow dots).

Table 21.2   Location of Data Available on C Stocks and Sedimentation Rates in Indonesian Mangroves

Observation	Site Name	Latitude	Longitude	Institution	Start Date
C stocks	Percut, Sumatra	03°43′31″ N	98°47′41″ E	CIFOR[a]	2016
	Jaring Halus, Sumatra	03°56′28″ N	98°47′41″ E	CIFOR	2017
	Sembilang, Sumatra	02°04′28″ S	104°28′09″ E	CIFOR	2012
	Banten Bay, Java	06°01′03″ S	106°11′41″ E	CIFOR	2015
	Cilacap, Java	07°43′25″ S	108°57′29″ E	CIFOR	2012
	Kubu Raya, Kalimantan	00°40′33″ S	109°21′41″ E	CIFOR	2012
	Tanjung Puting, Kalimantan	02°51′30″ S	111°42′02″ E	CIFOR	2012
	Mahakam Delta, Kalimantan	00°45′05″ S	117°35′02″ E	MOEF[b]/CIFOR	2015
	Bunaken, Sulawesi	01°17′17″ N	124°30′37″ E	CIFOR	2012
	Bintuni Bay, Papua	02°10′12″ S	133°32′09″ E	CIFOR	2012
	Teminabuan, Papua	01°37′24″ S	131°48′35″ E	CIFOR	2012
	Timika, Papua	04°51′41″ S	136°47′18″ E	CIFOR	2012
	Kaimana, Papua	03°40′06″ S	133°45′29″ E	CIFOR/CI[c]	2016
Sedimentation	Percut, Sumatra	03°43′31″ N	98°47′41″ E	CIFOR	2016
	Jaring Halus, Sumatra	03°56′28″ N	98°47′41″ E	CIFOR	2017
	Bengkalis, Sumatra	01°39′51″ N	101°45′03″ E	MMAF[d]	2016
	Banten Bay, Java	06°01′03″ S	106°11′41″ E	CIFOR	2015
	Porong, Java	07°31′57″ S	112°51′12″ E	MMAF	2010
	Perancak, Bali	08°23′35″ S	114°3742″ E	MMAF	2013
	Bintuni Bay, Papua	02°10′12″ S	133°32′09″ E	CIFOR	2015
	Kaimana, Papua	03°40′06″ S	133°45′29″ E	CIFOR/CI	2016

[a] CIFOR: Center for International Forestry Research.
[b] MOEF: Ministry of Environment and Forestry.
[c] CI: Conservation International.
[d] MMAF: Ministry of Marine Affairs and Fishery.

The results from the effluent of volcanic mud **eruptions** channeled through the Porong River, which ends at the northeastern coast of East Java, showed an extremely high accumulation rate of 4 mm year^{-1} in the first 2 years, which then increased even further to 10 mm year^{-1} in the following 4 years (Sidik et al. 2016). Such a rapid accumulation rate inhibited the growth of *Avicennia* sp. It is not clear, however, how much carbon and other nutrients have been deposited.

In island-type settings, such as that found in Ilha Grande, Brazil, the rate was approximately 1.7 mm year^{-1} for the past 100 years (Sanders et al. 2008). Their assessments were done using the same technique of ^{210}Pb dating models from a single sediment core. The dominating *Rhizophora mangle* was shown to provide a significant input of organic matter and stable subareal habitat. The islands may provide natural protection to the tourist coasts near the city of Rio de Janeiro.

However, not all island settings gain direct benefit from the sediment, e.g., on the islands of Korsea and Pohnpei, Federated States of Micronesia (FSM). Surface accretion rates in habitats naturally dominated by mangroves ranged from 2.9 to 20.8 mm year^{-1}, but mangroves' susceptibility to sea level rises has caused the fringe mangrove in Pohnpei to experience an elevation deficit of up to—1.30 mm year^{-1} (Krauss et al. 2010). Over a 6-year observation period, the greatest elevation deficit was—3.2 mm year^{-1}, but the island of Korsea registered the greatest elevation gain in the interior mangroves of 4.1 mm year^{-1} (Krauss et al. 2010).

Preliminary results contrasting fringe and interior mangroves in North Sumatra, Indonesia, show exactly the opposite trends. Interior mangroves gain accreted sediment of 3.7 mm year^{-1}, which is less than that for fringe mangroves of 5.6 mm year^{-1} in the past 75 years (Personal comm. 2017). The region has been extensively degraded mostly from abandoned shrimp ponds. At the same time, sediment inputs from the fresh water from the eroded upstream areas have been quite significant. Recent replanting activities using *Rhizophora stylosa* and *Avicennia alba* in the mudflats is yet to be evaluated against their survival rates, as this setting accreted more sediments than did the interior mangroves, with a rate of 4.3 mm year^{-1}.

The effects of disturbance regimes in the form of logging were observed in Bintuni Bay, Papua. In general, it was demonstrated that there was no significant difference in accretion rates between ages of stands (5 years against 15 years after logging). Likewise, differences were not demonstrated between fringe and interior mangroves in either age class. They were in the range of 0.87–1.29 mm year^{-1} in 5-year stand and 1.09–1.17 mm year^{-1} in 15-year stands (Murdiyarso et al. 2017). However, it is interesting to note that due to differences in carbon density, and perhaps the source of carbon, the carbon accumulation rates differed significantly with stand age. More recently logged (5-year stands) accumulated C of 1.07 Mg ha^{-1} year^{-1}, more than double that of 15-year stands of 0.44 Mg ha^{-1} year^{-1}. These numbers were shown in fringe mangroves, while the interior mangroves accumulated much less sediment as the rate of accumulation was only one-third of the fringe mangroves (Murdiyarso et al. 2017).

Assuming that inputs from the standing biomass are relatively high, such a poor carbon deposition rate must be largely due to logging practices, which tend to clear up woody debris and biomass residues to allow replanting to succeed in logged-over areas. In equatorial mangroves, litter production from standing mangrove forests could reach as much as 20.3–27.6 Mg ha^{-1} year^{-1}, depending on the species composition (Sukardjo et al. 2013).

Restoration of disturbed mangrove in Vietnam significantly caused higher vertical accretion rates of 1.01 cm year^{-1} in naturally colonized mangroves and 1.06 cm year^{-1} in outplanted mangroves (MacKenzie et al. 2016). These are an order of magnitude higher than other for places in the region. No significant difference was shown between fringe (1.06 ± 0.12 cm year^{-1}) and interior mangroves (0.99 ± 0.09 cm year^{-1}).

The vulnerability of low-lying coastal zones and small islands to sea level rises for the Indo-Pacific region remains a huge challenge, as under the pathway 6 scenario of the Intergovernmental Panel on Climate Change (IPCC 2013) they would be submerged in 2070 (Lovelock et al. 2015). Based on a global systematic review, Sasmito (2015) estimated that under the highest scenario (pathway 8.5), fringe mangroves can only survive up to 2055 and interior mangroves up to 2070, as they can only accrete sediment to a level of 29 cm and 41 cm, respectively. It is widely known that IPCC's worst-case scenario sea level rise will be in the range of 0.52–0.98 cm by the end of this century.

With the said estimate of vertical accretion, north Sumatran mangroves would be able to cope with a global sea level rise of 2.6–3.2 mm year^{-1} (Church et al. 2011), but not with a regional sea level rise of 4.2 ± 0.4 mm year^{-1} (NOAA 2014).

21.3 MANGROVE LAND-BASED DEVELOPMENT

21.3.1 Fish Industries

Aquaculture development was the main cause of mangrove conversions, especially during the period of 1997–2005. The officially recorded active shrimp pond area was about 0.65 Mha (MMAF 2013). Some of these ponds are now abandoned or unproductive. Table 21.3 shows that over a period of 6 years, the mean annual shrimp production was 351,838 t. This means that Indonesia's shrimp pond productivity was just below 2 t ha^{-1} year^{-1}. This is relatively low compared with the productivity of shrimp ponds in Thailand of 12 t ha^{-1} year^{-1} (where more advanced technology is

Table 21.3   Production of Various Types of Shrimp and Area of Active Ponds in Each Island of Indonesia

Island	Year	Production (Tons)					Total	Area (ha)
		Tiger	Vannamei	White	Rostris	Others		
Papua	2006	1,285	30	27			1,342	598
	2007	20		16		26	52	807
	2008	29		16		27	72	1,005
	2009	14					14	851
	2010	13		13		5	31	690
	2011	22	8	13			43	669
	Mean	231	19	17		19	259	770
Kalimantan	2006	8,577	2,237	4,016			14,830	144,254
	2007	10,136	1,790	5,029		3,758	20,713	107,574
	2008	9,901	6,692	6,287		5,016	27,896	197,787
	2009	6,832	2,992	4,923		4,589	19,336	213,089
	2010	16,430	9,018	5,585		7,230	38,263	211,323
	2011	15,487	5,272	5,218		6,540	32,517	287,553
	Mean	11,227	4,667	5,176		5,427	25,593	193,597
Sumatra	2006	74,594	124,213	2,717			201,524	130,589
	2007	68,768	142,204	3,546		825	215,343	133,021
	2008	69,372	149,470	7,424	77	3,737	230,080	126,137
	2009	66,840	103,375	6,747		1,120	178,082	144,374
	2010	53,592	87,324	3,102		882	144,900	128,044
	2011	72,295	24,956	62		235	97,548	124,939
	Mean	67,577	101,466	3,933	77	1,360	177,913	131,184
Maluku	2006	137		6			143	2,401
	2007	104		7			111	1,536
	2008	172		8			180	1,161
	2009	185		6			191	1,186
	2010	450	46	3			499	8,227
	2011	1,427	20				1,447	309
	Mean	413	33	6			429	2,470
Sulawesi	2006	31,058	2,522	979			34,559	144,412
	2007	21,164	4,603	140		556	26,463	134,130
	2008	22,404	5,154	719		3,495	31,772	125,875
	2009	19,215	4,312	235		5,756	29,518	145,272
	2010	24,294	12,861	538	2	6,285	43,980	152,843
	2011	28,616	21,721	426		3,378	54,141	153,677
	Mean	24,459	8,529	506	2	3,894	36,739	142,702
Java	2006	30,499	4,143	27,829	7		62,478	165,854
	2007	31,652	14,468	8,091		21,864	76,075	170,089
	2008	31,865	17,513	14,792		20,192	84,362	157,976
	2009	30,995	32,516	10,363		7,621	81,495	156,049
	2010	30,055	62,168	7,051		16,375	115,649	173,216
	2011	36,403	68,285	4,888	16	5,989	115,581	173,139
	Mean	31,912	33,182	12,169	12	14,408	89,273	170,442

(Continued)

Table 21.3 (*Continued*)   Production of Various Types of Shrimp and Area of Active Ponds in Each Island of Indonesia

Island	Year	Production (Tons)						Area (ha)
		Tiger	Vannamei	White	Rostris	Others	Total	
Nusa Tenggara	2006	1,701	8,506	159			10,366	8,927
	2007	1,270	16,904	168	42	115	18,499	8,769
	2008	1,189	29,820	2,898		61	33,968	8,312
	2009	483	27,776	91		35	28,385	8,918
	2010	687	35,160	71		27	35,945	8,515
	2011	2,626	4				2,630	8,935
	Mean	1,326	19,695	677	42	60	21,632	8,729
Total of mean							351,838	649,894

Source: MMAF (2013).

involved) and in Vietnam of 6 t ha^{-1} year^{-1} (Konkeo 1997; Portley 2016). These are the top two shrimp-producing countries, followed by Indonesia (FAO 2014), all at the cost of mangrove deforestation (Murdiyarso et al. 2015).

It was also reported that the revenue from shrimp exports approached USD 1.5 billion in 2013. This is almost 40% of the total revenues arising from the Indonesian fishery sector (MMAF 2014). Unless appropriate technology is adopted to meet the demands of the market, further conversions of mangroves will have serious consequences.

21.3.2 Palm Oil Industries

Oil palm has become the major export and revenue-generating commodity of Indonesia. In 2016, with 11.6 Mha of plantation area, Indonesia produced 33.5 Mt of crude palm oil (CPO), most of which (25.7 Mt) was exported (MoA 2016). A recent report indicates that the revenue generated from the export palm oil sector was USD 17.8 billion.[*] The business community that is organized under the Indonesian Oil Palm Association is planning to increase the production to 40 Mt by 2020, as global consumption is expected to grow to almost 60 Mt. It is not clear, however, where the plantations will be expanded or located.

A moratorium on oil palm expansion, which was expected from the new government, has been widely misunderstood. The intended regulation was already enacted in 2011 by the previous government, known as the 2-year forest moratorium. It was then renewed in 2013 for another 2 years to suspend the issuance of permits to convert primary forests and peatlands.

The new government followed this up by renewing the regulation when it expired in 2015. It was also revised in the context of those government institutions that were to ensure the coordination took place. However, the targets of the moratorium remain the same—primary forests and peatlands, as a result oil palm plantation on mangrove ecosystems are not prohibited.

It is interesting to note that during the period of 2000–2012, more mangroves were deforested for oil palm plantations after aquaculture in Indonesia (Richards and Friess 2016). These mainly occurred in Sumatra, where secondary or degraded mangroves are found. If the global demand for palm oil keeps growing, it is very likely that permits to convert mangroves will be issued, as peatlands are under the moratorium. Weak governance systems regarding mangroves makes these ecosystems more vulnerable, including those distributed in Kalimantan and Papua, which are in much better condition in terms of ecosystem functioning.

---

[*] http://www.sawit.or.id/sawit-konsisten-penyumbang-terbesar-devisa/

21.4 PROMOTING GOVERNANCE SYSTEM FOR MANGROVES

The level of understanding about mangrove potentials for climate change adaptation and mitigation among relevant stakeholders is improving. At the same time, the appetite to convert mangrove forests is also overwhelming. The weak governance system for mangroves is one of the reasons why the conflicting objectives were never resolved.

There is growing concern that ecosystem services would offer common ground for these conservation and development agendas to meet and do a deal. Financial mechanisms have to be explored while local perspectives are carefully considered.

Another entry point with which the national processes could link is the global development agenda, the SDGs. It is stipulated in SDG14 (Life below Water) and Target 14.2 that by 2020, coastal ecosystems (including mangroves) have to be sustainably managed to avoid significant adverse impacts. These measures could include mangrove restoration to strengthen their resilience and enhance their health and productivity.

Furthermore, the Paris Agreement, which has been ratified by most parties to the UNFCCC, offers huge opportunities for mangrove countries, including Indonesia, to pledge their contributions through the nationally determined contributions (NDCs) as stipulated in Article 4 of the Paris Agreement. Climate change mitigation and adaptation strategies that are fully described in Articles 5 and 6, respectively, have been considered in a more balances manner than before, and describe how countries can exercise land-use activities, including mangrove development and conservation.

Nationally, Indonesia should develop and improve policies and legal frameworks to make mangrove governance systems work. Overlapping authority and regulation is one of the greatest problems of mangrove management in Indonesia (Banjade et al. 2016). Among other legal frameworks are Law 27/2007 on the Management of Coastal Areas and Small Islands, Law 32/2009 on Environmental Protection and Management, Law 41/1999 on Forestry, and Law 5/1990 on Natural Resources Conservation. By considering and deriving those entities, there is a specific legal framework that embraces several government agencies to implement Presidential Regulation 73/2012 on the National Strategy for Mangrove Ecosystem Management.

Figure 21.4 shows the likely cycle involved in implementing the existing legal framework for mangrove management that engages relevant stakeholders at global, national, and subnational levels.

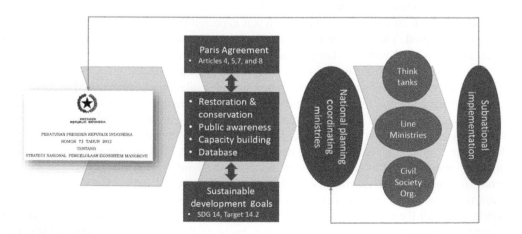

Figure 21.4   Legal framework that enables various government agencies to implement national and subnational agendas while considering the global frameworks such as those embodied by the SDGs and the Paris Agreement via which the sustainability of mangrove ecosystems is secured and their governance system is strengthened.

These legal frameworks were developed with sectoral interests and focus in mind. Thus, they are often confronted with challenges when implemented in a real-world situation involving stakeholders with wider interests. If the governance system for mangroves is to be strengthened, the national legal framework has to be implemented in an integrated manner in the context of meeting the national objectives. Global agendas, including the Paris Agreement and SDGs, which are usually **borderless,** may be adopted to promote the improvement of governance systems. Indonesia's Development Planning Agency, which happens to be the custodian of the SDGs can play a key role in facilitating the process. Other agencies having a similar mandate, including the National Focal Point to the UNFCCC, should be mobilized to play a similar role.

Feedback from society at large, including line ministries, practitioners, land managers, and academics may be used to improve the system to meet the new challenges as the ecosystems prevail and progress. The implementation at the subnational level should not be standalone and the monitoring and capacity development systems should already be in place beforehand.

21.5 CONCLUDING REMARKS

Mangroves are among the most carbon-rich ecosystems on Earth, per unit area. Their potential in addressing climate change mitigation and adaptation has been recognized. Being the world's largest home of mangroves, Indonesia needs to set priorities by means of which mangroves become prominently in focus.

The challenges from the fishery and palm oil development sectors are evident. The implications for the environment in general and climate change in particular are at odds with those for the economic growth objectives. Enhancing the benefits from ecosystem services rather than simply extracting the goods would potentially lead to the wise use of resources.

Mangrove ecosystems are unique in many ways as they are physically located in coastal zones that have strategic roles for development and conservation purposes. This entails many concerns ranging from the biophysical through to governance (institutional and legal framework) issues. Sustainable mangrove management will need the broad participation of stakeholders and strong governance.

27.3 CONCLUDING REMARKS

CHAPTER 22

The Fate of Blue Carbon in the Sacramento-San Joaquin Delta of California, USA

Judith Z. Drexler
U.S. Geological Survey, California Water Science Center

Shruti Khanna
California Department of Fish and Wildlife

David H. Schoellhamer
U.S. Geological Survey, California Water Science Center (Emeritus)

James Orlando
U.S. Geological Survey, California Water Science Center

CONTENTS

HIGHLIGHTS

1. The Sacramento-San Joaquin Delta of California was once a 1,900 km^2 wetland region, which stored ~150–210 Tg (1 Tg = 1 × 10^{12}g) of organic carbon (C).
2. Approximately 98% of the Delta land area has been converted to agriculture, resulting in the loss of ~83–100 Tg C.
3. The total area of restored wetlands in the Delta is projected to be 7,879 ha by 2020, which, after 100 years, will result in ~0.9 Tg C stored or roughly 1% of the original carbon sink.
4. To replenish the carbon sink, much larger areas must be restored; however, even small restorations will provide important co-benefits such as habitat for sensitive species.

22.1 INTRODUCTION: BACKGROUND AND RELEVANCE TO CALIFORNIA CARBON POLICY

It is difficult to imagine the size and physiognomy of the historic Sacramento-San Joaquin Delta of California (hereafter, the Delta) before ~98% of it was converted to agriculture. This inland delta, depicted in Figure 22.1a and b, is bounded by the Sacramento and San Joaquin Rivers. It is situated at the landward end of the San Francisco Estuary whose 163,000 km² (62,935 square miles) watershed is bordered by the Sierra Nevada to the east and the Coast Range and Cascades to the west (Cloern et al. 2011). Approximately 40% of California's freshwater flows pass through this region, providing drinking water to over 25 million people (California Resources Agency, 2016). The Delta was once dominated by 1,900 km² (734 sq. mi.) of herbaceous and wooded wetlands, mudflats, and channels (Atwater et al. 1979; Whipple et al. 2012). Figure 22.2a shows how channels were once unarmored and tree-lined, instead of leveed and diked as they are today.

Several deltaic systems have existed in the current location of the Delta through time (Shlemon and Begg 1975). The current form of the Delta took shape approximately 6,800 years ago (Drexler et al. 2009a), when sea level stabilized after a period of rapid rise, leading to the formation of many deltas around the world (Atwater et al. 1977). In much of the Delta, an extensive network of wetlands with highly organic soils (peatlands) began to develop at this time. Due to the mild Mediterranean climate and a nearly 10-month growing season (Atwater 1980; Western Regional Climate Center 2014), these peatlands sequestered great quantities of blue carbon (i.e., organic carbon stored in tidally influenced, coastal ecosystems) in their soils until the 1860s, when massive conversion to agriculture began and continued into the 1930s.

In this chapter, we will view this conversion as a case study on how land management and blue carbon stewardship have only recently been viewed as complementary goals. We will start with a history of blue carbon formation in the Delta. Next, we will chronicle the loss of over half of this carbon through conversion to agriculture and the ecological implications of this large-scale change in land use. We will conclude with a discussion about the challenges of restoring wetlands and increasing blue carbon storage in the Delta region.

22.2 THE HISTORICAL DELTA AND ITS CARBON SINK

The Delta has a semidiurnal, micro-tidal regime (a normal tidal range of ~1 m or less; Shlemon and Begg, 1975) and, throughout its history, it has largely been a tidal freshwater region, though the western periphery has long been an ecotone between fresh and slightly brackish conditions (Drexler et al. 2013). A historical ecology study of the Delta showed that prior to the mid-1800s, it consisted of three basic landscape types (Whipple et al. 2012). The central Delta was a broad expanse of tidal freshwater emergent wetlands surrounded by mudflats and sloughs. The northern Delta consisted largely of flooded basins lying adjacent to riparian forests of the Sacramento River and its tributaries, and the southern Delta contained abundant tributaries of the San Joaquin River, which formed a broad floodplain.

Studies of peat achenes (fruiting bodies of bulrushes (*Schoenoplectus* species)) and rhizomes have showed that plant communities in the Delta have not changed much from the past (Weir 1937; Goman and Wells 2000; Drexler et al. 2009a; Drexler 2011). Tidal freshwater emergent wetlands, in the form of herbaceous marshes, scrub-shrub communities, and a combination of these covered most of these three landscape types. Emergent macrophyte communities were dominated by bulrushes (*Schoenoplectus californicus, S. acutus, S. americanus,* and hybrids), cattails (*Typha angustifolia, T. latifolia, T. domingensis,* and hybrids), and common reed (*Phragmites australis*) (Atwater and Hedel 1976; Reed 2002). The scrub-shrub communities mainly contained willow (e.g., *Salix lasiolepis*), buttonbush (*Cephalanthus occidentalis*), and dogwood (*Cornus sericea*) (Drexler et al. 2009b). Areas below mean tidal level, which were relatively quiescent, contained aquatic species, including sago pondweed

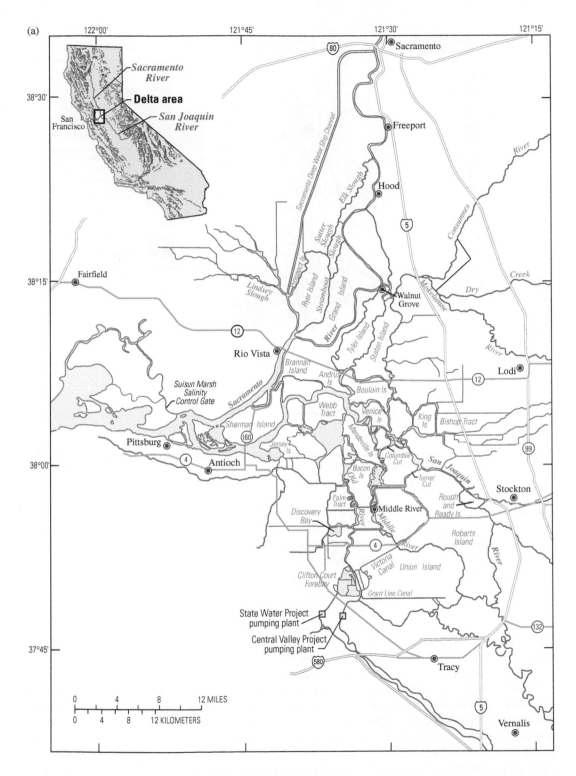

Figure 22.1 (See color insert following page 266.) (a) Map showing the location and configuration of the current Delta. (b) Land use in the historic and current Delta (Adapted from SFEI-ASC (2014).).

(Contiuned)

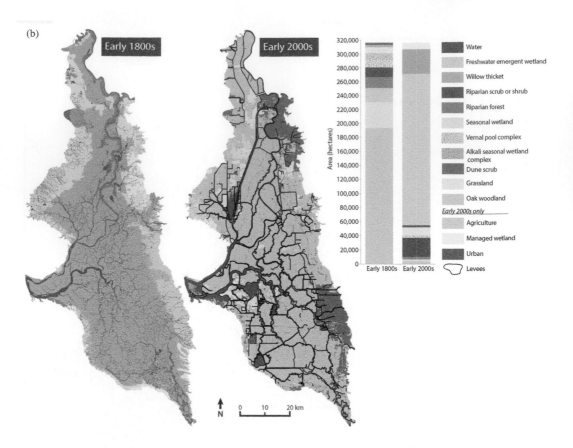

Figure 22.1 (CONTINUED)   (See color insert following page 266.) (a) Map showing the location and configuration of the current Delta. (b) Land use in the historic and current Delta (Adapted from SFEI-ASC (2014).).

(*Stuckenia pectinata*), coontail (*Ceratophyllum demersum*), and pennywort (*Hydrocotyle umbellata*) (Boyer and Sutula 2015). On natural levees and along the riparian corridors, there were large trees including oaks (*Quercus lobata, Q. agrifolia*) and cottonwoods (*Populus fremontii*). Due to the large expanses that were once wetland in the Delta, these plant communities likely formed huge complexes that graded from one type to another depending on hydroperiod, elevation, and local salinity.

Although productivity is unknown for historic wetlands of the Delta, recent measurements of aboveground peak live biomass in emergent marshes range from ~920 to 2,500 g m^{-2} year^{-1} (Miller and Fujii, 2010, 2011; Schile et al. 2014). Belowground peak live biomass has been shown to range from ~900 to 1,800 g m^{-2} year^{-1} (Miller and Fujii, 2010, 2011). Such a high level of belowground productivity is largely due to the robust root and rhizome systems of the various bulrush species.

The soils map in Figure 22.3 shows that high plant productivity over the millennia together with sediment from the greater watershed formed extensive peat soils (histosols) in the central Delta and mineral soils (mollisols) with extensive organic horizons in the northern and southern Delta (Goman and Wells 2000; Drexler 2011). The peat and organic horizons of mineral soils in the Delta are quite different from peats in temperate and boreal peatlands. These latter peatlands contain a large component of mosses and are composed almost entirely of organic matter (>95%; Shotyk 1996), whereas Delta peats have little if any moss component and a mean organic matter fraction by weight that is highly variable between sites, ranging from 40% to greater than 70%, depending on the hydrogeomorphic position of the wetland. Table 22.1 shows the peat characteristics in four sites in the Delta, which were chosen to represent the range of hydrogeomorphic conditions in the region.

(a)

(b)

Figure 22.2   (a) Undated photo of paddle wheel boat, San Joaquin No. 4, towing 4 barges. Trees lining unrein-
forced banks at approximately sea level suggest the time period before 1930 (Source: Center for
Sacramento History, California State Library Collection, 1977/028/005.) and (b) Looking up at the
levee and a container ship from Twitchell Island, a deeply subsided island in the western Delta.
The ship was traveling to Stockton, California along the deep ship channel of the San Joaquin
River, completed in 1933 (Crosby 1941) Credit: Drexler (2003).

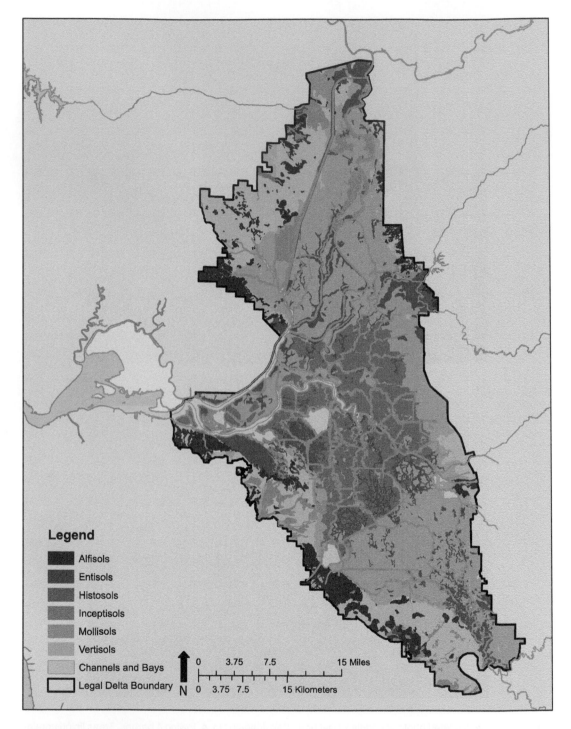

Figure 22.3 (See color insert following page 266.) Taxonomic soil order map of the Delta showing the central area covered by organic soils (histosols), and the large areas in the north and south covered by mineral soils including mollisols, which contain a thick organic surface horizon, and vertisols, which contain a high proportion of clay (Sprecher 2001). Data from Soil Survey Geographic Database (SSURGO) (Soil Survey Staff 2016).

Table 22.1 Characteristics of the Entire Peat Column from Four Extant Tidal Wetlands in the Sacramento-San Joaquin Delta

Core	Hydrogeomorphic Position	Length of Core (cm)	Basal Age of Core (Cal year BP)	Mean Bulk Density (g cm⁻³)	Mean % Organic Matter (sd)
Franks Wetland	Sheltered tributary	608	5550 (124)[a]	0.14 ± 0.08	72 ± 19
Browns Island	Confluence of Sac. and S.J. Rivers	922	6368 (125)[a]	0.31 ± 0.15	40 ± 18
Bacon Channel Island	Tributary of S.J. River	726	5830 (92)[a]	0.12 ± 0.05	76 ± 16
Tip of Mandeville Tip	Main channel of S.J. River	424	2340 (21)[b]	0.23 ± 0.11	41 ± 13

Data from Drexler et al. (2009b). Sac=Sacramento River and S.J.=San Joaquin River.
[a] Error in basal ages for all except Tip of Mandeville Tip is the total standard deviation as calculated by Heegaard et al. (2005).
[b] Error for Tip of Mandeville Tip=1 sigma.

Peat accumulation over the lifetime of the Delta (~6,800 years) resulted in an original peat layer that ranged from 2 to 15 m in thickness due to local differences in hydrogeomorphic conditions, plant community composition, and salinity (Dachnowski-Stokes 1936; Weir 1950; Atwater and Belknap 1980; Drexler et al. 2007; Drexler 2011). The entire volume of peat in the historical Delta has been estimated to be between 4.5 and 5.1 billion m³ (Mount and Twiss 2005; Deverel and Leighton 2010). A study of the accumulated peat layer in the four wetlands described in Table 22.1 showed that the organic carbon density in these wetlands was between 34.2 (sd = 7.2) and 41.0 (sd = 15.1) kg C m⁻³ (Drexler 2011).

Using these data, we have estimated that the original organic carbon sink in the Delta ranged from ~150 to 210 Tg (1 × 10¹²) C (or ~560–770 Tg CO_2e) as shown in Table 22.2. It is important to note that there are major uncertainties in this estimated range, including the accuracy of the two estimates for peat volume and the relevance of using mean organic carbon density measurements from the four remnant wetland sites studied by Drexler (2011) for the whole Delta region. Yet, despite such uncertainties, this 150 to 210 Tg C range (mean of ~180 Tg C or ~660 Tg CO_2) represents a reasonable first-order estimate of the original carbon sink in the Delta. The mean of this range represents nearly twice the energy-related CO_2 emissions in California in 2013 (353 Tg CO_2e; U.S. Energy Information Administration, 2015).

Table 22.2 Estimated Peat Volume, Carbon Density, Organic Carbon Sink, and Carbon Loss in the Delta

Estimated Total Historic Peat Volume in the Delta (m³)[a]	Mean Organic Carbon Density Range (kg C m⁻³) in Delta Peat[b]	Estimated Historic Organic Carbon Sink (Tg (1×10¹²g) C)	Estimated Peat Loss After Conversion to Agriculture (%)[a]	Estimated Carbon Loss After Conversion to Agriculture (Tg C)
4.5×10⁹	34.2–41.0	150–180	54	83–100
5.1×10⁹	34.2–41.0	170–210	49	86–100

[a] First estimate from a LiDAR-based elevation model (±30 cm vertical resolution; Deverel and Leighton (2010)), second from an elevation model constructed from the Shuttle Radar Topography Mission (SRTM) (±1 m vertical resolution; Mount and Twiss (2005)).
[b] Range in mean organic carbon density in Delta peat accumulated over the lifetime of four wetlands in the Delta from Drexler (2011).

22.3 SETTLEMENT AND CONVERSION TO AGRICULTURE

Several groups of Native Americans populated the Delta for at least 4,000 years prior to it being visited by Europeans (Prokopovich, 1985). Then on March 30, 1772, Father Juan Crespi, part of an exploration team headed by Captain Pedro Fages "ascended a pass to its highest point in order to make observations" (Mount Diablo, near Walnut Creek, CA; from Bolton, 1927). From his vantage point, Father Crespi determined that, "below the pass we beheld the estuary that we were following and saw that it was formed by two rivers (the Sacramento and San Joaquin at the head of Suisun Bay)" (from Bolton 1927). From this first visit by Europeans to approximately 1850, the Delta was mainly inhabited by Native Americans and fur trappers (Prokopovich 1985).

With the initiation of the Gold Rush period in 1849, farmers began settling the Delta region. Much of the initial conversion of wetlands was done in the western Delta, including Sherman and Twitchell Islands, which were first settled in the mid-1850s and early 1860s (Thompson 2006). Progress was dampened by multiple levee breaches due to floods and poor machinery, which made levee-building slow and laborious (Peatfield 1894; Thompson 2006). However, with Chinese labor and the invention of the clam dredge, progress accelerated. By the 1930s, approximately 97% of the Delta was drained and converted to agriculture (SFEI-ASC 2014). By then, the 1,400 km² region contained 1,800 km of levees and approximately 60 farmed islands and tracts, which, is largely the same configuration as the Delta of today shown in Figure 22.1a (Thompson 1957; Prokopovich 1985; Ingebritsen et al. 1999).

Throughout the Delta, in order to prepare their fields for cultivation, farmers dug drainage ditches to lower the water table, which was approximately at mean sea level (Gilbert 1917). Such large-scale dewatering of the landscape led to primary subsidence, or the lowering of the ground surface due to mechanical settling or loss of buoyant force within the peat (Everett 1983; Ewing and Vepraskas 2006). A description of early subsidence in the Delta by Walter Weir, a soil scientist at the University of California and subsidence pioneer, clearly states that settling or compaction was not an important process in the Delta (Weir 1950). Instead, subsequent to primary subsidence, the first major cause of secondary subsidence was the practice of burning to destroy weeds, kill seeds, eliminate pests, and liberate potash (Weir 1950). Such burning of fields was not done once but periodically to a depth of 7 to 13 centimeters per burn, which resulted in substantial loss of stored carbon in the soils. Other causes of this secondary wave of subsidence were shrinkage upon drying, wind and water erosion, anaerobic decomposition, dissolution of soil organic matter, and microbial oxidation (Weir 1950; Prokopovich 1985; Deverel and Rojstaczer 1996; Deverel and Leighton 2010).

Extensive research has shown that overall the majority of land-surface subsidence in the Delta has been caused by microbial oxidation, a process by which organic carbon in the peat is converted to carbon dioxide by microorganisms (Deverel and Rojstaczer 1996; Deverel and Leighton 2010). The combination of both primary and secondary subsidence has left the land-surface of much of the Delta from 1 to 8 m below mean sea level (Ingebritsen et al. 2000; Deverel and Leighton 2010). Figure 22.2b provides an example of what the severe subsidence of the western Delta looks like from the middle of a farmed island. Subsidence has also required continual inspection and expensive repair of the levee system following occasional levee blowouts and subsequent flooding of islands. Between the period from 1868 to 2011, there were 265 levee failures, with more than 100 since 1900 (Hopf 2011). For example, on a clear June day in 2004, a levee break on Upper Jones Tract, a 2,286 ha (5,648 acre) island about 10 miles west of Stockton, caused the loss of corn, wheat, alfalfa crops, displaced 270 residents, and ultimately cost approximately $90 million of state funds to repair (Bulwa et al. 2004; Deverel et al. 2016). In the deeply subsided western Delta, the mean annual probability of levee failure is ~7%, which translates into a probability of failure over a 25-year period of ~84% (base conditions from 2005, California Department of Water Resources (CDWR) 2009). Levee failures in this part of the Delta would pose a major risk to water quality and water supply, and cost into the billions of dollars to repair (CDWR 2009).

Land-surface subsidence continues to this day. The rate of land-surface subsidence in the Delta has decreased over time due to abandonment of peat burning, changes in land management, and ongoing reductions in the organic carbon content of surface soils (Deverel and Leighton 2010; Deverel et al. 2016). Maximum historic rates in the mid-20th century ranged from 2.8 to 11.7 cm year^{-1} (Weir 1950; CDWR 1980). In the late 1980s and 1990s, rates were found to be between 0.5 and 4 cm year^{-1} (Rojstaczer and Deverel 1993, 1995; Deverel and Rojstaczer 1996; Deverel et al. 1998). Current estimates are even lower (0.3 to 1.8 cm year^{-1}), demonstrating the reduction over time of secondary subsidence (Deverel et al. 2016). Current rates translate into an estimated loss of between 0.29 and 0.57 kg C m^{-2} year^{-1} from farmlands in the Delta (based on eddy covariance flux measurements in grazing lands and corn fields; Hatala et al. 2012; Knox et al. 2015).

Using estimates from both Deverel and Leighton (2010) and Mount and Twiss (2005) for the peat volume lost following conversion to agriculture (54% and 49%, respectively), we calculate the total loss of organic carbon from land-surface subsidence processes to be ~83–102 Tg C, which is shown in Table 22.2. This may be a conservative estimate because, in a study of the central and western Delta, Drexler et al. (2009a) showed that approximately two-thirds of the peat column had already been consumed in 2005.

Three factors continue to influence the rate of further carbon loss and land-surface subsidence in the Delta. The first is the elevation at which the artificial water table is maintained relative to the land surface to facilitate growing a variety of crops. This determines the depth to which the peat is oxidized (Rojstaczer and Deverel 1993; Drexler et al. 2009a). Other key factors that influence the rate of subsidence are the percent of organic carbon remaining in the peat and the thickness of the remaining peat layer, which are both highly variable across the Delta (Prokopovich 1985; Rojstaczer and Deverel 1995; Ingebritsen and Ikehara 1999; Deverel et al. 2016). Unless the artificial water table is raised to the point of soil inundation, secondary subsidence will continue until all the organic matter is depleted in the peat soils of the Delta.

The large-scale conversion of the Delta ecosystem to agriculture has led to other major problems besides subsidence. The use of the Delta by the State Water Project and the federal Central Valley Project for a water conveyance system, complete with large export pumps that supply water to agricultural and urban water users in the southern half of the state, has altered natural flow patterns and contributed to precipitous declines of native fish populations (Sommer et al. 2007; Delta Independent Science Board 2015). The loss of wetlands has led to a loss of outwelling detrital carbon in the form of plant litter, which was once a major part of the food web (Robinson et al. 2016). Finally, the influx of both invasive flora and fauna has also caused major changes in the food web, reduced habitat for native species, and disrupted natural ecosystem processes (Lopez et al. 2006, Lucas and Thompson 2012; Boyer and Sutula 2015). Collectively, these impacts have caused the collapse of the ecosystem, with multiple ecological consequences including dramatic declines in native fish populations (Sommer et al. 2007; Thomson et al. 2010). A small endemic fish, the Delta smelt (*Hypomesus transpacificus*), whose population serves as an index of overall ecosystem health, is currently on the brink of extinction, despite the protection provided for it under the federal Endangered Species Act (Brown et al. 2016).

22.4 RESTORATION OF DELTA WETLANDS AND CARBON STORAGE

The dire situation for native fishes and the desire to replenish key ecosystem services were the key motivating factors for early wetland restorations in the Delta. Much more recently, the need to reduce carbon emissions to 1990 levels by 2020 as mandated by The Global Warming Solutions Act of 2006 has spurred the entry of wetland restoration and rice cultivation projects as carbon offsets under the California Cap-and-Trade program (American Carbon Registry 2017; California Environmental Protection Agency 2017; CDFW 2017). This incorporation of wetland restoration into climate change mitigation was motivated at least partially by the need to rectify the environmental damage and major carbon loss sustained in the Delta region. The following section discusses the past, current, and future efforts for wetland restoration in the Delta and how such restoration may affect carbon storage in the Delta.

22.4.1 Restoration Projects

Small-scale efforts to restore wetlands in the Delta began in the 1960s with projects conducted by the U.S. Army Corps of Engineers (Atwater et al. 1979). By the mid-1990s, resource managers and policymakers realized that a landscape-scale effort was needed if major progress was to be made toward restoration of habitat and recovery of key ecosystem services. This led to the formation of a combined state and federal program called CALFED in 2000, which included a dedicated Ecosystem Restoration Program (ERP). The CALFED ERP focused its initial efforts on restoring habitats and functions in the Delta and Suisun Bay regions of the San Francisco Estuary, but little restoration was accomplished before the CALFED program ended. A subsequent major multi-agency habitat conservation plan, called the Bay-Delta Conservation Plan, also never came to fruition.

In 2006, Governor Schwarzenegger ordered a Blue Ribbon Task Force to study the needs for the region and develop a Delta Vision, complete with comprehensive recommendations for restoring ecosystem function and ensuring a reliable water supply (Governor's Delta Vision Blue Ribbon Task Force 2008). This led to the formation of the California Delta Stewardship Council (DSC), which completed its comprehensive Delta Plan in 2013 (DSC 2016b). This plan has the co-equal goals of water supply reliability and a healthy Delta ecosystem. Subsequently, the DSC's Delta Science Program (DSP) released a Delta Science Plan (DSC 2014) which provides the framework for conducting science through collaborative efforts aimed at guiding key actions and restoration efforts in the Delta. The major restoration goals in these plans include restoring habitat for fish, birds, and other native species in the Delta, improving ecosystem health and functionality, increasing hydrological connectivity, providing flood control, countering subsidence, and increasing carbon sequestration, and finally, providing diverse recreational activities (DSC 2016c).

Currently, the Sacramento-San Joaquin Delta Conservancy acts as the primary state agency for restoration projects in the Delta. The Delta Conservancy collaborates with multiple federal, state, and non-governmental agencies to implement a variety of projects supported by different funding agencies (Delta Conservancy 2016a,b). The newest plan for the Delta includes two projects, called California Waterfix and California EcoRestore, to safeguard the water delivery system while also restoring habitat in the region (California Natural Resources Agency (CNRA) 2016a,b). California Waterfix involves the construction of two huge tunnels to deliver water to the water projects, without passing through the rivers of the Delta, thus greatly reducing risks to water supply from levee breaks. California EcoRestore involves the restoration and protection of 12,141 ha (30,000 acres) of aquatic, sub-tidal, tidal, riparian, flood plain, and upland habitat (CNRA 2016b).

The data in Table 22.3 and the accompanying map in Figure 22.4 show that, through the various efforts described above, ~4,476 ha (11,060 acres) of wetlands have been restored or preserved, mostly along the periphery of the Delta. Future wetland restorations as part of EcoRestore total ~3,403 ha (8,409 acres). Summing both completed and planned restorations in Table 22.3 yields a grand total of ~7,879 ha (19,469 acres) of restored wetlands, which are slated to be in place by 2020.

If we assume this wetland area will accrete organic matter and sediment at a rate of 0.3 cm year^{-1} (recent accretion rate at Browns Island, from Callaway et al. 2012; lower end of range from Swanson et al. 2015) and at a mean carbon density of 0.038 g C cm^{-3}, taken from Table 22.2, this will result in a conservative estimate of carbon accumulation for all the restored wetlands of ~0.9 Tg C over the next 100 years. Clearly, even this conservative estimate may be inaccurate, because it is unknown how quickly the restored wetlands will establish and whether or not they

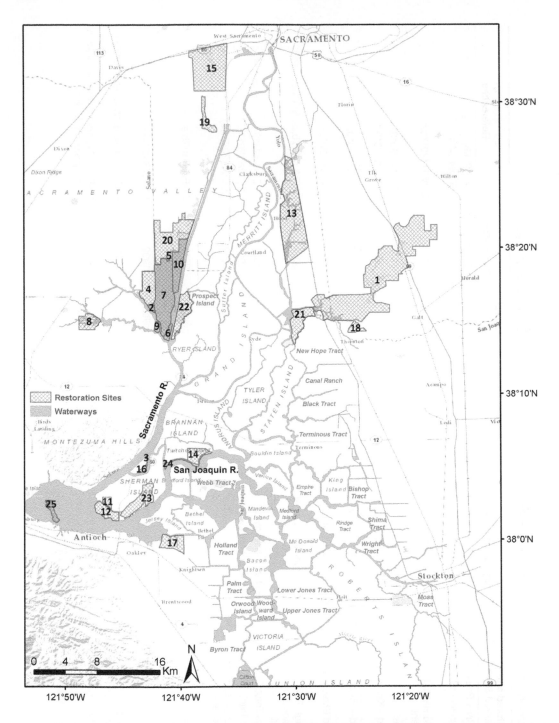

Figure 22.4 Location of wetland restoration projects in the Delta (1–15 completed; 16–25 planned as part of EcoRestore). Numbers refer to restoration projects described in Table 22.3.

will store carbon at the same rates as historic wetlands. Regardless, after a 100-year time frame, the total amount of carbon accumulated using the figures above represents roughly ~1% of the 83–100 Tg C lost from the Delta.

Table 22.3   Completed (1–15) and Planned (16–25) Wetland Restoration Projects in the Sacramento-San Joaquin Delta

No.	Restored and Preserved Sites	Purpose of Restoration	Total Hectares	Managed/Tidal	Year Completed	References
1	Cosumnes River Preserve	Ecosystem function, connectivity	420	Tidal	1998	CRP (2008)
2	Cache Slough Mitigation Area	Fish habitat	69	Tidal	1991	SFCWA (2013)
3	Decker Island Habitat Enhancement	Fish habitat, levee improvement	8.5 4.5	Managed, Tidal	2004	CDWR (2017b,c)
4	Liberty Farms	Habitat	688	Managed	2003	SFCWA (2013)
5	Liberty Island—Kerry Parcel	Habitat, connectivity	75	Tidal	2010	SFCWA (2013)
6	Liberty Island—Prospect, West	Fish habitat, connectivity	73	Tidal	1963	SFCWA (2013)
7	Liberty Island Ecological Reserve	Habitat for fish and birds, recreation	381	Tidal	1998	SFCWA (2013)
8	Lindsey Slough Tidal Habitat Restoration Calhoun Cut	Ecosystem function, connectivity	92	Tidal	2014	CNRA (2017b,c), SFCWA (2013)
9	Little Hastings Tract	Fish habitat, connectivity	65	Tidal	1992	SFCWA (2013)
10	Little Holland Tract	Fish habitat, connectivity	591	Tidal	1992	SFCWA (2013)
11	Sherman Island—Mayberry Farms Subsidence Reversal	Subsidence reversal, carbon sequestration	123	Managed	2010	CDWR (2017a,e)
12	Sherman Island—Whales Mouth Wetland Restoration	Habitat, flood control, carbon sequestration	243	Managed	2015	CNRA (2017b,e)
13	Stone Lake National Wildlife Refuge	Habitat connectivity, recreation	858	Controlled Tidal	1992	USFWS (2017)
14	Twitchell Island East End Habitat Restoration Project	Habitat, subsidence reversal	304	Managed	2013	CNRA (2017b,f), Miller et al. (2008)
15	Yolo Bypass Wildlife Area	Flood control, wildlife habitat, education, recreation, and agriculture	313 168	Tidal, managed	1997 and 2001	YBF (2017) personal communication, Jeffrey Stoddard, YBWA
	Total area restored in the Delta		**4,476**			

(Continued)

Table 22.3 (Continued) Completed (1–15) and Planned (16–25) Wetland Restoration Projects in the Sacramento-San Joaquin Delta

No.	EcoRestore Sites Planned for Restoration	Purpose of Restoration	Total Hectares	Managed/ Tidal	Expected completion	References
16	Decker Island Tidal Habitat Restoration	Habitat, connectivity	57	Tidal	Delayed	CNRA (2017b,d)
17	Dutch Slough Tidal Marsh Restoration	Habitat, recreation	227, 31	Tidal, Managed	2020	CNRA (2017b,m)
18	Grizzly Slough Restoration	Ecosystem function, recreation, flood control	162	Tidal	2018	CNRA (2017b,k)
19	Lower Putah Creek Realignment	Ecosystem function, fish passage, and habitat, recreation	210	Tidal	2019	CNRA (2017b,i)
20	Lower Yolo Restoration	Fish habitat, flood control	676	Tidal	2022	CNRA (2017b,l)
21	McCormack Williamson Tract	Ecosystem function, flood control, habitat, recreation	532	Tidal	2019	CNRA (2017b,j)
22	Prospect Island Tidal Habitat Restoration	Fish habitat, carbon sequestration	647	Tidal	2020	CDWR (2017d)
23	Sherman Island—Whale's Belly Wetland Restoration	Habitat	607	Managed	delayed	CNRA (2017b,n)
24	Twitchell Island—San Joaquin River Setback Levee	Flood (levee) protection, habitat	~6 to 16[a]	Tidal	2016	CNRA (2017b,g)
25	Winter Island Tidal Habitat Restoration	Fish habitat, primary productivity	238	Tidal	2019	CNRA (2017b,h)
	Total area planned for restoration		**3,403**			

Numbers refer to locations in Figure 22.4. The areas and completion dates of sites are based on published sources, which are subject to change depending on site-specific plans and restoration success. Some restorations are currently delayed, but were included in the total area planned because they are still expected to be completed.

[a] The higher value (16) was used for estimating total area planned for restoration.

It is important to note that this ~1 Tg estimate of regained carbon includes only the amount of carbon stored in the soil and does not imply that these wetlands will be long-term net carbon sinks. In fact, recent research suggests that the annual atmospheric carbon balance in Delta wetlands can vary greatly, resulting in different wetlands being a sink or a source or the same wetland varying between these two states (Knox et al. 2015; Anderson et al. 2016). Furthermore, once the global warming potential (GWP) of CH_4 over 100 years is counted in at 34 times that of CO_2 (Myhre et al. 2013) (or at 45 times CO_2 if one uses the sustained-flux GWP (Neubauer and Megonigal 2015)), the net effect makes wetlands into net GWP sources (Anderson et al. 2016). However, the data currently available are from studies conducted in impounded Delta marshes for only 1 or 2 years. Longer consecutive measurements of carbon fluxes are needed to definitively determine whether Delta wetlands have long-term net positive or negative GWPs.

It is important to note that all future wetland restorations in the Delta are subject to an array of challenges, which may affect their ultimate success. Some challenges, such as those spanning political, regulatory, and policy arenas, are largely outside the scope of this chapter. Here we will focus on several environmental issues, which stakeholders and practitioners will likely face as they move to complete the planned restorations in Table 22.3 and/or attempt to find even more lands to restore throughout the Delta region. The issues of particular concern include: (1) lack of suitable land at appropriate elevation ranges; (2) the reduction of available sediment to assist in wetland formation and vertical accretion; (3) the wide-scale infestation of invasive aquatic vegetation in subtidal waters, and (4) the impact of sea level rise on the future sustainability of restored wetlands. The remainder of the chapter is devoted to discussing each of these issues in turn.

22.4.2 Suitable Lands

Due to land-surface subsidence in the Delta, only some lands are within the proper elevation range for the growth of tidal freshwater wetland communities (~50 cm above and below mean sea level, Swanson et al. 2015). The following islands, which can be found in Figure 22.1a, have mean elevations less than 1 m below mean sea level as of 2005: Prospect Island, Grand Island, and Canal Ranch Tract in the north, Bishop Tract and neighboring islands in the east, Hotchkiss Tract in the west, and Roberts, Union, Rough and Ready, and Coney Islands in the south (Bates and Lund 2013). Of these only Union and Coney Islands have mean elevations of greater than −0.5 m MSL, assuring that these islands could support wetland plant communities. Of the islands above −1 m MSL, at least some portion of these lands is likely to be within the required elevation range. This examination of island elevations clearly demonstrates that, for most of the Delta, tidal wetland restoration is currently not possible.

The strategy that has been taken in deeply subsided areas of the Delta is to create impounded wetlands, which can then mitigate subsidence by accreting organic matter *in situ*, thereby gaining elevation (Deverel et al. 2014). It is assumed that such wetlands will continue to build up their elevations until they reach appropriate elevations for the islands to be breached to restore full tidal function. On Twitchell Island, two demonstration ponds with different water levels were restored to test how quickly impounded wetlands could increase their elevations. These permanently impounded wetlands, which were studied for over a decade, had mean elevation increases of approximately 4 cm year^{-1} (Miller and Fujii, 2011).

Bates and Lund (2013) used this mean accretion rate to calculate the potential elevation gains if such impoundments were established throughout the Delta. After 50 years, if all islands at elevations at −2.6 m MSL or higher were impounded, which is over half the Delta islands, elevations could be increased to at least −0.5 m MSL, the lowest limit for tidal wetland plant growth. In addition to elevation increases, Bates and Lund (2013) also calculated the probability of levee failure by year due to all causes including floods and earthquakes. For the islands at elevations of −2.6 m and higher, the probability of levee failure over a 50-year time period averaged 0.84 (range: 0.45–0.99).

Because the cost to repair levee breaks is extremely expensive (at least $100 million to $1 billion; CDWR, 2009), it is important to note that once levees break, they will likely not be repaired, causing the restored wetland to be converted to deeply flooded sub-tidal habitat.

Such a high probability of levee failure suggests that wetland restoration in the form of impoundments may not be a worthwhile investment in the Delta. Taken out of context, this might appear to be true, however, removing land from agricultural production and converting it back to wetlands has more benefits than simply increasing elevation relative to mean sea level. First of all, wetland restoration provides important habitat for a variety of birds, reptiles, and amphibians as well as provides detrital carbon for the aquatic food web. Even if each restored wetland may not reach its full trajectory to tidal wetland due to a levee break, a number of restoration sites, if managed with the relative risk of levee failures in mind, could serve as a series of temporary refugia for sensitive populations (c.f., Hermoso et al. 2013; van Hooidonk et al. 2013). Secondly, extensive research has shown that wetland impoundments as well as rice cultivation, which also requires impoundment, can slow or even stop subsidence, thus reducing risk of levee failure by decreasing the hydrostatic pressure on levees (Miller and Fujii 2011; Deverel et al. 2016). Finally, there may potentially be carbon benefits to wetland and rice impoundments relative to grazing lands and cultivated fields, however, in the few eddy covariance studies conducted over short time periods, results indicate that both impounded wetlands and rice likely have net positive global warming potentials due to their high methane emissions (Knox et al. 2013; Anderson et al. 2016).

22.4.3 Invasive Aquatic Vegetation

The San Francisco Estuary is the most invaded estuary in North America (Cohen and Carlton 1998). Because of this, the Delta, as the landward end of the Estuary, contains its share of alien species including fish, invertebrates, and aquatic vegetation (Cohen and Carlton 1998; Light et al. 2005; Moyle et al. 2010). Many of these species have colonized the Delta as a result of ship or recreational boat traffic (Cohen and Carlton 1995), and have been highly successful due to the spatial heterogeneity, the mild climate, and highly managed flows in the Delta (Gopal 1987; Sears et al. 2006; Santos et al. 2012). Several floating and submerged species of invasive aquatic vegetation (IAV) are particularly aggressive due to their phenotypic and morphological plasticity (Sakai et al. 2001).

Water hyacinth (*Eichhornia crassipes*) and water primrose (*Ludwigia* spp.), shown in Figure 22.5, comprise the major floating species of IAV in the Delta. Water hyacinth, a native of Brazil, has successfully invaded almost every region in the world suitable for its growth. It was first recorded in the Sacramento River in California by 1904 (Gopal 1987) and, by 1981, covered 506 hectares (Finlayson 1983). The annual cover and biomass of water hyacinth in the Delta varies due to climate and management. In recent years, the areal cover of water hyacinth has ranged from 200 hectares (June 2008) to 1,200 hectares (November 2014) (Khanna et al. 2015a,b).

Water hyacinth has many negative impacts on Delta channels and wetlands. It forms large monotypic floating mats, which reduce open-water habitat for other species and impact habitat quality for native fish species (Simenstad 1999). Due to the large amount of detritus produced under the water hyacinth mat, oxygen levels are low enough to stress fish and act as a barrier to accessing shallow habitat (Penfound and Earle 1948; Toft et al. 2003; Malik 2007). Water hyacinth can create permanent islands by depositing vast amounts of detritus and retarding water velocity (Toft 2003). In this way, water hyacinth may also block sediment from depositing on the wetland surface, interfering with natural vertical accretion processes in wetlands. Mats of water hyacinth also provide prime conditions for submerged IAV to colonize, particularly after water hyacinth is treated with herbicide and removed (Khanna et al. 2012).

Water primrose (*Ludwigia* spp.), in comparison to water hyacinth, has only recently invaded the Delta. It was first recorded in the Delta in 1949 (Light et al. 2005), and in recent years has shown exponential increase in cover (Cal-IPC 2006; Ustin et al. 2017). At least three species of the genus

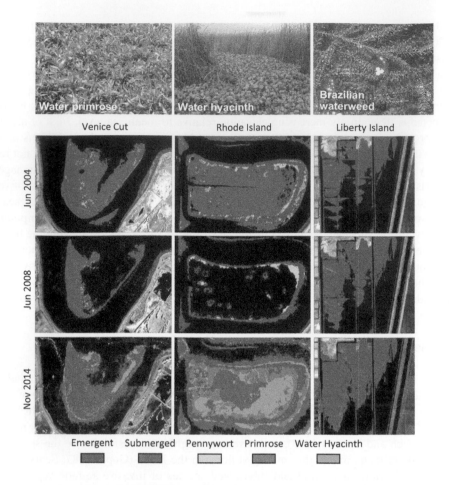

Figure 22.5   (See color insert following page 266.) Top panel: Dominant IAV species in the Delta. Credit: Shruti Khanna. Bottom panel: IAV distribution in the central (Venice Cut and Rhode Island) and northern Delta (Liberty Island) from 2004 to 2014. In June 2008, after a rigorous spray effort, submerged aquatic vegetation cover decreased in much of the Delta, including Rhode Island and Venice Cut. By 2014, however, cover had increased back to 2004 levels. Emergent=all emergent macrophytes, Submerged=all submerged species.

Ludwigia exist in the Delta (Rejmánková 1992; Cal-IPC 2006; Sears et al. 2006; Ustin et al. 2006; Boyer and Sutula 2015). *Ludwigia palustris* (water purslane) is native (Sears et al. 2006; Rejmánková; 1992; Boyer and Sutula 2015). *L. peploides* subsp. *peploides* and *L. grandiflora* subsp. *hexapetala* are known to be both alien and invasive (Rejmánková 1992; Dandelot et al. 2005; Sears et al. 2006; Thouvenot et al. 2013). *L. peploides* subsp. *montevidensis* has been documented in the Laguna system in Sonoma, CA, but there are no confirmed reports of it in the Delta (Sears et al. 2006).

Water primrose has both a creeping and floating habit and can invade patches of native macrophytes (Rejmánková, 1992). Between 2004 and 2014, water primrose cover has fluctuated from 270 hectares (June 2004) to more than 500 hectares (September 2015) (Khanna et al. 2015a,b; Ustin et al. 2015; 2016). It impacts the Delta ecosystem in much the same ways as water hyacinth by reducing water velocity and oxygen content in the water column, increasing sediment deposition in its vicinity, eliminating open-water habitat, and out-competing native vegetation (Dandelot et al. 2005; Sears et al. 2006; Thouvenot et al. 2013). Water primrose also has allelopathic properties, which actively discourage other native species from growing and give it a competitive edge (Dandelot et al. 2008).

Finally, there are also several submerged IAV in the Delta, including *Myriophyllum spicatum*, *Pomatogedon crispus,* and *Cabomba caroliniana*; however, none is as invasive in the Delta as Brazilian waterweed *(Egeria densa)* (Santos et al. 2012), which is shown in Figure 22.5. Brazilian waterweed first arrived in California around 1970 (Foschi and Liu 2002). Data collected between 2004 and 2014 showed that Brazilian waterweed grows as a monoculture or dominates biomass in 60–66% of all submerged aquatic vegetation patches in the Delta (Ustin et al. 2015; 2016). Brazilian waterweed has a very high growth rate as well as overwintering shoots, which are rooted at the bottom of the water column. It is capable of reproducing both from plant fragments and from seed, giving it a competitive advantage over other submerged aquatic vegetation (Santos et al. 2012). Brazilian waterweed acts as an ecosystem engineer (c.f., Jones et al. 1994a,b) by displacing native submerged plant species, supporting different heterotrophic pathways compared to native aquatic vegetation (Wilcock et al. 1999), decreasing dissolved oxygen in the water column (Getsinger and Dillon 1984), and reducing water flow velocity, causing suspended sediment to drop out of the water column (Wilcock et al. 1999; Champion and Tanner 2000; Hestir et al. 2013).

The annual distribution patterns of all submerged aquatic vegetation in the Delta depend on preceding winter and spring temperatures, runoff through the Delta, and the management actions from the previous year (Santos et al. 2009). Between 2004 and 2014, submerged aquatic vegetation as a whole covered from 2,080 hectares (June 2008) to more than 3,050 hectares (November 2014) (Khanna et al. 2015a,b). Overall, the area of the Delta colonized by submerged aquatic vegetation has increased by almost 700 hectares in the past decade, with a predominance of invasive species comprising this increase (Khanna et al. 2015a).

The IAV present in Delta channels and wetlands will continue to spread, unless there is a significant increase in management actions. The main agency responsible for IAV control, the California Department of Parks & Recreation-Division of Boating & Waterways (CDPR-DBW), has conceded that it cannot eradicate IAV, namely Brazilian waterweed and water hyacinth. Instead, CDPR-DBW is focused on controlling IAV, by means of herbicide application, in important navigational channels, marinas, and irrigation systems (CDPR-DBW 2017a,b).

IAV clearly represents a major challenge for new wetland restoration projects in the Delta, because the ubiquitous presence of plant fragments and seeds practically guarantees invasion. The fact that both submerged and floating IAV impede water flows, results not only in sediment falling out of the water column but also fine sediment settling on IAV leaves and stalks (Petticrew and Kalff 1992; Wilcock et al. 1999; Hestir et al. 2013), creating a sediment sink in IAV biomass. This sink has yet to be quantified but may at least partially explain the recent step decrease in turbidity in the Delta (Hestir et al. 2013). Such interference in natural inorganic sedimentation processes may reduce the sustainability of Delta wetlands, which rely on sediments in addition to organic matter to maintain their position in the tidal frame as sea level rises (Drexler 2011). Finally, although the carbon accumulation rates of sites with and without IAV in the Delta have yet to be studied, it is possible that invaded sites, particularly those invaded by the creeping form of water primrose, may not be storing as much carbon as macrophyte-dominated wetlands, which have much greater above- and belowground biomass. For all of these reasons, if wetland restoration is truly a priority for state and local officials, a greater emphasis is needed on IAV control to ensure the successful trajectory of wetland restoration projects into the future.

22.4.4 Sediment Availability

Hydraulic mining for gold during the period from 1852 to 1884 washed approximately 300 million tons of sediment into Sacramento Valley rivers (Gilbert 1917; Bouse et al. 2010). Much of this sediment pulse deposited in the rivers, their floodplains, and the Delta and bays of the San Francisco Estuary (Gilbert 1917). Since hydraulic mining was severely curtailed in 1884, several factors have decreased sediment supply from the San Francisco watershed (Schoellhamer et al.

2013). The hydraulic mining sediment pulse has diminished asymptotically and much of it now resides behind dams, on levee-protected flood plains, and in the Estuary (James 1999). During the 1900s, many dams that trap sediment were constructed in the watershed (Wright and Schoellhamer 2004). More than one-half of the banks of the lower Sacramento River were riprapped during the latter half of the 20th century to protect them from erosion. This results in decreased sediment transport in the river (U.S. Fish and Wildlife Service 2000). Flood control bypasses and settling basins built in the Sacramento River floodplain during the early 20th century further trap sediment and reduce downstream sediment supply (Singer et al. 2008). The net result of these factors is that sediment supply from the Sacramento River, the largest sediment source for the Estuary, decreased by about one-half from 1957 to 2001 despite no overall time trend in annual flow or flow variability during this period (Wright and Schoellhamer 2004).

Decreasing sediment supply has led to lower levels of suspended sediment in Delta channels. Total suspended-solids (TSSs) concentration in the Sacramento–San Joaquin River Delta decreased from 1975 to 1995 by 50% or 0.717 mg L^{-1} year^{-1} (Jassby et al. 2002). Hestir et al. (2013) found step decreases in Delta TSS of 27% in 1983 and 23% in 1998. These step decreases co-occurred with record high wet and dry season discharge following El Niño-driven high precipitation. Between 1998 and 2010, Delta TSS had a significantly decreasing trend of -0.64 mg L^{-1} year^{-1}. Total water discharge of the Sacramento Valley and mean TSS in the Sacramento River upstream from the Delta did not have significant trends during the 2000s, suggesting that neither flow nor river supply are responsible for the TSS decrease in the Delta. Sediment trapping by IAV may be the missing sediment sink responsible for the decrease in TSS in the early 21st century (Hestir et al. 2015).

Decreased TSS in the Delta decreases sediment supply to tidal wetlands, especially along the Sacramento River where inorganic sedimentation has historically been greatest (Reed 2002). Swanson et al. (2015) developed a wetland accretion model for the Delta and found that the magnitude of sea level rise over the next century was the primary driver of wetland surface elevation change and that sediment supply was the next most important controlling factor. Wetland restoration typically involves opening a diked area to tidal action and allowing sediment to deposit, which eventually results in plant colonization. Restored wetlands are often designed so that their initial elevations are slightly below target levels in order to capitalize on natural sedimentation processes and encourage the formation of tidal creeks (Williams and Orr 2002). In a newly restored wetland, the rate of deposition is proportional to TSS (Krone and Hu 2001), so the time required to create a wetland increases as TSS decreases. Therefore, if the rate of deposition is less than the rate of sea level rise, a vegetated wetland will never form. Thus, the current regime of decreased TSS in the Delta will likely affect restoration of subsided lands to tidal wetlands by (1) increasing the time needed to achieve colonization of wetland vegetation and (2) increasing the possibility that vertical accretion in restored wetlands will be incapable of keeping pace with sea level rise.

22.4.5 Sea-Level Rise and Salinity Incursion

The rate of sea level rise will have major implications for tidal wetland sustainability in the Delta, particularly for those sites that have yet to achieve target elevations. A committee of the National Research Council estimated sea level rise along the California coast south of Cape Mendocino to be 4–30 cm by 2030 (relative to 2000 levels), 12–61 cm by 2050, and 42–167 cm (17–66 in) by 2100 (Committee on Sea Level Rise in California, Oregon, and Washington, 2012). This means that tidal wetlands must be able to accrete at the same rate or faster than sea-level rise in order to maintain the elevation of the marsh platform within the tidal frame. Recent work has shown that marshes from the South Bay to the western boundary of the Delta accrete between 0.2 and 0.5 cm year^{-1} (Callaway et al. 2012). If tidal wetlands cannot keep pace through processes of organic accumulation and/or inorganic sedimentation, they become permanently inundated or drown, resulting in the loss of habitat, the cessation of carbon storage, and the transition to open water or mudflat habitat (Morris et al. 2002; Callaway 2007).

In addition to increased inundation, sea level rise will also lead to increased salinity incursions in the Delta, particularly on the western periphery. The high plant diversity in the Delta relative to other parts of the Estuary is associated with high productivity (Atwater and Hedel 1976; Drexler et al. 2009b). An increase in salinity would likely curtail productivity and alter plant community composition, which may ultimately have ramifications for estuarine food webs, carbon accumulation, and habitat value (Callaway et al. 2007).

Several researchers have applied mechanistic, vertical accretion models in the San Francisco Estuary in order to project how tidal wetlands will fare in the future (Orr et al. 2003; Stralberg et al. 2011; Deverel et al. 2014; Schile et al. 2014; Swanson et al. 2013; 2015). Of these studies, only two were focused on the Delta (Deverel et al. 2014; Swanson et al. 2015) and only one on determining the fate of the remaining tidal wetlands in the Delta (Swanson et al. 2015). Swanson et al. (2015) used a one-dimensional marsh surface elevation model, the Wetland Accretion Rate Model of Ecosystem Resilience (WARMER), to explore the impact of a broad suite of future conditions on a range of Delta tidal wetlands. A total of 450 simulations were conducted in order to cover the likely range of key model inputs including porosity values, initial elevations, organic and inorganic accumulation rates, and sea level rise under future conditions. As mentioned above, the results of these simulations showed that the magnitude of sea level rise over the next 100 years was the greatest driver of wetland elevation change and sediment supply as the second most important driver. Tidal wetlands faired relatively well until projected sea level rise increased to 133 cm and 179 cm by 2100. At these two highest levels, only 32% and 11%, respectively, of scenarios resulted in sustainable tidal wetlands. Tidal wetlands situated along main channels were slightly more resilient than tidal wetlands in tributaries. Overall, the results showed that upstream reaches of the Delta, where sea level rise will likely be attenuated, and tidal wetlands in main channels will be more resilient than other wetlands in the Delta.

The results from Swanson et al. (2015) are similar to other modeling studies in the San Francisco Estuary, which have showed that marsh sustainability relies strongly on the rate of sea level rise and the amount of available sediment (Orr et al. 2003; Stralberg et al. 2011; Swanson et al. 2013). It is important to note that each of these models has major uncertainties and can only project future scenarios, not predict future conditions. Nevertheless, it is clear that marshes with low sediment availability are highly vulnerable to drowning. This brings into question whether or not the organic component of vertical accretion, particularly belowground biomass, may be able to increase under heightened flooding, if sediment availability is not adequate for wetlands to keep pace with sea level rise. Although this is a major research topic on the east and gulf coasts (i.e., Nyman et al. 2006; Morris et al. 2013, Kirwan and Guntenspergen 2015; Watson et al. 2016) little research has been conducted to investigate the ability of dominant wetland plant species in the Delta to increase above-or belowground production under increased flooding. Such questions need to be addressed in order to gain a fuller understanding of the future fate of historic and restoring tidal wetlands in the Delta.

22.5 CONCLUSIONS

The Delta was once a vast store of blue carbon on the west coast of the United States. Subsequent to conversion of approximately 98% or 1,900 km² (734 sq. mi.) of historic wetland, floodplain, and riparian landscape to agriculture, ~100 Tg of the Delta blue carbon sink was lost, mainly as a result of microbial oxidation of organic soils. This has led to large-scale, land surface subsidence throughout the region. The massive change in land-use has also resulted in the collapse of the ecosystem, with dire consequences for native fishes. Restoration of tidal wetlands has been ongoing since the 1960s. The total area of wetland restoration is slated to reach ~7,879 ha by 2020. Approximately 0.9 Tg C will accumulate in these restored wetlands over 100 years. This represents roughly 1% of the carbon lost from the region. Although this amount of carbon is small compared to what has been lost, the co-benefits of wetland restoration, including slowing of land-surface subsidence, expansion of critical habitat for sensitive species, and support of the aquatic food web, are highly important in and of themselves. Much larger parcels of wetlands would need to be restored in order to replenish a meaningful portion of the Delta carbon sink. Additional restoration activities in the Delta may result from the recent entry of wetland restoration projects as carbon offsets under the California Cap-and-Trade program.

ACKNOWLEDGMENTS

We dedicate this chapter to Brian F. Atwater, our emeritus colleague at the USGS, whose passion for the Delta, evident in stellar geologic and ecological studies, lay much of the groundwork for the rest of us. We are grateful to Erik Loboschefsky for his help in providing the area of the Decker Island restoration. We appreciate the information on the Yolo Basin Wildlife Area provided by Martha Ozonoff, Robin Kulakow, and Jeffrey Stoddard. Lastly, we thank Sam Safron and Letitia Grenier for providing Figure 22.1b.

CHAPTER 23

Carbon Sequestration in Arid Blue Carbon Ecosystems

A Case Study from the United Arab Emirates

Lisa Schile-Beers and J. Patrick Megonigal
Smithsonian Institution

J. Boone Kauffman
Oregon State University

Stephen Crooks
Silvestrum Climate Associates, LLC

James W. Fourqurean
Florida International University

Justin Campbell and Bill Dougherty
Climate Change Research Group

Jane Glavan
Abu Dhabi Global Environmental Data Initiative

CONTENTS

HIGHLIGHTS

1. Carbon stocks were measured in the arid coastal ecosystems of the United Arab Emirates, focusing on traditional blue carbon ecosystems (natural and restored mangroves, salt marshes, and seagrass beds), and two ecosystems, coastal sabkha and microbial mats, that had not been studied previously in the context of blue carbon.
2. Mature mangroves contained the highest plant carbon pools; however, no detectible differences in soil carbon stocks were measured.
3. Microbial mats contained soil carbon pools comparable to vascular plant-dominated ecosystems and could arguably be recognized as a unique blue carbon ecosystem.
4. Compared to comparable ecosystems in temperate and tropical climates, arid blue carbon ecosystems had lower carbon stocks yet still provide the same ecosystem services.

23.1 INTRODUCTION

Vegetated coastal ecosystems produce and sequester significant amounts of organic carbon (Chmura et al. 2003; Duarte et al. 2005; Donato et al. 2011; McLeod et al. 2011; Fourqurean et al. 2012a), generating worldwide interest in the management, conservation, and restoration of mangroves, marshes, and seagrasses for the purpose of climate change mitigation (McLeod et al. 2011; Pendleton et al. 2012). The recent increase in attention to these blue carbon ecosystems has exposed considerable gaps in our understanding of carbon pools and sequestration rates in coastal environments. A major limitation is that field research has focused primarily on study sites located in humid regions at temperate and tropical latitudes (Chmura et al. 2003; Donato et al. 2011; Adame et al. 2012; Fourqurean et al. 2012a, Kauffman et al. 2014; Alongi et al. 2015), with relatively few studies in sub-humid or arid regions (Adame et al. 2012; Ezcurra et al. 2016) where differences in rainfall, evapotranspiration, and soil conditions could affect carbon storage. The limited range of climates examined makes it difficult to assess the potential for carbon-based ecosystem management across sites that vary tremendously across gradients of coastal climate, hydrology, geomorphology, and tide range (Sifleet et al. 2011), and limits our ability to generalize knowledge outside of warm humid regions. Perhaps the least-studied intertidal marine ecosystems occur in arid regions. For example, coastlines in the Arabian Gulf contain a mosaic of productive ecosystems, including mangroves, seagrasses, coral reefs, coastal sabkha (broad, flat inter- and supratidal salt flats lacking vascular plants), and cyanobacterial mats (hereafter microbial mats), among others, that provide food and habitat for diverse ecological communities and support over half a billion dollars in fisheries activities annually (Burt 2014). There is presently a dearth of research in arid tidal wetland and seagrass (herein blue carbon) ecosystems even though large investments have been made in creation, restoration, and protection activities by nations of the Arabian Peninsula (Aoki and Kugaprasatham 2009).

Our objective was to conduct the first comprehensive analysis of carbon stocks in coastal ecosystems of an arid region. The Abu Dhabi Blue Carbon Demonstration Project aimed to improve understanding of carbon capture and sequestration in an arid environment and contribute to this relatively new concept on a regional and international level. Additionally, the project endeavored to inform a science-based approach to making decisions through policies and appropriate management, in relation to sustainable ecosystem use and the preservation of their services for the current and future generations. The Abu Dhabi Global Environmental Data Initiative (AGEDI), a partnership initiative of the Environment Agency—Abu Dhabi (EAD) and the UN Environment, commissioned the project.

We quantified carbon stocks in salt marshes, seagrass beds, and natural and planted mangroves along 600 km of coastline of the United Arab Emirates (UAE) (Campbell et al. 2015; Schile et al. 2017). Further, we quantified carbon stocks in two coastal ecosystems—microbial mats and coastal sabkha—that are rarely studied in a carbon context. Based on our understanding of tidal inundation on carbon stocks in temperate and tropical wetland ecosystems, we hypothesized that carbon stocks in the intertidal ecosystems would decrease with decreasing flooding frequency, and be greater than carbon stocks of subtidal seagrass beds. Furthermore, we hypothesized that plant carbon pools in this arid environment would be lower than comparable ecosystems in humid tropical and subtropical systems due to lower primary production. We did not have a specific hypothesis for soil carbon pools because large pools occur in systems with both high and low net primary production and in regions of high and low temperature.

23.2 STUDY LOCATION AND HISTORY

We studied coastal and near-shore ecosystems within the UAE (Figure 23.1). Along the Arabian Gulf, air temperatures seasonally range from 12°C to >50°C, and water temperatures at the coastal margins range seasonally from 10°C to 36°C (EAD 2007; Piontkovski et al. 2012). Average annual rainfall is <100 mm and much less than evaporation rates of 1,000–2,000 mm (Evans et al. 1973). Salinity in the Arabian Gulf is high due to restricted tidal exchange and high rates of evaporation, reaching values >70 PSU in lagoons and other shallow waters during summer (EAD 2007). Localized areas of lower salinity are created by urban water outflows, through drainage networks (i.e., wadis) from mountain areas to the north and east, and increased water circulation following channel construction and dredging (Embabi 1993). Along the Gulf of Oman, average air temperature is 28°C and rainfall amounts range from 12 to 331 mm, averaging 138 mm in Khalba (Boer

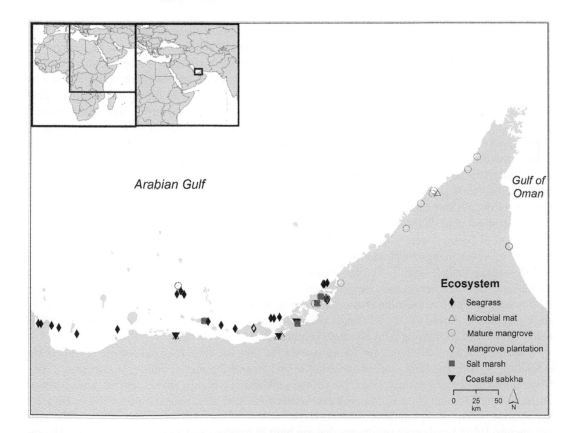

Figure 23.1 Site map of all reported intertidal sampling locations in Abu Dhabi.

1997). Tides are complex, driven by interfering standing waves across the Arabian Gulf, resulting in a mix of diurnal and semi-diurnal tides, with a spring range of approximately 2.5 m.

The geomorphology and soils of the Abu Dhabi shoreline hold a record of local, relative sea level change over the past 8,000 years (Evans et al. 2002). Set against a global trend in sea level rise, the position of the shore has varied in response to a balance between sea level, erosion, subsidence, and sedimentation. Along the coast, relatively rapid rising sea level pushed landwards to eventually truncate aeolian dunes sands some 4 km landward of the existing shore, marking the shoreline some 4,000 years ago (Evans 1964). This trend of shoreline retreat reversed at this point. Over recent millennia, during a period of relatively slow global sea level rise, accumulating aeolian sediments combined with evaporate accumulations within coastal sabkha deposits pushed the shoreline seaward.

The history of this seaward transgression and subsequent regression is held within the soils. Sedimentological analysis describes a sequence of Pleistocene sands, overlain by deposits of microbial mats, mangrove paleosols (buried soil horizons), and seagrass beds and shelly marine sands buried beneath a reversed sequence of mangrove soils and microbial mats, and capped by coastal sabkha deposits (Kenig et al. 1990). Observed in channel cuttings, buried soils can be tracked over several kilometers in extent. Examples of buried mangrove soils some 25 cm in thickness have been observed in Ras Ghanada, while in places more extensive buried microbial mat deposits have been logged up to 55 cm in thickness (Kenig et al. 1990). Evidence of such buried soils was observed in the present study.

23.3 ARID BLUE CARBON ECOSYSTEMS

Although plants in arid systems are adapted to survive under these conditions, they exist near the limit of their tolerance for extremes in temperature, rainfall, and salinity. Despite these potential limits (Noy-Meir 1973), low primary production does not necessarily prevent long-term accumulation of large soil carbon pools in blue carbon ecosystems, provided that soil carbon is preserved by development of anaerobic soil conditions. For example, mangrove forests that differ widely in plant biomass (e.g., 6.8–194.3 Mg ha^{-1}; Lovelock et al. 2005) nonetheless can form deep (7 m), organic (80% organic matter) soil profiles. Furthermore, low primary production in northern peatlands outweighs very slow decomposition rates over millennia, resulting in extremely carbon-rich soils despite that fact that plant production ranks near the bottom of terrestrial ecosystems (Frolking et al. 2001). On the other hand, labile carbon inputs are required to develop highly reducing anaerobic soil conditions, so there is a conceptual minimum rate of primary production required to accumulate significant soil carbon pools and plants in some arid regions might possibly fall below that rate (Megonigal et al. 2004).

A total of 58 sites were sampled across coastal UAE, with replication at each site: 18 natural mangroves; seven mangrove plantations; five salt marshes; five microbial mats; five coastal sabkha; and 18 seagrass beds (Figure 23.1). Intertidal sites were selected to represent a range of environmental settings (sheltered versus wave exposed), co-locating ecosystems when possible. Mangrove plantations were sampled to examine changes in carbon stock with age at Jubail (3-, 7- and 10-year-old sites) and Abu al Abyad (3-, 5-, 10-, and 15-year-old sites) Islands. These ecosystems are described below in order of decreasing tidal elevation (Figure 23.2).

Coastal sabkha ecosystems are found worldwide, but the largest expanses are in the UAE, extending 300 km long and more than 20 km inland in some places (Evans and Kirkham 2002). Coastal sabkha comprises the seaward part of the sabkha and typically floods several times per year during spring tides and when strong northerly winds drive seawater inland (Embabi 1993). They are hypersaline, inter-, or supratidal ecosystems largely devoid of vegetation (Kendall et al. 2002). Soils are unconsolidated carbonate sediments that are high in gypsum and anhydrite precipitates (Evans and Kirkham 2002).

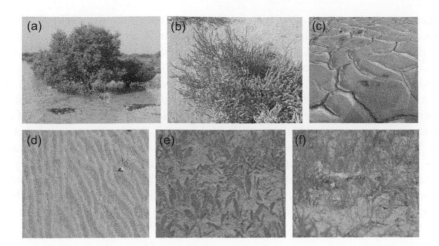

Figure 23.2   (See color insert following page 266.) Images of United Arab Emirate blue carbon ecosystems: (a) *Avicennia marina*, (b) *Arthrocnemum macrostachyum*, (c) microbial mat, (d) coastal sabkha, (e) *Halophila ovalis*, and (f) mix of *Halodule uninervis* (thin ribbons) and *Halophila stipulacea*.

Salt marshes are relatively limited in extent, occurring in patches along the fringe of sabkha, locally on sand veneers, and amongst higher intertidal areas of mangrove stands (Zahran and Al-Ansari 1999; Gul and Khan 2001; Khan et al. 2005). The salt marshes are almost exclusively dominated by the succulent, halophytic shrub *Arthrocnemum macrostachyum*. *Arthrocnemum macrostachyum* is found in tidal wetlands of arid and semi-arid climates in northeastern Africa, southeastern Asia, and the Mediterranean (Ghazanfar 1999; Serag 1999; Khan et al. 2005; Redondo-Gómez et al. 2010). Although *A. macrostachyum* is the dominant species, subdominant species also are found, such as *Halocnemum strobilaceum*, *Halopeplis perfoliata*, *Suaeda vermiculata*, *Salicornia europaea*, *Limonium axillare*, *Anabasis setifera* and *Salsola* spp. These species are typical of high salinity conditions and dryer, more aerated wetland soils.

Mangroves are found in scattered locations throughout the Emirate, particularly around the margins of lagoons and mud banks behind the barrier islands and on the outer islands. *Avicennia marina* is the only native mangrove species (Embabi 1993), though *Rhizophoraceae* was identified in charcoal fragments dating back to between 2,500 and 4,000 years ago (EAD 2007). Large areas of mangroves and other coastal ecosystems have been lost to recent and current coastal development (Embabi 1993), but mangrove planting has occurred since the 1960s, including revegetation along channels and degraded sites formerly occupied by mangroves (Saenger et al. 2004; Sheppard et al. 2010), with different aged stands co-ocurring. Recognizing the importance of mangroves, His Highness the late Sheikh Zayed Bin Sultan Al Nahyan initiated a program of mangrove planting to maintain and expand these forests. These mangrove plantation efforts provide a unique opportunity to investigate both rate of growth over time and differences across chronosequences. Currently, expansive engineering works involving excavation of coastal sabkha and microbial mats are ongoing on the mainland areas landward of Abu Al Abyad for the creation of mangrove plantations.

Along tidal margins of coastal sabkha where soils are consistently moist, large expanses of microbial mat are formed by stratified layers of cyanobacteria, colorless sulfur bacteria, purple sulfur bacteria, and sulfate-reducing bacteria (Kenig et al. 1990; Abed et al. 2007; Scherf and Rullkötter 2009). In sheltered locations, these organisms may form a thick 'leather-like' and moist mat, with a laminated fabric 5–20 cm in thickness, and can express different surface morphologies depending on location (Kendall and Skipwith 1968). Periodic storms bring sediments to the mats leading to layering of organic and non-organic sediment. Higher in the tidal frame with high evaporation, and in locations subject to more regular disturbance, the microbial film may only be a few millimeters in thickness, covering shelly sands (Kendall and Skipwith 1968). Globally, similar ecosystems are located near salt lakes and intertidal areas in Australia, the United States, and the Caribbean (Kendall and Skipwith 1968). Additionally, our study appears to be the first to quantify carbon stocks in microbial mat ecosystems.

Seagrass meadows are an extensive and important ecosystem in the Arabian Gulf. There are three species in the region, *Halodule uninervis, Halophila ovalis, Halophila stipulacea*. Although this represents a lower diversity compared to the eleven and seven species documented in the Red and Arabian Seas, respectively (Phillips 2003; Lipkin et al. 2003), the extent of this habitat is significant. Whereas only limited seagrass coverage is found in Kuwait and Iran, expansive areas of seagrass meadows are located between Qatar and the UAE. Within Abu Dhabi, an expansive complex of seagrass meadows extends around the islands and along the nearshore coastal plain. In sheltered locations, these meadows intermingle with microbial beds (*Hormophysa*). The seagrass beds support a commensurate population of dugongs (*Dugong dugon*) and green turtles.

23.4 STUDY DESIGN

23.4.1 Plant Carbon Pools

Plant carbon for mature mangrove trees were measured following methodology from Kauffman and Donato (2012). The goal was to use a standardized approach that previously had been implemented across multiple international studies. We modified the approach due to the UAE trees' smaller statue; all trees with stems >3 cm diameter at breast height (DBH; 1.3 m in height) were considered mature trees instead of restricting measurement to trees with DBH > 10 cm. At each mangrove site, carbon stocks were measured in six 7-m fixed radius circular plots placed 20 m apart along a 100 m transect. Trees taller than 1.3 m with a DBH < 3 cm were measured in a nested plot with a radius of 2 m. Seedlings, defined as individuals <1.3 m in height, were counted in the nested 2 m radius plot. Standing dead trees and downed woody debris were found at some Gulf of Oman sites and pools were quantified appropriately (Kauffman and Donato 2012). In the planted mangrove sites, five 2-m radius plots were established at 10 m intervals along a 40 m transect. When stands contained individuals <1.3 m tall, we measured the crown diameter and main stem diameter at 30–50 cm in height. In the 3-, 5-, and 10-year-old planted mangrove sites in Abu al Abyad, trees were planted in an evenly spaced grid; therefore, the plant density was calculated by measuring the average plant spacing and main stem and crown diameter of 50–75 trees.

Salt marsh transect length and plot spacing were the same as with the mature mangroves, although the plot radius ranged from 1 to 4 m depending on plant density. The height and elliptical crown area (perpendicular crown widths centered on the canopy) were measured on every plant rooted in each plot.

At each of the 18 seagrass survey locations, dive teams assessed seagrass cover along a 50 m transect and collected a shallow, large-diameter sediment core for quantifying seagrass biomass. The biomass samples were separated by species, and *H. uninervis* samples were further separated into above- and belowground components to allow comparison with other measures of abundance of this species throughout its global distribution.

Biomass for *A. marina* and *A. macrostachyum* were calculated using allometric equations (Schile et al. 2017). Global tree carbon percentages of 48% and 39% for above- and belowground biomass, respectively, were applied (Kauffman and Donato 2012). To examine differences in average annual carbon sequestered in planted mangrove trees, we divided total biomass for each stand by the number of years since plantation establishment. For *A. macrostachyum*, we calculated carbon content for aboveground biomass as $40.3 \pm 1.4\%C$ (Schile et al. 2017).

23.4.2 Soil Carbon Pools

At mangrove and salt marsh plots, undisturbed soil samples were collected following methodology from Kauffman and Donato (2012) using a 1-m long gouge auger with an open-face, semicylindrical chamber of 5.1 cm radius. Soils were cored to 3 m or until coarse marine sands or coral rubble representing the parent material was encountered. The soil core was divided into depth intervals of 0–15, 15–30, 30–50, 50–100, and >100 cm, or until refusal. Subsamples collected from center of each interval were analyzed for bulk density (dry mass per unit volume) and carbon concentration (organic and inorganic). If encountered, unique soil layers were sampled separately. The same soil sampling methodology was used within microbial mats and coastal sabkha; the number of plots sampled per transect varied from 3 to 6 plots spaced at 20 m intervals along a transect. Forty soil cores were collected from 18 distinct seagrass meadows. Soil cores were collected (in duplicate at most sites, in triplicate at a few sites) by driving a diver-operated piston core into the soils until a depth of 1 m or refusal was reached. The cores were subsampled at 3–9 cm intervals for the determination of dry bulk density, loss on ignition, and organic carbon content. We determined soil carbon stocks following methods outlined in Fourqurean et al. (2012a), which are designed to account for soils containing carbonates.

A variety of soil biogeochemical measurements, soil respiration, elevation, and tidal data were collected in selected mangroves, salt marsh, microbial mats, and sabkha to help characterize soil conditions. Redox potential (E_h) was estimated at each site (Megonigal and Rabenhorst 2013). At sites that had a shallow water table, soil pore-water was collected from corer boreholes at 5–10 cm below the surface and analyzed by the standard methods described in Keller et al. (2009). Pore-water measurements included salinity, pH, dissolved methane (CH_4), and concentrations of SO_4^{2-} and Cl^-. We quantified instantaneous CO_2 gas exchange rates in intertidal ecosystems to assist with ecosystem comparisons; more measurements were possible at higher respiration rates. In seagrass beds, water temperature, salinity, and water depth were recorded.

23.5 FINDINGS

Blue carbon ecosystems of a large arid region in which the dominant autotrophic inputs ranged from trees to mat-forming microbial assemblages exhibited a wide range in carbon stocks that are generally consistent with patterns expected from our understanding of humid and semi-humid systems. Natural mangroves had the largest plant and largest soil carbon pools, while seagrass beds had the lowest (Figure 23.2). Despite highly diverse carbon inputs in terms of both carbon quantity and quality, average soil carbon pools across intertidal ecosystems varied over a relatively narrow range of 80–156 Mg C ha^{-1}, and all were significantly larger than seagrass soil pools of 49 Mg C ha^{-1}. The largest soil carbon pools were ~50% higher in mature mangroves than coastal sabkha, which completely lacked plants (Figure 23.2). Despite having an autotrophic source of carbon input, seagrass bed soils contained a third of the carbon found in mangroves. Carbon sequestration in wood biomass accounted for 1–55% of total mangrove ecosystem carbon pools, compared to 2–51% in non-arid mangrove ecosystems (Donato et al. 2011) and 1–6% in UAE salt marshes. About 5% of salt marsh carbon was in plant biomass, compared to almost 30% in mangroves. Thus, trees are particularly important for carbon sequestration in this arid region.

Biomass and stand characteristics of natural mangroves varied widely across sites (Schile et al. 2017). Spatial variation in UAE mangrove carbon pools also was large, varying 10- and 20-fold in soil and tree carbon pools, respectively. Mangrove stands along the northern Arabian Gulf and the Gulf of Oman had the lowest salinity and supported larger trees, denser forests, deeper soils, and larger carbon pools than stands on the Arabian Gulf. Mangroves were the only ecosystem type to develop an organic-rich surface horizon, but this horizon was poorly developed compared to non-arid systems, ranging in thickness from <1 to 20 cm and containing 2–29% organic carbon in soil surface horizons. Factors that could favor higher soil carbon accumulation rates in some mangrove stands include: (i) longer periods of soil saturation due to lower elevations; (ii) higher vascular plant production rates, which increase carbon inputs and decrease decomposition by favoring microbial O_2 consumption and reduced soil conditions; and (iii) producing more recalcitrant plant litter that is more resistant to decay than seagrass plant litter. Understanding the relative contributions of these factors to regulating soil carbon sequestration rates is important because mangroves are regional hotspots of soil carbon, and have the most potential for significant carbon management. Additional factors such as landscape context (lagoonal versus wind/wave exposed), intertidal slope, tidal flushing, freshwater input, and nutrient availability all could affect plant production, resiliency, and carbon pools within this patchy coastal ecosystem.

Salt marshes dominated by *A. macrostachyum* had the second highest plant carbon pools and did not vary widely across sites, but they were nonetheless 20-fold lower than in mature mangroves and similar in size to planted mangroves (Schile et al. 2017). Salt marsh soil carbon pools were comparable to sabkha ecosystems, despite the absence of vascular plants. Both systems are relatively high in tidal elevations, which do not favor the low redox, anaerobic soil condition required to preserve carbon.

The relationship between dry bulk density (DBD) and the proportion of soil organic carbon followed a curve that has been reported by many in other wetland ecosystems (Figure 23.3). There is an inverse relationship between DBD and soil organic carbon, although the majority of points were

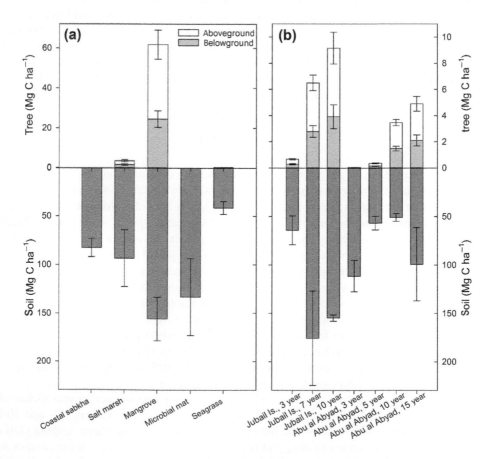

Figure 23.3 Average total carbon stock in Mg C ha⁻¹ of (a) each blue carbon ecosystem and (b) chronosequences of planted mangroves in Abu Dhabi (error bars = ±1 SE).

characterized by low organic carbon content and high DBD. DBD was the greatest when soil organic carbon is lowest and dropped off exponentially as the proportion of organic carbon increased. Mature mangroves were the only sites with soil organic carbon values greater than 10% (Figure 23.3).

An intriguing and somewhat unexpected aspect of this project was the high soil carbon storage potential of microbial mats, with organic carbon values ranging from 0.2% to 7.8%. These microorganisms are embedded in a matrix of extracellular polymeric substances (EPS). EPS are composed mostly of polysaccharides that constitute 50%–90% of mat carbon (Vu et al. 2009), and are transformed by heterotrophic bacteria from labile to refractory forms that accumulate as sediment organic matter (Braissant et al. 2009). Biofilms enhance organic matter preservation by increasing water holding capacity (Bhaskar and Bhosle 2005) and decreasing water infiltration rates (Vandevivere and Baveye 1992). Evidences that these features preserve sediment organic matter are reports that such ancient microbial mats contributed to the substantial production of fossil fuels in the region (Cardoso et al. 1978). We demonstrate that microbial mats store carbon at rates comparable to vegetated intertidal ecosystems, and argue that they meet the criteria of a blue carbon ecosystem. To our knowledge there are no formal policies that recognize ecosystem services that these systems may provide.

Carbon stocks in sabkha were entirely in soil pools and fell within the range of most mangrove and salt marsh sites. The fact that soil organic carbon was at approximately 1% of total soil weight suggests that allochthonous deposition of carbon is the source of recalcitrant soil carbon in these hyper-arid blue carbon ecosystems. Recalcitrant forms of plant tissue or deposition of black carbon (particulate matter caused by the incomplete combustion of fossil fuels or biomass) are potential sources (Schmidt and Noack 2000). We also observed buried organic-rich (9% organic carbon) soil surfaces up to 10 cm thick, indicating that rapid transitions to less-frequently inundated ecosystem types have occurred and remained in the soil (Evans et al. 1973; Kendall et al. 2002).

Carbon stored in the seagrass biomass of Abu Dhabi seagrass beds was relatively modest. There was no significant relationship between the abundance of seagrasses, as assessed either as percent cover or living plant biomass, with soil carbon stores (Campbell et al. 2015). This is likely a result of the very broad area of subtidal soils in the coastal zone that support ephemeral seagrass patches. The low soil carbon storage in our study resulted from a combination of both limited soil thickness and relatively low soil organic carbon concentrations, and our data suggest that seagrass carbon storage may be strongly influenced by site depositional and geomorphological characteristics that are independent of seagrass cover. We argue that research targeting poorly represented coastal habitats will likely highlight the need to conserve and restore these environments as viable strategies for the mitigation of anthropogenic carbon emissions. Thus, critical attention should be placed on preserving this important blue carbon ecosystem and preventing the negative impacts of development and oil exploration (Campbell et al. 2015).

23.6 GLOBAL COMPARISONS

Plant carbon stocks in mangrove, marsh, and seagrass ecosystems of this arid region were low, but not markedly so, compared to similar ecosystems in other climatic regions of the world (Schile et al. 2017). Most natural mangrove plant carbon stocks in the UAE were comparable to short stature mangroves from the Caribbean and Indopacific, with the exception of the two northern/Gulf of Oman sites that were more similar to medium stature stands in the Caribbean and Indopacific. Compared to mangrove soil carbon stocks in arid Baja California, Mexico, UAE mangrove carbon stocks were approximately half (Ezcurra et al. 2016). Salt marsh aboveground organic carbon was lower than *A. macrostachyum* (i.e., shrub-dominated) salt marshes in the Ebre delta, Spain (Curcó et al. 2002) and in Portugal (Neves et al. 2010), and also lower than herbaceous marshes in the northwest Atlantic coast of Canada, where cold temperatures limit growth (Frolking et al. 2001).

Salt marsh soil carbon stocks in the UAE were the lowest out of all the global data compiled (Curcó et al. 2002; Neves et al. 2010; Sifleet et al. 2011). Carbon stock data, particularly soil stocks, are sparse for salt marshes generally, and even more so for marshes dominated by woody species such as those in the UAE. This highlights a suprising need for more deep (1 m minimum) soil carbon data in tidal marshes. Compared to other seagrass systems for which C stocks have been estimated worldwide, lower carbon storage in Arabian Gulf beds was not unexpected considering the relatively small biomass of these particular seagrass species, which all share a small stature and short life span (Fourqurean et al. 2012b). However, the low biomass and small size of the UAE seagrasses cannot be explained completely as a function of the arid nature of the region, as dense beds of relatively large-bodied seagrasses are found along the shore of deserts in North Africa (Sghaier et al 2013) and Australia (Fourqurean et al. 2012b).

Although UAE plant biomass was moderate to low compared to other climatic regions, soil carbon stocks fell predominantly within the 25% quantile (245.3 Mg C ha^{-1}) of other sites globally. Two sites, Ras Al Kaimah and Khor Khalba South, however, fell within the lower end of the 50% quantile (560.7 Mg C ha^{-1}). Soil bulk density below the peat layers averaged 1.0 g m^{-3}, which is significantly larger than averages reported for semi-humid stands in Mexico (0.50 ± 0.06 g m^{-3}; Adame et al. 2013) and the Dominican Republic (0.39 ± 0.04 g m^{-3}; Kauffman et al. 2014). The deepest mangrove peat layers (if present) were approximately 15 cm thick, pools that are very small compared to mangroves in humid, tropical climates (Donato et al. 2012). Both salt marsh and seagrass soil carbon stocks were at the 1% quantile (Schile et al. 2017)). If blue carbon ecosystem soils preserve a constant fraction of the net primary production, as some carbon accretion models assume (Morris et al. 2012), then UAE soil carbon pools may have been expected to be higher than observed, falling within the 25%–50% quantile of global comparisons. These observations suggest that preservation of wetland plant biomass is generally lower in the UAE than in similar ecosystems elsewhere.

23.7 RESTORATION POTENTIAL

Biomass of planted mangroves increased linearly with stand age within the two regions, yet at a faster rate on Jubail Island (Schile et al. 2017). Stands on Jubail Island had greater biomass than those on Abu Al Abyad across all ages; the 3-year-old plantation biomass was 16 times as large and 10-year-old plantation biomass was twice as large. Yearly sequestration in total plant biomass was four times greater in the oldest than the youngest stands, but did not differ between the 7- and 10-year stands at Jubail Island or the 10- and 15-year stands at Abu al Abyad (Figure 23.2).

Mangrove ecosystem creation and restoration are growing practices in Abu Dhabi. We found that UAE mangrove ecosystems are critical for carbon sequestration compared to other blue carbon ecosystems, and report evidence that mangrove creation project designs may not consistently promote the successful establishment of mature mangrove forests. We found striking differences between two created mangrove chronosequences that are consistent with the tenet that flooding stress limits tree growth (Megonigal et al. 1997). Stands at Abu al Abyad were planted ~50 cm lower in the tidal frame than on Jubail Island, resulting in significantly longer hydroperiods and reduced growth up to 15-fold (Schile et al. 2017). Indeed, many of the seedlings at Abu al Abyad were covered in barnacles. Future planting and restoration efforts will require research to establish the optimum tidal elevations required to minimize inundation stress and maximize growth and carbon storage in this arid environment (Crooks et al. 2014). Excavation of coastal sabkha and microbial mats for the purpose of mangrove planting may not result in net greenhouse gas benefits. Thus, conserving natural mangroves is a more effective means to protect carbon stocks that restoring mangrove.

23.8 POTENTIAL FACTORS LIMITING CARBON STORAGE

Managing coastal ecosystems for carbon sequestration services in arid environments requires identifying the factors that account for their relatively small soil carbon pools. Coastal wetlands in the UAE could have been expected to support larger soil carbon stocks based on climatic factors alone; indeed, large soil carbon pools occur in comparable ecosystems elsewhere under conditions that suppress plant production (e.g., nutrient limitation) or enhance decomposition (e.g., tropical climates) (McKee et al. 2007). The interactions that control organic matter preservation are complex and may operate relatively inefficiently in these systems due to several factors. The soils in coastal UAE were often coarse-textured, carbon-poor, and weakly reducing. Course-textured soils generally have high water infiltration rates that allow hypoxic porewater (if present) to be rapidly replaced by O_2-rich tidal water or air. Given the circumneutral soil pH, E_h values >200 on many of the sites suggest that microbial activity is dominated by aerobic respiration (Megonigal and Rabenhorst 2013), with anaerobic respiration restricted to microsites. The O_2 consumption rate in these soils may also be limited by the low supply of labile organic carbon required for microbial respiration. Collectively, these traits are expected to inhibit development of the highly reducing, anoxic conditions found in sites with large soil carbon pools, and to favor nearly complete oxidation of organic carbon to CO_2 (Kenig et al. 1990). Indeed, soil organic matter concentrations in some study sites averaged 1% across all soil depths (Figure 23.4), which is comparable to upland forests and deserts below the A horizon. The relatively high soil carbon content of some microbial mats (up to 7.8% organic carbon) is evidence that that microbial biofilms can alter the redox environment of these course-textured soils, presumably by increasing the length of time a soil remains saturated (Vandevivere and Baveye 1992).

The highly dynamic and low-gradient UAE coastline is very sensitive to small changes in relative sea level (Kendall et al. 2002) that may prevent plant carbon inputs from persisting in one location long enough to accumulate large soil carbon pools (Craft et al. 2003). We observed buried organic-rich soil surfaces up to 10 cm thick, indicating rapid transitions to less-frequently inundated

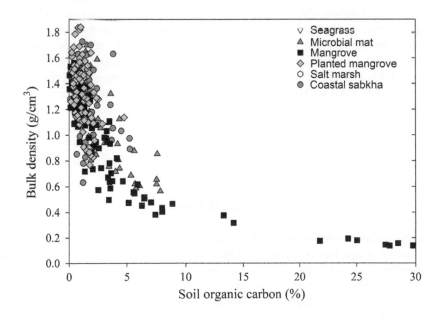

Figure 23.4   Bulk density versus organic carbon curve by blue carbon ecosystem in Abu Dhabi.

ecosystem types (Evans et al. 1973; Kendall et al. 2002). The ecosystems with the largest soil carbon pools—mangroves and microbial mats—had enriched soil carbon profiles < 30 cm deep, which contrasts with many comparable ecosystems in humid and semi-humid climates where soil profiles develop in the same location over thousands of years (Drexler et al. 2009).

23.9 MANAGEMENT APPLICATIONS AND POLICY IMPLICATIONS

Interest in mangrove, marsh, and seagrass carbon stocks and sequestration rates has grown rapidly since the publication of *Blue Carbon. A Rapid Response Assessment* (Nellemann et al. 2009), which highlighted the fact that these ecosystems are hot spots for carbon storage when compared to other terrestrial, aquatic, and marine ecosystems. This is the first study to recognize that other coastal ecosystems—sabkha and microbial mats—also support carbon stocks, and may therefore be important features in arid coastal landscapes. These data enabled identification of Ecologically or Biologically Significant Marine Areas (EBSA) under criteria developed by the Convention on Biological Diversity (CBD), and contributed toward RAMSAR designation for coastal sabkha as a distinct intertidal ecosystem because of its carbon storage, with potential application to other coastal blue carbon ecosystems.

Blue Carbon has been integrated into several UAE policy and planning documents including the Maritime Spatial Plan led by the Urban Planning Council (UPC), the National Biodiversity Strategic Action Plan (NBSAP), the Intended Nationally Determined Contributions (INDC) for the UAE as well as draft UAE and Abu Dhabi Climate Change Adaptation Plans. At the broadest level, two policy assessments were carried out for Abu Dhabi and the UAE with the purpose to build blue carbon ecosystem conservation into the daily business within the UAE, within the broader framework of environmental management. The policy assessments seek to provide those responsible for administering relevant activities with clear suggestions and policy direction to ensure that blue carbon ecosystems as well as associated ecosystems are managed in a way that accounts for their local and global environmental benefits. Through the stakeholders driven approach they identified five key components, including improving information management systems, sustainably managing blue carbon ecosystems, enhancing institutional coordination, engaging in and supporting international actions on blue carbon, and promoting public awareness of blue carbon ecosystem benefits.

The carbon values determined by this study form the basis of country-specific carbon stock values to aid the UAE inventory compilers in their calculations of carbon stock changes with management activities. The UAE is also one of the first countries to test and release the application of the 2013 Supplement to the 2006 IPCC Guidelines for National Greenhouse Gas Inventories for Wetlands for its GHG Inventory at the Abu Dhabi level (Hiraishi et al. 2014), providing feedback to the United Nations Framework Convention on Climate Change on the inclusion of wetlands within the national greenhouse gas inventories of other countries.

23.10 SUMMARY

In summary, carbon stocks of the coastal zones of Abu Dhabi are likely the highest stocks in the Emirate, and these coastal ecosystems provide a variety of services; climate change mitigation may be a subordinate service relative other functions such as supporting fish populations, biodiversity, and cultural values. Microbial mats have been identified as a potentially new blue carbon ecosystem, one specific to arid, high-salinity environments. Coastal sabkha has been identified as associated blue carbon ecosystems. Carbon stock levels are small compared with many regions of the world, a reflection of the arid climate and temperature range. As such, carbon stocks within Abu Dhabi blue carbon ecosystems fall below the range of those typically of interest to carbon finance framework, though blue carbon and associated ecosystem services have been highly included into conservation, management, and planning activities.

Across the landscape, based upon available survey maps, seagrass meadows hold the greatest quantity of carbon compared to other ecosystems; 7.9×10^6 Mg for mangroves 1.3×10^6 Mg for planted mangroves, 0.5×10^6 Mg for microbial mats, and salt marsh at 0.4×10^6 Mg. The extent of intertidal coastal sabkha is unknown. The total area, and hence the total carbon stock of seagrass meadows is unknown. Currently seagrass is mapped to 3 m below sea level but diver observations found seagrass to be widespread to depths of 14 m or more.

While relatively low rates of net primary productivity are likely due to low nutrient availability and plant stress, we hypothesize that small soil carbon pools reflect low carbon preservation due to weakly reducing soil conditions, which in turn reflect coarse soil texture, high hydraulic conductivity, and low moisture-holding capacity. The time period over which carbon has been accumulating may also explain the relatively small carbon pools. Currently, we do not have sufficient data to disentangle these factors. As in other regions of the world, blue carbon ecosystems of the UAE are the most carbon dense features of the landscape, providing ecosystem services other functions such as supporting fish, sea turtle, and dugong populations, biodiversity, and cultural values.

CHAPTER 24

Mikoko Pamoja

A Demonstrably Effective Community-Based Blue Carbon Project in Kenya

James G. Kairo and A. J. Hamza
Kenya Marine and Fisheries Research Institute

C. Wanjiru
Kenya Marine and Fisheries Research Institute
Kenyatta University

CONTENTS

HIGHLIGHTS

1. Mangrove and associated blue carbon ecosystems are highly productive ecosystems with potentials to mitigate climate change through carbon capture and storage
2. Mangroves in Kenya are threatened by human-induced stresses ranging from over exploitation of resources, conversion pressure, and pollution leading to a 18% loss in the forest from 1985 to 2010
3. Incentive-based schemes, such as Payment for Ecosystem Services, can help reverse these threats
4. Mikoko Pamoja, is the first community-type project in the world to restore and protect mangroves through sales of carbon credits
5. The project is regulating climate, helping communities, and conserving biodiversity

24.1 INTRODUCTION

Mangrove forests in Kenya provide a range of ecosystem goods and services. These **blue carbon ecosystems** are important nurseries and breeding grounds for many varieties of fish and other wildlife (Kimani et al. 1996; Huxham et al. 2004; Mirera & Moksnes 2015). They also play a key role in combating effects of rising sea levels, coastal erosion, and flooding from storm surges and tsunamis (Huxham et al. 2015). As an important source of renewable resources—notably fisheries and wood products—mangroves are important in coastal development (Kairo et al. 2001).

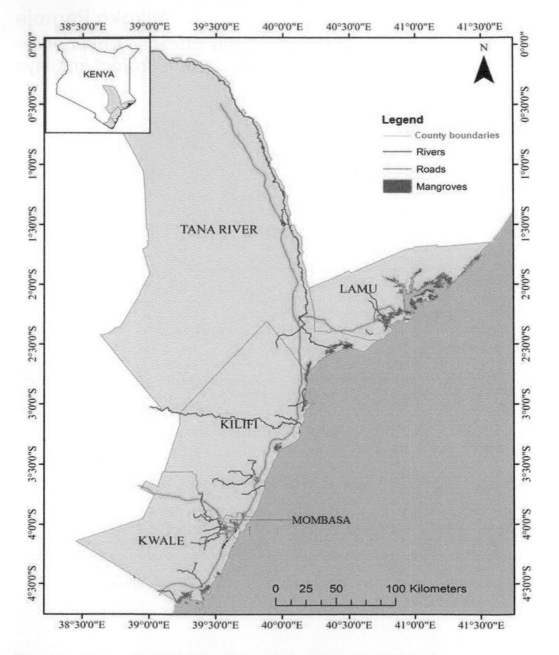

Figure 24.1   Mangrove distribution within the five counties along the coastal strip in Kenya.

There are 60,323 ha of mangrove forests distributed all along the 536 km Kenyan coastline as shown in Figure 24.1. These forests represent approximately 3% of the natural forest cover or less than 1% of the national land area in Kenya. As shown in Table 24.1 below, a large part of mangroves in Kenya (about 61%) occur in Lamu county, followed by Kwale (14), Kilifi (14), Mombasa (6), and Tana River (5) Counties.

There are nine mangrove species in Kenya as shown in Table 24.2 below. Two of the species, *Rhizophora mucronata* and *Ceriops tagal* are the most dominant and are represented in almost all mangrove formations. The rare species are *Heritiera littoralis* and *Xylocarpus moluccensis*. Major use of mangrove wood products include firewood, building poles, fencing and furniture.

Naturally growing mangroves exhibits horizontal distribution of species (or zonation). This is greatly influenced by levels of inundation, geomorphology, and the salinity (Tomlinson 1986). A typical zonation of mangrove in Kenya starts with *Sonneratia alba* on the seaward margin, followed by large *Avicennia* and *Rhizophora mucronata*. In the creeks, *Rhizophora-Avicennia* mix is the most dominant. *Avicennia* expresses a double zonation but mostly found in the landward side Knowledge of mangrove distribution across the intertidal area is important in their management.

The diversity of fauna within mangroves is high due to ample food resources and a wide range of microhabitats in the system, such as soil surface, permanent and temporary tidal pools, tree roots, trunks, and canopies (Bosire et al. 2016). In Kenya, these animals are represented by different phyla, ranging from protozoa and nematodes to molluscs, insects, crustaceans, birds, fish, and mammals. The main groups are molluscs, crustaceans, fish, and birds. Common groups of birds occurring in mangrove areas are: wading birds (herons, egrets, ibises), shore birds (plovers, sandpipers), floating, and diving birds (pelicans, cormorants, terns, gulls, kingfishers), birds of prey (fish eagle, osprey) and arboreal birds (bee-eaters, sunbirds). Further, mangrove forests receive thousands of migratory birds during winters every year (Huxham 2013).

Table 24.1 Main Mangrove Areas per County along the Kenya Coast

County	Forested Mangrove Area (ha)	%	Major Mangrove Areas in the County
Lamu	37,350	61	Northern Swamps, North-central swamps, Southern Swamps, Mongoni and Dodori Creek, Pate Island
Tana River	3,260	5	Kipini and Mto Tana
Kilifi	8,536	14	Ngomeni, Mida, Kilifi, Mtwapa
Mombasa	3,771	6	Tudor and Port Ritz creeks
Kwale	8,354	14	Gazi, Funzi, Vanga
Total	61,271	100	

Source: GoK (2017).

Table 24.2 Mangrove Species Found in Kenya and Their Uses

Species	Local Name	Main Use
Rhizophora mucronata	Mkoko	Poles, dye, firewood, fencing, charcoal
Bruguiera gymnorhiza	Muia	Poles, firewood, charcoal
Ceriops tagal	Mkandaa	Poles, firewood, charcoal
Sonneratia alba	Mlilana	Boat ribs, poles, firewood
Avicennia marina	Mchu	Firewood, poles
Lumnitzera racemosa	Kikandaa	Fencing poles, firewood
Xylocarpus granatum	Mkomafi	Furniture, poles, firewood
Xylocarpus moluccensis	Mkomafi dume	Fencing poles, firewood
Heritiera littoralis	Msikundazi	Timber, poles, boat mast

The principal groups of fish and crustacean associated with mangroves of Kenya are snappers, groupers, rabbit fish, grant, milkfish, mullet, terapons, carangids, shrimp, crabs, and oysters (Kimani et al. 1996; Mirera et al. 2010). The high biomass of fish, molluscs, and crabs that mangroves support has significant economic value to artisanal and commercial fisheries.

24.2 THREATS TO MANGROVES IN KENYA

In Kenya, mangroves are being lost and degraded due to a combination of human and natural factors, ranging from over-harvesting of wood products to conversion of mangrove land to other land uses, particularly for agriculture, pond aquaculture, and infrastructure development (Abuodha & Kairo 2001; Kirui et al. 2013). Conditions are worse in peri-urban areas, such as Mombasa, where mangroves are being cleared to pave ways for infrastructure and human settlement (Bosire et al. 2014). Less than half of the original mangrove forests in Kenya remain, and the current rate of loss (about 0.7% per year) is a major cause of concern (Kirui et al. 2013).

Climate change effects such as sea level rise, increased rainfall, and storm surges are expected to negatively impact the remaining mangrove areas in Kenya (Kairo & Bosire, 2016). Site-based responses to climate change effects such as the construction of hard civil structures are likely to exacerbate these effects. Of the predicted impacts to occur due to climate change, sea level rise is perhaps the greatest threat to the ecological integrity of mangroves and associated biological resources (Gilman et al. 2008). In addition to shortages of harvestable wood products, declines in fisheries and increased shoreline erosion, destruction of mangrove forests in Kenya release huge quantities of stored carbon into the atmosphere, contributing to global warming and other climate change trends (Lang'at et al. 2014). Fortunately, compensation for conservation and restoration can potentially help reverse these trends.

24.3 MARKETING MANGROVE SERVICES

Currently, the value of mangrove ecosystems in Kenya is captured mostly for provisioning services, such as wood products, capture fisheries, and some value-added cultural services such as ecotourism (Kairo et al. 2009). There is still a big gap in approaches for capturing the value of much of the regulating, supporting, and cultural services provided by mangroves. Payments for ecosystem service (PES) are emerging resource management tools that provide incentives for behavioral changes to increase the provision of ecosystem services, e.g., by discouraging losses and degradation of forests (Locatelli et al. 2014).

The Government of Kenya is pursuing market-based approaches to environmental protection, with a strategic focus on Ecosystem Services (ES), including biodiversity, carbon sequestration, food provision, recreation, and shoreline protection (GoK 2017). Specifically, the country's conservation strategy is to "identify the benefits of environmental services and to seek a system where beneficiaries of such services pay service providers." PES schemes are attractive because they reward those that supply or provide ES. However, the potential of forestry based PES schemes is hugely untapped in Kenya. Through the UK's Ecosystem Service for Poverty Alleviation (ESPA) funded projects in Kenya (www.espa.ac.uk), experience has been gained in facilitating the development and implementation of small-scale mangrove PES projects, with carbon credits supporting community development and mangrove conservation at Gazi Bay. This work led to the establishment of **Mikoko Pamoja**—the world's first community-based mangrove project funded by carbon credits (www.planvivo.org/project-network/mikoko-pamoja-kenya/)—Project timelines and achievements of Mikoko Pamoja are illustrated in Figure 24.2.

Mikoko Pamoja is being executed in Gazi Bay area of the southern coast of Kenya, about 55 km south of Mombasa in Kwale County. Figure 24.3 shows the project area. The bay is bordered by 620 ha

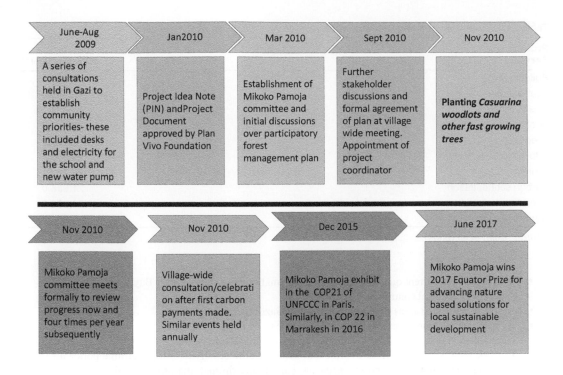

Figure 24.2   Mikoko Pamoja timeline and achievements.

of mangrove forests, which are heavily used by local people as fishing ground and source of wood for building and fuel (Dahdouh-Guebas et al. 2000). These forests have been extensively used and degraded—with large areas clear-felled in the 1970s and have not naturally recovered to-date. Selective logging of mangroves for commercial and subsistence use are still continuing in some stands and forest degradation continues through illegal harvesting of firewood and building poles. This has resulted in shortages of building poles and firewood, decreased fishery resources, and increased coastal erosion.

24.4 SOCIAL ECONOMIC CHARACTERISTICS OF GAZI BAY

There are approximately 6,000 residents in the two villages adjacent the bay, Gazi and Makongeni. The main ethnic group is the Swahili, while the Mijikenda of Bantu origin is the second commonest. Livelihoods are provided predominantly by fishing, farming, and tourism. Some 25% of households receive remittances from kin outside of the area, and around one-third of people are recent immigrants, who have come mostly to exploit the reef-based fisheries. Other sources of employment are provided by titanium mining and sugar cane farming companies. Historically, local communities have exploited mangrove forests for firewood, construction poles, and medicine. The net benefits of mangroves in Kenya have been estimated to have a value of over US$2,500 per hectare every year. Shoreline protection functions of mangroves is allocated the highest value when compared to other habitat functions provided by the ecosystem as demonstrated in Table 24.3. It is therefore vital that mangroves are protected for their benefits to community and the ecosystem functioning.

According to the Project Design Document (PDD) submitted to Plan Vivo (Huxham 2010), the overall objective of **Mikoko Pamoja** is to channel finance for the protection and restoration of mangrove ecosystems through the provision of and payments for quantifiable ecosystem services (PES). Specific objectives are:

Table 24.3 Valuation of Mangrove Ecosystem in Kenya

Product and Services	KES ha⁻¹year⁻¹
Building poles	30,659.5
Fuel wood	4,505.0
Onsite fisheries	9,612.7
Beekeeping	1,249.5
Integrated aquaculture	408.0
Education & Research	65,469.6
Tourism	782.0
Carbon sequestration	21,896.0
Shoreline protection	134,866.1
Total	269,448.3

Source: Kairo et al. (2009).

- To preserve the current quality and extent of the mangrove forests of Gazi Bay and of the services they provide to local communities
- To restore degraded areas of mangrove forest in Gazi Bay
- To raise income from forest resources, including carbon credits, for community benefit
- To establish alternative sources of timber and firewood in the Gazi area
- To work with the Kenya Forest Service and other government agencies to determine policy about engaging communities in land management, particularly through the provision of ecosystem services through international carbon offset markets

24.5 ELIGIBLE ACTIVITIES FOR CARBON FINANCING

Mikoko Pamoja activities are implemented through three distinct and interlinked project activities in Gazi Bay as shown in Figure 24.3 and Table 24.4.

- **Activity 1: Avoided deforestation and forest restoration.** This involves protection of existing natural *Rhizophora mucronata* forest over an area of 107 ha. The area has previously suffered from deforestation and forest degradation.
- **Activity 2: Reforestation and forest protection**. This has involved establishment of 10.0 ha of *Rhizophora mucronata* stand in formerly deforested area.
- **Activity 3: Reforestation and forest protection**. Replanting of a *Sonneratia alba* fringing forest of 40–70 m depth and 800 m length, along a wave-exposed beach. Mangrove wood was originally removed from parts of the area for industrial use, leaving open areas of sand, which have not regenerated naturally, and exposing adjacent agricultural field to erosion. As part of community commitments in the project, replanting of about 4,000 seedlings is undertaken in a succession of planting areas of about 0.4 ha every year. Cumulatively, over the 20 years contracting period, the activity would have replanted a total of 8 ha of mangroves.

24.6 CARBON BASELINE

Mangrove forests develop and sustain above- and belowground carbon pools. The latter constitutes a very long-term sink, with large amounts of carbon held in peat (Gress et al. 2017). Gazi Bay is among the best-studied mangrove system in the world (Kairo 2001), and there is detailed information on above- and belowground biomass carbon for different forest types (Tamooh et al. 2008;

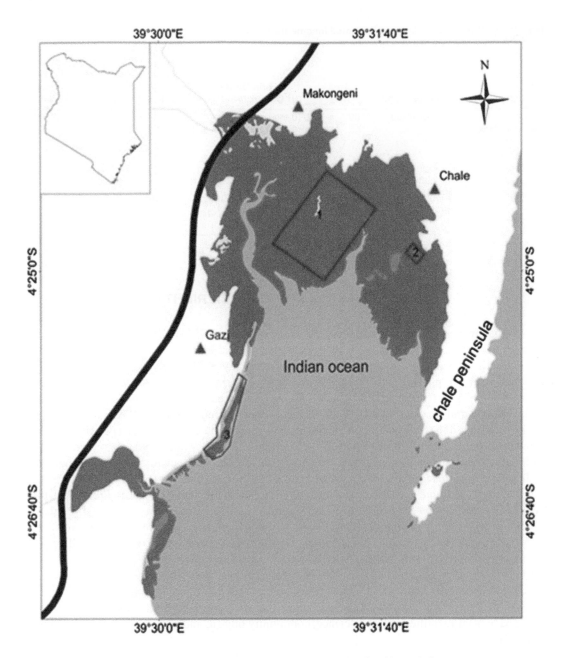

Figure 24.3 Map of Gazi showing Mikoko Pamoja activity areas within lined boundaries.

Cohen et al. 2013). The aboveground biomass of mangroves in Gazi is estimated at 250 t ha^{-1}, although this varies depending on age, species, and location in the intertidal area. Belowground biomass (to 60 cm depth) varies from 7.5 to 75 t ha^{-1}. This equates to approximately 155,000 t aboveground and 23,250 t belowground (178,250 t total) in the Gazi area. In addition, mangroves sequester around 1.5 t C ha^{-1} year^{-1} in accretion of new sediment, and approximately 5 t C ha^{-1} year^{-1} in new biomass (for a mature forest—rates are higher for young forests and plantations).

Table 24.4   Carbon Benefits and Projected Income from Mikoko Pamoja

Activity	Forest Type	Area (ha)	Carbon Benefit (t CO_2ha^{-1}year^{-1})	Total Annual Project Carbon Benefit From Initial Activities (t CO_2year^{-1})	Income ($)[a]
Avoided deforestation	Natural mixed	100	18 (based on mature forest, so conservative given this is a recovering forest)	1,800	10,800
Reforestation	*Rhizophora* plantation	7	29 (based on 12-year-old plantation)	203	1,218
Reforestation	New plantation (*Sonneratia*)	5 (after 5 years)	Four (but increasing to ~10 after 10 years)	20	120
			Total	2,023 tCO_2	$12,138

[a] Assumes a conservative price of $6 tonne^{-1} CO_2.

24.7 OWNERSHIP OF CARBON RIGHTS

All mangroves in Kenya are recognized under statutory laws as government reserve forests. Management of these forests is vested with the Kenya Forest Serve (KFS), either alone, or in partnership with Kenya Wildlife Service (KWS) whenever they occur within marine protected areas (MPA). Under the provisions of Forest Conservation and Management Act (2016), Community Forest Associations (CFAs) are designated specific forest areas that they could co-manage for the desired goods and services. Mikoko Pamoja is managed through GOGACOFA, a local CFA that has been involved in the development of participatory forest management plan (PFMP) for forests within and adjacent to Gazi Bay, and in the signing of Forest Management Agreement (FMA) with KFS. The management agreement enables Mikoko Pamoja to engage in sale of carbon credits from the designated mangrove area. Mikoko Pamoja is verified under the Plan Vivo System and Standards, a framework for supporting communities to manage their natural resources more sustainably with a view to generating climate, community, and biodiversity benefits through payments for environmental services; in this case carbon. Income from carbon credits, worth over US$12,500 each year, is used to fund continued mangrove conservation activities as well as priority projects chosen by communities, such as water and sanitation, health, and education (Abdalla et al. 2015). Communication of the value of blue carbon in both economic terms and in terms of ecosystem services provided by mangrove habitats was key to community uptake of Mikoko Pamoja. The building blocks of Mikoko Pamoja, which could be replicated to other mangrove areas, have been identified as good science, community buy-in, and government support (Abdalla et al. 2015).

24.8 GOVERNANCE STRUCTURE OF MIKOKO PAMOJA

Mikoko Pamoja is governed by a 13-member committee democratically elected from participating villages every two years during a village consultative meeting and in adherence to regional balance and gender equity (Figure 24.4). A paid project coordinator who also serves as a link between the group and the steering committee coordinates day-to-day project activities. The steering committee provides technical expertise in carbon accounting and socio-economic monitoring and also coordinates scientific and educational activities.

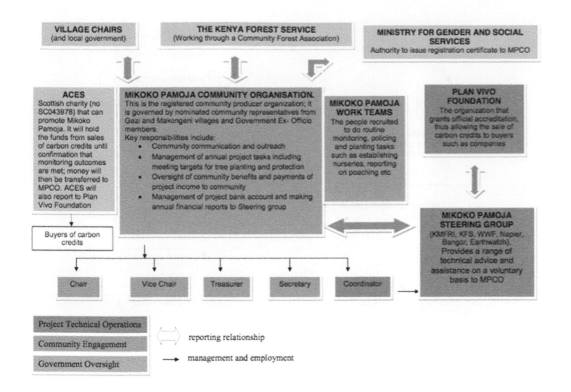

Figure 24.4   Mikoko Pamoja organization structure and governance.

24.9 SHARING BENEFITS FROM MIKOKO PAMOJA

Revenue generated through Mikoko Pamoja is divided such that 36% of the funds are utilized for project activities, including mangrove replanting, while 26% is for supporting community development activities as illustrated in Figure 24.5. Community projects to be supported through Mikoko Pamoja are decided annually through village-level consultative meetings. Mikoko Pamoja has improved education, water and sanitation systems, as well as the management of mangrove ecosystem in Gazi Bay. The purchased books and stationery for local schools have helped improve the education standards in the area. The achievements of the project are outlined in Table 24.5 below. Overall, Mikoko Pamoja is meeting the demand of 73% of 4,000 resident population by supplying water through water points or connecting water pipes directly into people's houses. The project is also supporting nature-based enterprises such as mangrove ecotourism and integrated aquaculture, leading to improved livelihood for the local people living in area.

24.10 CONCLUSION

Unsustainable exploitation of mangrove forests in Kenya has led to shortages of firewood and building materials, decline in fisheries, and increased shoreline erosion. Mikoko Pamoja is reversing this trend by attracting carbon finances and channeling them to the conservation and restoration of degraded mangrove areas as well as initiating community development projects in Gazi Bay. The

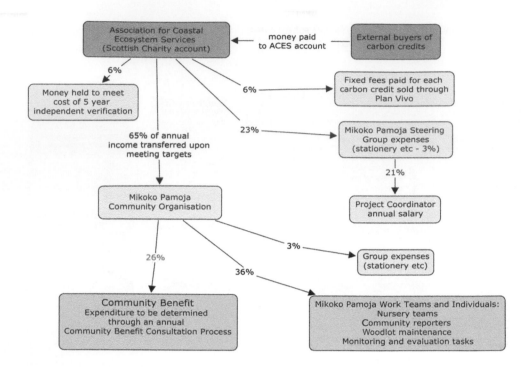

Figure 24.5 Benefit sharing scheme for Mikoko Pamoja.

project is supported by the village community, consisting largely of largely fishers whose livelihoods are connected to the health of the mangroves, with whom there is a clear payment arrangement for sold carbon credits. These payments partially cover dedicated staff time for the project, with the remaining funds being allocated to community projects and additional mangrove activities overseen by village leaders.

Success of Mikoko Pamoja stems from strong community support for the project, well established and ongoing scientific research on the mangroves in the area, knowledgeable government agencies interested in partnering with the local community on the project, and a supportive national policy that promotes participatory forest management.

One of the strengths of the project is the approach taken to reduce illegal harvesting of mangroves, by including the cultivation of fast-growing terrestrial forest plantations to serve as alterative wood sources and leakage control. The project has also established a mangrove boardwalk—a community-based ecotourism facility managed by the Gazi Women Group for recreation and educational activities. Recently, the project partnered with World Wide Fund for Nature (WWF) to promote energy saving stoves and solar lights that would further reduce community dependency on mangrove forests for wood. This is in addition to carrying out of integrated aquaculture as well as beekeeping (apiculture) in the mangrove areas as alternative livelihood activities. Mikoko Pamoja provides an excellent example of a **triple-win** situation in Kenya with benefits to community, climate, and biodiversity.

Table 24.5 Direct Impacts of the Mikoko Pamoja Project on Multiple Ecosystem Services and Public Health

Community Development Project	Input	Output	Outcome	Impact
Installation of clean water	Purchase of 2 PVC water storage tanks, two water pumps, piping, and creation of three water points	Two water sellers employed Generation of revenue from the sale of water Piped water into the community health center and primary schools	Clean water made available to about 70% of the community 47% reduction in the incidences of water-borne diseases Improved enrolment of students into local school Better learning environment	Improved Livelihood as a result of reduction of the price of water from KES 20- 3 per 20l jerican Reduced number man-hours lost in drawing water
Purchase of books and Renovating classrooms	Purchase of *ca.* 600 textbooks, roofing materials, and security doors for local primary schools	Classrooms renovated, increased number of books in a class	Reduction in the ratio of students sharing a textbook Better learning environment	10% increase in number of students joining national schools. Improved living standards as a result of improved education
Environment education	Community education on importance of mangroves in schools Engagement of students in marking environmental events Mangrove competitions (local and international) e.g., by Mangrove action project	~1,500 students from 15 schools educated on mangroves in 2016. Two international training efforts: 55 trained in 2016 and 13 teachers trained in 2017	Participation in more environment competitions and events from 1 event in a year to 3 Improved knowledge of the marine environment	Increase in mangrove conservation
Link with external communities	Three interested community conservation group visited and educated on mangroves (Big Ship in Mikindani, Madagascar and Mwikamba in Diani)	300 mangroves planted with Mwimkamba	Increased knowledge on importance of mangroves	Increase in areas under mangrove conservation
Woodlot	2,500 Casuarina planted BSc student recruited to monitor the woodlots	Data on the status of the forest gotten	Casuarinas currently providing building poles and firewood for the communities	Reduction in mangrove degradation
Livelihood	Employment	Four directly employed: project Co-ordinator, Assistant co-ordinator and 2 community scouts 32 indirectly employed in 2016 through planting and monitoring activities	Minimal cutting as a result of Increased surveillance in the project area	Increased conservation of the mangroves Improved livelihood

Table 24.5 Direct Impacts of the Mikoito Female Project on Multiple Ecosystem Services and Public Health

Community Development Project	Input	Output	Outcome	Impact

CHAPTER 25

Blue Carbon as a Tool to Support Coastal Management and Restoration
Bringing Wetlands to Market Case Study

Tonna-Marie Surgeon Rogers
Waquoit Bay National Estuarine Research Reserve

Kevin D. Kroeger and Meagan Eagle Gonneea
U.S. Geological Survey Woods Hole Coastal & Marine Science Center

Omar Abdul-Aziz
Ecological and Water Resources Engineering Laboratory
West Virginia University

Jianwu Tang
Marine Biological Laboratory

Serena Moseman-Valtierra
University of Rhode Island

CONTENTS

HIGHLIGHTS

1. A collaborative research approach involving substantial end user and stakeholder engagement was applied to great effect to guide broad, integrated investigation of the science, policy, and management of blue carbon and carbon markets as drivers for coastal wetland management and restoration.
2. Expanding awareness about blue carbon concepts among local, state, and federal agencies and the public was found to be effective at stimulating interest in taking climate-change mitigation action through wetland protection and restoration, as well as increasing understanding about the overall value of coastal wetlands and the services they provide.
3. Interest in blue carbon is high regardless of the potential for financial benefits through a carbon market. For end users such as the Massachusetts Division of Ecological Restoration and National Park Service Cape Cod National Seashore and many others, the ability to highlight the climate benefits of ecological restoration to stakeholders may have greater value than financial benefits from sale of carbon market credits.

25.1 INTRODUCTION TO THE BRINGING WETLANDS TO MARKET PROJECT

Inspired by new research on the interactions between coastal wetlands and global climate, a team of scientists worked with the Waquoit Bay National Estuarine Research Reserve (WBNERR) to engage with multidisciplinary partners to develop an innovative coastal blue carbon research project. Responding to a thrust within the National Estuarine Research Reserve System (NERRS) and WBNERR to prioritize climate change research, this new project focused on providing information and tools to support wetlands protection and restoration. The Bringing Wetlands to Market (BWM) project was initiated in 2011 to explore the connections between climate change, salt marshes, nitrogen loading, and coastal wetland restoration, and develop tools to incentivize investment in coastal wetland protection and restoration. The project is currently in its second phase. This chapter aims to share information about the project as a case study for managers, restoration practitioners, policymakers, and researchers to illustrate the thoughtful progression of using coastal blue carbon science to inform management and policy, as well as key lessons learned along the way. It covers the goals of the research, the project's integrated research plan, engagement of end users, and approach to implementation, as well as its overall impact on advancing blue carbon awareness and application locally and nationally.

25.2 WHY THIS WORK?

Seven key facts provided the impetus for initiating BWM: (i) salt marshes have tremendous value to society because of the suite of important ecosystem services that they provide, including carbon storage; (ii) globally and nationally coastal wetlands, and their carbon stores, are threatened by a number of factors such as development and sea level rise, and these ecosystems are being destroyed and/or degraded at an alarming rate; (iii) lack of adequate funding is limiting our ability to keep pace with restoration goals for coastal wetlands; (iv) carbon markets could serve as an important mechanism to secure resources for wetlands restoration, however, absence of a greenhouse gas (GHG) protocol to enable coastal wetlands to be considered in carbon markets (similar to what exists for forests) was a critical barrier; (v) additional research is needed to better understand the behavior and critical environmental drivers of GHGs and carbon storage in coastal wetlands; (vi) there is both a need and an opportunity to build awareness about the importance of coastal blue carbon and its climate mitigation benefits, as well as develop information and tools to enable managers and policymakers to use blue carbon to advance coastal protection and restoration, and (vii) local decision makers working in Massachusetts, where the study was conducted, were particularly concerned about the impact of anthropogenic nitrogen loading on water quality and coastal wetland health. Examining the impact of nitrogen loading was therefore included as a BWM objective to increase the project's relevance to the local community.

Against this background, the BWM project was launched with goals to provide science-based information and tools for coastal decision-makers, to inform strategies for wetlands protection and restoration, and for carbon and nitrogen management in coastal wetlands, and to develop policy frameworks and market-based mechanisms to reduce GHG fluxes and maximize carbon sequestration. Research goals were to develop and apply new techniques to quantify annual rates of GHG emission and carbon sequestration in salt marshes, to better understand controlling processes, and to enable prediction of fluxes across a range of environmental settings. The project also aimed to address key gaps hindering the incorporation of wetlands into carbon markets.

25.3 USING A DIFFERENT RESEARCH PARADIGM

Funding for BWM was secured from the NOAA NERRS Science Collaborative, which requires research teams to integrate potential end users of the science (i.e., those who could apply it in their work and decision-making) into the research process from concept to implementation. This collaborative approach contrasts with the more typical research paradigm in which research is conducted by scientists and published, but findings are either never directly shared with end users or are only disseminated to end users at the end of projects. While this has been the norm, social science research reveals some compelling findings, which suggest that a more effective approach may exist to bring science to bear on management and policy decisions. A study by Cash et al. (2003) found that decision-makers are more likely to apply science-based information in their work if they deem that information to be credible, legitimate, and salient. Also, moving from knowledge to action is enhanced where there is communication that is active, iterative, and inclusive between experts and decision makers. Thus, addressing needs that end users identify, and facilitating end user involvement in the research process, increases the likelihood that end users trust the science and have confidence in the results. This in turn increases the likelihood that the science and tools generated will be applied.

The Waquoit Bay NERR embraced this collaborative approach and worked with project partners to structure the BWM Team to include a combination of science investigators, tool developers, and end users working at the international, national, state, and local levels. Potential end users were thoughtfully identified and their information needs and interests assessed early and over the course of the project and utilized to shape the project's direction. A broad suite of potential end users was identified for the first phase of the project. End users included coastal managers, land managers, wetland restoration practitioners, scientists, policymakers, local officials with management responsibilities for both wetlands and for reducing nitrogen loading to coasts, stakeholders involved in climate change mitigation and adaptation efforts, and those concerned with wetlands conservation and restoration, as well as the NERRS. In the second phase of the project, a narrower group of end users was identified from this larger group to facilitate more detailed exploration of a potential carbon market opportunity for a wetland restoration project, focused particularly on tidal restoration.

At the outset of the project a plan was developed to facilitate and sustain engagement with end users over the course of the project. This plan identified critical junctures where it was important to seek end user input on the research plan and project outputs. End user engagement was carried out using collaborative learning—a social science methodology—as the backbone. Collaborative learning as described by Daniels and Walker (1996, 2001) emphasizes activities that encourage systems thinking, joint learning and open communication, and is appropriate for ecosystem system-based management work. Collaborative learning emphasizes that end users/decision-makers are not simply recipients of information from experts, but rather valuable contributors themselves of knowledge and expertise. Bringing together the full breadth of researcher and end user expertise in the research process strengthens the ability of groups to tackle complex environmental issues and solve problems (Feurt 2008). At the start of the project, BWM researchers and end users alike were introduced to what collaborative learning entailed to ensure that everyone had a common understanding. This collaborative framework was used in both phases of the BWM project and has helped to both shape objectives and strengthen project outcomes.

25.4 AN OVERVIEW OF BWM ACTIVITIES

The first phase of the project was implemented from 2011 to 2015 to achieve the following goals:

   i. Quantify carbon sinks and GHG fluxes (both production and consumption), in salt marshes, and assess the impact of anthropogenic nitrogen loading, sea level rise and climate on both carbon sequestration and net GHG emissions;

   ii. Develop a carbon sequestration and GHG emissions model for coastal wetlands (hereafter referred to as the BWM GHG model), to aid in identification of suitable carbon market projects, and to inform planning for coastal wetland conservation and restoration;

   iii. Develop a GHG offset methodology that is accepted by carbon registries, to enable coastal wetlands to be considered in carbon markets, as well as an associated guidebook, to provide information for managers interested in using the methodology;

   iv. Conduct a scoping economic assessment of the carbon sequestration and greenhouse benefits of tidal wetlands restoration to assess the financial relevance of blue carbon to land conservation and wetlands restoration decisions, and provide managers with an economic framework within which to consider potential market and non-market benefits of wetland restoration projects;

   v. Advance the ability of the NERRS to monitor the effects of climate change on coastal ecosystems, with Reserves serving as platforms to advance blue carbon research.

BWM Phase 1 had four main components as highlighted in Figure 25.1.

Phase 2 of the project is currently in progress and is scheduled to be completed in 2019. BWM 2 builds on the work done in Phase 1 and aims to expand blue carbon implementation by expanding

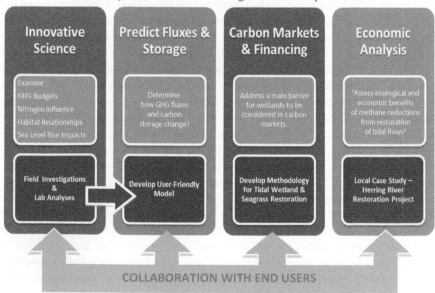

Figure 25.1 BWM Project schematic showing main components and linkages between them.

research and the BWM GHG Model development in wetlands with restricted or blocked tidal flows, and in wetlands that have been restored to salt marshes by restoring tidal flow. In addition, the BWM GHG Model will be expanded to provide predictions in tidal wetlands that occur under a broader range of geographic settings and conditions, by engaging an additional set of non-BWM (Tier 2) data providers.

Drawing on end user feedback and the findings of the economic assessment conducted in Phase 1, it became clear that while end users were learning about and becoming interested in the potential that carbon market financing presented, the transaction costs associated with actually bringing wetlands to market could be a barrier to this application. One way to lower these costs is to use accepted models to predict carbon storage and GHG emissions, thereby reducing the need for expensive data collection and monitoring. To address this need, the project team focused Phase 2 on expanding and generalizing the model created under Phase 1 to have application across a wider environmental range. Following on the work done in Phase 1, new techniques are being used to quantify the key fluxes, including vertical GHG emissions, lateral carbon fluxes, and carbon sequestration in several tidal marshes on Cape Cod, representing a range of restoration conditions. Additionally, to meet another end user-identified need, BWM 2 also aims to conduct a first of its kind feasibility assessment for a wetlands restoration carbon market project. This will be done by using the planned National Park Service Herring River Restoration Project on Cape Cod as a case study. Lastly, BWM 2 also focuses on continuing to build awareness about blue carbon among coastal decision-makers and local communities. This involves developing new communication products to better highlight the value of coastal wetlands, the ecosystem services they provide, and ways in which blue carbon can be used to support coastal restoration and climate adaptation and mitigation efforts. Sections 25.5 and 25.6 provide more information on the science questions investigated in both phases of the project and the tools developed through the project for the coastal management community.

25.5 A DEEPER DIVE INTO BWM RESEARCH—CONTEXT, OBJECTIVES, AND APPROACH

At the initiation of BWM, the potential for coastal wetland management as a climate-change mitigation measure was largely a theoretical concept. The term **blue carbon** had been coined only two years prior, and the state of knowledge was insufficient for regional or national scale accounting, or for prediction and verification of carbon market projects. In addition, there was, and still is, significant skepticism from the science community regarding the scale of potential climate change mitigation. Seven years later, as a result of increasing interest from stakeholders and federal agencies, and substantial effort from an increasing number of scientists who have engaged in synthesis, GHG inventory, and process investigations, it is clear that coastal vegetated ecosystems do, indeed, comprise a quantifiable and significant component of the U.S. GHG budget (EPA 2017, SOCCR2). The broader point regarding the concept of blue carbon is that, while the potential for GHG management at the regional or national scale is modest, the same is true for most of the large number of measures that will ultimately be taken to mitigate climate change. Further, among the measures available to society, few provide a cost-effective and sustainable means for the negative carbon emissions that will be a necessary part of our efforts to reduce the ultimate scale of global temperature increase. Coastal wetland management offers the potential for both anthropogenic emissions reduction (Pendleton et al. 2011; Kroeger et al. 2017), and negative emissions (Chmura et al. 2003; Bridgham et al. 2006) at a high rate per unit area. Thus, while the potential contribution of wetland management to national or global scale GHG reductions is limited, changes in emissions can be potent at the project scale, and widespread utilization for GHG management has significant potential to promote coastal wetland protection and restoration.

Within the BWM project, and a number of linked projects, a team of investigators endeavored to quantify, elucidate, and predict carbon and GHG cycling processes in tidal marshes, with a goal to provide the knowledge necessary for wise GHG management in coastal wetlands. The initial science team was comprised of Jim Tang (MBL), Serena Moseman-Valtierra (URI), Kevin Kroeger (USGS), and Omar Abdul-Aziz (WVU). The team expanded substantially over time, however, to include the broader science team identified in "Acknowledgments" section of this chapter. Here we will briefly highlight the primary objectives, approaches, and outcomes of the field science investigations undertaken.

25.5.1 Research Objectives

Broadly, the objectives were first to improve, or develop new, methods to more accurately quantify salt marsh carbon cycle processes, and to apply those techniques to measure annual process rates. Accurate, annual rate measurements are clearly necessary to quantify annual and long-term carbon storage, but further, they are a prerequisite to development of sufficient understanding of controls such that we can predict responses to environmental conditions. Environmental features and changes of interest, and that are likely to alter carbon storage and GHG fluxes, include anthropogenic nitrogen enrichment and change in salinity, elevation and vegetation zone, plant productivity, tidal range, climate change, water level and sea level rise, sediment exchanges, and hydrological management. Further goals for the field investigations were to produce the data necessary for development and testing of a predictive, end user-friendly model of marsh carbon cycle processes.

25.5.2 Approach

Design of the research program within BWM was organized around the concept of the Net Ecosystem Carbon Balance (NECB; Chapin et al. 2006): $NECB = NEP - RCH_4 - FL$. NEP refers to net ecosystem production, and is the net result of photosynthesis (gross primary production; GPP) and ecosystem respiration (plant respiration and microbial respiration), measured with the closed chamber method (Tang et al. 2008). Nighttime NEP is a measure of ecosystem respiration. Daytime GPP can be calculated as the difference between NEP and ecosystem respiration, which is extrapolated from nighttime respiration with a temperature response function (Tang et al. 2005; Tang et al. 2008). RCH_4 is the CH_4 flux measured simultaneously with NEP. In tidal wetlands, FL is the net lateral flux, through tidal aquatic export, of carbon. Important forms of carbon exchange at the study sites examined include dissolved inorganic carbon (DIC), dissolved organic carbon (DOC), particulate organic carbon (POC), and dissolved CH_4. Ultimately, the flux-based NECB measurements will be compared with long-term (decadal and century-scale) soil C sequestration rates measured using depth profiles of soil bulk density, and C content, with time-increments of vertical accretion based on naturally occurring radiochemical tracers, including [137]Cs (Delaune et al. 1978), and particularly through application of a modified version of the ^{210}Pb continuous rate of supply model (Appleby and Oldfield 1978). At the outset of the BWM project, and still today, simultaneous, annual measurements of the primary carbon cycle processes in tidal wetlands are rare. To our knowledge, the BWM project is the first to conduct simultaneous, annual rate measurements of all terms of the NECB at a single salt marsh site (Figure 25.2), though the complete analysis is yet to be published.

In the BWM 1 project, primary field investigations occurred across four polyhaline salt marsh sites, within and near the Waquoit Bay NERR, which comprised a 12-fold nitrogen-loading gradient. Additional data collections occurred, and are ongoing, at marsh sites within Narragansett Bay, Rhode Island, in part to expand the range of nitrogen loads. The driving questions and objectives of BWM 2 were an outgrowth of end user engagement in BWM 1, and are a good success story for collaborative learning. At our first stakeholder engagement workshop, a restoration ecologist and designated project end user from the Massachusetts Division of Ecological Restoration (DER) presented information on wetland restorations in the state. All restorations consisted of opening tidal restrictions, the majority of which were caused by undersized culverts under roads and railroads.

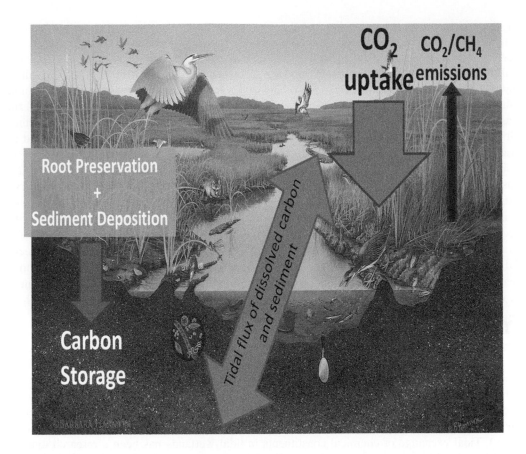

Figure 25.2 **(See color insert following page 266.)** Schematic diagram of the major carbon and GHG fluxes of interest in the BWM field research program. Net photosynthetic uptake of carbon dioxide supplies root biomass that contributes to soil carbon stock increase and soil elevation accretion. Tidal exchange with adjoining coastal waters carries lateral fluxes of respiratory and detrital carbon, as well as net import or export of sediment.

Those restorations did not build new wetland to promote new carbon storage, and thus initially the activity did not seem relevant to carbon markets. Further consideration, however, led to the concept that, since low-salinity wetlands generally have greater CH_4 emissions than saline wetlands, conversion of salt marshes to fresh or brackish wetlands and impoundments, as a result of tidal restriction, will commonly result in an increase in CH_4 emissions. Conversely, restoration of tidal flows, and therefore of more natural salinity and water level, will often reduce CH_4 emissions. Thus, in many cases, tidal restoration can be a method to reduce anthropogenic CH_4 emissions, and to engage in a new form of blue carbon management, with significant emissions reduction potential at the national scale (Kroeger et al. 2017). Field investigations within BWM 2, therefore, occurred at a series of tidally restricted, restored, and reference marsh sites on the Cape Cod Bay coast of Cape Cod, including sites within the National Park Service Herring River basin, to investigate the carbon stock and GHG flux consequences of tidal restriction and restoration.

25.5.3 Measurement Techniques

Vertical GHG Fluxes: Exchanges of CO_2 and CH_4 between salt marsh sites and the atmosphere were and are currently being measured to enable calculations of NEP, GPP, RCO_2, and RCH_4. In addition, nitrous oxide fluxes are measured to quantify their contribution to the net climate forcing (net climatic warming versus cooling) of the ecosystems under the range of conditions studied. Vertical GHG fluxes in the BWM project are measured using the closed chamber technique and laser absorption spectroscopy, using a CO_2 and CH_4 analyzer (G2301-F Fast CO_2, CH_4 and H_2O Analyzer, Picarro, Inc., Santa Clara, CA), and a N_2O analyzer (Los Gatos Research, Inc., Mountain View, CA). The sensors are connected via tubing to a portable chamber and a gas-tight air pump. Multiple permanent bases for the chambers have been installed in the soil at each study site. During measurements, transparent and shaded chambers are placed over soil and vegetation (Moseman-Valtierra et al. 2011; 2015), to measure net ecosystem exchange. Variations on this method are utilized for a range of conditions and purposes, including use of smaller, opaque chambers for respiration measurements where shrub and forest vegetation occur, and utilization of tall (>2 m) transparent chambers where *Phragmites* and *Typha* vegetation occur. Air inside the chambers is mixed using a battery-powered fan and is pumped through the instruments and back to the chamber in a closed loop. Chamber deployments are brief (typically ~4 to 7 min), to minimize chamber effects (e.g., Windham-Myers et al. 2018). Recently, project participants have additionally installed two eddy covariance systems for continuous, long-term measurement of gas exchange between a tidally restricted, freshened, *Phragmites australis*-invaded wetland within the Herring River (http://ameriflux.lbl.gov/sites/siteinfo/US-HRP), and in a reference, polyhaline, macrotidal, high marsh site within the Barnstable Great Marsh (http://ameriflux.lbl.gov/sites/siteinfo/US-BSM). Both are now part of the American carbon flux network (AmeriFlux, http://ameriflux.lbl.gov/).

Lateral Fluxes: Lateral fluxes or outwelling of fixed C and GHG from coastal wetlands to estuaries due to tidal water exchanges can be large relative to other terms in tidal wetland biogeochemical budgets (Gardner and Kjerfve 2006; Guo et al. 2009; Tong et al. 2010). Further, estimates of the lateral fluxes of carbon are integral to calculations of C sequestration in the marsh sites based on the NECB. Estimation of lateral fluxes is complex in part due to simultaneous contributions of dissolved carbon in marsh-derived materials carried by brackish and saline porewater and surface water exchange and of both estuarine and terrestrial dissolved carbon and GHGs (Kroeger et al. unpub.). Tidal exchange of chemical constituents in tidal wetlands has been a research topic for several decades, but methods employed generally allow only broad estimation of fluxes. To improve data and knowledge of this component of wetland carbon budgets, we developed new methods in the BWM project. Objectives were to measure annual rates of net lateral exchange of dissolved, respiratory CO_2 (DIC), detrital DOC, and living and detrital particulate carbon (POC), and sediment. The methods, still undergoing further development and described in more detail by Wang et al. (2016) and Chu et al. (in review), involve long-term (one year minimum) deployment of a Sontek-IQ acoustic Doppler flow meter, as well as water chemical and physical sensors (our primary device was a YSI EXO2 sonde; other devices included ProOceanus pCO2 sensors, a CHANOS automated DIC analyzer, and a Picarro CO_2 and CH_4 analyzer coupled to an air/water equilibrator) in a marsh creek with a reasonably identifiable contributing marsh basin. Hydrodynamic modeling is utilized to correct measured instantaneous water flows, based on modeled sheet flow outside of the creek, as well as to quantify contributions from fresh (terrestrial) groundwater. High-frequency estimation of carbon concentrations are achieved by regressing laboratory-measured concentrations on multiple predictive variables, including salinity, pH, temperature, day of year, and others. It is expected that the multiple regression models thus developed are best used for interpolation during sampled time-periods, and are not of utility across sites without site-specific data. The approach, however, does allow unprecedented frequency of carbon concentration estimations. Instantaneous fluxes are calculated as the product of instantaneous water flow and constituent concentration. Net flux, over a range of timescales, is integrated and normalized by contributing marsh area to estimate net flux per unit area of wetland. Source attribution for major carbon species (DOC, DIC, and POC) are supported by carbon stable isotope ratio measurements to distinguish between terrestrial, estuarine, and wetland sources (e.g., Bouillon et al. 2007). Analysis of ^{13}C in DOC, in fresh to saline samples, is conducted at a shared USGS/Woods Hole Oceanographic Institution IRMS facility.

Soil Carbon Accumulation: Investigation within the BWM project of soil carbon accumulation, and storage of a range of timescales, has been led by Meagan Gonneea of the USGS Woods Hole Coastal & Marine Science Center. Utilization of a 10-cm diameter piston core, to avoid compaction, 1 cm core sectioning, and use of the ^{210}Pb continuous rate of supply (CRS) model (Appleby and Oldfield 1978) have allowed improved accuracy and high-resolution reconstructions of marsh accretion and carbon storage during the past century (Gonneea et al. 2018). Examinations to date include response to sea level rise, as well as history of tidal restriction and restoration.

25.5.4 Model Development

Several process-based models, primarily for freshwater wetlands, are available for estimating fluxes of CO_2 and CH_4 (e.g., Cao et al. 1996; Walter and Heimann 2000; Zhang et al. 2002; St-Hilaire et al. 2010) and N_2O (e.g., Li et al. 1992; Parton et al. 2001; Hénault et al. 2005). Those models could be adapted for coastal wetlands. However, mechanistic GHG flux models are highly detailed, often over parameterized, involve intensive input data; and may not generate better predictions than that of a simpler model. Further, the model complexity requires specialized skills and training, which may limit application by end users. Through BWM Phase 1 a systematic data analytics and modeling framework was developed to first identify the dominant controls of wetland GHG fluxes and estimate their relative linkages with the hydro-climatic, sea level, biogeochemical, and ecological drivers. The dominant environmental drivers (e.g., photosynthetically active radiation, soil temperature, porewater salinity) were then used to develop power law-based emergent scaling models to predict the CO_2 and CH_4 fluxes, and estimate potential carbon storage of coastal salt marshes. A model was not developed for N_2O fluxes which had represented trace amounts (<1% of GHG flux) for the coastal wetlands in Waquoit Bay (BWM 1) and greater New England (BWM 2) areas. However, alongside expanding the models of vertical GHG fluxes across New England (Tier 1 data) and the broader U.S. East Coast (Tier 2 data), we will develop a new empirical model in BWM 2 to predict vertical, lateral, and soil fluxes based on their dominant environmental drivers, and relationships to each other. The model will be generalized in space and time to resolve heterogeneity of climatic, biogeochemical, and ecological processes across regions and seasons. The first step of model generalization will be to test the accuracy of predictions at new sites (reflecting larger process gradients) with the parameters estimated with Waquoit Bay data. The second step of model generalization will be to divide the overall data into several wetland regimes based on that of the dominant environmental drivers (e.g., climate, vegetation biomass and productivity, salinity). The model parameters will then be re-estimated for individual regimes and compared for similarity across different regimes. The third step is to develop dimensionless scaling models by defining a reference scaling point that contains information for both time and space for each wetland regime or site. Normalization of the response and predictor variables by the reference scaling point should ideally collapse the site and regime-specific relationships into a single generalized model. The most robust models for GHG and lateral fluxes will be identified based on prediction performance. Finally, the generalized models will be integrated into a user-friendly Excel spreadsheet tool by modifying the underlying Visual Basic code. Similar to the Excel model of BWM Phase 1, managers will be able to input site-specific climate and environmental variables and obtain the predicted potential wetland carbon storage under different scenarios (e.g., IPCC) by simply clicking **RUN** on the spreadsheet.

25.6 NEW TOOLS FOR COASTAL MANAGERS

The BWM project has developed several tools for the coastal management community to inform carbon management in coastal wetlands, climate adaptation and mitigation efforts, ecosystem service valuations as it pertains to carbon storage, development of potential blue carbon market projects, and public education on the value of coastal wetlands. In this section, we highlight the tools and their purpose.

25.6.1 A Model to Predict GHG Fluxes and Carbon Storage: BWM GHG Model

As introduced in Section 25.5, data from BWM field and lab investigations have been used to develop a novel, user-friendly model to predict CO_2 and CH_4 fluxes from coastal wetlands. The BWM GHG model is intended to serve as a planning tool and enables managers to determine whether a tidal wetland is acting as a source or sink of GHGs. The model is presented as a simple Excel spreadsheet and requires a small set of readily accessible inputs, which reduces the data collection burden on users. The BWM GHG model is currently being generalized for application across a wider latitudinal gradient and can be accessed here: www.waquoitbayreserve.org/research-monitoring/salt-marsh-carbon-project/

25.6.2 Methodology for Tidal Wetland Restoration Projects to Secure Carbon Financing

The BWM Project was instrumental in the development of the first methodology to enable coastal blue carbon projects to secure credits through verified carbon markets. Development of *Methodology for Tidal Wetland and Seagrass Restoration (VM0033)* was led by BWM team member, Restore America's Estuaries, which worked with many contributing partners to produce this new tool. The methodology was approved by the Verified Carbon Standard (VCS) in 2015 and is now available for application. This significant achievement has paved the way for project developers interested in supporting tidal wetland and seagrass restoration to now advance blue carbon market projects. The methodology can be accessed at: http://database.v-c-s.org/methodologies/methodology-tidal-wetland-and-seagrass-restoration-v10

Coupled with the development of the methodology, a guidebook—*Coastal Blue Carbon in Practice: A Manual for Using the VCS Methodology for Tidal Wetland and Seagrass Restoration*, was also produced for managers and project developers interested in utilizing the methodology. The guidebook also includes a section on grouping smaller restoration projects together to make projects of a large enough scale to be feasible as carbon market projects. The novel idea of aggregating projects for this purpose was raised by BWM end users in New England given the smaller size of many tidal restoration projects in the New England region. The guidebook can be accessed here: www.estuaries.org/images/rae_coastal_blue_carbon_methodology_web.pdf.

Following acceptance and publication of the methodology is available the next steps involved in moving from conceptual to **real** blue carbon market projects include (i) identifying potential tidal wetland and seagrass restoration projects that are of the right scale, (ii) identifying potential project developers, (iii) conducting assessments to determine if the projects would be good candidates for market application, and (iv) putting together the necessary documentation to advance these initiatives. Demonstration projects are now needed to show how these projects could work through the voluntary carbon market. In recognition of this need for demonstration projects, a key objective of BWM 2 is to conduct a first of its kind feasibility assessment for a blue carbon market project using the Herring River Restoration Project as a case study.

25.6.3 Herring River Carbon Project Feasibility Study

The Herring River Restoration Project at Cape Cod National Seashore is a 1,100 acre—planned tidal wetlands restoration effort—the largest in New England. The project presents an opportunity to achieve GHG reduction benefits, particularly for CH_4. The BWM Team is working with end users from the Herring River Restoration Committee (comprised of federal, state, and local partners), and the Friends of Herring River, to utilize the approved tidal wetland restoration methodology to explore the potential for gaining carbon credits for this project. The BWM research team is collecting data to support this analysis in response to end user requests.

The feasibility study is currently being led by Restore America's Estuaries and Terra Carbon Inc. in close collaboration with Herring River partners, BWM researchers, and the entire project team. The study will examine the following issues: (i) market feasibility—potential voluntary and regulatory markets that could provide funding for the project; (ii) technical feasibility—apply the methodology to estimate the carbon offset potential of the project; (iii) financial feasibility—estimate costs for both the restoration project itself and the carbon market project and potential revenue from carbon financing; (iv) legal feasibility—determining who would own the carbon credits and any legal agreements that might be necessary to develop; and lastly (v) organizational feasibility—the organizations that could be involved in the effort and their potential roles. These elements are intended to provide important information to end users that they can use in determining whether the Herring River Project would be a worthwhile carbon market project from the standpoint of both revenue gains and restoration of important wetland ecosystem services.

25.7 BUILDING AWARENESS ABOUT BLUE CARBON

One of the biggest challenges faced by the team at the outset of the project was lack of awareness about coastal blue carbon as a key wetland ecosystem service and its relevance to climate change mitigation. Given this, the project team focused on ramping up blue carbon education efforts on blue carbon ecosystem services. The Waquoit Bay Reserve worked on several fronts with team members to develop tailored communication products that addressed questions and issues end users were most interested in learning about. Some of these educational initiatives included developing communication materials such as videos explaining the research and its relevance to coastal management, fact sheets covering topics such as the importance of blue carbon, market and policy management applications, many presentations tailored for different audiences, as well as the previously described BWM tools. These products (available here: www.waquoitbayreserve.org/research-monitoring/salt-marsh-carbon-project) have since been viewed and accessed by thousands of interested individuals. Also, a blue carbon science curriculum for high school teachers was developed from a small science transfer project with educators in the NERRS and is now being used to bring blue carbon science to the classroom. The curriculum is available here: www.waquoitbayreserve.org/research-monitoring/salt-marsh-carbon-project/teachers/

Of special note is that all these products were not just developed for but **with** end users to increase their usefulness.

The team coupled the development of communication products with a variety of targeted educational opportunities for a range of stakeholders working at the local, state, and national levels. These included workshops, conferences, briefings, seminar presentations, one-on-one conversations, and field trips to research sites. As an example of the type of activities done, the BWM team organized and convened a regional blue carbon conference for New England in 2015 centered on the theme of "Capitalizing on Coastal Blue Carbon" and sharing science from BWM and other research blue carbon projects as well as BWM tools. This conference attracted close to 100 participants including resource managers, policymakers, scientists, state agencies, land management groups, and local officials at a time when blue carbon was still a new emerging concept for many managers in the region.

Another end user engagement strategy involved taking the BWM team "on the road" to hold blue carbon roadshow dialogues with key decision-makers working at the state level in Massachusetts and gathering their input on the research plan and BWM tools. Examples of organizations that were engaged and consulted in Massachusetts included: MA Executive Office of Energy and

Environmental Affairs, MA Division of Ecological Restoration, MA Office of Coastal Zone Management, The Nature Conservancy, the Cape Cod National Seashore, the U.S. Fish and Wildlife Service, the MA Department of Environmental Protection, non-profit organizations such as local land trusts, Association to Preserve Cape Cod, Friends of Herring River, and local officials such as elected town leaders and Conservation Commissions which are appointed municipal boards with management responsibilities for coastal wetlands.

The Waquoit Bay NERR also organized special webinars and workshops and participated in a number of science transfer projects to expose staff working at research reserves all across the country to blue carbon and what was being learned through BWM. At the start of the BWM project, of the 28 Reserves that comprised the NERRS at that time, only the Waquoit Bay Reserve was involved in blue carbon research. Today many more NERRS are either actively involved in some aspect of blue carbon research themselves or are supporting blue carbon research and related projects at their sites. This provides good examples of both the impact of the project's extensive stakeholder engagement strategy, and the expansion of interest in blue carbon that the project has helped to promote.

While the activities mentioned above do not represent the full breadth of the team's work to engage and educate stakeholders about blue carbon, they provide great examples of this work. A core ingredient of all these outreach efforts was maintaining a focus on what end users and stakeholders cared about and using their interests as primary touch points for communicating about the topic and its relevance to local communities. This proved to be an effective strategy for building awareness among a range of audiences.

25.8 KEY PROJECT OUTCOMES

The BWM project has had a significant impact on advancing coastal blue carbon research, by investigating and widely discussing the climate benefits of coastal wetlands and of improvements in management. Some of these outcomes are noted below for illustration:

- Produced new science approaches and results, as well as innovative tools to help managers and policymakers leverage blue carbon to support broader wetlands conservation and restoration goals
- Increased awareness of coastal blue carbon locally, nationally, and internationally
- Paved the way for coastal wetlands to be considered in carbon markets
- Motivated end users to begin incorporating blue carbon and GHG impacts into policy, management, and restoration initiatives and project evaluations (e.g., under the Global Warming Solutions Act, and as part of updating its climate action plan, Massachusetts investigated the inclusion of land use change as it pertains to wetlands on the State's GHG footprint).
- Identified, researched, and highlighted opportunities for methane emission reduction via restoration of tidal flows
- Advanced the first feasibility study for a tidal restoration carbon market project (Herring River Restoration Project Case Study)
- Increased use of the NERRS as platforms for blue carbon research
- Contributed to many follow-on projects and collaborative partnerships around coastal blue carbon research
- Trained a significant number of young scientists in coastal carbon and GHG management, research, and modeling
- Demonstrated the value of using a collaborative research approach which incorporates meaningful engagement of end users

ACKNOWLEDGMENTS

The BWM Project has been a collaboration success story and showcases the value of partnerships. Key contributing organizations and individuals include:

Jianwu (Jim) Tang, Faming Wang, Kate Morkeski, Joanna Carey—Marine Biological Laboratory; Serena Moseman-Valtierra, Rose Martin, Katelyn Szura—University of Rhode Island; Omar Abdul-Aziz and Khandker Ishtiaq, West Virginia University; Kevin Kroeger, Meagan Eagle Gonneea, Neil Ganju, John Pohlman, Adrian Mann, Wally Brooks, Sandy Brosnahan, Jen O'Keefe-Suttles, Michael Casso—United States Geological Survey; Steve Emmett-Mattox & Methodology Development Team, Restore America's Estuaries; Stephen Crooks, Silvestrum Inc.; Tom Walker, Manomet Center for Conservation Sciences; Tim Smith, Cape Cod National Seashore; Tim Purinton, MA Division of Ecological Restoration; Scott Settelmyer, Terra Carbon Inc.; Aleck Wang, Sophie Chu, Amanda Spivak, Katherine Hoering—Woods Hole Oceanographic Institution; Rebecca Roth, National Estuarine Research Reserve Association; members of the Herring River Restoration Committee and the Friends of Herring River; James Rassman, Joan Muller, Jordan Mora, Chris Weidman, Alison Leschen, Kate Harvey, Tonna-Marie Surgeon Rogers—Waquoit Bay National Estuarine Research Reserve, as well as a host of other contributors from the above mentioned organizations and others. Primary funding was provided by the NOAA NERRS Science Collaborative. Additional funding to Omar Abdul-Aziz was provided by a grant from the NSF CBET Environmental Sustainability Program (Award No. 1336911/1561941). Other major support has been provided by the USGS Coastal & Marine Geology Program, USGS Land Carbon Program, NSF, and NOAA Sea Grant. We thank Sandra Brosnahan and Lisamarie Windham-Myers for helpful reviews. Any use of trade, firm, or product names is for descriptive purposes only and does not imply endorsement by the U.S. Government.

CHAPTER 26

Tampa Bay Estuary Case Study
Identifying Blue Carbon Incentives to Further Bolster Future Critical Coastal Habitat Restoration and Management Efforts

Edward T. Sherwood, Holly S. Greening, and Gary E. Raulerson
Tampa Bay Estuary Program

Lindsey Sheehan, Dave Tomasko, and Doug Robison
Environmental Science Associates

Stephen Crooks
Silvestrum Climate Associates, LLC

Ryan P. Moyer and Kara R. Radabaugh
Fish & Wildlife Research Institute

Stefanie Simpson and Steve Emmett-Mattox
Restore America's Estuaries

Kimberly Yates
U.S. Geological Survey

CONTENTS

HIGHLIGHTS

1. Tampa Bay blue carbon ecosystems are currently being restored despite continuing urbanization within the watershed
2. Tampa Bay blue carbon ecosystems have been providing significant carbon sequestration benefits to the estuary over contemporary periods (carbon burial rates range from 0.71 Mg C ha^{-1} year^{-1} for salt marsh to 1.78 Mg C ha^{-1} year^{-1} for mature mangroves)
3. Continued urbanization and climate change stressors were simulated to potentially change the distribution and extent of blue carbon ecosystems in the future
4. Maintenance and potential future expansion of blue carbon ecosystem in the Tampa Bay estuary will require continuing improvements in bay water quality and allowance/planning for habitat migrations upslope from the developed coast

26.1 INTRODUCTION: TAMPA BAY, AN URBAN ESTUARY IN RECOVERY

Tampa Bay has supported a thriving coastal community since at least the mid-1800s. Since that time, coastal development patterns to expand industrial, commercial, and residential land uses within the watershed have impacted critical coastal habitats to varying degrees (Simon 1974). Beginning in the 1950s, the most significant impacts to emergent tidal and subtidal habitats started to occur. Large-scale dredge and fill activities to create shoreline residential and commercial development opportunities resulted in the burial or removal of many of the Bay's critical coastal habitats (LES and CE 1996). It was not until the late 1970s and early 1980s that more stringent development regulations and environmental standards were implemented to prevent further loss of Tampa Bay's coastal habitats and helped foster the recovery of bay water quality (Greening et al. 2014). Since that time, bay resource managers have had varying success in restoring Tampa Bay's coastal blue carbon ecosystems, which include seagrass meadows, mangrove forests, salt marshes, and salt barrens.

As depicted in Figure 26.1, in 2011 urban and suburban land uses dominated (>60% of the land area or ~180,614 ha) the upland and coastal land cover within 15 km of Tampa Bay's shoreline (SWFWMD 2011). An additional 13.8% (41,572 ha) has been developed for agricultural purposes within a 15-km shoreline buffer area. The coastal habitats contained within this area have been recognized for their importance to the life history of specific estuarine species guilds, as well as for the varied ecosystem services to which they contribute (LES and CE 1996; TAS 1999; Cicchetti and Greening 2011; Russell and Greening 2015). Consequently, Figure 26.1 also shows that the remaining natural uplands, freshwater and estuarine wetlands, and restored natural lands now only encompass about 25.6% (76,957 ha) of the 15 km buffer area along Tampa Bay's shoreline (as of 2011).

Despite these significant, historic conversions of coastal habitats to developed land uses, Tampa Bay continues to support a thriving estuarine ecosystem. In turn, the ecosystem services provided by the Bay's coastal habitats support and directly contribute toward a substantial regional economy (TBRPC 2015). About half of the regional employment is dependent upon the Bay itself, and in

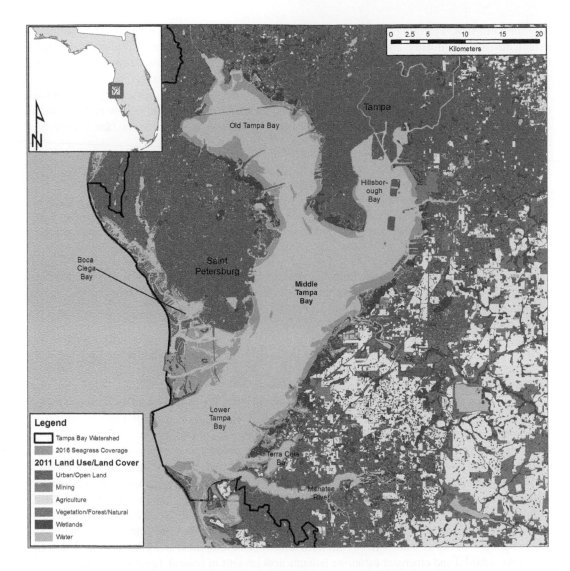

Figure 26.1 **(See color insert following page 266.)** Current (as of 2011 for emergent areas and 2016 for subtidal seagrass meadows), generalized land use, and land cover along coastal Tampa Bay. (Sources: SWFWMD 2011, 2016.)

total, 1 in 5 jobs in the region depends on a "healthy" Tampa Bay. In total, a healthy Tampa Bay contributes about $22 billion to the local regional economy in the form of jobs and commerce (TBRPC 2015). In addition, Russell and Greening (2015) estimated specific ecosystem service values for seagrass and estuarine wetland restoration efforts in Tampa Bay over the 1990–2008 period. Russell and Greening (2015) conservatively estimated that the habitats generate $24 million year^{-1} in denitrification and carbon sequestration ecosystem services that are not necessarily reflected in the regional economic estimates for a "healthy" Tampa Bay.

A growing regional economy and a recovering, healthy estuary and waterfront may create new pressures for the future management of the Bay's limited coastal habitats. Future development patterns and an anticipated doubling of the current population by 2050 could diminish both the economic and environmental integrity of Tampa Bay's coastal habitat resources. Not only does the expanding human populace pose a threat through both direct and indirect land development stressors, but new, emerging climate change stressors may also work synergistically to alter the extent, distribution, quality, and ecosystem services provided by the remaining critical coastal habitats in Tampa Bay. As such, bay managers have been looking to develop new incentives and policies to promote the protection and expansion of these important estuarine habitats.

Because Tampa Bay hosts three primary blue carbon ecosystems (seagrass, mangrove, and salt marsh), opportunities to encourage continued public and private investment in new restoration and conservation measures is being investigated. Recent research has focused on the blue carbon benefits of Tampa Bay habitats and has aimed to:

1. determine the greenhouse gas (GHG) sequestration benefits of existing coastal habitats and the climate change mitigation benefits derived from coastal habitat restoration efforts to date;
2. determine the anticipated changes in GHG emission and removal from coastal habitats in response to anticipated future land development, sea level rise (SLR), and climate change impacts; and,
3. identify future land management options that sustain and expand upon current coastal habitat extent and carbon sequestration potential.

Outcomes from this research are helping to guide future coastal restoration and habitat management objectives in Tampa Bay and will contribute to both local and global efforts to restore coastal ecosystems for the added incentive of mitigating future GHG impacts.

26.2 TAMPA BAY'S BLUE CARBON ECOSYSTEMS

Tampa Bay is located on Florida's west-central, Gulf of Mexico coast between $27.5°$ and $28.0°$ latitude, as depicted in Figure 26.1. The northern extent of the watershed borders between subtropical and temperate climates (Yates et al. 2011), though recent temperatures are trending toward a more subtropical climate (Martinez et al. 2012). As a result, a mosaic of subtidal and emergent estuarine wetlands tolerant to mild subtropical climates is present. These habitats (seagrass meadows, mangrove forests, salt marshes, and salt barrens) are collectively referred to as blue carbon ecosystems throughout this chapter. Recently, Raabe et al. 2012 demonstrated that Tampa Bay emergent blue carbon ecosystems are now trending toward subtropical-dominant mangrove forest habitats. Raabe et al. (2012) hypothesized that salt marsh to mangrove forest conversion over the 19th to 20th centuries was due to three primary drivers: global climate change, hydrologic alterations to emergent tidal wetlands, and the changes to freshwater inflows from continued landscape development and urbanization. A description of the subtidal and emergent estuarine habitats now present in coastal Tampa Bay follows.

26.2.1 Subtidal Blue Carbon Ecosystems

26.2.1.1 Seagrass Meadows

Five seagrass species occur in Tampa Bay; however, three species are the most predominant. Stable seagrass beds in the higher salinity regions of Tampa Bay—typically toward the mouth of the Bay and adjacent to the Gulf of Mexico—are primarily composed of *Thalassia testudinum* (turtle grass). Turtle grass is the largest seagrass species, with long strap-shaped leaves and robust rhizomes, which can provide significant carbon storage and GHG sequestration potential (McLeod et al. 2011; Fourqurean et al. 2012; Greiner et al. 2013). In the more dynamic salinity regions of the Bay—typically adjacent to the headwaters and ranging down to Middle Tampa Bay—*Halodule wrightii* (shoal grass) is the dominant species. It has flat, narrow leaves and a shallow root system. Shoal grass can tolerate more frequent exposure from low tides than other Tampa Bay seagrass species, and thus usually occupies shallow, fringing areas adjacent to more dense turtle grass beds. Shoal grass may also dominate the deep-water edge of Tampa Bay seagrass meadows, depending upon the region. The third dominant species, *Syringodium filiforme* (manatee grass), is usually found in the higher salinity regions of Tampa Bay, in association with turtle grass, but typically toward the deeper extent of these meadows. Manatee grass can be distinguished by its long, cylindrical leaves. Two other minor species occur in patchy, ephemeral distributions primarily in the upper, lower salinity portions of Tampa Bay primarily adjacent to shoal grass-dominated meadows: *Ruppia maritima* (widgeon grass) and *Halophila engelmannii* (star grass).

Globally, subtidal seagrass beds are highly valued coastal habitats that are experiencing rapid decline (Waycott et al. 2009; Unsworth et al. 2015). In Tampa Bay, historic declines in seagrass have also been documented (Greening et al. 2014), though recent efforts to restore Bay water quality have resulted in a significant recovery of these habitats (Greening et al. 2014; Sherwood et al. 2015; 2017). Contemporary restoration efforts continue to focus on maintaining or improving water quality conditions in Tampa Bay so that adequate light reaches shallow (<2 m), subtidal flats located throughout the Bay (Yates et al. 2011; Greening et al. 2014). Of the seagrass species present, turtle and shoal grass probably form the greatest complex of seagrass meadows that possess inherent blue carbon benefits in Tampa Bay. Both species can form dense, stable, climax seagrass meadows if water quality and available light are conducive to a particular species' growth, reproduction, and rhizome expansion. Additionally, carbonate-rich sediments in Tampa Bay may also provide a mechanism for enhanced blue carbon benefits from seagrass meadows. Chemically driven carbon sequestration through the "bicarbonate pathway" (Smith 1981; Burdidge and Zimmerman 2002; Tomasko et al. 2016) may account for the disparity between high production rates and low soil-carbon content for seagrass meadows in carbonate-rich, subtropical environments.

26.2.2 Emergent Tidal Wetland Blue Carbon Ecosystems

As is the case for seagrass meadows, a mosaic of emergent tidal wetland habitats consisting of mangrove, salt marsh, and salt barren species occur in specific salinity gradients and intertidal elevations throughout Tampa Bay proper and the tidal tributaries flowing into the Bay (LES & CE 1996; PBS&J 2010; Cicchetti and Greening 2011; Yates et al. 2011). These emergent habitats typically occur along the Bay's low-energy, natural, shallow-sloping, intertidal shelves that have not been converted to alternative land uses as a result of Tampa Bay's historic suburban and urban development patterns. They can also fringe the more developed areas throughout the bay, including bulkhead and seawalled shorelines—though the ecosystem function and service value of these altered shoreline habitats are presumed to be reduced.

26.2.2.1 Mangrove Forests

Mangrove forests are the dominant emergent tidal wetland in Tampa Bay, consisting of four primary species. At the lowest elevation usually along the fringing intertidal/shoreline zone, red mangroves (*Rhizophora mangle*) are typical. Black mangroves (*Avicennia germinans*), white mangroves (*Laguncularia racemosa*), and buttonwood (*Conocarpus erectus*) typically follow, in that order, upslope along the intertidal zone (PBS&J 2010; Yates et al. 2011). Mangrove forests produce, sequester, and export large pools of organic carbon (Odum and McIvor 1990), and as such are an important global blue carbon ecosystem (Donato et al. 2011; Murdiyarso et al. 2015).

Some of the more successful emergent tidal wetland restoration projects in Tampa Bay have involved the creation of mangrove forests, though the planting of salt marsh was generally a precursor to mangrove succession. The functional equivalence of restored mangrove habitats in terms of soil-carbon storage and accumulation, however, has not yet approached natural systems (Osland et al. 2012; Gonneea 2016). Nonetheless, mangrove forests and soils collectively represent a significant carbon stock within the Tampa Bay estuary.

26.2.2.2 Salt Marshes

26.2.2.2.1 Polyhaline Salt Marshes

Polyhaline salt marshes may occur seaward of the fringing mangrove coast in Tampa Bay proper where salinities generally remain above 20 psu throughout the rainy season (nominally June–September). Smooth cordgrass (*Spartina alterniflora*) is the common species in these circumstances; however, at higher shoreline elevations around seasonal high tide levels other species may occur, particularly around beach and dune formations in the lower part of Tampa Bay. In these regions, saltmeadow cordgrass (*Spartina patens*), saltgrass (*Distichlis spicata*), saltwort (*Batis maritima*) and salt jointgrass (*Paspalum vaginatum*) may also be present. For Tampa Bay habitat management purposes, polyhaline salt marshes are typically included as a component of mangrove forests due to the difficulty in partitioning these habitats from mapping and land use cover analyses. Further, tidal wetland restoration practices in Tampa Bay have evolved to include careful grading and planting of pioneering salt marsh species (e.g., *Spartina* spp.) to encourage recruitment and retention of mangrove seedlings in order to revegetate and restore intertidal sites to a climax mangrove forest condition (Henningsen et al. 2003; Ries 2009; Osland et al. 2012).

26.2.2.2.2 Meso-Oligohaline Salt Marshes

Distinct meso-oligohaline salt marsh plant communities are found predominantly in Tampa Bay tidal rivers and creeks where freshwater inflow reduces salinities year-round to between 0.5 and 20 psu (PBS&J 2010). In the upper tidal reaches, oligohaline marshes persist where salinities typically range between 0.5 and 5 psu. Within the upper tidal reaches, the plant community exhibits a mixture of true marine plants and typical freshwater taxa such as cattails (*Typha domingensis*) and sawgrass (*Cladium jamaicense*) that tolerate low-salt concentrations. The predominant plant species of oligohaline marshes include black needlerush (*Juncus roemerianus*), leather fern (*Acrostichum danaeifolium*), cattails, sawgrass, bulrush (*Scirpus robustus*), and spider lily (*Hymenocallis palmeri*). A transition to mesohaline salt marshes typically occurs in the lower tidal reaches where black needlerush and leather fern tend to dominate and mangrove species start to invade the mesohaline salt marsh habitats. Ecologically, these low salinity tidal reach marshes have been identified as critical nursery habitats to commercially and recreationally important fisheries (Wessel and Dixon 2016). In addition, the organic-rich sediments have been recognized as important carbon sinks that provide substantial blue carbon benefits (McLeod et al. 2011).

26.2.2.3 Salt Barrens

Tampa Bay salt barrens are typically located slightly upslope of mangrove forest or salt marsh habitats at somewhat higher elevations where tidal inundation periodically occurs due to spring tides. As a result, soils are typically hypersaline and seasonal expansion of low-growing, succulent, salt-tolerant vegetation occurs during the rainy season (June-September). Annual glasswort (*Salicornia bigelovii*), perennial glasswort (*Salicornia virginica*), key grass (*Monoanthochloe littoralis*), sea lavender (*Limonium carolinianum*), samphire (*Blutaparon vermiculare*), and sea purslane (*Sesuvium portulacastrum*) are common plant species that occupy Tampa Bay salt barren habitats. Due to their low structural complexity and apparent lack of numerous fauna, salt barrens are often assumed to have low ecological value; however, these tidally inundated habitats can be important wading bird forage areas and are occupied by fiddler crab species (*Uca* spp.) that actively process sediment deposits. As such, the inherent blue carbon benefits elicited by this habitat type are not well known, though recent research by Moyer et al. (2016) and Gonneea (2016) suggest that salt barrens provide some above- and below-ground carbon sequestration benefits in the Tampa Bay estuary.

26.3 CURRENT BLUE CARBON ECOSYSTEM STATUS AND TRENDS

26.3.1 Areal Extent

As shown in Figure 26.2, Tampa Bay blue carbon ecosystems declined in coverage between the 1900s and early 1990s primarily due to historic land development activities that have resulted in the expansion of suburban and urban areas throughout the watershed. Since the 1980s, public land conservation and acquisition, as well as discrete habitat restoration activities undertaken primarily by local and state agencies, but also by non-profit and private entities, has led to modest gains in emergent coastal habitat coverage, as depicted in Figure 26.2. In addition, a significant and concerted effort to curtail nitrogen load pollution over the 1980-present period has led to documented improvements in Tampa Bay water quality and a significant expansion of subtidal seagrass habitats (Greening et al. 2014; Sherwood et al. 2017). As a result, seagrass extent in Tampa Bay is now commensurate to 1950s coverage levels—a baywide restoration goal set by the community through the Tampa Bay Estuary Program (TBEP) and the Tampa Bay Nitrogen Management Consortium (Greening and Janicki 2006; Greening et al. 2011, 2014; Gross and Hagy 2017). Proportionally, recovery of seagrass coverage in Tampa Bay has far exceeded other emergent blue carbon ecosystem coverage gains over the contemporary period (1990–2016), as depicted in Figure 26.2.

26.3.2 Current Blue Carbon Ecosystem Benefits

Tampa Bay-specific blue carbon ecosystem stock and accumulation estimates for mangrove, salt marsh, and salt barren habitats were recently developed by Gonneea (2016), Moyer

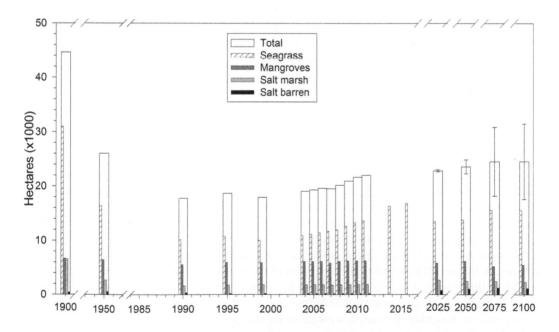

Figure 26.2 Estimated coverage of Tampa Bay blue carbon ecosystems prior to rapid population growth and urbanization in the watershed (ca. 1900 and 1950; adapted from LES & CE 1996), over the contemporary data record (1990–2015), and projected into the future utilizing a habitat evolution model (2025–2100; adapted from Sheehan and Crooks (2016), Sources: SWFWMD, TBEP).

Table 26.1   Comparison of Recent Tampa Bay Habitat Carbon Pools Relative to Other Published Values

Habitat	Location	Citation	Core Depth (cm)	Belowground C (Mg C·ha⁻¹)	Aboveground C (Mg C·ha⁻¹)	Total Carbon (Mg C·ha⁻¹)
Mangrove	Global	Alongi (2014)	62–3800			956
	Tampa Bay, FL	Radabaugh et al. (2018)	13–50	93	41	134
	Merritt Island, FL	Doughty et al. (2015)	30	67	55	122
Salt Marsh	Global	Alongi (2014)	12–50			593
	Tampa Bay, FL	Radabaugh et al. (2018)	4–50	62	4	66
	Merritt Island, FL	Doughty et al. 2015	30	52	8	61
Salt Barren	Tampa Bay, FL	Radabaugh et al. (2018)	4–50	26	1	27

Source: Adapted from Moyer et al. (2016) and Radabaugh et al. (2018).

et al. (2016), and Radabaugh et al. (2017). Radabaugh et al. (2017) expanded upon the available literature to estimate aboveground biomass, carbon and nitrogen content for 18 species of Tampa Bay salt marsh plants. Allometric equations and elemental content estimates developed by Radabaugh et al. (2017) highlighted the need for refined local data because salt marsh plant growth and elemental composition can vary according to position within the marsh, seasonal climate (temperate vs. sub-tropical), and/or other regionally unique conditions (e.g., hydrologic alterations). Moyer et al. (2016) and Radabaugh et al. (2018) expanded upon this work to develop estimates for both vegetative and soil-carbon stocks for salt marshes, mangroves, and salt barrens in Tampa Bay. Moyer's et al. (2016) results further emphasized site-specific carbon stock estimates for blue carbon ecosystems in Tampa Bay. Table 26.1 summarizes the divergence in carbon stock estimates for Florida habitats relative to other assessments for functionally similar habitats. Lastly, Gonneea (2016) found that mangrove and salt marsh ecosystems appeared to be keeping pace with recent regional SLR trends (accreting at 2.9–3.5 mm year⁻¹ relative to NOAA St. Petersburg, FL Tide station SLR trend of 2.66 mm year⁻¹ over the 1947–2015 period*), and are burying carbon at rates between 0.71 Mg C ha⁻¹ year⁻¹ (salt marsh) to 1.78 Mg C ha⁻¹ year⁻¹ (mature mangroves) over contemporary periods (1950–present).

Additionally, Sheehan and Crooks (2016) recently developed a tool to estimate the GHG sequestration benefits from ongoing coastal habitat restoration activities within the Tampa Bay estuary. Habitat restoration projects from 2006 to 2015 ($N = 254$) were utilized in tool development and included the following activities: debris removal, erosion control and grading, hydrologic restoration, invasive species control, land acquisition, mechanical thinning, prescribed burning, reef construction and vegetation establishment. Dependent upon the activity, net GHG emissions or sequestration was estimated for the initial restoration activity. Utilizing this tool and projecting the GHG sequestration benefits of newly restored habitats until 2050 (an approximate 35- to 45-year crediting period) results in 0.53 million tonnes CO_2 eq. of total sequestration. However, this total is offset by initial losses of vegetation (−0.24 million tonnes CO_2 eq.) due to specific restoration activities that are sometimes necessary to establish native habitats (e.g., invasive species removal, mechanical thinning, and prescribed burns) and CH_4 emissions (−0.17 million tonnes CO_2 eq.) from newly created mangrove or freshwater marsh ecosystems. As a result, the net GHG sequestration benefits of these projects until 2050 is only estimated to be 0.12 million tonnes CO_2 eq. These results are contingent upon the new habitats being maintained in the future despite potential SLR impacts, and they highlight that the initial restoration activity and the resulting habitats that establish on-site can influence the potential GHG sequestration benefits estimated over a given time period.

*NOAA Sea Level Trends: https://tidesandcurrents.noaa.gov/sltrends/sltrends_station.shtml?stnid=8726520.

26.4 PROJECTED CHANGES TO BLUE CARBON ECOSYSTEMS

26.4.1 Areal Extent

Glick and Clough (2006), Sherwood and Greening (2013) and Geselbracht et al. (2013) developed initial estimates of coastal habitat changes anticipated from varying future SLR scenarios. Although each study's focus was slightly different, all applied variations of the Sea Level Affecting Marshes Model (SLAMM*) and all showed general increases in mangrove areal extent dependent upon the SLR scenario simulated. Sherwood and Greening (2013) reported a likely reduction in blue carbon ecosystem areal extent at higher SLR scenarios with an overall greater proportional shift to mangrove habitats by 2100 regardless of SLR scenario. As a result, Sherwood and Greening (2013) called for a more pragmatic approach to future coastal habitat management in Tampa Bay that considered the implications of both impending coastal urbanization and climate change impacts that may alter the composition of blue carbon ecosystems in Tampa Bay in the future.

More recently, Sheehan and Crooks (2016) utilized an updated, GIS-based, habitat evolution model that extends implementation of the SLAMM to include considerations for seagrass habitat migration in response to SLR scenarios in Tampa Bay. Sheehan and Crooks (2016) developed modeled changes for blue carbon ecosystems based on two regionally specific SLR scenarios as recommended by the TBCSAP (2015) (i.e., NOAA intermediate low, 0.6 m by 2100; and NOAA intermediate high, 1.3 m by 2100), two accretion rate assumptions dependent upon blue carbon ecosystem (low range: 1.6–3.75 mm year^{-1}; high range: 3.0–5.0 mm year^{-1}), and two future management alternatives (protect developed land versus allowing habitats to migrate into undeveloped and "softly" developed lands). Sheehan and Crooks (2016) highlighted that:

1. Total blue carbon ecosystem extent could potentially be maintained in the future despite emergent coastal habitat losses that result from limited migration potential into currently developed lands, if a significant expansion of subtidal seagrass habitats into newly inundated coastal areas occurs, as depicted in Figures 26.2–26.4a;
2. The simulated expansion and future maintenance of subtidal seagrass habitats will be highly dependent upon maintaining adequate water quality in newly inundated areas, as well as in the areas where seagrass currently exist but that would become deeper with future SLR;
3. Emergent blue carbon ecosystems (mangroves and salt marshes) were more likely to persist into the future if the ecosystems responded to future SLR with higher accretion rates; lower accretion rate assumptions resulted in a mangrove tipping point toward the latter part of the century when significant mangrove extent losses were modeled to occur in response to SLR overwhelming the intertidal elevations where these habitats occurred; and,
4. Maintaining and expanding emergent blue carbon ecosystems to any great extent in the future would require the future sacrifice of uplands regardless of their current development status (i.e., setting aside conservation areas and/or refugia and/or providing migration corridors within both undeveloped and "softly" developed uplands).

26.4.2 Sequestration Value and Benefits

The simulated changes in blue carbon ecosystems developed by Sheehan and Crooks (2016) were further utilized to develop estimates of net GHG fluxes and sequestration by 2100 for all projected land conversions under six model scenarios in Tampa Bay, as summarized in Table 26.2. Sheehan and Crooks (2016) applied the IPCC (2006) GHG accounting framework and values from Gonneea (2016) and Moyer et al. (2016) to estimate land use-specific changes in biomass C stock, additional

---

*Sea Level Affecting Marshes Model: http://warrenpinnacle.com/prof/SLAMM/.

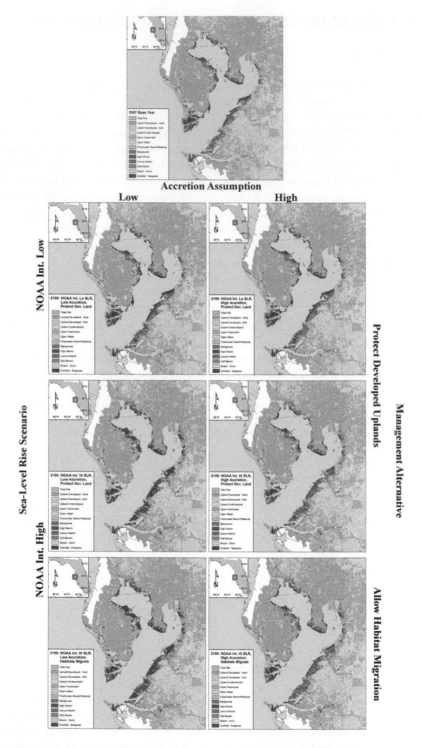

Figure 26.3   **(See color insert following page 266.)** Comparison of 2007 blue carbon ecosystem extent and distribution to modeled changes in ecosystems relative to two SLR scenarios (low versus high), two accretion assumptions (low vs. high), and two management alternatives (protect currently developed land vs. habitats allowed to migrate to undeveloped and "softly" developed lands) (Adapted from Crooks and Sheehan (2016), Sources: SWFWMD, TBEP.).

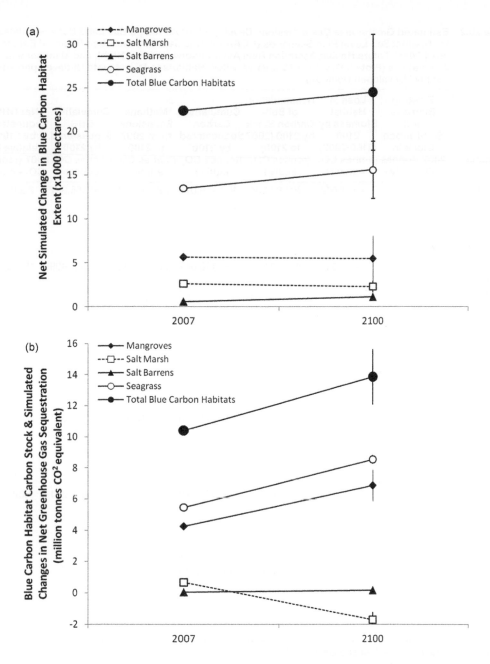

Figure 26.4 (a) Average change in blue carbon ecosystem extent over the 2007 and 2100 modeled period under the six scenarios presented in Figure 26.3. (b) Current C stock (modeled in 2007) and average estimated net total change in GHG sequestration potential for 2100 (adapted from Crooks and Sheehan 2016). Error bars for 2100 period represent the standard deviation of the average of 6 model simulations, as depicted in Figure 26.3. Dotted lines indicate the simulated average is reduced from 2007 estimates.

soil C sequestration generated from future land use extent, soil C emissions from converted habitats, and CH_4 emissions from future habitat extent to develop a net GHG sequestration estimate relative to a 2007 modeled base year. When accounting for all potential future land-use conversions, Tampa Bay GHG sequestration potential was estimated to increase on average by 2.1 ± 0.5 million tonnes

Table 26.2 Estimated Greenhouse Gas Emissions Developed for All Tampa Bay Land Uses Simulated for Distinct Sea-Level Rise Scenarios (NOAA Intermediate Low and High Regional Estimates, as of 2016), Future Habitat Accretion Rate Assumptions (Low vs. High), and Management Alternatives (Protect Developed Lands vs. Allow Habitats to Migrate into Undeveloped and "Softly" Developed Uplands)

Scenario (Run)	Existing Biomass and Soil-Carbon Stock in 2007 (tonnes CO$_2$ equiv.)	Loss in Habitat Biomass by 2100 (2100-2007) (tonnes CO$_2$ equiv.)	Accumulation of Soil-Carbon Stock by 2100 (2007 to 2100) (tonnes CO$_2$ equiv.)	Cumulative Carbon Sequestered by 2100 (tonnes CO$_2$ equiv.)	Methane Emissions from 2007 to 2100 (tonnes CO$_2$ equiv.)	Cumulative GHG Sequestered by 2100 (tonnes CO$_2$ equiv.)	Net GHG Sequestered by 2100 relative to 2007 (tonnes CO$_2$ equiv.)
Run 1-low slr, low accretion, protect developed lands		−161,042	30,682,234	102,458,079	−28,302,683	74,155,397	2,218,509
Run 2-low slr, high accretion, protect developed lands		−157,558	30,719,337	102,498,667	−28,391,168	74,107,499	2,170,611
Run 3-high slr, low accretion, protect developed lands	71,936,887	−1,535,591	29,997,627	100,398,923	−26,983,877	73,415,046	1,478,159
Run 4-high slr, high accretion, protect developed lands		−352,169	30,697,049	102,281,768	−27,964,916	74,316,851	2,379,964
Run 5-high slr, low accretion, habitats allowed to migrate		−1,373,153	30,019,445	100,583,179	−27,023,918	73,559,261	1,622,374
Run 6-high slr, high accretion, habitats allowed to migrate		−191,194	30,908,125	102,653,818	−28,004,019	74,649,799	2,712,912

Source: Adapted from Sheehan and Crooks (2016).

CO$_2$ eq. (±1 S.D.) by 2100. The average is equivalent to eliminating CO$_2$ emissions from >9.6 × 10^6 L of consumed gasoline **every** year between 2007 and 2100 (USEPA 2016).

Specifically for Tampa Bay blue carbon ecosystems, Figure 26.4b shows that, on average, additional net positive GHG sequestration was estimated for all blue carbon ecosystems under the six scenarios except for salt marsh—a result of direct loss in salt marsh extent and conversions to brackish marsh habitats that emit higher methane levels in the future. Figure 26.4a shows that although on average the total mangrove extent was estimated to be reduced by 2100, mangrove habitats were still expected to provide net positive GHG sequestration potential in the future due to continued soil C accumulation, as depicted in Figure 26.4b. Collectively, seagrass and mangrove habitats are projected to provide the greatest GHG sequestration potential for the future, and compensate for significant loss and emissions projected by salt marsh habitats by 2100, as depicted in Figure 26.4b. In total, blue carbon ecosystems were estimated to provide a net positive GHG sequestration potential of 3.5 ± 1.8 million tonnes CO_2 eq. (± 1 S.D.) by 2100. The average is equivalent to eliminating CO_2 emissions from ~16.1×10^6 L of consumed gasoline **every** year between 2007 and 2100 (USEPA 2016).

26.5 INCORPORATING BLUE CARBON BENEFITS INTO TAMPA BAY'S COASTAL HABITAT RESTORATION AND PROTECTION STRATEGY

Over recent decades, the Tampa Bay environmental resource community has been largely successful in increasing the extent of important coastal habitats within the estuary, despite continued coastal development and an expanding human populace within the watershed (Cicchetti and Greening 2011; Greening et al. 2014; Sherwood et al. 2017). The challenge for future bay management efforts is to sustain this momentum with the looming pressures of additional coastal development and future climate change impacts. The existing ecosystem benefits and services garnered from contemporary management efforts have been well documented for Tampa Bay coastal habitats and also provide significant regional economic benefits (Greening et al. 2014; Russell and Greening 2015; TBRPC 2015). Further, the current 72 million tonnes CO_2 eq. carbon stock of all land uses has the potential to provide additional GHG sequestration ecosystem benefits, if properly managed into the future. If improperly managed and important blue carbon ecosystems are completely lost, then a significant C pool equal to the one-year emissions of about 21 coal-fired power plants (USEPA 2016) would be expected to further exacerbate the USA's contributions to global climate change.

However, managing Tampa Bay coastal habitats for their inherent blue carbon benefits alone may not be entirely compatible with existing management objectives. For example, maximizing blue carbon benefits by focusing restoration efforts only toward those habitat types with high GHG sequestration rates (i.e., mangroves and/or seagrasses) could preclude inclusion of other habitat types with lower sequestration rates (salt barrens, oligohaline habitat) into restoration plans that may provide important ecosystem function or other ecosystem benefits in the future (e.g., neglecting the fishery production benefits of restoring meso- and oligohaline nursery salt marsh habitats because of greater CH_4 emissions when compared to mangrove or polyhaline salt marsh habitats). Also, exotic plant species removals [e.g., Brazilian pepper (*Schinus terebinthifolius*) and other woody invasives] that support restoration and recovery activities of native coastal habitat species can temporarily increase CO_2 emissions by removing mature vegetation and temporarily exposing sediment C pools (see 26.3.2). Thus, future Tampa Bay coastal habitat restoration initiatives must carefully weigh and determine any added blue carbon benefits that may be provided through different management and restoration practices.

Regardless, knowledge of the blue carbon benefits resulting from implementation of a refined Tampa Bay coastal habitat restoration and protection strategy could provide additional rationale and incentive to invest in future restoration efforts in a watershed limited by undeveloped lands, as depicted in Figure 26.1. Determining the added GHG sequestration benefits that these habitats provide, in addition to the more traditionally understood ecosystem service values and benefits for fisheries and wildlife, nutrient reduction and coastal protection values, could further drive investment into restoration activities on managed public lands as well as privately held undeveloped lands. Leveraging additional investment capital within the region for new restoration projects focused on capitalizing on the blue carbon benefits of these habitats may be a challenge, though. The size and scope of future restoration activities in the Tampa Bay watershed is limited due to existing development. So, the costs associated with initially scoping a blue carbon ecosystem restoration project for entrance into a voluntary C market can be prohibitive. The voluntary C markets are currently the only mechanism available in the USA to potentially help finance restoration activities through GHG accounting. As such, the TBEP has begun to investigate the potential for a "grouped" watershed restoration activity whereby GHG accounting could be collectively implemented across both public and private land management interests and the resulting voluntary C market incentives invested and shared among partnering entities (Simpson 2016). In this approach, restoration and GHG sequestration accounting would be pooled across multiple restoration projects that may be too small a scale to initiate a C financing program on their own.

The potential for voluntary C market financing of blue carbon ecosystem restoration in Tampa Bay would be an additional resource management tool that would help sustain momentum for the Bay's continued ecosystem recovery. The added monetary incentives could influence alternative land management practices within the watershed, especially for areas that could become highly vulnerable to tidal inundation in the future. The projected changes in coastal habitats anticipated from future SLR, as referenced in earlier sections, highlights the potential and need for blue carbon ecosystem restoration opportunities along vulnerable coastal areas in the Bay. If higher SLR projections are realized, creating more space for landward habitat migration will be necessary along the undeveloped and "softly" developed coast to maintain the current balance of ecosystem types. As depicted in Figure 26.3, public and private land owners in vulnerable areas could also capitalize on the added incentives of increasing net GHG sequestration through the voluntary C market by accounting for and allowing blue carbon ecosystem migration and establishment to occur along the periphery of the Bay. The same could be true for newly inundated subtidal areas that will eventually become held in the public interest. Public entities who become owners of newly inundated subtidal areas that establish new seagrass meadows could capitalize on the GHG sequestration potential of these habitats. This scenario would be highly dependent on maintaining adequate water quality and ensuring that seagrass expand into these newly inundated areas in the future. This, in turn, could provide additional incentive to invest in watershed restoration activities that continue to improve water quality throughout the Bay—the overarching management paradigm of the TBEP and its partners since the 1990s that has led to positive seagrass recovery in Tampa Bay (Greening et al. 2014). These examples highlight that establishing future subtidal and upland refugia is a viable management mechanism for blue carbon ecosystem restoration and protection and that public and private land owners in these potential refugia areas could capitalize on the GHG sequestration potential of sustaining these habitats in these vulnerable areas through a collaborative, "grouped" approach (Simpson 2016).

Other ecosystem benefits may also arise from continued blue carbon ecosystem protection and restoration in Tampa Bay. For example, water column pH throughout Tampa Bay has been increasing since the mid-1980s, coincident with the expansion of seagrass meadows observed over the 1990–2016 period and contrary to global ocean acidification (OA) trends (Sherwood et al. 2015; 2017). The role that seagrass may play in buffering coastal OA processes within the Tampa Bay estuary is currently being investigated. Yates et al. (2016) has initially found that seagrass meadows in parts of Tampa Bay are capable of increasing daytime pH values by 0.5 units, consistent with the expectation that inorganic carbon is assimilated through photosynthesis. In addition, Yates et al. (2016) found that seagrass meadows increased, at least locally, carbonate saturation rates in the water column, consistent with the findings of Manzello et al. (2012) and suggesting that the mechanisms involved in the bicarbonate pathway outlined by Burdige and Zimmerman (2002) could be occurring in Tampa Bay seagrass meadows. Thus, continued expansion of an important blue carbon ecosystem in Tampa Bay could mitigate global and coastal OA processes that are a direct result of increasing, global atmospheric CO_2 concentrations, and allow the Bay to become a regionally significant OA refugia along the Gulf of Mexico coast for estuarine species, including commercially and recreationally important fisheries susceptible to OA impacts. This further adds support for the economic and ecologic benefit of continued blue carbon ecosystem restoration in Tampa Bay.

In summary, future Tampa Bay coastal habitat protection and restoration initiatives must consider the impending pressures of ongoing development and climate change in activities that promote long-term habitat persistence, succession, and migration. Resource managers now have an added tool to engage public and private landowners in potential C market financing incentives that would provide the needed space for Tampa Bay blue carbon ecosystems to thrive in the future. Currently, the barriers to attempt this management approach are centered on the limited-scale of the remaining, restorable lands within the watershed and the prohibitively high costs to register, monitor, and verify the individual, small-scale projects. A solution to this could be a collaborative, "grouped" project approach throughout the Tampa Bay watershed. A collaborative restoration approach has already worked for the recovery of seagrass in Tampa Bay (Greening et al. 2014), now the resource management community will need to implement additional strategies, such as C market financing, to ensure that all of Tampa Bay's blue carbon ecosystems, including seagrass meadows, mangroves, salt marshes and salt barrens, are maintained and allowed to expand in the future.

CHAPTER 27

Blue Carbon Opportunities for Mangroves of Mexico

Maria Fernanda Adame
Griffith University

Richard Birdsey
Woods Hole Research Center

Jorge Herrera Silveira
CINVESTAV-IPN Unidad Merida

CONTENTS

HIGHLIGHTS

1. Mexico has the fourth largest area of mangroves in the world, comprising a range of climatic and geomorphological settings from the subtropical arid to the humid tropical.
2. Mangrove deforestation rates were high until 2015, when rates stabilized close to zero. However, mangrove degradation is increasing.
3. In recent decades, there has been an improvement in the number and quality of studies of the C cycle of mangroves in Mexico, supported by federal and international programs, including a federal program to monitor changes in mangrove area.
4. Mexico has the legal framework, federal and international support, and the institutional arrangements to benefit from financial mechanisms to reduce greenhouse gas emissions. However, implementation at the local scale, lack of scientific information in some locations, and corruption can hamper such efforts.
5. Improving scientific knowledge and communication, involving the local community, and strengthening current programs for payment of ecosystem services and early REDD+ projects could promote conservation and restoration of the large areas of mangroves in Mexico.

Mexico has the fourth largest area of mangroves in the world with 775,557 ha covering much of the country's coastline (National Commission on Biodiversity, CONABIO 2015; Figure 27.1). In Mexico, mangroves are dominated by four species: red mangrove (*Rhizophora mangle*), white

Figure 27.1   (See color insert following page 266.) Mangrove extension and regions on the basis of climate, hydrology, geologic and geomorphologic characteristics. Map and regionalization are
from CONABIO (2009). Key locations for studies of mangroves are: (a) arid mangroves in Bahia
Magdalena, Baja California; (b) riverine mangroves in La Encrucijada Biosphere Reserve, Chiapas;
(c) karstic mangroves in Celestun, Yucatan; and (d) karstic mangroves in Sian Kaan, the Mexican
Caribbean. Photos from MF Adame.

mangrove (*Laguncularia racemosa*), black mangrove (*Avicennia germinans*), and buttonwood mangrove (*Conocarpus erectus*). The mangroves in Mexico are highly variable in structure and function, occupying a range of sites from the subtropical arid north coast, to the tropical and humid

Table 27.1 Mangrove Regions of Mexico

Region	Climate	Annual Mean Precipitation (mm)	Mean Min T (°C)	Mean Max T (°C)	Tidal Range (m)	Mangrove Characteristics
I) North Pacific	Dry	100–300	18	22	0.5–3.5	Estuarine mangroves. Presence of scrub hypersaline mangroves
II) Central Pacific	Hot sub-humid	1,000–2,000	22	26	0.5–1.2	Estuarine and riverine
III) South Pacific	Hot humid and sub-humid	1,000–4,000	22	26	0.5–1.2	Mostly riverine. Tallest mangroves in the country
IV) Gulf of Mexico	Temperate sub-humid to humid	600–4,000	10	22	0.3–0.4	Estuarine and riverine
V) Yucatan Peninsula	Hot sub-humid	1,000–2,000	22	26	0.0–0.05	Karstic substrate, groundwater. Presence of scrub mangroves and mangroves associated to freshwater springs

Precipitation and temperature values from National Water Commission (CAN)-National Institute of Geography and Statistics (INEGI). Tidal range values from National Autonomous University of Mexico (UNAM), Geophysics Institute.

far south. Mangroves in Mexico can be classified in five regions according to climatic, hydrologic, geologic, and geomorphologic characteristics (CONABIO 2009; Table 27.1, Figure 27.1).

Until recently, Mexico had high deforestation rates. From 1970 to 2005, Mexico lost 10% of its mangrove cover, with some regions losing up to 1% of their mangroves every year (Ramírez-García et al. 1998; Valderrama et al. 2014). Agriculture, cattle grazing, shrimp farming, urbanization, and tourist development are the most common causes of mangrove loss (Ramírez-García et al. 1998; Ruiz-Luna et al. 2008). In recent years, deforestation has decreased with recent estimates reaching a national mean of zero losses from 2010 to 2015 (CONABIO 2016).

In 2007, all mangroves in Mexico were declared fully protected under the innovative law 60 TER (General Wildlife Law). This comprehensive law prohibits the removal, fill, transplant, cut or any other activity that affects the integrity or the hydrological flux of the mangroves, its ecosystem, and its area of influence. Despite the legal protection, mangrove deforestation and degradation is still a serious problem in some areas. However, there is now a wider national appreciation of the values and ecosystem services that mangroves provide (López-Medellín et al. 2011), including their role in sequestering carbon (C).

27.1 CARBON CYCLE SCIENCE IN MEXICO

In recent decades, there has been an improvement in the number and quality of studies of the C cycle of mangroves in Mexico, with more than 200 studies published (Herrera-Silveira et al. 2015). Numerous national programs and some international organizations have provided funding for such studies. For example, the long-term monitoring program of the National Commission on Biodiversity (CONABIO) has financed more than 25 projects on the study of mangroves in the country since 2008.

Mangrove research at large spatial scales in Mexico has been facilitated by the release of maps of mangrove area throughout the country in 2009 by CONABIO. These maps are readily available to download for free (http://www.conabio.gob.mx/informacion/gis/) and are reassessed every 5 years. These maps along with improved satellite image analyses and field data have facilitated accurate estimations of C content, losses and gains at large spatial scales (e.g., Valderrama et al. 2014; Adame et al. 2015).

Data about the C content of different ecosystem components, and how carbon is affected by disturbance are needed to understand the potential for mangroves to store and sequester carbon. First, we need to know how much C is stored in the live biomass, the soil, and the dead wood accumulated on the forest floor. Second, we need to know how much C is sequestered as trees grow and soil accretes through time. Third, we need to know how much C is lost to land-use change and other disturbances. In the following paragraphs, we review the information available on C stocks and sequestration of mangroves in Mexico.

The C stored in the biomass of live trees can be estimated from aboveground measurements, i.e., measurements of the diameter and height of the trees. In Mexico, aboveground biomass of mangroves is generally well studied (e.g., Flores-Verdugo et al. 1987; Zaldívar-Jiménez et al. 2004). However, mangroves from the arid temperate North and Central Pacific Coast are poorly documented compared to the rest of the country (Herrera-Silveira et al. 2015). This gap is noteworthy, as mangroves in the North and Central Pacific coast comprise one-fourth of the mangroves of the country (CONABIO 2009).

Information on C stored in the soil is less available compared to information of C in trees. Recent studies in the Gulf of Mexico, South Pacific Coast, and the Yucatan Peninsula have provided detailed and valuable information on C stocks of mangroves that include the soil component (e.g., Adame et al. 2013, 2015; Kauffman et al. 2015). With this newly available information, a better understanding of C stored in mangroves in Mexico has developed (Figure 27.2). Research has shown that unique mangroves in freshwater springs of Yucatan have high potential for storing C. We also know that scrub mangroves, which are typically only 40 cm tall, can store as much C as a tropical rainforest (Adame et al. 2013, Gutierrez-Mendoza and Herrera-Silveira 2015). We have learned that flooding is associated with the potential of mangroves to store C in the Yucatan peninsula (Cerón-Bretón et al. 2011), and that riverine mangroves in the Pacific coast can store C for centuries (Adame and Fry 2016).

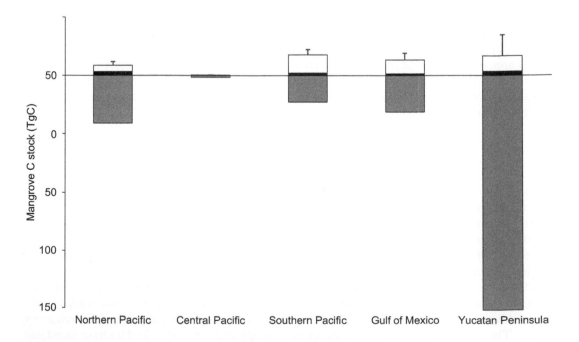

Figure 27.2 C stocks Mexico of different kinds of mangroves in Mexico. Data from Adame et al. (2013, 2015); Kauffman et al. (2015); Caamal et al. (2015).

Soil C has different sources; it can be derived from the ocean, the land, or the mangroves themselves. In Mexico, there have been two studies that have revealed the primary sources of C in mangroves within a coastal lagoon and a river mouth (Eagle et al. 2004; Adame and Fry 2016). The results show that most or the entire C stored in mangroves is from autochthonous origin, i.e., from the local mangrove production.

Sequestration of C refers to the rate at which C is incorporated and retained in the forests. Sequestration rates of mangroves are more difficult to measure than C stocks. Estimation of sequestration rates requires long-term measurements of C movement throughout the system; for example, how much C is incorporated into the trees as they grow? How much C is deposited on the forest floor as litterfall? How much of the litter is exported to the ocean as the tide floods the forest or from storm runoff? How much of the litter is decomposed and released back to the atmosphere? Production rates of aboveground biomass are generally well quantified in mangrove forest of Mexico because this information can be based on repeated tree measurements (e.g., Flores-Verdugo et al. 1987; Adame et al. 2013; Mendoza-Morales et al. 2015). But rates of root production, C burial, decomposition, and import and export during tidal inundation are very scarce (e.g., Flores-Verdugo et al. 1987; Adame et al. 2014; Camacho-Rico and Herrera-Silveira 2015), with most studies taking place in a limited region, mostly involving mangroves of the Yucatan Peninsula (Herrera-Silveira et al. 2015).

To estimate the potential for sequestering C by mangroves over the long term, the most important factor is to understand how much C is accreted and preserved in the soil. In the South Pacific Coast of Mexico, it has been estimated that mangrove soils accumulate between 1 and 5 tonne of C per hectare, every year (Adame et al. 2015). And arid mangroves in Baja California are estimated to accumulate between 0.1 and 2.6 tonne of C per hectare every year (Ezcurra et al. 2016). Mangroves in Campeche, in the Yucatan Peninsula accrete 3–4 mm every year (Lynch et al. 1989), which roughly corresponds to sequestration rate of one tonne of C per hectare per year. This information is not available for other regions in Mexico.

The loss of C in mangroves resulting from disturbances is often estimated by assuming that a proportion of the existing C in each of the different ecosystem components is emitted to the atmosphere or transferred to another carbon pool following disturbance. In Mexico, C loss has been estimated at 370 MgC per hectare as a result of converting mangrove forest to cattle pasture in Pantanos de Centla in the Gulf of Mexico (Kauffman et al. 2015). This value is in the upper end compared to the global estimated average loss of 112–392 MgC ha^{-1} cleared for other uses, depending on how deeply the clearing affects stored C in the soil (Donato et al. 2011). The emissions of methane, a powerful greenhouse gas, also need to be measured and monitored on intact and disturbed mangroves, especially those associated with freshwater springs. In coastal lagoons surrounded by mangroves in Yucatan, methane emissions were associated with low salinity and high nutrient pollution (Chuang et al. 2016), and in arid mangroves, the production of methane is below detection levels (Giani et al. 1996).

Detailed information on C fluxes is still scarce for disturbed and intact mangroves of the country. However, the recent National Intensive Carbon Monitoring Network aims to fill this information gap by measuring daily CO_2 fluxes between forest and atmosphere (Vargas et al. 2012). The Network includes different ecosystems throughout the country, such as temperate forests, dry and wet rainforests, and a mangrove forest in the South Pacific Coast in La Encrucijada Biosphere Reserve.

Currently, numerous projects are tackling the knowledge gaps on C stocks and sequestration of Mexican mangroves. Unfortunately, much of the information is not readily available either to the scientific community or to the general public. There have been efforts to make government project reports freely available online, but most of these reports are in Spanish and are not peer-reviewed. There is a clear need to encourage scientists and government organizations to make available this information to advance in the science of C for mangroves in Mexico. There is also a lack of communication between scientists, governmental organizations, and NGOs (Lopez-Medellin et al. 2011). In 2004, the Mexican Carbon Program (PMC, *Programa Mexicano del Carbono*) was established to coordinate and share information on the C cycles of ecosystems in Mexico. This program has promoted the dissemination of information on C in different ecosystems including mangroves (Vargas et al. 2012).

Finally, standardization of protocols for measuring and monitoring C stocks and fluxes is required to establish information at the national scale. Currently, there is a published handbook to guide field measurements based the IPCC guidelines for monitoring mangroves (Kauffman and Donato 2012), adapted for Mexico (Kauffmann et al. 2015). However, standardization at the national level, for example, between the National Commission on Forestry and the National Commission for Biodiversity, has yet to be achieved.

27.2 NATIONAL POLICY

In Mexico, 54% of mangroves are within Protected Natural Areas while the rest are legally protected by federal law (CONABIO 2009). At the national level, there is strong support to protect and restore mangroves, and to improve monitoring and assessment, as part of the National Climate Change Adaptation Strategy backed by the National Law on Climate Change (*Ley General de Cambio Climatico*).

Federal funding has being provided for mangrove restoration projects in the Pacific Coast and in the Yucatan Peninsula (Toledo et al. 2001; Herrera-Silveira et al. 2008). The price for restoring mangroves as a climate change strategy is relatively low, with costs per hectare in the Yucatan Peninsula estimated at US$ 5077 (Herrera et al. 2008). Mangrove restoration has had social benefits. For example, in Yucatan, the restoration of a mangrove forest in 2007 employed 251 fisherman and their wives through the Temporal Employment Program (PET, *Programa de Empleo Temporal*; Mexican Government 2014; Herrera-Silveira et al. 2008), which supports marginalized communities during periods of economic hardship, during tropical storms, and fishing bans (Figure 27.3). Additionally, the community benefited from the program's efforts to enhance capacity and the appreciation they acquired of mangrove forests after the experience (Herrera-Silveira and Adame, pers. comm).

Other programs in Mexico such as Payment for Ecosystem Services for forest management (CONAFOR) benefited 1,448 communities and 683,000 ha of forested land throughout Mexico during 2003–2007 (Corbera et al. 2009) (US$25.7 per mangrove ha; US$ 82,3336 spent in 2012). This program has supported the restoration of mangroves in Mexico. Finally, Early action projects under the international program REDD+ is already financing conservation, restoration, and improved management projects of mangroves in the Yucatan Peninsula and the Central Pacific Coast (Corbera et al. 2009; CONAFOR 2014).

However, national commitments and federal legal protection have not always had a strong influence at the local level, and the available funding is minuscule compared to that allocated to other interests. For example, Yucatan has one of the highest rates of mangrove deforestation in the country, mainly due to tourist developments. Yucatan has also suffered from coastal erosion caused from sea level rise, inadequate coastal constructions, and mangrove destruction (Meyer-Arendt 1993). Every year, the government of Yucatan spends US$158,000 nourishing the beach with sand to sustain the tourist industry (SEDUMA 2011), the equivalent of restoring 45 ha of mangroves every year. Thus, legal and financial support exists from the federal government from pro-poor livelihood, conservation agendas, and national climate-change adaptation policies. But at the state level, vested interests in local priorities, such as beachfront protection in Yucatan, are where most funds are spent.

Figure 27.3   (a) Workshop on the value of carbon sequestration in mangroves at La Encrucijada, Chiapas Mexico; (b) Measurement of C stocks in trees of *Rhizophora mangle* in La Encrucijada, (c) Mangrove restoration in Sisal, Yucatan Mexico. Photos from MF Adame (a, b) and JA Herrera-Silveira (c).

27.3 OPPORTUNITIES

Mexico has the legal framework, federal support, and the institutional arrangements to benefit from financial mechanisms to reduce greenhouse gas emissions. However, implementation at the local scale, lack of scientific information in some locations, and corruption can hamper such efforts. For example, in January 2016, a stretch of 58 ha of mangroves was destroyed in Cancun for the construction of a tourist development named Malecon Tajamar. Despite being legally protected, the government claims that the permits for the destruction of the mangroves were released before the national ban. There was outcry from the local, national, and international community and the development has been momentarily stopped. By destroying the mangrove ecosystem, there was a release to the atmosphere of 133,000 tonne of CO_2, worth US $1 million in the California Market (US$15 per tonne of CO_2). It is still unknown whether the land will be restored to mangroves, but participation in a C financial mechanism could provide the support to do so.

The case of Tajamar exemplifies not only the threats that mangroves face in Mexico, but also the opportunity to strengthen public policy and practice of mangrove conservation and restoration. Currently, there has been some success of restoration projects conducted by CONAFOR, The Marine Secretary, and various NGOs, universities and research centers, with mangroves partially recovering within the first 5–10 years (Herrera-Silveira et al. 2008; Toledo et al. 2009). Carbon financial incentives might increase such restoration programs. For instance, in Sian Kaan Biosphere Reserve, Yucatan, mangroves are being restored through financial support from Sweden as a way of mitigating their C emissions (CONAFOR, F. Camacho, pers comm.). Mexico has the opportunity to establish itself as a pioneer in mangrove restoration as a green strategy to mitigate and adapt to climate change (Zaldívar-Jiménez et al. 2010; Teutli-Hernandez and Herrera-Silveira, 2016).

Mangrove experts in the country have identified 81 mangrove forests that have been prioritized for conservation and restoration purposes (CONABIO 2009). Other regions that would benefit from C markets or international climate mitigation programs are those where the mangroves might expand as sea level rises in the North Pacific coast (López-Medellín et al. 2011) and the Yucatan Peninsula (Adame et al. 2013).

Fortunately, there is a growing support for mangrove restoration and conservation in the country, reinforced by the high level of concern in Mexico about the impacts of climate change (Pew Research Centre 2015). Environmental tourism based on mangroves has also increased in the past decades. In the Yucatan, about 500 members within 14 localities provide ecotourism activities such as boat trips and bird watching (García de Fuentes et al. 2011). Such groups have established themselves as guardians of the environment; they also increase the appreciation of the mangroves, which permeates throughout the community and to the tourists (García de Fuentes et al. 2011).

Mexico has an opportunity to benefit from C markets by protecting and restoring the large areas of existing mangroves. The large C stocks and sequestration potential of mangroves, along with the institutional support, legal framework, national policies, and scientific knowledge could make Mexico an ideal setting for engagement in C markets and other international programs. The financial incentives could hopefully raise environmental awareness, promote social equality, and increase the services that mangroves provide not only to Mexicans, but to the rest of the world.

CHAPTER 28

Blue Carbon Futures
Moving Forward on Terra Firma

Lisamarie Windham-Myers
U.S. Geological Survey

Stephen Crooks
Silvestrum Climate Associates, LLC

Tiffany Troxler
Florida International University

CONTENTS

HIGHLIGHTS

1. Maintaining coastal carbon sequestration and storage services is economically valuable in providing a potentially long-term contribution toward climate resilience, both in terms of adaptation and mitigation.

2. The volumetric accumulation of coastal carbon stocks is unique from other terrestrial and aquatic processes, and inconsistent use of terminology is holding back understanding of the range, magnitude, and processes critical to this carbon sink.

3. Documenting net greenhouse gas (GHG) benefits of coastal ecosystem management needs integrated models that quantitatively incorporate geomorphic, biogeochemical, atmospheric, and hydrologic exchanges to account for both carbon accumulation and loss, across a range of timescales.

4. A community effort is necessary to explore similarities among coastal ecosystems to determine the drivers and scale of true variability, to prioritize specific wetland management options, and develop the most effective monitoring approaches.

5. While there are further scientific aspects of blue carbon to be explored, there is sufficient knowledge and experience to advance demonstration projects across a range of systems and conditions, which can inform policy development and scaled implementation.

28.1 INTRODUCTION

The last 15 years have seen tremendous advancement in the recognition of blue carbon ecosystems (BCEs) as a component of the global carbon cycle and as a management opportunity to partially reduce atmospheric GHG levels while improve human and environmental sustainability. The current vulnerability of coastal lands to increasing sea level rise, storm surge and development pressures have generated various intervention opportunities to protect and enhance ecosystem services through conservation and restoration (Chapters 15–20). Despite the wide range of wetland types and threats globally, an emerging appreciation of the basic principles BCE function are being used to initiate management on most temperate and tropical coastlines (Chapters 21–27). This attention shines a light on uncertainties and gaps in science, policy, and practice. The science community has increased investment in experimentation, monitoring, and modeling of coastal carbon fluxes (Chapters 3–14), conservation groups and agencies have started including blue carbon as a project element (Chapter 2) and policy developers are considering mechanisms to include coastal ecosystems in a post-Paris Climate Agreement context (Chapter 1). Many of the chapters herein illustrate how far we have come in mapping, monitoring, and managing BCEs. Many chapters point the way forward in reducing uncertainties empirically and with models. And many, also, show how far we still have to go.

The scientific response to understanding BCE carbon and GHG fluxes has been exponential, with researchers rapidly generating critical reviews of data availability and gaps (e.g., Chmura et al. 2003; Duarte et al. 2005; Ouyang and Lee 2014), developing conceptual and process-based models (e.g., Kirwan and Megonigal 2013; Morris et al. 2016), and predicting regional sensitivity to global change (e.g., Thorne et al. 2015; Ward et al. 2016). For example, large synthetic datasets have been compiled of published data for soil and stocks in seagrasses, mangroves and marshes (Fourqurean et al. 2012; Holmquist et al. 2018; Rovai et al. 2018; Sanderman et al. 2018), which improve our knowledge of both the size and the distribution of these large C pools. Synthetic datasets of gaseous and hydrologic fluxes have also been compiled (e.g., Najjar et al. 2018; Windham-Myers, Cai, et al. 2018) and demonstrate a large range of emissions and exchanges, both positive and negative, across sites and years.

Substantial progress also has been made in the arenas of policy and practice. Along with terrestrial ecosystems, BCEs are recognized as potent management opportunities in combatting climate change in the Paris Climate Agreement (https://unfccc.int/process-and-meetings/the-paris-agreement/the-paris-agreement). Specifically, ocean components are recognized within the Preamble of the Agreement in the context of Ecosystem Integrity, with Articles 4 and 5 providing that Parties should promote sustainable management and "take action to conserve and enhance, as appropriate, sinks and reservoirs of greenhouse gases." This inclusion provides an internationally recognized basis for marine protection policies, and encourages expansion of science and action to preserve and restore BCEs (e.g., International Partnership on Blue Carbon, https://bluecarbonpartnership.org/). In parallel, grass-root efforts among scientists, practitioners, and policy developers are promoting the expansion of information-sharing networks and data portals—such as the Coastal Carbon Research Coordination Network (http://serc.si.edu/coastalCarbon), the Sustainable Wetlands Adaptation and Mitigation Program (http://www.cifor.org/swamp/) and the Ocean Wealth program (http://oceanwealth.org). Further, the building of tools, such as GHG quantification methodologies (e.g., http://verra.org/method-ologies/), are connecting conservation and restoration of coastal BCEs to the voluntary carbon market (Emmer et al. 2015; 2018a,b).

While the BCE research community continues to conduct both basic and applied studies to reduce uncertainty, especially in poorly represented global regions undergoing rapid changes, this does not preclude action on today's immediate opportunities. Blue carbon management concepts are generally sound (e.g., Orr et al. 2003) and some pressing reasons to begin implementation of sizable, scalable, and repeatable on-the-ground actions include:

1. **Emerging policy recognition**—Policy interest in blue carbon opportunities and productive policy-science interactions are relatively recent, and time is required to build products and frameworks for decision support.
2. **Recent scientific focus on management**—Scientific research has traditionally focused on less-disturbed ("pristine") settings, and studies of management actions through BCE implementation provides the necessary data to model impacts of human activity on carbons and emissions from inhabited coastlines.
3. **Timescale of implementation**—Large-scale interventions, once identified, may take years to decades to implement, but accelerating sea level rise places an imperative on near-term action.
4. **Opportunities for avoided loss**—The greatest human-induced BCE emissions are occurring in countries for whom wetland loss rates are higher, and where coastal sustainability is an important component of their national strategy for both economic development and increased standard of living of their citizens.

28.2 SCIENCE ADVANCES THROUGH SYNTHESES

Growing advances for the BCE scientific community support to synthesize knowledge from multiple disciplines and from disparate and historic datasets. From data analyses to shared experiences in implementing BCE management, these empirical syntheses are a "low hanging fruit" to generate conceptual and numerical models that enhance our ability to quantify past, present, and future coastal carbon dynamics. Below we review five recent North American efforts to survey what is known about carbon fluxes in coastal environments, with an emphasis on sustainability and responses of BCEs carbon storage capacity.

a. Coastal Carbon Synthesis (CCARS, https://www.us-ocb.org/coastal-carbon-synthesis-ccars/): The Ocean Carbon and Biogeochemistry (OCB) and North American Carbon Program (NACP) supported a series of workshops over an eight-year period, which sought to assess the state of knowledge on key carbon fluxes at the land/ocean margin for North America. Divided across five regions of North America (Pacific Coast, Atlantic Coast, Gulf of Mexico, Arctic Coast, and Laurentian Great Lakes), modeled and measured C fluxes were compiled and updated for inter-comparison within and between regions. Benway et al (2016) reports the consensus of 62 scientists across North American governmental, academic, and nonprofit organizations. Data gaps and modeling limitations for carbon fluxes were notable for tidal wetlands and seagrass meadows, and especially those on Pacific and Arctic coastlines. Key recommendations from the CCARS data synthesis activities included additional use of satellite observations, and coordination of an all-inclusive carbon flux measurement campaign with universally established protocols across a small set of representative estuarine and tidal wetland systems (using a typology approach) by regions. Of particular relevance to BCE management, the science plan calls for increased carbon flux measurements in degraded or drowned coastal ecosystems to better understand policy implications of coastal ecosystem change (Benway et al. 2016).

b. Second State of the Carbon Cycle Report (SOCCR-2, https://www.carboncyclescience.us/state-carbon-cycle-report-soccr): The NACP and USGCRP have coordinated an interagency and cross-sector assessment of North American carbon cycle processes, stocks, fluxes, and interactions with global-scale carbon budgets and climate change impacts in managed and unmanaged systems. Carbon stocks and fluxes assessed included soils, water (including near-coastal oceans), vegetation, aquatic-terrestrial interfaces (e.g., coasts, estuaries, wetlands, inland waters), human settlements, agriculture, and forestry. One decade after the initial SOCCR report from 2007, this assessment notably has added a dedicated chapter on tidal wetlands and estuaries, which provides updated synthetic assessments of BCE stocks and fluxes, including methane emissions (Windham-Myers et al. 2018). A review of North American mapping efforts illustrate that areal extent is greatest for seagrass, then mangrove, then marsh, despite a likelihood of underrepresentation of ecosystems on the freshwater end of the tidal spectrum. These North American BCEs today likely contain roughly 1.9 Pg of organic carbon in the top meter of soils, alone. Rates of key carbon fluxes—such as C burial and net ecosystem exchange of CO_2 (NEE) or methane—show strong variability between years and sites, indicating a need to improve modeling and observations to reduce uncertainty. When considered at a continental scale, upscaled carbon budgets for tidal wetlands suggested that roughly 20% of NEE is stored annually in soil C burial, with the remaining 80% of NEE exported laterally to adjoining estuarine habitats, although these fractions varied strongly among sites.

c. NASA-USGS Blue Carbon Monitoring System (BCMS, https://water.usgs.gov/nrp/blue-carbon/nasa-blue-cms/): From 2014 to 2018, a team of 18 scientists across U.S. institutions assessed, and sought to reduce the uncertainty in U.S. coastal carbon fluxes reported in the IPCC National Greenhouse Gas Inventory (NGGI). As seen in Chapter 16, coastal wetlands were included for the first time in the 2017 report, with an estimated 8.6 MT of CO_2 sequestered due to land management. The BCMS synthesis across all U.S. coasts supported this assessment but quantified uncertainty in a spatially explicit manner, identifying key trends as well as bias and error associated with national-scale approaches to soil C stock, aboveground biomass C stocks, and mapped habitat. Holmquist et al. (2018a) found a surprisingly narrow distribution of soil carbon densities both at the surface and downcore to 1 m, among a representative set of 1,900+ in tidal wetland soil core locations across U.S. coasts (0.027 g C $cm^{-3}+-0.0001$ SE). Further, with a universally applicable algorithm at 30 m scale using remotely satellite data from LandSat and the National Agricultural Imaging Program, Byrd et al (2018) found a surprisingly similar distribution of tidal marsh aboveground biomass carbon stocks at validated sites located in six states across the U.S. coastline (Massachusetts, Maryland, Florida, Louisiana, California and Washington). Ultimately these two products (soil and biomass maps) illustrate that C stock accounting at continental scales is fairly well constrained, and not a major point of uncertainty; C stocks in the top meter of tidal wetland soil in the continental United States can be parsimoniously estimated at 270 Mg ha^{-1} and C stocks in aboveground vegetation are roughly 1.85 Mg ha^{-1}. Agency-provided national scale products, however, were found to have significant biases for C accounting, with overestimation of soil C stocks (USDA-NRCS Soil Survey Geographic Database) and underestimation of tidally influenced wetlands (USFWS National Wetlands Inventory). Uncertainty assessments across all GHG accounting components point to the particular need for (i) improved assessment of methane emissions, and (ii) improved modeling of the fate of eroded soil carbon and rate of carbon dioxide emissions following wetland conversion to open water (Holmquist et al. 2018b). Numerous chapters in this book show that increasing empirical data and metadata sets, finer-scaled mapping, and model development will greatly improve the spatial extrapolation of field data into mapped products.

d. NASA Wetland-Estuary Transports and CARbon Budgets (WETCARB, https://cce.nasa.gov/cgi-bin/cce/cce_profile.pl?project_group_id=3165): From 2015 to 2017, a team of 12 PIs performed a landscape-scale synthesis effort to establish a baseline of coastal carbon fluxes within 139 watersheds of the CONUS, including input from rivers, exchange with the atmosphere and open ocean, burial, photosynthesis, and respiration. The datasets synthesized thus far indicate strong variability in these fluxes along the U.S. Atlantic coastline (Najjar et al. 2018) and a great paucity of gas flux data available to validate the accounting approach. In contrast to tidal wetlands, estuarine components of the land/ocean margin were generally found to be net heterotrophic. Considering BCEs within this landscape context (i.e., considering both terrestrial inputs and estuarine outputs, see Chapter 4), appears to be a significant factor in predicting the long-term potential of their carbon sink.

e. National Academy of Sciences, Engineering and Medicine Carbon Dioxide Removal (CDR) and Reliable Sequestration Research Study (http://nas-sites.org/dels/studies/cdr/): In 2017–2018, a critical review and comparison of atmospheric CDR approaches, for achieving negative emissions included coastal "blue carbon" as part of a CDR portfolio of approaches. A broad assessment of opportunities and challenges to implementation of BCE approaches identified significant benefits, but also significant risks to the long-term viability of the removals. When compared with terrestrial and engineering (i.e., direct air capture, geologic sequestration) approaches to removal of carbon dioxide from the atmosphere, BCEs have a smaller annual flux and potential capacity. These estimates of flux and capacity did not consider **avoided loss** as a means to reduce future atmospheric carbon dioxide emissions (i.e., the study focused only on active, negative emissions). While smaller than rates of terrestrial and engineered CDR approaches, management and restoration of BCE and coastal adaptation integrating BCE function was found to comprise a relatively low-cost, technologically ready CDR approach that can be applied immediately. Demonstration and monitoring projects were encouraged where knowledge gaps exist, especially because of their value beyond C sequestration (i.e., ecosystem services).

28.3 TOWARD TERRA FIRMA: FOUR OPPORTUNITIES TO ADVANCE BCE SCIENCE, POLICY, AND PRACTICE

Because C fluxes in coastal systems are sensitive to biological, chemical, and physical drivers, all of which are extremely dynamic over timescales and over scales of time and space, the questions remaining are largely about how effectively we can reduce uncertainty, and especially bias (systematic error), in our understanding and quantification of BCE ecosystem services.

28.3.1 Terminology

One first step toward reducing uncertainty is clarity in terminology, timeframes, and signage conventions. Scientists across disciplines have not yet converged on a framework to communicate key fluxes at appropriate scales (but see Barr et al. 2014). We suggest that terminology identify **carbon burial** as the most inclusive and liberal measurement, including all timeframes (from days to millennia) and carbon sources (autochthonous and allochthonous), whereas **carbon sequestration** represents a more conservative measurement, referring only to autochthonous inputs and long-term preservation (>50 years to millennia). Akin to the reconciling of carbon budget concepts in Chapin et al (2006), this qualified terminology is a first but necessary step for rectifying different measurements. While these separate terms address the primary source of communication errors among practitioners of carbon accounting, we suggest that ultimately a quantified terminology is more accurate, with a specific timescale used for a specific sink term. Further, the convention of using negative signage to indicate flux direction to or from the atmosphere has worked sufficiently for terrestrial fluxes (negative C emissions indicate removals from the atmosphere). In BCEs, the importance of hydrologic exchanges with both the atmosphere and with wetland soil pools in BCEs requires clarity, and thus the positive or negative signage chosen must be clearly stated in relation to the specific exchange in question. Clarification of the direction of fluxes is critical to process intercomparison and reducing uncertainty in conceptual and process-based model development.

28.3.2 Fate of Eroded Soil Carbon

While rapid oxygenation of wetland organic soils through drainage likely leads to immediate gaseous carbon dioxide emissions through microbial oxidation, the rate and fate of soil C eroded into aquatic systems is less well quantified, both physically and biogeochemically (e.g., Ward et al. 2016). BCE gain and loss are historic landscape processes along coastal margins, and thus the increasing loss of wetland acreage, and associated emissions from soil erosion, is a fundamentally important metric for understanding long-term C burial under accelerated conditions of global change (Lovelock et al. 2017). Rather than projecting emissions from substrate quality alone (e.g., lignin content), microbial activity during transport and deposition in coastal sedimentary environments is likely the largest driver of emission rates from BCE eroded soil carbon—with microbial responses to conditions such as temperature, sediment texture and disturbance dynamics, and water body stratification (e.g., warm, shallow, and oxygenated extensive shelves versus cold, dark, oxygen-limited high latitude fjords). Monitoring of tidal fluxes of particulate carbon show seasonal variation (Bergamaschi et al. 2012) and sedimentary redistribution (Ensign and Currin 2017). The current inability to sufficiently track aqueous phases of carbon (gaseous, particulate, and dissolved), especially in episodic environments, make this important flux difficult to constrain and therefore difficult to model relative to changes in coastal management. For the purpose of carbon markets, updates that will be released under the Voluntary Carbon Standard protocol (VCS; Needelman et al 2018) will include guidance on calculating emissions of mobilized wetland carbon based upon geomorphology of depositional environment as a near-term approach for filling that knowledge gap (Emmer et al. 2018).

28.3.3 Modeling Sustainability Thresholds (Tipping Points)

The projection of BCE carbon storage and sequestration services is critical to carbon accounting, as management decisions and market-based crediting are calculated as baseline C budgets versus projected C budgets. Both the model used and the timeframe used for baseline and projected budgets are critical to evaluating carbon benefits. Modeling sustainability has come a long way from **bathtub** approaches, with more process-based models that incorporate the ecogeomorphic feedbacks that allow marshes to sustain their relative elevation to sea level rise. Many one-dimensional process-based models (Morris et al. 2002; Stralberg et al. 2011; Thorne et al. 2014; Byrd et al. 2016; Clough et al. 2016; Mogensen and Rogers 2018) have been applied at landscape scales and show different sensitivities to input variables (e.g., suspended sediment concentration, sea level rise, vegetation productivity). A simple conceptual model (Kirwan and Megongial 2013) illustrates how positive accretion responses to sea level rise can persist and be aided by sediment supply, up to a point. The looming predictive advance of process-based models will be determining the **tipping point** of a given marsh based on key conditions such as initial relative elevation in the tidal frame (Z^*), relative sea level rise, sediment supply, and net autochthonous productivity (see Chapter 6). Regardless of these improvements, empirically based conceptual models still provide sufficient support for implementation and monitoring (e.g., Orr et al. 2003).

Advanced modeling and monitoring will require improved mapping of BCE extents and conditions (e.g., relative elevation, salinity class, vegetation density), including changes in the upper estuary where freshwater dominates (e.g., Krauss et al. 2018). The increasing patchiness of drowning marshes and mangroves can be observed with fine-scale (e.g., sub-meter) image classification, a characterization often referred to as the Unvegetated:Vegetated Ratio or Fraction Green Vegetation (UVVR, Ganju et al. 2017; FGV, Byrd et al. 2018). For tidal wetlands, sufficient elevation mapping is critical (e.g., accurate to less than 10cm), and elevations appear to be greatly improved by reducing the positive bias associated with vegetation (Buffington et al. 2016). While relative elevation is critical in nearly all model projections, a community-driven and cross-model effort to assess sensitivity of BCE resilience (e.g., rates of accretion/erosion/expansion) to known edaphic drivers across gradients (climate, salinity, relative elevation, relative sea level rise) is needed. We currently lack intertidal and subtidal models that integrate physical and biogeochemical processes sufficiently to link stock and flux measurements at annual or seasonal timescales.

28.3.4 The "Salinity Curtain" for Methane Emissions

For stable wetlands, not undergoing a conversion to open water or upland land uses, net methane emissions appear to play the largest role in determining whether the GHG radiative balance is positive or negative (Holmquist et al. 2018b). Currently, a compilation of field gas flux data, primarily from static chambers, identify a threshold at 18 parts per thousand salinity, whereby BCEs with salinities from half-strength to full seawater are negligible in their methane emissions; BCEs with salinities below this threshold have significantly higher and more variable methane emissions (Poffenbarger et al. 2011, Figure 20.2). This distinction sets a boundary whereby higher salinity wetlands are more favorable for reducing GHG emissions than fresher water wetlands. However, salinity intrusion can promote soil C release and reduce productivity (e.g., Chambers et al. 2015), thus while potentially reducing methane emissions, the salinity may reduce the C sink potential of a site. Across other wetlands globally, process-based models are emerging that indicate alternative drivers of methane emission (e.g., Yang et al. 2017), such as productivity and temperature. Integrating the unique presence of abundant marine-based sulfate in tidal wetlands with these process-based models may allow greater predictions of methane fluxes toward future climate and vegetation assemblages (e.g., *Phragmites*, Martin et al. 2017). Further, the variable responses among brackish and freshwater wetlands seen in Poffenbarger et al (2011) are likely also a function of the differing methods of data collection and upscaling. Eddy-covariance approaches, where measured, have consistently shown lower annual methane emission estimates than static chamber approaches (Windham-Myers et al. 2018; see Figure 20.2). Moving toward improved models, more reporting of site metadata, and improved means of upscaling and cross-validation of site-specific datasets can improve quantification and prediction of methane fluxes. It remains possible that the salinity curtain will shift substantially downward, such that lower salinity marshes, those fresher than half-strength salinity, will be poised for lower methane emissions than predicted from a regression alone. We propose the likelihood that "a little bit of sulfate can go a long way" toward reducing methane emissions from BCEs.

28.4 NEXT STEPS: SIX GLOBALLY APPLICABLE PATHWAYS

Coastal areas, both urban and natural, are actively experiencing the impacts of climate change and management decisions. Even as scientific progress continues to reduce uncertainty, the time is now to move forward with data collection and exchange between BCE scientists, practitioners, and policy analysts. Coastal planners and policy makers will continue to need improved projections and demonstrated evidence of C capacity as well as guidance on how to reduce risks and ensure permanence of BCEs and their carbon stocks in the face of uncertainty. In a recent paper, Bai et al (2018) summarized the findings of a scientific steering committee for the IPCC Cities and Climate Change Science Conference (https//citiesipcc.org). These findings, calling for long-term cross-disciplinary studies, mirror priorities for natural systems, and the majority of coastal systems that are a mosaic of urban and natural elements. To conclude this volume, we place our findings in the context of those same six priorities.

28.4.1 Expand (and Integrate) Observations

Moving forward with demonstration projects that include quantification of key stocks, and provide supporting information to model lateral and atmospheric fluxes, can help answer remaining questions and will provide evidence and guidance for other implementation projects. While the scientific accounting has not necessarily found full closure for carbon flux budgets within BCEs (e.g., Najjar et al. 2018), neither have carbon fluxes found full closure in other terrestrial ecosystems credited with natural climate mitigation (e.g., forested lands; see also SOCCR-2, 2018). The stock gain-loss approach used in other terrestrial systems accounts for soil C stock change sufficiently to validation BCE implementation under multiple accredited standards (e.g., Verified Carbon Standard, American Carbon Registry). Expanding observations through a community intercomparison of datasets is the next major step forward, allowing cross-site and cross-system evaluations of model assumptions and method bias. Intercomparison can also provide clarification of the gaps and necessary metadata required to generate both (i) model insight, and (ii) allow spatially explicit accounting. A major improvement from Morris et al (2016) has been the use of a **mixing model** to examine organic and inorganic contributions and their **packing densities** in tidal wetland soils, thus illuminating geomorphic differences between karstic (marine) and sedimentary (deltaic) environments (Breithaupt et al. 2014). This type of first-principles approaches to assessing carbon storage complements machine-learning (e.g., Sanderman et al. 2018) or geomorphic assessments (e.g., Rovai et al. 2018), providing insight to physical versus chemical versus microbial mechanisms of preservation or loss.

28.4.2 Understand Climate Interactions

Climate interactions with BCEs include both direct drivers of carbon processes (e.g., sea level rise, Chapter 10; temperature Chapter 11), and indirect effects on land and water management (e.g., operations). Understanding the relative responses of BCEs to these impacts is key to model development, but we caution that a full understanding and quantification will always be elusive. Rather than focus on the complexity of responses, we point out that years of monitoring data have illustrated some dominant principles of resilience or vulnerability in BCEs throughout the world. Key factors of resilience include adequate sediment supplies, limited nitrogen loading, and pre-existing elevation capital (e.g., Orr et al. 2003; Crooks et al. 2014). Monitoring landscape changes is a significant form of empirical support for climate drivers and remote sensing is among the most useful ways (Chapters 12–14) to document ecosystem responses and assess climate sensitivities.

28.4.3 Study Informal Settlements

Social drivers are a dominant cause of BCE loss. More than 40% of world's population lives within 100 km of the coast (Neumann et al. 2015). Urban areas are sprawling, and by the year 2050 it is estimates that 3 billion people will be living in unregulated areas (e.g., slums) with no mainstream governance, on unzoned land (Bai et al. 2018). These areas are highly vulnerable to climate change, populated by displaced people from rural areas. While urban development is not the topic of this volume, we call attention to the fact that natural spaces are becoming fewer, less remote, and more dissected by human infrastructure. A significant challenge for resource management is the ongoing uncoordinated conversions of BCEs for aquaculture or resource extraction that often occur quickly, transitioning large-scale coastal ecosystems from "pristine" to "degraded" conditions in a few years. Carbon management approaches require a quantified baseline of land conversion, but with these rapid changes, baseline projections are highly inaccurate. The Mahakam Delta, South Sulawesi, is a typical example. In a fairly isolated setting, with low-density populations, the delta supported a largely pristine mangrove ecosystem isolated from heavy human disturbance until 1980 (Sidik, 2008). By 2000, 47%, and then 2010, 75%, of mangroves had been converted to 21,000 ha of shrimp ponds

(Rahman et al. 2017). The initiation and pace of land-use conversion overwhelms the capacity of agencies and communities to develop a response. Many other coastal areas are similarly threatened. The Zambezi delta currently exists in a similarly pristine state, adjacent to terrestrial uplands being heavily deforested (https://rainforests.mongabay.com/deforestation/archive/Mozambique.htm). With depletion of terrestrial wood sources, attention will likely turn to mangrove timber. Outside views of carbon management projects might consider the carbon accounting or financial arrangements to be the more complex aspect; in actuality, the BCE futures described here hinge on sustainable human resource use. The finance component is often is applied to aid transition (i.e., funding of support services) but the future of these BCE's is tied directly to a workable improvement in human well-being.

28.4.4 Harness Disruptive Technology

Growing empowerment of individuals through communications technology is a global institutional disruption that may be a key in developing local economies based upon sustainable practices, a core need for long-term BCE management. Emergence of business cluster enterprises based around sustainability and connected to external markets would help communities improve livelihoods. Coupled to this, establishment of sustainability accreditation to products that are recognized and valued in consumer markets could help shift production approaches.

Perhaps one of the largest potential disrupters may be driven by the financial sector. The surge in sales of green bonds (and subclasses of blue bonds, climate resilience bonds, etc.), reflect the amount of available capital seeking stable investment, and financial institutions (pension and reinsurance companies) are recognizing the fiduciary responsibility of reducing risk brought on by climate changes. From inception in 2011, the green bond market grew to $150 billion issuance in 2015 and a projected issuance of $250 billion in 2018 (www.climatebonds.net). Early beneficiaries of climate bonds have been wind farms, a relatively simple investment option with quantifiable climate benefits. Of relevance to BCEs, work is ongoing to establish the standards under which financing might be applied to support coastal resilience (www.CPICfinance.com). Such a standard may include procedures developed under the voluntary carbon market for conservation and restoration of BCEs (Emmer et al. 2015; 2018a,b).

Opportunities to include wetlands under bonds may materialize in a number of ways. The U.S. state of California, for example, offers financial support to cities to reduce their climate vulnerability. As part of this funding, credit is gained for the inclusion of wetlands as natural infrastructure, and in some cases for the carbon sequestration benefits specifically (http://resources.ca.gov/climate/natural-working-lands/). Green bond standards might also call for conservation or restoration of wetlands as compensatory measures for GHG emissions or ecological impacts, but these requirements have yet to be set.

28.4.5 Support Transformation

Transforming the tools and scales of coastal ecosystem decision making may also be a means of BCE management, as suggested for cities (Bai et al. 2018). Bold strategies to foster early demonstration of alternative approaches and good practice are necessary to accelerate scaling up of successful action. It is true for coastal settings, just as for terrestrial settings, that it takes many years to develop the skills that link science and practice toward enacting and scaling-up management projects. For coastal wetlands, good practice in restoration exists, which can be coupled with experience in terrestrial carbon project development, but there is a need to connect and apply this learning globally (Crooks et al. 2014). Other barriers for blue carbon projects include the challenge in finding lands of sufficient size and scalability. There is a need to remove complexity and bring down verification costs, so that small-scale landowners can join regional collective actions (called grouped projected under the Verified Carbon Standard) and take part in landscape-scale carbon management.

Carbon financing has been promoted as a transformative tool for the past 2–3 decades, but even in the land sector, representing 25% of anthropogenic emissions (10–12 $PgCO_2e$/year), there has been very little success in drawing soil management under carbon financing. In their review of carbon markets for soil management, Von Unger and Emmer (2018) document that only 20 soil carbon projects, globally, on grasslands and agriculture have been registered for the sale of credits, despite the relatively simple land-use opportunities presented. In the absence of compliance market leadership, today, due largely to soil carbon methodology development within the voluntary market, robust methodologies and procedures are available and may stimulate implementation for almost any project category, including those of BCEs.

Beyond carbon finance, the inclusion of BCEs in Nationally Determined Contributions for the Paris Agreement will be critical for many countries in linking management of these ecosystems to climate resilience, both in terms of mitigation and adaptation. Sources of financing may be derived via multilateral and bilateral national agreements, or carbon taxes, adaptation support, debt for nature swap and green bonds. As such, blue carbon projects integrate critical land decisions and cannot succeed for long in the absence of broader policy-level transformation, addressing strategic planning, zoning, land tenure, investment.

28.4.6 Recognize the Global Sustainability Context

Coastal areas are inherently difficult places to manage, involving complex, multi-scaled, dynamic, and interconnected natural and social systems. While loss of BCEs is often a response to local drivers, national-scale impacts are often driven by global pressures, such as food production. If BCEs are to be managed for sustainability, mechanisms must be found to accommodate the pressures from population growth, changing demands in population preferences, and global environmental changes. Tackling these challenges requires whole system level analysis, linking strategic planning across land and water, and from river to coast. There will be a need for intergovernmental cooperation to tackle cross-boundary issues, and for connections to be built between government silos. It is less common now, but still not uncommon, to find that different government agencies might have different plans for the same parcel of land, ranging from preservation to resource extraction (e.g., Chapters 21 and 24). Though difficult to tackle there is an imperative need to seek solutions in the nexus between food provision, energy production, access to markets, environmental protection, all in the context of resilience and risk reduction.

28.5 CLOSING NOTE

We are optimistic that knowledge of BCE function and structure is growing, and useful for management of multiple coastal ecosystem services. This edited collection of 28 chapters is a snapshot of the current drivers and conditions of BCEs at local to global scales. The chapters provide a reference on carbon dioxide and sinks and sources within coastal lands, and makes significant strides in reducing knowledge gaps for the science, policy development, and management practices in BCEs. From theory to case studies, there is a wealth of information available for stakeholders to explore and apply as needed; further, there is abundant opportunity critique and compare the information herein to point forward in other directions where improvements can be made. We expect that in another 10 years the BCE community will have evolved even further toward broader understanding of climate mitigation and adaptation opportunities on coastlines, and more quantitative predictions of carbon sequestration and burial across the BCE continuum. The growing interest and findings are likely to support coastal management toward restoration, rehabilitation, creation, and conservation of BCE and accelerate the use of nature-based approaches for building resilience of coastlines globally.

ACKNOWLEDGMENTS

We recognize the support of the NASA Carbon Monitoring Program (NNH14AY671) and the USGS LandCarbon Program. We appreciate the detailed review by Isa Woo (USGS).

References

Abdalla S., Kairo J., Huxham M., Ruzowitzsky L. (2015). 2014–2015 Plan Vivo Annual Report Mikoko Pamoja, (December).

Abdalla S., Kairo J.G., Huxham M. (2014). 2013–2014 Plan Vivo Annual Report: Mikoko Pamoja. Plan Vivo. http://planvivo.org/docs/2013-14_Annual-Report-Mikoko-Pamoja_Published.pdf.

Abed R.M., Kohls K., De Beer D. (2007). Effect of salinity changes on the bacterial diversity, photosynthesis and oxygen consumption of cyanobacterial mats from an intertidal flat of the Arabian Gulf. *Environmental Microbiology, 9*: 1384–1392.

Abuodha P.A.W., Kairo J.G. (2001). Human-induced stresses on mangrove swamps along the Kenyan coast. *Hydrobiologia, 458*(1): 255–265. doi: 10.1023/A:1013130916811.

Adam P. (2002). Saltmarshes in a time of change. *Environmental Conservation, 29*: 39–61.

Adame B., et al. (2019). Mangroves of Mexico, In: *A Blue Carbon Primer: The State of Coastal Wetland Carbon Science, Practice, and Policy*, Lisamarie W.-M., Stephen C., and Tiffany G.T. (eds), CRC Press: Boca Raton, London.

Adame M.F., et al. (2013). Carbon stocks of tropical coastal wetlands within the karstic landscape of the mexican caribbean, H. Y. H. Chen, (ed.). *PLoS One, 8*(2): e56569. doi: 10.1371/journal.pone.0056569.

Adame M.F., et al. (2013). Drivers of mangrove litterfall within a karstic region affected by frequent hurricanes. *Biotropica, 45*(2): 147–154.

Adame M.F., et al. (2014). Root biomass and production of mangroves surrounding a karstic oligotrophic coastal lagoon. *Wetlands, 34*(3): 479–488.

Adame M.F., et al. (2015). Carbon stocks and soil sequestration rates of tropical riverine wetlands. *Biogeosciences, 12*(12): 3805–3818.

Adame M.F., Fry B. (2016). Source and stability of soil carbon in mangrove and freshwater wetlands of the Mexican Pacific coast. *Wetlands Ecology and Management*, 1–9. doi: 10.1007/s11273-015-9475-6.

Adame M.F., Kauffman J.B., Medina I., Gamboa J.N., Torres O., Caamal J., Reza M., Herrera-Silveira J.A. (2012). Carbon stocks of tropical coastal wetlands within the Karstic landscape of the Mexican Caribbean. *PloS One, 8*: e56569.

Adame M.F., Lovelock C.E. (2011). Carbon and nutrient exchange of mangrove forests with the coastal ocean. *Hydrobiologia, 663*: 23–50.

Adame M.F., Reef R., Grinham A., Glen G., Lovelock C.E. (2012). Nutrient exchange of extensive cyanobacterial mats in an arid subtropical wetland. *Marine and Freshwater Research, 63*: 457–467.

Adams M.P., Saunders M.I., Maxwell P.S., Tuazon D., Roelfsema C.M., Callaghan D.P., Leon J., Grinham A.R., O'Brien K.R. (2015). Prioritizing localized management actions for seagrass conservation and restoration using a species distribution model. *Aquatic Conservation, 26*(4): 639–659.

Adger W.N. (2000). Social and ecological resilience: are they related? *Progress in Human Geography, 24*(3): 347–364.

Aich S., McVoy C.W., Dreschel T.W., Santamaria F. (2013). Estimating soil subsidence and carbon loss in the Everglades Agricultural Area, Florida, using geospatial techniques. *Agriculture, Ecosystems and Environment*. doi: 10.1016/j.agee.2013.03.017.

Akanle T., Appleton A., Kulovesi K., Recio E., Schulz A., Willetts L. (2011). Daily Web Coverage from UN Climate Change Conference June 2011, Earth Negotiations Bulletin. www.iisd.ca/climate/sb34.

Alizad K., Hagen S.C., Morris J.T., Medeiros S.C., Bilskie M.V., Weishampel J.F. (2016). Coastal wetland response to sea level rise in a fluvial estuarine system. *Earth's Future, 4*. doi: 10.1002/2016EF000385.

Alizad K., Scott C.H., Morris J.T., Bacopoulos P., Bilskie M.V., Weishampel J.F., Medeiros S.C. (2016). A coupled, two-dimensional hydrodynamic-marsh model with biological feedback. *Ecological Modeling, 327*: 29–43.

Allison M.A., Kolker A.S., Meselhe E.A. (2014). Water and Sediment Dynamics through the Wetlands and Coastal Water Bodies of Large Rivers. In: *Biogeochemical Dynamics at Major River-Coastal Interfaces: Linkages with Global Change*, Bianchi T., Allison M., Cai W.-J. (eds). Cambridge University Press: New York, Chapter 2, pp. 21–54.

Allison M.A., Yuill B.T., Tornqvist T., Amelung F., Dixon T., Erkens G., Stuurman R., Jones C., Milne G., Steckler M., Syvitski J., Teatini P. (2016). Coastal subsidence: global risks and mitigation. 2016. *EOS—Transactions of the American Geophysical Union, 97*: 22–27.

Allison S.D. (2014). Modeling adaptations of carbon use efficiency in microbial communities. *Frontiers in Microbiology*, 5: 571.

Allison S.D., Wallenstein M.D., Bradford M.A. (2010). Soil-carbon response to warming dependent on microbial physiology. *Nature Geosciences*, 3: 336–340.

Alongi D., Murdiyarso D., Fourqurean J., Kauffman J., Hutahaean A., Crooks S., Lovelock C., Howard J., Herr D., Fortes M. (2016). Indonesia's blue carbon: a globally significant and vulnerable sink for seagrass and mangrove carbon. *Wetlands Ecology and Management*, 24(1): 1–11.

Alongi D.M. (1998). *Coastal Ecosystem Processes*, CRC Press: New York.

Alongi D.M. (2002). Present state and future of the world's mangrove forests. *Environmental Conservation*, 29(03): 331–349.

Alongi D.M. (2008). Mangrove forests: resilience, protection from tsunamis, and responses to global climate change. *Estuarine, Coastal and Shelf Science*, 76: 1–13.

Alongi D.M. (2014). Carbon cycling and storage in mangrove forests, *Annual Review of Marine Science*, 6: 195–219.

Alongi D.M., Trott L.A., Undu M.C., Tirendi F. (2008). Benthic microbial metabolism in seagrass meadows along a carbonate gradient in Sulawesi, Indonesia. *Marine Ecology Progress Series*, 51: 141–152.

American Carbon Registry (2017). Restoration of California deltaic and coastal wetlands. Winrock International. http://americancarbonregistry.org/carbon-accounting/standards-methodologies/restoration-of-california-deltaic-and-coastal-wetlands (accessed June 30, 2017).

Amran M.A. (2017). Mapping seagrass condition using Google Earth Imagery. *Journal of Engineering Science and Technology Review*, 10(1): 18–23.

Anderson F.E., Bergamaschi B., Sturtevant C., Knox S., Hastings L., Windham-Myers L., Detto M., Hestir E.L., Drexler J., Miller R.L., Matthes J.H., Verfaillie J., Baldocchi D., Snyder R.L., Fujii R. (2016). Variation of energy and carbon fluxes from a restored temperature freshwater wetland and implications for carbon market verification protocols. *Journal of Geophysical Research: Biogeosciences.* doi: 10.1002/2015/JG003083.

Aoki C., Kugaprasatham S. (2009). *Support for Environmental Management of the Iraqi Marshlands*, United Nations Environment Programme, UNEP Publication DTI/1171/JP: Nairobi (Kenya).

Apostolaki E.T., Holmer M., Marbá N., Karakassis I. (2011). Epiphyte dynamics and carbon metabolism in a nutrient enriched Mediterranean seagrass (*Posidonia oceanica*) ecosystem. *Journal of Sea Research*, 66: 135–142.

Apple J.K., Del Giorgio P.A. (2007). Organic substrate quality as the link between bacterioplankton carbon demand and growth efficiency in a temperate salt-marsh estuary. *The ISME Journal*, 1(8): 729–742.

Apple J.K., Del Giorgio P.A., Kemp W.M. (2006). Temperature regulation of bacterial production, respiration, and growth efficiency in a temperate salt-marsh estuary. *Aquatic Microbial Ecology*, 43: 243–254.

Appleby P.G., Oldfield F. (1978). The calculation of lead-210 dates assuming a constant rate of supply of unsupported ^{210}Pb to the sediment. *Catena*, 5: 1–8.

Appleby P.G., Oldfield F. (1983). The assessment of ^{210}Pb data from sites with varying sediment accumulation rates. *Hydrobiologia*, 103: 29–35.

Ardón, M., Morse J.L., Colman B.P., Bernhardt E.S. (2013). Drought-induced saltwater incursion leads to increased wetland nitrogen export. *Global Change Biology*, 19: 2976–2985.

Arkema K.K., Guannel G., Verutes G., Wood S.A., Guerry A., Ruckelshaus M., Kareiva P., Lacayo M., Silver J.M. (2013). Coastal habitats shield people and property from sea-level rise and storms. *Nature Climate Change*, 3: 913–918.

Armitage A.R., Highfield W.E., Brody S.D., Louchouarn P. (2015). The contribution of mangrove expansion to salt marsh loss on the Texas Gulf Coast. *PloS One*, 10(5): e0125404.

Armstrong M.E. (1989). Modern carbonate soil production and its relation to bottom variability, Graham's Harbour. *MSc Dissertation*, Univ. Cincinnati: Ohio, San Salvador, Bahamas

Arneth A., Sitch S., Pongratz J., Stocker B.D., Ciais P., Poulter B., Bayer A.D., Bondeau A., Calle L., Chini L.P. Gasser T., Fader M., Friedlingstein P., Kato E., Li W., Lindeskog M., Nabel J.E.M.S., Pugh T.A.M., Robertson E., Viovy N., Zaehle S. (2017). Historic carbon dioxide emissions caused by land use changes are possibly larger than assumed. *Nature Geoscience*, 10: 79–84.

Arnold T., Mealey C., Leahey H., Miller A.W., Hall-Spencer J.M., Milazzo M., Maers K. (2012). Ocean acidification and the loss of phenolic substances in marine plants. *PLoS One*, 7(4): e35107.

Arnosti C. (2011). Microbial extracellular enzymes and the marine carbon cycle. *Annual Review of Marine Science*, *3*: 401–425.

Arnosti C. (2014). Patterns of microbially driven carbon cycling in the ocean: links between extracellular enzymes and microbial communities. *Advances in Oceanography*. doi: 10.1155/2014/706082.

Arrieta J.M., Mayol E., Hansman R.L., Herndl G.J., Dittmar T., Duarte C.M. (2015). Dilution limits dissolved organic carbon utilization in the deep ocean. *Science*, *348*(6232): 331–333.

Asner G., Mascaro J., Anderson C., Knapp D., Martin R., Kennedy-Bowdoin T., van Breugel M., Davies S., Hall J., Muller-Landau H., Potvin C., Sousa W., Wright J., Bermingham E. (2013). High-fidelity national carbon mapping for resource management and REDD+. *Carbon Balance and Management*, *8*: 7.

Astor M. (2017). *2017 Hurricane Season really is More Intense than Normal*, New York Times. (September 19, 2017).

Atkinson G., Bateman Mourato I., S. (2012). Recent advances in the valuation of ecosystem services and biodiversity. *Oxford Review of Economic Policy*, *38*: 22–47.

Attermeyer K., Tittel J., Allgaier M., Frindte K., Wurzbacher C., Hilt S., Kamjunke N., Grossart H.P. (2015). Effects of light and autochthonous carbon additions on microbial turnover of allochthonous organic carbon and community composition. *Microbial Ecology*, *69*: 361–371.

Atwater B.F. (1979). Ancient processes at the site of southern San Francisco Bay: Movement of the crust and changes in sea level. In: *San Francisco Bay: The urbanized estuary*, Conomos T.J. (ed.), Pacific Division, American Association for the Advancement of Science: San Francisco, CA, pp. 31–45.

Atwater B.F. (1980). Attempts to correlate late Quaternary climatic records between San Francisco Bay, the Sacramento-San Joaquin Delta, and the Mokelumne River, California. *PhD Dissertation*. University of Delaware: Newark, DE, USA.

Atwater B.F., Belknap D.F. (1980). Tidal-wetland deposits of the Sacramento-San Joaquin Delta, California. *Pacific Coast Paleogeography Symposium*, *4*: 89–103.

Atwater B.F., Conard S.G., Dowden J.N., Hedel C.W., MacDonald R.L. (1979). History, landforms, and vegetation of the estuary's tidal marshes. *Proceedings of the Fifty-Eighth Annual Meeting of the Pacific Division/American Association for the Advancement of Science*. San Francisco Bay – The Urbanized Estuary, San Francisco State University, California Academy of Sciences: San Francisco, San Francisco, CA, USA, pp. 347–85.

Atwater B.F., Hedel C.W., Helley E.J. (1977). *Late Quaternary Depositional History, Holocene Sea-Level Changes, and Vertical Crust Movement, Southern San Francisco Bay*, U.S. Geological Survey Professional Paper 1014, United States Government Printing Office: California, Washington, D.C.

Atwood T.B., Connolly R.M., Almahasheer H., Carnell P.E., Duarte C.M., Ewers Lewis C.J., Irigoien X., Kelleway J.J., Lavery P.S., Macreadie P.I., Serrano O., Sanders C.J., Santos I., Steven A.D., Lovelock C.E. Global patterns in mangrove soil carbon stocks and losses. *Nature Climate Change*. doi: 10.1038/NCLIMATE3326.

Atwood T.B., Connolly R.M., Ritchie E.G., Lovelock C.E., Heithaus M.R., Hays G.C., Macreadie P.I. (2015). Predators help protect carbon stocks in blue carbon ecosystems. *Nature Climate Change*, *5*(12): 1038–1045.

Aukland L., Moura P., Costa, Brown S. (2003). A conceptual framework and its application for addressing leakage: the case of avoided deforestation. *Climate Policy*, *3*: 123–136.

Bahlmann E., Weinberg I., Lavrič J.V., Eckhardt T., Michaelis W., Santos R., Seifert R. (2015). Tidal controls on trace gas dynamics in a seagrass meadow of the Ria Formosa lagoon (southern Portugal). *Biogeosciences*, *12*: 1683–1696.

Bai X., Dawson R.J., Ürge-Vorsatz D., Delgado G.C., Barau A.S., Dhakal S., Dodman D., Leonardsen L., Masson-Delmotte V., Roberts D. (2018). Six research priorities for cities and climate change. *Nature*, *555*(7694): 23–25.

Bakirman T., Gumusay M.U., Tuney I. (2016). Mapping of the seagrass cover along the Mediterranean coast of Turkey using Landsat 8 OLI images. *The International Archives of the Photogrammetry, Remote Sensing and Spatial Information Sciences*, XLI-B8: 1103–1105.

Baldocchi D. (2014). Measuring fluxes of trace gases and energy between ecosystems and the atmosphere – the state and future of the eddy covariance method. *Global Change Biology*, *20*: 3600–3609.

Baldocchi D., Falge E., Gu L.H., Olson R., Hollinger D., Running S., Anthoni P., Bernhofer C., Davis K., Evans R., Fuentes J., Goldstein A., Katul G., Law B., Lee X.H., Malhi Y., Meyers T., Munger W., Oechel W., K.T.P.U., Pilegaard K., Schmid H.P., Valentini R., Verma S., Vesala T., Wilson

K., Wofsy S. (2001). FLUXNET: a new tool to study the temporal and spatial variability of ecosystem-scale carbon dioxide, water vapour and energy flux densities. *Bulletin of the American Meteorological Society*, *82*: 2415–2434.

Baldwin A.H., Egnotovich M.S., Clarke E. (2001). Hydrologic changes and vegetation of tidal freshwater marshes: field, greenhouse, and seed-bank experiments. *Wetlands*, *21*: 519–531.

Ball D., Soto-Berelov M., Young P. (2014). Historical seagrass mapping in Port Phillip Bay, Australia. *Journal of Coastal Conservation*, *18*: 257–272.

Ball D.F. (1964). Loss-on-ignition as an estimate of organic matter and organic carbon in non-calcareous soils. *Journal of Soil Science*, *15*: 84–92.

Banjade M.R., Liswanti N., Herawati T., Mwangi E. (2016). *Assessment of Natural Resource Governance Including Land and Forest Tenure in Coastal Mangrove Forests of Indonesia*, CIFOR, Tetra Tech and USAID: Bogor, Indonesia.

Barbier E., Hacker S., Kennedy C., Koch E., Stier A., Silliman B. (2011). The value of estuarine and coastal ecosystem services. *Ecological Monographs*, *81*: 169–193.

Barbier E.B. (2006). Natural barriers to natural disasters: replanting mangroves after the tsunami. *Frontiers in Ecology and the Environment*, *4*: 124–131.

Barbier E.B. (2015). Valuing the storm protection service of estuarine and coastal ecosystems. *Ecosystem Services*, *11*: 32–38.

Barbier E.B. (2017). Marine ecosystem services. *Current Biology 27*: R431–R510.

Barbier E.B., Georgiou I.Y., Enchelmeyer B., Reed D.J. (2013). The value of wetlands in protecting southeast louisiana from hurricane storm surges. *Plos One*, *8*(3), e58715.

Barillé, L., Robin M., Harin N., Bargain A., Launeau P. (2010). Increase in seagrass distribution at Bourgneuf Bay (France) detected by spatial remote sensing. *Aquatic Botany*, *92*: 185–194.

Barlow N.L., Long A.J., Saher M.H., Gehrels W.R., Garnett M.H., Scaife R.G. (2014). Salt-marsh reconstructions of relative sea-level change in the North Atlantic during the last 2000 years. *Quaternary Science Reviews*, *99*: 1–16.

Barr J.G., Troxler T.G., Najjar R.G. (2014). Understanding coastal carbon cycling by linking top-down and bottom-up approaches. *Eos, Transactions American Geophysical Union*, *95*(35): 315–315.

Barrell J., Grant J., Hanson A., Mahoney M. (2015). Evaluating the complementarity of acoustic and satellite remote sensing for seagrass landscape mapping. *International Journal of Remote Sensing*. doi: 10.1080/01431161.2015.1076208.

Barrón, C., Duarte C.M. (2009). Dissolved organic matter release in a *Posidonia oceanica* meadow. *Marine Ecology Progress Series*, *374*: 75–84.

Barrón, C., Duarte C.M., Frankignoulle M., Borges A.V. (2006). Organic carbon metabolism and carbonate dynamics in a Mediterranean seagrass (*Posidonia Oceanica*) meadow. *Estuaries and Coasts*, *29*: 417–428.

Barrón, C., Marbà N., Terrados J., Kennedy H., Duarte C.M. (2004). Community metabolism and carbon budget along a gradient of seagrass (*Cymodocea nodosa*) colonization. *Limnology and Oceanography*, *49*: 1642–1651.

Bartlett K.B., Bartlett D.S., Harriss R.C., Sebacher D.I. (1987). Methane emissions along a salt marsh salinity gradient. *Biogeochemistry*, *4*: 183–202.

Basso D. (2012). Carbonate production by calcareous red algae and global change. *Geodiversitas*, *34*: 13–33.

Bates M.E., Lund J.R. (2013). Delta subsidence reversal, levee failure, and aquatic habitat—a cautionary tale. *San Francisco Estuary Water Sci.* *11*(1). http://escholarship.org/uc/item/9pp3n639.

Batllori-Sampedro E., SEDUMA, Secretaría de Desarrollo Urbano y Medio Ambiente (2011). Gestión para la rehabilitación y recuperación de playas en la franja costera del Estado de Yucatan. Foro Erosión costera.

Bauer J., Cai W.-J., Raymond P., Bianchi T., Hopkinson C. S., Regnier P. (2013). The changing carbon cycle of the coastal ocean. *Nature*, *493*: 27–36. doi: 10.1038/nature 12857.

Baumstark R., Duffey R., Pu R. (2016). Mapping seagrass and colonized hard bottom in Springs Coast, Florida using WorldView-2 satellite imagery. *Estuarine, Coastal and Shelf Science*, *181*: 83–92.

Bayen S. (2012). Occurrence, bioavailability and toxic effects of trace metals and organic contaminants in mangrove ecosystems: a review. *Environment International*, *48*: 84–101.

Bayrak M., Marafa L. (2016). Ten Years of REDD+: A Critical Review of the Impact of REDD+ on Forest-Dependent Communities, 8 Sustainability 620.

Bayraktarov E., Saunders M.I., Abdullah S., Mills M., Beher J., Possingham H.P., Mumby P.J., Lovelock C.E. (2016). The cost and feasibility of marine coastal restoration. *Ecological Applications*, *26*: 1055–1074. doi: 10.1890/15-1077.

Beal E.J., House C.H., Orphan V.J. (2009). Manganese- and iron-dependent marine methane oxidation. *Science*, *325*: 184–187.

Beaumont N., Jones L., Garbutt A., Hansom J., Toberman M. (2014). The value of carbon sequestration and storage in coastal habitats. *Estuarine Coastal and Shelf Science*, *137*: 32–34.

Beck M.W., Murrell M.C., Hagy J.D. III (2015). Improving estimates of ecosystem metabolism by reducing effects of tidal advection in dissolved oxygen time series. *Limnology and Oceanography: Methods*, *13*: 731–745.

Beer S., Koch E. (1996). Photosynthesis of marine macroalgae and seagrasses in globally changing CO sub (2) environments. *Marine Ecology Progress Series*, *141*: 199–204.

Belshe E.F., Mateo M.A., Gillis L., Zimmer M., Teichberg M. (2017). Muddy waters: unintentional consequences of blue carbon research obscure our understanding of organic carbon dynamics in seagrass ecosystems. *Frontiers in Marine Science*, *4*: 125.

Benner R., Amon R.M. (2015). The size-reactivity continuum of major bioelements in the ocean. *Annual Review of Marine Science*, *7*: 185–205.

Benner R., Fogel M., Sprague E. (1991). Diagenesis of belowground biomass of *Spartina alterniflora* in salt-marsh sediments. *Limnology and Oceanography*, *36*: 1358–1374.

Benway H., Alin S., Boyer E., Cai W.-J., Coble P., Cross J., Friedrichs M., Goñi, M., Griffith P., Herrmann M., Lohrenz S., Mathis J., McKinley G., Najjar R., Pilskaln C., Siedlecki S., Smith R. (2016). A Science Plan for Carbon Cycle Research in North American Waters. Report of the Coastal CARbon Synthesis (CCARS), North American Carbon Program, p. 84. doi: 10.1575/1912/7777.

Bergamaschi B.A., Krabbenhoft D.P., Aiken G.R., Patino E., Rumbold D.G., Orem W.H. (2012). Tidally driven export of dissolved organic carbon, total mercury, and methylmercury from a mangrove-dominated estuary. *Environmental science and technology*, *46*(3): 1371–1378.

Bernal B., Megonigal J.P., Mozdzer T.M. (2017). Effects of tidal marsh native and invasive plant species on deep soil organic matter decomposition. *Global Change Biology*, *23*(5): 2104–2116.

Bertness M.D., Ewanchuk P.J. (2002). Latitude and climate-driven variation in the strength and nature of biological interactions in New England salt marshes. *Oecologia*, *132*: 392–401.

Bhaskar P., Bhosle N.B. (2005). Microbial extracellular polymeric substances in marine biogeochemical processes. *Current Science 88*: 45–53.

Bianchi T. (2011). The role of terrestrially derived organic carbon in the coastal ocean: a changing paradigm and the priming effect. *Proceedings of the National Academy of Sciences of the United States of America*, *108*(49): 19473–19481.

Bianchi T., Galler J., Allison M. (2007). Hydrodynamic sorting and transport of terrestrially derived organic carbon in sediments of the Mississippi and Atchafalaya Rivers. *Estuarine Coastal and Shelf Science*, *73*: 211–222.

Bianchi T.S. (2006). *Biogeochemistry of Estuaries*, Oxford University Press: Oxford, UK.

Bianchi T.S., Allison M.A., Zhao J., Li X., Comeaux R.S., Feagin R.A., Wasantha Kulawardhana R. (2013). Historical reconstruction of mangrove expansion in the Gulf of Mexico: linking climate change with carbon sequestration in coastal wetlands. *Estuarine and Coastal Shelf Science*, *119*: 7–16.

Bianchi T.S., Canuel E.A. (2011). *Chemical Biomarkers in Aquatic Ecosystems*, Princeton University Press: Princeton, NJ.

Bianchi T.S., Lambert C.D., Santschi P.H., Guo L. (1997). Sources and transport of land-derived particulate and dissolved organic matter in the Gulf of Mexico (Texas shelf/slope): the use of lignin-phenols and loliolides as biomarkers. *Organic Geochemistry*, *27*: 65–78.

Bianchi T.S., Thornton D.C.O., Yvon-Lewis S.A., King G.M., Eglinton T.I., Shields M.R., Ward N.D., Curtis J. (2015). Positive priming of terrestrially-derived dissolved organic matter in a freshwater microcosm system. *Geophysical Research Letters*, *42*(13): 5460–5467. doi: 10.1002/2015GL064765.

Bilkovic D.M., Mitchell M.M., La M.K., Peyre Toft J.D. (eds) (2017). *Living Shorelines: The Science and Management of Nature-Based Coastal Protection*, CRC Press: Boca Raton, p. 519.

Bird E.C.F. (1969). *Coasts*, MIT Press: Cambridge. p. 246.

Bird E.C.F. (1971). Mangroves as land builders. *The Victorian Naturalist*, *88*: 189–197.

Blair N.E., Aller R.C. (2012). The fate of terrestrial organic carbon in the marine environment. *Annual Review of Marine Science*, 4: 401–423.

Blanchet M., Pringault O., Panagiotopoulos C., Lefèvre D., Charrière B., Ghiglione J., Fernandez C., Aparicio F.L., Marrasé C., Catala P., Oriol L., Caparros J., Joux F. (2016). When riverine dissolved organic matter (DOM) meets labile DOM in coastal waters: changes in bacterial community activity and composition. *Aquatic Sciences*, 79(1): 1–17.

Blum L.K. (1993). *Sartina alterniflora* root dynamics in a Virginia marsh. *Marine Ecology Progress Series*, 102: 169–178.

Blum L.K., Christian R.R. (2004). Belowground production and decomposition along a tidal gradient in a Virginia salt marsh. In: *Ecogeomorphology of Tidal Marshes*, Fagherazzi S., Marani M., Blum L. (eds), AGU: Washington D.C., pp. 47–75. doi: 10.1029/CE059p0047.

Blum L.K., Christian R.R. (2004). Belowground production and decomposition along a tidal gradient in a Virginia salt marsh. In: *The Ecogeomorphology of Tidal Marshes*, Fagherazzi S., Marani M., Blum L.K. (eds), American Geophysical Union: Washington D.C., pp. 47–75. doi: 10.1029/CE059p0047.

Bodelier P.L.E., Frenzel P., Drake H.L., Hurek T., Küsel K., Lovell C., Megonigal P., Reinhold-Hurek B., Sorrell B. (2006). Ecological aspects of microbes and microbial communitites inhabiting the rhizosphere of wetland plants. In: Verhoeven J.T.A., Beltman B., Bobbink R., Whigham D.F. (eds), *Ecological Studies, Wetlands and Natural Resource Management*. Springer Verlag: Berlin, Heidelberg, *vol. 190*.

Böer B. (1997). An introduction to the climate of the United Arab Emirates. *Journal of Arid Environments*, 35: 3–16.

Boesch D.F., Josselyn M.N., Mehta A.J., Morris J.T., Nuttle W.K., Simenstad C.A., Swift D.J. (1994). Scientific assessment of coastal wetland loss, restoration and management in Louisiana. *Journal of Coastal Research*, i: 103.

Bolton H.E. (1927). *Fray Juan Crespi: Missionary Explorer of the Pacific Coast, 1769–1774*, University of California Press, p. 402.

Bonacorsi M., Pergent-Martini C., Bréand N., Pergent G. (2013). Is *Posidonia oceanica* regressing a general feature in the Mediterranean Sea? *Mediterranean Marine Science*, 14(1): 193–203.

Boone D.R., Whitman W.B., Rouviere P. (1993). *Diversity and Taxonomy of Methanogens*, Ferry J.G. (ed.), Chapman and Hall: United Kingdom, pp. 35–80.

Borowitzka M., Lavery P., Van Keulen M. (2006). *Epiphytes of seagrass*. In: *Seagrasses: Biology, Ecology and Conservation*, Larkum A., Orth R.J., Duarte C.M. (eds), Springer: Dordrecht, pp. 441–461.

Borum J., Raun A.L., Hasler-Sheetal H., Pedersen M.Ø., Pedersen O., Holmer M. (2014). Eelgrass fairy rings: sulfide as inhibiting agent. *Marine Biology*, 161: 351–358.

Bosence D. (1989). Biogenic carbonate production in Florida Bay. *Bulletin of Marine Science*, 44: 419–433.

Bosire J.O., et al. (2014). Mangroves in peril: unprecedented degradation rates of peri-urban mangroves in Kenya. *Biogeosciences*, 11(10): 2623–2634. doi: 10.5194/bg-11-2623-2014.

Bosire J.O., Kaino J.J., Olagoke A.O., Mwihaki L.M., Ogendi G.M., Kairo J.G., Macharia D. (2014). Mangroves in peril: unprecedented degradation rates of peri-urban mangroves in Kenya. *Biogeosciences*, 11(10): 2623–2634.

Bosire J.O., Lang'at J.K., Kirui B., Kairo J.G., Mwihaki L.M., Hamza A.J. (2016). Mangroves in Kenya. In: *Mangroves of the Western Indian Ocean: Status and Management*, Bosire J.O., Mangora M.M., Bandeira S., Ratsimbazafy R., C. Appadoo, J.G. Kairo (eds), WIOMSA: Zanzibar, Town, pp. 15–31.

Boström, C., Baden S., Bockelmann A.-C., Dromph K., Fredriksen S., Gustafsson C., Krause-Jensen D.T., Möller, Nielsen S.L., Olesen B., Olsen J., Pihl L., Rinde E. (2014). Distribution, structure and function of Nordic eelgrass (*Zostera marina*) ecosystems: implications for coastal management and conservation. *Aquatic Conservation: Marine and Freshwater Ecosystems*, 24: 410–434.

Bouillon S., Connolly R., Lee S. (2008). Organic matter exchange and cycling in mangrove ecosystems: recent insights from stable isotope studies. *Journal of Sea Research*, 59: 44–58.

Bouma T.J., van Belzen J., Balke T., Zhu Z., Airoldi L., Blight A.J., Davies A.J., Galvan C., Hawkins S.J., Hoggart S.P.G., Lara J.L., Losada I.J., Maza M., Ondiviela B., Skov M.W., Strain E.M., Thompson R.C., Yang S., Zanuttigh B., Zhang L., Herman P.M.J. (2014). Identifying knowledge gaps hampering application of intertidal habitats in coastal protection: opportunities & steps to take. *Coastal Engineering*, 87: 147–157.

Boumans R.M.J., Day J.W. Jr. (1993). High precision measurements of sediment elevation in shallow coastal areas using a sedimentation-erosion table. *Estuaries*, 16: 375–380.

Bouse R., Fuller C.C., Luoma S., Hornberger M.I., Jaffe B.E., Smith R.E. (2010). Mercury-contaminated hydraulic mining debris in San Francisco Bay. *San Francisco Estuary and Watershed Science*, 8, 1–28.

Bowman G.T., Delfino J.J. (1980). Sediment oxygen demand techniques: a review and comparison of laboratory and *in situ* systems. *Water Research*, 14(5): 491–499.

Boyer K., Sutula M. (2015). Factors Controlling Submersed and Floating Macrophytes in the Sacramento-San Joaquin Delta. Southern California Coastal Water Research Project. Technical Report No. 870. Costa Mesa, CA.

Bradley P., Morris J.T. (1990). Influence of oxygen and sulfide concentration on nitrogen uptake kinetics in *Spartina alterniflora*. *Ecology*, 71: 282–287.

Braissant O., Decho A.W., Przekop K.M., Gallagher K.L., Glunk C., Dupraz C., Visscher P.T. (2009). Characteristics and turnover of exopolymeric substances in a hypersaline microbial mat. *FEMS microbiology ecology*, 67: 293–307.

Brand L.A., Smith L.M., Takekawa J.Y., Athearn N.D., Taylor K., Shellenbarger G.G., Schoellhamer D.H., Spenst R. (2012). Trajectory of early tidal marsh restoration: elevation, sedimentation and colonization of breached salt ponds in the northern San Francisco Bay. *Ecological Engineering*, 42: 19–29.

Breithaupt J.L., Smoak J.M., Smith T.J., Sanders C.J., Hoare A. (2012). Organic carbon burial rates in mangrove sediments: strengthening the global budget. *Global Biogeochemical Cycles*, 26(3).

Brevik E.C., Homburg J.A. (2004). A 5000 year record of carbon sequestration from a coastal lagoon and wetland complex, Southern California, USA. *Catena*, 57: 221–232.

Brew D., Williams P.B. (2010). Predicting the impact of large-scale tidal wetland restoration on morphodynamics and habitat evolution in South San Francisco Bay. *Journal of Coastal Research*, 26(5): pp. 912–924. (September, 2010).

Bricker S.B., Clement C.G., Pirhalla D.E., Orlando S.P., Farrow D.R.G. (1999). *National Estuarine Eutrophication Assessment: Effects of Nutrient Enrichment in the Nation's Estuaries, NOAA, National Ocean Service, Special Projects Office and the National Centers for Coastal Ocean Science*, Silver Spring: MD, p. 71.

Bricker-Urso S., Nixon S.W., Cochran J.K., Hirschberg D.J., Hunt C. (1989). Accretion rates and sediment accumulation in Rhodes Island salt marshes. *Estuaries*, 12: 300–317.

Bridgham S., Megonigal J., Keller J., Bliss N., Trettin C. (2006). The carbon balance of North American wetlands. *Wetlands* 26: 889–916.

Bridgham S.D., Cadillo-Quiroz H., Keller J.K., Zhuang Q. (2013). Methane emissions from wetlands: biogeochemical, microbial, and modeling perspectives from local to global scales. *Global Change Biology*, 19: 1325–1346.

Bridgham S.D., Moore T.R., Richardson C.J., Roulet N.T. (2014). Errors in greenhouse forcing and soil carbon sequestration estimates in freshwater wetlands: a comment on Mitsch et al. (2013). *Landscape Ecology*, 29: 1481–1485.

Bridgham S.D., Ye R. (2013). Organic matter mineralization and decomposition. In: *Methods in Biogeochemistry in Wetlands*, DeLaune R.D., Reddy K.R., Richardson C.J., Megonigal J.P. (eds), Soil Science Society of America: Madison, WI, SSSA Book Series, no. 10, pp. 385–406.

Brinson M.M. (1993). *A Hydrogeomorphic Classification for Wetlands*, East Carolina University: Greenville, NC.

Brinson M.M., Christian R.R., Blum L.K. (1995). Multiple states in the sea-level induced transition from terrestrial forest to estuary. *Estuaries*, 18: 648–659.

Brown B., Fadillah R., Nurdin Y., Soulsby I., Ahmad R. (2014). CASE STUDY: community based ecological mangrove rehabilitation (CBEMR) in indonesia. From small (12–33 ha) to medium scales (400 ha) with pathways for adoption at larger scales (>5000 ha). *SAPIEN. Surveys and Perspectives Integrating Environment and Society*, (7.2).

Brown K.M. (2005). On modern seagrasses as carbonate soil producers in shallow cool-water marine environments, South Australia. *PhD Dissertation*, University of Adelaide.

Brown L.R., Komoroske L.M., Wagner R.W., Morgan-King T., May J.T., Connon R.E. et al. (2016). Coupled downscaled climate models and ecophysiological metrics forecast habitat compression for an endangered estuarine fish. *PLoS One*, 11(1): e0146724. doi: 10.1371/journal.pone.0146724.

Buffington K.J., Dugger B.D., Thorne K.M., Takekawa J.Y. (2016). Statistical correction of lidar-derived digital elevation models with multispectral airborne imagery in tidal marshes. *Remote Sensing of Environment*, 186: 616–625.

Buffington K.J., Dugger B.D., Thorne K.M., Takekawa J.Y. (2016). Statistical correction of lidar-derived digital elevation models with multispectral airborne imagery in tidal marshes. *Remote Sensing of Environment, 186*: 616–625.

Bullock A.L., Sutton-Grier A.E., Megonigal J.P. (2013). Anaerobic metabolism in tidal freshwater wetlands: III. Temperature regulation of iron cycling. *Estuaries and Coasts, 36*: 482–490.

Bulwa D., Hallissy E., Lucas G., Fagan K. (2004). Deluge in the Delta – 'It was like an ocean'/Levee gives way – farmland inundated, San Francisco Chronicle. (June 4).

Burdige D.J. (2006). *Geochemistry of Marine Sediments.* Princeton Univ. Press: Princeton, NJ, p. 630.

Burdige D.J., Hu X., Zimmerman R.C. (2010). The widespread occurrence of coupled carbonate dissolution/reprecipitation in surface soils on the Bahamas Bank. *American Journal of Science, 310*: 492–521.

Burdige D.J., Zimmerman R.C. (2002). Impacts of seagrass density on carbonate dissolution in Bahamian sediments. *Limnology and Oceanography, 47*: 1751–1763.

Burdige D.J., Zimmerman R.C., Hu X. (2008). Rates of carbonate dissolution in permeable soils estimated from pore-water profiles: the role of sea grasses. *Limnology and Oceanography, 53*: 549–565.

Burgin A.J., Hamilton S.K. (2007). Have we overemphasized the role of denitrification in aquatic ecosystems? A review of nitrate removal pathways. *Frontiers in Ecology and the Environment, 5*: 89–96.

Burgin A.J., Hamilton S.K., Gardner W.S., McCarthy M.J. (2013). Nitrate reduction, denitrification, and dissimilatory nitrate reduction to ammonium in wetland sediments. In: *Methods in Biogeochemistry in Wetlands*, DeLaune R.D., Reddy R., K., Richardson C.J., Megonigal J.P. (eds), Soil Science Society of America: Madison, WI, SSSA Book Series, no. 10, pp. 519–538.

Burt J.A. (2014). The environmental costs of coastal urbanization in the Arabian Gulf. *City 18*: 760–770.

Buth G.J.C., Voesenek L.A.C.J. (1987). Decomposition of standing and fallen litter of halophytes in a Dutch salt marsh. In: *Geobotany 11: Vegetation Between Land and Sea*, Huiskes A.H.L., Blom C.W.P.M., Rozema J. (eds), Dr. Junk W. Pub.: Dordrecht, pp. 146–165.

Butzeck C., Eschenbach A., Gröngröft A., Hansen K., Nolte S., Jensen K. (2015). Sediment deposition and accretion rates in tidal marshes are highly variable along estuarine salinity and flooding gradients. *Estuaries and Coasts, 38*: 434–450.

Byrd K.B., Ballanti L., Thomas N., Nguyen D., Holmquist J.R., Simard M., Windham-Myers L. (2018). A remote sensing-based model of tidal marsh aboveground carbon stocks for the conterminous United States. *ISPRS Journal of Photogrammetry and Remote Sensing, 139*: 255–271.

Byrd K.B., J.L. O'Connell Di Tommaso S., Kelly M. (2014). Evaluation of sensor types and environmental controls on mapping biomass of coastal marsh emergent vegetation. *Remote Sensing of Environment, 149*: 166–180.

Byrd K.B., Windham-Myers L., Leeuw T., Downing B., Morris J.T., Ferner M.C. (2016). Forecasting tidal marsh elevation and habitat change through fusion of Earth observations and a process model. *Ecosphere 7*, e01582. doi: 10.1002/ecs2.1582.

Byrd K.B. et al. (2019). Status of Tidal Marsh Mapping for Blue Carbon Inventories. In: *A Blue Carbon Primer: The State of Coastal Wetland Carbon Science, Practice, and Policy*, Lisamarie W.-M., Stephen C., and Tiffany G.T. (eds), CRC Press: Boca Raton, London.

Caamal-Sosa J.P., Zaldívar A., Adame M.F., Teutli C., Andueza M.T., Pérez R., Herrera-Silveira J.A. (2015). Almacenes de carbon en diferentes tipos ecológicos de manglares en un escenario cárstico. In: Paz F., Wong J. (eds). *Estado Actual del Conocimiento del ciclo de Carbon y sus Interacciones en México: Síntesis a 2014*, Texcoco, Edo de Mex: Mexico. p. 642.

Cadol D., Engelhardt K., Elmore A., Sanders G. (2014). Elevation-dependent surface elevation gain in a tidal freshwater marsh and implications for marsh persistence. *Limnology and Oceanography, 59*: 1065–1080.

Cahoon D.R., Guntenspergen G.R. (2010). Climate change, sea-level rise, and coastal wetlands. *National Wetlands Newsletter, 32*: 8–12.

Cahoon D.R., Lynch J.C. (1997). Vertical accretion and shallow subsidence in a mangrove forest of southwestern Florida, USA. *Mangroves and Salt Marshes, 1*(3): 173–86.

Cahoon D.R., Lynch J.C., Hensel P., Boumans R., Perez B.C., Segura B., Day J.W. (2002). High-precision measurements of wetland sediment elevation: I. recent improvements to the sedimentation-erosion table. *Journal of Sedimentary Research 72*: 730–733.

Cahoon D.R., Lynch J.C., Perez B.C., Segura B., Holland R.D., Stelly C., Stephenson G., Hensel P. (2002b). High-precision measurements of wetland sediment elevation: II. The rod surface elevation table. *Journal of Sedimentary Research, 72*: 734–739.

Cahoon D.R., Reed D.J., Day J.W. (1995). Estimating shallow subsidence in midrotidal salt marshes of the southeastern United States: Kaye and Barghoorn revisited. *Marine Geology, 128*: 1–9.

Cahoon D.R., Reed D.J., Kolker A.S., Brinson M.M., Stevenson J.C., Riggs S.R., Christian R., Voss C., Kunz D. (2009). Coastal wetland sustainability. In: *Coastal Sensitivity to Sea-Level Rise: A Focus on the Mid-Atlantic Region. A Report by the U.S. Climate Change Program and the Subcommittee on Global Change Research*, Titus J.G., Anderson K.E., Cahoon D.R., Gesch D.B., Gill S.K., Guitierrez B.T., Thieler E.R., Williams S.J. (eds), U.S. Environmental Protection Agency: Washington D.C., pp. 57–72.

Cahoon D.R., Turner R.E. (1989). Accretion and canal impacts in a rapidly subsiding wetland: II. Feldspar marker horizon technique. *Estuaries, 12*: 260–268.

Cahoon D.R., White D.A., Lynch J.C. (2011). Sediment infilling and wetland formation dynamics in an active crevasse splay of the Mississippi River delta. *Geomorphology, 131*: 57–68.

Cai W.J. (2011). Estuarine and coastal ocean carbon paradox: CO_2 sinks or sites of terrestrial carbon incineration? *Annual Review of Marine Science, 3*: 123–145.

California Department of Fish and Wildlife (CDFW) (2016). Water and Watershed Conservation. www.wildlife.ca.gov/Conservation/Watersheds (accessed March 15, 2016).

California Department of Parks & Recreation-Division of Boating & Waterways (CDPR-DBW) (2017a). Submersed aquatic vegetation. http://dbw.parks.ca.gov/?page_id=28994 (accessed July 24, 2017).

California Department of Water Resources (CDWR) (2009). Delta Risk Management Strategy: Final Phase 1 Report. www.water.ca.gov/floodsafe/fessro/levees/drms/phase1_information.cfm (accessed March 9, 2016).

California Environmental Protection Agency (2014). Assembly Bill 32: Global Warming Solutions Act, www.arb.ca.gov/cc/ab32/ab32.htm (accessed March 28, 2016).

California Environmental Protection Agency (2017). Air Resources Board, Cap-and-Trade Program, www.arb.ca.gov/cc/capandtrade/capandtrade.htm (accessed June 29, 2017).

California Natural Resources Agency (CNRA) (2016a). History of Water Project Conveyance in the Delta. Water Conveyance History Fact Sheet. www.californiawaterfix.com/resources/outreach-materials/ (accessed March 2, 2016).

Cal-IPC (2006). *California Invasive Plant Inventory*. Cal-IPC Publication, Berkeley, California.

Callaway J., Nyman J.A., DeLaune R.D. (1996). Sediment accretion in coastal wetlands: a review and simulation model of processes. *Current Topics in Wetland Biogeochemistry, 2*: 2–23.

Callaway J.C. (2005). The challenge of restoring functioning salt marsh ecosystems. *Journal of Coastal Research Special Issue, 40*: 24–36.

Callaway J.C., Borgnis E.L., Turner R.E., Milan C.S. (2012). Carbon sequestration and sediment accretion in San Francisco Bay tidal wetlands. *Estuaries and Coasts, 35*: 1163–1181. doi: 10.1007/s12237-012-9508-9.

Callaway J.C., Cahoon D.R., Lynch J.C. (2013). The surface elevation table – marker horizon method for measuring wetland accretion and elevation dynamics. In: *Methods in Biogeochemistry of Wetlands*, DeLaune R.D., Reddy K.R., Richardson C.J., Megonigal P. (eds), Soil Science Society of America, SSSA Book Series, no. 10, *vol. 78*, no. 3, pp. 901–917.

Callaway J.C., DeLaune R.D., Patrick W.H. Jr. (1996). Chernobyl ^{137}Cs used to determine sediment accretion rates at selected northern European coastal wetlands. *Limnology and Oceanography, 41*: 444–450.

Callaway J.C., Parker V.T., Vasey M.C., Schile L.M. (2007). Emerging issues for the restoration of tidal marsh ecosystems in the context of predicted climate change. *Madroño 54*(3): 234–248.

Callaway J.C., Parker V.T., Vasey M.C., Schile L.M., Herbert E.R. (2011). Tidal wetland restoration in San Francisco Bay: history and current issues. *San Francisco Estuary and Watershed Science, 9*: 3.

Campbell J., Lacey E., Decker R., Crooks S., Fourqurean J. (2015). Carbon storage in seagrass beds of Abu Dhabi, United Arab Emirates. *Estuaries and Coasts, 38*: 242–251.

Campbell J.E., Fourqurean J.W. (2013). Effects of in situ CO_2 enrichment on the structural and chemical characteristics of the seagrass *Thalassia testudinum*. *Marine Biology, 160*: 1465–1475.

Camuffo D., Sturaro G. (2003). Sixty-cm submersion of Venice discovered thanks to Canaletto's paintings. *Climatic Change, 58*: 333–343.

CAN, National Water Commission (*Comisión Nacional del Agua*). (2016). http://smn.cna.gob.mx/index.php?option=com_content&view=article&id=103&Itemid=80 (accessed January 27).

Canals M., Ballesteros E. (1997). Production of carbonate particles by phytobenthic communities on the Mallorca-Menorca shelf, northwestern Mediterranean Sea. *Deep-Sea Research Part II, 44*: 611–629.

Cao M., Marshall S., Gregson K. (1996). Global carbon exchange and methane emissions from natural wet-lands: application of a processes based model. *Journal of Geophysical Research 101*: 14399–14414.

Cardoso J., Watts C., Maxwell J., Goodfellow R., Eglinton G., Golubic S. (1978). A biogeochemical study of the Abu Dhabi microbial mats: a simplified ecosystem. *Chemical Geology, 23*: 273–291.

Carey A.E., Oliver F.W. (1918). *Tidal Lands: A Study of Shore Problems*, Blackie and Son Limited: London.

Cartaxana P., Catarino F. (1997). Allocation of nitrogen and carbon in an estuarine salt marsh in Portugal. *Journal of Coastal Conservation, 3*: 27–34.

Cash D.W., William C.C., Frank A., Nancy M.D., Noelle E., David H.G., Jill J., Ronald B.M. (2003). Knowledge Systems for Sustainable Development, *Proceedings of the National Academy of Sciences of the United States of America, 100*(14): 8086–8091.

Cavanaugh K.C., Kellner J.R., Forde A.J., Gruner D.S., Parker J.D., Rodriguez W., Feller I.C. (2014). Poleward expansion of mangroves is a threshold response to decreased frequency of extreme cold events. *Proceedings of the National Academy of Sciences of the United States of America, 111*: 723–727.

Cavanaugh K.C., Kellner J.R., Forde A.J., Gruner D.S., Parker J.D., Rodriguez W., Feller I.C. (2014). Reply to Giri and Long: Freeze-mediated expansion of mangroves does not depend on whether expansion is emergence or reemergence. *Proceedings of the National Academy of Sciences of the United States of America, 111*(15): E1449.

Cazenave A., Llovel W. (2010). Contemporary sea level rise. *Annual Review of Marine Science, 2*: 145–173.

CBD (2010a). Decision X/29. Coastal, Marine Biodiversity. UNEP/CBD/COP/DEC/X/29.

CBD (2012). Decision XI/21. Biodiversity, climate change: integrating biodiversity considerations into climate-change related activities. UNEP/CBD/COP/DEC/XI/21.

CBD (2012a). Decision XI/16. Ecosystem restoration. UNEP/CBD/COP/DEC/XI/16.

CBD (2014). Decision XII/20. Biodiversity, climate change, disaster risk reduction UNEP/CBD/COP/DEC/XII/20.

CBD (2016). Decision XIII/5. Ecosystem restoration: short-term action plan. UNEP/CBD/COP/DEC/XIII/5.

CBD (2016a). Decision XIII/4. Biodiversity, climate change. UNEP/CBD/COP/DEC/XIII/4.

CBD (Convention on Biological Diversity) (2010). Decision X/33. Biodiversity, climate change. UNEP/CBD/COP/DEC/X/33.

CBD/SBSTA (2016). Recommendation XX/10. Biodiversity, climate change. UNEP/CBD/SBSTA/REC/XX/10.

CCA (2017). *Análisis de las Oportunidades Para la Integración del Concepto de Carbono Azul en la Política Pública Mexicana*, Comisión para la Cooperación Ambiental: Montreal: Canadá.

CDFW (2017). The Wetlands Restoration for Greenhouse Gas Reduction Grant Program. www.wildlife.ca.gov/Conservation/Watersheds/Greenhouse-Gas-Reduction (accessed June 30, 2017).

CDPR-DBW (2017b). Floating aquatic vegetation. http://dbw.parks.ca.gov/?page_id=28995 (accessed July 24, 2017).

CDWR (2016). FloodSAFE Environmental Stewardship and Statewide Resources Office (FESSRO). www.water.ca.gov/floodsafe/fessro/ (accessed March 16, 2016).

CDWR (2017a). Managing & Protecting California's Water Resources. www.water.ca.gov/ (accessed July 17, 2017).

CDWR (2017b). FloodSAFE Environmental Stewardship and Statewide Resources Office (FESSRO). www.water.ca.gov/floodsafe/fessro/ (accessed July 17, 2017).

CDWR (2017c). Delta Ecosystem Enhancement: Decker Island Habitat Development/ Levee Improvement Project. www.water.ca.gov/floodsafe/fessro/restoration/Decker_Island.cfm (accessed July 17, 2017).

CDWR (2017d). Prospect Island Tidal Habitat Restoration Project: Component of the Fish Restoration Program (FRP). www.dwr.water.ca.gov/environmentalservices/docs/frpa/Prospect_Island_Fact_Sheet_9-12-14.pdf (accessed July 17, 2017).

CDWR (2017e). Mayberry Farms Subsidence Reversal and Carbon Sequestration Project. www.water.ca.gov/deltainit/docs/mayberry_factsheet.pdf (accessed July 17, 2017).

CEC (2016). *North America's Blue Carbon: Assessing Seagrass, Salt Marsh and Mangrove Distribuition and Carbon Sinks*, Commission for Environmental Cooperation: Montreal, Canada.

Cerón-Bretón, J.G., et al. (2011). Determination of carbon sequestration rate in soil of a mangrove forest in Campeche, Mexico. *WSEAS Transactions on Environment and Development, 7*(2): pp.54–64.

Chakravortty S., Shah S., Chowdhury A.S. (2014). Application of Spectral Unmixing Algorithm on Hyperspectral Data for Mangrove Species Classification. In: *Applied Algorithms*, Springer: Cham, 223–236

Chambers L., Osborne T., Reddy K. (2013). Effect of salinity-altering pulsing events on soil organic carbon loss along an intertidal wetland gradient: a laboratory experiment. *Biogeochemistry, 115*: 363–383.

Chambers L.G., Davis S.E., Troxler T., Boyer J.N., Downey-Wall A., Scinto L.J. (2014). Biogeochemical effects of simulated sea level rise on carbon loss in an Everglades mangrove peat soil. *Hydrobiologia, 726*: 195–211.

Chambers R.M., Meyerson L.A., Saltonstall K. (1999). Expansion of *Phragmites australis* into tidal wetlands of North America. *Aquatic Botany, 64*(3): 261–273.

Champion P.D., Tanner C.C. (2000). Seasonality of macrophytes and interaction with flow in a New Zealand lowland stream. *Hydrobiologia, 441*: 1–12.

Chan K.M.A., Satterfield T., Goldstein J. (2012b). Rethinkinge cosystem services to better address and navigate cultural values. *Ecological Economics, 74*: 8–18.

Chapin F.S., et al. (2006). Reconciling carbon-cycle concepts, terminology, and methods. *Ecosystems 9*(7): 1041–1050.

Chapin F.S., Matson P.A., Mooney H.A. (2011). *Principles of Terrestrial Ecosystem Ecology*, 2nd edn, Springer-Verlag: New York, USA.

Chapin F.S., Woodwell G.M., Randerson J.T., Rastetter E.B., Lovett G.M., Baldocchi D.D., Clark D.A., Harmon M.E., Schimel D.S., Valentini R., Wirth C. (2006). Reconciling carbon-cycle concepts, terminology, and methods. *Ecosystems, 9*(7): 1041–1050.

Chapman V.J. (1944). 1939 Cambridge University Expedition to Jamaica. Part 1. A study of the botanical processes concerned in the development of the Jamaican shoreline. *The Journal of the Linnaean Society London, Botany, 52*: 40–447.

Chapman V.J. (1960). *Salt Marshes and Salt Deserts of the World (No. 581.9095 C43)*, London.

Charles H., Dukes J.S. (2009). Effects of warming and altered precipitation on plant and nutrient dynamics of a New England salt marsh. *Ecological Applications, 19*: 1758–1773.

Chassereau J.E., Bell J.M., Torres R. (2011). A comparison of GPS and lidar salt marsh DEMs. *Earth Surface Processes and Landforms, 36*: 1770–1775.

Chefaoui R.M., Assis J., Duarte C.M., Serrão E.A. (2016). Large-scale prediction of seagrass distribution integrating landscape metrics and environmental factors: the case of *Cymodocea nodosa* (Mediterranean-Atlantic). *Estuaries and Coasts, 39*(1): 123–137.

Chen C.-F., Lau V.-K., Chang N.-B., Son N.-T., Tong P.-H.-S., Chiang S.-H. (2016). Multi-temporal change detection of seagrass beds using integrated Landsat TM/ETM+/OLI imageries in Cam Ranh Bay, Vietnam. *Ecological Informatics, 35*: 43–54.

Chen G.C., Tam N.F.Y., Ye Y. (2010). Summer fluxes of atmospheric gases N_2O, CH_4 and CO_2 from mangrove soil in South China. *Science of the Total Environment, 408*: 2761–2767.

Chen G.C., Tama N.F.Y., Wong Y.S., Yed Y. (2011). Effect of wastewater discharge on greenhouse gas fluxes from mangrove soils. *Atmospheric Environment, 45*: 1110–1115.

Chen X., Zhang X., Church J.A., Watson C.S., King M.A., Monselesan D., Legresy B., Harig C. (2017). The increasing rate of global mean sea-level rise during 1993–2014. *Nature Climate Change, 7*: 492–495.

Chen Y., Chen G., Ye Y. (2015). Coastal vegetation invasion increases greenhouse gas emission from wetland soils but also increases soil carbon accumulation. *Science of the Total Environment, 526*: 19–28.

Chile (2016). Submission on the relevance of the ocean in the global response to climate change. Submission to the UNFCCC. http://www4.unfccc.int/Submissions/Lists/OSPSubmission Upload/39_279_131200401965740037-Submission%20on%20NDC%20and%20oceans%20Chile.pdf.

Chmura G.L. (2013). What do we need to assess the sustainability of the tidal salt marsh carbon sink? *Ocean and Coastal Management, 83*: 25–31.

Chmura G.L., Anisfeld S.C., Cahoon D.R., Lynch J.C. (2003). Global carbon sequestration in tidal, saline wetland soils. *Global Biogeochemical Cycles 17*(4): 1111. doi: 10.1029/2002GB001917.

Chmura G.L., Kellman L., Guntenspergen G.R. (2011). The greenhouse gas flux and potential global warming feedbacks of a northern macrotidal and microtidal salt marsh. *Environmental Research Letters, 6*. doi: 10.1088/1748–9326/1086/1084/044016.

Chmura G.L., Kellman L., van Ardenne L., Buntenspergen G.R. (2016). Greenhouse gas fluxes from salt marshes exposed to chronic nutrient enrichment. *PLOS One, 11*: e0149937.

Christian R.R., Mazzilli S. (2007). Defining the coast and sentinel ecosystems for coastal observations of global change. *Hydrobiologia, 577*: 55–70. (See Hydrobiologia 583: 385 for erratum on order of authorship from original. It should read Mazzilli, Christian).

Christian R.R., Stasavich L., Thomas C., Brinson M.M. (2000). Reference is a moving target in sea-level controlled wetlands. In: *Concepts and Controversies in Tidal Marsh Ecology,* Weinstein M.P., Kreeger D.A. (eds) Kluwer Press: The Netherlands, pp. 805–825.

Chu S.N., Wang Z.A., Gonneea M., Kroeger K.D., Ganju N.K. (2018). Assessing inorganic carbon export from intertidal salt marshes using direct, high-frequency measurements. *Geochimica et Cosmochimica Acta.*

Chuang P.-C., Young M.B., Dale A.W., Miller L.G., Herrera-Silveira J.A., Paytan A. (2016). Methane and sulfate dynamics in sediments from mangrove-dominated tropical coastal lagoons, Yucatán, Mexico. *Biogeosciences, 13*: 2981–3001.

Chuang P.C., Young M.B., Dale A.W., Miller L.G., Herrera-Silveira J.A., Paytan A. (2017). Methane fluxes from tropical coastal lagoons surrounded by mangroves, Yucatán, Mexico. *Biogeosciences 122*: 1156–1174.

Chung I., Beardall J., Mehta S., Sahoo D., Stojkovic S. (2011). Using marine macroalgae for carbon sequestration: a critical appraisal. *Journal of Applied Phycology, 23*: 877–86.

Church J.A., White N.J. and Arblaster J.M. (2013). Coauthors, 2013: Sea level change. *Climate Change 2013: The Physical Science Basis,* 1137–1216.

Church J.A., White N.J. (2006). A 20th century acceleration in global sea-level rise. *Geophysical Research Letters, 33*(1).

Church J.A., White N.J. (2011). Sea-level rise from the late 19th to the early 21st century. *Surveys in Geophysics, 32*(4–5): 585–602.

Church J.A., White N.J., Konikow L.F., Domingues C.M., Cogley J.G., Rignot E., Gregory J.M., van den Broeke M.R., Monaghan A.J., Velicogna I. (2011). Revisiting the earth's sea level and energy budgets from 1961 to 2008. *Geophysical Research Letters, 38*: L18601. doi: 10.1029/2011GL048794.

Ciais P., Sabine C., Bala G., Bopp L., Brovkin V., Canadell J., Chhabra A., DeFries R., Galloway J., Heimann M., Jones C., Le Quéré C., Myneni R.B., Piao S., Thornton P. (2013). Carbon and Other Biogeochemical Cycles. In: *Climate Change 2013: The Physical Science Basis. Contribution of Working Group I to the Fifth Assessment Report of the Intergovernmental Panel on Climate Change,* Stocker T.F., Qin D., Plattner G.-K., Tignor M., Allen S.K., Boschung J., Nauels A., Xia Y., Bex V., Midgley P.M. (eds), Cambridge University Press: Cambridge, New York, UK, USA.

Cicchetti G., Greening H. (2011). Estuarine biotope mosaics and habitat management goals: an application in Tampa Bay, FL, USA. *Estuaries and Coasts, 34*: 1278-1292. doi: 10.1007/s12237-011-9408-4.

Cicin-Sain B., et al. (2016). *Towards a Strategic Action Roadmap on Oceans and Climate: 2016 to 2021,* USGlobal A., Ocean Forum: Washington D.C.

Cintron G., Lugo A.E., Pool D.J., Morris G. (1978). Mangroves of arid environments in Puerto Rico and adjacent islands. *Biotropica, 10*: 110–121.

Clark P.U., Mitrovica J.X., Milne G.A., Tamisiea M.E. (2002). Sea-Level Fingerprinting as a Direct Test for the Source of Global Meltwater Pulse IA. *Science, 295*: 2438–2441.

Clewell A.F., Aronson J. (2013). *Ecological Restoration: Principles, Values, and Structure of an Emerging Profession,* Society of Ecological Restoration, p. 303.

Climate Focus (2011). *Blue Carbon Policy Options Assessment,* Linden Trust for Conservation: Washington, DC.

Cloern J.E. (2001). Our evolving conceptual model of the coastal eutrophication problem. *Marine Ecology Progress Series, 210*: 223–253.

Cloern J.E., Abreu P.C., Carstensen J., Chauvaud L., Elmgren R., Grall J., Xu J. (2016). Human activities and climate variability drive fast-paced change across the world's estuarine–coastal ecosystems. *Global Change Biology, 22*(2): 513–529.

Cloern J.E., Knowles N., Brown L.R., Cayan D., Dettinger M.D., Morgan T.L., et al. (2011). Projected evolution of California's San Francisco Bay–Delta-River system in a century of climate change. *PLoS One, 6*(9): e24465. doi: 10.1371/journal.pone.0024465.

CNRA (2017b). California EcoRestore: A Stronger Delta Ecosystem. http://resources.ca.gov/ecorestore/ (accessed July 17, 2017).

CNRA (2017c). Lindsey Slough Tidal Marsh Restoration. http://resources.ca.gov/docs/ecorestore/projects/Lindsey_Slough.pdf (accessed July 17, 2017).

CNRA (2017d). Decker Island Tidal Habitat Restoration. http://resources.ca.gov/docs/ecorestore/projects/Decker_Island_Tidal_Habitat_Restoration.pdf (accessed July 17, 2017).

CNRA (2017e). Sherman Island – Whale's Mouth Wetland Restoration. http://resources.ca.gov/docs/ecore-store/projects/Sherman_Island-_Whale's_Mouth_Wetland.pdf (accessed July 17, 2017).

CNRA (2017f). Twitchell Island – East End Wetland Restoration. http://resources.ca.gov/docs/ecorestore/proj-ects/Twitchell_Island-_East_End_Wetland.pdf (accessed July 17, 2017).

CNRA (2017g). Twitchell Island – San Joaquin River Setback Levee. http://resources.ca.gov/docs/ecorestore/projects/Twitchell_Island-_SJ_River_Setback_Levee.pdf (accessed July 17, 2017).

CNRA (2017h). Winter Island Tidal Habitat Restoration Project. http://resources.ca.gov/docs/ecorestore/proj-ects/Winter_Island_Tidal_Habitat_Restoration.pdf (accessed July 17, 2017).

CNRA (2017i). Lower Putah Creek Realignment Project. http://resources.ca.gov/docs/ecorestore/projects/Lower_Putah_Creek_Realignment.pdf (accessed July 17, 2017).

CNRA (2017j). McCormack-Williamson Tract Restoration Project. http://resources.ca.gov/docs/ecorestore/projects/McCormack_Williamson_Tract_Project.pdf (accessed July 17, 2017).

CNRA (2017k). Grizzly Slough Floodplain Project. http://resources.ca.gov/docs/ecorestore/projects/Grizzly_Slough_Floodplain_Project.pdf (accessed July 17, 2017).

CNRA (2017l). Lower Yolo Restoration Project. http://resources.ca.gov/docs/ecorestore/projects/Lower_Yolo_Restoration.pdf (accessed July 17, 2017).

CNRA (2017m). Dutch Slough Tidal Habitat Restoration Project. http://resources.ca.gov/docs/ecorestore/proj-ects/Dutch_Slough_Tidal_Marsh_Restoration.pdf (accessed July 17, 2017).

CNRA (2017n). Sherman Island – Belly Wetland Restoration. http://resources.ca.gov/docs/ecorestore/projects/Sherman_Island_Belly_Wetland_Restoration.pdf (accessed July 17, 2017).

Cohen A.N., Carlton J.T. (1995). Biological study. Non-indigenous aquatic species in a United States estu-ary: A case study of the biological invasions of the San Francisco Bay and Delta. A report for the United States Fish and Wildlife Service, Washington D.C. and the National Seagrant College Program, Connecticut Seagrant, NTIS Report Number PB96-166525.

Cohen A.N., Carlton J.T. (1998). Accelerating invasion rate in a highly invaded estuary. *Science 279*: 555–558.

Cohen R., Kaino J., Okello J.A., Bosire J.O., Kairo J.G., Huxham M., Mencuccini M. (2013). Propagating uncertainty to estimates of above-ground biomass for Kenyan mangroves: a scaling procedure from tree to landscape level. *Forest Ecology and Management, 310*: 968–982.

Coldren G.C., Barreto C., Wykoff D., Morrssey E., Langley J.A., Feller I.C., Chapman S.K. (2016). Chronic warming stimulates growth of marsh grasses more than mangroves in a coastal wetland ecotone. *Ecology, 97*: 3167–3175.

Cole J.J., Prairie Y.T., Caraco N.F., McDowell W.H., Tranvik L.J., Stricgl R.G., Duarte C.M., Koterlainen P., Downing J.A., Melack J.M. (2007). Plumbing the global carbon cycle: integrating inland waters into the terrestrial carbon budget. *Ecosystems, 10*(1): 172–185.

Colgan C.S., Beck M.W., Narayan S. (2017). *Financing Natural Infrastructure for Coastal Flood Damage Reduction*. Lloyd's Tercentenary Research Foundation: London.

Collier C.J., Adams M.P., Langlois L., Waycott M., O'Brien K.R., Maxwell P.S., McKenzie L. (2016). Thresholds for morphological response to light reduction for four tropical seagrass species. *Ecological Indicators, 67*: 358–366.

Collins D.S., Avidis A., Allison P.A., Johnson H.D., Hills J., Piggott M.D., Hassan M.H., Damit A.R. (2017). Tidal dynamics and mangrove carbon sequestration during the Oligo-Miocene in the South China Seas. *Nature Communications.* doi: 10.1038/ncomms15698.

Colonnello G., Medina E. (1998). Vegetation changes induced by dam construction in a tropical estuary: the case of the Mánamo river, Orinoco Delta (Venezuela). *Plant Ecology, 139*(2): 145–154.

Comeaux R.S., Allison M.A., Bianchi T.S. (2012). Mangrove expansion in the Gulf of Mexico with climate change: implications for wetland health and resistance to rising sea levels. *Estuarine Coastal and Shelf Science, 96*: 81–95.

Comeaux R.S., Allison M.A., Bianchi T.S. (2012). Mangrove expansion in the Gulf of Mexico with climate change: implications for wetland health and resistance to rising sea levels. *Estuarine Coastal and Shelf Science, 96*: 81–95.

Committee on Sea Level Rise in California, Oregon, and Washington (2012). *Sea Level Rise for the Coasts of California, Oregon, and Washington: Past, Present, and Future. Board on Earth Sciences and Resources and Oceans Studies Board, Division on Earth and Life Sciences*, National Research Council of the National Academies: Washington, D.C., p. 260.

CONABIO (2013). *Manglares de México:Actualizacion y Exploracion de los datos del Sistema de monitoreo 1970/1980–2015*, National Comission on Biodiversity: Mexico City, Mexico, p. 128.

CONABIO, (2009). *Manglares de México: Extensión y distribución*, 2nd edn, National Comission on Biodiversity: Mexico City, Mexico.

CONABIO. Mangrove projects. (2016). www.conabio.gob.mx/institucion/cgi-bin/consulta_proy2.cgi (accessed January 2016).

Congalton R.G., Green K. (2009). *Assessing the accuracy of remotely sensed data, principles and practices*, 2nd edn, CRC Press, Taylor and Francis Group: Baca Raton.

Conley D.J., Paerl H.W., Howarth R.W., Boesch D.F., Seitzinger S.P., Havens K.E., Likens G.E. (2009). Controlling eutrophication: nitrogen and phosphorus. *Science*, *323*(5917): 1014–1015.

Connor R.F., Chmura G.L., Beecher C.B. (2001). Carbon accumulation in Bay of Fundy salt marshes: implications for restoration of reclaimed marshes. *Global Biogeochemical Cycles*, *15*(4): 943–954.

Corbera E., Soberanis C.G., Brown K. (2009). Institutional dimensions of Payments for Ecosystem Services: an analysis of Mexico's carbon forestry programme. *Ecological Economics*, *68*(3): pp. 743–761. doi: 10.1016/j.ecolecon.2008.06.008.

Corlett H., Jones B. (2007). Epiphyte communities on *Thalassia testudinum* from Grand Cayman, British West Indies: their composition, structure, and contribution to lagoonal soils. *Sedimentary Geology*, *194*: 245–262.

Costanza R., d'Arge R., de Groot R., Farber S., Grasso M., Hannon B., Limburg K., Naeem S., O'Neill R.V., Paruelo J., Raskin R.G., Sutton P., van den Belt M. (1997). The value of the world's ecosystem services and natural capital. *Nature*, *387*: 253–260.

Costanza R., de Groot R., Sutton P., van der Ploeg S., Anderson S.J., Kubiszewski I., Farber S., Turner R.K. (2014). Changes in the global value of ecosystem services. *Global Environmental Change*, *26*: 152–158. doi: 10.1016/J.GLOENVCHA.2014.04.002.

Costanza R., Perez-Maqueo O., Luisa Martinez M., Sutton P., Anderson S.J., Mulder K. (2008). The value of coastal wetlands for hurricane protection. *Ambio*, *37*: 241–248.

Cosumnes River Preserve (2008). Cosumnes River Preserve Management Plan. http://cosumnes.org/documents/managementplan.pdf (accessed July 17, (2017).

Council on Climate Preparedness and Resilience Climate and Natural Resources Working Group (2014). Priority Agenda for Enhancing the Climate Resilience of America's Natural Resources. The White House, www.whitehouse.gov/sites/default/files/docs/enhancing_climate_resilience_of_americas_natural_resources.pdf.

Couvillion B.R., Steyer G.D., Wang H., Beck H.J., Rybczyk J.M. (2013). Forecasting the effects of coastal protection and restoration projects on wetland morphology in coastal louisiana under multiple environmental uncertainty scenarios. *Journal of Coastal Research*, SI 67: 29–50.

Cox T.E., Nash M., Gazeau F. et al. (2017). Effects of in situ CO_2 enrichment on *Posidonia oceanica* epiphytic community composition and mineralogy. *Marine Biology*, *164*: 103. doi: 10.1007/s00227-017-3136-7.

Craft C. (2007). Freshwater input structures soil properties, vertical accretion, and nutrient accumulation of Georgia and US tidal marshes. *Limnology and Oceanography*, *52*(3): 1220–1230.

Craft C. (2013). Emergent macrophyte biomass production. In: *Methods in Biogeochemistry of Wetlands*, DeLaune R.D., Reddy K.R., Richardson C.J., Megonigal J.P. (eds), Soil Science Society of America: Madison, Wisconsin, USA, pp. 137–153.

Craft C., Broome S., Campbell C. (2002). Fifteen years of vegetation and soil development after brackish water marsh creation. *Restoration Ecology*, *10*: 248–258.

Craft C., Megonigal P., Broome S., Stevenson J., Freese R., Cornell J., Zheng L., Sacco J. (2003). The pace of ecosystem development of constructed Spartina alterniflora marshes. *Ecological Applications*, *13*(5): 1417–1432.

Craft C., Pennings S., Clough J., Park R., Ehman J. (2009). SLR and ecosystem services: a response to Kirwan and Guntenspergen. *Frontiers in Ecology and the Environment*, *7*(3): 127–128.

Craft C., Reader J., Sacco J.N., Broome S.W. (1999). Twenty-five years of ecosystem development of constructed *Spartina alterniflora* (Loisel) marshes. *Ecological Applications*, *9*: 1405–1419.

Craft C.B. (2012). Tidal freshwater forest accretion does not keep pace with sea level rise. *Global Change Biology*, *18*(12): 3615–3623.

Craft C.B., Richardson C.J. (1993). Peat accretion and N, P, and organic C accumulation in nutrient- enriched and unenriched Everglades peatlands. *Ecological Applications*, *3*: 446–458.

Craft C.B., Seneca E.D., Broome S.W. (1991). Loss on ignition and Kjeldahl digestion for estimating organic carbon and total nitrogen in estuarine marsh soils: calibration with dry combustion. *Estuaries*, *14*: 175–179.

Crooks S., Emmer I., von Under M., Brown B., Orr M.K., Murdiyarso D. (2014). *Guiding Principles for Delivering Coastal Wetland Carbon Projects.* Report by United Nations Environment Program and Center for International Forestry Research.

Crooks S., Findsen J., Igusky K., Orr M.K., Brew D. (2009). Greenhouse Gas Mitigation Typology Issues Paper: Tidal Wetlands Restoration. Report by PWA and SAIC to the California Climate Action Reserve.

Crooks S., Herr D., Tamelander J., Laffoley D., Vanderver J. (2011). World Bank Env. Dept Paper 121.

Crooks S., Herr D., Tamelander J., Laffoley D., Vandever J. (2011). Mitigating Climate Change through Restoration and Management of Coastal Wetlands and Near-shore Marine Ecosystems: Challenges and Opportunities. Environment Department Paper 121, World Bank: Washington, D.C.

Crooks S., Turner R.K. (1999). Coastal zone management: sustaining estuarine natural resources. *Advances in Ecological Research*, *29*: 241–291.

Crooks S., von Unger M., Schile L., Allen C., Whisnant R. (2017). Understanding Strategic Blue Carbon Opportunities across East Asia. Report by Silvestrum Climate Associates for Partnerships in Environmental Management for the Seas of East Asia, Conservation International, The Nature Conservancy, with support from the Global Environment Facility, United Nations Development Program.

Crooks S., von Unger M., Schile L., Allen C., Whisnant R. (2017). *Understanding Strategic Blue Carbon Opportunities across East Asia. Report by Silvestrum Climate Associates for Partnerships in Environmental Management for the Seas of East Asia*, Conservation International, The Nature Conservancy, with support from the Global Environment Facility, United Nations Development Program.

Crooks[a] et al. (in review). *Nature Climate Change.*

Crooks[b] et al. (submitted). *Wetlands.*

Crosby S.W. (1941). *Soil Survey: the Sacramento-San Joaquin Delta*, Series 1935, no. 21, United States Department of Agriculture, Bureau of Plant Industry, In cooperation with the University of California Agricultural Experiment Station, Washington D.C.

Culver S.J., Leorri E., Mallinson D.J., Corbett D.R., Shazili N.A.M. (2015). Recent coastal evolution and sea-level rise, Setiu Wetland, Peninsular Malaysia. *Palaeogeography, Palaeoclimatology, Palaeoecology*, *417*: 406–421.

Curcó, A., Ibàñez C., Day J.W., Prat N. (2002). Net primary production and decomposition of salt marshes of the Ebre Delta (Catalonia, Spain). *Estuaries*, *25*: 309–324.

Currin C.A., Davis J., Malhotra A. (2017). Response of Salt Marshes to Wave Energy Provides Guidance for Successful Living Shoreline Implementation. In: *Living Shorelines: The Science and Management of Nature-Based Coastal Protection*, Bilkovic D.M., Mitchell M.M., La Peyre M.K., Toft J.D. (eds), CRC Press, Boca Raton, p. 519.

Curtin P.D. (1981). African enterprise in the mangrove trade: the case of Lamu. *African Economic History*, (10): 23–33.

Cyronak T., Santos I.R., McMahon A., Eyre B.D. (2013). Carbon cycling hysteresis in permeable carbonate sands over a diel cycle: implications for ocean acidification. *Limnology and Oceanography*, *58*: 131–143.

D'Angelo E.M., Reddy K.R. (1999). Regulators of heterotrophic microbial potentials in wetland soils. *Soil Biology and Biochemistry*, *31*: 815–830.

Daby D. (2003). Effects of seagrass bed removal for tourism purposes in a Mauritian bay. *Environmental Pollution*, *125*(3): 313–324.

Dachnowski-Stokes A.P. (1936). *Peat Land in the Pacific Coast States in Relation to Land and Water Resources*, U.S. Department of Agriculture: Washington, DC, USA, p. 248. (Miscellaneous publication).

Dahdouh-Guebas F., Jayatissa L.P., Di Nitto D., Bosire J.O., Lo Seen D., Koedam N. (2005). How effective were mangroves as a defence against the recent tsunami? *Current Biology, 15*: R443–R447.

Dahdouh-Guebas F., Mathenge C., Kairo J.G., Koedam N. (2000). Utilization of mangrove wood product around Mida Creek (Kenya) amongst subsistence and commercial users. *Economic Botany*, *54*(4): 513–527.

Dahl T.E., Stedman S.M. (2013). Status and trends of wetlands in the coastal watersheds of the Conterminous United States 2004 to 2009. U.S. Department of the Interior, Fish and Wildlife Service and National Oceanic and Atmospheric Administration, National Marine Fisheries Service.

Dahl T.E., Stedman S.M. (2013). *Status and Trends of Wetlands in the Coastal Watersheds of the Conterminous United States 2004 to 2009*, U.S. Department of the Interior, Fish and Wildlife Service, National Oceanic and Atmospheric Administration, National Marine Fisheries Service: Arlington, VA, p. 46.

Dame R.F., Chrzanowski T., Bildstein K., Kjerfve B., Mckellar H., Nelson D., Spurrier J., Stancyk S., Stevenson H., Vernberg J., Zingmark R. (1986). The outwelling hypothesis and North Inlet, South Carolina. *Marine Ecology Progress Series*, *33*: 217–229.

Dame R.F., Koepfler E., Gregory L. (2002). Benthic-Pelagic coupling in marsh-estuarine ecosystems. In: *Concepts and Controversies in Tidal Marsh Ecology*, Weinstein M.P., Kreeger D.A. (eds), Kluwer Academic Publisher: New York, pp. 369–390, 859.

Dame R.F., Spurrier J., Williams T.M., Kjerfve B., Zingmark R.G., Wolaver T.G., Chrzanowski T.H., McKellar H.N., Vernberg F.J. (1991). Annual material processing by a salt marsh-estuarine basin in South Carolina, USA. *Marine Ecology Progress Series*, *72*: 153–166.

Dandelot S., Robles C., Pech N., Cazaubon A., Verlaque R. (2008). Allelopathic potential of two invasive alien *Ludwigia* spp. *Aquatic Botany*, *88*: 311–316.

Dandelot S., Verlaque R., Dutartre A., Cazaubon A. (2005). Ecological, dynamic and taxonomic problems due to Ludwigia (Onagraceae) in France. *Hydrobiologia*, *551*: 131–136.

Danger M., Cornut J., Chauvet E., Chavez P., Elger A., Lecerf A. (2013). Benthic algae stimulate leaf litter decomposition in detritus-based headwater streams: a case of aquatic priming effect? *Ecology*, *94*: 1604–1613.

Daniel T.C., Muhar A., Arnberger A., et al. (2012).Contributions of cultural services to the ecosystem services agenda. *Proceedings of the National Academy of Sciences of the United States of America*, *109*: 8812–8819.

Daniels S.E., Walker G.B. (1996). Collaborative learning: improving public deliberation in ecosystem-based management. *Environmental Impact Assessment Review*, *16*: 71–102.

Daniels S.E., Walker G.B. (2001). *Working through Environmental Conflict: The Collaborative Learning Approach*, Prager Published: Westport, CT.

Danielsen F., Sørensen M.K., Olwig M.F., Selvam V., Parish F., Burgess N.D., Hiraishi T., Karunagaran V.M., Rasmussen M.S., Hansen L.B. (2005). The Asian tsunami: a protective role for coastal vegetation. *Science*, *310*: 643.

Darby F.A., Turner R.E. (2008). Below- and aboveground biomass of *Spartina alterniflora*: response to nutrient addition in a louisiana salt marsh. *Estuaries and Coasts*, *31*: 326–334.

Das S., Lyla P., Khan S.A. (2006). Marine microbial diversity and ecology: importance and future perspectives. *Current Science*, *90*: 1325–1335.

Davis C.A. (1910). Salt marsh formation near Boston and its geological significance. *Economic Geology*, *5*: 623–639.

Davis J. (1940). The ecology and geologic role of mangroves in Florida. *Papers from Tortugas Laboratory*, *32*: 307–412.

Dawes C.J., Andorfer J., Rose C., Uranowski C., Ehringer N. (1997). Regrowth of the seagrass Thalassia testudinum into propeller scars. *Aquatic Botany*, *59*(1): 139–155.

Dawes C.J., Phillips R.C., Morrison G. (2004). *Seagrass Communities of the Gulf Coast of Florida: Status and Ecology*, Florida Fish and Wildlife Conservation Commission, Fish and Wildlife Research Institute, Tampa Bay Estuary Program: St. Petersburg, FL. http://myfwc.com/research/habitat/seagrasses/publications/communities-gulf-coast/.

Day J., Kemp W., Yanez-Arancibia A., Crump B. (2013). *Estuarine Ecology*, 2nd edn, Wiley Blackwell: New York, p. 568.

Day J.W., Christian R.R., Boesch D.M., Yáñez-Arancibia A., Morris J.T., Twilley R.R., Naylor L., Schaffner L. (2008). Consequences of climate change on the ecogeomorphology of coastal wetlands. *Estuaries and Coasts*, *31*(3): 477–491.

Day J.W., Hopkinson C.S., Conner W. (1982). An analysis of environmental factors regulating community metabolism and fisheries production in a *Louisiana* estuary. In: *Estuarine Comparisons*, Kennedy W. (ed.), Academic Press, pp. 121–138.

Day J.W., Hopkinson C.S., Loesch H. (1977). Modeling man and nature in southern Louisiana. In: *Ecosystem Modeling in Theory and Practice: An Introduction with Case Histories*, Hall C.A.S., Day J.W. (eds), Wiley Interscience: New York, pp. 381–393.

Day J.W., Shaffer G.P., Britsch L.D., Reed D.J., Hawes S.R., Cahoon D. (2000). Pattern and process of land loss in the Mississippi Delta: a spatial and temporal analysis of wetland habitat change. *Estuaries*, *23*: 425–438.

Day J.W., Smith W., Wagner P., Stowe W. (1973). *Community Structure and Carbon Budget of a Salt Marsh and Shallow Bay Estuarine System in Louisiana*, LSU Center for Wetland Resources Publ. LSU-SG-72-04, pp. 1–80.

de Fouw J., Govers L.L., van de Koppel J., van Belzen J., Dorigo W., Sidi Cheikh M.A., Christianen M.J.A., van der Reijden K.J., van der Geest M., Piersma T., Smolders A.J.P., Olff H., Lamers L.P.M., van Gils J.A., van der Heide T. (2016). Drought, mutualism breakdown, and landscape-scale degradation of seagrass beds. *Current Biology*, *26*: 1051–1056.

Deegan L.A., Johnson D.S., Warren R.S., Peterson B.J., Fleeger J.W., Fagherazzi S., Wollheim W.M. (2012). Coastal eutrophication as a driver of salt marsh loss. *Nature*, *490*(7420): 388–392.

Dekker A., Brando V., Anstee J. (2006). Remote sensing of seagrass ecosystems: use of spaceborne and airborne sensors. In: *Seagrasses: Biology, Ecology and Conservation*, Larkum A.W.D., Orth R.J., Duarte C.M. (eds), Springer: The Netherlands, Dordrecht, pp. 347–359.

Delaune R.D., Patrick W.H., Jr., Buresh R.J. (1978). Sedimentation rates determined by ^{137}Cs dating in a rapidly accreting salt marsh. *Nature*, *275*: 532–533.

DeLaune R.D., Sasser C.E., Evers-Hebert E., White J.R., Roberts H.H. (2016). Influence of the Wax Lake Delta sediment diversion on aboveground plant productivity and carbon storage in deltaic island and mainland coastal marshes. *Estuarine, Coastal and Shelf Science*, *177*: 83–89.

DeLaune R.D., Smith C.J., Patrick W.H., Roberts H.H. (1987). Rejuvenated marsh and bay-bottom accretion on the rapidly subsiding coastal plain of U.S. Gulf coast: a second-order effect of the emerging Atchafalaya delta. Estuarine, *Coastal and Shelf Science*, *25*: 381–389.

DeLaune R.D., White J.R. (2012). Will coastal wetlands continue to sequester carbon in response to an increase in global sea level? A case study of the rapidly subsiding Mississippi river deltaic plain. *Climatic Change*, *110*: 297–314.

Delta Conservancy (2016a). Restoration Projects in the Delta and Suisun Marsh. http://deltaconservancy. ca.gov/restoration-projects-delta-and-suisun-marsh/ (accessed March 14, 2016.

Delta Conservancy (2016b). Restoration Partners and Programs. http://deltaconservancy.ca.gov/restoration-partners-and-programs/ (accessed March 14, 2016).

Delta Independent Science Board (2015). Flows and fishes in the Sacramento-San Joaquin Delta: Research needs in support of adaptive management. Lund, J. (chair). http://deltacouncil.ca.gov/sites/default/files/2015/09/2015-9-29-15-0929-Final-Fishes-and-Flows-in-the-Delta.pdf (accessed March 10, 2016).

Delta Stewardship Council (2016a). The Delta Plan. http://deltacouncil.ca.gov/delta-plan-0 (accessed March 7, 2016).

den Hertog C., Kuo J. (2006). Taxonomy and biogeography of seagrasses. In: *Seagrasses: Biology, Ecology and Conservation*, Larkum A.W.D., Orth R.J., Duarte C.M. (eds), Springer: The Netherlands, Dordrecht, pp. 1–23.

Deng X., Zhan Y., Wang F., Ma W., Ren Z., Chen X., Lv X. (2016). Soil organic carbon of an intensively reclaimed region in China: current status and carbon sequestration potential. *Science of the Total Environment*, *565*: 539–546.

Dennison W.C., Orth R.J., Moore K.A., Stevenson J.C., Carter V., Kollar S., Bergstrom P.W., Batiuk R.A. (1993). Assessing water quality with submersed aquatic vegetation. *BioScience*, *43*: 86–94.

Deverel S.J., Ingrum T., Leighton D. (2016). Present day oxidative subsidence of organic soils and mitigation in the Sacramento-San Joaquin Delta, California, USA. *Hydrogeology Journal 24*: 569–586.

Deverel S.J., Ingrum T., Lucero C., Drexler J.Z. (2014). Impounded marshes on subsided islands: simulated vertical accretion, processes, and effects, Sacramento-San Joaquin Delta, CA USA. *San Francisco Estuary and Watershed Science*, *12*(2): jmie_sfews_12893. Retrieved from: https://escholarship.org/uc/item/0qm0w92c (accessed March 2, 2016).

Deverel S.J., Leighton D.A. (2010). Historic, recent, and future subsidence, Sacramento-San Joaquin Delta, California, USA. *San Francisco Estuary and Watershed Science, 8*(2): jmie_sfews_11016. http://eschol-arship.org/uc/item/7xd4x0xw (accessed March 2, 2016).

Deverel S.J., Rojstaczer S. (1996). Subsidence of agricultural lands in the Sacramento–San Joaquin Delta, California: role of aqueous and gaseous carbon fluxes. *Water Resources Research, 8*: 2359–2367.

Deverel S.J., Wang B., Rojstaczer S. (1998). Subsidence in the Sacramento–San Joaquin Delta. In: *Land subsidence: case studies and current research. Proceedings of the Dr. Joseph F. Poland Subsidence Symposium*, Borchers J.W. (ed.), Association of Engineering Geologists, Star Pub. Co.: Belmont, CA, Book Series, no. 8.

Deverel S.J., Rojstaczer S (1996). Subsidence of agricultural lands in the Sacramento–San Joaquin Delta, California: role of aqueous and gaseous carbon fluxes. *Water Resources Research, 32*(8): 2359–2367.

DeVêvre O.C., Horwáth W.R. (2000). Decomposition of rice straw and microbial use efficiency under different soil temperatures and moistures. *Soil Biology and Biochemistry, 32*: 1773–1785.

Diaz-Almela E., Marbà N., Duarte C.M. (2007). Consequences of Mediterranean warming events in seagrass (Posidonia oceanica) flowering records. *Global change biology, 13*(1): 224–235.

Dickson A.G., Sabine C.L., Christian J.R. (2007). Guide to best practices for ocean CO_2 measurements. *PICES Special Publication, 3.*

Dierssen H.M., Chlus A., Russell B. (2015). Hyperspectral discrimination of floating mats of seagrass wrack and the macroalgae *Sargassum* in coastal waters of Greater Florida Bay using airborne remote sensing. *Remote Sensing of Environment, 167*: 247–258.

Dierssen H.M., Zimmerman R.C. (2003). Ocean color remote sensing of seagrass and bathymetry in the Bahamas Banks by high-resolution airborne imagery. *Limnology & Oceanography, 48*(1–2): 444–455.

Dijkstra P., Thomas S.C., Heinrich P.L., Koch G.W., Schwartz E., Hungate B.A. (2011). Effect of temperature on metabolic activity of intact microbial communities: evidence for altered metabolic pathway activity but not for increased maintenance respiration and reduced carbon use efficiency. *Soil Biology and Biochemistry, 43*: 2023–2031.

Dittmar T., Hertkorn N., Kattner G., Lara R.J. (2006). Mangroves, a major source of dissolved organic carbon to the oceans. *Global Biogeochemical Cycles, 20*(1).

Dokka R.K. (2011). The role of deep processes in late 20th century subsidence of New Orleans and coastal areas of southern Louisiana and Mississippi. *Journal of Geophysical Research: Solid Earth, 116*: B06403. doi: 06410.01029/02010jb008008.

Domingues C.M., Church J.A., White N.J., Gleckler P.J., Wijffels S.E., Barker P.M., Dunn J.R. (2008). Improved estimates of upper-ocean warming and multi-decadal sea-level rise. *Nature, 453*: 1090–1093.

Donato D.C., Kauffman J.B., Murdiyarso D., Kurnianto S., Stidham M., Kanninen M. (2011). Mangroves among the most carbon-rich forests in the tropics. *Nature Geoscience, 4*: 293–297. doi: 10.1038/ngeo1123.

Doughty C.L., Langley J.A., Walker W.S., Feller I.C., Schaub R., Chapman S.K. (2016). Mangrove range expansion rapidly increases coastal wetland carbon storage. *Estuaries and Coasts, 39*(2): 385–396.

Downie A.-L., von Numers M., Boström C. (2013). Influence of model selection on the predicted distribution fo the seagrass *Zostera marina. Estuarine, Coastal and Shelf Science, 121–122*: 8–19.

Doyle T.W., Krauss K.W., Conner W.H., From A.S. (2010). Predicting the retreat and migration of tidal forest along the northern Gulf of Mexico under sea-level rise. *Forest Ecology and Management, 259*: 770–777.

Drake L.A., Dobbs F.C., Zimmerman R.C. (2003). Effects of epiphyte load on optical properties and photosynthetic potential of the seagrasses *Thalassia testudinum* Banks ex König and *Zostera marina* L. *Limnology and Oceanography, 48*(1–2): 456–463.

Drexler J.Z. (2011). Peat formation processes through the millennia in tidal marshes of the Sacramento-San Joaquin Delta, California, USA. *Estuaries and Coasts*, doi: 10.1007/s12237-011-9393-7.

Drexler J.Z., de Fontaine C.S., Brown T.A. (2009). Peat accretion histories during the past 6,000 years in marshes of the Sacramento-San Joaquin Delta of California, USA. *Estuaries and Coasts, 32*: 871–892.

Drexler J.Z., de Fontaine C.S., Deverel S.J. (2009). The legacy of wetland drainage on the remaining peat in the Sacramento–San Joaquin Delta, California, USA. *Wetlands, 29*: 372–386.

Drexler J.Z., de Fontaine C.S., Knifong D.L. (2007). *Age Determination of the Remaining Peat in the Sacramento-San Joaquin Delta*, U.S. Geological Survey: California, USA, p. 2. (Open file report 2007-1303).

Drexler J.Z., et al. (2019). The fate of blue carbon in the Sacramento-San Joaquin Delta of California, USA. In: *A Blue Carbon Primer: The State of Coastal Wetland Carbon Science, Practice, and Policy*, Lisamarie W.-M., Stephen C., and Tiffany G.T. (eds), CRC Press: Boca Raton, London.

Drexler J.Z., Paces J.B., Alpers C.N., Windham-Myers L., Neymark L., Taylor H.E. (2013). $^{234}U/^{238}U$ and ^{87}Sr in peat as useful tracers of paleosalinity in the Sacramento–San Joaquin Delta of California. *Applied Geochemistry, 40*, 164–179.

Drupp P.S., De Carlo E.H., Mackenzie F.T. (2016). Porewater CO_2-carbonic acid system chemistry in permeable carbonate reef sands. *Marine Chemistry, 185*: 48–64.

Duarte B., Reboreda R., Caçador I. (2008). Seasonal variation of extracellular enzymatic activity (EEA) and its influence on metal speciation in a polluted salt marsh. *Chemosphere, 73*(7): 1056–1063.

Duarte C., Middleburg J., Caraco N. (2005). Major role of marine vegetation on the oceanic carbon cycle. *Biogeosciences, 2*: 1–8.

Duarte C.M. (1995). Submerged aquatic vegetation in relation to different nutrient regimes. *Ophelia, 41*: 87–112.

Duarte C.M. (2002). The future of seagrass meadows. *Environmental conservation, 29*: 192–206.

Duarte C.M., Borum J., Short F.T., Walker D.I. (2008). Seagrass ecosystems: their global status and prospects. In: *Aquatic Ecosystems: Trends and Global Prospects*, Polunin N.V.C. (ed.), Cambridge University Press: Cambridge, pp. 281–294.

Duarte C.M., Cebrian J. (1996). The fate of marine autotrophic production. *Limnology and Oceanography, 41*(8): 1758–1766.

Duarte C.M., Fourqurean J.W., Krause-Jensen D., Olesen B. (2005). Dynamics of seagrass stability and change. In: *Seagrasses: Biology, Ecology and Conservation*, Larkum A.W.D., Orth R. J., Duarte C.M. (eds.), Springer, The Netherlands, pp. 271–294.

Duarte C.M., Kennedy H., Marbà N., Hendriks I. (2013). Assessing the capacity of seagrass meadows for carbon burial: current limitations and future strategies. *Ocean & Coastal Management, 83*: 32–38.

Duarte C.M., Losada I.J., Hendriks I.E., Mazarrasa I., Marbà N. (2013). The role of coastal plant communities for climate change mitigation and adaptation. *Nature Climate Change, 3*(11): 961–968.

Duarte C.M., Marba N., gacia E., Fourqurean J.W., Beggins J., Barron C., Apostolaki E.T. (2010). Seagrass community metabolism: assessing the carbon sink capacity of seagrass meadows. *Global Biogeochemical Cycles, 24*: GB4032.

Duarte C.M., Middelburg J.J., Caraco N. (2004). Major role of marine vegetation on the oceanic carbon cycle. *Biogeosciences discussions, 1*(1): 659–679.

Duarte. C.M., Hendriks I.E., Moore T.S. et al. (2013). Is ocean acidification an open-ocean syndrome? Understanding anthropogenic impacts on seawater pH. *Estuaries and Coasts, 36*: 221–236.

Dunfield P., Knowles R., Dumont R., Moore T.R. (1993). Methane production and consumption in temperate and subarctic peat soils: responses to temperature and pH. *Soil Biology and Biochemistry, 25*: 231–326.

Dutton A., Lambeck K. (2012). Ice volume and sea level during the last interglacial. *Science, 337*: 216–219.

EAD (2007). *Marine Environment and Resources of Abu Dhabi. Environmental Agency*, Abu Dhabi and Motivate publishing: Abu Dhabi.

Eagle M.G., Paytan A., Herrera-Silveira J. (2004). Tracing organic matter sources and carbon burial in mangrove sediments over the past 160 years. *Estuarine, Coastal and Shelf Science, 61*(2): 211–227.

Egler F. (1952). Southeast saline Everglades vegetation, Florida: and tis management. *Vegetatio, 3*: 213–265.

El-Kammar M.M. Abuassy E.M.A., Wali A.M.A., Abu El-Ezz A.R. (2014). The possible origin of hydrocarbon generation sourced from an evaporative environment: a comparative analog of recent and older environments. *Petroleum Science and Technology, 33*(1): 51–61.

Ellison A.M., Farnsworth E.J. (1996). Anthropogenic disturbance of Caribbean mangrove ecosystems: past impacts, present trends, and future predictions. *Biotropica 28*(4): 549–565.

Ellison J.C., Stoddart D.R. (1991). Mangrove ecosystem collapse during predicted sea-level rise: holocene analogues and implications. *Journal of Coastal Research*, pp.151–165.

Embabi N.S. (1993). Environmental aspects of geographical distribution of mangrove in the United Arab Emirates. In: Lieth H., Al Masoom A. (eds), *Towards the Rational Use of High Salinity Tolerant Plants*, Kluwer Academic: Netherlands, *vol. 1*, pp. 45–58.

Emmer I.M., Needelman B.A., Emmett-Mattox S., Crooks S., Megonigal J.P., Myers D., Oreska M.P.J., McGlathery K.J., Shoch D. (2015a). *Methodology for Tidal Wetland and Seagrass Restoration. VCS Methodology VM0033, v 1.0*, Verified Carbon Standard: Washington D.C.

Emmer I.M., Needelman B.A., Emmett-Mattox S., Crooks S., Megonigal J.P., Myers D., Oreska M.P.J., McGlathery K.J. (2018a). *Estimation of Baseline Carbon Stock Changes and Greenhouse Gas Emissions in Tidal Wetland Restoration and Conservation Project Activities (BL-TW). VCS Module, v 1.0.* Verra (Verified Carbon Standard): Washington, D.C.

Emmer I.M., Needelman B.A., Emmett-Mattox S., Crooks S., Megonigal J.P., Myers D., Oreska M.P.J., McGlathery K.J. (2018b). *Methods for Monitoring of Carbon Stock Changes and Greenhouse Gas Emissions and Removals in Tidal Wetland Restoration and Conservation Project Activities (M-TW). VCS Module, v 1.0,* Verra (Verified Carbon Standard): Washington, D.C.

Emmer I.M., von Unger M., Needelman B.A., Crooks S., Emmett-Mattox S. (2015b). *Coastal Blue Carbon in Practice: A Manual for Using the VCS Methodology for Tidal Wetland and Seagrass Restoration,* Restore America's Estuaries: Arlington, VA.

Engle D.V. (2011). Estimating the provision of ecosystem services by Gulf of Mexico coastal wetlands. *Wetlands, 31*: 179–193.

Enríquez S., Schubert N. (2014). Direct contribution of the seagrass *Thalassia testudinum* to lime mud production. *Nature Communications, 5*: 3835. doi: 10.1038/ncomms4835.

Ensign S.H., Currin C. (2017). Geomorphic implications of particle movement by water surface tension in a salt marsh. *Wetlands, 37*(2): pp.245–256.

Environmental Protection Agency (2017). www.epa.gov/ghgemissions/inventory-us-greenhouse-gas-emissions-and-sinks-1990-2015.

Enwright N.M., Griffith K.T., Osland M.J. (2016). Barriers to and opportunities for landward migration of coastal wetlands with sea-level rise. *Frontiers in Ecology and the Environment, 14*(6): 307–316.

Epple C., García Rangel S., Jenkins M., Guth M. (2016). Managing ecosystems in the context of climate change mitigation: A review of current knowledge and recommendations to support ecosystem-based mitigation actions that look beyond terrestrial. Technical Series No.86. Secretariat of the Convention on Biological Diversity, Montreal.

Erftemeijer P.L., Lewis R.R.R. (2006). Environmental impacts of dredging on seagrasses: a review. *Marine Pollution Bulletin, 52*(12): 1553–1572.

Erisman J.W., Sutton M.A., Galloway J., Klimont Z., Winiwarter W. (2008). How a century of ammonia synthesis changed the world. *Nature Geoscience, 1*(10): 636–639.

Evans G., Kendall C.G., Skipwith P.A. (1964). Origin of Coastal Flats, The Sabkha, of the Trucial Coast, Persian Gulf. *Nature, 202*: 759–761.

Evans G., Kirkham A. (2002). The Abu Dhabi Sabkha, In: H.-J. Barth, B. Böer, (eds), *Sabkha Ecosystems, Volume I: The Arabian Peninsula and Adjacent Countries,* Kluwer Academic Publishers: Dordrecht, pp. 7–20,

Evans G., Murray J., Biggs H., Bate R., Bush P. (1973). *The Oceanography, Ecology, Sedimentology and Geomorphology of Parts of the Trucial Coast Barrier Island Complex, Persian Gulf,* The Persian Gulf: Springer, pp. 233–277

Eyre B.D., Maher D.T., Squire P. (2013). Quantity and quality of organic matter (detritus) drives N_2 effluxes (net denitrification) across seasons, benthic habitats, and estuaries. *Global Biogeochemical Cycles, 27*: 1083–1095.

Ezcurra P., Ezcurra E., Garcillán P.P., Costa M.T., Aburto-Oropeza O. (2016). Coastal landforms and accumulation of mangrove peat increase carbon sequestration and storage. *Proceedings of the National Academy of Sciences of the United States of America, 113*: 4404–4409.

Ezer T., Corlett W.B. (2012). Is sea level rise accelerating in the Chesapeake Bay? A demonstration of a novel new approach for analyzing sea level data, *Geophysical Research Letters, 39*: L19605. doi: 10.1029/2012GL053435.

Fabricius K.E., De'ath G., Humphrey C., Zagorskis I., Schaffelke B. (2013). Intra-annual variation in turbidity in response to terrestrial runoff on near-shore coral reefs of the Great Barrier Reef. *Estuarine, Coastal and Shelf Science, 116*: 57–65.

Fagherazzi S., Kirwan M.L., Mudd S.M., Guntenspergen G.R., Temmerman S., D'Alpaos A., van de Koppel J., Craft C., Rybczyk J., Reyes E., Clough J. (2012). Numerical models of salt marsh evolution: ecological, geomorphic, and climatic factors. *Reviews of Geophysics, 50*: RG1002.

Fagherazzi S., Marani M., Blum L.K. (eds) (2004). *The Ecogeomorphology of Tidal Marshes.* American Geophysical Union Monograph: Washington D.C., p. 268.

Fagherazzi S., Mariotti G., Wiberg P., McGlathery K. (2013). Marsh collapse does not require sea level rise. *Oceanography, 26*: 70–77 doi: 10.5670/oceanog.2013.47.

Fagherazzi S., Wiberg P.L. (2009). Importance of wind conditions, fetch, and water levels on wave-generated shear stresses in shallow intertidal basins. *Journal of Geophysical Research: Earth Surface*, *114*(F3).

Ferguson A.J., Gruber R., Potts J., Wright A., Welsh D.T., Scanes P. (2017). Oxygen and carbon metabolism of *Zostera muelleri* across a depth gradient – Implications for resilience and blue carbon. *Estuarine Coastal and Shelf Science*, *187*: 216–230.

Ferrario Beck F.M.W., Storlazzi C.D., Micheli F., Shepard C.C., Airoldi L. (2014). The effectiveness of coral reefs for coastal hazard risk reduction and adaptation. *Nature Communications*, *5*: 3794. doi: 10.1038/ncomms4794.

Feurt C. (2008). *Collaborative Learning Guide for Ecosystem Management*, 5th edn. dune.une.edu.

Field C.R., Bayard T.S., Gjerdrum C., Hill J.M., Meiman S., Elphick C.S. (2017). High-resolution tide projections reveal extinction threshold in response to sea-level rise. *Global Change Biology*, *23*: 2058–2070.

Finlayson B.J. (1983). Water hyacinth: threat to the Delta? *Outdoor California*, 44, 10–14.

Firestone M.K., Davidson E.A. (1989). Microbiological basis of NO and N_2O production and consumption in soil. In: *Exchange of Trace Gases Between Terrestrial Ecosystems and the Atmosphere*, Andreae M.O., Schimel D.S. (eds), Wiley: Chichester, UK, pp. 7–21.

Fish U.S., and Wildlife Service (2000). *Impacts of Riprapping to Ecosystem Functioning*, Lower Sacramento River: California. www.fws.gov/sacramento/ES_Species/Accounts/Fish/Documents/Riprap_Effects_2004_revision.pdf (accessed on March 14, 2016).

Fish U.S., Wildlife Service (2014). National Wetlands Inventory website. www.fws.gov/wetlands/. Washington, D.C.: U.S. Department of the Interior, Fish and Wildlife Service.

Fisher J., Acreman M.C. (2004). Wetland nutrient removal: a review of the evidence. *Hydrology and Earth System Sciences*, *8*: 673–685.

Flores-Verdugo F., Day, J.W.J., Briseño-Duenas R. (1987). Structure, litterfall, decomposition, and detritus dynamics of mangroves in a Mexican coastal lagoon with an ephemeral inlet. *Marine Ecology Progress Series*, *35*: 83–90.

Foden J., Brazier D.P. (2007). Angiosperms (seagrass) within the EU water framework directive: a UK perspective. *Marine Pollution Bulletin*, *55*(1): 181–195.

Food and Agricultural Organization (FAO) (2007). The world's mangroves 1980–2005.

Food and Agricultural Organization (FAO) (2016). Valuing Coastal Ecosystems as Economic Assets: The importance of mangroves for food security and livelihoods among communities in Kilifi Country and the Tana Delta, Kenya. United Nations, www.fao.org/3/a-i5689e.pdf.

Food and Agriculture Organization of the United Nations [FAO] (2014). *Global Aquaculture Production (FishStat) Dataset*, Rome: FAO.

Food and Agriculture Organization of the United Nations [FAO] (2007). The World's Mangroves 1980–2005. FAO Forestry Paper No. 153. Rome: FAO.

Forbrich I., Giblin A. (2015). Marsh-atmosphere CO_2 exchange in a New England salt marsh. *Journal of Geophysical Research: Biogeosciences*, *120*. doi: 10.1002/2015JG003044.

Forbrich I., Giblin A., Hopkinson C. S. (2018). Carbon storage in a New England salt marsh: a comparison of estimates using eddy covariance measurements and long-term estimates from sediment cores. *Limnology and Oceanography*, *123*(3): 867–878.

Forest Trends (2014). *Turning Over a New Leaf: State of the Forest Carbon Markets 2014*, p. 110. www.forest-trends.org/documents/files/doc_4770.pdf.

Fortes M.D. (2013). A review: biodiversity, distribution and conservation of Philippine seagrasses. *Philippine Journal of Science*, *142*: 95–111.

Foschi P., Liu H. (2002). *Active Learning for Classifying a Spectrally Variable Subject*, Citeseer: Niagara Falls, Canada, pp. 115–124, in 2nd International Workshop on Pattern Recognition for Remote Sensing (PRRS 2002).

Fourqueran J.W., et al. (2012). Seagrass ecosystems as a globally significant carbon stock. *Nature Geoscience*, *5*: 505–509. doi: 10.1038/ngeo1477.

Fourqurean J., Johnson B., Kauffman J.B., Kennedy H., Lovelock C., Alongi D.M., Cifuentes M., Copertino M., Crooks S., Duarte C., Fortes M., Howard J., Hutahaean A., Kairo J., Marbà N., Morris J., Murdiyarso D., Pidgeon E., Ralph P., Saintilan N., Serrano O. (2014). Field sampling of soil carbon pools in coastal ecosystems. In: *Coastal Blue Carbon: Methods for Assessing Carbon Stocks and Emissions Factors in Mangroves, Tidal Salt Marshes, and Seagrass Meadows*, Howard J., Hoyt S., Isensee K., Pidgeon E., Telszewski M. (eds), Conservation International, Intergovernmental Oceanographic Commission of UNESCO, International Union for Conservation of Nature: Arlington, Virginia, USA, pp. 39–66.

Fourqurean J.W., Duarte C.M., Kennedy H., Marbà N., Holmer M., Mateo M.A., Apostolaki E.T., Kendrick G.A., Krause-Jensen D., McGlathery K.J., Serrano O. (2012). Seagrass ecosystems as a globally significant carbon stock. *Nature Geoscience*, 5: 505–509.

Fourqurean J.W., Kendrick G.A., Collins L.S., Chambers R.M., Vanderklift M.A. (2012b). Carbon and nutrient storage in subtropical seagrass meadows: examples from Florida Bay and Shark Bay. *Marine and Freshwater Research*, 63: 967–983.

Frankignoulle M., Distèche A. (1984). CO_2 chemistry in the water column above a *Posidonia* seagrass bed and related air-sea exchanges. *Oceanologica Acta*, 7: 209–219.

Franklin M.J., Wiebe W.J., Whitman W.B. (1988). Populations of methanogenic bacteria in a Georgia salt marsh. *Applied and Environmental Microbiology*, 54: 1151–1157.

Frankovich T.A., Zieman J.C. (1994). Total epiphyte and epiphytic carbonate production on *Thalassia testudinum* across Florida Bay. *Bulletin of Marine Science*, 54: 679–695.

Fraser M.W., Kendrick G.A., Statton J., Hovey R.K., Zavala-Perez A., Walker D.I. (2014). Extreme climate events lower resilience of foundation seagrass at edge of biogeographical range. *Journal of Ecology*, 102: 1528–1536.

French J.R., Spencer T., Murray A.L., Arnold N.S. (1995). Geostatistical analysis of sediment deposition in two small tidal wetlands, Norfolk, U.K. *Journal of Coastal Research*, 11: 308–321.

French J.R., Spencer T., Murray A.L., Arnold N.S. (1995). Geostatistical analysis of sediment deposition in two small tidal wetlands, Norfolk, U.K. *Journal of Coastal Research*, 11: 308–321.

Frey R.W., Basan P. (1978). Coastal salt marshes. In: *Coastal Sedimentary Environments*, Davis R. (ed.), Springer: NY. pp. 101–169.

Frey S.D., Lee J., Melillo J.M., Six J. (2013). The temperature response of soil microbial efficiency and its feedback to climate. *Nature Climate Change*, 3: 395–398.

Friedrichs C.T., Perry J.E. (2001). Tidal salt marsh morphodynamics: a synthesis. *Journal of Coastal Research*, 27: 7–37.

Friess D.A. (2016). Ecosystem services and disservices of mangrove forests: insights from historical colonial observations. *Forests*, 7(9): 183.

Friess D.A., Webb E.L. (2014). Variability in mangrove change estimates and implications for the assessment of ecosystem service provision. *Global ecology and biogeography*, 23(7): 715–725.

Frolking S., Roulet N.T., Moore T.R., Richard P.J., Lavoie M., Muller S.D. (2001). Modeling northern peatland decomposition and peat accumulation. *Ecosystems 4*: 479–498.

Fuhrman J.A. (1999). Marine viruses and their biogeochemical and ecological effects. *Nature*, 399: 541–548.

Fujita K. (2002). Soilary process in seagrass beds at Ishigaki-jima, Ryukyu Islands. *MSc Dissertation*, Tohoku University Sendai.

Fung I., John J., Lerner J., Matthews E., Prather M., Steele L.P., Fraser P.J. (1991). Three-dimensional model synthesis of the global methane cycle. *Journal of Geophysical Research*, 96: 2156–2202.

Fussel H.M. (2007). Adaptation planning for climate change: concepts, assessment approaches, and key lessons. *Sustainability Science 2*: 265–275.

Fyfe S.K. (2003). Spatial and temporal variation in spectral reflectance: are seagrass species spectrally distinct? *Limnology & Oceanography*, 48(1–2): 464–479.

Ganju N.K., Defne Z., Kirwan M.L., Fagherazzi S., D'Alpaos A., Carniello L. (2017). Spatially integrative metrics reveal hidden vulnerability of microtidal salt marshes. *Nature communications*, 8: 14156.

Ganong W.F. (1903). The vegetation of the Bay of Fundy salt and diked marshes: an ecological study. *Botanical Gazette*, 36(3): 161–186.

García de Fuentes A., et al. (2011). La costa de Yucatán en la perspectiva del desarrollo turístico. *Comisión Nacional para el Conocimiento y Uso de la Biodiversidad, Colección Corredor Biológico Mesoamericano, México, Serie Conocimientos, 9.*

Gardner L.R., Kjerfve B. (2006). Tidal fluxes of nutrients and suspended sediments at the North Inlet – Winyah Bay National Estuarine Research Reserve. *Estuarine, Coastal and Shelf Science*, 70: 682–692.

Gattuso J.-P., Pichon M., Frankignoulle M. (1995). Biological control of air-sea CO_2 fluxes: effect of photosynthetic and calcifying marine organisms and ecosystems. *Marine Ecology Progress Series*, 129: 307–312.

Gedan K.B., Altieri A.H., Bertness M.D. (2011). Uncertain future of New England salt marshes. *Marine Ecology and Progress Series*, *434*: 229–237.

Gedan K.B., Bertness M.D. (2009). Experimental warming causes rapid loss of plant diversity in New England salt marshes. *Ecology Letters*, *12*: 842–848.

Gedan K.B., Kirwan M.L., Wolanski E., Barbier E.B., Silliman B.R. (2011). The present and future role of coastal wetland vegetation in protecting shorelines: answering recent challenges to the paradigm. *Climatic Change*, *106*: 7–29.

Gedan K.B., Silliman B.R., Bertness M.D. (2009). Centuries of human-driven change in salt marsh ecosystems. *Annual Review of Marine Science*, *1*: 117–141.

GEF Blue Forest Project (2016). Blue Carbon on the Side: List of relevant Side Events at UNFCCC COP22. www.gefblueforests.org/blue-carbon-on-the-side-list-of-relevant-side-events-at-unfccc-cop22/.

Gehrels W.R., Shennan I. (2015). Sea level in time and space: revolutions and inconvenient truths. *Journal of Quarternary Science*, *30* (2): 131–143.

Geneletti D., Zardo L. (2015). Ecosystem-based adaptation in cities: an analyses of European urban climate adaptation plans. *Land Use Policy*, *50*: 38–47.

Gesch D.B. (2009). Analysis of lidar elevation data for improved identification and delineation of lands vulnerable to sea-level rise. *Journal of Coastal Research*, *53*: 49–58.

Getsinger K.D., Dillon C.R. (1984). Quiescence, growth and senescence of *Egeria densa* in Lake Marion. *Aquatic Botany*, *20*: 329–338.

Geyer K.M., Dijkstra P., Sinsabaugh R., Frey S.D. (2019). Clarifying the interpretation of carbon use efficiency in soil through methods comparison. *Soil Biology and Biochemistry*, 128: 79–88.

Ghazanfar S. (1999). Coastal vegetation of Oman. *Estuarine, Coastal and Shelf Science*, *49*: 21–27.

Ghosh S., Mishra D.R., Gitelson A.A. (2016). Long-term monitoring of biophysical characteristics of tidal wetlands in the northern Gulf of Mexico —A methodological approach using MODIS. *Remote Sensing of Environment*, *173*: 39–58.

Giani L., Bashan Y., Holguin G., Strangmann A. (1996). Characteristics and methanogenesis of the Balandra lagoon mangrove soils, Baja California Sur, Mexico. *Geoderma*, *72*: 149–160.

Gilbert G.K. (1917). Hydraulic mining debris in the Sierra Nevada. U.S. Geological Survey Professional Paper 105. http://pubs.usgs.gov/pp/0105/report.pdf.

Gilhespy S.L., Anthony S., Cardenas L., Chadwick D., del Pradod A., Lie C., Misselbrooka T., Reesf R.M., Salas W., Sanz-Cobena A., Smith P., Tilston E.L., Topp C.F.E., Vetter S., Yeluripati J.B. (2014). First 20 years of DNDC (DeNitrification DeComposition): model evolution. *Ecological Modelling*, *292*: 51–62.

Gilman E.L., Ellison J., Duke N.C., Field C. (2008). Threats to mangroves from climate change and adaptation options: a review. *Aquatic Botany*, *89*: 237–250.

Gilman E.L., Ellison J., Duke N.C., Field C. (2008). Threats to mangroves from climate change and adaptation options: a review. *Aquatic Botany*, *89*(2): 237–250. doi: 10.1016/j.aquabot.2007.12.009.

Ginsburg R.N., Lowenstam H.A. (1958). The influence of marine bottom communities on the depositional environment of soils. *Journal of Geology*, *66*: 310–318.

Giosan L., Syvitski J., Constantinescu S., Day J. (2014). Climate change: protect the world's deltas. *Nature*, *516*(7529): 31–33.

Giri C., et al. (2019). Mapping and monitoring of mangrove forests of the world using Remote Sensing. In: *A Blue Carbon Primer: The State of Coastal Wetland Carbon Science, Practice, and Policy*, Lisamarie W.-M., Stephen C., and Tiffany G.T. (eds), CRC Press: Boca Raton, London.

Giri C., Long J. (2016). Is the Geographic Range of Mangrove Forests in the Conterminous United States Really Expanding? *Sensors*, *16*: 2010.

Giri C., Long J., Abbas S., Murali R.M., Qamer F.M., Pengra B., Thau D. (2014). Distribution and dynamics of mangrove forests of South Asia. *Journal of Environmental Management*, 148, 101–111.

Giri C., Long J., Tieszen L. (2011a). Mapping and monitoring Louisiana's mangroves in the aftermath of the 2010 Gulf of Mexico oil spill. *Journal of Coastal Research*, 27, 1059–1064.

Giri C., Muhlhausen J. (2008). Mangrove forest distributions and dynamics in Madagascar (1975–2005). *Sensors*, *8*: 2104–2117.

Giri C., Ochieng E., Tieszen L.L., et al. (2011). Status and distribution of mangrove forests of the world using earth observation satellite data. *Global Ecology and Biogeography*, *20*: 154–59. doi: 10.1111/j.1466–8238.2010.00584.x.

Giri C., Pengra B., Zhu Z.L., Singh A., Tieszen L.L. (2007). Monitoring mangrove forest dynamics of the Sundarbans in Bangladesh and India using multi-temporal satellite data from 1973 to 2000. *Estuarine Coastal and Shelf Science*, *73*, 91–100.

Giri C., Zhu Z., Tieszen L.L., Singh A., Gillette S., Kelmelis J.A. (2008). Mangrove forest distributions and dynamics (1975–2005) of the tsunami-affected region of Asia. *Journal of Biogeography*, *35*, 519–528.

Giri C.P., Long G. (2014). Mangrove reemergence in the northernmost range limit of eastern Florida. *Proceedings of the National Academy of Sciences of the United States of America*, *111*: E1447–E1448.

Gitelson A.A. (2004). Wide dynamic range vegetation index for remote quantification of biophysical characteristics of vegetation. *Journal of Plant Physiology*, *161*: 165–173.

Gittman R.K., Peterson C.H., Currin C.A., Fodrie F.J., Piehler M.F., Bruno J.F. (2016b). Living shorelines can enhance the nursery role of threatened estuarine habitats. *Ecological Applications*, *26*: 249–263. doi: 10.1890/14–0716.1

Gittman R.K., Popowich A.M., Bruno J.F., Peterson C.H. (2014). Marshes with and without sills protect estuarine shorelines from erosion better than bulkheads during a Category 1 hurricane. *Ocean and Coastal Management*, *102*: 94–102.

Gittman R.K., Scyphers S.B., Smith C.S., Neylan I.P., Grabowski J.H. (2016a). Ecological consequences of shoreline hardening: a meta-analysis. *Bioscience*, *66*: 763–773.

Glaser P.H., Volin J.C., Givnish T.J., Hansen B.C.S., Stricker C.A. (2012). Carbon and sediment accumulation in the Everglades (USA) during the past 4000 years: rates, drivers, and sources of error. *Journal of Geophysical Research: Biogeosciences*, *117*: G03026. doi: 03010.01029/02011jg001821.

Goals Project (2015). *The Baylands and Climate Change: What We Can Do. Baylands Ecosystem Habitat Goals Science Update 2015*, San Francisco Bay Area Wetlands Ecosystem Goals Project, California State Coastal Conservancy: Oakland, California.

GoK G.K. (2017). *National Mangrove Ecosystem Management Plan*, Nairobi, Kenya.

Golden J.S., Virdin J., Nowacek D., Halpin P., Bennear L., Patil P.G. (2017). Making sure the blue economy is green. *Nature Ecology and Evolution*, *1*: 17. doi: 10.1038/s41559-016-0017.

Goman M., Wells L. (2000). Trends in river flow affecting the northeastern reach of the San Francisco Bay Estuary over the past 7,000 years. *Quaternary Research*, *54*: 206–217.

Goni M.A., Ruttenberg K.C., Eglinton T.I. (1998). A reassessment of the sources and importance of land-derived organic matter in surface sediments from the Gulf of Mexico. *Geochimica et Cosmochimica Acta*, *62*: 3055–3075.

Gonneea M. (2016). Tampa Bay carbon burial rates across mangrove and salt marsh ecosystems. In: Sheehan L., Crooks S. (eds), *Tampa Bay Blue Carbon Assessment: Summary of Findings*, TBEP Technical Report #07-16: St. Petersburg, FL, Appendix D, pp. 159–209. www.tbeptech.org/TBEP_TECH_PUBS/2016/TBEP_07_16_Tampa-Bay-Blue-Carbon-Assessment-Report-FINAL_June16-2016.pdf.

Gontikaki E., Thornton B., Huvenne V.A., White U. (2013). Negative priming effect on organic matter mineralisation in NE Atlantic slope sediments. *PLoS One*, *8*: e67722.

Gonzalez Trilla G., Pratolongo P., Beget M.E., Kandus P., Marcovecchio J., Di Bella C. (2013). Relating Biophysical Parameters of Coastal Marshes to Hyperspectral Reflectance Data in the Bahia Blanca Estuary, Argentina. *Journal of Coastal Research*, *29*: 231–238.

Gopal B. (1987). *Water Hyacinth*, Amsterdam: Elsevier.

Gosselink J.G., Hatton R., Hopkinson C.S. (1984). Relationship of organic carbon and mineral content to bulk density in Louisiana marsh soils. *Soil Science*, *137*: 177–180.

Gough L., Grace J.B. (1998). Effects of flooding, salinity, and herbivory on coastal plant communities, Louisiana, United States. *Oecologia*, *117*: 527–535.

Governor's Delta Vision Blue Ribbon Task Force (2008). http://deltavision.ca.gov/BlueRibbonTaskForce/FinalVision/Delta_Vision_Final.pdf (accessed March 15, 2016).

Gray A.J., Mogg R.J. (2001). Climate impacts on pioneer saltmarsh plants. *Climate Research*, *18*: 105–112.

Grech A., Chartrand-Miller K., Erftemeijer P., Fonseca M., McKenzie L., Rasheed M., Taylor H., Coles R. (2012). A comparison of threats, vulnerabilities and management approaches in global seagrass bioregions. *Environmental Research Letters*, *7*: 024006.

Grech A., Wolter J., Coles R., McKenzie L., Rasheed M., Thomas C., Waycott M., Hanert E. (2016). Spatial patterns of seagrass dispersal and settlement. *Diversity and Distributions*, *22*: 1150–1162.

Green E.P., Short F.T. (eds) (2003). *World Atlas of Seagrasses*, University of California Press: Berkeley, CA, p. 310.

Greening H., Janicki A. (2006). Toward reversal of eutrophic conditions in a subtropical estuary: water quality and seagrass response to nitrogen load reductions in Tampa Bay, Florida, USA. *Environmental Management*, 38(2): 163-178. doi: 10.1007/s00267-005-0079-4.

Greening H., Janicki A., Sherwood E.T., Pribble R., Johansson J.O.R. (2014). Ecosystem responses to long-term nutrient management in an urban estuary: Tampa Bay, Florida, USA. *Estuarine Coastal and Shelf Science*, 151: A1–A16. doi: 10.1016/j.ecss.2014.10.003.

Greening H.S., Cross L.C., Sherwood E.T. (2011). A multiscale approach to seagrass recovery in Tampa Bay, Florida. *Ecological Restoration*, 29(1–2): 82–93.

Greiner J.T., McGlathery K.J., Gunnell J., McKee B.A. (2013). Seagrass restoration enhances "blue carbon" sequestration in coastal waters. *PLoS One*, 8(8): e72469. doi: 10.1371/journal.pone.0072469.

Greiner J.T., Wilkinson G.M., McGlathery K.J., Emery K.A. (2016). Sources of sediment carbon sequestered in restored seagrass meadows. *Marine Ecology Progress Series*, 551: 95–105. doi: 10.1371/journal.pone.0072469.

Gress S.K., Huxham M., Kairo J.G., Mugi L.M., Briers R.A. (2017). Evaluating, predicting and mapping belowground carbon stores in Kenyan mangroves. *Global Change Biology*, 23(1): 224–234. doi: 10.1111/gcb.13438.

Grimsditch G., Alder J., Nakamura T., Kenchington R., Tamelander J. (2013). The blue carbon special edition–Introduction and overview. *Ocean and Coastal Management*, 83: 1–4.

Griscom B.W., Adams J., Ellis P.W., Houghton R.A., Lomax G., Miteva D.A., Schlesinger W.H., Shoch D., Siikamäki J.V., Smith P., Woodbury P., Zganjar C., Blackman A., Campari J., Conant R.T., Delgado C., Elias P., Gopalakrishna T., Hamsik M.R., Herrero M., Kiesecker J., Landis E., Laestadius L., Leavitt S.M., Minnemeyer S., Polasky S., Potapov P., Putz F.E., Sanderman J., Silvius M., Wollenberg E., Fargione J. (2017). Natural climate solutions. *Proceedings of the National Academy of Sciences of the United States of America*, 114(44): 11645–11650. doi: 10.1073/pnas.1710465114.

Gross C., Hagy J.D. (2017). Attributes of successful actions to restore lakes and estuaries degraded by nutrient pollution. *Journal of Environmental Management*, 187: 122–136. doi: 10.1016/j.jenvman.2016.11.018.

Gross M.F., Hardisky M.A., Klemas V., Wolf P.L. (1987). Quantification of biomass of the marsh grass *Spartina alterniflora* loisel using Landsat Thematic Mapper imagery. *Photogrammetric Engineering and Remote Sensing*, 53: 1577–1583.

Guenet B., Danger M., Abbadie L., Lacroix G. (2010). Priming effect: bridging the gap between terrestrial and aquatic ecology. *Ecology*, 91: 2850–2861.

Guenet B., Danger M., Harrault L., Allard B., Jauset-Alcala M., Bardoux G., Benest D., Abbadie L. (2014). Fast mineralization of land-born C in inland waters: first experimental evidences of aquatic priming effect. *Hydrobiologia*, 721: 35–44.

Gul B., Khan M.A. (2001). Seasonal seed bank patterns of an Arthrocnemum macrostachyum (Chenopodiaceae) community along a coastal marsh inundation gradient on the Arabian Sea near Karachi, Pakistan. *Pak J Bot*, 33: 305–314.

Guo H.Q., Noormets A., Zhao B., Chen J.Q., Sun G., Gu Y.J., Li B., Chen J.K. (2009). Tidal effects on net ecosystem exchange of carbon in an estuarine wetland. *Agricultural and Forest Meteorology*, 149: 1820–1828.

Gupta V., Smemo K.A., Yavitt J.B., Fowle D., Branfireun B., Basiliko N. (2013). Stable isotopes reveal widespread anaerobic methane oxidation across latitude and peatland type. *Environmental Science & Technology*, 47: 8273–8279.

Gutiérrez-Mendoza J., Herrera-Silveira J.A. (2015). Almacenes de carbon en manglares de tipo chaparro en un scenario cárstico. In: Paz F., Wong J. (eds), *Estado Actual del Conocimiento del Ciclo de Carbon y sus Interacciones en México: Síntesis a 2014*, Texcoco, Edo de Mex: Mexico. p. 642.

Hachani M.A., Ziadi B., Langar H., Sami D.A., Turki S., Aleya L. (2016). The mapping of the *Posidonia oceanica* (L.) Delile barrier reef meadow in the southeastern Gulf of Tunis (Tunisia). *Journal of African Earth Sciences*, 121: 358–364.

Hagerty S.B., van Groenigen K.J., Allison S.D., Hungate B.A., Schwartz E., Koch G.W., Kolka R.K., Dijkstra P. (2014). Accelerated microbial turnover but constant growth efficiency with warming in soil. *Nature Climate Change*, 4: 903–906.

Haines E.B. (1977). The origins of detritus in Georgia salt marsh estuaries. *Oikos*, 29: 254–260.

Hall M.O., Furman B.T., Merello M., Durako M.J. (2016). Recurrence of *Thalassia testudinum* seagrass die-off in Florida Bay, USA: initial observations. *Marine Ecology Progress Series*, 560: 243–249.

Hallock P., Cottey T.L., Forward L.B., Halas J. (1986). Population biology and soil production of *Archaias angulates, Foraminiferida*, in Largo Sound, Florida, USA. *Journal of Foraminiferal Research*, *16*: 1–8.

Hamana M., Komatsu T. (2016). Real-time classification of seagrass meadows on flat bottom with bathymetric data measured by a narrow multibeam sonar system. *Remote Sensing*, 8(96)1–14.

Hamilton S.E., Casey D. (2016). Creation of a high spatio-temporal resolution global database of continuous mangrove forest cover for the 21st century (CGMFC-21). *Global Ecology and Biogeography*, *25*(6): 729–738.

Hamilton S.E., Lovette J. (2015). Ecuador's mangrove forest carbon stocks: a spatiotemporal analysis of living carbon holdings and their depletion since the advent of commercial aquaculture. *PloS One*, *10*(3): e0118880.

Hammer J., Pivo G. (2017). The triple bottom line and sustainable economic development: theory and practice. *Economic Development Quarterly*. doi: 10.1177/0891242416674808.

Hannam P.M., Liao Z., Davis S.J., Oppenheimer M. (2015). Developing country finance in a post-2020 global climate agreement. *Nature Climate Change*. doi: 10.1038/NCLIMATE2731.

Hansen J.C.R., Reidenbach M.A. (2013). Seasonal growth and senescence of a Zostera marina seagrass meadow alters wave-dominated flow and sediment suspension within a coastal bay. *Estuaries and Coasts*, *36*: 1099–1114.

Hanson R.S., Hanson T.E. (1996). Methanotrophic bacteria. *Microbiological Reviews*, *60*: 439–471.

Haq B.U., Hardenbol J., Vail P.R. (1987). Chronology of fluctuating sea levels since the Triassic. *Science*, *235*: 1156–1167.

Hardisky M.A., Daiber F.C., Roman C.T., Klemas V. (1984). Remote sensing of biomass and annual net aerial primary productivity of a salt marsh. *Remote Sensing of Environment*, *16*: 91–106.

Hardisky M.A., Klemas V., Michael Smart R. (1983a). The influence of soil salinity, growth form, and leaf moisture on the spectral radiance of *Spartina alterniflora* canopies. *Photogrammetric Engineering and Remote Sensing*, *49*: 77–83.

Hardisky M.A., Michael R., Smart, Klemas V. (1983b). Growth response and spectral characteristics of a short *Spartina alterniflora* salt marsh irrigated with freshwater and sewage effluent. *Remote Sensing of Environment*, *13*: 57–67.

Harris R.J., Milbrandt E.C., Everham E.M., Bovard B.D. (2010). The Effects of Reduced Tidal Flushing on Mangrove Structure and Function Across a Disturbance Gradient. *Estuaries and Coasts*, *33* (5): 1176–1185.

Harrison P.G. (1989). Detrital processing in seagrass systems: a review of factors affecting decay rates, remineralization and detritivory. *Aquatic Botany*, *35*(3–4): 263–288.

Hartmann D.L., Klein Tank A.M.G., Rusticucci M., Alexander L.V., Brönnimann S., Charabi Y., Dentener F.J., Dlugokencky E.J., Easterling D.R., Kaplan A., Soden B.J., Thorne P.W., Wild M., Zhai P.M. (2013). Observations: Atmosphere and Surface. In: *Climate Change 2013: The Physical Science Basis. Contribution of Working Group I to the Fifth Assessment Report of the Intergovernmental Panel on Climate Change*, Stocker T.F., Qin D., Plattner G.-K., Tignor M., Allen S.K., Boschung J., Nauels A., Xia Y., Bex V., Midgley P.M. (eds), Cambridge University Press: Cambridge, New York, UK, USA.

Hashim M., Yahya N.N., Ahmad S., Komatsu T., Misbari S., Reba M.N. (2014). Determination of seagrass biomass at Merambong Shoal in Straits of Johor using satellite remote sensing technique. *Malayan Nature Journal*, *66*(1–2): 20–37.

Hatala J.A., Detto M., Sonnentag O., Deverel S.J., Verfaillie J., Baldocchi D.D. (2012). Greenhouse gas (CO_2, CH_4, H_2O) fluxes from drained and flooded agricultural peatlands in the Sacramento-San Joaquin Delta. *Agricultural, Ecosystems, and Environment*, *150*: 1–18.

Hatton R.S., DeLaune R.D., Patrick W.H. Jr. (1983). Sedimentation, accretion, and subsidence in marshes of Barataria Basin, Louisiana. *Limnology and Oceanography*, *28*: 494–502.

He Y., Widney S., Ruan M., Herbert E., Li X., Craft C. (2016). Accumulation of soil carbon drives denitrification potential and lab incubated gas production along a chornosequence of salt marsh development. *Estuarine Coastal and Shelf Science*, *172*: 72–80.

Heck K., Carruthers T., Duarte C.M., et al. (2008). Trophic transfers from seagrass meadows subsidize diverse marine and terrestrial consumers. *Ecosystems*, *11*: 1198. doi: 10.1007/s10021-008-9155-y.

Hedges J., Keil R. (1995). Sedimentary organic matter preservation: an assessment and speculative synthesis. *Marine Chemistry*, *49*: 81–115 doi: 10.1016/0304–4203(95)00008-F.

Hedges J.I., Mann D.C. (1979). The characterization of plant tissues by their lignin oxidation products. *Geochimica et Cosmochimica Acta, 43*: 1809–1818.

Hedley J., Russell B., Randolph K., Dierssen H. (2016). A physics-based method for the remote sensing of seagrasses. *Remote Sensing of Environment, 174*: 134–147.

Heegaard E., Birks H.J.B., Telford R.J. (2005). Relationships between calibrated ages and depth in stratigraphical sequences: an estimation procedure by mixed-effect regression. *The Holocene, 15*: 612–618.

Hénault C., Bizouard F., Laville P., Gabrielle B., Nicoullaud B., Germon J.C., Cellier P. 2005. Predicting in situ soil N_2O emission using NOE algorithm and soil database. *Global Change Biology, 11*: 115–127 doi: 10.1111/j.1365–2486.2004.00879.x.

Hendriks I.E., Duarte C.M., Olsen Y.S. et al. (2014). Biological mechanisms supporting adaptation to ocean acidification in coastal ecosystems. *Estuarine Coastal and Shelf Science, 152*: A1–A8.

Hendriks I.E., Olsen Y.S., Duarte C.M. (2017). Light availability and temperature, not increased CO_2, will structure future meadows of *posidonia oceanica*. *Aquatic Botany, 139*: 32–36.

Hendriks I.E., Sintes T., Bouma T.J., Duarte C.M. (2008). Experimental assessment and modeling evaluation of the effects of the seagrass *Posidonia oceanica* on flow and particle trapping. *Marine Ecology Progress Series, 356*: 163–173.

Henningsen B. (2005, October). The maturation and future of habitat restoration programs for the Tampa Bay estuarine ecosystem. In: *Proceedings, Tampa Bay area scientific information symposium, BASIS, vol. 4*, pp. 27–30. http://www.tbeptech.org/BASIS/BASIS4/BASIS4.pdf#page=173.

Herbert E.R., Boon P., Burgin A.J., Neubauer S.C., Franklin R.B., Ardón M., Hopfensperger K.N., Lamers L.P., Gell P. (2015). A global perspective on wetland salinization: ecological consequences of a growing threat to freshwater wetlands. *Ecosphere, 6*(10): 1–43.

Hermoso V., Ward D.P., Kennard M.J. (2013). Prioritizing refugia for freshwater biodiversity conservation in highly seasonal ecosystems. *Diversity and Distributions, 19*: 1031–1042.

Herr D., Agardy T., Benzaken D., Hicks F., Howard J., Landis E., Soles A., Vegh T. (2015). *Coastal "Blue" Carbon. A Revised Guide to Supporting Coastal Wetland Programs and Projects Using Climate Finance and other Financial Mechanisms*, IUCN: Gland, Switzerland.

Herr D., Himes-Cornell A., Laffoley D. (2016). *National Blue Carbon Policy Assessment Framework: Towards Effective Management of Coastal Carbon Ecosystems*. IUCN: Gland, Switzerland.

Herr D., Landis E. (2016). *Coastal Blue Carbon Ecosystems. Opportunities for Nationally Determined Contributions. Policy Brief. Gland*, IUCN, TNC: Switzerland, Washington, D.C., USA.

Herr D., Pidgeon E., Laffoley D. (eds) (2011). *Blue Carbon Policy Framework: Based on the First Workshop of the International Blue Carbon Policy Working Group*, IUCN: Gland, Switzerland; CI Arlington, USA.

Herr D., Pidgeon E., Laffoley D. (eds) (2012). *Blue Carbon Policy Framework 2.0: Based on the First Workshop of the International Blue Carbon Policy Working Group*, IUCN: Gland, Switzerland; CI Arlington, USA.

Herr D., Trines E., Howard J., Silvius M., Pidgeon E. (2014). *Keep It Fresh or Salty. An Introductory Guide to Financing Wetland Carbon Programs and Projects*, IUCN, CI, WI: Gland, Switzerland.

Herr D., von Unger M., Laffoley D., McGivern A. (2017). Pathways for implementation of blue carbon initiatives. *Aquatic Conservation. 27*(51): 116–129.

Herrera-Silveira J.A., et al. (2012). *Rehabilitación de Manglares en el estado de Yucatán Sometidos a Diferentes Condiciones Hidrológicas y Nivel de Impacto: el Caso de Celestún y Progreso*, Centro de Investigación y Estudios Avanzados, IPN, SNIB-CONABIO: México D. F.

Herrera-Silveira J.A., et al. (2015). Carbon dynamics (stocks and fluxes) of mangroves in Mexico. (In press).

Hestir E.L., Schoellhamer D.H., Greenberg J., Morgan-King T.L., Ustin S.L. (2015). The effect of submerged aquatic vegetation expansion on a declining turbidity trend in the Sacramento-San Joaquin River Delta, *Estuaries and Coasts*. doi: 10.1007/s12237-015-0055-z.

Hestir E.L., Schoellhamer D.H., Morgan-King T., Ustin S.L. (2013). A step decrease in sediment concentration in a highly modified tidal river delta following the 1983 El Niño floods. *Marine Geology, 345*: 304–313.

Hill V.J., Zimmerman R.C., Bissett W.P., Dierssen H., Kohler D.D.R. (2014). Evaluating light availability, seagrass biomass, and productivity using hyperspectral airborne remote sensing in Saint Joseph's Bay, Florida. *Estuaries and Coasts, 37*(6): 1467–1489.

Himes-Cornell A., Pendleton L., Atiyah P. (2018). Valuing ecosystem services from blue forests: a systematic review of the valuation of salt marshes, seagrass beds and mangrove forests. *Ecosystem Services. 30*: 36–48.

Hinrichs K.-U., Boetius A. (2002). The anaerobic oxidation of methane: new insights in microbial ecology and biogeochemistry. In: *Ocean Marine Systems*, Wefer G., Billett D., Hebbeln D., Jørgensen B.B., Schlüter M., van Weering T.C.E. (eds), Springer-Verlag: Berlin, pp. 457–477.

Hinson A.L., Feagin R.A., Eriksson M., Najjar R.G., Herrmann M., Bianchi T.S., Kemp M., Hutchings J.A., Crooks S., Boutton T. (2017). The spatial distribution of soil organic carbon in tidal wetland soils of the continental United States. *Global Change Biology*. (In press. CoBluCarb database also available on request from NASA or authors, as project # NNX14AM37G).

Hiraishi T., Krug T., Tanabe K., Srivastava N., Baasansuren J., Fukuda M., Troxler T.G. (2014). *2013 Supplement to the 2006). IPCC Guidelines for National Greenhouse Gas Inventories: Wetlands*, IPCC: Switzerland.

Hladik C., Alber M. (2012). Accuracy assessment and correction of a LIDAR-derived salt marsh digital elevation model. *Remote Sensing of Environment, 121*: 224–235.

Hladik C., Alber M. (2014). Classification of salt marsh vegetation using edaphic and remote sensing-derived variables. *Estuarine Coastal and Shelf Science, 141*: 47–57.

Hodson R.E., Christian R.R., Maccubbin A.E. (1984). Lignocellulose and lignin in the salt marsh grass *Spartina alterniflora*: initial concentrations and short-term, post-depositional changes in detrital matter. *Marine Biology, 81*: 1–7.

Hoekstra J.M., Molnar J.L., Jennings M., Revenga C., Spalding M.D., Boucher T.M., Robertson J.C., Heibel T.J., Ellison K. (2010). *The Atlas of Global Conservation: Changes, Challenges, and Opportunities to Make a Difference*, Molnar J.L. (ed.), University of California Press: Berkeley.

Holm G.O. Jr., Perez B.C., McWhorter D.E., Krauss K.W., Johnson D.J., Raynie R.C., Killebrew C.J. (2016). Ecosystem level methane fluxes from tidal freshwater and brackish marshes of the Mississippi River Delta: Implications for coastal wetland carbon projects. *Wetlands, 36*: 401–413. doi: 10.1007/s13157-016-0746-7.

Holmquist J.R., Windham-Myers L., Bliss N., Crooks S., Morris J., Megonigal J.P., Troxler T., Weller D., Callaway J., Drexler J., Ferner M.C., Gonneea M., Kroeger K., Schile-Beers L., Woo I., Buffington K., Breithaupt J., Boyd B.M., Brown L.N., Dix N., Hice L., Horton B., MacDonald G.M., Moyer R.P., Reay W., Shaw T., Smith E., Smoak J., Sommerfield C., Thorne K., Velinsky D., Watson E., Grimes K.W., Woodrey M. (2018a). Accuracy and precision of soil carbon estimates for tidal wetlands in the conterminous United States. *Nature Scientific Reports, 8*(1): 9478.

Holmquist J., Windham-Myers L., Bernal B., Byrd K.B., Crooks S., Gonneea M.E., Herold N., Knox S.H., Kroeger K.D., McCombs J., Megonigal J.P., Meng L., Morris J.T., Sutton-Grier A.E., Troxler T.G., Weller D.E. (2018b). Uncertainty in United States Coastal Wetland Greenhouse Gas Inventorying. ERL105350 2018. https://doi.org/10.1088/1748-9326/113157.

Hopf F. (2011). Levee failures in the Sacramento-San Joaquin River Delta: Characteristics and Perspectives, *PhD Thesis*, Texas A&M University, College Station: Texas, p. 392. http://hdl.handle.net/1969.1/ETD-TAMU-2011-12-10691.

Hopkinson C.S. (1988). Patterns of organic carbon exchange between coastal ecosystems: the mass balance approach in salt marsh ecosystems. In: *Coastal-Offshore Ecosystem Interactions*, Jansson B.-O. (ed.), Springer-Verlag: Lecture Notes on Coastal and Estuarine Studies: Berlin, Heidelberg, *vol. 22*, pp. 122–154.

Hopkinson C.S. (1992). A comparison of ecosystem dynamics in freshwater wetlands. *Estuaries, 15*: 549–562.

Hopkinson C.S., Cai W.-J., Hu X. (2012). Carbon sequestration in wetland dominated coastal systems – a global sink of rapidly diminishing magnitude. *Current Opinion in Environmental Sustainability, 4*: 1–9. doi: 10.1016 /j.cosust.2012.03.005.

Hopkinson C.S., Morris J.T., Fagherazzi S., Raymond P. (2016). Loss for Gain: marsh cannibalization provides critical sediment to maintain salt marsh accretion. (In preparation).

Hopkinson C.S., Vallino J. (2005). The relationships among man's activities in watersheds and estuaries: a model of runoff effects on patterns of estuarine community metabolism. *Estuaries, 18*: 598–621.

Horton B.P., Rahmstorf S., Engelhart S.E., Kemp A.C. (2014). Expert assessment of sea-level rise by AD 2100 and AD 2300. *Quaternary Science Reviews, 84*: 1–6.

Hossain M.S., Bujang J.S., Zakaria M.H., Hashim M. (2014). The application of remote sensing to seagrass ecosystems: an overview and future research prospects. *International Journal of Remote Sensing, 36*(1): 61–113.

Hossain M.S., Bujang J.S., Zakaria M.H., Hashim M. (2015a). Assessment of the impact of Landsat 7 Scan Line Corrector data gaps on Sungai Pulai Estuary seagrass mapping. *Applied Geomatics, 7*: 189–202.

Hossain M.S., Bujang J.S., Zakaria M.H., Hashim M. (2015b). Application of Landsat images to seagrass areal cover change analysis for Lawas, Terengganu and Kelantan of Malaysia. *Continental Shelf Research*, *110*: 124–148.

Howard J., Sutton-Grier A., Herr D., Kleypas J., Landis E., Mcleod E., Pidgeon E., Simpson S. (2017). Clarifying the role of coastal and marine systems in climate mitigation. *Frontiers in Ecology and the Environment*, *15*(1): 42–50. doi: 10.1002/fee.1451.

Howard J.L., Creed J.C., Aguiar M.V.P., Fouqurean J.W. (2017). CO_2 released by carbonate sediment production in some coastal areas may offset the benefits of seagrass "Blue Carbon" storage. *Limnology and Oceanography*. doi: 10.1002/lno.10621,5 2017.

Howard R.J., Day R.H., Krauss K.W., From A.S., Allain L., Cormier N. (2016). Hydrologic restoration in a dynamic subtropical mangrove-to-marsh ecotone. *Restoration Ecology*, *25*(3): 471–482.

Howes B., Dacey J., Teal J. (1985). Annual carbon mineralization and belowground production of Spartina alterniflora in a New England salt marsh. *Ecology*, *66*: 595–605.

Hu X., Burdige D.J. (2007). Enriched stable carbon isotopes in the pore waters of carbonate soils dominated by seagrasses: evidence for coupled carbonate dissolution and reprecipitation. *Geochimica et Cosmochimica Acta*, *71*: 129–144.

Hu Z., Lee J.W., Chandran K., Kim S., Khanal S.K. (2012). Nitrous oxide (N_2O) emission from aquaculture: a review. *Environmental Science and Technology*, *46* (12): 6470–6480.

Hu Z., Lee J.W., Chandran K., Kim S., Sharma K., Brotto A.C., Khanal S.K. (2013). Nitrogen transformations in intensive aquaculture system and its implication to climate change through nitrous oxide emission. *Bioresource Technology*, *130*: 314–320.

Huang S., Pollack H.N., Shen P.-Y (2000). Temperature trends over the past five centuries reconstructed from borehole temperatures. *Nature*, *403*: 756–758.

Humm H.J. (1964). Epiphytes of the sea grass Thalassia testudinum, in Florida. *Bulletin of Marine Science of the Gulf and Caribbean*, *14*: 306–341.

Huwyler F., Käppeli J., Serafimova K., Swanson E., Tobin J. (2014). *Conservation Finance: Moving Beyond Donor Funding Toward an Investor-Driven Approach*, Credit Suisse, WWF, McKinsey & Company: Zurich, Switzerland.

Huxham M. (2010). Plan Vivo Project Idea Note Project Title: Mikoko Pamoja Mangrove restoration in Gazi Bay, Kenya, (March).

Huxham M. (2013). MIKOKO PAMOJA Mangrove conservation for community benefit, *Plan Vivo PDD*, 44(August): 1–38.

Huxham M. (2013). Plan vivo project design document. Mikoko Pamoja: Mangrove conservation for community benefit. http://planvivo.org/docs/Mikoko-Pamoja-PDD_published.pdf.

Huxham M., Emerton L., Kairo J., Munyi F., Abdirizak H., Muriuki T., ... Briers R.A. (2015). Applying Climate Compatible Development and economic valuation to coastal management: a case study of Kenya's mangrove forests. *Journal of Environmental Management*, *157*: 168–181. doi: 10.1016/j.jenvman.2015.04.018.

Huxham M., Kimani E., Augley J. (2004). Mangrove fish: a comparison of community structure between forested and cleared habitats. *Estuarine, Coastal and Shelf Science*, *60*(4): 637–647. doi: 10.1016/j.ecss.2004.03.003.

ICBP (International Blue Carbon Partnership) (2016). http://bluecarbonpartnership.org/.

IDDRI (2015). Pathways to deep decarbonization 2015 report – executive summary, SDSN-IDDRI. http://www.iddri.org/Themes/DDPP_EXESUM.pdf.

IFAD (2001). *Assessment of Rural Poverty: Western and Central Africa*. International Fund for Agricultural Development: Rome, Italy, *vol. 130*.

IISD (2016).Oceans Action Day at COP 22. http://enb.iisd.org/climate/cop22/oceans-action-day/.

Ilman M., Dargusch P., Dart P., Onrizal KC (2016). A historical analysis of the drivers of loss and degradation of Indonesia's mangroves. *Land Use Policy*, *54*: 448–459.

Indarto G.B., Murharjanti P., Khatarina J., Pulungan I., Ivalerina F., Rahman J., Prana M.N., Resosudarmo I.A.P., Muharrom E. (2012). The Context of REDD+ in Indonesia: Drivers, agents and institutions, Working Paper 92. CIFOR, BOGOR, Indonesia, www.cifor.org/publications/pdf_files/WPapers/WP92Resosudarmo.pdf.

Indonesian Minister of Forestry (2008). The Implementation of Demonstration Activities on Reduction of Emission from Deforestation and Degradation.

Indonesian Ministry of National Development Planning/ National Development Planning Agency (2010). 2010–2014 National Medium-Term Development Plan (RPJMN 2010–2014).

Ingebritsen S.E., Ikehara M.E. (1999). Sacramento-San Joaquin Delta: the sinking heart of the state. In: Galloway D., Jones D.R., Ingebritsen S.E. (eds) *Land Subsidence in the United States. Circular 1182*, U.S. Geological Survey: Reston, VA, USA, pp. 83–94.

Ingebritsen S.E., Ikehara M.E., Galloway D.L., Jones D.R. (2000). *Delta Subsidence in California: the Sinking Heart of the State*, U.S. Geological Survey FS-005-00.

Inglett K.S., Chanton J.P., Inglett P.W. (2013a). Methanogenesis and methane oxidation in wetland soils. In: *Methods in Biogeochemistry in Wetlands. Soil Science Society of America*, DeLaune R.D., Reddy K.R., Richardson C.J., Megonigal J.P. (eds), Madison: WI, SSSA Book Series, no. 10, pp. 407–426.

Inglett K.S., Ogram A.V., Reddy K.R. (2013b). Ammonium oxidation in wetland soils. In: *Methods in Biogeochemistry in Wetlands*, DeLaune R.D., Reddy K.R., Richardson C.J., Megonigal J.P. (eds), Soil Science Society of America Madison: WI, SSSA Book Series, no. 10, pp. 485–502.

Inglett P.W., Kana T.M., An S. (2013c). Denitrification measurements using membrane inlet mass spectrometry. In: *Methods in Biogeochemistry in Wetlands*, DeLaune R.D., Reddy K.R., Richardson C.J., Megonigal J.P. (eds), Soil Science Society of America: Madison, WI, SSSA Book Series, no. 10, pp. 503–516.

Interagency Working Group on Social Cost of Carbon (2015). Technical Support Document: Technical Update of the Social Cost of Carbon for Regulatory Impact Analysis Under Executive Order 12866. Available at www.whitehouse.gov/sites/default/files/omb/inforeg/scc-tsd-final-july-2015.pdf (accessed May 2, 2016).

Intergovernmental Panel on Climate Change (2014). Fifth Assessment Report (AR5). Available at www.ipcc. ch/report/ar5/ (accessed May 2, 2016).

IPCC (2003). *Intergovernmental Panel on Climate Change Good Practice Guidance for Land Use, Land-Use Change and Forestry*, Penman J., Gytarsky M., Hiraishi T., Krug T., Kruger D., Pipatti R., Buendia L., Miwa K., Ngara T., Tanabe K., Wagner F. (eds), IPCC: Japan.

IPCC (2006). 2006 IPCC Guidelines for National Greenhouse Gas Inventories, Volume 4: Agriculture, Forestry and Other Land Use. In: *Prepared by the National Greenhouse Gas Inventories Programme*, Eggleston, H.S., Buendia, L., Miwa, K., Ngara, T., Tanabe, K. (eds), IGES: Japan.

IPCC (2013). Annex III: Glossary. *Climate Change 2013: The Physical Science Basis. Contribution of Working Group I to the Fifth Assessment Report of the Intergovernmental Panel on Climate Change*, Planton S., Stocker T.F., Qin D., Plattner G.-K., Tignor M., Allen S.K., Boschung J., Nauels A., Xia Y., Bex V., Midgley P.M. (eds), Cambridge University Press: Cambridge, New York, UK, USA.

IPCC (2014). *2013 Supplement to the 2006 IPCC Guidelines for National Greenhouse Gas Inventories: Wetlands*, Hiraishi T., Krug T., Tanabe K., Srivastava N., Baasansuren J., Fukuda M., Troxler T.G. (eds). IPCC: Switzerland.

IPCC (2014). Climate Change 2014: Synthesis Report. In: *Contribution of Working Groups I, II and III to the Fifth Assessment Report of the Intergovernmental Panel on Climate Change*, Pachauri R.K., Meyer L.A. (eds), IPCC: Geneva, Switzerland, p. 151.

IPCC (2014). In: Climate Change 2014: Mitigation of Climate Change. *Contribution of Working Group III to the Fifth Assessment Report of the Intergovernmental Panel on Climate Change*. Edenhofer O., Pichs-Madruga R., Sokona Y., Farahani E., Kadner S., Seyboth K., Adler A., Baum I., Brunner S., Eickemeier P., Kriemann B., Savolainen J., Schlömer S., von Stechow C., Zwickel T., Minx J.C. (eds), Cambridge University Press: Cambridge, New York, UK, USA.

Irvine I.C., Vivanco L., Bentley P.N., Martiny J.B.H. (2012). The effect of nitrogen enrichment on C1-cycling microorganisms and methane flux in salt marsh sediments. *Frontiers in Microbiology*, *3*. doi: 10.3389/fmicb.2012.00090.

Isla F.I., Bujaleski G.G. (2008). Coastal geology and morphology of Patagonia and the Fuegian Archipelago. In: *Late Cenozoic of Patagonia and Tierra del Fuego*, pp. 227–239.

IUCN (2017). The Blue Foundations to Sustainable Development – Climate Change Adaptation and Mitigation Actions. Information brief. www.iucn.org/sites/dev/files/content/documents/2017/the_blue_foundations_of_sustainable_development_and_climate_action_final.pdf.

James A. (1999). Time and the persistence of alluvium: river engineering, fluvial geomorphology, and mining sediment in California. *Geomorphology*, *31*: 265–290.

James N.P., Bone Y., Brown K.M., Cheshire A. (2009). Calcareous epiphyte production in cool-water carbonate seagrass depositional environments-southern Australia. *Perspectives in Carbonate Geology, International Association of Sedimentology Special Publication*, *41*: 123–148.

Jassby A.D., Cloern J.E., Cole B.E. (2002). Annual primary production: patterns and mechanisms of change in a nutrient-rich tidal ecosystem. *Limnology and Oceanography*, *47*(3): 698–712.

Jensen H.S., McGlathery K.J., Marino R., Howarth R.W. (1998). Forms and availability of soil phosphorus in carbonate sand of Bermuda seagrass beds. *Limnology and Oceanography*, *43*: 799–810.

Jensen H.S., Nielsen O.I., Koch M.S., de Vicente I. (2009). Phosphorus release with carbonate dissolution coupled to sulfide oxidation in Florida Bay seagrass soils. *Limnology and Oceanography*, *54*: 1753–1764.

Jerath M., Bhat M., Rivera-Monroy V.H., Castañeda-Moya E., Simard M., Twilley R.R. (2016). The role of economic, policy, and ecological factors in estimating the value of carbon stocks in Everglades mangrove forests, south Florida, USA. *Environmental Science and Policy*, *66*: 160–169.

Jevrejeva S., Moore J.C., Grinsted A., Woodworth P.L. (2008). Recent global sea level acceleration started over 200 years ago? *Geophysical Research Letters*, *35*(8).

Jha C.S., Rodda S.R., Thumaty K.C., Raha A.K., Dadhwal V.K. (2014). Eddy covariance based methane flux in Sundarbans mangroves, India. *Journal of Earth System Science*, *123*: 1089–1096.

Jickells T.D., Andrews J.E., Parkes D.J. (2016). Direct and indirect effects of estuarine reclamation on nutrient and metal fluxes in the global coastal zone. *Aquatic Geochemistry*, *22*(4): 337–348.

Jimenez J.A., Lugo A.E., Cintron G. (1985). Tree mortality in mangrove forests. *Biotropica*, 177–185.

Johannessen S.C., Macdonald R.W. (2016). Geoengineering with seagrasses: is credit due where credit is given? *Environmental Research Letters*, *11*: 113001.

Johnson G.C., Wijffels S.E. (2011). Ocean density change contributions to sea level rise. *Oceanography*, *24*(2): 112–21.

Jones C.G., Lawton J.H., Shachak M. (1994a). Organisms as ecosystem engineers. *Oikos*, *69*: 373–386.

Jones C.G., Lawton J.H., Shachak M. (1994b). Positive and negative effects of organisms as physical ecosystem engineers. *Ecology*, *78*: 1946–1957.

Joosten H., Couwenberg J., von Unger M., Emmer I. (2016). Peatlands, forests and the climate architecture: setting incentives through markets and enhanced accounting. *Climate Change*, *14*. Umweltbundesamt, Germany.

Joshi I.D., D'Sa E.J., Osburn C.L., Bianchi T.S., Ko D., Oviedo-Vargas D., Arellano A.R., Ward N.D. (2017). Assessing chromophoric dissolved organic matter (CDOM) distribution, stocks, and fluxes in Apalachicola Bay using combined field, VIIRS ocean color, and model observations. *Remote Sensing of Environment*, *191*: 359–372.

Joye S.B., Hollibaugh J.T. (1995). Influence of sulfide inhibition of nitrification on nitrogen regeneration in sediments. *Science*, *270*: 623–625.

Kairo J., Dahdouh-Guebas F., Bosire J., Koedam N. (2001). Restoration and management of mangrove systems – a lesson for and from East African region. *Soth Africa Journal of Botany*, *67*: 383–389.

Kairo J.G. (2001). *Ecology and Restoration of Mangrove Systems in Kenya*. University of Brussels (VUB): Belgium.

Kairo J.G., Bosire J.O. (2016). Emerging and Crosscutting issues. In: Bosire J.O., Mangora M.M., Bandeira S., Rajkaran A., Ratsimbazafy R., Appadoo C., Kairo J.G. (eds), *Mangroves of the Western Indian Ocean: Status and Management*, Zanzibar, Town: WIOMSA, pp. 136–151.

Kairo J.G., et al. (2019). Mikoko Pamoja – a demonstrably effective community-based blue carbon project in Kenya. In: *A Blue Carbon Primer: The State of Coastal Wetland Carbon Science, Practice, and Policy*, Lisamarie W.-M., Stephen C., and Tiffany G.T. (eds), CRC Press: Boca Raton, London.

Kairo J.G., Wanjiru C., Ochiewo J. (2009). Net Pay: economic analysis of a replanted mangrove plantation in Kenya. *Journal of Sustainable Forestry*, *28*(3–5): 395–414. doi: 10.1080/10549810902791523.

Kakuta S., Takeuchi W., Prathep A. (2016). Seaweed and seagrass mapping in Thailand measured using Landsat 8 optical and textual image properties. *Journal of Marine Science and Technology*, *24*(6): 1155–1160.

Kaldy J.E. (2006). Carbon, nitrogen, phosphorus and heavy metal budgets: how large is the eelgrass (*Zostera marina* L.) sink in a temperate estuary? *Marine Pollution Bulletin*, *52*: 332–356.

Kaldy J.E. (2012). Influence of light, temperature and salinity on dissolved organic carbon exudation rates in *Zostera marina* L. *Aquatic Biosystems*, *8*: 19.

Kao J.T., Titus J.E., Zhu W.-X. (2003). Differential nitrogen and phosphorus retention by five wetland plant species. *Wetlands*, *23*: 979–987.

Kathilankal J.C., Mozdzer T., Fuentes J., D'Ordorico P., McGlathery K., Zieman J. (2008). Tidal influences on carbon assimilation by a salt marsh. *Environmental Research Letters*, *3*(044010): 6. doi: 10.1088/1748–9326/3/4/044010.

Kathiresan K. (2012). Importance of mangrove ecosystem. *International Journal of Marine Science*, 2(1).

Kathiresan K., Rajendran N. (2005). Coastal mangrove forests mitigated tsunami. *Estuarine, Coastal and Shelf Science*, 65: 601–606.

Kauffman J.B., Donato D.C. (2012). *Protocols for the measurement, monitoring and reporting of structure, biomass and carbon stocks in mangrove forests*, CIFOR: Bogor, Indonesia. Working paper 86.

Kauffman J.B., Donato D.C., Adame M.F. (2012). *Protocolo para la medición, monitoreo y reporte de la estructura, biomasa y reservas de carbono de los manglares*, CIFOR: Bogor Indonesia. Working paper 117.

Kauffman J.B., Heider C., Norfolk J., Payton F. (2014). Carbon stocks of intact mangroves and carbon emissions arising from their conversion in the Dominican Republic. *Ecological Applications*, 24(3): 518–527.

Kauffman J.B., Trejo H.H., Garcia M.D.C.J., Heider C., Contreras W.M. (2016). Carbon stocks of mangroves and losses arising from their conversion to cattle pastures in the Pantanos de Centla, Mexico. *Wetlands Ecology and Management*, 24(2): 203–216.

Kearney M.S., Stutzer D., Turpie K., Stevenson J.C. (2009). The effects of tidal inundation on the reflectance characteristics of coastal marsh vegetation. *Journal of Coastal Research*, 25: 1177–1186.

Kearney M.S., Turner R.E. (2016). Microtidal marshes: can these widespread and fragile marshes survive increasing climate–sea level variability and human action? *Journal of Coastal Research*, 32(3): 686–99.

Keller J.K., Wolf A.A., Weisenhorn P.B., Drake B.G., Megonigal J.P. (2009). Elevated CO_2 affects porewater chemistry in a brackish marsh. *Biogeochemistry*, 96: 101–117.

Kelleway J.J., Saintilan N., Macreadie P.I., Skilbeck C.G., Zawadzki A., Ralph P.J. (2016). Seventy years of continuous encroachment substantially increases 'blue carbon' capacity as mangroves replace intertidal salt marshes. *Global Change Biology*, 22: 1097–1109.

Kelley C.A., Chanton J.P., Bebout B.M. (2015). Rates and pathways of methanogenesis in hypersaline environments as determined by ^{13}C-labeling. *Biogeochemistry*, 126: 1–13.

Kemp A.C., Horton B., Donnelly J., Mann M., Vermeer M., Rahmstorf S. (2011). Climate related sea-level variations over the past two millennia. *Proceedings of the National Academy of Sciences of the United States of America*, 108: 11017–11022.

Kemp A.C., Kegel J.J., Culver S.J., Barber D.C., Mallinson D.J., Leorri E., Bernhardt C.E., Cahill N., Riggs S.R., Woodson A.L., Mulligan R.P. (2017). Extended late Holocene relative sea-level histories for North Carolina, USA. *Quaternary Science Reviews*, 160: 13–30.

Kemp W.M., Smith E., Marvin-DiPasquale M., Boynton W. (1997). Organic carbon balance and net ecosystem metabolism in Chesapeake Bay. *MEPS*, 150: 229–248.

Kendall C.G.S.C., Alsharhan A.S., Cohen A. (2002). The Holocene Tidal Flat Complex of the Arabian Gulf Coast of Abu Dhabi. In: *Sabkha Ecosystems, Volume I: The Arabian Peninsula and Adjacent Countries*, Barth H.-J., Böer B. (eds), Kluwer Academic Publishers: Dordrecht, pp. 21–36.

Kendall C.G.S.C., Skipwith P.A. (1968). Recent microbial mats of Persian Gulf lagoon. *Journal of Sedimentary Petrology*, 38: 1040–1058.

Kenig F., Huc A.Y., Purser B.H., Oudin J. (1990). Sedimentation, distribution and diagenesis of organic matter in a recent carbonate environment, Abu Dhabi, U.A.E. *Organic Geochemistry*, 16: 735–747.

Kennedy H., Beggins J., Duarte C.M., Fourqurean J.W., Holmer M., Marbà N., Middelburg J.J. (2010). Seagrass sediments as a global carbon sink: isotopic constraints. *Global Biogeochemical Cycles*, 24: 1–8.

Kerr A.M., Baird A.H. (2007). Natural barriers to natural disasters. *BioScience*, 57, 102–103.

Kerwin M.L., Megonigal J.P. (2013). Tidal wetland stability in the face of human impacts and sea-level rise. *Nature*. 504: 53–60. doi: 10.1038/nature12856.

Keuskamp J.A., Dingemans B.J., Lehtinen T., Sarneel J.M., Hefting M.M. (2013). Tea Bag Index: a novel approach to collect uniform decomposition data across ecosystems. *Methods in Ecology and Evolution*, 4(11): 1070–1075.

Khan M.A., Ungar I.A., Showalter A.M. (2005). Salt stimulation and tolerance in an intertidal stem-succulent halophyte. *Journal of plant nutrition*, 28: 1365–1374.

Khanna S., Bellvert J., Shapiro K., Ustin S.L. (2015a). *Dynamic Changes in an Estuary Over the Past 11 Years*, AGU Fall Meeting: San Francisco, California, USA.

Khanna S., Bellvert J., Shapiro K., Ustin S.L. (2015b). *Invasions in State of the Estuary 2015: Status and Trends Updates on 33 Indicators of Ecosystem Health*, The San Francisco Estuary Partnership: Oakland, California, USA.

Khanna S., Santos M.J., Hestir E.L., Ustin S.L. (2012). Plant community dynamics relative to the changing distribution of a highly invasive species, Eichhornia crassipes: a remote sensing perspective. *Biological Invasions, 14*: 717–733.

Khanna S., Santos M.J., Ustin S.L., Koltunov A., Kokaly R.F., Roberts D.A. (2013). Detection of Salt Marsh Vegetation Stress and Recovery after the Deepwater Horizon Oil Spill in Barataria Bay, Gulf of Mexico Using AVIRIS Data. *Plos One, 8*.

Kim K., Choi J.-K., Ryu J.-H., Jeong H.J., Lee K., Park M.G., Kim K.Y. (2015). Observation of typhoon-induced seagrass die-off using remote sensing. *Estuarine, Coastal and Shelf Science, 154*: 111–121.

Kimani E.N., Mwatha G.K., Wakwabi E.O., Ntiba J.M., Okoth B.K. (1996). Fishes of a shallow tropical mangrove estuary, Gazi, Kenya. *Marine and Freshwater Research, 47*(7): 857–868. doi: 10.1071/MF9960857.

Kimura M., Jia Z.-J., Nakayama N., Asakawa S. (2008). Ecology of viruses in soils. Past, present and future perspectives. *Soil Science and Plant Nutrition, 54*: 1–32.

King G.M., Klug M., Lovley D. (1983). Metabolism of acetate, methanol, and methylated amines in intertidal sediments of Lowes Cove, Maine. *Applied and Environmental Microbiology, 45*: 1848–1853.

King L. (2012). Notes From The Field: Including mangrove forests in REDD+, CDKN 2012. http://cdkn.org/wp-content/uploads/2012/12/Notes-from-the-field-Lesley-King-1.pdf.

Kirchman D.L., Mazzella L., Alberte R.S., Mitchell R. (1984). Epiphytic bacterial production on *Zostera marina. Marine Ecology Progress Series, 15*: 117–123.

Kirchner G., Ehlers H. (1998). Sediment geochronology in changing coastal environments: potentials and limitations of the Cs-137 and Pb-210 methods. *Journal of Coastal Research, 14*: 483–492.

Kirui K.B., Kairo J.G., Bosire J., Viergever K.M., Rudra S., Huxham M., Briers R. (2013). Mapping of mangrove forest land cover change along the Kenya coastline using Landsat imagery. *Ocean and Coastal Management, 83*: 19–24.

Kirwan M., Megonigal J.P. (2013). Tidal wetland stability in the face of human impacts and sea-level rise. *Nature.* doi: 10.1038/nature12856.

Kirwan M., Mudd S. (2012). Response of salt-marsh carbon accumulation to climate change. *Nature, 489*(7417): 550–553.

Kirwan M., Temmerman S., Skeehan E.E., Guntenspergen G.R., Fagherazzi S. (2016). Overestimation of marsh vulnerability to sea level rise. *Nature Climate Change, 6*: 253–260.

Kirwan M.L., Walters D.C., Reay W., Carr J.A. (2016b). Sea level driven marsh expansion in a coupled model of marsh erosion and migration. *Geophysical Research Letters, 43*, 4366–4373.

Kirwan M.L., Blum L.K. (2011). Enhanced decomposition offsets enhanced productivity and soil carbon accumulation in coastal wetlands responding to climate change. *Biogeosciences, 8*: 987–993.

Kirwan M.L., et al. (2010). Limits on the adaptability of coastal marshes to rising sea level. *Geophysical Research Letters, 37*: L23401.

Kirwan M.L., Guntenspergen G.R. (2010). Influence of tidal range on the stability of coastal marshland. *Journal of Geophysical Research-Earth Surface, 115*: F02009. doi: 02010.01029/02009JF001400.

Kirwan M.L., Guntenspergen G.R. (2015). Response of Plant Productivity to Experimental Flooding in a Stable and a Submerging Marsh. *Ecosystems, 18*: 903–913. doi: 10.1007/s10021-015-9870-0.

Kirwan M.L., Guntenspergen G.R., D'Alpaos A., Morris J.T., Mudd S.M., Temmerman S. (2010). Limits on the adaptability of coastal marshes to rising sea level. *Journal of Geophysical Research, 37*: L23401. doi: 10.1029/2010GL045489.

Kirwan M.L., Guntenspergen G.R., Langley J.A. (2014). Temperature sensitivity of organic-matter decay in tidal marshes. *Biogeosciences, 11*: 4801–4808.

Kirwan M.L., Guntenspergen G.R., Morris J.T. (2009). Latitudinal trends in *Spartina alterniflora* productivity and the response of coastal marshes to global change. *Global Change Biology, 15*: 1982–1989.

Kirwan M.L., Megonigal J.P. (2013). Tidal wetland stability in the face of human impacts and sea-level rise. *Nature, 504*(7478): 53–60.

Kirwan M.L., Mudd S.M. (2012). Response of salt-marsh carbon accumulation to climate change. *Nature, 489*: 550–553.

Kirwan M.L., Temmerman S., Skeehan E.E., Guntenspergen G.R., Fagherazzi S. (2016). Overestimation of marsh vulnerability to sea level rise. *Nature Climate Change, 6*: 253–260.

Knittel K., Boetius A. (2009). Anaerobic oxidation of methane: progress with an unknown process. *Annual Review of Microbiology, 63*: 311–334.

Knox S.H., Sturtevant C., Matthes J.H., Koteen L., Verfaillie J., Baldocchi D. (2015). Agricultural peatland restoration: effects of land-use change on greenhouse gas (CO_2 and CH_4) fluxes in the Sacramento-San Joaquin Delta. *Global Change Biology, 21*: 750–765.

Koch E.W., Ackerman J.D., Verduin J., van Keulen M. (2006). Fluid dynamics in seagrass ecology – from molecules to ecosystems. In: *Seagrasses: Biology, Ecology, and Conservation*, Larkum A.W.D, Orth R.J., Duarte C.M. (eds), Springer, 193–225.

Koedsin W., Intararuang W., Ritchie R.J., Huete A. (2016). An integrated field and remote sensing method for mapping seagrass species, cover, and biomass in Southern Thailand. *Remote Sensing, 8*(292): 1–18.

Konkeo H. (1997). Comparison of intensive shrimp farming systems in Indonesia, Philippines, Taiwan and Thailand. *Aquaculture Research, 28*: 789–96.

Kopp R.E. (2013). Does the mid-Atlantic United States sea level acceleration hot spot reflect ocean dynamic variability? *Geophysical Research Letters, 40*: 3981–3985.

Kopp R.E., Simons F.J., Mitrovica J.X., Maloof A.C., Oppenheimer M. (2009). Probabilistic assessment of sea level during the last interglacial stage. *Nature, 462*: 863–868.

Kopp R.E., Simons F.J., Mitrovica J.X., Maloof A.C., Oppenheimer M. (2013). A probabilistic assessment of sea level variations within the last interglacial stage. *Geophysical Journal International, 193*: 711–716.

Krause E., Wichels A., Gimenez L., Gerdts G. (2013). Marine fungi may benefit from ocean acidification. *Aquatic Microbial Ecology, 69*: 59–67.

Krauss, K.W., Cormier, N., Osland M.J., Kirwan M.L., Stagg C.L., Nestlerode J.A., Russell M.J., From A.S., Spivak A.C., Dantin D.D., Harvey J.E., Almario A.E. (2017). Created mangrove wetlands store belowground carbon and surface elevation change enables them to adjust to sea-level rise. *Science Reports, 7*: 1030.

Krauss K.W., Cahoon D.R., Allen J.A., et al. (2010). Surface elevation change and susceptibility of different mangrove zones to sea-level rise on Pacific high islands of Micronesia. *Ecosystems, 13*: 129–143. doi: 10.1007/s10021-009-9307-8.

Krauss K.W., Duberstein J.A., Doyle T.W., Conner W.H., Day R.H., Inabinette L.W., Whitbeck J.L. (2009). Site condition, structure, and growth of bald cypress along tidal/non-tidal salinity gradients. *Wetlands, 29*(2): 505–519.

Krauss K.W., Holm G.O., Perez B.C., McWhorter D.E., Cormier N., Moss R.F., Johnson D.J., Neubauer S.C., Raynie R.C. (2016). Component greenhouse gas fluxes and radiative balance from two deltaic marshes in Louisiana: pairing chamber techniques and eddy covariance. *Journal of Geophysical Research: Biogeosciences, 121*: 1503–1521.

Krauss K.W., McKee K.L., Lovelock C.E., Cahoon D.R., Saintilan N., Reef R., Chen L. (2014). How mangrove forests adjust to rising sea level. *New Phytologist, 202*(1): 19–34.

Krauss K.W., Noe G.B., Duberstein J.A., Conner W.H., Stagg C.L., Cormier N., Jones M.C., Bernhardt C.E., Graeme Lockaby B., From A.S., Doyle T.W. (2018). The role of the upper tidal estuary in wetland blue carbon storage and flux. *Global Biogeochemical Cycles, 32*(5): pp.817–839.

Krauss K.W., Whitbeck J.L. (2012). Soil greenhouse gas fluxes during wetland forest retreat along the lower Savannah River, Georgia (USA). *Wetlands, 32*: 73–81.

Kreuzwieser J., Buchholz J., Rennenberg H. (2003). Emissions of methane and nitrous oxide by Australian mangrove ecosystems. *Plant Biology, 5*: 423–431.

Kristensen E., Bouillon S., Dittmar T., Marchand C. (2008). Organic carbon dynamics in mangrove ecosystems: a review. *Aquatic Botany, 89*(2): 201–219.

Kroeger K.D. www.nacarbon.org/meeting_ab_presentations/2017/2017_Mar30_AM_Kroeger_177.pdf.

Kroeger K.D., Crooks S., Moseman-Valtierra S., Tang J. (2017). Restoring tides to reduce methane emissions as a new and potent Blue Carbon intervention. *Scientific Reports, 7*: 11914. doi: 10.1038/s41598-017-12138-4.

Krone R.B. (1962). *Flume Studies of the Transport of Sediment in Estuarial Shoaling Processes: Final Report*, Hydraulic Engineering Laboratory. University of California: Berkeley.

Krone R.B. (1987). A method for simulating historic marsh elevations. In: *Coastal Sediments '87*, Krause N.C. (ed.), American Society of Civil Engineers: New York, pp. 316–323.

Krone R.B., Hu G. (2001). Restoration of subsided sites and calculation of historic marsh elevations. *Journal of Coastal Research, 27*: 162–169.

Ku T.C.W., Walter L.M., Coleman M.L., Blake R.E., Martini A.M. (1999). Coupling between sulfur cycling and syndepositional carbonate dissolution: evidence from oxygen and sulfur isotope composition of pore water sulfate, South Florida Platform, U.S.A. *Geochimica et Cosmochimica Acta, 63*: 2529–2546.

Kulawardhana R.W., Feagin R.A., Popescu S.C., Boutton T.W., Yeager K.M., Bianchi T.S. (2015). The role of elevation, relative sea level history, and vegetation transition in determining carbon distribution in *Spartina alterniflora* dominated salt marshes. *Estuarine, Coastal and Shelf Science, 154*: 48–57.

Kulawardhana R.W., Popescu S.C., Feagin R.A. (2014). Fusion of lidar and multispectral data to quantify salt marsh carbon stocks. *Remote Sensing of the Environment, 154*: 345–357.

La Viña, A., de Leon A., Barrer R. (2016). History and future of REDD+ in the UNFCCC: issues and challenges. In: *Research Handbook on REDD+ and International Law*, Voigt C. (ed.), Elgaronline.

Laanbroek H.J. (2010). Methane emission from natural wetlands: interplay between emergent macrophytes and soil microbial processes. *A Mini-Review. Annals of Botany, 105*: 141–153.

Laffoley D.d'A., Grimsditch G. (eds) (2009). *The Management of Natural Coastal Carbon Sinks*, IUCN: Gland, Switzerland.

Lal R. (2016). Beyond COP 21: potential and challenges of the "4 per thousand" initiative. *Journal of Soil and Water Conservation, 71*(1): 20A–25A.

Lal R., Lorenz K., Hüttl R.F., Scheider D.U., von Braun J. (eds) (2012). *Recarbonizing the Biosphere: Ecosystems and the Global Carbon Cycle*. IASS Potsdam and Springer: London.

Land L.S. (1970). Carbonate mud: production by epibiont growth on *Thalassia testudinum*. *Journal of Sedimentary Petrology, 40*: 1361–1363.

Lane R.R., Mack S.K., Day J.W., et al. (2016). Fate of Soil Organic Carbon During Wetland Loss. *Wetlands, 36*: 1167–1181. doi: 10.1007/s13157-016-0834-8.

Lang'at J.K.S., Kairo J.G. (2013). *Conservation and Management of Mangrove Forests in Kenya*, Kenya Marine and Fisheries Research Institute: Mombasa, Kenya.

Lang'at J.K.S., Kairo J.G., Mencuccini M., Bouillon S., Skov M.W., Waldron S., Huxham M. (2014). Rapid losses of surface elevation following tree girdling and cutting in tropical mangroves. *PLoS One, 9*(9): 1–8. doi: 10.1371/journal.pone.0107868.

Langley J.A., Megonigal J.P. (2010). Ecosystem response to elevated CO_2 levels limited by nitrogen-induced plant species shift. *Nature, 466*: 96–99.

Larkin D.J., Lishawa S.C., Tuchman N.C. (2012). Appropriation of nitrogen by the invasive cattail *Typha × glauca*. *Aquatic Botany, 100*: 62–66.

Lau W.W.Y. (2013). Beyond carbon: conceptualizing payments for ecosystem services in blue forests on carbon and other marine and coastal ecosystem services. *Ocean and Coastal Management, 83*: 5–14.

Lavery P.S., Mateo M.-Á., Serrano O., Rozaimi M. (2013). Variability in the carbon storage of seagrass habitats and its implications for global estimates of blue carbon ecosystem service. *PLoS One, 8*(9): e73748.

Le Quéré, C., Andrew R.M., Canadell J.G., Sitch S., Korsbakken J.I., Peters G.P., Manning A.C., Boden T.A., Tans P.P., Houghton R.A., Keeling R.F., Alin S., Andrews O.D., Anthoni P., Barbero L., Bopp L., Chevallier F., Chini L.P., Ciais P., Currie K., Delire C., Doney S.C., Friedlingstein P., Gkritzalis T., Harris I., Hauck J., Haverd V., Hoppema M., Klein Goldewijk K., Jain A.K., Kato E., Körtzinger A. Landschützer P., Lefèvre N., Lenton A., Lienert S., Lombardozzi D., Melton J.R., Metzl N., Millero F., Monteiro P.M.S., Munro D.R., Nabel J.E.M.S., Nakaoka S.-I., O'Brien K., Olsen A., Omar A.M., Ono T., Ierrot D., Poulter B., Rödenbeck C., Salisbury J., Schuster U., Schwinger J., Séférian R., Skjelvan I., Stocker B.D., Sutton A.J., Takahashi T., Tian H., Tilbrook B., van der Laan-Luijkx I.T., van der Werf G.R., Viovy N., Walker A.P., Wiltshire A.J., Zaehle S. (2016). Global carbon budget. *Earth System Science Data, 8*: 605–649.

Lefebvre A., Thompson C.E.L., Collins K.J., Amos C.L. (2009). Use of a high-resolution profiling sonar and a towed video camera to map a *Zostera marina* bed, Solent, UK. *Estuarine Coastal and Shelf Science, 82*: 323–334.

Leonard S. (2015). Analysis. Paris Agreement: Not perfect, but the best we could get. CIFOR Blog. http://blog.cifor.org/38995/paris-agreement-not-perfect-but-the-best-we-could-get?fnl=en&utm_source=January+2016&utm_campaign=NEWS+UPDATE+English+v2&utm_medium=email.

Leonardi N., Ganju N.K., Fagherazzi S. (2016). A linear relationship between wave power and erosion determines salt-marsh resilience to violent storms and hurricanes. *Proceedings of the National Academy of Sciences of the United States of America, 113*(1): 64–68.

Leorri E., Mulligan R., Mallinson D., Cearreta A. (2011a). Sea-level rise and local tidal range changes in coastal embayments: an added complexity in developing reliable sea-level index points. *Journal of Integrated Coastal Zone Management, 11*: 307–314.

Leorri E., Woodroffe S., Mitra S. (2011b). Biomarkers from mangrove sediments (Seychelles, Indian Ocean): assessing their ability to reconstruct former sea levels. *11th International Estuarine Biogeochemistry Symposium*, 22.

LES (Lewis Environmental Services, Inc.), and CE (Coastal Environmental, Inc.) (1996). *Setting Priorities for Tampa Bay Habitat Protection and Restoration: Restoring the Balance.* TBEP Technical Report #09-95: St. Petersburg, FL. www.tbeptech.org/TBEP_TECH_PUBS/1995/TBEP_09-95Restoring-Balance.pdf.

Lewis E., Wallace D.W.R. (1998). Program Developed for CO_2 System Calculations. ORNL/CDIAC-105. *Carbon Dioxide Information Analysis Center*, Oak Ridge National Laboratory, U.S. Department of Energy: Oak Ridge, Tennessee.

Lewis R.R. (2005). Ecological engineering for successful management and restoration of mangrove forests. *Ecological engineering, 24*(4): 403–418.

Lewis R.R., III, Brown B. (2014). *Ecological Mangrove Rehabilitation: A Field Manual for Practitioners*, Mangrove Action Project, Canadian International Development Agency, OXFAM.

Lewis R.R., Milbrandt E.C., Brown B., Krauss K.W., Rovai A.S., Beever J.W., Flynn L.L. (2016). Stress in mangrove forests: early detection and preemptive rehabilitation are essential for future successful worldwide mangrove forest management. *Marine pollution bulletin, 109*(2): 764–771.

Lewis S.L., Maslin M.A. (2015). Defining the anthropocene. *Nature, 519*: 171–180.

Ley General de Cambio Climatico. www.diputados.gob.mx/LeyesBiblio/pdf/LGCC_130515.pdf.

Li C., Frolking S., Frolking T.A. (1992). A model of nitrous oxide evolution from soil driven by rainfall events: 1. Model structure and sensitivity. *Journal of Geophysical Research, 97*: 9759–9776.

Li J.W., Wang G.S., Allison S.D., Mayes M.A., Luo Y.Q. (2014). Soil carbon sensitivity to temperature and carbon use efficiency compared across microbial-ecosystem models of varying complexity. *Biogeochemistry, 119*: 67–84.

Light T., Grosholz E.D., Moyle P.B. (2005). *Delta Ecological Survey (Phase I): Nonindigenous Aquatic Species in the Sacramento–San Joaquin Delta, a Literature Review. Final Report for Agreement DCN 1113322J011*, US Fish and Wildlife Service: Stockton, CA.

Lindeman R. (1942). The trophic-dynamic aspect of ecology. *Ecology, 23*: 399–418.

Lipkin Y., Beer S., Zakai D. (2003). The seagrasses of the Eastern Mediterranean and the Red Sea. In: Green E.P., Short F. T. (eds), *World Atlas of Seagrasses*, University of California Press: Berkeley, Chapter 5, p. 298.

Locatelli T., et al. (2014). Turning the tide: how blue carbon and Payments for Ecosystem Services (PES) might help save mangrove forests. *Ambio, 43*(8): 981–995.

Loebl M., van Beusekom J.E.E., Reise K. (2006). Is spread of the neophyte *Spartina anglica* recently enhanced by increasing temperatures? *Aquatic Ecology, 40*: 315–324.

Lohnis F. (1926). Nitrogen availability of green manures. *Soil Science, 22*: 253–290.

Long J., Napton D., Giri C., Graesser J. (2013b). A Mapping and Monitoring Assessment of the Philippines' Mangrove *Forests* from 1990 to 2010. *Journal of Coastal Research*, 260–271.

Lopez C.B., Cloern J.E., Schraga T.S., Little A.J., Lucas L.V., Thompson J.K., Burau J.R. (2006). Ecological values of shallow-water habitats: implications for the restoration of disturbed ecosystems. *Ecosystems, 9*: 422–440.

López N.I., Duarte C.M., Vallespinós F., Romero J., Alcoverro T. (1995). Bacterial activity in NW Mediterranean seagrass (Posidonia oceanica) sediments. *Journal of Experimental Marine Biology and Ecology, 187*(1): 39–49.

López-Medellín, X., et al. (2011). Oceanographic anomalies and sea-level rise drive mangroves inland in the Pacific coast of Mexico. *Journal of Vegetation Science, 22*(1): 143–151.

López-Merino L., Colás-Ruiz N.R., Adame M.F., Serrano O., Martínez Cortizas A., Mateo M.A. (2017). A six thousand-year record of climate and land-use change from Mediterranean seagrass mats. *Journal of Ecology.*

Lotze H.K., Lenihan H.S., Bourque B.J., Bradbury R.H., Cooke R.G., et al. (2006). Depletion, degradation, and recovery potential of estuaries and coastal seas. *Science, 312*: 1806–1809.

Lovelock C.E., Atwood T., Baldock J., Duarte C.M., Hickey S., Lavery P.S., Masque P., Macreadie P.I., Ricart A.M., Serrano O., Steven A. (2017). Assessing the risk of carbon dioxide emissions from blue carbon ecosystems. *Frontiers in Ecology and the Environment, 15*(5): 257–265.

Lovelock C.E., Cahoon D.R., Friess D.A., Guntenspergen G.R., Krauss K.W., Reef R., Rogers L., Saunders M., Sidik F., Swales A., Saintilan N., Thuyen L.X., Tran T. (2015). Vulnerability of Indo pacific mangrove forests to sea level rise. *Nature, 526*: 559–563. doi: 10.1038/nature15538.

Lovelock C.E., Feller I.C., McKee K.L., Thompson R. (2005). Variation in Mangrove Forest Structure and Sediment Characteristics in Bocas del Toro, Panama. *Caribbean Journal of Science*, *41*: 456–464.

Lovelock C.E., Reuss R.W., Feller I.C. (2011). CO_2 efflux from cleared mangrove peat. *PLoS One*, *6*(6): e21279.

Lu W.Z., Xiao J.F., Liu F., Zhang Y., Liu C.A., Lin G.H. (2017). Contrasting ecosystem CO_2 fluxes of inland and coastal wetlands: a meta-analysis of eddy covariance data. *Global Change Biology*, *23*: 118–1198.

Lucas L.V., Thompson J.K. (2012). Changing restoration rules: exotic bivalves interact with residence time and depth to control phytoplankton productivity. *Ecosphere*, *3*(12): 117. www.esajournals.org/doi/pdf/10.1890/ES12-00251.1.

Lynch J.C., et al. (1989). Recent Accretion in Mangrove Ecosystems Based on 137 Cs and 210 Pb S. *Estuaries*, *12*(4): 284–299.

Lynch J.C., Hensel P., Cahoon D.R. (2015). The surface elevation table and marker horizon technique: A protocol for monitoring wetland elevation dynamics. In: *Natural Resource Report NPS/NCBN/NRR—2015/1078*, National Park Service: Fort Collins, Colorado. Available at: www.nature.nps.gov/publications/nrpm/.

Lyons M., Roelfsema C., Kovacs E., Samper-Villarreal J., Saunders M., Maxwell P., Phinn S. (2015). Rapid monitoring of seagrass biomass using a simple linear modelling approach, in the field and from space. *Marine Ecology Progress Series*, *530*: 1–14.

MacKenzie L.J., Finkbeiner M.A., Kirkman H. (2001). Methods for mapping seagrass distribution. In: *Global Seagrass Research Methods*, Short F.T., Coles R.G. (eds), Elsevier Science: Amsterdam, pp. 101–121.

MacKenzie R.A., Foulk P.B., Klump J.V., Weckerly K., Purbopuspito J., Murdiyarso D., Donato D.C., Nam V.N. (2016). Sedimentation and belowground carbon accumulation rates in mangrove forests that differ in diversity and land use: a tale of two mangroves. *Wetlands Ecology and Management*. doi: 10.1007/s11273-016-9481-3.

MacPherson T.A., Cahoon L.B., Mallin M.A. (2007). Water column oxygen demand and sediment oxygen flux: patterns of oxygen depletion in tidal creeks. *Hydrobiologia*, *586*(1): 235–248.

Macreadie P.I., Nielsen D.A., Kelleway J.J., Atwood T.B., Seymour J.R., Petrou K., Connolly R.M., Tomson A.C.G., Trevathan-Tackett S.M., Ralph P.J. (2017). Can we manage coastal ecosystems to sequester more blue carbon? *Frontiers in Ecology and the Environment*, *15*(4): 206–213.

Macreadie P.I., Trevathan-Tackett S.M., Skilbeck C.G., Sanderman J., Curlevski N., Jacobsen G., Seymour J.R. (2015). Losses and recovery of organic carbon from a seagrass ecosystem following disturbance. *Proceedings of the Royal Society B*, *282*: 20151537.

Maeda K., Spor A., Edel-Hermann V., Heraud C., Breuil M.-C., Bizouard F., Toyoda S., Yoshida N., Steinberg C., Philippot L. (2015). N_2O production, a widespread trait in fungi. *Scientific Reports*, *5*: 9697.

Maher D.T., Santos I.R., Golsby-Smith L., Gleeson J., Eyre B.D. (2013). Groundwater-derived dissolved inorganic and organic carbon exports from a mangrove tidal creek: the missing mangrove carbon sink? *Limnology and Oceanography*, *58*(2): 475–488.

Malik A. (2007). Environmental challenge *vis à vis* opportunity: the case of water hyacinth. *Environment International*, *33*: 122–138.

Mallinson D.J., Culver S.J., Corbett D.R., Parham P.R., Shazili N.A.R., Rosnan Y. (2014). Holocene coastal response to monsoons and sea-level changes in north-east peninsular Malaysia. *Journal of Asian Earth Science*, *91*: 194–205.

Manzello D.P., Enochs I.C., Melo N., Gledhill D.K., Johns E.M. (2012). Ocean acidification refugia of the Florida Reef Tract. *PLoS One*, *7*(7): e41715. doi: 10.1371/journal.pone.0041715.

Manzoni S., Taylor P., Richter A., Porporato A., Ågren G.I. (2012). Environmental and stoichiometric controls on microbial carbon-use efficiency in soils. *New Phytologist*, *196*: 79–91.

Marbà, N., Arias-Ortiz A., Masqué, P., Kendrick G.A., Mazarrasa I., Bastyan G.R., Duarte C.M. (2015). Impact of seagrass loss and subsequent revegetation on carbon sequestration and stocks. *Journal of Ecology*, *103*(2): 296–302.

Marbà, N., Holmer M., Gacia E., Barrón C. (2006). Seagrass beds and coastal biogeochemistry. In: *Seagrasses: Biology, Ecology and Conservation*. Larkum A.W.D., Orth R.J., Duarte C.M. (eds), Springer: Dordrecht, pp. 135–157, 159–192.

Marchio D.A., Savarese M. Jr., Bovard B., Mistch W.J. (2016). Carbon sequestration and sedimentation in mangrove swamps influenced by hydro-geomorphic conditions and urbanization in Southwest Florida. *Forest*, *7*(116). doi: 10.3390/f7060116.

Margono B.A., Potapov P.V., Turubanova S., Stolle F., Hansen M.C. (2014). Primary forest cover loss in Indonesia over 2000–2012. *Nature Climate Change*, *4*: 730–35.

Marín-Spiotta E., Gruley K.E., Crawford J., Atkinson E.E., Miesel J.R., Greene S., Spencer R.G.M. (2014). Paradigm shifts in soil organic matter research affect interpretations of aquatic carbon cycling: transcending disciplinary and ecosystem boundaries. *Biogeochemistry*, *117*(2–3): 279–297.

Mariotti G., Fagherazzi S. (2013). Critical width of tidal flats triggers marsh collapse in the absence of sea-level rise. *Proceedings of the National Academy of Sciences of the United States of America*. doi: 10.1073/pnas.1219600110.

Mariotti G., Fagherazzi S., Wiberg P., McGlathery K., Carniello L., Defina A. (2010). Influence of storm surges and sea level on shallow tidal basin erosive processes. *JGR*, *115*: C11012. doi: 10.1029/2009JC005892.

Martin A., Landis E., Bryson C., Lynaugh S., Mongeau A., Lutz S. (2016). *Blue Carbon – Nationally Determined Contributions Inventory. Appendix to: Coastal Blue Carbon Ecosystems. Opportunities for Nationally Determined Contributions*, GRIDArendal: Norway.

Martinez C.J., Maleski J.J., Miller M.F. (2012). Trends in precipitation and temperature in Florida, USA. *Journal of Hydrology*, *452–453*: 259-281. doi: 10.1016/j.jhydrol.2012.05.066.

Mateo M.A., Cebrián J., Dunton K., Mutchler T. (2006). Carbon flux in seagrass ecosystems. Seagrasses: biology, ecology and conservation. In: *Seagrasses: Biology, Ecology and Conservation*, Larkum A.W.D., Orth R.J., Duarte C.M. (eds), Springer: Dordrecht, Netherlands, pp. 159–192.

Maxwell P.S., Eklöf J.S., van Katwijk M.M., O'Brien K.R., de la Torre-Castro M., Boström C., Bouma T.J., Krause-Jensen D., Unsworth R.K.F., van Tussenbroek B.I., van der Heide T. (2016). The fundamental role of ecological feedback mechanisms for the adaptive management of seagrass ecosystems – a review. *Biological Reviews*, *2016*: 1–18.

Mazarrasa I., Marbà N., Lovelock C.E., et al. (2015). Seagrass meadows as a globally significant carbonate reservoir. *Biogeosciences*, *12*: 4993–5003.

McClure A., Liu X., Hines E., Ferner M.C. (2016). Evaluation of Error Reduction Techniques on a LIDAR-Derived Salt Marsh Digital Elevation Model. *Journal of Coastal Research*, *32*: 424–433.

McDonald S., Pringle J.M., Prenzler P.D., Bishop A.G., Robards K. (2007). Bioavailability of dissolved organic carbon and fulvic acid from an Australian floodplain river and billabong. *Marine and Freshwater Research*, *58*(2): 222–231.

McEwin A.R.M. (2014). *Organic Shrimp Certification and Carbon Financing: An Assessment for the Mangroves and Markets Project in Ca Mau Province, Vietnam*, SNV World: Vietnam.

McGlathery K., Shoch D. (2015). Methodology for Tidal Wetland and Seagrass Restoration, *VM0033 Version 1.0 Verified Carbon Standard*, University of Virginia: Washington, D.C.

McIvor A., Spencer T., Möller I., Spalding M. (2013). *The response of mangrove soil surface elevation to sea level rise*. The Nature Conservancy and Wetlands International, Natural Coastal Protection Series: Report, 3.

McKee K.L. (2011). Biophysical controls on accretion and elevation change in Caribbean mangrove ecosystems. *Estuarine, Coastal and Shelf Science*, *91*(4): 475–483.

McKee K.L., Cahoon D.R., Feller I.C. (2007). Caribbean mangroves adjust to rising sea level through biotic controls on change in soil elevation. *Global Ecology and Biogeography*, *16*: 545–556.

McKee K.L., Feller I.C., Popp M., Wanek W. (2002). Mangrove isotopic ($\delta^{15}N$ and $\delta^{13}C$) fractionation across a nitrogen vs. phosphorus limitation gradient. *Ecology*, *83*: 1065–1075.

McKee K.L., Mendelssohn I.A., Materne M.D. (2004). Acute salt marsh dieback in the Mississippi River deltaic plain: a drought-induced phenomenon? *Global Ecology and Biogeography*, *13*: 65–73.

McKee K.L., Patrick W.H. Jr. (1988). The relationship of smooth cordgrass (*Spartina alterniflora*) to tidal datums: a review. *Estuaries*, *11*: 143–151.

McKinnell S.M., Crawford W.R. (2007). The 18.6-year lunar nodal cycle and surface temperature variability in the northeast Pacific. *Journal of Geophysical Research: Oceans*, *112*(C2).

Mcleod E.W., Chmura G.L., Bouillon S., Salm R., Björk M., Duarte C.M., Lovelock C.E., Schlesinger W.H., Silliman B.R. (2011). A blueprint for blue carbon: toward an improved understanding of the role of vegetated coastal habitats in sequestering CO_2. *Frontiers in Ecology and the Environment*, *9*(10): 552–560.

Mcowen C., Weatherdon L.V., Bochove J., Sullivan E., Blyth S., Zockler C., Stanwell-Smith D., Kingston N., Martin C.S., Spalding M., Fletcher S. (2017). A global map of saltmarshes. *Biodiversity Data Journal*, *5*: e11764. doi: 10.3897/BDJ.5.e11764. http://data.unep-wcmc.org/datasets/43 (v.4).

Megonigal J.P. (1996). *Methane Production and Oxidation in a Future Climate*, Duke University: Durham, NC, p. 151.

Megonigal J.P., Conner W.H., Kroeger S., Sharitz R.R. (1997). Aboveground production in southeastern floodplain forests: a test of the subsidy–stress hypothesis. *Ecology*, *78*: 370–384.

Megonigal J.P., Hines M.E., Visscher P.T. (2004). Anaerobic metabolism: linkages to trace gases and aerobic processes. In: *Biogeochemistry*, Schlesinger W.H. (ed.), Elsevier-Pergamon, Oxford, UK, pp. 317–424.

Megonigal J.P., Mines M.E., Visscher P.T. (2005). Linkages to trace gases and aerobic processes. *Biogeochemistry*, *8*: 350–362.

Megonigal J.P., Rabenhorst M. (2013). Reduction-oxidation potential and oxygen. In: DeLaune R.D., Reddy K.R., Richardson C.J., Megonigal J.P. (eds), *Methods in Biogeochemistry of Wetlands. Soil Science Society of America*, Madison, WI, pp. 37–51

Megonigal J.P., Schlesinger W.H. (1997). Enhanced CH_4 emissions from a wetland soil exposed to elevated CO_2. *Biogeochemistry*, *37*: 77–88.

Megonigal J.P., Schlesinger W.H. (2002). Methane-limited methanotrophy in tidal freshwater swamps. *Global Biogeochemical Cycles*, *16*: 1088.

Melton J.R., Wania R., Hodson E.L., Poulter B., Ringeval B., Spahni R., Bohn T., Avis C.A., Beerling D.J., Chen G., Eliseev A.V., Denisov S.N., Hopcroft P.O., Lettenmaier D.P., Riley W.J., Singarayer J.S., Subin Z.M., Tian H., Zürcher S., Brovkin V., van Bodegom P.M., Kleinen T., Yu Z.C., Kaplan J.O. (2013). Present state of global wetland extent and wetland methane modelling: conclusions from a model intercomparison project (WETCHIMP). *Biogeosciences*, *10*: 753–758.

Mendelssohn I.A., Morris J.T. (2000). Eco-physiological controls on the productivity of *Spartina alterniflora* Loisel. In: *Concepts and Controversies in Tidal Marsh Ecology*, Weinstein M.P., Kreeger D.A. (eds), Kluwer Academic Publishers: Boston, MA, pp. 59–80.

Mendoza-Morales A., González-Sansón, G., Aguilar-Betancourt C. (2015). Producción espacial y temporal de hojarasca del manglar en la laguna Barra de Navidad, Jalisco, México. *Revista de Biología Tropical*, *64*(1): 275–289.

Meng L., Hess P.G.M., Mahowald N.M., Yavitt J.B., Riley W.J., Subin Z.M., Lawrence D.M., Swensen S.C., Jauhiainen J., Fuka D.R. (2012). Sensitivity of wetland methane emissions to model assumptions: application and model testing against site observations. *Biogeosciences*, *9*: 2793–2819.

Merbold L., Eugster W., Stieger J., Zahniser M., Nelson D., Buchmann N. (2014). Greenhouse gas budget (CO_2, CH_4 and N_2O) of intensively managed grassland following restoration. *Global Change Biology*, *20*: 1913–1928.

Methodology V. (2015). VM0033: Methodology for tidal wetland and sea grass restoration.

Meyer-Arendt K.J. (1993). Recreational urbanization and shoreline modification along the North Coast of Yucatan, In: *Coasts at the Millenium*, Wilson 1980: Mexico, pp. 20–25.

Mikoko Plan P., Vivo, www.planvivo.org/docs/Mikoko-Pamoja-Annual-Report-2014-2015-web.pdf.

Milan C.S., Swenson E.M., Turner R.E., Lee J.M. (1995). Assessment of the ^{137}Cs method for estimating sediment accumulation rates: louisiana salt marshes. *Journal of Coastal Research*, *11*: 296–307.

Millennium Ecosystem Assessment (2005). *Ecosystems and Human Well-Being: Wetlands and Water Synthesis*, World Resources Institute: Washington, D.C.

Miller K.G., Kominz M.A., Browning J.V., Wright J.D., Mountain G.S., Katz M.E., Sugarman P.J., Cramer B.S., Christie-Blick N., Pekar S.F. (2005). The Phanerozoic record of global sea-level change. *Science*, *310*: 1293–1298.

Miller R.L., Fram M., Fujii R., Wheeler G. (2008). Subsidence reversal in a re-established wetland in the Sacramento-San Joaquin Delta, California, USA. *San Francisco Estuary Water Sci*, *6*(3). http://escholarship.org/uc/item/5j76502x.

Miller R.L., Fujii R. (2010). Plant community, primary productivity, and environmental conditions following wetland re-establishment in the Sacramento-San Joaquin Delta, California. *Wetland Ecology and Management*, *18*: 1–6.

Miller R.L., Fujii R. (2011). Re-establishing marshes can turn a current carbon source into a carbon sink in the Sacramento-San Joaquin Delta of California, USA. In: *River Deltas: Types, Structure, and Ecology*, Contreras, D.A. (ed.), Nova Science Publishers, Inc.: New York, pp. 1–34

Mills M., Leon J.X., Saunders M.I., Bell J., Liu Y., O'Mara J., Lovelock C.E., Tulloch V.J. (2016). Reconciling development and conservation under coastal squeeze from rising sea-level. *Conservation Letters*, *9*(5): 361–368.

Milne G., Gehrels W.R., Hughes C.W., Tamisiea M.E. (2009). Identifying the causes of sea-level change. *Nature Geosciences*, 2: 471–478.

Milne G., Shennan I. (2007). Sea level studies: isostasy. *Encyclopedia of Quaternary Science,* Amsterdam: Elsevier, pp. 3043–3051.

Ministry of Agriculture [MoA] (2016). *Indonesian Estate Crops Statistics – Oil Palm 2014–2016*, Directorate General of Estate Crops, Ministry of Agriculture: Jakarta, Indonesia.

Ministry of Forestry of the Republic of Indonesia [MoF] (2013). *Digital Land Cover and Land Use Map of Indonesia for Years 2000, 2003, 2006 and 2009*, Spatial Planning Agency, Ministry of Forestry of the Republic of Indonesia. Jakarta.

Ministry of Marine Affairs and Fishery of the Republic of Indonesia [MMAF] (2014). *Export of Fishery Products*, Ministry of Marine Affairs and Fishery of the Republic of Indonesia: Jakarta. http://statistik.kkp.go.id/index.php/statistik/c/430/0/0/0/0/Volume-dan-Nilai-Ekspor-Menurut-Komoditi-per-Provinsi-HS-2012 (accessed June 22, 2017).

Ministry of Marine Affairs and Fishery of the Republic of Indonesia [MMAF] (2013). *Shrimp Farms: Area and Production During 2006–2012*. Center for Statistical Data and Information, Ministry of Marine Affairs and Fishery of the Republic of Indonesia: Jakarta.

Mirera D.O., Moksnes P.O. (2014). Comparative performance of wild juvenile mud crab (Scylla serrata) in different culture systems in East Africa: effect of shelter, crab size and stocking density. *Aquaculture International*, 23(1): 155–173. doi: 10.1007/s10499-014-9805-3.

Mirera O.D., Kairo J.G., Kimani E., N., Waweru F.K. (2010). A comparison in fish assemblages in mangrove forests and on intertidal flats at Ungwana bay, Kenya. *African Journal of Aqautic Science*, 35(2): 165–171.

Misbari S., Hashim M. (2016A). Change detection of submerged seagrass biomass in shallow coastal water. *Remote Sensing*, 8(200): 1–29.

Misbari S., Hashim M. (2016B). *Light Penetration Ability Assessment of Satellite Band for Seagrass Detection Using Landsat 8 OLI Satellite Data*, ICCSA 2016, Springer: Cham. doi: 10.1007/978-3-319–42111-7_21.

Mishra D.R., Cho H.J., Ghosh S., Fox A., Downs C., Merani P.B.T., Kirui P., Jackson N., Mishra S. (2012). Post-spill state of the marsh: remote estimation of the ecological impact of the Gulf of Mexico oil spill on Louisiana Salt Marshes. *Remote Sensing of Environment*, 118: 176–185.

Mishra D.R., Ghosh S. (2015). Using moderate-resolution satellite sensors for monitoring the biophysical parameters and phenology of tidal marshes. In: Tiner R.W., Lang M.W., Klemas V.V. (eds), *Remote Sensing of Wetlands: Applications and Advances*, CRC Press: Boca Raton, Chapter 14, pp. 283–314

Mitchell D.H., Ian L.T., Michael A.K., Jason H.M., Peter J.M., Kristen D.S., Matthew S.P., Joshua A.S., David J.H., Andrew D.S. (2017). Extreme coastal erosion enhanced by anomalous extratropical storm wave direction. *Scientific Reports*, 7(1). doi: 10.1038/s41598-017-05792-1.

Mitsch W.J., Bernal B., Nahlik A.M., Mander U., Zhang L., Anderson C.J., Jørgensen S.E., Brix H. (2013). Wetlands, carbon, and climate change. *Landscape Ecology*, 28: 583–597.

Mitsch W.J., Gosselink J.G. (1986). *Wetlands*, Van Nostrand Reinhold: New York, USA.

Mizuno K., Asada A., Matsumoto Y., Sugimoto K., Fujii T., Yamamuro M., Fortes M.D., Sarceda M., Jimenez L.A. (2017). A simple and efficient method for making a high-resolution seagrass map and quantification of dugong feeding trail distribution: a field test at Mayo Bay, Philippines. *Ecological Informatics*, 38: 89–94.

Mo Y., Momen B., Kearney M.S. (2015). Quantifying moderate resolution remote sensing phenology of Louisiana coastal marshes. *Ecological Modelling*, 312: 191–199.

Mogensen L.A., Rogers K. (2018). Validation and Comparison of a Model of the Effect of Sea-Level Rise on Coastal Wetlands. *Scientific Reports*, 8(1): 1369.

Mojica K.D.A., Huisman J., Wilhelm S.W., Brussaard C.P.D. (2016). Latitudinal variation in virus-induced mortality of phytoplankton across the North Atlantic Ocean. *The ISME Journal*, 10: 500–513.

Möller I., Kudella M., Rupprecht F., Spencer T., Paul M., van Wesenbeeck B.K., Wolters G., Jensen K., Bouma T.J., Miranda-Lange M., Schimmels S. (2014). Wave attenuation over coastal salt marshes under storm surge conditions. *Nature Geoscience*, 7: 727–731.

Moore K.A., Orth R.J., Wilcox D.J. (2009). Assessment of the abundance of submersed aquatic vegetation (SAV) communities in the Chesapeake Bay and its use in SAV management. In: *Remote Sensing and Geospatial Technologies for Coastal Ecosystem Assessment and Management, Lecture Notes in Geoinformation and Cartography*, Yang X. (ed.), Springer-Verlag: Berlin, pp. 233–257.

Moran M.A., Hodson R.E. (1990). Contributions of degrading *Spartina alterniflora* lignocellulose to the dissolved organic carbon pool of a salt marsh. Marine ecology progress series. *Oldendorf, 62*(1): 161–168.

Moran M.A., Wicks R.J., Hodson R.E. (1991). Export of dissolved organic matter from a mangrove swamp ecosystem: evidence from natural fluorescence, dissolved lignin phenols, and bacterial secondary production. *Marine Ecology Progress Series, 76*: 175–184.

Morelli G., Gasparon M., Fierro D., Hu W.P., Zawadzki A. (2012). Historical trends in trace metal and sediment accumulation in intertidal sediments of Moreton Bay, southeast Queensland, Australia. *Chemical Geology, 300*: 152–164.

Moriarty D.J.W., Iverson R.L., Pollard P.C. (1986). Exudation of organic carbon by the seagrass *Halodule wrightii* Aschers and its effect on bacterial growth in the sediment. *Journal of Experimental Marine Biology and Ecology, 96*: 115–126.

Morris J.T. (1984). Effects of oxygen and salinity on ammonium uptake by *Spartina alterniflora* Loisel. And *Spartina patens* (Aiton) Muhl. *Journal of Experimental Marine Biology and Ecology, 78*: 87–98.

Morris J.T. (2016). Marsh equilibrium theory. In: *4th International Conference on Invasive Spartiana, ICI-Spartina 2014*, University of Rennes press: Rennes, France, pp. 67–71.

Morris J.T., Barber D.C., Callaway J.C., Chambers R., Hagen S.C., Hopkinson C.S., Johnson B.J., Megonigal P., Neubauer S.C., Troxler T., Wigand C. (2016). Contributions of organic and inorganic matter to sediment volume and accretion in tidal wetlands at steady state. *Earth's Future, 4*: 110–121. doi: 10.1002/2015EF000334.

Morris J.T., Edwards J., Crooks S., Reyes E. (2012). Assessment of carbon sequestration potential in coastal wetlands. In: *Recarbonization of the Biosphere: Ecosystems and the Global Carbon Cycle*, Lal R., et al. (eds), Springer, pp. 517–531. doi: 10.1007/978–94–007–4159-1_24.

Morris J.T., Haskin B. (1990). A 5-yr record of aerial primary production and stand characteristics of *Spartina alterniflora. Ecology, 71*: 2209–2217.

Morris J.T., Kjerfve B., Dean J.M. (1990). Dependence of estuarine productivity on anomalies in mean sea level. *Limnology and Oceanography, 35*: 926–930.

Morris J.T., Sundareshwar P.V., Nietch C.T., Kjerfve B., Cahoon D.R. (2002). Responses of coastal wetlands to rising sea level *Ecology, 83*(10): 2869–2877.

Morris J.T., Sundberg K., Hopkinson C.S. (2013). Salt marsh primary production and its responses to relative sea level and nutrients in estuaries at Plum Island, Massachusetts, and North Inlet, South Carolina, USA. *Oceanography, 26*(3): 78–84. doi: 10.5670/ oceanog.2013.48.

Morris R.L., Konlechner T.M., Ghisalberti M., Swearer S.E. (2018). From grey to green: efficacy of eco-engineering solutions for nature-based coastal defence. *Global Change Biology*, 1–16. doi: 10.1111/gcb.14063.

Moseman-Valtierra S., Gonzalez R., Kroeger K.D., Tang J., Chao W.-C., Crusius J., Bratton J., Green A., Shelton J. (2011). Short-term nitrogen additions can shift a coastal wetland from a sink to a source of N_2O. *Atmospheric Environment, 45*: 4390–4397.

Moseman-Valtierra S., Tang J., Martin R., Egan K., Mora J., Morkeski K., Carey J.C., Kroeger K.D. (2016). Carbon dioxide fluxes reflect plant zonation and below ground biomass in a coastal marsh. *Ecosphere, 7*: e01560. doi: 10.1002/ecs2.1560.

Mount J., Twiss R. (2005). Subsidence, sea level rise, seismicity in the Sacramento-San Joaquin Delta. *San Francisco Estuary Water Sci, 3*(1). http://repositories.cdlib.org/jmie/sfews/vol3/iss1/art5.

Mount R.E., Bricher P.J. (2008). Estuarine, Coastal and Marine (ECM) National Habitat Mapping Project, ECM National Habitat Map Series User Guide Version 1 February 2008. Spatial Science Group, School of Geography and Environmental Studies, University of Tasmania. Report to the Department of Climate Change and the National Land and Water Resources Audit, Canberra, ACT. ECM National Habitat Map Series User Guide.

Moyer R.P., Radabaugh K.R., Powell C.E., Bociu I., Chappel A.R., Clark B.C., Crooks S., Emmett-Mattox S. (2016). Quantifying carbon stocks for natural and restored mangroves, salt marshes and salt barrens in Tampa Bay. In: Sheehan L., Crooks S. (eds). *Tampa Bay Blue Carbon Assessment: Summary of Findings*, TBEP Technical Report #07-16: St. Petersburg, FL, Appendix C, pp. 119–158. www.tbeptech.org/TBEP_TECH_PUBS/2016/TBEP_07_16_Tampa-Bay-Blue-Carbon-Assessment-Report-FINAL_June16-2016.pdf.

Moyle P.B., Lund J.R., Bennett W.A., Fleenor W.E. (2010). Habitat variability and complexity in the upper San Francisco Estuary. *San Francisco Estuary and Watershed Science, 8*(3). jmie_sfews_11019. http://escholarship.org/uc/item/0kf0d32x.

Moyle P.B., Manfree A.D., Fiedler P.L (eds) (2014). *Suisun Marsh: Ecological History and Possible Futures*, University of California Press: Berkeley and Los Angeles, California, USA.

Mozdzer T.J., Megonigal J.P. (2013). Increased methane emissions by an introduced *Phragmites australis* lineage under global change. *Wetlands*. doi: 10.1007/s13157-013-0417-x.

Mudd S.M. (2011). The life and death of salt marshes in response to anthropogenic disturbance of sediment supply. *Geology*, *39*(5): 511–512.

Mudd S.M., D'Alpaos A., Morris J.T. (2010). How does vegetation affect sedimentation on tidal marshes? Investigating particle capture and hydrodynamic controls on biologically mediated sedimentation. *Journal of Geophysical Research-Earth Surface*, *115*: F03029. doi: 10.1029/2009JF001566.

Mudd S.M., Fagherazzi S., Morris J.T., Furbish D. (2004). Flow, sedimentation, and biomass production on a vegetated salt marsh in South Carolina: toward a predictive model of marsh morphologic and ecologic evolution. In: *Ecogeomorphology of Tidal Marshes*, Fagherazzi S., Marani A., Blum L. (eds), AGU: Washington D.C., pp. 165–287. doi: 10.1029/CE059p0165.

Mudd S.M., Howell S.M., Morris J.T. (2009). Impact of dynamic feedbacks between sedimentation, sea-level rise, and biomass production on near-surface marsh stratigraphy and carbon accumulation. *Estuarine, Coastal and Shelf Science*, *82*: 377–389.

Mudge B. (1858). The salt marsh formations of Lynn. *Essex Institute Proceedings*, 2: 117–119.

Muehllehner N., Langdon C., Venti A., Kadko D. (2016). Dynamics of carbonate chemistry, production, and calcification of the Florida Reef Tract (2009–2010): evidence for seasonal dissolution. *Global Biogeochemical Cycles*, *30*. doi: 10.1002/2015GB005327.

Muehlstein L.K., Porter D., Short F.T. (1991). *Labyrinthula zosterae* sp. nov., the causative agent of wasting disease of eelgrass, *Zostera marina*. *Mycologia*, *83*(2): 180–191.

Mueller P., Hager R.N., Meschter J.E., Mozdzer T.H., Langley J.A., Jensen K., Megonigal J.P. (2016). Complex invader-ecosystem interactions and seasonality mediate the impact of non-native *Phragmites* on CH_4 emissions. *Biological Invasions*. doi: 10.1007/s10530-016-1093-6.

Mueller P., Jensen K., Megonigal P. (2015). Plants mediate soil organic matter decomposition in response to sea level rise. *Global Change Biology*, doi: 10.1111/gcb.13082.

Murdiyarso D., et al. (2019). The nexus between conservation and development in Indonesian mangroves. In: *A Blue Carbon Primer: The State of Coastal Wetland Carbon Science, Practice, and Policy*, Lisamarie W.-M., Stephen C., and Tiffany G.T. (eds), CRC Press: Boca Raton, London.

Murdiyarso D., Purbopuspito J., Kauffman J.B., Warren M.W., Sasmito S.D., Donato D.C., Manuri S., Krisnawati H., Taberima S., Kurnianto S. (2015). The potential of Indonesian mangrove forests for global change mitigation. *Nature Climate Change*, *5*: 8–11. doi: 10.1038/NCLIMATE2734.

Murdiyarso D., Sasmito S., Silanpaa M. (2017). Biomass recovery and sediment maintenance in a long-term logging rotation in mangrove forests. (In preparation).

Murray B., Vegh T. (2012). *Incorporating Blue Carbon as a Mitigation Action Under the United Nations Framework Convention on Climate Change*. Technical Issues to Address. Report NI R 12–05, Nicholas Institute for Environmental Policy Solutions, Duke University: Durham, NC.

Murray B.C., Pendleton L., Jenkins W.A., Sifleet S. (2011). *Green Payments for Blue Carbon: Economic Incentives for Protecting Threatened Coastal Habitats*. Report NI R 11-04, Nicholas Institute for Environmental Policy Solutions, Duke University: Durham.

Murray R.H., Erler D.V., Eyre B.D. (2015). Nitrous oxide fluxes in estuarine environments: response to global change. *Global Change Biology*, *21*: 3219–3245.

Mustin K., Sutherland W.J., Gill J.A. (2007). The complexity of predicting climate-induced ecological impacts. *Climate Research*, *35*: 165–175.

Mutanga O., Adam E., Cho M.A. (2012). High density biomass estimation for wetland vegetation using WorldView-2 imagery and random forest regression algorithm. *International Journal of Applied Earth Observation and Geoinformation*, *18*: 399–406.

Mutanga O., Skidmore A.K. (2004). Narrow band vegetation indices overcome the saturation problem in biomass estimation. *International Journal of Remote Sensing*, *25*: 3999–4014.

Myhre G., Shindell D., Bréon F.-M., Collins W., Fuglestvedt J., Huang J., Koch D., Lamarque J.-F., Lee D., Mendoza B., Nakajima T., Robock A., Stephens G., Takemura T., Zhang H. (2013). Anthropogenic and natural radiative forcing. In: *Climate Change 2013: The Physical Science Basis. Contribution of*

Working Group I to the Fifth Assessment Report of the Intergovernmental Panel on Climate Change, Stocker T.F., Qin D., Plattner G.-K., Tignor M., Allen S.K., Boschung J., Nauels A., Xia Y., Bex V., Midgley P.M. (eds), Cambridge University Press: Cambridge, New York, UK, USA.

Myint S.W., Giri C.P., Le W., Zhu Z.L., Gillette S.C. (2008). Identifying mangrove species and their surrounding land use and land cover classes using an object-oriented approach with a lacunarity spatial measure. *Giscience and Remote Sensing, 45*: 188–208.

Nahlik A.M., Fennessy M.S. (2016). Carbon storage in US wetlands. *Nature Communications, 7*: doi: 10.1038/ncomms13835.

Najjar R.G., Herrmann M., Alexander R., Boyer E.W., Burdige D.J., Butman D., Cai W.J., Canuel E.A., Chen R.F., Friedrichs M.A., Feagin R.A. (2018). Carbon budget of tidal wetlands, estuaries, and shelf waters of eastern North America. *Global Biogeochemical Cycles, 32*(3): 389–416.

Narayan S., Beck M.W., Reguero B.G., Losada I.J., van Wesenbeeck B., Pontee N., Sanchirico J.N., Ingram J.C., Lange G.M., Burks-Copes K.A. (2016a). The effectiveness, costs and coastal protection benefits of natural and nature-based defences. *Plos One, 11*(5), e0154735.

Narayan S., Beck M.W., Reguero B.G., Losada I.J., van Wesenbeeck B., Pontee N., Sanchirico J.N., Ingram J.C., Lange G.M., Burks-Copes K.A. (2016). The effectiveness. costs and coastal protection benefits of natural and nature-based defences. *PLoS One, 11*.

Narayan S., Beck M.W., Wilson P., Thomas C., Guerrero A., Shepard C.C., Reguero B.G., Franco G., Ingram C.J., Trespalacios D. (2016b). *Coastal Wetlands and Flood Damage Reduction: Using Risk Industry-based Models to Assess Natural Defenses in the Northeastern USA*, Lloyd's Tercentenary Research Foundation: London.

National Council on Climate Change (2010). Setting a course for Indonesia's green growth.

National Ocean and Atmospheric Administration (NOAA) (2014). Seagrass distribution off California. Available at: https://data.noaa.gov/dataset/seagrass-distribution-off-california (accessed May 9, 2017).

National Oceanic and Atmospheric Administration [NOAA] (2014). Regional mean sea level research (Andaman Sea). www.star.nesdis.noaa.gov/ (accessed October 26, 2016).

National Oceanic and Atmospheric Administration, National Climate Data Center (2014). Climate Data Online: Freeze/Frost Occurrence Data for California. www.ncdc.noaa.gov/cdo-web/ (accessed April 4, 2014).

Natural Resources Conservation Service (2016). U.S. Soil Survey Geographic (SSURGO) Database.

Needelman B.A., Emmer I.M., Emmett-Mattox S., Crooks S., Megonigal J.P., Myers D., Oreska M.P.J., McGlathery K. (2018). The science and policy of the Verified Carbon Standard Methodology for Tidal Wetland and Seagrass Restoration. *Estuaries and Coasts*, 1–13.

Needleman B.A., Crooks S., Shumway C.A., Titus J.G., Takacs R., Hawkes J.E. (2012). Restore-adapt-mitigate: responding to climate change through coastal habitat restoration. Report by *Restore Americas Estuaries*.

Nelleman C., Corcoran E., Duarte C.M., Valdes L., DeYoung C., Fonseca L., Grimsditch G. (2009). *Blue Carbon: A Rapid Response Assessment*, United Nations Environment Programme, GRID-Arendal: Arendal, Norway. www.grida.no.

Nellemann C., Corcoran E. (2009). *Blue Carbon: the Role of Healthy Oceans in Binding Carbon: A Rapid Response Assessment*, UNEP/Earthprint.

Nellemann C., Corcoran E., Duarte C., Vales L., Fonseca C., Grimsditch G. (2009). *Blue Carbon – the Role of Healthy Oceans in Binding Carbon*, GRID-Arendal: United Nations Environment Programme.

Nelsen J.E., Ginsburg R.N. (1986). Calcium carbonate production by epibionts on *Thalassia* in Florida Bay. *Journal of Sedimentary Petrology, 56*: 622–628.

Nepf H.M., Vivoni E.R. (2000). Flow structure in depth-limited, vegetated flow. *Journal of Geophysical Research-Oceans, 105*: 28547–28557.

Neubauer S.C. (2014). On the challenge of modeling the net radiative forcing of wetlands: reconsidering, (Mitsch, et al. 2013). *Landscape Ecology, 29*: 571–577.

Neubauer S.C., Franklin R.B., Berrier D.J. (2013). Saltwater intrusion into tidal freshwater marshes alters the biogeochemical processing of organic carbon. *Biogeosciences, 10*: 10685–10720.

Neubauer S.C., Givler K., Valentine S., Megonigal J.P. (2005). Seasonal patterns and plant-mediated controls of subsurface wetland biogeochemistry. *Ecology, 86*: 3334–3344.

Neubauer S.C., Megonigal J.P. (2015). Moving beyond global warming potentials to quantify the climatic role of ecosystems. *Ecosystems, 18*: 1000–1013. doi: 10.1007/s10021-015-9879-4.

Neubauer S.C., Miller W.D., Anderson I.C. (2001). Carbon cycling in a tidal freshwater marsh ecosystem: a carbon gas flux study. *Marine Ecology Progress Series, 199*: 13–30.

Neumann B., Vafeidis A.T., Zimmerman J., Nicholls R.J. (2015). Future coastal population growth and exposure to sea-level rise and coastal flooding: a global assessment. *PLoS One, 10*: e0118571.

Neumann B., Vafeidis A.T., Zimmermann J., Nicholls R.J. (2015). Future Coastal Population Growth and Exposure to Sea-Level Rise and Coastal Flooding – A Global Assessment. *PLoS One, 10*(3): e0118571. doi: 10.1371/journal.pone.0118571.

Neves J., Simões M., Ferreira L., Madeira M., Gazarini L. (2010). Comparison of biomass and nutrient dynamics between an invasive and a native species in a Mediterranean saltmarsh. *Wetlands, 30*: 817–826.

Nixon S. (1982). Nutrient dynamics, primary production and fisheries yield of lagoons. *Oceanologica Acta, 5*: 357–371.

NOAA Office for Coastal Management (2014). National 2010 Coastal Change Analysis Program Accuracy Assessment.

NOAA Office for Coastal Management (2015). NOAA Coastal Change Analysis Program (C-CAP) Regional Land Cover Database, Charleston, SC. Data collected 1995-present. Data accessed at www.coast.noaa. gov/digitalcoast/data/ccapregional.

NOAA Office for Coastal Management (2016). National Coastal Land Cover Change Summary Report 1996–2010.

Nóbrega G.N., Ferreira T.O., Neto M.S., Queiroz H.M., Artur A.G., Mendonça E.D.S., Silv E.D.O., Oterof X.L. (2016). Edaphic factors controlling summer (rainy season) greenhouse gas emissions (CO_2 and CH_4) from semiarid mangrove soils (NE-Brazil). *Science of the Total Environment, 542*: 685–693.

Noe G.B., Hupp C.R., Bernhardt C.E., Krauss K.W. (2016). Contemporary deposition and long-term accumulation of sediment and nutrients by tidal freshwater forested wetlands impacted by sea level rise. *Estuaries and Coasts, 39*(4): 1006–1019.

Nolte S., Koppenaal E.C., Esselink P., Dijkema K.S., Schuerch M., De Groot A.V., Bakker J.P., Temmerman S. (2013). Measuring sedimentation in tidal marshes: a review on methods and their applicability in biogeomorphological studies. *Journal of Coastal Conservation, 17*: 301–325.

Nordlund L.M., Koch E.W., Barbier E.B., Creed J.C. (2016). Seagrass ecosystem services and their variability across genera and geographical regions. *PLoS One, 11*(10): e0163091.

Nowicki B.L. (1994). The effect of temperature, oxygen, salinity, and nutrient enrichment on estuarine denitrification rates measured with a modified nitrogen gas flux technique. *Estuarine, Coastal and Shelf Science, 38*(2): 137–156.

Nowicki R.J., Thomson J.A., Burkholder D.A., Fourqurean J.W., Heithaus M.R. (2017). Predicting seagrass recovery times and their implications following an extreme climate event. *Marine Ecology Progress Series, 567*: 79–93.

Noy-Meir I. (1973). Desert ecosystems: environment and producers. *Annual Review of Ecology and Systematics, 4*(1): 25–51.

Nyman J.A., DeLaune R.D., Patrick W.H. Jr. (1990). Wetland soil formation in the rapidly subsiding Mississippi River Deltaic Plain: Mineral and organic matter relationships. *Estuarine, Coastal and Shelf Science, 31*: 57–69.

Nyman J.A., DeLaune R.D., Roberts H.H., Patrick W.H. Jr. (1993). Relationship between vegetation and soil formation in a rapidly submerging coastal marsh. *Marine Ecology Progress Series, 96*: 269–279.

Nyman J.A., Walters R.J., Delaune R.D., Patrick W.H. Jr. (2006). Marsh vertical accretion via vegetative growth. *Estuarine, Coastal and Shelf Science, 69*: 370–380.

O'Connell J.L., Byrd K.B., Kelly M. (2014). Remotely-Sensed Indicators of N-Related Biomass Allocation in *Schoenoplectus acutus*. *Plos One, 9*: e90870.

O'Connell J.L., Byrd K.B., Kelly M. (2015). A hybrid model for mapping relative differences in belowground biomass and root: shoot ratios using spectral reflectance, foliar N and plant biophysical data within coastal marsh. *Remote Sensing, 7*: 16480–16503.

Obiria M. (2016). Protecting mangroves, Kenya's fishermen net cash – and more fish. Thomson Reuters Foundation, http://news.trust.org/item/20160927120410-sr9t5/.

Odum E.P. (1968). A research challenge: evaluating the productivity of coastal and estuarine water. In: *Proceedings of the Second Sea Grant Conference*, University of Rhode Island, p. 64.

Odum E.P. (1971). *Fundamentals of Ecology*. W.B. Saunders Co., p. 574.

Odum E.P., de la Cruz A.A. (1967). Particulate organic detritus in a Georgia salt marsh-estuarine ecosystem. In: *Estuaries*, Lauff G.H. (ed.), American Association for the Advancement of Science: Washington D.C., pp. 383–388.

Odum H.T., Odum E.P. (1955). Trophic structure and productivity of a windward coral reef community on Eniwetok. *Ecological Monographs*, 25: 291–320.

Odum W.E. (1988). Comparative ecology of tidal freshwater and salt marshes. *Annual Review of Ecology and Systematics*, 19(1): 147–176.

Odum W.E., Fisher J.S., Pickral J.C. (1979). Factors controlling the flux of particulate organic carbon from estuarine wetlands. *Ecological Processes in Coastal and Marine Systems*, Springer: Boston, MA, pp. 69–80.

Odum W.E., McIvor C.C. (1990). Mangroves. In: *Ecosystems of Florida*, Myers R.L., Ewel J.J. (eds), UCF Press: Orlando, pp. 517–548.

Odum W.E., Zieman J.C., Heald E.J. (1973). The importance of vascular plant detritus to estuaries. In: *Coastal Marsh and Estuary Symposium*, Chabreck R.H. (ed.), LSU: Baton Rouge, Louisiana, pp. 91–135.

Ólafsson E., Carlström S., Ndaro S.G. (2000). Meiobenthos of hypersaline tropical mangrove sediment in relation to spring tide inundation. *Hydrobiologia*, 426: 57–64.

Olander J., Ebeling J. (2011). Building Forest Carbon Projects: Step-by-Step Overview and Guide. In: *Building Forest Carbon Projects*, Ebeling J., Olander J. (eds), Forest Trends: Washington, D.C. www.foresttrends. org/documents/files/doc_2555.pdf (accessed November 27, 2014).

Olesen B., Sand-Jensen K. (1994). Biomass-density patterns in the temperate seagrass *Zostera marina*. *Marine Ecology Progress Series*, 109: 283–291.

Ollinger S.V. (2011). Sources of variability in canopy reflectance and the convergent properties of plants. *New Phytologist*, 189: 375–394.

Ooi J.L.S., van Niel K.P., Kendrick G.A., Holmes K.W. (2014). Spatial structure of seagrass suggests that size-dependent plant traits have a strong influence on the distribution and maintenance of tropical multispecies meadows. *PLoS One*, 9(1): e86782.

Opsahl S., Benner R. (1993). Decomposition of senescent blades of the seagrass *Halodule wrightii* in a subtropical lagoon. *Marine Ecology Progress Series*, 94: 191.

Oremland R.S. (1975). Methane production in shallow-water, tropical marine sediments. *Applied Microbiology*, 30: 602–608.

Oremland R.S., Kiene R.P., Mathrani I., Whiticar M.J., Boone D.R. (1989). Description of an estuarine methylotrophic methanogen which grows on dimethyl sulfide. *Applied and Environmental Microbiology*, 55: 994–1002.

Oreska M.P.J., et al. (2019). Seagrass mapping: A survey of the recent seagrass distribution literature. In: *A Blue Carbon Primer: The State of Coastal Wetland Carbon Science, Practice, and Policy*, Lisamarie W.-M., Stephen C., and Tiffany G.T. (eds), CRC Press: Boca Raton, London.

Oreska M.P.J., McGlathery K., Emmer I.M., Needelman B.A., Emmett-Mattox S., Crooks S., Megonigal J.P., Myers D. (2018). Comment on 'Geoengineering with seagrasses: is credit due where credit is given?'. *Environmental Research Letters*. doi: 10.1088/1748–9326/aaae72 (accepted).

Oreska M.P.J., McGlathery K.J., Porter J.H. (2017). Seagrass blue carbon spatial patterns at the meadow-scale. *PLoS One*, 12(4): e0176630.

Orr M., Crooks S., Williams P.B. (2003). Issues in San Francisco Estuary tidal restoration: will restored tidal marshes be sustainable? *San Francisco Estuary and Watershed Science*, 1(1): 108–142. http://escholarship.org/uc/item/8hj3d20t.

Orson R.A., Warren R.S., Niering W.A. (1998). Interpreting sea level rise and rates of vertical marsh accretion in a southern New England tidal salt marsh. *Estuarine, Coastal and Shelf Science*, 47: 419–429.

Orth R.J., Carruthers T.J., Dennison W.C., Duarte C.M., Fourqurean J.W., Heck K.L., Hughes A.R., Kendrick G.A., Kenworthy W.J., Olyarnik S., Short F.T., Williams S.L. (2006a). A global crisis for seagrass ecosystems. *Bioscience*, 56(12): 987–996.

Orth R.J., Dennison W.C., Lefcheck J.S., Gurbisz C., Hannam M., Keisman J., Landry J.B., Moore K.A., Murphy R.R., Patrick C.J., Testa J., Weller D.E., Wilcox D.J. (2017). Submersed aquatic vegetation in Chesapeake Bay: sentinel species in a changing world. *Bioscience*, 67 (8): 698–712

Orth R.J., Luckenbach M.L., Marion S.R., Moore K.A., Wilcox D.J. (2006b). Seagrass recovery in the Delmarva Coastal Bays, USA. *Aquatic Botany*, 84: 26–36.

Orth R.J., Marion S.R., Moore K.A., Wilcox D.J. (2010). Eelgrass (*Zostera marina* L.) in the Chesapeake Bay Region of Mid-Atlantic Coast of the USA: challenges in Conservation and Restoration. *Estuaries and Coasts, 33*: 139–150.

Orth R.J., Moore K.A. (1983). Chesapeake Bay: an unprecedented decline in submerged aquatic vegetation. *Science, 222*(4619): 51–53.

Orth R.J., Moore K.A. (1984). Distribution and abundance of submerged aquatic vegetation in Chesapeake Bay: an historical perspective. *Estuaries, 7*(4): 531–540.

Orth R.J., Moore K.A., Marion S.R., Wilcox D.J., Parrish D.B. (2012). Seed addition facilitates eelgrass recovery in a coastal bay system. *Marine Ecology Progress Series, 448*: 177–195.

Orth R.J., Wilcox D.J., Whiting J.R., Kenne A.K., Nagey L., Smith E.R. (2016). 2015 Distribution of Submerged Aquatic Vegetation in the Chesapeake Bay and Coastal Bays. VIMS Special Scientific Report Number 159. Final report to EPA, Chesapeake Bay Program, Annapolis, MD. Grant No CB96321901-0, http://www.vims.edu/bio/sav/sav15.

Osburn C.L., Mikan M.P., Etheridge J.R., Burchell M.R., Birgand F. (2015). Seasonal variation in the quality of dissolved and particulate organic matter exchanged between a salt marsh and its adjacent estuary. *Journal of Geophysical Research: Biogeosciences, 120*(7): 1430–1449.

Osgood D.T., Zieman J.C. (1993). Factors controlling aboveground *Spartina alterniflora* (smooth cordgrass) tissue element composition and production in different-age barrier island marshes. *Estuaries, 16*: 815–826.

Osland M.J., Enwright N., Day R.H., Doyle T.W. (2013). Winter climate change and coastal wetland foundation species: salt marshes versus mangrove forests in the southeastern US. *Global Change Biology, 19*: 1482–1494.

Osland M.J., Spivak A.C., Nestlerode J.A., Lessmann J.M., Almario A.E., Heitmuller P.T., Russell M.J., Krauss K.W., Alvarez F., Dantin D.D. (2012). Ecosystem development after mangrove wetland creation: plant– soil change across a 20-year chronosequence. *Ecosystems, 15*(5): 848–866. doi: 10.1007/s10021-012-9551-1.

Ouyang X., Lee S.Y. (2014). Updated estimates of carbon accumulation rates in coastal marsh sediments. *Biogeosciences, 11*: 5057–5071.

Ouyang X., Lee S.Y., Connolly R.M. (2017). The role of root decomposition in global mangrove and saltmarsh carbon budgets. *Earth-Science Reviews, 166*: 53–63.

Pace N. (2017). Permitting a Living Shoreline: A Look at the Legal Framework Governing Living Shoreline Projects At The Federal, State, And Local Level. In: *Living Shorelines: The Science and Management of Nature-Based Coastal Protection*, D.M. Bilkovic, M.M. Mitchell, M.K. La Peyre, J.D. Toft (eds). CRC Press: Boca Raton, p. 519.

Pan Z., Fernandez-Diaz J.C., Glennie C.L., Starek M. (2014). Shallow water seagrass observed by high resolution full waveform bathymetric LiDAR. *Geoscience and Remote Sensing Symposium (IGARSS)*. doi: 10.1109/IGARSS.2014.6946682.

Parrish C.E., Rogers J.N., Calder B.R. (2014). Assessment of waveform features for lidar uncertainty modeling in a coastal salt marsh environment. *IEEE Geoscience and Remote Sensing Letters, 11*: 569–573.

Parton W.J., Holland E.A., Del S.J., Grosso Hartman M.D., Martin R.E., Mosier A.R., Ojima D.S., Schimel D.S. (2001). Generalized model for NOx and N_2O emissions from soils. *Journal of Geophysical Research, 106*: 17403–17419.

Patrick W.H. Jr., DeLaune R.D. (1990). Subsidence, accretion, and sea level rise in south San Francisco Bay marshes. *Limnology and Oceanography, 35*: 1389–1395.

Patriquin D.G. (1972). Carbonate mud production by epibionts on *Thalassia*: an estimate based on leaf growth rate data. *Journal of Sedimentary Petrology, 42*: 687–689.

Patterson C.S., Mendelssohn I.A. (1991). A comparison of physicochemical variables across plant zones in a mangal salt-marsh community in Louisiana. *Wetlands, 11*: 139–161.

Paul M., Lefebvre A., Manca E., Amos C.L. (2011). An acoustic method for the remote measurement of seagrass metrics. *Estuarine, Coastal and Shelf Science, 93*: 68–79.

Paulose N.E., Dilipan E., Thangaradjou T. (2012). Integrating Indian remote sensing multi-spectral satellite and field data to estimate seagrass cover change in the Andaman and Nicobar Islands, India. *Ocean Science Journal, 48*(2): 1–9.

PBS & J (2010). *Tampa Bay Estuary Program Habitat Master Plan Update*, TBEP Technical Report #06-09: St. Petersburg, FL. www.tbeptech.org/TBEP_TECH_PUBS/2009/TBEP_06_09_Habitat_Master_Plan_Update_Report_July_2010.pdf.

PCAST (1998). *Teaming with Life: Investing in Science to Understand and Use America's Living Capital*, Executive Office of the President: Washington, D.C.

PCAST (2011). Report to the President Sustaining Environmental Capital: Protecting Society and the Economy.

Pe'eri S., Ru Morrison J., Short F., Mathieson A., Lippmann T. (2016). Eelgrass and macroalgal mapping to develop nutrient criteria in New Hampshire's estuaries using hyperspectral imagery. *Journal of Coastal Research, 76*: 209–218.

Pearce D.W., Turner R.K. (1990). *Economics of Natural Resources and the Environment*. The John Hopkins University Press: Baltimore.

Peatfield J.J. (1894). Dredging on the Pacific Coast. *Overland Monthly, 24*(2nd series 141): 315–327.

Peirano A., Damasso V., Montefalcone M., Morri C., Bianchi C.N. (2005). Effects of climate, invasive species and anthropogenic impacts on the growth of seagrass *Posidonia oceanica* (L.) Delile in Liguria (NW Mediterranean Sea). *Marine Pollution Bulletin, 50*: 817–822.

Pendleton L., Donato D.C., Murray B.C., Crooks S., Jenkins W.A., Sifleet S., Craft C., Fourqurean J.W., Kauffman J.B., Marba N., Megonigal P., Pidgeon E., Herr D., Gordon D., Baldera A. (2012). Estimating global "blue carbon" emissions from conversion and degradation of vegetated coastal ecosystems. *PloS One, 7*(9): e43542.

Pendleton L., Murray B.C., Gordon D., Cooley D., Vegh T. (2014). 17. Harnessing the financial value of coastal 'blue'carbon. In: *Valuing Ecosystem Services: Methodological Issues and Case Studies*, p. 361.

Pendleton L.H., Sutton-Grier A.E., Gordon D.R., Murray B.C., Victor B.E., Griffis R.B., Lechuga J.A.V., Giri C. (2013). Considering "Coastal Carbon" in Existing US Federal Statutes and Policies. *Coastal Management, 41*: 439–456.

Penfound W.T., Earle T.T. (1948). The biology of water hyacinth. *Ecological Monographs, 18*: 447–472.

Perillo G.M.E., Wolanski E., Cahoon D.R., Brinson M.M. (2009). *Coastal Wetlands – An Integrated Ecosystem Approach*, 1st edn, Elsevier: Amsterdam.

Perleman M.A. (2017). Bonds move disaster finance from recovery to resilience. Conservation Finance network. www.conservationfinancenetwork.org/2017/08/23/bonds-move-disaster-finance-from-recovery-to-resilience.

Perry C.L., Mendelssohn I.A. (2009). Ecosystem effects of expanding populations of *Avicennia germinans* in a Louisiana salt marsh. *Wetlands, 29*: 396–406.

Perry C.T., Beavington-Penney S.J. (2005). Epiphytic calcium carbonate production and facies development within sub-tropical seagrass beds, Inhaca Island, Mozambique. *Sedimentary Geology, 174*: 161–176.

Perry C.T., Salter M.A., Harborne A.R., Crowley S.F., Jelks H.L., Wilson R.W. (2011). Fish as major carbonate mud producers and missing components of the tropical carbonate factory. *Proceedings of the National Academy of Sciences, 108*: 3865–3869.

Peters G.P., Le Quéré C., Andrew R.M., Canadell J.G., Friedlingstein P., Ilyina T., Jackson R.B., Joos F., Korsbakken J.I., McKinley G.A., Sitch S., Tans P. (2017). Towards real-time verification of CO_2 emissions. *Nature Climate Change*. doi: 10.1038/s41558-017-0013–9, INRMM-MiD: 14477845.

Pethick J.S., Crooks S. (2000). Development of a coastal vulnerability index, a geomorphological perspective. *Environmental Conservation, 27*(4): 359–367.

Petticrew E.L., Kalff J. (1992). Water flow and clay retention in submerged macrophyte beds. *Canadian Journal of Fisheries and Aquatic Sciences, 49*: 2483–2489.

Phelan N., Shaw A., Baylis A. (2011). *The Extent of Saltmarsh in England and Wales: 2006–2009*. Environment Agency: Bristol, UK.

Phillips R.C. (2003). The seagrasses of the Arabian Gulf and Arabian Region, In: Green E.P., F. Short T. (eds), *World Atlas of Seagrasses*, University of California Press: Berkeley, Chapter 6, p. 298.

Phinn S.R. (1998). A framework for selecting appropriate remotely sensed data dimensions for environmental monitoring and management. *International Journal of Remote Sensing, 19*: 3457–3463.

Phinn S.R., Stow D., Zedler J. (1996). Monitoring wetland habitat restoration in southern California using airborne multispectral video data. *Restoration Ecology, 4*: 412–422.

Pidgeon E. (2009). Carbon Sequestration by Coastal Marine Habitats: Important Missing Sinks. In: *The Management of Natural Coastal Carbon Sinks*, Laffoley D.d'A., Grimsditch G. (eds), IUCN: Gland, Switzerland, pp. 47–51.

Piontkovski S.A., Al-Gheilani H.M., Jupp B.P., Al-Azri A.R., Al-Hashmi K.A. (2012). Interannual changes in the Sea of Oman ecosystem. *Open Marine Biology Journal*, *6*: 38–52.

Plan Vivo Foundation (2015). *Mikoko Pamoja*, Kenya, Edinburgh, UK. www.planvivo.org/project-network/mikoko-pamoja-kenya/.

Pleasant M.M., Gray S.A., Lepczyk C., Fernandes A., Hunter N., Ford D. (2014). Managing cultural ecosystem services. *Ecosystem Services*, *8*: 141–147.

Plummer L.N., Busenberg E. (1982). The solubilities of calcite, aragonite and vaterite in CO_2-H_2O solutions between 0 and 90°C, and an evaluation of the aqueous model for the system $CaCO_3$-CO_2-H_2O. *Geochimica et Cosmochimica Acta*, *46*: 1011–1040.

Poffenbarger H.J., Needelman B.A., Megonigal J.P. (2011). Salinity influence on methane emissions from tidal marshes. *Wetlands*, *31*(5): 831–842.

Poland J.F., Ireland R.L. (1988). *Land Subsidence in the Santa Clara Valley, California, as of 1982*. U.S. Geological Survey Professional Paper 497-F. U.S. Government Printing Office: Washington, D.C.

Pomeroy L.R., Wiegert R.G. (1981). *The Ecology of a Salt Marsh*. Springer-Verlag: NY, p. 271.

Poretsky R., Sun S., Mou X., Moran M. (2010). Transporter genes expressed by coastal bacterioplankton in response to dissolved organic carbon. *Environmental Microbiology*, *12*: 616–627.

Portley N. (2016). *SFP Report on the Shrimp Sector: Asian Farmed Shrimp Trade and Sustainability*, Sustainable Fisheries Partnership Foundation, Honolulup, p. 22. http://cmsdevelopment.sustainablefish. org.s3.amazonaws.com/2016/04/07/Asian%20shrimp%20summary%20report-65b964a4.pdf (accessed 22 June 2017).

Prahl F.G., Coble P.G. (1994). Input and Behavior of Dissolved Organic Carbon in the Columbia River Estuary. In: *Changes in Fluxes in Estuaries: Implications from Science and Management (ECSA22/ERF Symp., Plymouth, England)*, Dyer K.R., Orth R.J. (eds), Olsen and Olsen: Denmark, pp. 451–457.

Primavera J. (2006). Overcoming the impacts of aquaculture on the coastal zone. *Ocean & Coastal Management*, *49*(9): 531–545.

Primavera J.H. (2005). Mangroves, fishponds, and the quest for sustainability. *Science*, *310*(5745): pp. 57–59.

Primavera J.H., Savaris J.D., Bajoyo B., Coching J.D., Curnick D.J., Golbeque R., Guzman A.T., Henderin J.Q., Joven R.V., Loma R.A., Koldewey H.J. (2012). *Manual for Community Based Mangrove Rehabilitation*, Zoological Society of London: London.

PRISM Climate Group (2017). Oregon State University. http://prism.oregonstate.edu (accessed April 26).

Prokopovich N.P. (1985). Subsidence of peat in California and Florida. *Bulletin of the Association of Engineering Geologists*, *22*: 395–420.

Pu R., Bell S. (2017). Mapping seagrass coverage and spatial patterns with high spatial resolution IKONOS imagery. *International Journal of Applied Earth Observation and Geoinformation*, *54*: 145–158.

Pugh D.T. (ed.) (1987). *Tides, Surges, and Mean Sea Level*, John Wiley & Sons.

Purvaja R., Ramesh R. (2001). Natural and anthropogenic methane emission from coastal wetlands of South India. *Environmental Management*, *27* (4): 547–557.

Purvaja R., Ramesh R., Frenzel P. (2004). Plant-mediated methane emission from an Indian mangrove. *Global Change Biology*, *10*: 1825–1834.

Purvaja R., Ramesh R., Shalini A., Rixen T. (2008). Biogeochemistry of nitrogen in seagrass and oceanic systems. *Memoir Geological Society of India*, *73*: 435–460.

Raabe E.A., Roy L.C., McIvor C.C. (2012). Tampa Bay coastal wetlands: nineteenth to twentieth century tidal marsh-to-mangrove conversion. *Estuaries and Coasts*, *35*(5): 1145–1162. doi: 10.1007/s12237-012-9503-1.

Radabaugh K.R., Moyer R.P., Chappel A.C., Powell C.E., Bociu I., Clark B.C., Smoak J.M. (2018). Coastal blue carbon assessment of mangroves, salt marshes, and salt barrens in Tampa Bay, Florida, USA. *Estuaries and Coasts*, *41*: 1496–1510. doi: 10.1007/s12237-017-0362-7.

Radabaugh K.R., Powell C.E., Bociu I., Clark B.C., Moyer R.P. (2017). Plant size metrics and organic carbon content of Florida salt marsh vegetation. *Wetlands Ecology and Management*, *25*(4): 443-455. doi: 10.1007/s11273-016-9527-6.

Rahman F., Dragoni D., Didan K., Barreto-Munoz A., Hutabarat J. (2013). Detecting large scale conversion of mangroves to aquaculture with change point and mixed-pixel analyses of high-fidelity MODIS data. *Remote Sensing of the Environment*, *130*, doi: 10.1016/j.rse.2012.11.014.

Rahmstorf S. (2007). A semi-empirical approach to projecting future sea-level rise. *Science*, *315*: 368–370.

Rahmstorf S., Foster G., Cazenave A. (2012). Comparing climate projections to observations up to 2011. *Environmental Research Letters*, 7: 044035. doi: 10.1088/1748–9326/7/4/044035.

Raich J.W., Schlesinger W.H. (1992). The global carbon dioxide flux in soil respiration and its relationship to vegetation and climate. *Tellus B*, *44*(2): 81–99.

Ramírez-García, P., López-Blanco J., Ocaña, D. (1998). Mangrove vegetation assessment in the Santiago River Mouth, Mexico, by means of supervised classification using Landsat TM imagery. *Forest Ecology and Management*, *105*(1–3): 217–229.

Ramsar (2012). Resolution XI.14. Climate change and wetlands: implications for the Ramsar Convention on Wetlands.

Ramsar (2015). Resolution XII.11. Peatlands, climate change and wise use: Implications for the Ramsar Convention.

Ramsar (2018). Resolution X.24. Climate change and wetlands.

Ramsar Convention Secretariat (2016). *An Introduction to the Ramsar Convention on Wetlands*, (previously The Ramsar Convention Manual), 7th edn, Ramsar Convention Secretariat: Gland, Switzerland.

Ramsar Convention Secretariat (2016a). *The Fourth Ramsar Strategic Plan 2016–2024*. *Ramsar Handbooks for the Wise use of Wetlands*, 5th edn, Ramsar Convention Secretariat: Gland, Switzerland, *vol. 2*.

Ramsar/STRP (2016). Scientific and Technical Review Panel (STRP) Work Plan 2016–2018. As approved at the 52nd Meeting of the Standing Committee (June 13–17) of the Ramsar Convention on Wetlands.

Ramsey E. III, Rangoonwala A., Chi Z., Jones C.E., Bannister T. (2014). Marsh Dieback, loss, and recovery mapped with satellite optical, airborne polarimetric radar, and field data. *Remote Sensing of Environment*, *152*: 364–374.

Rannik U., Haapanala S., Shurpali N.J., Mammarella I., Lind S., Hyvonen N., Peltola O., Zahniser M., Martikainen P.J., Vesala T. (2015). Intercomparison of fast response commercial gas analysers for nitrous oxide flux measurements under field conditions. *Biogeosciences*, *12*: 415–432.

Ratliff K.A., Braswella A.E., Marani M. (2015). Spatial response of coastal marshes to increased atmospheric CO_2. *Proceedings of the National Academy of Sciences*, *112*: 15580–15584.

Ravit B., Ehrenfeld J.G., Haggblom M.M. (2003). A comparison of sediment microbial communities associated with *Phragmites australis* and *Spartina alterniflora* in two brackish wetlands of New Jersey. *Estuaries and Coasts*, *26*(2): 465–474.

Raymond P., Bauer J. (2001). Use of 14C and 13C natural abundances as a tool for evaluating freshwater, estuarine and coastal organic matter sources and cycling. *Organic Geochemistry*, *32*: 469–485.

Record S., Charney N.D., Zakaria R.M., Ellison A.M. (2013). Projecting global mangrove species and community distributions under climate change. *Ecosphere*, *4*: 34. doi: 10.1890/es12–00296.1.

Redfield A.C. (1967). The ontogeny of a salt marsh estuary. In: *Estuaries*, Lauff G. (ed.). AAAS, Pub. No. 83.

Redfield A.C. (1972). Development of a New England salt marsh. *Ecological Monographs*, *42*: 201–237 doi: 10.2307/1942263.

Redfield A.C., Rubin M. (1962). The age of salt marsh peat and its relation to recent changes in sea level at Barnstable, Massachusetts. *Proceedings of the National Academy of Sciences of the United States of America*, *48*: 1728–1735.

Redondo-Gómez S., Mateos-Naranjo E, Figueroa M.E., Davy A.J. (2010). Salt stimulation of growth and photosynthesis in an extreme halophyte, Arthrocnemum macrostachyum. *Plant Biology*, *12*: 79–87.

Reeburgh W.S. (2007). Oceanic methane biogeochemistry. *Chemical Reviews*, *107*: 486–513.

Reed D. (1995). The response of coastal marshes to SLR: survival or submergence? *Earth Surface Processes and Landforms*, *20*: 39–48. doi: 10.1002/esp.3290200105.

Reed D. (2002). SLR and coastal marsh sustainability: geological and ecological factors in the Mississippi delta plain. *Geomorphology*, *48*: 233–243.

Reed D.J. (1989). Patterns of sediment deposition in subsiding coastal salt marshes, Terrebonne Bay, Louisiana: the role of winter storms. *Estuaries*, *12*: 222–227.

Reed D.J. (2002). Sea-level rise and coastal marsh sustainability: geological and ecological factors in the Mississippi delta plain. *Geomorphology*, *48*: 233–243.

Reed D.J. (2002). Understanding tidal marsh sedimentation in the Sacramento–San Joaquin Delta, California. *Journal of Coastal Research SI*, *36*: 605–611.

Reef R., Spencer T., Moller I., Lovelock C.E., Christie E.K., Mcivor A.L., Evans B.R., Tempest J.A. (2017). The effects of elevated CO_2 and eutrophication on surface elevation gain in a European salt marsh. *Global Change Biology, 23*, 881–890.

Reide Corbett D., Walsh J.P. (2015). ^{210}Lead and ^{137}Cesium: Establishing a chronology for the last century. *Handbook of Sea-Level Research*, In: Shennan I., Long A.J., Horton B.P. (eds), Wiley: West Sussex, pp. 361–372.

Rejmánková, E. (1992). Ecology of creeping macrophytes with special reference to Ludwigia peploides (H.B.K.) Raven. *Aquatic Botany, 43*: 283–299.

Renaud F.G., Sudmeier-Rieux K., Estrella M. (2013). *The Role of Ecosystems in Disaster Risk Reduction*, United Nations University Press: Paris. p. 512.

Richard G.A. (1978). Seasonal and environmental variations in sediment accretion in a Long Island salt marsh. *Estuaries, 1*: 29–35.

Richards D.R., Friess D.A. (2016). Rates and drivers of mangrove deforestation in Southeast Asia, 2000–2012. *Proceedings of the National Academy of Sciences of the United States of America, 113*: 344–349

Richards T.A., Jones M.D., Leonard G., Bass D. (2012). Marine fungi: their ecology and molecular diversity. *Annual Review of Marine Science, 4*: 495–522.

Rickels W., Dovern J., Hoffmann J., Quaas M.F., Schmidt J.O., Visbeck M. (2016). Indicators for monitoring sustainable development goals, an application to oceanic development in the European Union, *Earth's Future, 4*: 252–267. doi: 10.1002/2016EF000353.

Ries T. (2009). Restored habitats: Lessons learned and the importance of management. In: *Proceedings of the 5th Tampa Bay Area Scientific Information Symposium (BASIS 5)*, Cooper S.T. (eds), St. Petersburg, FL, pp. 195–212. www.tbeptech.org/BASIS/BASIS5/BASIS5.pdf#page=209 (accessed October 20–23).

Ritchie J.C., McHenry J.R. (1990). Application of radioactive fallout cesium-137 for measuring soil erosion and sediment accumulation rates and patterns: a review. *Journal of Environmental Quality, 19*: 215–233.

Rivera-Monroy V.H., Castaneda-Moya E., Barr J.G., Engel V., Fuentes J.D., Troxler T.G., Twilley R.R., Bouillon S., Smith T.J., O'Halloran T.L. (2013). Current methods to evaluate net primary production and carbon budgets in mangrove forests. In: *Methods in Biogeochemistry of Wetlands*, DeLaune R.D., Reddy K.R., Richardson C.J., Megonigal J.P. (eds), Soil Science Society of America: Madison, Wisconsin, USA, pp. 243–288.

Robblee M.B., Barber T.R., Carlson P.R. Jr., Durako M.J., Fourqurean J.W., Muehlstein L.K., Zieman J.C. (1991). Mass mortality of the tropical seagrass *Thalassia testudinum* in Florida Bay (USA). *Marine Ecology Progress Series*, 297–299.

Robinson A., Safran S.M., Beagle J., Grenier J.L., Grossinger R.M., Spotswood E., Dusterhoff S.D., Richey A. (2016). *A Delta Renewed: A Guide to Science-Based Ecological Restoration in the Sacramento-San Joaquin Delta*. Delta Landscapes Project. Prepared for the California Department of Fish and Wildlife and Ecosystem Restoration Program. A Report of SFEI-ASC's Resilient Landscapes Program. SFEI Contribution No. 799. San Francisco Estuary Institute – Aquatic Science Center: Richmond, CA.

Rocchio J. (2005). *North American Arid West Freshwater Marsh Ecological System Ecological Integrity Assessment*. Colorado Natural Heritage Program: Fort Collins.

Roelfsema C., Kovacs E.M., Saunders M.I., Phinn S., Lyons M., Maxwell P. (2013). Challenges of remote sensing for quantifying changes in large complex seagrass environments. *Estuarine, Coastal and Shelf Science, 133*: 161–171.

Roelfsema C.M., Lyons M., Kovacs E.M., Maxwell P., Saunders M.I., Samper-Villarreal J., Phinn S.R. (2014). Multi-temporal mapping of seagrass cover, species and biomass: a semi-automated object based image analysis approach. *Remote Sensing of Environment, 150*: 172–187.

Rogers J.N., Parrish C.E., Ward L.G., Burdick D.M. (2015). Evaluation of field-measured vertical obscuration and full waveform lidar to assess salt marsh vegetation biophysical parameters. *Remote Sensing of Environment, 156*: 264–275.

Rogers S.-M., et al. (2019). Blue Carbon as a Tool to Support Coastal Management and Restoration: Bringing Wetlands to Market Case Study. In: *A Blue Carbon Primer: The State of Coastal Wetland Carbon Science, Practice, and Policy*, Lisamarie W.-M., Stephen C., and Tiffany G.T. (eds), CRC Press: Boca Raton, London.

Rogers T.-M. (2015). Bringing Wetlands to Market: Nitrogen and Coastal Blue Carbon. *A Final Report Submitted to the National Estuarine Research Reserve System Science Collaborative*, August 25.

Rojstaczer S., Deverel S.J. (1993). Time dependence in atmospheric carbon inputs from drainage of organic soils. *Geophysical Research Letters, 20*: 1383–86.

Romero J. (1988). Epifitos de las hojas de Posidonia oceanica: variaciones estacionales y batimetricas de bio-masa en la pradera de las islas Medes (Girona). *Oecologica Aquatica*, 9: 19–25.

Roner M., D'Alpaos A., Ghinassi M., Marani M., Silvestri S., Franceschinis E., Realdon N. (2016). Spatial variation of salt-marsh organic and inorganic deposition and organic carbon accumulation: inferences from the Venice lagoon, Italy. *Advances in Water Resources*, 93: 276–287.

Rovai A., Twilley R., Castañeda-Moya E., Riul P., Cifuentes-Jara, M., Manrow-Villalobos M., Horta P., Simonassi J., Alessandra L.F., Pagliosa P. (2018). Global controls on carbon storage in mangrove soils. *Nature Climate Change*, 8. doi: 10.1038/s41558-018-0162-5.

Roy E.D., White J.D. (2013). Measurements of nitrogen mineralization potential in wetland soils In: *Methods in Biogeochemistry in Wetlands*, DeLaune R.D., Reddy K.R., Richardson C.J., Megonigal J.P. (eds), Soil Science Society of America: Madison, WI, SSSA Book Series, no. 10, pp. 465–472.

Rude P.D., Aller R.C. (1991). Fluorine mobility during early diagenesis of carbonate soil: an indicator of min-eral transformations. *Geochimica et Cosmochimica Acta*, 55: 2491–2509.

Ruengsorn C., Mekumpun S., Ichimi K., Yamaguchi H., Tada K. (2015). Development of mapping tech-niques for small seagrass meadows: a case study of *Zostera marina* and *Halodule pinifolia*. *Plankton & Benthos Research*, 10(2): 81–90.

Ruiz-Luna A., Acosta-Velázquez J., Berlanga-Robles C.A. (2008). On the reliability of the data of the extent of mangroves: a case study in Mexico. *Ocean and Coastal Management*, 51(4): 342–351.

Runting R.K., Lovelock C.E., Beyer H.L., Rhodes J.R. (2016). Costs and opportunities for preserving coastal wetlands under sea level rise. *Conservation Letters*, 10(1), 49–57.

Runting R.K., Wilson K.A., Rhodes J.R. (2013). Does more mean less? The value of information for conserva-tion planning under sea level rise. *Global Change Biology*, 19: 352–363.

Russell B.D., Connell S.D., Uthicke S., Muehllehner N., Fabricius K.E., Hall-Spencer J.M. (2013). Future seagrass beds: can increased productivity lead to increased carbon storage? *Marine Pollution Bulletin*, 73(2): 463–469.

Russell M., Greening H. (2015). Estimating benefits in a recovering estuary: Tampa Bay, Florida. *Estuaries and Coasts*, 38(1): S9–S18. doi: 10.1007/s12237-013-9662-8.

Sacramento – San Joaquin Delta Conservancy (2017). Restoration Projects in the Delta and Suisun Marsh. http://deltaconservancy.ca.gov/restoration-projects-delta-and-suisun-marsh/ (accessed July 17, 2017).

Saenger P., Blasco F., Yousseff A.M.M., Loughland R.A. (2004). Mangroves of the United Arab Emirates with particular emphasis on those of Abu Dhabi Emirate. In: Loughland R.A., Muhairi F.S.A., Fadel S.S., Mehdi A.M.A., Hellyer P. (eds), *Marine Atlas of Abu Dhabi*. Emirates Heritage Club: Abu Dhabi, pp. 58–69

Saenger P., Snedaker S.C. (1993). Pantropical trends in mangrove above-ground biomass and annual litterfall. *Oecologia*, 96: 293–299.

Sagawa T., Komatsu T. (2015). Simulation of seagrass bed mapping by satellite images based on radiative transfer model. *Ocean Science Journal*, 50(2): 335–342.

SAGE (Systems Approach to Geomorphic Engineering). (2015). Natural and Structural Measures for Shoreline Stabilization. www.SAGEcoast.org.

Saintilan N., Wilson N.C., Rogers K., Rajkaran A., Krauss K.W. (2014). Mangrove expansion and salt marsh decline at mangrove poleward limits. *Global Change Biology*, 20(1): 147–157.

Sakai A.K., Allendorf F.W., Holt J.S., Lodge D.M., Molofsky J., With K.A., Baughman S., Cabin R.J., Cohen J.E., Ellstrand N.C., McCauley D.E., P. O'Neil Parker I.M., Thompson J.N., Weller S.G. (2001). The Population Biology of Invasive Species. *Annual Review of Ecology and Systematics*, 32: 305–332.

San Francisco Estuary Institute-Aquatic Science Center (SFEI-ASC) (2014). *A Delta Transformed: Ecological Functions, Spatial Metrics, and Landscape Change in the Sacramento-San Joaquin Delta*, Prepared for the California Department of Fish and Wildlife and Ecosystem Restoration Program. A Report of SFEI-ASC's Resilient Landscapes Program, Publication #729, San Francisco Estuary Institute-Aquatic Science Center: Richmond, CA.

Sanderman J., Hengl T., Fiske G., Solvik K., Adame M.F., Benson L., Bukoski J.J., Carnell P., Cifuentes-Jara M., Donato D., Duncan C. (2018). A global map of mangrove forest soil carbon at 30 m spatial resolu-tion. *Environmental Research Letters*, 13(5): 055002.

Sanders C.J., Smoak J.M., Naidu A.S., Patchineelam S.R. (2008). Recent sediment accumulation in a man-grove forest and its relevance to local sea-level rise (Ilha Grande, Brazil). *Journal of Coastal Research*, 24(2): 533–6.

Sandifer P.A., Sutton-Grier A.E. (2014). Connecting stressors, ocean ecosystem services, and human health. *Natural Resources Forum, 38*(3): 157–167.

Sandifer P.A., Sutton-Grier A.E., Ward B.P. (2015). Exploring connections among nature, biodiversity, ecosystem services, and human health and well-being: opportunities to enhance health and biodiversity conservation. *Ecosystem Services, 12*: 1–15.

Santos M.J., Hestir E.L., Khanna S., Ustin S.L. (2012). Image spectroscopy and stable isotopes elucidate functional dissimilarity between native and nonnative plant species in the aquatic environment. *New Phytologist, 193*: 683–695.

Santos M.J., Khanna S., Hestir E.L., Andrew M.E., Rajapakse S.S., Greenberg J.A., Anderson L.W.J., Ustin S.L. (2009). Use of Hyperspectral Remote Sensing to Evaluate Efficacy of Aquatic Plant Management in the Sacramento-San Joaquin River Delta, California. *Invasive Plant Science and Management, 2*: 216–229.

Sasmito S.D., Murdiyarso D., Friess D.A., Kurnianto S. (2015). Can mangroves keep pace with contemporary sea level rise? A global data review. *Wetlands Ecology and Management, 24*, 263–278.

Savaresi A. (2013). REDD+ and human rights: addressing synergies between international regimes, *Ecology and Society, 18*: 5.

Scales I.R., Friess D.A., Glass L., Ravaoarinorotsihoarana L.A. (2017). Rural livelihoods and mangrove degradation in southwestern Madagascar: lime production as an emerging threat. *Oryx, 1–5.*

Schaefer M., Goldman E., Bartuska A.M., Sutton-Grier A., Lubchenco J. (2015). Nature as capital: advancing and incorporating ecosystem services in United States federal policies and programs. *Proceedings of the National Academy of Sciences of the United States of America, 112*: 7383–7389.

Schalles J.F., Hladik C.M., Lynes A.A., Pennings S.C. (2013). Landscape Estimates of Habitat Types, Plant Biomass, and Invertebrate Densities in a Georgia Salt Marsh. *Oceanography, 26*: 88–97.

Schelske C., Odum E.P. (1962). Mechanisms maintaining high productivity in Georgia estuaries. *Proceedings of the Gulf and Caribbean Fisheries Institute, 14*: 75–80.

Scherf A.-K., Rullkötter J. (2009). Biogeochemistry of high salinity microbial mats–Part 1: lipid composition of microbial mats across intertidal flats of Abu Dhabi, United Arab Emirates. *Organic Geochemistry, 40*: 1018–1028.

Schile L., Kauffman J.B., Crooks S., Fourqurean J.W., Glavan J., Megonigal J.P. (2017). Limits on Carbon Sequestration in Arid Blue Carbon Ecosystems. *Ecological Applications, 27*(3): 859–874.

Schile L.M., Callaway J.C., Morris J.T., Stralberg D., Parker V.T., Kelly M. (2014). Modeling tidal wetland distribution with sea-level rise: evaluating the role of vegetation, sediment, and upland habitat in marsh resiliency. *PLoS One, 9*(2): e88760. doi: 10.1371/journal.pone.0088760.

Schleussner C.F., Rogelj J., Schaeffer M., Lissner T., Licker R., Fischer E.M., et al. (2016). Science and policy characteristics of the Paris Agreement temperature goal. *Nature Climate Change.* doi: 10.1038/nclimate3096.

Schmidt M., Torn M.S., Abiven S., Dittmar T., Guggenberger G., Janssens I.A., Kleber M., Kögel-Knabner I., Lehmann J., Manning D.A.C., Nannipieri P., Rasse D.P., Weinerand S. Trumbore S.E. (2011). Persistence of soil organic matter as an ecosystem property. *Nature, 478*: 49–56.

Schmidt M.W., Noack A.G. (2000). Black carbon in soils and sediments: analysis, distribution, implications, and current challenges. *Global Biogeochemical Cycles, 14*: 777–793.

Schoellhamer D.H., Wright S.A., Drexler J.Z. (2013). Adjustment of the San Francisco estuary and watershed to reducing sediment supply in the 20th century. *Marine Geology, 345*: 63–71.

Scholle P.A. (1978). Carbonate rock constituents, textures, cements, and porosities. *American Association of Petroleum Geologists, 27*: 241.

Schubauer J.P., Hopkinson C.S. (1984). Above- and belowground emergent macrophyte production and turnover in a coastal marsh ecosystem, Georgia. *Limnology and Oceanography, 29*: 1052–1065.

Schutz H., Schroder P., Rennenberg H. (1991). In: *Role of Plants in Regulating Methane Flux to the Atmosphere.* T.E.H., Sharkey, H.M. (eds), Academic Press: California, pp. 29–63.

Scott D.S., Medioli F.S. (1978). Vertical zonations of marsh foraminifera as accurate indicators of former sea-levels. *Nature, 272*: 528–31.

SeagrassNet (2010). Available at: http://www.seagrassnet.org (accessed May 15, 2017).

Sears A.L.W., Meisler J., Verdone L.N. (2006). Invasive Ludwigia management plan. Sonoma County Ludwigia Task Force. www.goldridgercd.org/pdfs/ludwigia_man_plan.pdf (accessed March 15, 2016).

Segarra K.E.A., Comerford C., Slaughter J., Joye S.B. (2013). Impact of electron acceptor availability on the anaerobic oxidation of methane in coastal freshwater and brackish wetland sediments. *Geochimica et Cosmochimica Acta, 115*: 15–30.

Semesi I.S., Beer S., Björk M. (2009). Seagrass photosynthesis controls rates of calcification and photosynthesis of calcareous macroalgae in a tropical seagrass meadow. *Marine Ecology Progress Series, 382*: 41–47. doi: 10.3354/meps07973.

Serag M. (1999). Ecology of Four Succulent Halophytes in the Mediterranean Coast of Damietta Egypt. *Estuarine, Coastal and Shelf Science, 49*: 29–36.

Serrano O., Mateo M.A., Renom P., Julià R. (2012). Characterization of soils beneath a *Posidonia oceanica* meadow. *Geoderma, 185*: 26–36.

Serrano O., Ruhon R., Lavery P.S., Kendrick G.A., Hickey S., Masqué, P., Duarte C.M. (2016). Impact of mooring activities on carbon stocks in seagrass meadows. *Scientific Reports, 6*.

Service R.F. (2007). Delta blues, California style. *Science, 317*: 442–445.

Servicio Mareográfico Nacional, National Autonomous University of Mexico, Geophysics Institute. (2016). www.mareografico.unam.mx:8080/Mareografico/Pages/masInformacion/Informacion.jsp (accessed January 27).

Sghaier Y.R., Zakhama-Sraieb R.Y.M., Charfi-Cheikhrouha F. (2013). Patterns of shallow seagrass (*Posidonia oceanica*) growth and flowering along the Tunisian coast. *Aquatic Botany. 104*: 185–192.

Sheehan L., Crooks S. (2016). *Tampa Bay Blue Carbon Assessment: Summary of Findings*. TBEP Technical Report #07-16. St. Petersburg, FL. http://www.tbeptech.org/TBEP_TECH_PUBS/2016/TBEP_07_16_Tampa-Bay-Blue-Carbon-Assessment-Report-FINAL_June16-2016.pdf.

Sheng Y.P., Lapetina A., G.M.a. (2012). The reduction of storm surge by vegetation canopies: three-dimensional simulations, *Geophysical Research Letters, 39*.

Shennan I., Long A.J., Horton B.P. (eds) (2015). *Handbook of Sea-Level Research*, John Wiley & Sons.

Shepard C.C., Crain C.M., Beck M.W. (2011). The protective role of coastal marshes: a systematic review and meta-analysis. *PLoS One, 6*, doi: 10.1371/journal.pone.0027374.

Sheppard C., Al-Husiani M., Al-Jamali F., Al-Yamani F., Baldwin R., Bishop J., Benzoni F., Dutrieux E., Dulvy N.K., Durvasula S.R.V. (2010). The Gulf: a young sea in decline. *Marine Pollution Bulletin, 60*: 13–38.

Sherwood E.T., et al. (2019). Tampa Bay Estuary Case Study: Identifying Blue Carbon Incentives to Further Bolster Future Critical Coastal Habitat Restoration and Management Efforts. In: *A Blue Carbon Primer: The State of Coastal Wetland Carbon Science, Practice, and Policy*, Lisamarie W.-M., Stephen C., and Tiffany G.T. (eds), CRC Press: Boca Raton, London.

Sherwood E.T., Greening H.S. (2013). Potential impacts and management implications of climate change on Tampa Bay estuary critical coastal habitats. *Environmental Management, 53*(2): 401–415. doi: 10.1007/s00267-013-0179-5.

Sherwood E.T., Greening H.S., Janicki A., Karlen D. (2015). Tampa Bay estuary: monitoring long-term recovery through regional partnerships. *Regional Studies in Marine Science, 4*: 1–11. doi: 10.1016/j.rsma.2015.05.005.

Sherwood E.T., Greening H.S., Johansson J.O.R., Kaufman K., Raulerson G.E. (2017). Tampa Bay (Florida, USA): Documenting seagrass recovery since the 1980's and reviewing the benefits. *Southeastern Geographer, 57*(3): 294–319. doi: 10.1353/sgo.2017.0026.

Shiah F.K., Ducklow H.W. (1995). Multiscale variability in bacterioplankton abundance, production, and specific growth rate in a temperate salt-marsh tidal creek. *Limnology and Oceanography, 40*(1): 55–66.

Shields M.R., Bianchi T.S., Y. Gélinas Allison M.A., Twilley R.R. (2016). Enhanced terrestrial carbon preservation promoted by reactive iron in deltaic sediments. *Geophysical Research Letters, 43*(3). doi: 10.1002/2015GL067388.

Shlemon R.J., Begg E.L. (1975). Late quaternary evolution of the Sacramento–San Joaquin Delta, California. In: *Quaternary Studies*, Suggate R.P., Creswell M.M. (eds), The Royal Society of New Zealand, pp. 259–266.

Short F., Carruthers T., Dennison W., Waycott M. (2007). Global seagrass distribution and diversity: a bioregional model. *Journal of Experimental Marine Biology and Ecology, 350*: 3–20.

Short F.T., Coles R., Fortes M.D., Victor S., Salik M., Isnain I., Andrew J., Seno A. (2014). Monitoring in the Western Pacific region shows evidence of seagrass decline in line with global trends. *Marine Pollution Bulletin, 83*: 408–416.

Short F.T., Coles R.G., Pergent-Martini C. (2001). Global Seagrass Distribution. In: *Global Seagrass Research Methods*, Short F.T., Coles R.G. (eds), Elsevier Science: Amsterdam, pp. 5–30.

Short F.T., Duarte C.M. (2001). Methods for measurement of seagrass growth and production. In: *Global Seagrass Research Methods*, Short F.T., Coles R.G. (eds), Elsevier Science: Amsterdam, pp. 155–182.

Short F.T., Polidoro B., Livingstone S.R., Carpenter K.E., Bandeira S., Bujang J.S., Calumpong H.P., Carruthers T.J.B., Coles R.G., Dennison W.C., Erftemeijer P.L.A., Fortes M.D., Freeman A.S., Jagtap T.G., Kamal A.H.M., Kendrick G.A., Kenworthy W.J., La Nafie Y.A., Nasution I.M., Orth R.J., Prathep A., Sanciangco J.C., van Tussenbroek B., Vergara S.G., Waycott M., Zieman J.C. (2011). Extinction risk assessment of the world's seagrass species. *Biological Conservation, 144*: 1961–1971.

Short F.T., Wyllie-Echeverria S. (1996). Natural and human-*induced* disturbance of seagrasses. *Environmental Conservation, 23*(01): 17–27.

Shotyk W. (1996). Natural and anthropogenic enrichments of As, Cu, Pb, Sb, and Zn in ombrotrophic vs. minerotrophic peat bog profiles, Jura Mountains, Switzerland. *Water, Air, and Soil Pollution, 90*, 375–405.

Shoun H., Kim D.-H., Uchiyama H., Sugiyama J. (1992). Denitrification by fungi. *FEMS Microbiology Letters, 94*: 277–282.

Sidik F., Neil D., Lovelock C.E. (2016). Effect of high sedimentation rates on surface sediment dynamics and mangrove growth in the Porong River, Indonesia. *Marine Pollution Bulletin*. doi: 10.1016/j.marpolbul.2016.02.048.

Sidik F., Supriyanto B., Krisnawati H., Muttaqin M.Z. (2018). Mangrove conservation for climate change mitigation in Indonesia. *Wiley Interdisciplinary Reviews: Climate Change*, e529.

Sifleet S., Pendleton L., Murray B. (2011). State of the science on coastal blue carbon. A Summary for Policy Makers. Nicholas Institute Report: 11–06.

Siikamäki J., Sanchirico J.N., Jardine S., McLaughlin D., Morris D. (2013). Blue carbon: coastal ecosystems, their carbon storage, and potential for reducing emissions. *Environment: Science and Policy for Sustainable Development, 55*: 14–29.

Simenstad C., Toft J., Higgins H., Cordell J., Orr M., Williams P., Grimaldo L., Hymanson Z., Reed D. (1999). Preliminary Results from the Sacramento-San Joaquin Delta Breached Levee Wetland Study (BREACH). *IEP Newsletter, 12*: 15–20.

Simon J.J. (1974). Tampa Bay estuarine system – A synopsis. *Florida Scientist, 37*(4): 217–244.

Simpson S. (2016). Analysis and recommendation for grouping blue carbon projects in Tampa Bay. In: Sheehan L., Crooks S. (eds), *Tampa Bay Blue Carbon Assessment: Summary of Findings*, TBEP Technical Report #07-16: St. Petersburg, FL, Appendix L, pp. 319–332. www.tbeptech.org/TBEP_TECH_PUBS/2016/TBEP_07_16_Tampa-Bay-Blue-Carbon-Assessment-Report-FINAL_June16-2016.pdf.

Singer M.B., Aalto R., James L.A. (2008). Status of the lower Sacramento Valley flood-control system within the context of its natural geomorphic setting. *Natural Hazards Review, 9*(3): 104–115.

Sinsabaugh R.L., Manzoni S., Moorhead D.L., Richter A. (2013). Carbon use efficiency of microbial communities: stoichiometry, methodology and modelling. *Ecology Letters, 16*: 930–939.

Slonecker E.T., Jones D.K., Pellerin B.A. (2016). The new Landsat 8 potential for remote sensing of colored dissolved organic matter (CDOM). *Marine Pollution Bulletin, 107*(2): 518–527.

Small C., Nicholls R.J. (2003). A global analysis of human settlement in coastal zones. *Journal of Coastal Research*, 584–599.

Smith C.J., DeLaune R.D., Patrick W.H. Jr. (1983). Nitrous oxide emission from Gulf Coast wetlands. *Geochimica et Cosmochimica Acta, 47*: 1805–1814.

Smith S.V., Atkinson M.J. (1983). Mass balance of carbon and phosphorus in Shark Bay, Western Australia. *Limnology and Oceanography, 28*: 625–639.

Smith S.V., Hollibaugh J. (1993). Coastal metabolism and the oceanic carbon balance. *Reviews of Geophysics, 31*: 75–89.

SNV World (2014). MAM Mangroves & Markets: protecting the ecosystem (full documentary). Running time: 7:06 mins, Link at: www.snv.org/project/mangroves-and-markets.

Soil Survey Staff, Natural Resources Conservation Service, United States Department of Agriculture (2016). Web Soil Survey. http://websoilsurvey.nrcs.usda.gov/ (accessed March 16, 2016).

Sommer T., Armor C., Baxter R., Breuer R., Brown L., Chotkowski M., et al. (2007). The collapse of pelagic fishes in the upper San Francisco Estuary. *Fisheries, 32*: 270–277. doi: 10.1577/15488446(2007)32[270:TCOPFI]2.0.CO;2.

Sorokin D.Y., Makarova K.S., Abbas B., Ferrer M., Golyshin P.N., Galinski E.A., Ciordia S., CarmenMena M., Merkel A.Y., Wolf Y., van Loosdrecht M.C.M., Koonin E.V. (2017). Discovery of extremely halophilic, methyl-reducing euryarchaea provides insights into the evolutionary origin of methanogenesis. *Nature Microbiology, 2*: 17081.

Spalding M., Taylor M., Ravilious C., Short F., Green E. (2003). Global overview: The distribution and stuatus of seagrasses. In: *World Atlas of Seagrasses*, Green E.P., Short F.T. (eds), University of California Press: Berkeley, pp. 5–26.

Spalding M.D., McIvor A.L., Beck M.W., Koch E.W., Möller I., Reed D.J., Wesenbeeck B.K. (2014). Coastal ecosystems: a critical element of risk reduction. *Conservation Letters*, 7(3): 293–301.

Spalding M.D., Ruffo S., Lacambra C., Meliane I., Hale L.Z., Shepard C.C., Beck M.W. (2014). The role of ecosystems in coastal protection: adapting to climate change and coastal hazards. *Ocean and Coastal Management*, 90: 50–57.

Spencer T.M., Schuerch M., Nicholls R.J., Hinkel J., Lincke D., Vafeidis A.T., Reef R., McFadden L., Brown S. (2016). Global coastal wetland change under sea-level rise and related stresses: the DIVA Wetland Change Model. *Global and Planetary Change*, 139: 15–30.

Sprecher S.W. (2001). Basic concepts of soil science. In: *Wetland Soils: Genesis, Hydrology, Landscapes, and Classification*, Richardson, J.L., Vepraskas, M.J. (eds), CRC Press: Boca Raton, pp. 3–18.

Stagg C.L., Mendelssohn I.A. (2011). Controls on resilience and stability in a sediment-subsidized salt marsh. *Ecological Applications*, 21: 1731–1744.

Stagg C.L., Schoolmaster D.R., Krauss K.W., Cormier N., Conner W.H. (2017). Causal mechanisms of soil organic matter decomposition: deconstructing salinity and flooding impacts in coastal wetlands. *Ecology*. doi: 10.1002/ecy.1890.

Stanley D.J., Warne A.G. (1998). Nile Delta in its destruction phase. *Journal of Coastal Research*, 14: 794–825.

State and Federal Contractors Water Agency (2016). SFCWA: State and Federal Contractors Water Agency. www.sfcwa.org/ (accessed March 16, 2016).

State and Federal Contractors Water Agency (SFCWA) (2013). Final Lower Yolo Ranch Restoration Long-term Management Plan, Yolo County, California. www.sfcwa.org/wp-content/uploads/Lower-Yolo-Restoration-Management-Plan.pdf (accessed July 17, 2017).

Stearns L.A., MacCreary D. (1957). The case of the vanishing brick dust: contribution to knowledge of marsh development. *Mosquito News*, 17: 303–304.

Stedmon C., Nelson N.B. (2015). The optical properties of DOM in the ocean. In: *Biogeochemistry of Marine Dissolved Organic Matter*, Hansell DA, Carlson CA (eds). Elsevier Science, pp. 481–508.

Steers J.A. (1938). The rate of sedimentation on salt marshes on Scolt Heat Island, Norfolk. *Geological Magazine*, 75: 26–39.

Steinweg J.M., Plante A.F., Conant R.T., Paul E.A., Tanaka D.L. (2008). Patterns of substrate utilization during long-term incubations at different temperatures. *Soil Biology and Biochemistry*, 40: 2722–2728.

Stephens F.C., Louchard E.M., Reid R.P., Maffione R.A. (2003). Effects of microalgal communities on reflectance spectra of carbonate sediments in subtidal optically shallow marine environments. *Limnology & Oceanography*, 48(1–2): 535–546.

Stevens P.W., Fox S.L., Montague C.L. (2006). The interplay between mangroves and saltmarshes at the transition between temperate and subtropical climate in Florida. *Wetlands Ecology and Management*, 14(5): 435–444.

St-Hilaire F., Wu J., Roulet N.T., Frolking S.E., Lafleur P.M., Humphreys E.R., Arora V. (2010). McGill wetland model: evaluation of a peatland carbon simulator developed for global assessments. *Biogeosciences*, 7: 3517–3530, doi: 10.5194/bg-7-3517-2010.

Stockman K.W., Ginsburg R.N., Shinn E.A. (1967). The production of lime mud by algae in South Florida. *Journal of Sedimentary Petrology*, 37: 633–648.

Stokes B., Wike R., Steward R. (2015). *Global Concern about Climate Change, Broad Support for Limiting Emissions*, Pew Research Centre: Washington, DC.

Stralberg D., Brennan M., Callaway J.C., Wood J.K., Schile L.M., Jongsomjit D., Kelly M., Parker V.T., Crooks S. (2011). Evaluating tidal marsh sustainability in the face of sea-level rise: a hybrid modeling approach applied in San Francisco Bay. *PLoS One*, 6(11), e27388.

Streck C., Keenlyside P., von Unger M. (2016). The Paris agreement: a new beginning. *Journal for European Environmental and Planning Law*, 13, 3–29.

Subair L., Rahman S., Sampara S., Mustamin H. (2017). Implementation of Coastal Area Management Legislation in South Sulawesi. *ADRI International Journal of Law and Social Science*, 1(1): 42–51.

Sukardjo S., Alongi D.M., Kusmana C. (2013). Rapid litter production and accumulation in Bornean mangrove forests. *Ecosphere*, 4(7): 79, doi: 10.1890/ES13-00145.1.

Sutton-Grier A.E., Keller J.K., Koch R., Gilmour C., Megonigal J.P. (2011). Electron donors and acceptors influence anaerobic soil organic matter mineralization in tidal marshes. *Soil Biology and Biochemistry*, *43*: 1576–1583.

Sutton-Grier A.E., Moore A. (2016). Leveraging Carbon Services of Coastal Ecosystems for Habitat Protection and Restoration. *Coastal Management*, *44*: 259–277.

Sutton-Grier A.E., Moore A.K., Wiley P.C., Edwards P.E. (2014). Incorporating ecosystem services into the implementation of existing U.S. natural resource management regulations: the case for carbon sequestration and storage. *Marine Policy*, *43*: 246–253.

Sutton-Grier A.E., Wowk K., Bamford H. (2015). Future of our coasts: the potential for natural and hybrid infrastructure to enhance the resilience of our coastal communities, economies and ecosystems. *Environmental Science and Policy*, *51*: 137–148.

Swanson K.M., Drexler J.Z., Fuller C., Schoellhamer D.H. (2015). Modeling tidal freshwater marsh sustainability in the Sacramento-San Joaquin Delta under a broad suite of potential future scenarios. *San Francisco Estuary Water Sci.*, *13*(1). http://escholarship.org/uc/item/7x65r0tf.

Swanson K.M., Drexler J.Z., Schoellhamer D.H., Thorne K.M., Casazza M.L., Overton C.T., Callaway J.C., Takekawa J.Y. (2014). Wetland Accretion Rate Model of Ecosystem Resilience (WARMER) and its application to habitat sustainability for endangered species in the San Francisco Estuary. *Estuaries and Coasts*, *37*: 476–492. doi: 10.1007/s12237-013-9694-0.

SWFWMD (Southwest Florida Water Management District) (2011). *Land Use/Cover GIS Shapefiles*, Brooksville, FL. http://data-swfwmd.opendata.arcgis.com/datasets/f325a3417c92444d9cba838154d6fa0d_11.

SWFWMD (Southwest Florida Water Management District) (2016). *Southwest Florida Seagrass Coverage GIS Shapefiles*, Brooksville, FL. http://data-swfwmd.opendata.arcgis.com/datasets/f0ecff0cf0de491685f8fb074adb278b_20.

Syrbe R.U., Walz U. (2012). Spatial indicators for the assessment of ecosystem services: providing, benefitting and connecting areas and landscape metrics. *Ecological Indicators*, *21*: 80–88.

Tamooh F., Huxham M., Karachi M., Mencuccini M., Kairo J.G., Kirui B. (2008). Below-ground root yield and distribution in natural and replanted mangrove forests at Gazi bay, Kenya. *Forest Ecology and Management*, *256*(6): 1290–1297. http://doi.org/10.1016/j.foreco.2008.06.026.

Tang J., Baldocchi D.D., Xu L. (2005). Tree photosynthesis modulates soil respiration on a diurnal time scale. *Global Change Biology*, *11*: 1298–1304.

Tang J., Bolstad P.V., Desai A.R., Martin J.G., Cook B.D., Davis K.J., Carey E.V. (2008). Ecosystem respiration and its components in an old-growth forest in the Great Lakes region of the United States. *Agricultural and Forest Meteorology*, *148*: 171–185.

TAS (Tampa Audubon Society) (1999). *No Crayfish, No Ibis? The Importance of Freshwater Wetlands for Coastal-Nesting White Ibis in Tampa Bay*, TBEP Technical Report #11-99: St. Petersburg, FL. www.tbeptech.org/TBEP_TECH_PUBS/1999/TBEP_11_99NoIbisNoCrayfish.pdf.

TBCSAP (Tampa Bay Climate Science Advisory Panel) (2015). *Recommended Projection of Sea Level Rise in the Tampa Bay Region*, St. Petersburg, FL.

TBRPC (Tampa Bay Regional Planning Council) (2014). *Economic Valuation of Tampa Bay*, TBEP Technical Report #04-14: , St. Petersburg, FL. www.tbeptech.org/TBEP_TECH_PUBS/2014/TBEP_04_14_%20FinalReport_Economic_Valuation_of_Tampa_Bay_Estuary.pdf.

Teal J. (1957). Community metabolism in a temperate cold spring. *Ecological Monographs*, *27*: 283–302.

Teal J.M. (1962). Energy flow in the salt marsh ecosystem of Georgia. *Ecology*, *43*: 614–624.

Telesca L., Belluscio A., Criscoli A., Ardizzone G., Apostolaki E.T., Fraschetti S., Gristina M., Knittweis L., Martin C.S., Pergent G., Alagna A., Badalamenti F., Garofalo G., Gerakaris V., Pace M.L., Pergent-Martini C., Salomidi M. (2015). Seagrass meadows (*Posidonia oceanica*) distribution and trajectories of change. *Scientific Reports*, *5*: 12505.

Temmerman S., Meire P., Bouma T.J., Herman P.M., Ysebaert T., De Vriend H.J. (2013). Ecosystem-based coastal defence in the face of global change. *Nature*, *504*(7478): 79–83.

Tessler Z.D., Vorosmarty C.J., Grossberg M., Gladkova I., Aizenman H., Syvitski J.P.M., Foufoula-Georgiou E. (2015). Profiling risk and sustainability in coastal deltas of the world. *Science*, *349*: 638–643.

Teutli-Hernández C., Herrera Silveira J.A. (2016). Estrategias de la restauración de manglares de México: el caso Yucatán, In: *Experiencias Mexicanas en la Restauración de Ecosistemas*, CONABIO: UNAM.

The REDD Desk (2017). Indonesia. http://theredddesk.org/countries/indonesia, Global Canopy Program (in).

Thenkabail P.S., Enclona E.A., Ashton M.S., Van B., Der Meer (2004). Accuracy assessments of hyperspectral waveband performance for vegetation analysis applications. *Remote Sensing of Environment*, *91*: 354–376.

Thiele T., Gerber L.R. (2017). Innovative financing for the High Seas. *Aquatic Conservation: Marine and Freshwater Ecosystems*, *27*(S1): 89–99. doi: 10.1002/aqc.2794.

Thilenius J.F. (1990). Plant succession on earthquake uplifted coastal wetlands, Copper River Delta, Alaska. *Northwest Science*, *64*: 259–262.

Thom R., Southard S., Borde A. (2014). Climate-linked mechanisms driving spatial and temporal variation in eelgrass (*Zostera marina* L.) growth and assemblage structure in Pacific Northwest estuaries, U.S.A. *Journal of Coastal Research*, *68*: 1–11.

Thomas N., Lucas R., Bunting P., Hardy A., Rosenqvist A., Simard M. (2017). Distribution and drivers of global mangrove forest change, 1996–2010. *PLoS One*, *12*: e0179302. doi: 10.1371/journal.pone.0179302.

Thomas S. (2014). Blue carbon: knowledge gaps, critical issues, and novel approaches. *Ecological Economics*, *107*: 22–38.

Thomas S., Ridd P.V. (2004). Review of methods to measure short time scale sediment accumulation. *Marine Geology*, *207*: 95–114.

Thompson B.S., Clubbe C.P., Primavera J.H., Curnick D., Koldewey H.J. (2014). Locally assessing the economic viability of blue carbon: a case study from Panay Island, the Philippines. *Ecosystem Services*, *8*: 128–140.

Thompson B.S., Primavera J.H., Friess D.A. (2017). Governance and implementation challenges for mangrove forest Payments for Ecosystem Services (PES): empirical evidence from the Philippines. *Ecosystem Services*, *23*: 146–155.

Thompson J. (1957). The settlement geography of the Sacramento-San Joaquin Delta, California. *PhD Dissertation*, Stanford University, Stanford, CA, USA.

Thompson J. (2006). Early reclamation and abandonment of the central Sacramento-San Joaquin Delta. Sacramento History: *Journal of the Sacramento County Historical Society*, *VI*(1–4): 41–72.

Thomson J.R., Kimmerer W.J., Brown L.R., Newman K.B., Mac Nally R.M., Bennett W.A., et al. (2010). Bayesian changepoint analysis of abundance trends for pelagic fishes in the upper San Francisco Estuary. *Ecological Applications*, *20*: 181–198, doi: 10.1890/090998.1.

Thorne K., MacDonald G., Guntenspergen G., Ambrose R., Buffington K., Dugger B., Freeman C., Janousek C., Brown L., Rosencranz J., Holmquist J. (2018). US Pacific coastal wetland resilience and vulnerability to sea-level rise. *Science Advances*, *4*(2): eaao3270.

Thorne K.M., Dugger B.D., Buffington K.J., Freeman C.M., Janousek C.N., Powelson K.W., Gutenspergen G.R., Takekawa J.Y. (2015). *Marshes to Mudflats—Effects of Sea-Level Rise on Tidal Marshes Along a Latitudinal Gradient in the Pacific Northwest*, no. 2015-1204, US Geological Survey.

Thorne K.M., Elliott-Fisk D.L., Wylie G.D., Perry W.M., Takekawa J.Y. (2014). Importance of biogeomorphic and spatial properties in assessing a tidal salt marsh vulnerability to sea-level rise. *Estuaries and coasts*, *37*(4): 941–951.

Thouvenot L., Haury J., Thiebaut G. (2013). A success story: water primroses, aquatic plant pests. *Aquatic Conservation: Marine and Freshwater Ecosystems*, *23*: 790–803.

Tobias C., Neubauer S.C. (2009). Salt marsh biogeochemistry – an overview. In: *Coastal Wetalnds: An Integrated Approach*, Perillo G., Wolanski E., Cahoon D., Brinson M. (eds), Elsevier: Amsterdam, pp. 445–492.

Tobias V.D., Williamson M.F., Nyman J.A. (2014). A comparison of the elemental composition of leaf tissue of *Spartina patens* and *Spartina alternifora* in Louisiana's coastal marshes. *Journal of Plant Nutrition*, *37*: 1327–1344.

Toft J.D., Simenstad C.A., Cordell J.R., Grimaldo L.F. (2003). The effects of introduced water hyacinth on habitat structure, invertebrate assemblages, and fish diets. *Estuaries*, *26*: 746–758.

Tokoro T., Hosokawa S., Miyoshi E. et al. (2014). Net uptake of atmospheric CO_2 by coastal submerged aquatic vegetation. *Global Change Biology*, *20*: 1873–1884.

Toledo G., Rojas A., Bashan Y. (2001). Monitoring of black mangrove restoration with nursery-reared seedlings on an arid coastal lagoon. *Hydrobiologia*, *444*: 101–109.

Tomasko D., Crooks S., Robison D. (2016). Carbon sequestration estimates of seagrass meadows in Tampa Bay. In: *Proceedings, Tampa Bay Area Scientific Information Symposium, BASIS 6*, Burke M. (eds), St. Petersburg, FL, pp. 259–272. www.tbeptech.org/BASIS/BASIS6/BASIS6_Proceedings_FINAL.pdf (accessed September 28–30, 2015).

Tomlinson P.B. (1986). *The Botany of Mangroves*, Cambridge University Press: Cambridge, UK, p. 419.

Tonelli M., Fagherazzi S., Petti M. (2010). Modeling wave impact on salt marsh boundaries. *Journal of Geophysical Research: Oceans*, *115*(C9).

Tong C., Wang W.Q., Huang J.F., Gauci V., Zhang L.H., Zeng C.S. (2012). Invasive alien plants increase CH_4 emissions from a subtropical tidal estuarine wetland. *Biogeochemistry*, *111*: 677–693.

Tong C., Wang W.Q., Zeng C.S., Marrs R. (2010). Methane (CH_4) emission from a tidal marsh in the Min River estuary, southeast China. *Journal of Environmental Science and Health Part a- Toxic/Hazardous Substances and Environmental Engineering*, *45*: 506–516.

Törnqvist T.E., Rosenheim B.E., Hu P., Fernandez A.B. (2015). Radiocarbon dating and calibration. *Handbook of Sea-Level Research*, In: Shennan I., Long A.J., Horton B.P. (eds), Wiley-Blackwell: Hoboken, NJ, pp. 347–360.

Törnqvist T.E., Wallace D.J., Storms J.E.A., Wallinga J., Van Dam R.L., Blaauw M., Derksen M.S., Klerks C.J.W., Meijneken C., Snijders E.M.A. (2008). Mississippi Delta subsidence primarily caused by compaction of Holocene strata. *Nature Geoscience*, *1*: 173–176.

Torres-Pulliza D., Wilson J.R., Darmawan A., Campbell S.J., Andréfouët S. (2013). Ecoregional scale seagrass mapping: a tool to support resilient MPA network design in the Coral Triangle. *Ocean & Coastal Management*, *80*: 55–64.

Towler E., Rajagopalan B., Gilleland E., Summers R.S., Yates D., Katz R.W. (2010). Modeling hydrologic and water quality extremes in a changing climate: a statistical approach based on extreme value theory. *Water Resources Research*, *46*: W11504. doi: 10.1029/2009WR008876.

Trefry J.H., Metz S., Nelsen T.A., Trocine T.P., Eadie B.A. (1994). Transport and fate of particulate organic carbon by the Mississippi River and its fate in the Gulf of Mexico. *Estuaries*, *17*(4): 839–849.

Troche-Souza C., Rodríguez-Zúñiga M.T., Velázquez-Salazar S., Valderrama-Landeros L., Villeda-Chávez E., Alcántara-Maya A., Vázquez-Balderas B., Cruz-López M.I., Ressl R. (2016). *Manglares de México: extensión, distribución y monitoreo (1970/1980–2015)*. *CONABIO*, Comisión Nacional para el Conocimiento y Uso de la Biodiversidad: México, D.F.

Troxler T.G., et al. Integrated carbon budget models for the Everglades terrestrial-coastal-oceanic gradient: current status and needs for inter-site comparisons. *Oceanography*, *26*: 98–107. doi: 10.5670/oceanog.2013.51.

Troxler T.G., Starr G., Boyer J.N., Fuentes J.D., Jaffe R., et al. (in press). Carbon Cycling in the Florida Coastal Everglades Social-Ecological System Across Scales. In: *The Coastal Everglades: The Dynamics of Socio-Ecological Transformations in the South Florida Landscape*. Childers D.L., Ogden L., Gaiser E. (eds), Chapter 6.

Tschakert P., Sagoe R. (2009). Mental models (concept maps): understanding the causes and consequences of climate change. *Participatory Learning and Action, Community-Based Adaptation to Climate Change*, *60*: 154–159.

Tsompanoglou K., Croudace I.W., Birch H., Collins M. (2011). Geochemical and radiochronological evidence of North Sea storm surges in salt marsh cores from the Wash embayment (UK). *Holocene*, *21*: 225–236.

Tucker C.J. (1977). Asymptotic nature of grass canopy spectral reflectance. *Applied Optics*, *16*: 1151–1156.

Tucker C.J. (1979). Red and photographic infrared linear combinations for monitoring vegetation. *Remote Sensing of Environment*, *8*: 127–150.

Turk D., Yates K.K., Vega-Rodriguez M. et al. (2015). Community metabolism in shallow coral reef and seagrass ecosystems, lower Florida Keys. *Marine Ecology Progress Series*, *538*: 35–52.

Turner R. (2004). Coastal wetland subsidence arising from local hydrologic manipulations. *Estuaries and Coasts*, *27*(2): 265–272.

Turner R.E. (1991). Tide gauge records, water level rise, and subsidence in the Northern Gulf of Mexico. *Estuaries*, *14*: 139–147.

Turner R.E., Howes B.L., Teal J.M., et al. (2009). Salt marshes and eutrophication: an unsustainable outcome. *Limnology and Oceanography*, *54*: 1634–1642.

Turner R.E., Howes B.L., Teal J.M., Milan C.S., Swenson E.M., Tonerb D.D.G. (2009). Salt marshes and eutrophication: an unsustainable outcome. *Limnology and Oceanography*, *54*(5): 1634–1642.

Turner R.E., Swenson E.M., Milan C.S. (2000). Organic and inorganic contributions to vertical accretion in salt marsh sediments. In: *Concepts and Controversies in Tidal Marsh Ecology*, Weinstein M.P., Kreeger D.A. (eds), Kluwer Academic Press: Dordrecht, The Netherlands, pp. 583–595.

Turner R.E., Swenson E.M., Milan C.S., Lee J.M., Oswald T.A. (2004). Below-ground biomass in healthy and impaired salt marshes. *Ecological Research*, *19*: 29–35.

Turner R.K., Jones T. (1991). *Market and Intervention Failures: Four Case Studies*, Earthscan: Wetlands, London.

Turpie K.R., Klemas V.V., Byrd K., Kelly M., Jo Y.-H. (2015). Prospective HyspIRI global observations of tidal wetlands. *Remote Sensing of Environment*, *167*: 206–217.

Twilley R.R., Chen R., Hargis T. (1992). Carbon sinks in mangroves and their implication to carbon budget of tropical ecosystems. *Water, Air and Soil Pollution*, *64*: 265–288.

Twilley R.R., Pozo M., Garcia V.H., RiveraMonroy V., Bodero R.Z.A. (1997). Litter dynamics in riverine mangrove forests in the Guayas River estuary, Ecuador. *Oecologia*, *111*: 109–122.

Tzortziou M., Neale P.J., Megonigal J.P., Pow C.L., Butterworth M. (2011). Spatial gradients in dissolved carbon due to tidal marsh outwelling into a Chesapeake Bay estuary. *Marine Ecology Progress Series*, *426*: 41–56.

Tzortziou M., Neale P.J., Osburn C.L., Megonigal J.P., Maie N., Jaffé R. (2008). Tidal marshes as a source of optically and chemically distinctive colored dissolved organic matter in the Chesapeake Bay. *Limnology and Oceanography*, *53*(1): 148.

Tzortziou M., Osburn C.L., Neale P.J. (2007). Photobleaching of dissolved organic material from a tidal marsh-estuarine system of the Chesapeake Bay. *Photochemistry and Photobiology*, *83*(4): 782–792.

U.S. Energy Information Administration (2015). Energy-related carbon dioxide emissions at the state level, 2000–2013. www.eia.gov/environment/emissions/state/analysis/pdf/stateanalysis.pdf/ (accessed on April 1, 2016).

Uhrin A.V., Townsend P.A. (2016). Improved seagrass mapping using linear spectral unmixing of aerial photographs. *Estuarine, Coastal and Shelf Science*, *171*: 11–22.

UN Habitat (2010). *The Changwon Declaration: Cities Responding to Climate Change*, UN Habitat – Government of Changwon: Korea.

UNDP (2016). *The 2016 BIOFIN Workbook: Mobilizing Resources for Biodiversity and Sustainable Development*, The Biodiversity Finance Initiative.

UNEP (2005). *After the Tsunami: Rapid Environmental Assessment Report*. Nairobi, Kenya. February 22.

UNESCO, (2009). World Water Assessment Programme. Water in a Changing World. The United Nations World Water Development Report 329.

UNFCCC (2011a). Provisional Agenda and Annotations. Subsidiary Body for Scientific and Technological Advice, Thirty-Fourth Session. http:// unfccc.int/resource/docs/2011/sbsta/eng/01a01.pdf.

UNFCCC (2011b). Decision 2/CMP.7. Land use, land-use change, forestry.

UNFCCC (2011c). Decision 12/CP.17. Guidance on systems for providing information on how safeguards are addressed, respected, modalities relating to forest reference emission levels, forest reference levels as referred to in decision 1/CP.16.

UNFCCC (2012c). Update on Developments in Research Activities Relevant to the Needs of the Convention, Including on the Long-Term Global Goal; and Information on Technical and Scientific Aspects of Emissions and Removals of all Greenhouse Gases from Coastal and Marine Ecosystems, Submissions from Regional and International Climate Change Research Programmes and Organizations. Subsidiary Body for Scientific and Technological Advice, Thirty-Sixth Session. http://unfccc.int/resource/docs/2012/sbsta/eng/misc03.pdf.

UNFCCC (2013). Warsaw Framework for REDD-plus. http://unfccc.int/land_use_and_climate_change/redd/items/8180.php.

UNFCCC (2013a). UNFCCC Workshop on technical and scientific aspects of ecosystems with high-carbon reservoirs not covered by other agenda items under the Convention. http://unfccc.int/science/workshops_meetings/items/7797.php.

UNFCCC (2016). Decisions adopted by COP 22 and CMP 12 and CMA 1 http://unfccc.int/2860.php.

UNFCCC (2016a). FOCUS: Mitigation – NAMAs, Nationally Appropriate Mitigation Actions http://unfccc.int/focus/mitigation/items/7172.php.

UNFCCC (2016b). NAMA Registry. http://unfccc.int/cooperation_support/nama/items/7476.php.

UNFCCC (United Nations Framework Convention on Climate Change) (2011). SBSTA 34 Dialogue on Development of Research Activities Relevant to the needs of the Convention. http://unfccc.int/methods_and_science/research_and_systematic_observation/items/6044.php.

UNFCCC/SBSTA (2012a). Report of the Subsidiary Body for Scientific and Technological Advice on its thirty-seventh session, held in Doha from 26 November to 2 December (2012).

UNFCCC/SBSTA (2014). 40th session. Report on the workshop on technical and scientific aspects of eco-systems with high-carbon reservoirs not covered by other agenda items under the Convention FCCC/SBSTA/2014/INF.1.

UNFCCC/SBSTA (2015). Report of the Subsidiary Body for Scientific and Technological Advice on its forty-third session, held in Paris from 1 to 4 December 2015 FCCC/SBSTA/2015/5.

UNFCCC/SBSTA (2016). Report of the Subsidiary Body for Scientific and Technological Advice on its forty-fourth session, held in Bonn from 16 to 26 May 2016. FCCC/SBSTA/2016/2.

UNFCCC/SBSTA (Subsidiary Body for Scientific and Technological Advice) (2012). Summary Report by the SBSTA Chair. SBSTA 36 Research Dialogue.

United Nations Environment Programme [UNEP] (2014). *The Importance of Mangroves to People: A Call to Action*, van Bochove J., Sullivan E., Nakamura T. (eds), United Nations Environment Programme World Conservation Monitoring Centre: Cambridge. p. 128.

United Nations Environment Programme-World Conservation Monitoring Centre [UNEP-WCMC], Short F.T. (2016). Global Distribution of Seagrasses (version 4.0). Available at: http://data.unep-wcmc.org/datasets/7 (accessed May 9, 2017).

United States Fish and Wildlife Service (2017). *Stone Lakes: National Wildlife Refuge*. www.fws.gov/refuge/Stone_Lakes/about.html (accessed July 17, 2017).

Unsworth R.K.F., Collier C.J., Henderson G.M., McKenzie L.J. (2012). Tropical seagrass meadows modify seawater carbon chemistry: implications for coral reefs impacted by ocean acidification. *Environmental Research Letters*, 7: 024026. doi: 10.1088/1748-9326/7/2/024026.

Unsworth R.K.F., Collier C.J., Waycott M., Mckenzie L.J., Cullen-Unsworth L.C. (2015). A framework for the resilience of seagrass ecosystems. *Marine Pollution Bulletin*, 100(1): 34-46. doi: 10.1016/j.marpolbul.2015.08.016 .

US Army Corps of Engineers Coastal Risk Reduction and Resilience (USACE 2013). www.corpsclimate.us/docs/USACE_Coastal_Risk_Reduction_final_CWTS_2013-3.pdf.

US DOE (United States Department of Energy) (2017). *Research Priorities to Incorporate Terrestrial-Aquatic Interfaces in Earth System Models*, US Department of Energy, Office of Science, Office of Biological and Environmental Research. p. 96. tes.science.energy.gov.

US Forest Service. http://www.nrs.fs.fed.us/fia/data-tools/state-reports/glossary/default.asp.

USEPA (United States Environmental Protection Agency) (2016). *Greenhouse Gas Equivalencies Calculator*. Online resource: www.epa.gov/energy/greenhouse-gas-equivalencies-calculator.

U.S. Environmental Protection Agency (2017). Inventory of US greenhouse gas emissions and sinks: 1990–2015. Environmental Protection Agency 2017. www.epa.gov/ghgemissions/inventory-us-greenhouse-gas-emissions-and-sinks-1990-2015.

USGCRP (2017). *Climate Science Special Report: Fourth National Climate Assessment, Volume I*, Wuebbles D.J., Fahey D.W., Hibbard K.A., Dokken D.J., Stewart B.C., Maycock T.K. (eds). Global U.S. Change Research Program: Washington, DC, USA, p. 470. doi: 10.7930/J0J964J6.

Ustin S.L., Greenberg J.A., Hestir E.L., Khanna S., Santos M.J., Andrew M.E., Whiting M., Rajapakse S., Lay M. (2006). *Mapping Invasive Plant Species in the Sacramento-San Joaquin Delta Using Hyperspectral Imagery*, California Department of Boating and Waterways: Sacramento, CA.

Ustin S.L., Khanna S., Bellvert J., Boyer J.D., Shapiro K. (2016). Impact of drought on Submerged Aquatic Vegetation (SAV) and Floating Aquatic Vegetation (FAV) using AVIRIS-NG airborne imagery. *California Department of Fish and Wildlife*, p. 40. (June, 2015).

Ustin S.L., Khanna S., Lay M., Shapiro K., Ghajarnia N. (2017). *Enhancement of Delta Smelt (Hypomesus Transpacificus) Habitat through Adaptive Management of Invasive Aquatic Weeds in the Sacramento-San Joaquin Delta*. Report submitted to California Department of Fish and Wildlife, p. 40. (June, 2017).

Ustin S.L., Santos M.J., Hestir E.L., Khanna S., Casas A., Greenberg J. (2015). Developing the capacity to monitor climate change impacts in Mediterranean estuaries. *Evolutionary Ecology Research*, 16: 529–550.

Valderrama L., et al. (2014). Evaluation of mangrove cover changes in Mexico during the 1970–2005 Period. *Wetlands*, 34(4): 747–758.

Valenzuela E.I., Prieto-Davó A., López-Lozano N.E., Hernández-Eligio A., Vega-Alvarado L., Juárez K., García-González A.S. López M.G., Cervantes F.J. (2017). Anaerobic methane oxidation driven by microbial reduction of natural organic matter in a tropical wetland. *Applied and Environmental Microbiology*, 83: e00645–e00617.

Valiela I. (1995). *Marine Ecological Processes*, 2nd edn, Springer: New York.

Valiela I. (2006). *Global Coastal Change*, Oxford, U.K. Blackwell Publishing.

Valiela I., Bowen J.L., York J.K. (2001). Mangrove Forests: one of the World's Threatened Major Tropical Environments: at least 35% of the area of mangrove forests has been lost in the past two decades, losses that exceed those for tropical rain forests and coral reefs, two other well-known threatened environments. *Bioscience, 51*(10): 807–815.

Valiela I., Cole M.L. (2002). Comparative evidence that salt marshes and mangroves may protect seagrass meadows from land-derived nitrogen loads. *Ecosystems, 5*: 92–102.

Valle M., Chust G., del Campo A., Wisz M.S., Olsen S.M., Garmendia J.M., Borja Á. (2014). Projecting future distribution of the seagrass *Zostera noltii* under global warming and sea level rise. *Biological Conservation, 170*: 74–85.

Valle M., Palà V., Lafon V., Dehouck A., Garmendia J.M., Borja Á., Chust G. (2015). Mapping estuarine habitats using airborne hyperspectral imagery, with special focus on seagrass meadows. *Estuarine, Coastal and Shelf Science, 164*: 433–442.

Valle M., van Katwijk M.M., de Jong D.J., Bouma T.J., Schipper A.M., Chust G., Benito B.M., Garmendia J.M., Borja Á. (2013). Comparing the performance of species distribution models of *Zostera marina*: implications for conservation. *Journal of Sea Research, 83*: 56–64.

Vallino J.J., Hopkinson C.S., Garritt R.H. (2005). Estimating estuarine gross production, community respiration and ent ecosystem production: a nonlinear inverse technique. *Ecological Modelling, 187*: 281–296.

Van de Broek M., Temmerman S., Merckx R., Govers G. (2016). Controls on soil organic carbon stocks in tidal marshes along an estuarine salinity gradient. *Biogeosciences, 13*: 6611–6624.

van der Heide T., van Nes E.H., Geerling G.W., Smolders A.J.P., Bouma T.J., van Katwijk M.M. (2007). Positive feedbacks in seagrass ecosystems: implications for success in conservation and restoration. *Ecosystems, 10*: 1311–1322.

van Hooidonk R., Maynard J.A., Planes S. (2013). Temporary refugia for coral reefs in a warming world. *Nature Climate Change, 3*: 508–511.

Van Katwijk M.M., Bos A.R., De Jonge V.N., Hanssen L.S.A.M., Hermus D.C.R., De Jong D.J. (2009). Guidelines for seagrass restoration: importance of habitat selection and donor population, spreading of risks, and ecosystem engineering effects. *Marine Pollution Bulletin, 58*(2): 179–188.

Van Oudenhoven A.P.E., Petz K., Alkemade R., Hein L., de R.S., Groot. (2012). Framework for systematic indicator selection to assess effects of land management on ecosystem services. *Ecological Indicators, 21*: 110–122.

Vandevivere P., Baveye P. (1992). Effect of bacterial extracellular polymers on the saturated hydraulic conductivity of sand columns. *Applied and Environmental Microbiology, 58*: 1690–1698.

Vann C.D., Megonigal J.P. (2003). Elevated CO_2 and water depth regulation of methane emissions: comparison of woody and non-woody wetland plant species. *Biogeochemistry, 63*: 117–134.

Vargas R., et al. (2012). Opportunities for advancing carbon cycle science in Mexico: toward a continental scale understanding. *Environmental Science and Policy, 21*: 84–93.

Vasilijevic A., Miskovic N., Vukic Z., Mandic F. (2014). Monitoring of seagrass by lightweight AUV: a *Posidonia oceanica* case study surrounding Murter Island of Croatia. *2014 22nd Mediterranean Conference of Control and Automation (MED)*. doi: 10.1109/MED.2014.6961465.

VCS (2018). VCS Project Database – Sectoral Scope 14. http://vcsprojectdatabase.org/.

Vegh T., Jungwiwattanaporn M., Pendleton L., Murray B.C. (2014). *Mangrove Ecosystem Services Valuation: State of the Literature*, Durham, NC, USA, Duke University.

Verhoeven J.T., et al. (2006). Regional and global concerns over wetlands and water quality. *Trends in Ecology & Evolution, 21*: 96–103.

Verified Carbon Standard (2017). *Agriculture, Forestry, and Other Land Use (AFOLU) Requirements. VCS Version 3 Requirements Document*, Verified Carbon Standard: Washington, D.C.

Vermaat J.E., Thampanya U. (2006). Mangroves mitigate tsunami damage: a further response. *Estuarine, Coastal and Shelf Science, 69*: 1–3.

Vermeer M., Rahmstorf S. (2009). Global sea level linked to global temperature. *Proceedings of the National Academy of Sciences of the United States of America. 106*: 21527–21532.

Verra (2018). Standards for a Sustainable Future. http://verra.org/.

Villamagna A.M., Angermeier P.L., Bennett E.M. (2013). Capacity, pressure, demand, and flow: a conceptual framework for analyzing ecosystem service provision and delivery. *Ecological Complexity, 15*: 114–121.

Vizza C., West W.E., Jones S.E., Hart J.A., Lamberti G.A. (2017). Regulators of coastal wetland methane production and responses to simulated global change. *Biogeosciences*, *14*: 431–446.

Vo Q.T., Kuenzer C., Vo Q.M., Moder F., Oppelt N. (2012). Review of valuation methods for mangrove ecosystem services. *Ecological Indicators*, *23*: 431–446.

Von Unger M., Emmer I.E. (in prep). Carbon Market Incentives to Conserve, Restore and Enhance Soil Carbon. Draft report by Silvestrum Climate Associates to The Nature Conservancy.

Vonk J.A., Middelburg J.J., Stapel J., Bouma T.J. (2008). Dissolved organic nitrogen uptake by seagrasses. *Limnology and Oceanography*, *53*: 542–548.

Vorosmarty C.J., McIntyre P.B., Gessner M.O., Dudgeon D., Prusevich A., Green P., Glidden S., Bunn S.E., Sullivan C.A., Liermann C.R., Davies P.M. (2010). Global threats to human water security and river biodiversity. *Nature*, *467*: 555–561.

Vu B., Chen M., Crawford R.J., Ivanova E.P. (2009). Bacterial extracellular polysaccharides involved in biofilm formation. *Molecules*, *14*: 2535–2554.

Walker D.I., Kendrick G.A., McComb A.J. (2006). Decline and recovery of seagrass ecosystems—the dynamics of change. In: *Seagrasses: Biology, Ecology and Conservation*, Larkum A.W.D., Orth R.J., Duarte C.M., Springer: Dordrecht, The Netherlands, pp. 551–565.

Walker D.I., Woelkerling W.J. (1988). Quantitative study of soil contribution by epiphytic coralline red algae in seagrass meadows in Shark Bay, Western Australia. *Marine Ecology Progress Series*, *43*: 71–77.

Walker M., Johnsen S., Rasmussen S.O., Popp T., Steffensen J.P., Gibbard P., Hoek W., Lowe J., Andrews J., Björck S., Cwynar L.C. (2009). Formal definition and dating of the GSSP (Global Stratotype Section and Point) for the base of the Holocene using the Greenland NGRIP ice core, and selected auxiliary records. *Journal of Quaternary Science*, *24*(1): 3–17.

Walter B.P., Heimann M. (2000). A process-based, climate-sensitive model to derive methane emissions from natural wetlands: application to five wetland sites, sensitivity to model parameters, and climate. *Global Biogeochemical Cycles*, *14*: 745– 765.

Walters B.B., Rönnbäck P., Kovacs J.M., Crona B., Hussain S.A., Badola R., Primavera J.H., Barbier E., Dahdouh-Guebas F. (2008). Ethnobiology, socio-economics and management of mangrove forests: a review. *Aquatic Botany*, *89* (2): 220–236.

Wamsley T.V., Cialone M.A., Smith J.M., Atkinson J.H., Rosati J.D. (2010). The potential of wetlands in reducing storm surge. *Ocean Engineering*, *37*: 59–68.

Wamsley T.V., Cialone M.A., Smith J.M., Ebersole B.A., Grzegorzewski A.S. (2009). Influence of landscape restoration and degradation on storm surge and waves in southern Louisiana. *Natural Hazards*, *51*: 207–224.

Wang Q.K., Wang S.L., He T.X., Liu L., Wu J.B. (2014). Response of organic carbon mineralization and microbial community to leaf litter and nutrient additions in subtropical forest soils. *Soil Biology and Biochemistry*, *71*: 13–20.

Wang S., Di Iorio D., Cai W.-J., Hopkinson C. S. (2017). Inorganic carbon and oxygen dynamics in a marsh-dominated estuary. *Limnology and Oceanography*. doi: 10.1002/lno.10614.

Wang Y. (2012). Detecting Vegetation Recovery Patterns After Hurricanes in South Florida Using NDVI Time Series.

Wang Z.A., Cai W.-J. (2004). Carbon dioxide degassing and inorganic carbon export from a marsh-dominated estuary (the Duplin River): a marsh CO_2 pump. *Limnology and Oceanography*, *49*: 341–354.

Wang Z.A., Kroeger K.D., Ganju N.K., Chu S.N., Gonneea M.E. (2016). Salt marshes as an important source of inorganic carbon to the coastal ocean. *Limnology and Oceanography*, *61*: 1916–1931. doi: 10.1002/lno.10347.

Wankel S.D., Ziebis W., Buchwald C., Charoenpong C., de Beer D., Dentinger J., Xu Z., Zengler K. (2017). Evidence for fungal and chemodenitrification based N_2O flux from nitrogen impacted coastal sediments. *Nature Communications*, *8*: 15595.

Ward N.D., Bianchi T.S., Medeiros P.M., Seidel M., Richey J.E., Keil R.G., Sawakuchi H.O. (2017). Where carbon goes when water flows: carbon cycling across the aquatic continuum. *Frontiers in Marine Science*, *4*: 7. doi: 10.3389/fmars.2017.00007.

Ward N.D., Bianchi T.S., Medeiros P.M., Seidel M., Richey J.E., Keil R.G., Sawakuchi H.O. (2017). Where carbon goes when water flows: carbon cycling across the aquatic continuum. *Frontiers in Marine Science*, *4*: 7.

Ward N.D., Bianchi T.S., Sawakuchi H.O., Maynard W.G., Cunha A.C., Brito D.C., Neu V., de Matos Valerio A., da Silva R., Krusche A.V., Richey J.E., Keil R.G. (2016). The reactivity of plant-derived organic matter and the potential importance of priming effects along the lower Amazon River. *Journal of Geophysical Research-Biogeosciences*, *121*: 1522–1539.

Watson E.B., Wigand C., Davey E.W., Andrews H.M., Bishop J., Raposa K.B. (2016). Wetland loss patterns and inundation-productivity relationships prognosticate widespread salt marsh loss for southern New England. *Estuaries and Coasts*. doi: 10.1007/s12237-016-0069-1.

Waycott M., Collier C., McMahon K., Ralph P., McKenzie L., Udy J., Grech A. (2007). *Vulnerability of Seagrasses in the Great Barrier Reef to Climate Change*, Great Barrier Reef Marine Park Authority and Australian Greenhouse Office, pp. 193–236.

Waycott M., Duarte C.M., Carruthers T.J.B., Orth R.J., Dennison W.C., Olyarnik S., Calladine A., Fourqurean J.W., Heck K.L. Jr., Hughes A.R., Kendrick G.A., Kensorthy W.J., Short F.T., Williams S.L. (2009). Accelerating loss of seagrass across the globe threatens coastal ecosystems. *Proceedings of the National Academy of Sciences*, *106*(30): 12377–12381. doi: 10.1073/pnas.090562010.

Waycott M., et al. (2009). Accelerating loss of seagrasses across the globe threatens coastal ecosystems. *Proceedings of the National Academy of Sciences of the United States of America*, *106*(30): 12377–12381. doi: 10.1073/pnas.0905620106.

WCED [World commission on Environment and Development] (1987). *Our Common Future*, Oxford University Press: Oxford.

Webb E.L., Friess D.A., Krauss K.W., Cahoon D.R., Guntenspergen G.R., Phelps J. (2013). A global standard for monitoring coastal wetland vulnerability to accelerated sea-level rise. *Nature Climate Change*, *3*: 458–465.

Webb J.R., Santos I.R., Tait D.R., Sippo J.Z., Macdonald B.C., Robson B., Maher D.T. (2016). Divergent drivers of carbon dioxide and methane dynamics in an agricultural coastal floodplain: post-flood hydrological and biological drivers. *Chemical Geology*, *440*: 313–325.

Webster T.J.M., Parnankape M.A. (1977). Sedimentation of organic matter in St Margerete Bay, Nova Scotia. *Journal of the Canadian Fisheries Research Board*, *32*: 1399–1407.

Weigman A.R.H., Day J., et al. (2017). Modeling impacts of sea-level rise, oil price, and management strategy on the costs of sustaining Mississippi delta marshes with hydraulic dredging. *Science of the Total Environment*. doi: 10.1016/j.scitotenv.2017.09.314.

Weir W.W. (1937). *Subsidence of Peat Land on the Sacramento-San Joaquin Delta of California. Volume B*, Trans. 6th Commission, International Society of Soil Science.

Weir W.W. (1950). Subsidence of peat lands of the Sacramento-San Joaquin Delta, California. *Hilgardia*, *20*: 37–56.

Welsh D., Bartoli M., Nizzoli D., Castaldelli G., Riou S.A., Viaroli P. (2000). Denitrification, nitrogen fixation, community primary productivity and inorganic-N and oxygen fluxes in an intertidal Zostera noltii meadow. *Marine Ecology Progress Series*, *208*: 65–77.

Wessel M., Dixon K. (2016). *Southwest Florida Tidal Creeks Nutrient Study: Final Report to the Sarasota Bay Estuary Program*. Prepared by Janicki Environmental Inc., Mote Marine Laboratory. TBEP Technical Report #02-16: St. Petersburg, FL. www.tbeptech.org/TBEP_TECH_PUBS/2016/TBEP_02_16_SW_FL_Tidal_Creeks_Final_Draft_Report_160121.pdf.

Western Regional Climate Center (2014). Historical climate information. www.wrcc.dri.edu/index.html (accessed April 1, 2014).

Weston N.B. (2014). Declining sediments and rising seas: an unfortunate convergence for tidal wetlands. *Estuaries and Coasts*, *37*: 1–23. doi: 10.1007/s12237-013-9654-8.

Weston N.B., Dixon R.E., Joye S.B. (2006). Ramifications of increased salinity in tidal freshwater sediments: geochemistry and microbial pathways of organic matter mineralization. *Journal of Geophysical Research-Biogeosciences*, *111*: G01009.

Weston N.B., Giblin A.E., Banta G.T., Hopkinson C.S., Tucker J. (2010). The Effects of Varying Salinity on Ammonium Exchange in Estuarine Sediments of the Parker River, Massachusetts. *Estuaries and Coasts*, *33*: 985–1003.

Weston N.B., Neubauer S.C., Velinsky D.J., Vile M.A. (2014). Net ecosystem exchange and greenhouse gas balance of tidal marshes along an estuarine salinity gradient. *Biogeochemistry*, *120*: 163–189.

Weston N.B., Vile M., Neubauer S., Velinksy D. (2011). Accelerated microbial organic matter mineralization flowing salt-water intrusion into tidal freshwater marsh soils. *Biogeochemistry*, *102*: 135–151.

Weston N.B., Vile M.A., Neubauer S.C., Velinsky D.J. (2011). Accelerated microbial organic matter mineralization following salt-water intrusion into tidal freshwater marsh soils. *Biogeochemistry, 102*(1–3): 135–151.

Weston N.B., Vile M.A., Neubauer S.C., Velinsky D.J. (2011). Accelerated microbial organic matter mineralization following salt-water intrusion into tidal freshwater marsh soils. *Biogeochemistry, 102*: 135–151.

Whipple A.A., Grossinger R.M., Rankin D., Stanford B., Askevold R.A. (2012). *Sacramento-San Joaquin Delta Historical Ecological Investigation: Exploring Pattern and Process.* Prepared for the California Department of Fish and Game and Ecosystem Restoration Program. San Francisco Estuary Institute-Aquatic Science Center: Richmond, CA. www.sfei.org/DeltaHEStudy (accessed February 2, 2016).

Whisnant R., Reyes A. (2015). Blue Economy for Business in East Asia: Towards an Integrated Understanding of Blue Economy. *Partnership in Environmental Management for the Seas of East Asia (PEMSEA),* Quezon City, Philippines, p. 69. www.pemsea.org/sites/default/files/PEMSEA%20Blue%20Economy%20 Report%2011.10.15.pdf.

White House NSTC [National Science and Technology Council] (2015). *Ecosystem-Services Assessment: Research Needs for Coastal Green Infrastructure,* U.S. Government: Washington D.C.

White House Office of Management and Budget, Council on Environmental Quality, and Office of Science and Technology Policy (2015). Memorandum for Executive Departments and Agencies on Incorporating Ecosystem Services into Federal Decision Making. The White House, www.whitehouse.gov/sites/ default/files/omb/memoranda/2016/m-16-01.pdf.

Whittle A., Gallego-Sala A. (2016). Vulnerability of the peatland carbon sink to sea-level rise. Scientific Reports 6.

Wicaksono P., Hafizt M. (2013). Mapping seagrass from space: addressing the complexity of seagrass LAI mapping. *European Journal of Remote Sensing, 46*(1): 18–39.

Wieder W.R., Bonan G.B., Allison S.D. (2013). Global soil carbon projections are improved by modelling microbial processes. *Nature Climate Change, 3*: 909–912.

Wigand C., Brennan P., Stolt M., Holt M., Ryba S. (2009). Soil respiration rates in coastal marshes subject to increasing watershed nitrogen loads in southern New England, USA. *Wetlands, 29*(3): 952–963.

Wilcock R.J., Champion P.D., Nagels J.W., Croker G.F. (1999). The influence of aquatic macrophytes on the hydraulic and physico-chemical properties of a New Zealand lowland stream. *Hydrobiologia, 416*: 203–214.

Williams M., Zalasiewicz J., Waters C.N., Edgeworth M., Bennett C., Barnosky A.D., Ellis E.C., Ellis M.A., Cearreta A., Haff P.K., Ivar do Sul J.A. (2016). The Anthropocene: a conspicuous stratigraphical signal of anthropogenic changes in production and consumption across the biosphere. *Earth's Future, 4*(3): 34–53.

Williams P.B., Faber P. (2001). Salt Marsh Restoration Experience in the San Francisco Bay Estuary. *Journal of Coastal Research. 27*: 203–211.

Williams P.B., Orr M.K. (2002). Physical evolution of restored breached levee salt marshes in the San Francisco Bay estuary. *Restoration Ecology, 10*: 527–542.

Williams P.B., Orr M.K. (2002). Physical evolution of restored breached levee salt marshes in Wilson J.O., Buchsbaum R., Valiela I., Swain T. (1986). Decomposition in salt marsh ecosystems: phenolic dynamics during decay of litter of *Spartina alterniflora, Marine Ecology Progress Series, 29*: 177–187.

Windham-Myers L., Bergamaschi B., Anderson F., Knox S., Miller R., Fujii R. (2018). Potential for negative emissions of greenhouse gases (CO_2, CH_4 and N_2O) through coastal peatland re-establishment: novel insights from high frequency flux data at meter and kilometer scales. *Environmental Research Letters, 13*(4), p.045005. doi: 10.1088/1748-9326/aaae74.

Windham-Myers L., Cai W.-J. Alin S.R., Andersson A., Crosswell J., Dunton K.H., Hernandez-Ayon J.M., Herrmann M., Hinson A.L., Hopkinson C.S., Howard J., Hu X., Knox S.H., Kroeger K., Lagomasino D., Megonigal P., Najjar R.G., Paulsen M.-L., Peteet D., Pidgeon E., Schäfer K.V.R., Tzortziou M., Wang Z.A., Watson E.B. (2018). Chapter 15: Tidal wetlands and estuaries. In: *Second State of the Carbon Cycle Report (SOCCR2): A Sustained Assessment Report,* Cavallaro N., Shrestha G., Birdsey R., Mayes M.A., Najjar R.G., Reed S.C., Romero-Lankao P., Zhu Z. (eds), U.S. Global Change Research Program: Washington, DC, pp. 596–648. https://doi.org/10.7930/SOCCR2.2018.Ch15.

Winters G., Edelist D., Shem-Tov R., Beer S., Rilov G. (2016). A low cost field-survey method for mapping seagrasses and their potential threats: an example from the northern Gulf of Aqaba, Red Sea. *Aquatic Conservation: Marine and Freshwater Ecosystems, 27*(2): 324–339.

Witte S., Giana L. (2016). Greenhouse gas emissions and balance of marshes at the Southern North Sea Coast. *Wetlands*, *36*: 121–132.

Wolf A.A., Drake B.G., Erickson J.E., Megonigal J.P. (2007). An oxygen-mediated positive feedback between elevated CO_2 and soil organic matter decomposition in a simulated anaerobic wetland. *Global Change Biology*, *13*: 2036–2044.

Wolff S., Schulp C.E.J., Verburg P. (2015). Mapping ecosystem services demand: a review of current research and future perspectives. *Ecological Indicators*. doi: 10.1016/j.ecolind.2015.03.016.

Wolf-Gladrow D.A., Zeebe R.E., Klaas C., Körtzinger A., Dickson A.G. (2007). Total alkalinity: the explicit conservative expression and its application to biogeochemical processes. *Marine Chemistry*, *106*: 287–300.

Woodroffe C.D. (2002). *Coasts: Form, Process and Evolution*, University press: Cambridge, p. 623.

Woodroffe C.D., Rogers K., McKee K.L., Lovelock C.E., Mendelssohn I.A., Saintilan N. (2016). Mangrove Sedimentation and Response to Relative Sea-Level Rise. *Annual Review of Marine Science*, *8*: 243–266.

Woodroffe S.A., Long A.J., Milne G.A., Bryant C.L., Thomas A.L. (2015). New constraints on late Holocene eustatic sea-level changes from Mahé, Seychelles. *Quaternary Science Reviews*, *115*: 1–16.

Woodwell G.M., Houghton R., Hall C., Whitney D., Moll R., Juers D. (1979). The Flax Pond ecosystem study: the annual metabolism and nutrient budgets of a salt marsh. In: *Ecological Processes in Coastal Environments*, Jefferies R., Davy A. (eds), Blackwell Scientific: Oxford, pp. 491–511.

Woodworth P.L., Gehrels W.R., Nerem R.S. (2011). Nineteenth and twentieth century changes in sea level. *Oceanography*, *24*(2): 80–93.

Wöppelmann G., Marcos M. (2016). Vertical land motion as a key to understanding sea level change and variability. *Reviews of Geophysics*, *54*(1): 64–92.

Wöppelmann G., Pouvreau N., Coulomb A., Simon B., Woodworth P.L. (2008). Tide gauge datum continuity at Brest since, 1711: France's longest sea-level record. *Geophysical Research Letters*, *35*(22).

Wöppelmann G., Miguez B.M., Bouin M.-N., Altamini Z. (2007). Geocentric sea-level trend estimates from GPS analyses at relevant tide gauges world-wide. *Global and Planetary Change*, *57*: 396–406.

World Bank, Ecofys and Vivid Economics (2016). *State and Trends of Carbon Pricing 2016*, World Bank, Washington, D.C. (accessed October).

Wozniak A.S., Roman C.T., Wainright S.C., McKinney R.A., James-Pirri M.J. (2006). Monitoring food web changes in tide-restored salt marshes: a carbon stable isotope approach. *Estuaries and Coasts*, *29*(4): 568–578.

Wright H.E. Jr. (1991). Coring tips. *Journal of Paleolimnology*, *6*: 37–49.

Wright S.A., Schoellhamer D.H. (2004). Trends in the Sediment Yield of the Sacramento River, California, 1957–2001. *San Francisco Estuary and Watershed Science*. *2*(2). http://repositories.cdlib.org/jmie/sfews/vol2/iss2/art2.

Wylie L., Sutton-Grier A.E., Moore A. (2016). Keys to successful blue carbon projects: lessons learned from global case studies. *Marine Policy*, *65*: 76–84.

Yaakub S.M., Lim R.L.F., Lim W.L., Todd P.A. (2013). The diversity and distribution of seagrass in Singapore. *Nature in Singapore*, *6*: 105–111.

Yamamuro M., Hiratsuka J.-I., Ishitobi Y., Hosokawa S., Nakamura Y. (2006). Ecosystem shift resulting from the loss of eelgrass and other submerged aquatic vegetation in two estuarine lagoons, Lake Nakaumi and Lake Shinji, Japan. *Journal of Oceanography*, *62*: 551–558.

Yang W.H., McNicol G., Teh Y.A., Estera-Molina K., Wood T.E., Silver W.L. (2017). Evaluating the classical versus an emerging conceptual model of peatland methane dynamics. *Global Biogeochemical Cycles*, *31*(9): 1435–1453.

Yates K.K., Greening H., Morrison G. (2011). *Integrating Science and Resource Management in Tampa Bay*, United States Geological Survey Circular 1348: Florida. http://pubs.usgs.gov/circ/1348.

Yates K.K., Halley R.B. (2006). Diurnal variation in rates of calcification and carbonate soil dissolution in Florida Bay. *Estuaries and Coasts*, *29*: 24–39.

Yates K.K., Moyer R.P., Moore C., Tomasko D., Smiley N., Torres-Garcia L., Powell C.E., Chappel A.R., Bociu I. (2016). Ocean acidification buffering effects of seagrass in Tampa Bay. In: *Proceedings, Tampa Bay Area Scientific Information Symposium, BASIS 6*, Burke M. (eds), St. Petersburg, FL, pp. 273–284. www.tbeptech.org/BASIS/BASIS6/BASIS6_Proceedings_FINAL.pdf (accessed September 28–30, 2015).

Yokoyama Y., Esat T.M. (2011). Global climate and sea level: enduring variability and rapid fluctuations over the past 150,000 years. *Oceanography*, *24*(2): 54–69.

Yolo Basin Foundation (2017). About the Yolo Bypass Wildlife Area. http://yolobasin.org/yolobypasswild-lifearea/ (accessed July 17, 2017).

Yu K., Hiscox A., DeLaune R.D. (2013). Greenhouse gas emission by static chamber and eddy flux methods. In: *Methods in Biogeochemistry in Wetlands*, DeLaune R.D., Reddy K.R., Richardson C.J., Megonigal J.P. (eds), Soil Science Society of America: Madison, WI, SSSA Book Series, no. 10, pp. 427–437.

Yu Z., Loisel J., Brosseau D., Beilman D., Hunt S. (2010). Global peatland dynamics since the Last glacial maximum. *Geophysical Research Letters, 37*(13): 10.1029/2010GL043584.

Yvon-Durocher G., Allen A., Montoya J., Trimmer M., Woodward G. (2010). The temperature dependence of the carbon cycle in aquatic systems. *Advances in Ecological Research, 43*. doi: 10.1016/S0065–2504(10)43007-X.

Yvon-Durocher Y., Allen A.P., Bastviken D., Conrad R., Gudasz C., St-Pierre A., Thanh-Duc N., del Giorgio P.A. (2014). Methane fluxes show consistent temperature dependence across microbial to ecosystem scales. *Nature, 507*: 488–491.

Zahran M., Al-Ansari F. (1999). The Ecology of Al-Samaliah Island, UAE. *Estuarine, Coastal and Shelf Science, 49*: 11–19.

Zaldívar-Jiménez A., et al. (2004). Estructura y productividad de los manglares en la reserva de la biosfera Ría Celestún, Yucatán, México. *Maderas y Bosques, Número esp*, 23–25.

Zaldívar-Jiménez M.A., et al. (2010). Conceptual framework for mangrove restoration in the Yucatán Peninsula. *Ecological Restoration, 28*(3): 333–342.

Zarfl C., Lumsdon A.E., Berlekamp J., Tydecks L., Tockner K. (2015). A global boom in hydropower dam construction. *Aquatic Sciences, 77*(1): 161–170.

Zeebe R.E., Wolf-Gladrow D.A. (2001). *CO_2 in Seawater: Equilibrium, Kinetics, Isotopes*, Elsevier.

Zhang K., Li Y., Liu H., Rhome J., Forbes C. (2013). Transition of the coastal and estuarine storm tide model into an operational storm surge forecast model: a case study of the Florida coast. *Weather and Forecasting, 28*, 1019–1037.

Zhang K., Liu H., Li Y., Xu H., Shen J., Rhome J., Smith T.J. III, (2012).: The role of mangroves in attenuating storm surges. *Estuarine, Coastal, and Shelf Science*, (102–103): 11–23.

Zhang K., Xiao C., Shen J. (2008).: Comparison of the CEST and SLOSH models for storm surge flooding. *Journal of Coastal Research, 24*: 489–499.

Zhang M., Ustin S.L., Rejmankova E., Sanderson E.W. (1997). Monitoring Pacific Coast salt marshes using remote sensing. *Ecological Applications, 7*: 1039–1053.

Zhang Y., Li C., Trettin C.C., Li H., Sun G. (2002). An integrated model of soil, hydrology, and vegetation for carbon dynamics in wetland ecosystems. *Global Biogeochemical Cycles, 16*: X-1–X-18. doi: 10.1029/2001GB001838.

Zhuang G.-C., Elling F.J., Nigro L.M., Samarkin V., Joye S.B., Teske A., Hinrichs K.-U. (2016). Multiple evidence for methylotrophic methanogenesis as the dominant methanogenic pathway in hypersaline sediments from the Orca Basin, Gulf of Mexico. *Geochimica et Cosmochimica Acta, 187*: 1–20.

Ziegler S., Benner R. (1999). Dissolved organic carbon cycling in a subtropical seagrass-dominated lagoon. *Marine Ecology Progress Series, 180*: 149–160.

Zimmerman R.C. (2003). A biooptical model of irradiance distribution and photosynthesis in seagrass canopies. *Limnology & Oceanography, 48*(1–2): 568–585.

Zimmerman R.C. (2006). Light and photosynthesis in seagrass meadows. In: *Seagrasses: Biology, Ecology and Conservation*, Larkum A.W.D., Orth R.J., Duarte C.M., Springer: Dordrecht, The Netherlands, pp. 303–321.

Zimmerman R.C., Dekker A.G. (2006). Aquatic optics: basic concepts for understanding how light affects seagrasses and makes them measurable from space. In: *Seagrasses: Biology, Ecology and Conservation*, Larkum A.W.D., Orth R.J., Duarte C.M., Springer: Dordrecht, The Netherlands, pp. 295–301.

Zimmerman R.C., Hill V.J., Gallegos C.L. (2015). Predicting effects of ocean warming, acidification, and water quality on Chesapeake region eelgrass. *Limnology & Oceanography, 60*: 1781–1804.

Zimmerman R.C., Hill V.J., Jinuntuya M. et al. (2017). Experimental impacts of climate warming and ocean carbonation on eelgrass *Zostera marina*. *Marine Ecology progress Series, 566*: 1–15.

Zinder S.H. (1993). *Physiological Ecology of Methanogens*, Ferry J.G. (ed.), Chapman and Hall: United Kingdom, pp. 128–206.

Zostera Experimental Network [ZEN] (2017). Available at: www.zenscience.org (accessed May15, 2017).

Zweig C.L., Burgess M.A., Percival H.F., Kitchens W.M. (2015). Use of Unmanned Aircraft Systems to Delineate Fine-Scale Wetland Vegetation Communities. *Wetlands, 35*: 303–309.

Index

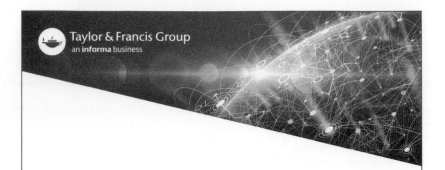

Taylor & Francis eBooks

www.taylorfrancis.com

A single destination for eBooks from Taylor & Francis with increased functionality and an improved user experience to meet the needs of our customers.

90,000+ eBooks of award-winning academic content in Humanities, Social Science, Science, Technology, Engineering, and Medical written by a global network of editors and authors.

TAYLOR & FRANCIS EBOOKS OFFERS:

A streamlined experience for our library customers

A single point of discovery for all of our eBook content

Improved search and discovery of content at both book and chapter level

REQUEST A FREE TRIAL
support@taylorfrancis.com

 Routledge
Taylor & Francis Group

 CRC Press
Taylor & Francis Group

Printed and bound by CPI Group (UK) Ltd, Croydon, CR0 4YY

01/11/2024

01782603-0011